CYTOGENETICS, EVOLUTION, BIOSTATISTICS AND PLANT BREEDING

CYTOGENETICS, EVOLUTION, BIOSTATISTICS AND PLANT BREEDING

For the Students of B.Sc. and M.Sc.

Dr. R.S. SHUKLA
M.Sc., Ph.D.
Head, Department of Botany
D.A.V. College, Kanpur (U.P.)

Dr. P.S. CHANDEL
M.Sc., Ph.D.
Department of Botany
Jai Narain Degree College (K.K. College)
Lucknow (U.P.)

S. CHAND
PUBLISHING

S Chand And Company Limited
(ISO 9001 Certified Company)

S Chand And Company Limited
(ISO 9001 Certified Company)
Head Office: D-92, Sector–2, Noida – 201301, U.P. (India), Ph. 91-120-4682700
Registered Office: A-27, 2nd Floor, Mohan Co-operative Industrial Estate, New Delhi – 110 044, Phone: 011-49731800
www.**schandpublishing.com;** e-mail: **info@schandpublishing.com**

Marketing Offices:

Chennai : Ph: 23632120; chennai@schandpublishing.com
Guwahati : Ph: 2738811, 2735640; guwahati@schandpublishing.com
Hyderabad : Ph: 40186018; hyderabad@schandpublishing.com
Jalandhar : Ph: 4645630; jalandhar@schandpublishing.com
Kolkata : Ph: 23357458, 23353914; kolkata@schandpublishing.com
Lucknow : Ph: 4003633; lucknow@schandpublishing.com
Mumbai : Ph: 25000297; mumbai@schandpublishing.com
Patna : Ph: 2260011; patna@schandpublishing.com

First Edition 1974
Subsequent Editions and Reprints 1977, 80, 88, 89, 92, 93, 96, 98, 2002, 2004
Thoroughly Revised & Enlarged Edition 2007

Fifth Revised Edition 2013
Reprints 2018, 2019, 2020, 2021, 2022

Reprint 2023

ISBN: 978-81-219-1270-9 **Product Code:** H6CEB68BOTN10ENAE13O

PRINTED IN INDIA

By Vikas Publishing House Private Limited, Plot 20/4, Site-IV, Industrial Area Sahibabad, Ghaziabad – 201 010 and Published by S Chand And Company Limited, A-27, 2nd Floor, Mohan Co-operative Industrial Estate, New Delhi – 110 044.

PREFACE TO THE FIFTH EDITION

In this edition, efforts have been made to make the text attractive and up-to-date. Many additions, alterations and corrections have been made in the text. Important additions made in this edition are as follows:

1. In the Chapter 1 a comparative account of procaryotic cells (bacteria & cyanobacteria) and eukaryotic cell have been added and special mention is made of the recently discovered group of very interesting organisms which are tentatively placed in a group named *Archaea.*
2. In Chapter 9, more details have been added to Procaryotic Genome especially bacterial genome and plasmids.
3. Chapter 22 on Mutations has been revised thoroughly. Classification of mutations on different basis have been discussed.

Gene or point mutations, effect of gene mutation on trait, gene mutations at molecular level : Base pair substitution–transitions and transversion, and Frame Shifts involving deletion, insertion or addition and inversions have been discussed.

Under types of mutations and their molecular basis, spontaneous mutations and induced mutations including chemical, physical mutagenesis have been discussed. In the detection of mutations, Methods of detecting macromutations and micromutations in plants have also been added.

With this, we feel that the present edition will meet the requirements of students in a better way. Letters of appreciation, suggestions and comments have been received from the students and teachers of several universities and colleges from different states of India. We sincerely thanks them all, further suggestions and comments are welcome.

We are thankful to the Management Team and the Editorial Department of S. Chand & Company Ltd. for all help and support in the publication of this book.

Dr. R.S. SHUKLA
Dr. P.S. CHANDEL

PREFACE TO THE FIRST EDITION

The present book has been written to cover up the courses on Cytogenetics and Evolution prescribed for the biology and agriculture students of different Indian Universities. The level of discussion is suitable for the under-graduate students, and for the graduate students it provides the indispensable knowledge without which the most recent advancements over a wide front cannot be understood. The subject matter of the book is represented in the way which is best suited to the needs of our average Indian students. The book presents the fundamental ideas of the subject in simple and concise form and it gives to the students a proper background of the subject and introduces them to the wide body of factual informations revealed by modem researches of the last two decades.

The special features of the book are as follows:

1. It has been written in simple and fluent language.
2. The subject-matter is systematically arranged and it is fully illustrated by easy diagrams.
3. Each topic is discussed in the light of recent researches that have greatly illuminated many of the earlier discoveries on certain cytogenetical phenomena by revealing the molecular basis of those phenomena.

It is hoped that the teachers and students will find this book suitable.

The obligations both general and specific, that we have in the preparation of this book are numerous.

While preparing this book, we have followed generally accepted views of standard books, Research Journals, periodicals and magazines, hence we do not claim for any originality of the approach. The writers of any such book as this draw-upon a body of knowledge built up through the efforts of many workers over many years. Our indebtedness to the authors and scientists, whose works we have taken, is acknowledged with pleasure.

Although every care has been taken in making the book free from error yet the recurrence of errors is possible. For those mistakes, and for any other short comings, we alone are responsible. We shall be highly grateful to those teachers and students who point out the mistakes of the book and also send to us their criticism and healthy suggestions for the improvement of this book.

We sincerely express our thanks to our friends who have offered valuable suggestions for the improvement of original manuscript.

We are specially indebted to our friend Dr. A.K. Misra of C.M. College, Darbhanga (Bihar) for kindly going through some parts of the original manuscript and giving suggestions.

AUTHORS

CONTENTS

CYTOLOGY

GENETICS

BIOSTATISTICS

PLANT BREEDING

1 CHAPTER The Cell

All the living bodies are made up of protoplasm. The fabric of life (protoplasm) is seldom found in large masses but is found in the form of discrete structures. These minute morphological bodies are known as the ***cells***. The cells are small structural units of the body just as the bricks are the units of structure of a house. The cell can be defined as "*a unit of biological activity delimited by a semi-permeable membrane and capable of self-reproduction in a medium free of other living systems*." The cell is an independent functional unit in all the living organisms. In the unicellular organism the single celled body is highly differentiated physiologically and it performs all the vital activities, such as nutrition, respiration, excretion, reproduction and so on. In the multicellular organisms the life activities are accomplished by highly specialised groups of innumerable cells. Thus the vital processes of the body are the sum of the co-ordinated functions of its cells.

Historical Accounts

Though the speculation about cell can be traced back to the age of Greek philosophers, the science of cells began to develop only after the discovery of compound microscope by Jansens at the end of the sixteenth century. The first treatise in microscopy was given by Robert Hooke, the curator of experiments for the Royal Society [Fig. 1.1 (*a*)]. He published a purely descriptive account of the cell, a hitherto unexplored field of life science. He observed plant cells for the first time in 1665 using the newly invented primitive model of compound microscope. Hooke observed a thin section of cork under the microscope and found that the bottle cork was not a homogeneous compact material but consisted of a great many small empty boxes or cavities bounded by definite walls [Fig. 1.1 (*b*)]. Hooke applied the name **cells** to those boxes (Latin *cellula* = a little room). He published his observations in his book 'The Micrographia'. Later he observed

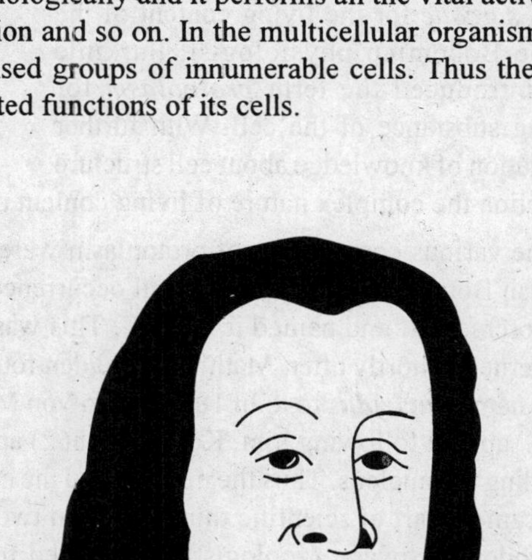

Fig. 1.1 (*a*) Robert Hooke (1635–1723), the architect of the compound microscope and discoverer of plant cell.

the cells in fresh vegetable material and found that the cavities of the living cells were filled with "juice" but he could not notice the mystery of protoplasm.

After Robert Hooke, a great curiosity developed in scientists and more details about cells became available. Anton Van Leeuwenhoek, the famous Dutch microscopist, observed for the first time green coloured bodies in the cells of plant. These are now known as chloroplasts. Two contemporaries, an English physician Nehemiah Grew and an Italian Professor Marcello Malphigi, continued their observations on the internal structure of plants and described in great detail the different types of cells in stem but they also could not notice the importance of protoplasm. Corti (1772) and Fontana (1781) emphasised the presence of semi-liquid substance in the plant cells and Dutrochet (1824) noticed it in the animal cells. In 1835, a French scientist Dujardin proposed the term *sarcode* for the living content of the cell. The Bohemian physiologist, Purkinje (1839) introduced the term *protoplasm* for the living substance of the cell. With further accumulation of knowledge about cell structure and function the complex nature of living content of the cell became more clear.

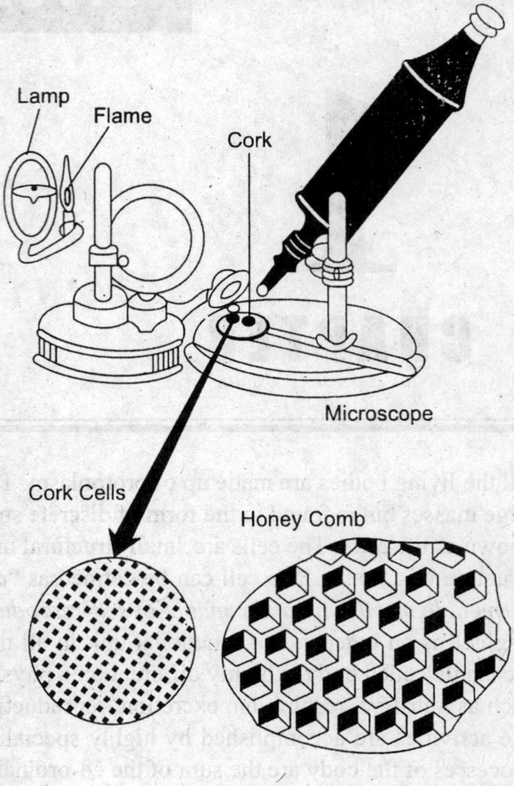

Fig. 1.1 (*b*) Hooke's diagram of cork cells.

The various components of protoplasm were recognised one by one. In 1831 Robert Brown, an English Botanist, noticed the general occurrence of a definite spherical body in each of the skin cells of orchid leaf and named it 'areola'. This was later designated as ***nucleus*** (Latin word which means kernel). Shortly after, Mathias Schleiden found a small spherical body inside the nucleus and gave the name ***nucleolus*** to it. In 1846, Hugo Von Mohl introduced a distinction between protoplasm and cell sap and following him, Kolliker (1862) applied the name **cytoplasm** to the living material surrounding the nucleus. Thus the nucleus and the cytoplasm together form the protoplasm. All these facts became a part of scientific thinking when two German biologists, named Schleiden (botanist) and Theodore Schwann (zoologist), formulated their well known "*Cell Theory*" (1938–39). The theory at present suggests that—

(*i*) all the living organisms are composed of cells and cell products,

(*ii*) there are considerable similarities in the chemical constitution and metabolic activities of all the cells, and

(*iii*) the total activities of an organism is the sum of the activities and interactions of its independent cell units.

Virchow (1855) stated the fact that all the cells in the body arise by divisions of pre-existing ones. Mechanism of cell division was established by the end of 1870. In the year 1861, Schultze also supported the cell theory and put forth the *protoplasm doctrine*. Hanstein (1880) introduced the term *protoplast* to the organised discrete mass of protoplasm and stressed that the term protoplast be used instead of the term cell. By continuous observations more parts of the cell came to light. These parts included the vacuoles, mitochondria, Golgi bodies, etc.

The cell theory was severely criticised and the concepts were considerably refined by some eminent scientists. The discovery of virus particles in the recent years has raised a problem before the biologists as to whether these particles are living or non-living. A virus can be defined as a self-reproducing biological unit which does not possess a finite semi-permeable membrane and is capable of self-reproduction only when it is present within a living cell. Outside the living cell, it is just like non-living particle. If they are regarded as living then the cell theory cannot be applied to them because their bodies are in the form of very minute particles and are not cellular. The viruses are formed of nucleic acid and proteins, the key compounds invariably present in all the cells (Fig. 1.2). These particles have been interpreted in various ways by different biologists. According to them—

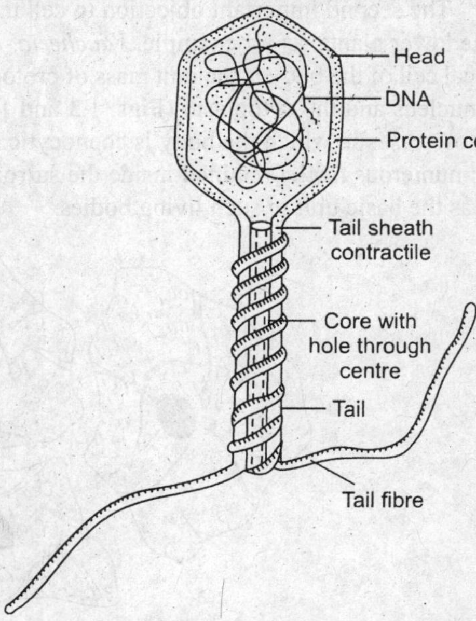

Fig. 1.2 A bacterial virus.

(*a*) Viruses are living chemicals.

(*b*) Viruses evolved from cellular forms through degeneration and they became reduced to particle size through parasitism.

(*c*) Viruses are the primitive organisms which could not reach to cellular state.

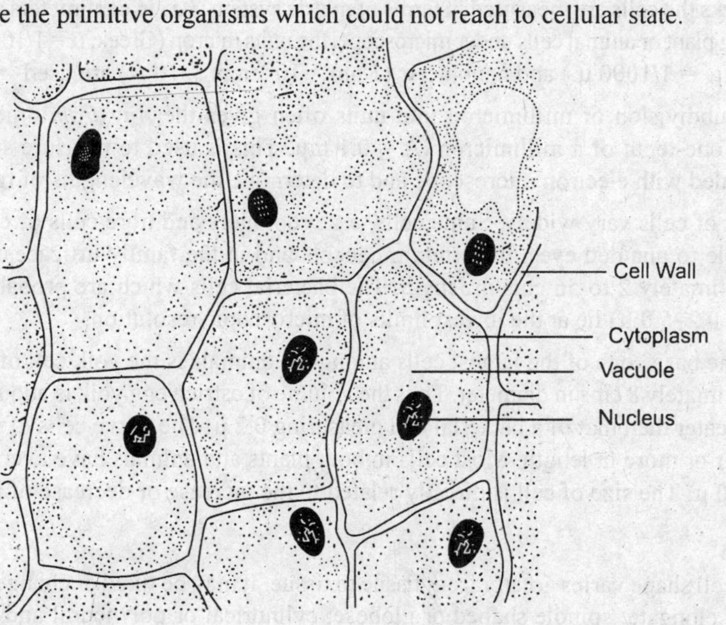

Fig. 1.3 Onion Cells.

Besides viruses, there also exist some forms that appear intermediate between viruses and cells. These systems, called **mycoplasma** and **rickettsias** appear to have membranes but are incapable of growing outside living cells. It is likely, however, that they have evolved from more independent bacteria which have been converted to parasitism and have lost some of their biochemical potentialities.

The second important objection to cell theory is that it does not explain coenocytic bodies of some lower plants, as for example, *Vaucheria, Mucor, Rhizopus,* etc. According to the cell theory, a typical cell of the body is the unit mass of protoplasm consisting of two distinct components namely the nucleus and the cytoplasm (Figs. 1.3 and 1.4). This definition, however, does not seem proper in those cases in which the body is coenocytic. The coenocytic body is a mass of living substances with numerous nuclei retained inside the surrounding wall. Thus it would be difficult to define the cell as the basic unit in such living bodies.

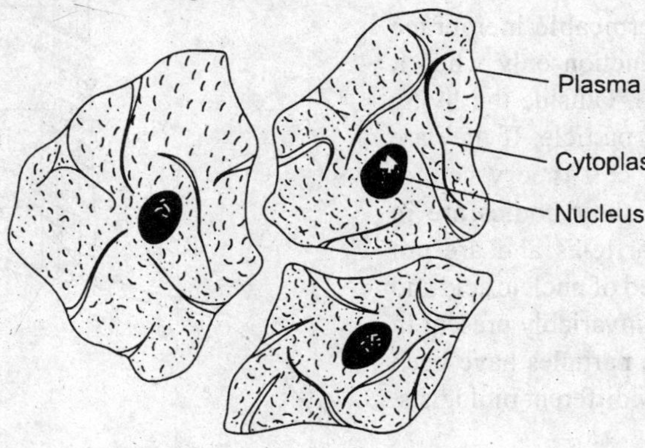

Fig. 1.4 Cells from inner lining of human cheek.

Cell Size

In scientific works the cells are measured in terms of metric system, *i.e.*, in centimeters and millimeters. When measuring plant or animal cells under microscope, the units micron (Greek, $\mu = 1/1000$ or 0.001 mm) *millimicron* ($m\mu = 1/1000\ \mu$) and nanometer (1 nm = 10^{-6} mm = 10 Å) are used.

Further subdivision of millimicron into units often gives the *Angstrom* unit (Å). This unit corresponds to one-tenth of a millimicron (Å = 0.1 mμ. This is used to measure sub-microscopic structures revealed with electron microscope and to determine the wavelengths of radiations.

The sizes of cells vary widely. Some cells, *e.g.*, root hairs and fibre cells of cotton and other plants are visible to unaided eyes. In certain monocots and in the family urticaceae the fibre cells measure approximately 2 to 56 cms in length. The bacterial cells which are probably the smallest (approximately 0.2–5.0 μ) lie at the lowest limits of microscopic resolution.

On volume basis, one of the largest cells among the animals is the yolk cell of the ostrich egg which is approximately 8 cms in diameter. Thus the volume of ostrich egg yolk is about fifteen million billion times greater than that of a bacterial cell measuring 0.2 μ. The nerve cells of some mammals measure a meter or more in length. Most cells in both plants and animals have average dimensions from 0.5 μ to 20 μ. The size of cell is directly related to the increase or decrease of its content.

Cell Shape

In general, the cell shape varies greatly from tissue to tissue. It may be fixed or changeable; spherical or oval; flat or elongate; spindle shaped or globose; cylindrical or polyhedral and so forth. These different shapes of the cells are related to the particular functions they perform. Mechanical force also determines the shapes of cells to some extent, *e.g.*, the cells from undifferentiated tissue are polyhedral in shape because of their mutual pressure. Regardless of their shapes, all the cells tend to be spherical if they are freed from various restraining factors. The shape of the newly developed cell is also dependent upon the mother cell, plane of cell division and times of division the mother cell has undergone. On maturity the cells generally become surrounded by a rigid or semi-rigid non-living membranes which also stabilize their shapes.

STRUCTURE OF THE CELL

Before the discovery of electron microscope, most of our knowledge regarding the internal structure of cell was based on the observations made with the help of light microscope. In recent years the electron microscope has been used in the study of cell. The term "ultrastructures" refers to the fact that the cell structures accounted for have been analysed with the help of electron microscope. The electron microscope makes use of a beam of light and is not in principle very different from the light microscope. For electron microscopic studies the tissues are fixed in osmium oxide (OsO_4), formaldehyde solutions etc., dehydrated, embedded in liquid plastics (usually the mixture of butyl methyl methacrylate) and sectioned with ultratome (section thickness about 100 Å). The electron microscope magnifies the image upto 3,00,000 or more times the actual size of the object on a fluorescent plate resembling that of a television tube. Photographs are made for further studies. One drawback with this apparatus is that it is difficult to examine materials in living state.

The detailed general reviews on the electron microscopy of plant cells have been given by Muihlethaler (1957), Buvat (1958) and many other workers.

There are two well recognized basic plans of cellular organization. These are:

(*i*) Simple prokaryotic plan.

(*ii*) Complex eukaryotic plan.

STRUCTURE OF PROKARYOTES

The organisms in which nucleus is not bounded by a definite nuclear membrane are called *prokaryotes*. Among the prokaryotes are the viruses, Pleuropnemonia like organisms (PPLO), bacteria and blue-green algae. Prokaryotes are very small (generally from 0.5 μ to 5 μ) and exhibit no differentiation into nucleus and cytoplasm. In the eukaryotic organisms the nucleus is bounded by a definite nuclear membrane. Prokaryotic and eukaryotic cells differ fundamentally in following points:

A comparison of Prokaryotes (bacteria and Cyanobacteria) and Eukaryotes

	Prokaryotes	Eukaryotes
1.	0.5μ–5μ; unicellular	5–100μ; unicellular or multecellutar
2.	No differentiation into nucleus and cytoplasm, i.e. cytoplasmic and nuclear materials remain intermingled.	Nucleus and cytoplasm are distinct and a definite nuclear membrane bounds the nucleus.
3.	Cytoplasmic organelles are a few or none. Clearly defined membrane bound organelles such as mitochondria, golgibodies, endoplasmic reticulum, lysosomes, centrioles and chloroplasts are lacking. Enzymatic functions of mitochondria are carried out by plasma membrane which become infolded at several points to form mesosomes.	In eukaryotic cells membrane limited organelles such as plastids, mitochondria, endoplasmic reticulum, golgi complex, lysosomes and centrioles are present which perform definite functions. Mesosomes are absent in eukaryotic cells.
4.	Genetic information is located in a linear or circular naked DNA which lacks basic proteins called *histones*. DNA occurs as nucleoid.	Linear DNA molecules are organised in chromosomes. The chromosomes are founded in a nucleus. In eukaryotic cells, the DNA of chromosome is complexed with basic proteins or histones.

	Prokaryotes	Eukaryotes
5.	Spindle apparatus is not formed during cell division in prokaryotes and nucleolus is lacking. The replication of DNA takes place when the circular DNA is attached to plasma membrane which also pulls the two strands apart. Mitosis and meiosis are absent.	Spindle apparatus develops in the cell during cell division and chromatids are pulled to opposite poles by spindle fibres. Nucleolus present in the nucleus of eukaryotes plays some role in spindle formation. Mitosis and meiosis are of common occurrence.
6.	In prokaryotes, photosynthetic pigments, when present are associated with free lamellae and pigmented plates are not enclosed in a membrane.	In eukaryotes, pigmented plates are enclosed in distinct membrane to from plastids.
7.	Flagella and cilia in prokaryotic ceils do not have 9 peripheral and two central fibrils (9 + 2 configuration).	In eukaryotic cells flagella and cilia exhibit a definite structural plan, having 9 peripheral and 2 central fibrils.
8.	Synthesis of RNA and proteins occur in the same compartment.	Synthesis and processing of RNA occur in the nucleus and protein synthesis takes place in the cytoplasm.
9.	Cell wall of prokaryotes contains amino sugars, muramic acid, and other carbohydrates.	In eukaryotes, the cell wall does not contain these substances.

Mention of a special group of recently discovered living organisms in 1977 may be made here which is named *Archaea*. This group is distinct from bacteria and eukaryotes. The living cells, therefore, are now divided into (1) *Bacteria*, (ii) *Archaea*, (iii) *Eucarya*. Such a classification was substantiated by the discovery of the genome of an archaeon namely *Methanococcus jannaschii* which was fully sequenced (i.e. nucleotide sequence fully determined) and compared with the nucleotide sequences of the genomes of bacteria and eukaryotes. By 1996 the nucleotides of genomes of three other organisms (*Haemophilus influenzae, Mycoplasma genitalium* and *Saccharomyces cerevisiae*) were fully determined and the sequences of genome of *Methanococcus jannaschii* was the fourth to be fully described. In the year 1997, the complete DNA sequence of the genome of *Escherichia* coli has also come to light. A comparison of the archaeon genome with those of other four organisms revealed that some genes of the archaeon were similar to those of bacteria, some genes resembled those of eukaryotes and about 56% of the archaeon genes were completely new to science. It has also been pointed out that the archaeal genetic system is more like a eukaryotic system, *Methanococcus jannaschii* was discovered in 1982 from a hot spot during an expedition to deep sea, 3 kilometers beneath the Pacific ocean under the leadership of *Holger jannasch*. The species name of the organism is after the name of expedition leader. The special or unique features of *Methanococcus jannaschii* are:

(i) It inhabited they region of ocean where temperature ranges from 48°C to 94°C and at pressures of more than 200 atmosphere (some members of Archaea grow at salt concentration 10 times higher than the salt concentration of sea water)

(ii) It is perfectly anaerobic and is killed if exposed to oxygen.

(iii) It is an autotroph and utilizing CO_2, nitrogen and hydrogen it produces methane and so it is *methanogen*.

(iv) Some of the Achaean genes resemble those of bacteria (Prokaryota), some genes resemble those of eukaryotes and about 56% genes are new to science. It has also been pointed out that the archaeal system is close to eukaryotic system. These archaeic features of *Methanococcus* have forced biologists to place it in a new domain.

Archaea, distinct from bacteria (prokaryota) and eukaryotes. Scientists now believe that all the modern living creatures originated from common thermophilic archaean creatures which gave rise to three major ancestral stocks; *pròkaryotes* (bacteria), *archaea* and *eukarya* more than 3.6 billion years ago.

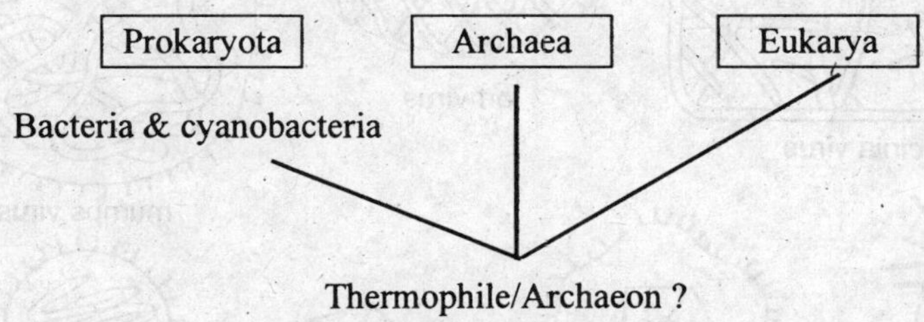

VIRUSES

Virus is a Latin word which means poison. Adolf Mayer described for the first time a disease of tobacco plant. Dimitri Ivanowski, a Russian botanist in 1892 demonstrated experimentally that sap of mosaic tobacco plant was capable of inducing the mosaic disease in healthy tobacco plants after it had been passed through bacteria proof filter. This indicated that the infective agent was smaller than any known bacteria and so he was the first to give clear cut evidence of virus. Beizerinck, a Dutch microbiologist in the year 1898 confirmed the observation of Ivanowski and named the infectious fluid obtained from the diseased tobacco plant as *'contagium vivum fluidium'* and referred it as *virus*. Stanley (1935) isolated a crystalline protein from the diseased tobacco leaves. Bawden and Pirie (1937) established that viruses are nucleoproteins. Marcham (1949) isolated tobacco mosaic virus (TMV).

Viruses comprise a unique group of infectious agents which are characterised by their small size, simple composition and parasitic mode of life.

Sizes and Shapes of Viruses

Viruses are much smaller than bacteria and their size is variable. The larger viruses may be about 300 Å in diameter, *i.e.*, they may be as large as some of the small bacteria. The majority of the viruses are about 200 Å in diameter.

Viruses occur mainly in the following three shapes:

1. *Spherical or Polyhedral,* as for example, Polio virus, Adeno viruses and Herpes viruses (Fig. 1.5).

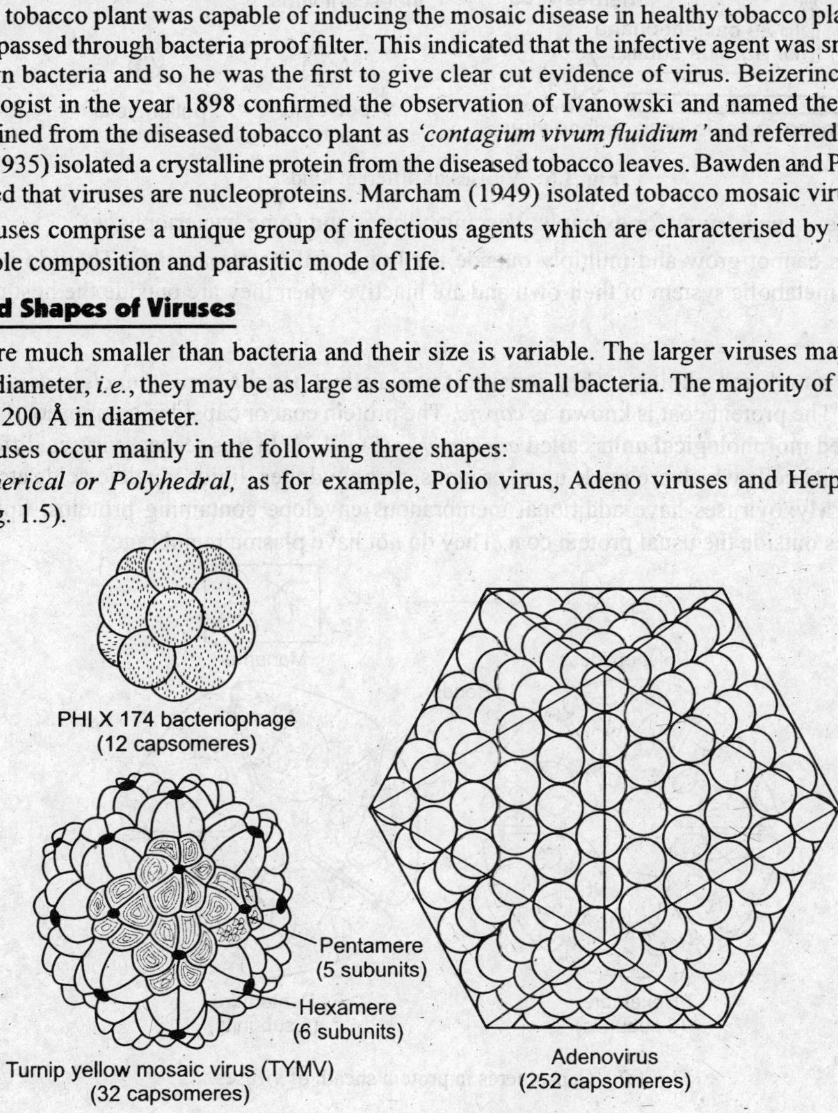

Fig. 1.5 Spherical or polyhedral viruses.

2. *Helical or Cylindrical,* as for example, Tobacco mosaic virus. Influenza virus, etc., (Fig. 1.6).

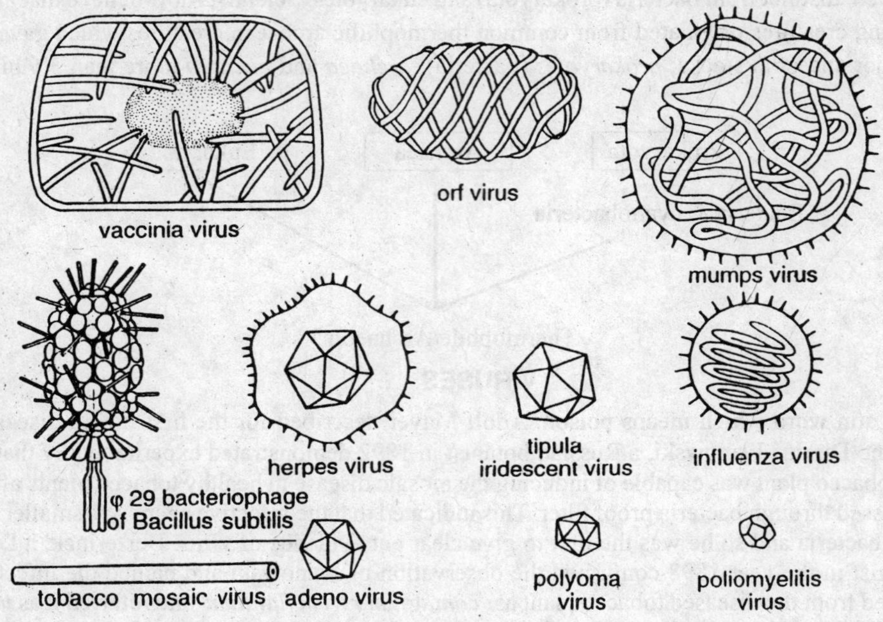

Fig. 1.6 Viruses of different kinds.

3. *Complex symmetry,* as for example, Vaccinia viruses and some bacteriophages.

Viruses cannot grow and multiply outside the living cell (total parasite). They do not have independent metabolic system of their own and are inactive when they are outside the host cell.

Structure

Viruses have simple morphology. They consist of two distinct parts: a *core of nucleic acid* and the *protein coat.* The protein coat is known as *capsid.* The protein coat or capsid is composed of several closely packed morphological units called *capsomeres* (Fig. 1.7). In this respect viruses differ from typical bacterial cells which are made up of proteins, carbohydrates, lipids, nucleic acids etc. Some viruses, *e.g.* Myxoviruses have additional membranous envelope containing proteins, lipids and carbohydrates outside the usual protein coat. They do not have plasma membrane.

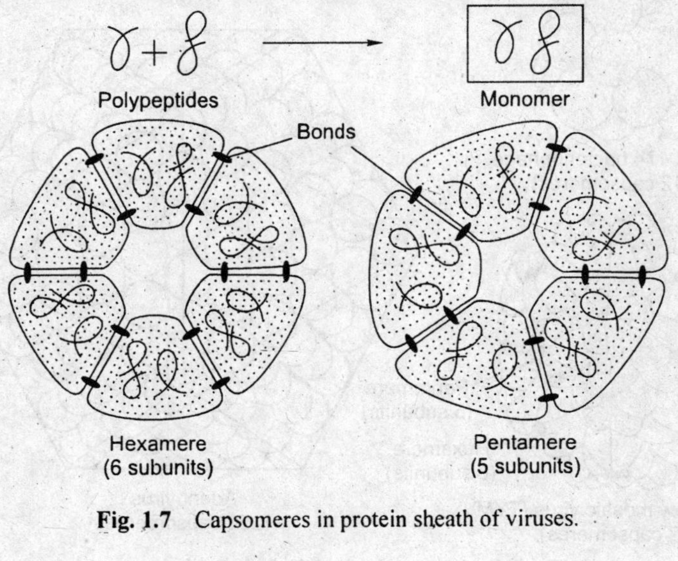

Fig. 1.7 Capsomeres in protein sheath of viruses.

Viruses lack cytoplasm and thus cell organelles such as mitochondria, golgibodies, ribosomes and lysosomes as well as enzyme systems are absent.

Viruses usually have either DNA or RNA whereas the typical cell contains both DNA and RNA. Certain animal viruses, *e.g.* Rous sarcoma virus (RSV) have both DNA and RNA. Thus with respect to nucleic acid, the viruses are of three types:

1. RNA virus,
2. DNA virus, and
3. DNA-RNA virus.

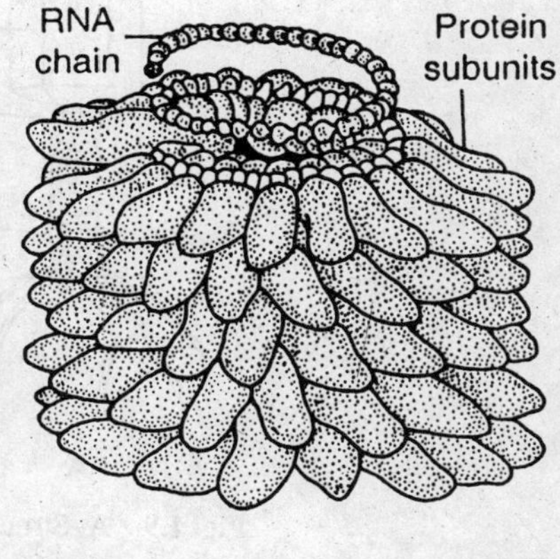

Fig. 1.8 Tobacco mosaic virus (TMV).

Plant viruses contain only RNA (Fig. 1.8). Most bacteriophages (viruses infecting bacteria) contain only DNA (some bacteriophages like MS-2, F_2 coliphages, R-17 are the exceptions which contain RNA). Animal viruses contain RNA or DNA or rarely both DNA-RNA. The nucleic acid component of viruses may be composed of either single stranded or double stranded RNA or double stranded DNA or single stranded DNA. Table 1.1 shows the type of nucleic acid and the number of strand(s) in different viruses:

Table 1.1

	Viruses	*Nucleic acid*	*Number of strands*
1.	Pox virus	DNA	2
2.	Coliphages T-even series (T_2, T_4, T_6)	DNA	2
3.	Coliphage T-odd series (T_3, T_7)	DNA	2
4.	Coliphage Lambda (γ)virus	DNA	2
5.	Adeno viruses	DNA	2
6.	Papilloma virus	DNA	2
7.	Polyoma virus SV–40	DNA	2
8.	Herpes viruses	DNA	2
9.	Bacteriophages ϕ X174	DNA	1
10.	M13	DNA	1
11.	Reo virus	RNA	2
12.	Wound tumour virus	RNA	2
13.	Rice dwarf virus	RNA	2
14.	Tobacco mosaic virus (TMV)	RNA	1
15.	Influenza virus	RNA	1
16.	Poliomyelitis virus	RNA	1
17.	Bacteriophage MS-2	RNA	1
18.	Bacterial virus F_2	RNA	1
19.	Avian Leukemia virus	RNA	1

BACTERIOPHAGE

Bacteriophages are the viruses which multiply within bacteria. Twort (1915) and d'Herelle (1917) discovered them and named them.

Brenner and his associates studied the ultra-structure of bacteriophages. Most of the bacteriophages have complex symmetry. Bacteriophages of T-even series (T_2, T_4, T_6) which grow in the colon bacterium *Escherichia coli* are tadpole shaped consisting of a head and a tail. (Fig. 1.9). Besides tailed form, there are some *tailless forms* and *filamentous forms* or tubular forms.

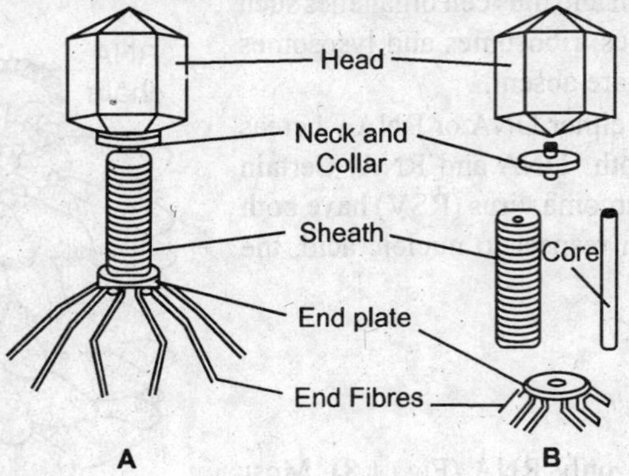

Fig. 1.9 A–Structure and B–Parts of T_4 bacteriophage.

The head is pyramidal prolate icosahedran made up of about 2,000 protein sub-units or capsomeres and is packed with DNA. The tail is cylindrical which varies in diameter and structure. The tail is composed of a helical *core tube*, 80 Å in diameter surrounded by a protein sheath. Core tube is connected with the head by a disc called *collar* at the upper end and with a hexagonal plate at the lower end (*end plate* or *tail plate*) (Fig. 1.9B).

The base plate has six corners, each with a spike. Each of the six corners of base plate is provided with a long *tail fibre* which is about 1300 Å in length. These are meant for attachment of bacteriophage on the surface of bacterial cell.

MYCOPLASMA

Mycoplasmas are tiny, virus-like non-motile prokaryotic organisms. They are the simplest and smallest known organisms capable of growing on cell free laboratory culture media containing sterols. Mycoplasmas lack definite cell wall and are characteristically delimited by a distinct, soft, flexible trilaminar unit membrane of lipoprotein. Because of the soft and plastic membrane they can assume a variety of shapes and sizes (Pleomorphic). The electron micrographs have revealed that mycoplasmas are small spherical bodies or branched worm-like structures which are smaller than bacteria but are within the size range of large viruses.

They can pass easily through bacterial filter and can be squeezed through filter pores smaller than their actual diameter (Morowitz 1972). They are highly resistant to many antibiotics.

The unit membrane of mycoplasma surrounds the cytoplasm which is packed with ribosomes, fibrilar DNA, one or more electron dense areas, and some empty vacuoles (Fig. 1.10).

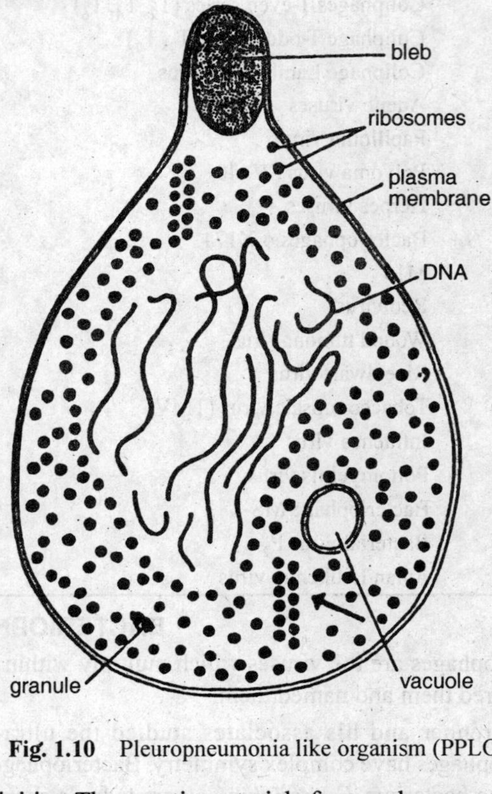

Fig. 1.10 Pleuropneumonia like organism (PPLO).

They are capable of performing all life activities. The genetic material of mycoplasma consists of a naked circular chromosome of fibrilar DNA which is about 3 nm in thickness and 1000 × 10^6

daltons in weight (1 dalton = 1/16 the atomic weight of oxygen = 1.650×10^{-24} g). This replicates in the same way as does the bacterial chromosome. The ribosomes are generally 72 S and are about 14 nm in diameter.

Reproduction in mycoplasma has not yet been conclusively proved. They can reproduce by binary fission, by formation of spores (elementary bodies), by filamentous growth and by budding.

Structure of Pleuropneumonia like Organism (PPLO)

The Pleuropneumonia like organisms (PPLO) are bacteria like organisms and show structural organization somewhat intermediate between those of viruses and bacteria. They are more complex than the primitive self-duplicating system from which life presumably arose. Nevertheless, they may be more close to the minimum size and complexity required for independent cellular life and reproduction under present day condition (Morowitz, 1966). They are filterable like viruses but unlike viruses they can grow in cell free non-living medium.

Mycoplasma, a group of widely studied PPLO, measure nearly 0.1 μ to 0.25 μ in diameter. They differ from bacteria in lacking cell wall and mesosomes and exhibiting very simple organization.

The PPLOs possess a thick plasma membrane of about 75 Å thickness and a hyaloplasm (Fig. 1.10). In hyaloplasm there occurs a single free DNA molecule, numerous ribosomes, some granular structures and empty spaces of unknown significance. The vacuole like structures presumably arise due to infoldings of the plasma membrane.

The PPLOs are free living and have all materials and enzymes required in the completion of some biological processes as DNA replication, protein synthesis, anaerobic break-down of sugar, ATP synthesis and so on.

These primitive micro-organisms unlike viruses do not require host cells for their multiplication. They multiply by binary fission, budding, fragmentation etc.

BACTERIAL CELL

The bacteria are unicellular, achlorophyllous prokaryotic micro-organisms leading either saprophytic or parasitic mode of life. An average bacterial cell may measure 1.25 μ in diameter. The smallest among the known bacteria is *Dialister pneumosintes* (0.15 to 0.3 μ in length) and the largest bacterium is *Spirillum volutans* (13 to 15 μ in length). The bacterial cell may be (*i*) spherical (**coccus**), (*ii*) rod shaped or cylindrical (**bacillus**), and (*iii*) spiral (**spirillum**) or ***spirochetes***. The spherical bacteria (cocci) may occur either singly (micro or **monococci**) or in pairs (**diplococci**) or in a group of 4 (**tetracocci**) or in chains (**streptococci**) or in irregular bunches (**staphylococci**). Rod shaped bacteria occur usually singly but may occasionally be found in pairs (**diplobacilli**) or in chains (**streptobacilli**).

Structure. The bacterial cells show a typical prokaryotic organization consisting of the following parts (Fig. 1.11):

1. Outer covering,
2. Cytoplasm, and
3. Nucleoid.

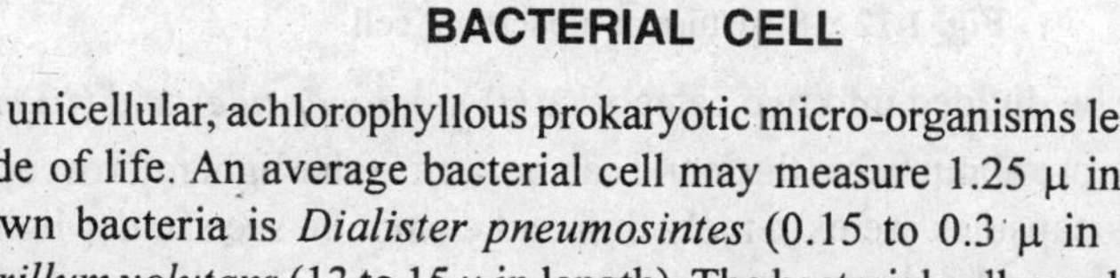

Fig. 1.11 A generalized structure of a bacterium.

1. Outer covering of the cell. The outer covering in most of the bacterial cells consists of the following layers:

(*a*) Capsule,
(*b*) Cell wall, and
(*c*) Plasma membrane.

(*a*) **Capsule.** In some bacteria there may be a slimy layer outside the cell wall which is called capsule. The slimy capsule is secreted by protoplasm and is made up of di or polysaccharides. The polysaccharides of most bacterial capsules contain more than one type of sugar residues and so they are heteropolysaccharides such as D-glucose, D-galactose, D-Mannose, D-glucouronic and L-rhamnose residues etc. In some bacteria the polysaccharides do contain single sugar residue and so they are homopolysaccharides such as polyfructose.

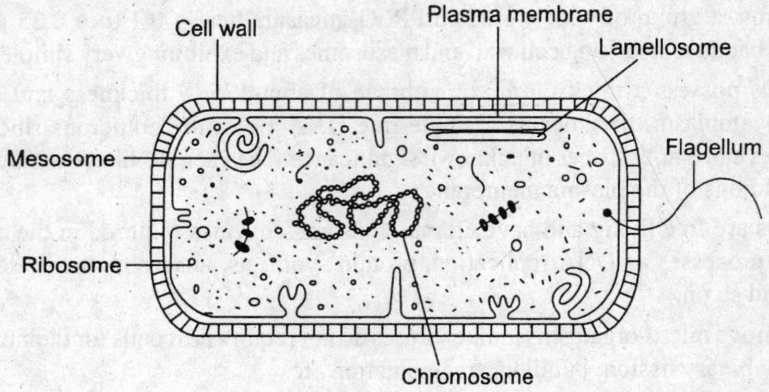

Fig. 1.12 Structure of a bacterial cell.

Capsules may be divided into two categories (*i*) *macrocapsules* and (*ii*) *microcapsules.* Macrocapsules are about 0.2 m thick and can be seen under light microscope after special staining. Microcapsules are extremely thin and cannot be seen under light microscope. They can, however, be demonstrated immunologically. The capsule layer serves as a protective covering for bacterial protoplasm. In pathogenic bacteria capsule provides enough resistance against drugs and phagocytes.

Appreciable number of bacterial species are motile and they are provided with flagella. Bacterial flagella are about 100 to 200 Å thick and show varied length. They develop from basal granules lying below the cell wall. The bacterial flagella, unlike multistranded flagella of eukaryotes, are made up of special type of proteins (Fig 1.12).

In addition to flagella, there occur some hair-like or peg-like outgrowths on the surface of some bacterial cells. These are called pili or fimbriae. They are composed of helically arranged units of special protein called pilin. Pilin helps in the attachment of bacterial cells to some other objects, and in some they act as conjugation channels through which DNA of donor cell moves into the female cell.

(*b*) **Cell wall.** Cell wall is rigid structure, 10 nm thick which is present just below the capsule and outside the plasma membrane. The cell wall maintains the shape of the cell and protects the cell from high osmotic pressure gradient. The chemical composition of bacterial cell wall is different from that of higher plant cell. It does not contain cellulose. The main structural components of almost all bacterial cell walls are peptidoglycans or mucopeptides consisting of acetylglucosamine and acetyl muramic acid molecules linked alternately in chain. In certain bacteria, *Teichoic* acid molecules are covalently linked to peptidoglucan. Other compounds found in the cell wall are proteins, polysaccharides, lipids, certain inorganic salts, phosphorous and aminoacids, diaminopimallic acid (a substance found only in bacteria and blue-green algae). The polysaccharides of bacterial cell wall consist of different sugars such as glucose, galactose and mannose or corresponding aminosugars depending upon bacteria. The electron microscopy of bacterial cell wall reveals that it is composed of granular units of 50 to 140 nm diameter arranged in regular hexagonal or rectangular pattern.

(*c*) **Plasma membrane.** Beneath the cell wall there lies a thin living membrane, the plasma membrane which forms the outer boundary of cytoplasm. The plasma membrane is 75 Å thick unit membrane and is composed chiefly of proteins and phospholipids. Some amount of carbohydrate, DNA and RNA have also been reported from plasma membrane but it still needs substantial proof. The lipids in the prokaryotic plasma membrane are polar lipids which may be glycophosphates or glycolipids. Small amount of quinone, Co-enzyme-Q, vitamin K and carotenoids may also be found in the bacterial plasma membrane.

This membrane is a selective barrier to the surrounding medium. The plasma membranes of bacteria contain enzymes involved in oxidation of metabolites or respiratory chain and many multienzyme complexes. It performs the functions including oxidative phosphorylation which are usually done by mitochondria in eukaryotic cells. Bacterial plasma membrane contains many specific transport systems for compounds of the sugars, aminoacids, mineral ions etc. The plasma membrane is capable of not only transporting materials by simple diffusion but is also involved in active transport against concentration gradient.

At certain places membrane may be infolded to form whorls of convoluted membranes called **mesosomes**. The mesosomes perform several important metabolic activities such as respiration, secretion, etc. They are thought to increase the surface area for transporting systems of the cells. They are also sites of DNA replication enzymes and nucleoid separation.

2. Cytoplasm. The cytoplasm of bacterial cell is dense and colloidal substance containing a variety of organic compounds such as proteins, glycogens and lipids. Besides, granulose (a polymer of glucose), volutin (polymetaphosphate), polybeta hydroxibutyric acid and elemental sulphur may also be found in the form of granules. The majority of cytoplasmic organelles such as endoplasmic reticulum, golgi complex, plastids, mitochondria, lysosomes and centrioles are lacking in the bacterial cytoplasm. Photosynthetic bacteria contain a pigment called **bacteriochlorophyll** which is somewhat different from the chlorophylls found in eukaryotes. The photosynthetic pigments and enzymes are generally found in the groundplasm associated with the lamellae, tubules or vesicles called chromatophores. The bacterial cells contain ribosome particles of about 25 nm diameter which exist in the cytoplasm in free state or in the form of polysomes but are not attached to the membrane.

Ribosome particles constitute upto 30% of the bacterial weight and they are the sites for protein synthesis.

3. Nucleoid. Nuclear region or nucleoid contains genophore which is single circular double stranded molecule of deoxyribonucleic acid (DNA). DNA molecule of bacterium is about 1mm long (10^6 nm) and contains all the genetic information. One molecule is sufficient to code for about 2000 to 3000 different proteins. DNA molecule is folded and packed within the nuclear region. It lies free in the cytoplasm and is not bounded by nuclear membrane. DNA molecule of bacterial cell appears to be free and not complexed with histone but contains polyamines (non-histone type) which may be linked to some of its phosphate groups. Under certain conditions bacterial cell may contain two or more DNA molecules, because of replication of original DNA.

Evidence from studies of *Bacillus subtilis* indicates that the circular DNA may be attached to the plasma membrane *via* mesosome. So, it is possible that mesosome plays an important role in DNA replication by providing a mechanism for unwinding of DNA double helix as well as energy. In general, during division of the bacterial cell DNA molecule replicates and the two molecules go to different daughter cells.

CYANOBACTERIA (BLUE GREEN ALGAE)

Blue-green algae are the most primitive organisms in the plant kingdom and show typical prokaryotic organization (Fig. 1.13). A typical cell of blue-green algae is composed of the following components:

1. Outer covering.
2. Cytoplasm.
3. Nucleic material.

1. Outer covering. The outer covering of cell includes :

(*a*) Mucilaginous layer

(*b*) Cell wall

(*c*) Innermost plasma membrane

(*a*) *Mucilaginous layer.* Mucilaginous sheath is the outermost layer covering the cell wall. In some cases the mucilaginous layer is very conspicuous and forms mucilaginous sheath but in others it may be inconspicuous. It protects the cell from the injurious factors of the environment.

(*b*) *Cell wall.* Just below the mucilaginous layer is present cell wall. Electron microscopy has revealed that the cell wall is relatively complex structure. The cell wall is 2 or 3-layered and the inner layer lies in between outerwall layer and plasma membrane. The cell wall is formed of polysaccharides and mucopeptides.

(*c*) *Plasma membrane.* The plasma membrane is selectively permeable living membrane enclosing the cytoplasm and is lipoproteinic in nature.

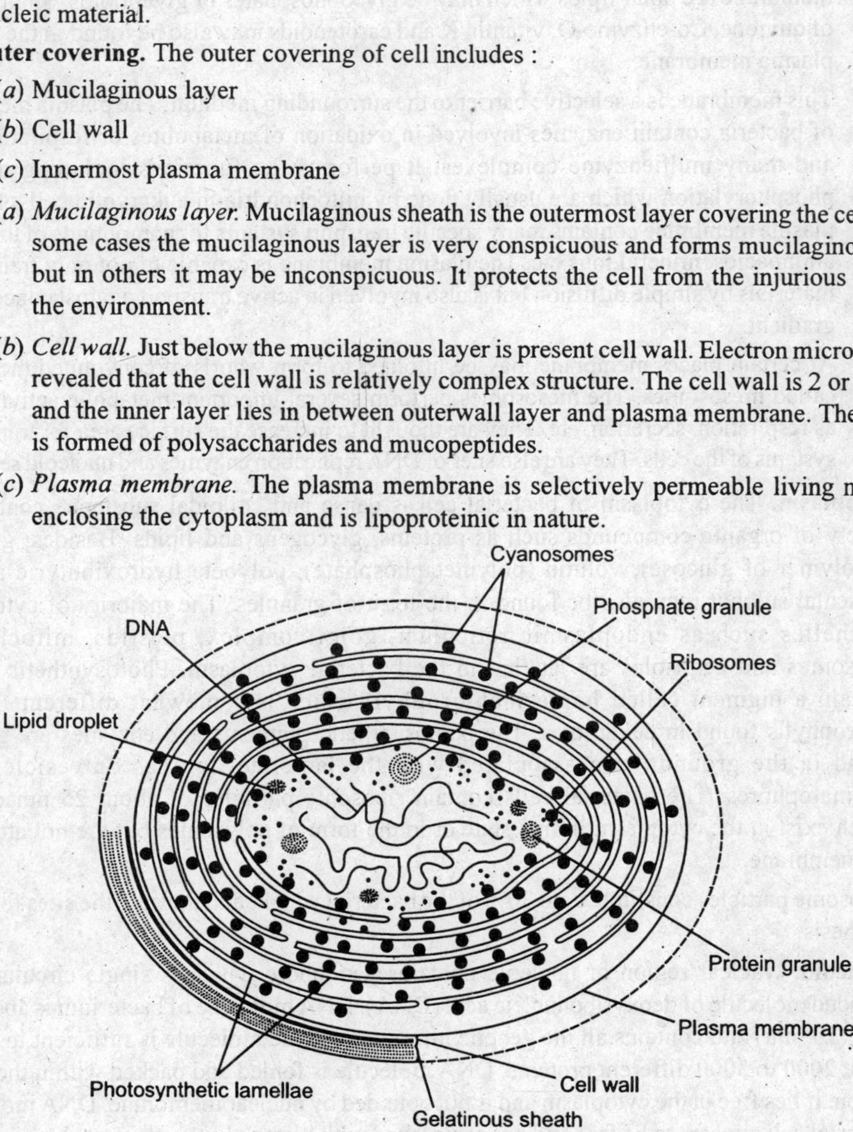

Fig. 1.13 Diagram of Cyanobacterial cell.

2. **Cytoplasm.** Below the plasma membrane is seen the groundplasm which contains structures of different shapes and functions. In the peripheral region of cytoplasm are located lamellae which contain pigments (Fig. 1.14). Fine structure study has made it clear that the pigmented lamellae are not organised into plastid. Lamellae or membranes are derived from plasma membrane. The pigments in lamellae include chlorophylls, carotenes, xanthophylls, c-phycoerythrin and c-phycocyanin, the last two are characteristically found in blue-green algae only. In addition to lamellae, several membrane bound vesicles may also be seen in the cytoplasm and they may sometimes be stacked in layers. Besides, ribosomes may be found scattered in the groundplasm.
3. **Nuclear material.** The nucleoplasm or DNA containing region is centrally located in the cell and shows a fibrillar structure. Nucleoplasm is Feulgen-positive but is not organised into a nucleus, *i.e.*, there is no nuclear boundary and no nucleolus. During division the nucleoplasmic material dispersed throughout the cell divides into two and no spindle apparatus participates in this process.

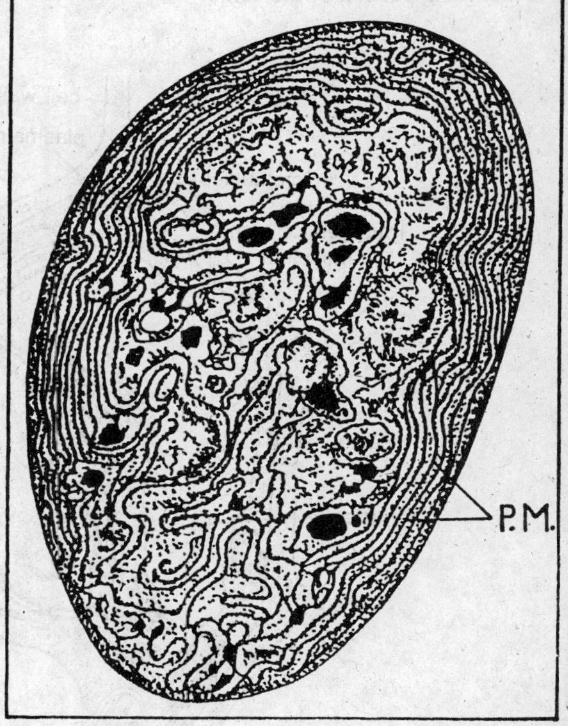

Fig. 1.14 Electron micrograph of *Anabaena* cell.

EUKARYOTIC CELLS

The eukaryotic cells are found in most of the algae, fungi, protozoa and all the higher plants and animals. Irrespective of their variable shapes and sizes, these cells display the following fundamental components (Fig 1.15):

1. Cell wall and Plasmalemma or cytoplasmic membrane
2. Cytoplasm
3. Nucleus

The eukaryotic cells contain extensive internal membrane systems such as endoplasmic reticulum (E.R.), as well as organelles surrounded by membranes, as for example, the mitochondria, plastids, lysosomes etc. These systems are incapable of internal movement and contain, in addition, a number of energy transducing systems such as mitotic apparatus, multi-stranded flagella and so on. The nuclei are delimited by nuclear membranes and contain nucleoli and more than one chromosome made up of DNA and histones.

The Cell Wall and Plasma Membrane

There are two important structural components of the surface boundary of the cell: (*i*) the outer layer, the ***glycocalyx***, and (*ii*) inner component, ***the plasma membrane*** or ***cytoplasmic membrane***. The glycocalyx is the general name applied to any structure which lies outside the plasma membrane. Glycocalyx contains polysaccharides which are associated sometimes with some other substances as aminoacids, lipids, lignins, proteins, nucleotides, calcium phosphate and so on. Glycocalyx of plants and bacteria are familiarly known as cell wall. Plasma membrane, the outer boundary of cytoplasm is the living portion of the cell.

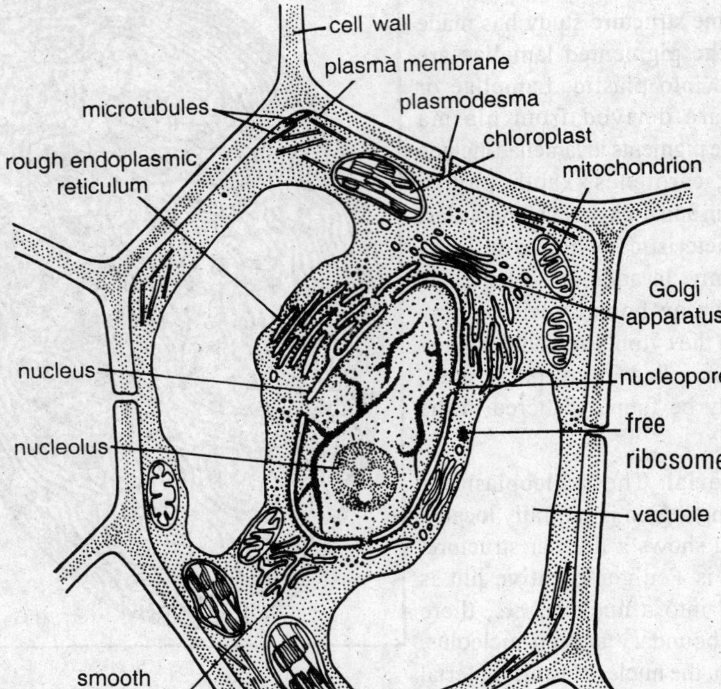

Fig. 1.15 A typical eukaryotic plant cell.

The Cell Wall. The plant cell is always surrounded by a cell wall. The cell wall which differs from plasma membrane in being non-living is the additional layer synthesized by the living cytoplasm just on the external surface of plasma membrane. In most of the plant cells, the cell wall is made up of cellulose, hemicellulose and pectin. In many fungi the wall is formed of fibrous polysaccharides and chitin. In animal cells, the non-living cell wall is lacking, but in some a thin layer of mucopolysaccharide and cuticle is deposited just outside the plasma-membrane in the form of pellicle.

The cell wall is perforated at one or more places through which the cytoplasm of a cell is connected with that of adjacent cells. These cytoplasmic bridges interconnecting any two adjacent cells are termed as *plasmodesmata.* These are of great physiological importance because they allow flow of cytoplasmic materials from one cell to the other. Between the walls of any two adjacent cells intercellular substancc is deposited in the form of *middle lamella* (Fig. 1.15). In plants the middle lamella is formed of calcium pectate. The intercellular matrix cements the cells together cell wall to cell wall or pellicle to pellicle as the case may be. The middle lamella and outer cell wall or pellicle are emphasized as non-living because of the fact that they do not play important role in the life of cell.

The cell wall or pellicle can be removed from a number of plant and animal cells by microsurgery and other techniques. The naked cells thus obtained after the removal of cell wall or pellicle are surrounded by plasma membrane only and are more liable to mechanical injuries. Though they remain alive yet they cannot retain their original shape and size.

In plants the cell wall can be differentiated into primary, secondary and sometimes tertiary layers. The secondary layer is deposited on the inner face of the primary membrane and the tertiary layer below the secondary one (Fig. 1.16).

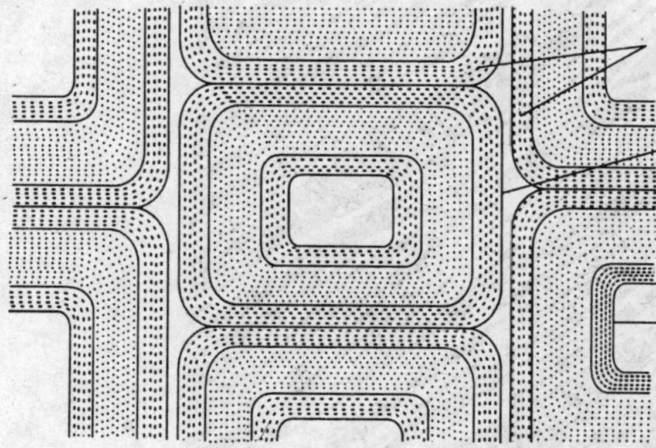

Fig. 1.16 Gross-structure of plant cell wall.

Electron microscopy has revealed that the primary cell wall is composed of sub-microscopic strands of cellulose which in turn can be split into even finer threads by means of ultrasonic irradiations. These microfibrils of the primary wall layer are interwoven to form a very dense network (Fig. 1.17).

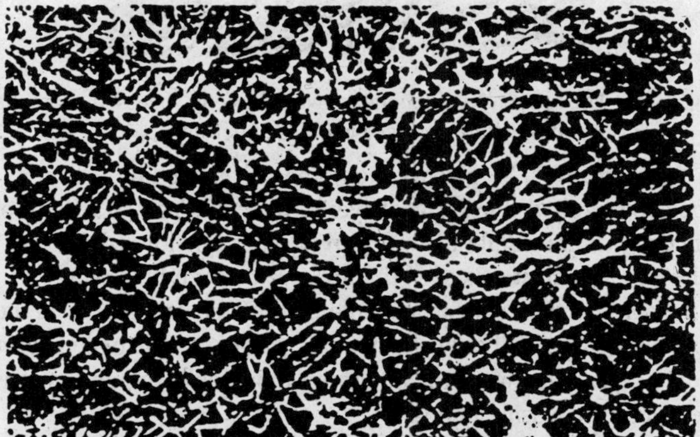

Fig. 1.17 Ultrastructure of primary cell wall showing dense network of fine threads (microfibrils).

The microfibrils of cellulose in the secondary cell wall tend to lie in parallel fashion forming successive lamellae instead of forming a network. In the successive layers of secondary wall the orientation of the parallel microfibrils is changed. At certain places the microfibrils of secondary wall materials are not deposited, resulting thereby depressions or perforations. Such characteristic perforations in the secondary walls are known as pits. There are microcapillary spaces between the microfibrils of cellulose in the secondary wall. In these microcapillary spaces lignin, cutin, suberine, hemicellulose, minerals and some other wall materials are deposited which make the secondary wall tensile and sometimes impermeable to water and gases (Fig. 1.18). In general, the structure and the chemical constitution of cell walls will vary according to the physiology of plant tissue in which they are found . The cell wall provides rigidity and support to the cells. It also protects the protoplasm from mechanical injury and maintains the characteristic shape of cell.

Fig. 1.18 Ultrastructure of secondary wall showing successive lamellae of parallely arranged cellulose microfibrils.

The Plasma Membrane. The plasma membrane is exceedingly thin and not visible as a separate layer. It appears merely as the surface layer of the cytoplasm. Earlier biologists doubted its appearance. Existence of plasma membrane was proved when phenol-red dye was microinjected into the cytoplasm. In the experiment it was seen that the dye coloured the whole of the cytoplasm and did not pass out of the cell. Further, when some colourless cells were placed in the dye, it was found that the dye could colour the cell wall only and not the cytoplasm. The two observations suggest that there exists a boundary layer in between the outer non-living cell wall and internal cytoplasm. This is called plasma membrane or plasmalemma.

The plasma membrane is a living semi-permeable membrane covering the cell cytoplasm. It provides mechanical supports and definite shape to the cell. It delimits the protoplast from the outside environment and due to its semi-permeable nature it transmits necessary material into the cell and checks the entry of undesirable substances into the cell. It is found to contain small pores through which exchange of molecules occurs.

THE CYTOPLASM

It is defined as the extranuclear part of the protoplast of a cell. The cytoplasm is a heterogeneous phase system consisting of several types of granular, fibrillar, lamellated subcellular living bodies called the **organoids** or **organelles, inclusions** and **vacuoles** suspended in a clear groundplasm (Fig. 1.15). The various components of the cytoplasm are described under the following heads:

(*a*) Groundplasm or cytoplasmic matrix,

(*b*) Organoids or organelles,

(*c*) Inclusions or Ergastic substances, and (*d*) Vacuoles,

Groundplasm or Cytoplasmic matrix

In light microscopy, the term groundplasm refers to the liquid fraction of cytoplasm in which other components of the protoplast are suspended.

In ultrastructure, ground substance or cytoplasmic matrix is defined as the viscous, homogeneous, clear and transparent liquid part of the cytoplasm. It has unusual property of being capable of both viscous flow like a liquid and elastic deformations like a solid. The cytoplasmic matrix near the outer membrane tends to be dense like solid and is generally referred to as *ectoplasm* whereas the cytoplasmic matrix in the interior of the cell, generally in fluid state, is referred to as *endoplasm*. The ground plasm contains a great variety of inorganic salts and ions, as well as carbohydrates, proteins, fats and many other organic substances which are beyond the resolution limit of electron microscope. This homogeneous ground substance still remains a challenge because the structural world of the atoms of cytoplasm still remains unknown and needs exploration.

The Organoids or Organelles

A variety of living bodies of definite structures and functions are seen suspended in the cytoplasm which are known as ***organoids or organelles.*** These organelles are the main sites for the various cytoplasmic activities. The organelles are of two types : some concerned with the chemical works or metabolism of the cytoplasm and the others concerned with the mechanical works. The organelles concerned with the chemical works of the cytoplasm are as under:

(*i*) Plastids (only in the plant cells)

(*ii*) Mitochondria

(*iii*) Golgi complex

(*iv*) Ribosomes

(*v*) Endoplasmic reticulum

(*vi*) Lysosomes, Microbodies and Peroxysomes.

The other kinds of organelles concerned with the mechanical works of cytoplasm are:

(*i*) Microtubules

(*ii*) Centrosomes (not seen in the cells of higher plants)

(*iii*) Flagella and Cilia.

The Plastids. These are coloured or colourless cytoplasmic bodies present in all the plant cells except fungi and prokaryotes. It is only on account of the presence of these plastids the plant cells are variously coloured. On colour basis, these plastids may be grouped into the following types:

(*i*) leucoplasts or leukoplasts (colourless)

(*ii*) chloroplasts (green)

(*iii*) chromoplasts (coloured other than green).

The Chromoplasts are responsible for yellow, pink, red colours of the flowers, fruits and young leaves of plants. Leucoplasts are sometimes called amyloplasts if they store starch. Amyloplasts are found chiefly in the cells of storage organs, as in potato tubers.

The plastids are generally coloured owing to the presence of several kinds of colouring substances called pigments. Some important pigments and their respective colours are listed below:

	Pigment	*Colour*
1.	Chlorophylls	Green
2.	Carotenes	Yellow or red
3.	Xanthophylls	Yellow, orange or brown

4.	Phycoerythrin	Red
5.	Haematochrom	Red
6.	Phycocyanin	Blue green

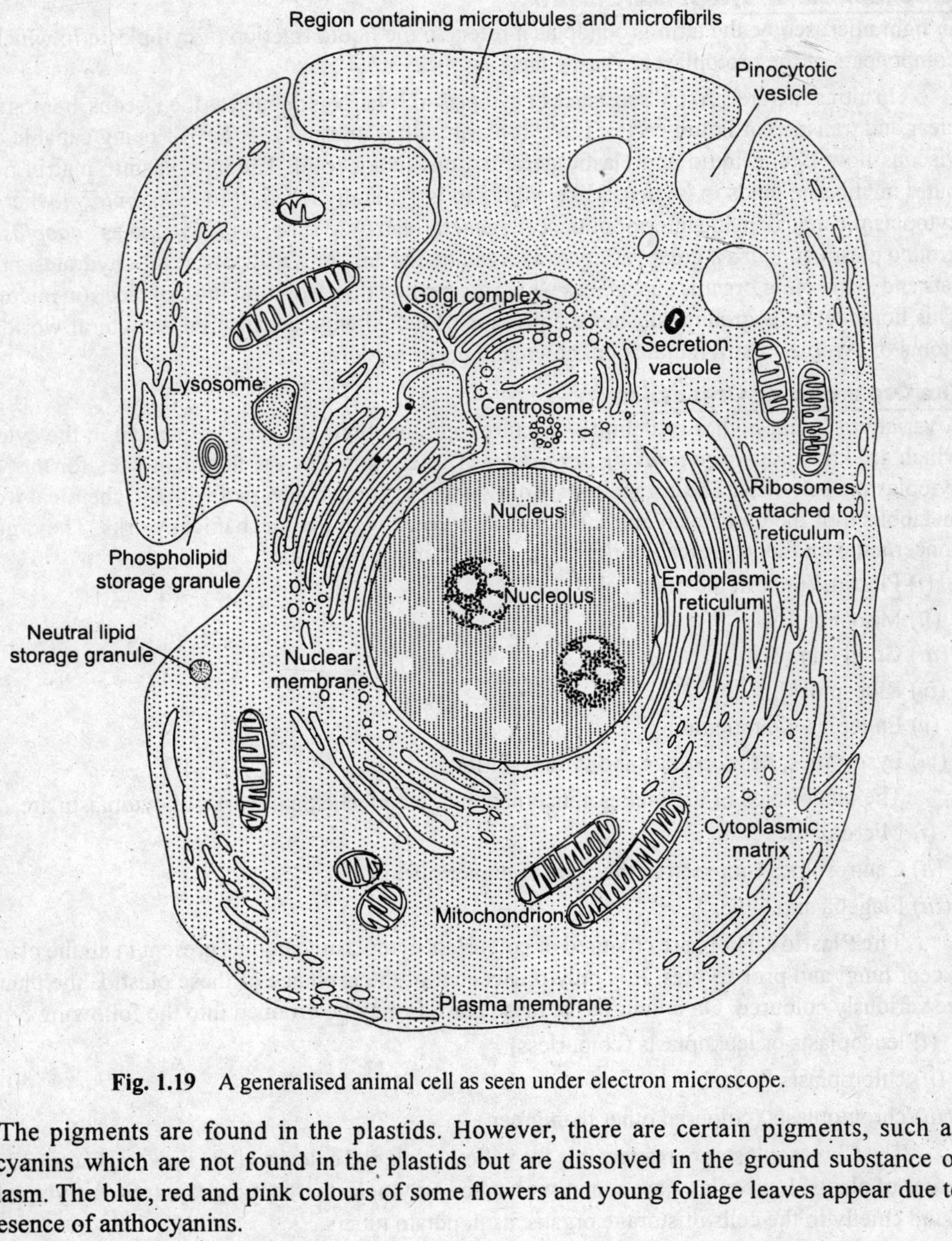

Fig. 1.19 A generalised animal cell as seen under electron microscope.

The pigments are found in the plastids. However, there are certain pigments, such as anthocyanins which are not found in the plastids but are dissolved in the ground substance of cytoplasm. The blue, red and pink colours of some flowers and young foliage leaves appear due to the presence of anthocyanins.

Chloroplast is the most important among all the plastids and is the chief site of photosynthesis in plant cells. The chloroplasts are of various shapes and sizes. In higher plants they are generally shaped like a biconvex lens and are 4 to 6 μ in diameter and 2 to 3 μ in thickness. Their number in a cell varies from plant to plant and tissue to tissue. Juvenile cells usually lack chloroplasts but contain sub-microscopic bodies called proplastids from which chloroplasts develop as the cells mature. The chloroplast develops only in the cells exposed to light.

Mitochondria. There are several synonymous terms for mitochondria such as chondriochonds, chondriomes, mitosomes, chondriosomes and so on. It was Kolliker who observed granule-like structures in the muscle cells of insects in the year 1880. Flemming (1882) named them *fila* and later on Altmann (1890) named them *bioplasts.* In 1897, Benda demonstrated similar objects in cells and assigned the name *mitochondria* to them. Lewis and Lewis (1914) demonstrated the possibility that mitochondria are concerned with some metabolic activity of the cell and Hogeboom (1948) has shown that the mitochondria are the chief sites for the cellular respiration. The presence of mitochondria in the plant cell was first detected by F. Meves in 1904 in *Nymphaea.*

The mitochondria are virtually present in all the aerobic cells. They are absent in bacteria, other prokaryotic cells and mature red blood cells of the multicellular organisms.

The mitochondria are bounded by two layers of unit membranes. The inner layer is invaginated to form finger-like, plate-like or sac-like plates in the lumen of mitochondrion. These folds or plates are called cristae. The space between the outer and inner membranes and the central space is filled with viscous mitochondrical matrix which contains oxidative enzymes and coenzymes.

The mitochondria are colourless bodies widely distributed in the ground substance of the cytoplasm. These are easily differentiated from other cytoplasmic components by staining process. They are selectively stained by a special stain *Janus green.* The mitochondria are of different shapes. They may be fibrillar, spherical, rod shaped and oval and they may change from one form to another depending upon the physiological conditions of the cells. Mitochondria measure usually from 0.5 μ to 1.0 μ in diam. and reach a length up to 40 μ. They contain numerous enzymes which take part in the oxidative steps of Krebs' cycle in respiratory process. The high energy phosphate compounds such as Adenosine diphosphate (ADP) and Adenosine triphosphate (ATP) are also synthesized and stored in the mitochondria. These phosphate compounds after breakdown liberate tremendous amount of energy which is required in the completion of many chemical processes of the cytoplasm. This is why mitochondria are regarded as "power house" of the cell. They are the principal but not the only sites of oxidation since the oxidation of some compounds also takes place in the ground substance of cytoplasm with the help of enzymes present therein (Fig. 1.19).

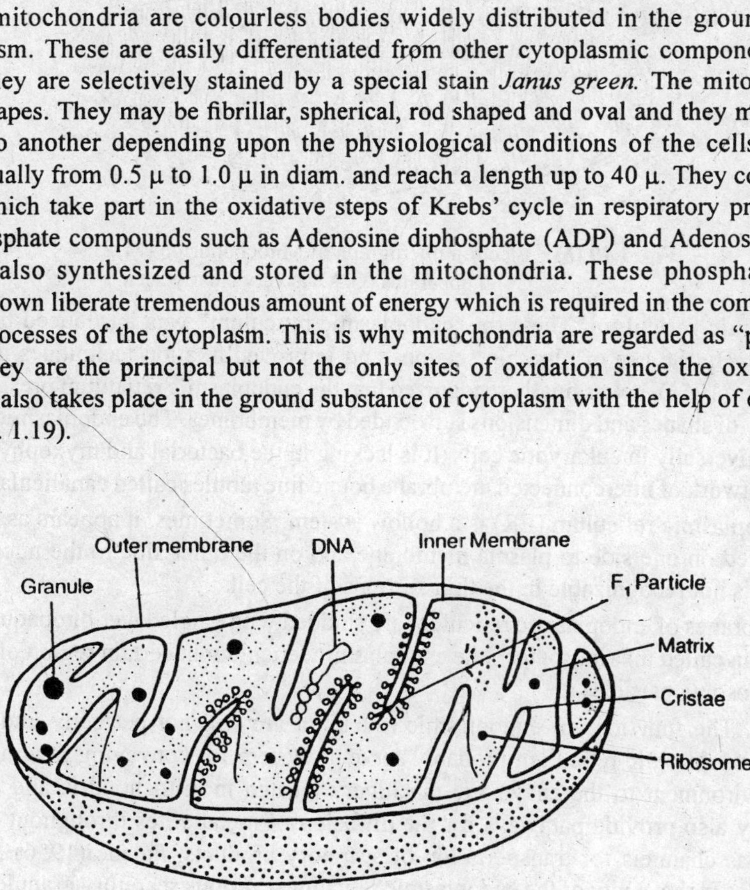

Fig. 1.19 (*a*) Mitochondrion cut open to show the inner structures.

Golgi complex. This cytoplasmic organelle is named after its discoverer Golgi. The structure was discovered in 1898. The golgi bodies are also called lipochondria. For several years there was considerable disagreement about the existence of that organelle. Most of the early biologists believed that it was an artifact of fixation or staining procedures. Studies with phase contrast microscope in the early 1940s also indicated the existence of golgi bodies. The study of electron micrographs of thin sections of cells in 1950s finally proved beyond doubt the existence of golgi bodies in all the cells of eukaryotes. The golgi apparatus does not exist in the prokaryotes.

The electron microscopic studies have revealed that this organelle consists of a series of compactly grouped smooth contoured membrane limited vesicles of variable shapes and dimensions and variable number of small vacuoles [Fig. 1.19 (*b*)]. The membranes of golgi complex are of lipoproteins. The functions of golgi complex are storage of proteins and enzymes and secretion of many important materials including cell wall materials.

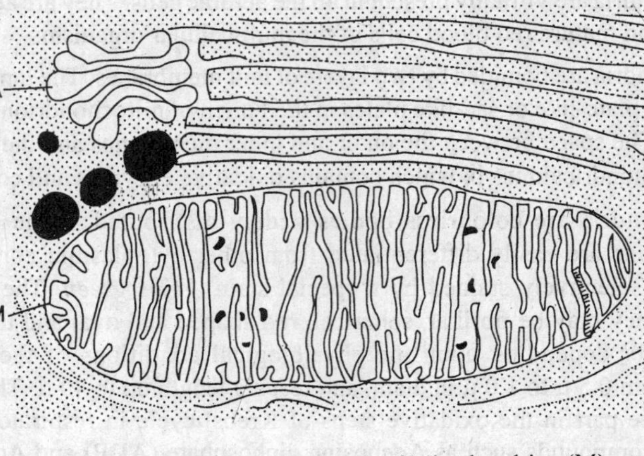

Fig. 1.19 (*b*) Electron micrograph of Mitochondrion (M). Golgi apparatus (G.A.) is very clearly seen.

Endoplasmic reticulum. The term "endoplasmic reticulum" was introduced by Porter and Kallman (1952). By the use of ultrathin sections and improved fixation techniques developed by Palade and Porter (1954), it was finally recognized that the endoplasmic reticulum presented cavities of a great variety of shapes and dimensions surrounded by membranes. The endoplasmic reticulum is found almost universally in eukaryotic cells. It is lacking in the bacterial and myxophycean cells. It is a system or network of interconnected membrane bound fine tubules called canaliculae (Fig. 1.20). Obviously, endoplasmic reticulum (ER) is a hollow system. Sometimes, it appears as a continuous system, connected on one side to plasma membrane and on the other side to the nuclear envelop. This continuity is not recognizable in the thin sections of the cell.

The membranes of endoplasmic reticulum are studded with single layer of opaque particles of ribonucleoprotein called the ribosomes. The endoplasmic reticulum in certain region of the cell may be devoid of ribosome particles.

Function. The functions of endoplasmic reticulum are still not perfectly understood. The canaliculae of endoplasmic reticulum probably serve as system for conveying raw materials from cytoplasmic environment to the enzymatic machinery located in mitochondria and elsewhere in the cell and they also provide pathways for the diffusion of metabolites throughout the cell. The canaliculae act as channels for transportation of secretory products. Palade (1956) has observed secretory granules in the cavity of the endoplasmic reticulum. Various secretory granules of granular endoplasmic reticulum are transported to other organelles as follows:

Granular ER → agranular ER → cisternae of golgibody → secretory granules

The endoplarmic reticulum acts as ultrastructural skeletal framework in the cell. The endoplasmic reticulum provides increased surface for the various enzymatic activities. It contains many enzymes which perform various metabolic reactions.

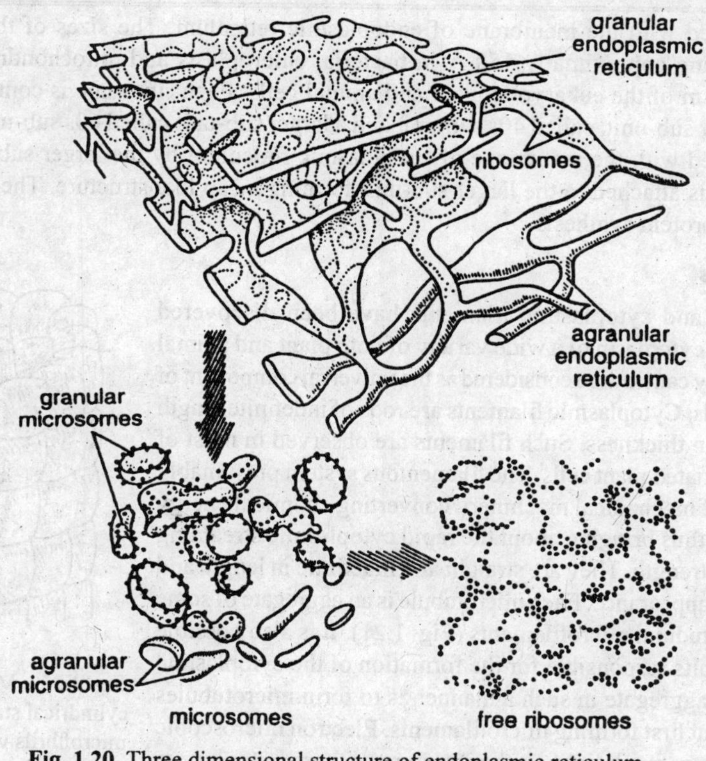

Fig. 1.20 Three dimensional structure of endoplasmic reticulum, microsomes and ribosomes.

Microsomes. The microsomes are special type of organelles in the cytoplasm which are described as small vesicles bounded by thin surface membranes of lipoproteins impregnated with small ribosome particles. These are supposed to have developed when canaliculae of the ergastoplasm are broken into small spheres. The microsomes lie scattered in the cytoplasm. These RNA rich bodies are the chief sites for protein synthesis and their membranes are involved in the steroid synthesis.

Lysosomes, Microbodies and Peroxysomes. In the cytoplasmic matrix of animal cells are found variously shaped bodies usually bounded by a single surface membrane and containing hydrolytic enzymes. These are called *lysosomes* (Fig. 1.19). Lysosomes were first reported by de Duve in 1955. In 1964, P.Matile demonstrated the occurrence of lysosomes in the fungus *Neurospora.* The lysosomes originate from golgi complex. The lysosomes are lytic in nature and are involved in the digestion of intracellular substances. The function of lysosome membrane is to separate the hydrolytic enzymes from other parts of the cell thus protecting the cell from self-digestion. When the cells become dead the lysosomes release their enzymes which rapidly digest the cell. The lysosomes of plant cells are membrane bound storage granules containing a variety of hydrolytic enzymes and they comprise of sphaerosomes, aleurone grains and vacuoles.

Microbodies. In the ground plasm of many cells are found spherical or ovoid membrane bound bodies of variable size, 0.2–1.5 mμ in diameter which are called *microbodies.* These bodies are surrounded by single unit membrane and contain dense or crystalline materials, the matrix. Recently they have been found to contain enzymes particularly catalases, oxidases and enzymes for hydrogen peroxide metabolism.

Ribosomes. In the cytoplasm of cell there occur particles composed of ribonucleic acid (RNA) and proteins. These are called *Ribosomes.* These particles occur freely in the cytoplasm and also remain attached with the membrane of endoplasmic reticulum. The sizes of the ribosomes vary somewhat, being approximatly 150 Å in bacteria, chloroplasts and mitochondria and 140-200 Å in the cytoplasm of the eukaryotic plant and animal cells. Each ribosome is composed of two sub-units; a smaller sub-unit, called 40s sub-unit and a larger sub-unit called 60s sub-unit. The ribosomes remain attached with the membrane of endoplasmic reticulum by the larger sub-unit. The smaller subunit remains attached to the larger sub-unit forming a cap like structure. The ribosomes are the chief sites of protein synthesis.

Microtubules

Microtubules and cytoplasmic filaments have been discovered recently in the cytoplasm of a wide variety of both plant and animal cells and so they can now be considered as the universal component of eukaryotic cells. Cytoplasmic filaments are rods of indefinite length and 40–50 Å in thickness. Such filaments are observed in most of the undifferentiated plant cells. The filamentous system presumably forms a part of mechanical machinery converting chemical energy into work and thus bringing about the rapid cytoplasmic streaming and provides strength. They are straight and indefinite in length and have a hollow appearance. Each microtubule is an aggregate of some 10 to 14 longitudinal microfilaments (Fig. 1.21). It is also possible that the sub-units responsible for the formation of the cytoplasmic filaments can aggregate in such a manner as to form microtubules directly without first forming microfilaments. Electron microscopic study has thrown much light on the structure of these cytoplasmic tubules. They are different from other tubular structures, such as endoplasmic reticulum, golgi bodies. Microtubules are about 200 Å to 270 Å in external diameter having an electron dense wall of some 50 to 70 Å thickness. They are found in rotating condition (R. Jarosch, 1956 and K. Kuroda). They are present in nuclear spindle, the kinetosomes and the cilia.

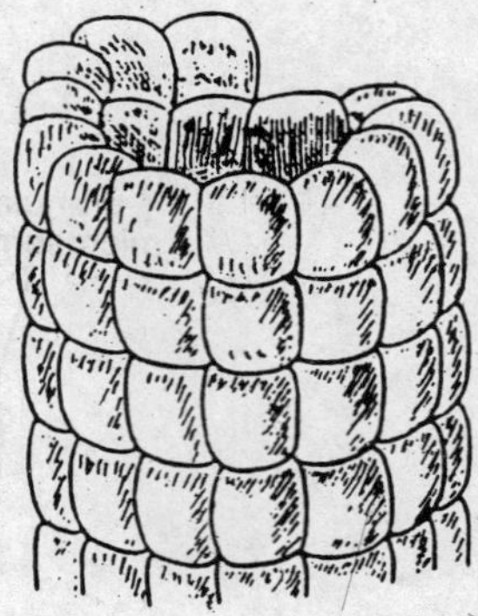

Fig. 1.21 Microtubule, a hollow cylindrical structure formed by 13 microfibrils which themselves are linear arrays of globular proteins.

Thus, like filaments, microtubules seem to be involved with the machinery of motion. These tubules also play important role in maintaining the shape of the cell. Thus like filaments, microtubules form the part of cytoskeleton in the cytoplasmic matrix. The other functions of these tubules are yet to be discovered.

Centrosome

In the cells of some lower plants and all animals there occurs radiating structure in the cytoplasm very close to the nucleus. That is termed *centrosome*. The term *centrosome* was coined by T. Boveri in 1888. It consists of radiating system, the aster or astral rays, and a pair of granules, the centrioles or diplosomes (Fig. 1.19). The centrioles are independent self-duplicating bodies, 300—500 mμ in length and 150 mμ to 160 mμ in diameter. They form 90° angle with

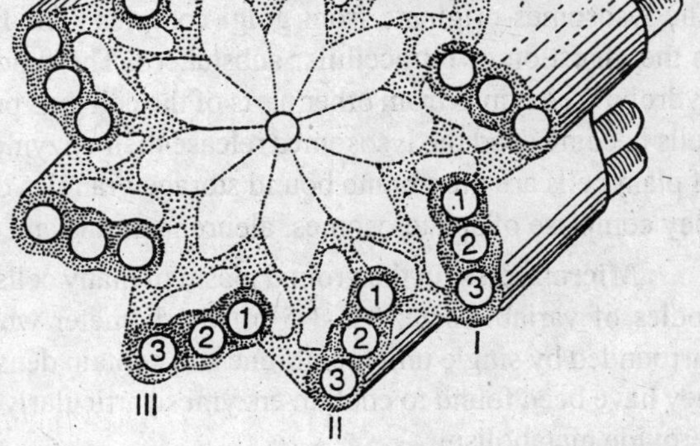

Fig. 1.22 A cross-section of a centriole showing 9 triplets. Where 1, 2, 3 represent sub-fibrils A, B and C respectively.

each other. The ultrastructure of a cell reveals that the centriole is a small barrel shaped cylinder made up of evenly spaced nine longitudinally oriented triplets of rodlets or fibrils embedded in dense amorphous matrix. One end of centriole appears to be closed while the other end is open. The innermost unit of each triplet is designated subfibrils *A* and the other two subfibril *B* and *C*. All the three subfibrils have a dense rim and a less dense centre giving them hollow appearance. In fact, it has not been established whether they are tubules or solid fibres with a dense cortex and light interior. Subfibrils *A* of the nine triplets are uniformly spaced on the circumference of a circle about 150 mμ in diameter and each triplet is inclined so that a line through the centres of its sub-units make an angle of about 30° with a tangent to this circle at the midpoint of subfibril *A*. The arrangement of fibrils in the wall of centriole thus resembles the set of the propellants of a pyrotechnic pinwheel. Subfibril *A* is provided with two short diverging arms. One of the two arms, directed inward along a radius, has free end pointing toward the centre of the centriole and the other arm, directed outward connects *A* with subfibril *C* of the next triplet. The successive triplets are thus linked together, subfibril *A* to *C*, around the circumference of centriole by a series of slender linear densities (Fig. 1.22).

The interior of the centriole is usually filled with a homogeneous cytoplasm of low density but may contain one or more small dense granules. Sometimes fibrous arms or microtubules also radiate from them. In some cell types two or more pericentriolar structures or satellites are deployed around the centriole. These take various forms.

Recently, the presence of DNA and RNA has been noticed in centrioles. Probably the information for the synthesis of centrioles is obtained from a tiny DNA containing unit with weight of about 2×10^{-16} gm.

Function. Centrosome performs mechanical function in the cell. Before the onset of nuclear division the centrosome divides into two. Two centrosomes migrate to the two opposite poles of the nucleus where they are involved in the formation of mitotic apparatus and direct the separation of chromosomes during nuclear division.

Flagella and Cilia

The cilia and flagella are rapidly beating and contractile filamentous processes emerging out from the cytoplasm. Recently they have been examined with eletron microscope and it has been found that cilia and flagella throughout the plant and animal kingdoms show identical structures. Cilia have been defined as long cylindrical processes tapered at the tip and composed of an axial filament complex embedded in a matrix and enclosed in a ciliary membrane which is continuous at the base with cell membrane proper. Though there is no fundamental structural difference between cilia and flagella yet both can be distinguished from each other by the following features:

(*i*) The flagella are few in number in each cell but the cilia are numerous in number per cell.

(*ii*) The flagella are long in proportion to the size of the cell whereas the cilia are small. The shaft of the cilium is 0.2–0.25 μ in diameter and 5 to 10 μ long. Flagella range from this length upward to 150 μ or more.

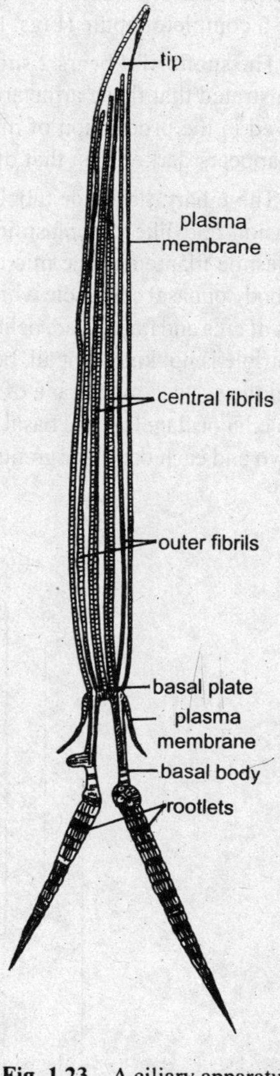

Fig. 1.23 A ciliary apparatus.

(*iii*) The flagella beat independently and exhibit undulatory motion while the cilia tend to beat in coordinated rhythms and move in a sweeping or pendular stroke.

The flagella and cilia have 'axial-fibre-complex' consisting of a constant number of internal fibrils or microtubules, that is 11, out of which two are in the centre and nine evenly spaced microfibrils remain arranged around the two central fibrils (9 + 2). These eleven fibrils remain embedded in the matrix of liquid consistency and enclosed in a unit membrane of 90 Å thickness (Fig. 1.23). The two central fibrils of the axial-fibre complex are in singlet state. They are approximately circular in cross- section, 240 Å in diameter and about 360 Å from centre to centre. They are enveloped by central sheath which shows spiral composition in longitudinal section (Fig. 1.24). They have tubular appearance having denser outer region and light central core. The nine peripheral fibrils differ from the two central fibrils in being doublets (*i.e.*, each composed of two subfibrils). They measure approximately 38 Å by 260 Å in cross-section. One subfibril of each doublet is designated as subfibril *A* and the other as subfibril *B*. The subfibril *A* has a circular profile while the subfibril *B* is crescentric and shares a sector of the wall of subfibril *A*. The curving septum between the two is thus a part of the wall of subfibril *A* and is convex toward subfibril *B*. Subfibril *B* independent of subfibril *A* would not be a complete tubule (Figs. 1.24 and 1.25).

The subfibril *A* bears 2 short arms that project toward subfibril *B* of the next doublet. Gibbons demonstrated that these arms are composed of proteins which he named dynein, that appears to be involved in the production of mechanical work. The interior of subfibril *A* in the cilia of some cell types appears darker than that of subfibril *B*.

The ciliary fibrils or tubules and the cytoplasmic microtubules are similar in appearance and dimension and like cytoplasmic microtubules ciliary microtubules are composed of some 10–14 cytoplasmic filaments. The microtubular complex in both cilia and flagella terminates at its base in a basal body or basal corpuscle which is a hollow cylinder with the same structure as a centriole. The basal bodies of cilia and flagella are believed to arise by reduplication of centrioles. The manner of reduplication of centriole is not known at all, but like the reduplication manner of chromosomes, the centrioles move toward the surface membrane of the cell and form the basal bodies or kinetosomes which in turn give rise to cilia or flagella. The basal granules situated at the base of cilia or flagella are also able to divide into two and each daughter granule is able to produce a new cilium or flagellum as the case may be.

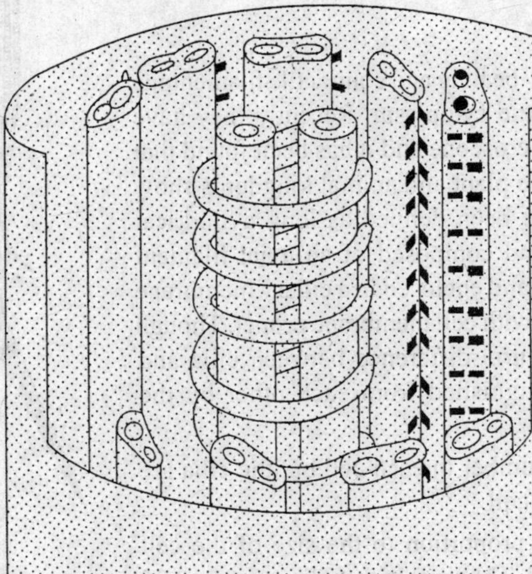

Fig. 1.24 A piece of flagellum cut open at different levels.

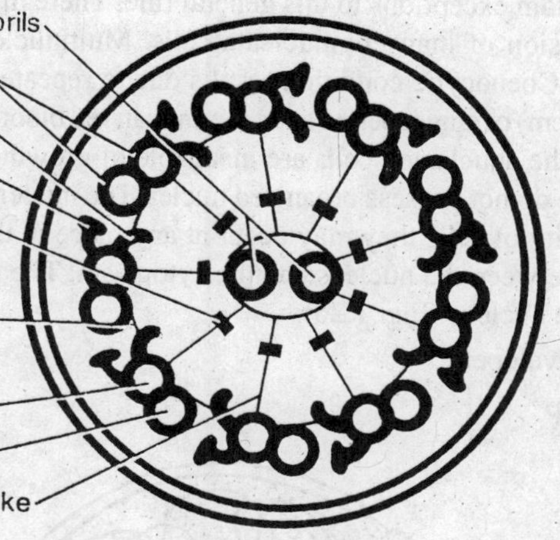

Fig. 1.25 A cross-section of cilium.

The architecture of the cilium is intimately related to the ciliary beat. The direction of ciliary beat is perpendicular to a line joining the two central tubules.

Ergastic Substances or Cytoplasmic Inclusions

In the ground substance of the cytoplasm are seen many non-living bodies called cytoplasmic inclusions. These substances are formed as a result of metabolism and are accumulated in the form of granules or crystals as for example, calcium carbonate, calcium oxalate, starch grains, proteins, pigments, tannin, resins and oil drops.

Vacuole

In the ground substance of the cytoplasm may be seen one or more vacuoles in addition to the living components. It is a debatable matter whether a vacuole should be treated as a cell organelle. The vacuoles have vacuolar membranes or tonoplasts and are filled with vacuolar sap. It is still an open question whether or not vacuoles are delimited by a membrane or a cytoplasmic interface. Some vacuoles seem to represent smooth walled vesicles formed by endoplasmic reticulum and have fibrillar extensions, *i.e.*, the endoplasmic reticulum contributes to the membranes of vacuoles. This view was questioned on the ground that vacuoles lack any continuity with either endoplasmic reticulum or nuclear membrane. The second school of thought holds that the vacuoles are derived from golgi apparatus. The vacuolar development in the cells of barley shoot apex supports the second view (Marinos, 1963). The vacuoles gain in volume by simple fusion of smaller ones.

Mollenhauer and his associates (1961) on the other hand suggest a different type of vacuolar development in the cells of root cap. The golgi apparatus of these cells has been found to produce small vesicles which serve to transport the material to the peripheri of the cell. The material of the vacuole after disposition goes into the composition of cell wall while the vacuolar membrane fuses with plasma membrane. The vacuoles are of common occurrence in the cells of both plants and animals. In some primitive unicellular organisms the vacuoles contain food particles, hence they are called food vacuoles. These cavities have got special property of contraction and expansion. They pump the excess amount of water and waste products out of cytoplasm and thus maintain the definite internal pressure in the cell. The liquid portion of the vacuole, the so called vacuolar sap, is never a living substance. It is a watery solution of soluble wastes including pigments, sugars, uric acid etc.

NUCLEUS
(The controlling centre of cell)

The protoplast of eukaryotic cell contains prominent spherical or subspherical body called the nucleus (Pl.—nuclei) which controls all the life activities. As a rule, there is only one nucleus in one cell, but there are certain exceptions to this general rule. There may be many nuclei in **syncytium** which is formed by fusion of many uninucleate cells. Multinucleate cells (**coenocytes**) are commonly found in plants. Coenocytic condition results due to repeated nuclear division without cytokinesis. Sieve cells (phloem) of angiosperms and mammalian red blood cells lose their nuclei during the process of maturation. The anucleated cells are incapable of growth and division. The cells of bacteria and blue-green algae do not possess organised nuclei. The nuclei in some cells may be seen in the centre but in the majority of cells they may occur at any place in the cytoplasm. There is but little optical differentiation between the nucleus and the cytoplasm. The following are the principal components of the interphase nucleus (Fig. 1.26):

(*a*) nuclear envelope

(*b*) nuclear sap

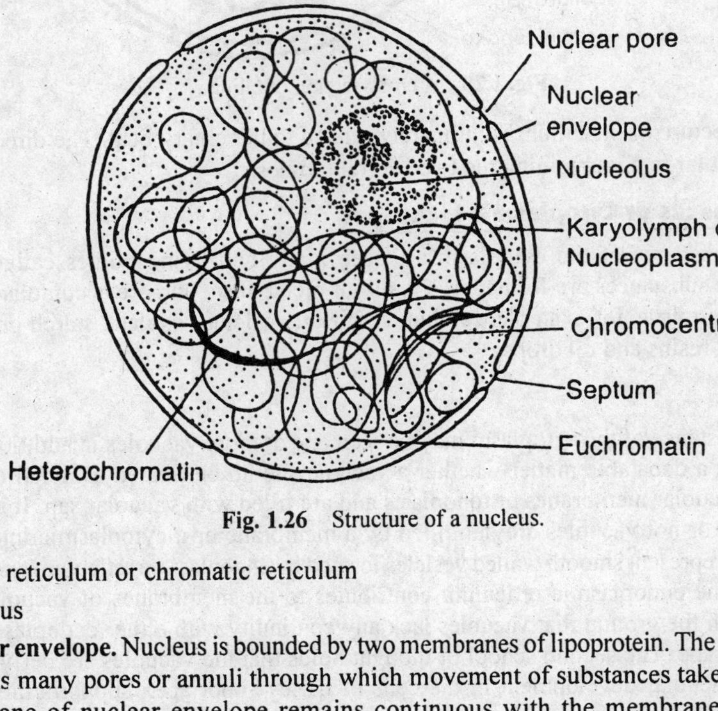

Fig. 1.26 Structure of a nucleus.

(*c*) nuclear reticulum or chromatic reticulum

(*d*) nucleolus

(*a*) **Nuclear envelope.** Nucleus is bounded by two membranes of lipoprotein. The nuclear membrane contains many pores or annuli through which movement of substances takes place. The outer membrane of nuclear envelope remains continuous with the membranes of endoplasmic reticulum.

(*b*) **Nuclear sap.** The space bounded by nuclear envelope is filled by the groundplasm, known as *nucleoplasm, nuclear sap* or *karyolymph.* The nucleoplasm contains dissolved sugars, proteins, nucleotides, nucleic acids, energy rich compounds and a variety of enzymes.

(*c*) **Nuclear reticulum or chromatic reticulum.** The nucleoplasm contains thread like elongated structures of chromatin called chromatin fibres. The chromatin fibres later on differentiate to form thread-like structures of definite shapes and sizes which are called chromosomes. The chromatin fibres in the nucleus of a growing cell form a network type structure that is called chromatic reticulum. Chromatin is fundamentally nucleoprotein complex (Nucleic acids and proteins).

(*d*) **Nucleolus.** In the nucleoplasm there occurs a conspicuous body of spherical shape attached to a particular chromosome on a definite location. This is called nucleolus. Nucleolus is composed of a large amount of ribosomal proteins and ribosomal RNA and thus provides the raw materials for the biogenesis of ribosomes.

Functions of different parts of the cells

The different components of the cell perform various biological functions which are summarized in Table 1.2.

Table 1.2: Functions of Different Components of a Cell

Cell Component	*Functions*
Cell wall	(*i*) It protects the protoplast from injuries.
	(*ii*) Provides definite shape to the cell.
	(*iii*) It provides rigidity and strength to the cell.
Plasma membrane	It is a differentially permeable membrane through which translocation of materials occurs.
Cytoplasm	It is the executive centre of the cellular functions.
Hyaloplasm or Matrix	It contains substrates and enzymes for the biochemical processes.
Cytoplasmic organelles:	
(*a*) *Plastids*. (in plant cells only)	(*i*) Photosynthesis
	(*ii*) Storage of pigments, starch and some other cellular products.
(*b*) *Mitochondria*	(*i*) Completion of Krebs' cycle.
	(*ii*) Synthesis of ATP—the energy rich compound.
	(*iii*) Electron transport chain.
	(*iv*) β-oxidation of fatty acids.
(*c*) *Golgi complex*	(*i*) storage of synthetic proteins and enzymes.
	(*ii*) Secretion of certain substances.
	(*iii*) Formation of excretory vacuoles and lysosomes.
(*d*) *Endoplasmic reticulum*	(*i*) Acts as ultrastructural skeletal framework in the cell
	(*ii*) Provides increased surface for the various enzymatic activities which normally occur at or across membrane surface.
	(*iii*) Transports extracellular and intracellular molecules.
	(*iv*) Forms nuclear envelope at telophase.
(*e*) *Microsomes*	The ribosome particles attached to the membranes of endoplasmic reticulum and microsomes are the main sites for protein synthesis in the cell.
(*f*) *Ribosomes*	These are the sites for protein synthesis in the cell.
(*g*) *Lysosome*	(*i*) Acts as the store house for a number of hydrolysing enzymes which help in the digestion of intracellular particles.
	(*ii*) Destroys bacteria and other foreign bodies inside the cell.
(*h*) *Microbodies*	These contain enzymes for hydrogen peroxide metabolism, purine metabolism, gluconeogenesis and photorespiration.
(*i*) *Microtubules*	(*i*) Transport ions, water and small molecules of various substances.
	(*ii*) Form structural units of centrioles, basal granules, cilia and flagella and mitotic spindles.
	(*iii*) Help in cytoplasmic movement.
(*j*) *Centrosome*	The centrioles form the spindle and have significant role in the movement of chromosomes during nuclear division.
(*k*) *Flagella and Cilia*	These are extracellular projections which help in the movement of cell.
Vacuoles	(*i*) Act as osmoregulatory structures in the cell.
	(*ii*) Act as storage depots for excess water, waste products, pigments, etc.
	(*iii*) Act as excretory structures.

Cell Component	Functions
Nucleus	It is the controlling centre for all vital functions of the cell.
(*a*) *Nuclear envelope*	It provides active continuity of ground plasm between nucleus and cytoplasm.
(*b*) *Nucleoplasm*	It contains materials for the synthesis of nucleic acids.
(*c*) *Chromatic reticulum*	These are the bearers of hereditary instructions and are concerned with the regulations of cellular processes.
(*d*) *Nucleolus*	(*i*) It disappears during nuclear division and forms matrix sheath around the chromosomes.
	(*ii*) Acts as a store house for ribosomal proteins and ribosomal RNA.

Difference between Plant Cell and Animal Cell

Plant cell	Animal cell
1. Presence of cellulosic cell wall.	No cell wall of cellulose present.
2. Presence of large vacuoles.	Vacuoles in animal cells are small.
3. Plastids present in most cells.	Plastids absent in animal cells excepting some protozoa (*Euglenae*).
4. Centrioles absent in the plant cells excepting a few algae.	Centrioles are common in animal cells.

From the detailed study of different components of protoplasm it is evident that each structure in the cells has to perform special vital activity. Thus the cell is actually an exceedingly complex factory in which a large number of chemical processes are going on simultaneously. If the cytoplasm by virtue of its different organelles is regarded as executive centre for metabolic processes, the nucleus by virtue of its genes can be regarded as the regulating or controlling centre of the cell.

QUESTIONS

1. What do you understand by a cell? Give well labelled diagram of a plant cell based on electron micrograph.
2. Justify the statement: "The use of electron microscope has helped in a better understanding of the structure of the cell".
3. Describe the structure of Plastids.
4. Describe the structure of mitochondria.
5. Give an account of the structure of Plasma membrane and explain how the adjacent cells in multicellular plants are held together.
6. Give an account of the structure of endoplasmic reticulum, microsome and golgibody.
7. Give an account of the structure of the nucleus and discuss the functions of different components of nucleus.
8. Give an account of the morphology of chromosomes and discuss that in all the cellular organelles the chromosome is the structure most likely to serve as genetic material.
9. Write short notes on:
   (*a*) Prokaryotic nucleoids
   (*b*) Nucleolus
   (*c*) Nuclear membrane
   (*d*) Chromatids
10. Discuss the structure of centrosome, flagella and cilia in the light of recent electron microscopic studies and point out the differences between flagella and cilia.

2 CHAPTER Cell Membrane

The protoplasm of the cell is bounded by an exceedingly thin elastic, semi-permeable living limiting membrane called the ***plasmalemma*** (J.Q. Plowe, 1931) or ***Plasma membrane*** or ***Cytoplasmic membrane*** or ***cell membrane*** (Nageli and C. Cramer, 1855). A plasma membrane forms the outer limiting boundary of most animal cells. In plant cell and bacterial cell the plasma membrane lies between the cell wall and cytoplasm. Many botanists prefer the term plasmalemma to plasma membrane. The earlier biologists doubted its appearance. Existence of plasma membrane was proved when phenol-red dye was injected into the cytoplasm. In the experiment it was found that the dye coloured the whole of cytoplasm and did not pass out of the cell. Further when some colourless cells were placed in the dye, only the cell wall was coloured and not the cytoplasm. These observations suggested the existence of a boundary layer between the outer non-living cell wall and internal cytoplasm. W. Pfeffer (1890) established that the plasma-membrane could not be differentiated from the inner plasma with the help of light microscope. It can, however, be easily studied under electron microscope due to better resolution and it appears as a *unit membrane* of 75 to 100 Å thickness. Now it is an accepted view that all the unit membranes have the same structure. The unit membrane forms not only the plasma membrane, but also the limiting layer of most of the cell organelles. Such a view was initially put forth by Robertson in 1960.

In broad sense, the term cell membrane refers to the limiting membrane of the cell organelles like plastid, mitochondria, lysosomes and the other membranous systems like endoplasmic reticulum, Golgicomplex and nuclear envelopes. The structure of plasma membrane has not been worked out in every type of cell because of the problem of technique for isolation. The electron microscope has been quite helpful in study of cell membranes.

The detailed structure of plasma membrane could be studied only after the improvement of micro-technique for isolation of pure and intact plasma membrane from the living cell and their subsequent electron microscopy. The studies made by E. Gorter and F. Grendel (1925) on mammalian red blood cells and the myelin shealth of the nerve fibre have provided bulk of information regarding the structure and functions of the plasma membrane. The recent observations on the plasma membranes of many other cell types are similar to the findings on the membranes of these two cells and hence applicable to all others cell types. The red blood cells are treated to remove all protoplasmic contents except plasma membrane. The residue of such treatment is called a 'red cell ghost'. Mitchell (1959) isolated and analysed the plasma membrane of a bacterium by removing the cell wall through enzyme treatment and then by bursting the cell in hypotonic solution.

Ponder and others have raised objections on the grounds that there is no direct evidence for the ghost to be a true likeness of plasma membrane of living cells. Some workers have suggested that the ghost is simply an artifact of technique involved in the isolation of plasma membrane. Nevertheless, the studies on red blood cells, nerve cells and some other cells have provided basic information about the plasma membrane.

CHEMICAL COMPOSITION OF PLASMA MEMBRANE

The chemical composition of plasma membranes differs in different cells. The membranes are composed mainly of proteins, phospholipids and a small percentage (1–5%) of oligosaccharides. In general, plasma membranes contain about 60% protein and 40% lipids by dry weight but there is wide variation in the protein lipid ratio of different cell membranes. Oligosaccharides may be found attached to either lipids (glycolipids) or proteins (glycoproteins). Carbohydrates usually form 1–10% of the total dry weight. In the membranes of some cell organelles the presence of nucleic acids, DNA and RNA has also been reported but they are presumed to be contaminations.

Proteins of the Plasma Membranes

Plasma membranes contain three different classes of protein : structural proteins, enzymes and carrier proteins.

The structural proteins form the mechanical structure of the membrane. They have little catalytic activity and are extremely hydrophilic in nature. The plasma membrane consists largely of structural proteins, the average molecular weight being 3×10^4. The structural proteins of plasma membrane have been classified into two main groups:

1. The peripheral or extrinsic proteins.
2. Intrinsic or integral proteins.

1. ***The peripheral proteins*** are weakly associated with the membrane surface and may be separated by mild treatments. They are soluble in aqueous solutions and are usually free from phospholipids. The most common examples of peripheral proteins are spectrin found in erythrocytes, ATPase and Cytochrome c found in mitochondria and acetylcholinesterase, etc.
2. ***Integral or Intrinsic proteins*** are associated strongly with the membrane and represent more than 70% of the structural proteins. They are usually associated with the lipid extending all the way through it and require drastic procedures for isolation such as treatment with hydrophobic bond breaking agents like detergents, organic solvents, chaotropic agents, etc. They are usually insoluble in aqueous solution. The most common examples of integral proteins are rhodopsin found in rod cells of retina and cytochrome oxidase found in mitochondrial membrane.

Enzymes of Plasma Membranes

Enzymes form the major part or component of many membranes and are catalytic proteins. The membranes of endoplasmic reliculum, mitochondria, and other membranes contain many enzymes. The structure of enzymes varies from membrane to membrane. About 30 enzymes have been reported from the plasma membrane, the most common being 5–nucleotidase, Mg^{++} ATPase, Na^{+}–K^{+} activated Mg^{++} ATPase, alkaline phosphatase, adenylcyclase, acid phosphomonoesterase and RNAase. Some of the enzymes have specific localizations, *e.g.*, alkaline phosphatase are more abundant in the bile capillaries while disaccharides are present in the microvilli of intestine.

Carrier Proteins

Carrier proteins are permeases which transport substances across the membrane against concentration gradient. The carrier proteins are more or less similar to structural proteins in molecular weight.

Lipid fraction of Plasma Membrane

The lipid fraction constitutes about 20–79% of the plasma membrane. The main lipid constituents of cell membrane are phospholipids, cholesterol and galactolipids in variable proportions in different cells. Phospholipid components are of two kinds:

(*i*) ***Neutral phospholipids,*** such as phosphatidicholine, phosphatidylethanolamine and sphingomyline which tend to pack tightly in two layers.

(*ii*) ***Acidic phospholipids,*** such as phosphatidyl inocetol, phosphatidyl serine, cardiolipid, phosphatidyl glycerol, sulpholipids which are negatively charged and associated with proteins by way of lipid protein interactions (Maddy, 1966).

The lipids of plasma membrane are polar lipids which contain hydrophilic heads (glycerol) and two hydrophobic tails (fatty acids). The hydrophobic and hydrophilic regions may be bridged by glycerol moiety, a sphinganine derivative or homologue of a sterol. Plasma membranes of different cells consist of a variety of fatty acids in their lipid layer, *e.g.*, red blood cells of mammal contain fatty acids such as palmitic, stearic, oleic, linoleic and arachidonic acids.

Carbohydrate fraction of Plasma Membranes

According to Bell (1962) the polysaccharides are present in the outer protein layer of the plasma membrane which provide some stability to the lipoprotein complex. The common oligosaccharides which remain bounded in plasma membrane of red blood cells and liver cells of mammals are hexose, hexosamine, Fucose, sialic acid.

MEMBRANE MODELS

In order to explain physical and biological features of cell membranes many hypotheses have been proposed. These hypotheses have been discussed under the following categories:

(*A*) Lamellar Models.

(*B*) Micellar or Sub-unit Models.

(A) LAMELLAR MODELS

(I) Lipid bilayer Hypothesis

Overton (1895) suggested that the cell membranes contain lipids. This conclusion was based on the fact that fat solvents dissolved the membrane easily and fat soluble substances passed easily through the cell membranes. Some workers like Hober (1910) and Fricke (1925) found low electrical conductivity in the cell membrane which indicated the presence of lipid layer. Gorter and Grendel (1925) were the first to suggest a possible structure of the cell membrane. On the basis of studies of cell membranes of erythrocytes they showed that the lipid extracted from the red cell ghost spread in area twice that of a simple molecular film. So it was thought that the membrane consisted of double layers of lipid molecules, the polar hydrophilic groups of the molecules being situated on the outside and hydrophobic ends standing at right angles to the surface are oriented inwardly (Fig. 2.1).

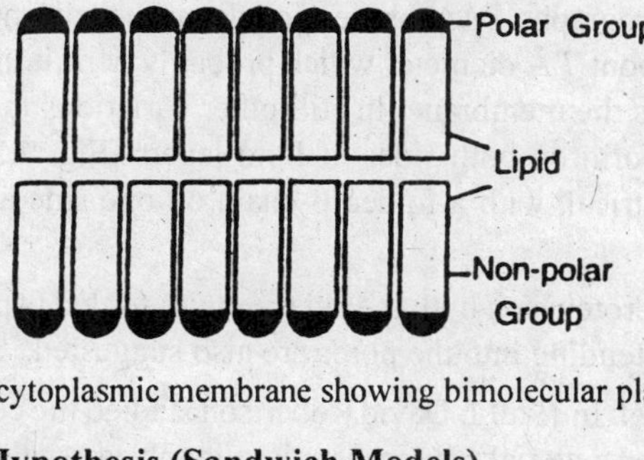

Fig. 2.1 A piece of cytoplasmic membrane showing bimolecular plate of a phospholipid.

(II) Protein-lipid-protein Hypothesis (Sandwich Models)

(*i*) **Danielli and Davson Model.** Harvey and Coley (1931) and Danielli and Harvey (1935) studied surface tension of cell membrane and on the basis of their observation they pointed out the existence of protein molecules adsorbed on the surface of lipid droplets which reduce the surface tension of droplets. This conclusion led James Danielli and Hugh Davson in 1935 to suggest bimolecular leaflet model of cell membrane. Danielli and Davson model was the first attempt to describe membrane structure in terms of molecules and to relate the structure to biological and chemical properties.

According to bimolecular model of Danielli and Davson, plasma membrane consists of two layers of phospholipid molecules (a bimolecular leaflet) in which phospholipid molecules are arranged in such a way that hydrophilic heads of the phospholipid molecules face outside and hydrophobic non-polar lipid chains are associated in the inner region of leaflet. The hypothesis also suggested that the polar ends of lipid molecules are associated with monomolecular layer of globular proteins. The plasma membrane would thus consist of a double layer of phospholipid molecules sandwiched between two essentially continuous layers of protein (Fig. 2.2).

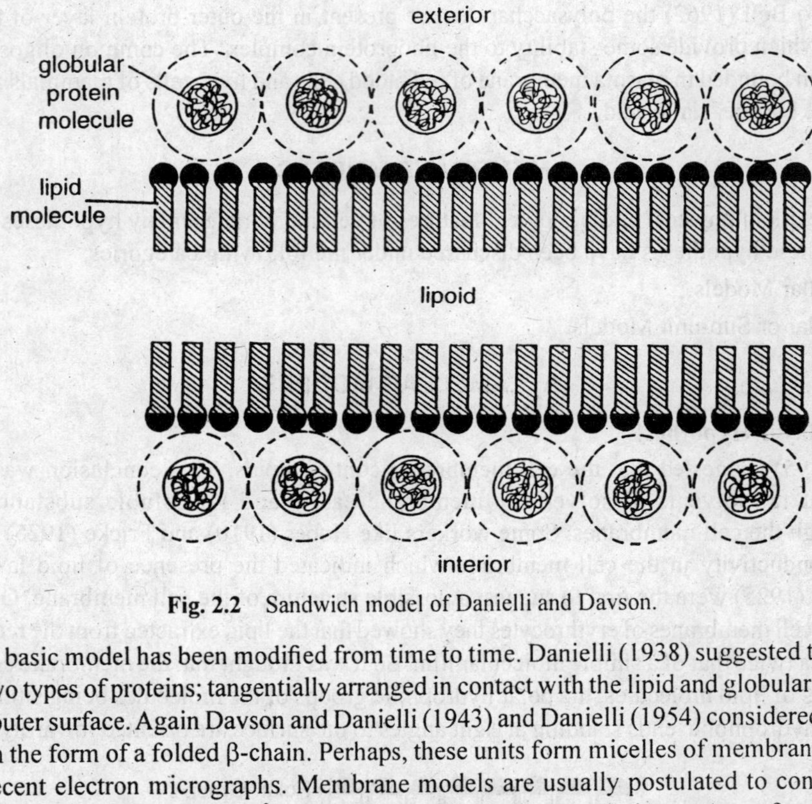

Fig. 2.2 Sandwich model of Danielli and Davson.

This basic model has been modified from time to time. Danielli (1938) suggested the presence of two types of proteins; tangentially arranged in contact with the lipid and globular proteins on the outer surface. Again Davson and Danielli (1943) and Danielli (1954) considered proteins to be in the form of a folded β-chain. Perhaps, these units form micelles of membranes indicated in recent electron micrographs. Membrane models are usually postulated to contain protein lined polar pores of about 7 Å diameter which probably permit the passage of small ions and water molecules across the membrane. In still other variations the proteins are thought to be in coiled or globular form on both sides of lipid layers [Fig. 2.3 (A), (B), (C)] or they are thought to be asymmetrical, with a folded β-chain on one side and globular proteins on the other [Fig. 2.3 (D)].

Models with globular proteins on both the sides or with folded β-proteins on both the surfaces and helical proteins extending into the pores are also suggested.

(*ii*) **Unit Membrane Model.** In 1950 J. David Robertson studied the cell membranes from electron micrographs of sectioned material. The preparations involved usual fixing in solutions of osmium tetraoxide and potassium permanganate ($KMnO_4$), and dehydrating in solvents such as acetone before sectioning. In late 1950s Robertson summarised a large number of ultra-structural data obtained by him and some other workers and concluded that the plasma membrane and the membranes of all cell organalles were similar in structure. Although the similarity is not resolved by light microscopy, it is clearly seen in electron micrographs.

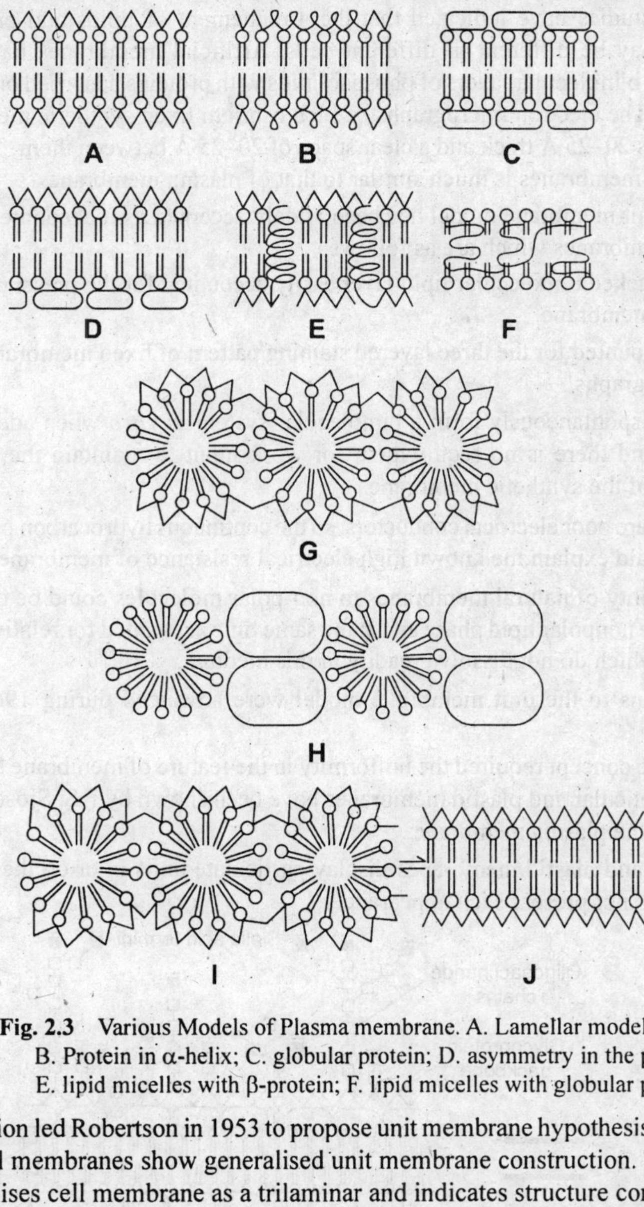

Fig. 2.3 Various Models of Plasma membrane. A. Lamellar model; B. Protein in α-helix; C. globular protein; D. asymmetry in the protein; E. lipid micelles with β-protein; F. lipid micelles with globular protein.

This conclusion led Robertson in 1953 to propose unit membrane hypothesis according to which all biological membranes show generalised unit membrane construction. The unit membrane model visualises cell membrane as a trilaminar and indicates structure consisting of two dark osmiophilic layers separated by a light osmiophilic layer. The physical appearance of this trilaminar model has led to the term **unit membrane**. The unit membrane concept implies a trilaminar appearance with a bimolecular lipid layer between two protein layers (Fig. 2.4). Each dense osmiophilic band is made up of protein (20 Å) and the polar groups of phospholipids (5 Å) and is thus 25 Å thick. The clear Osmiophilic zone 35 Å in thickness is a bimolecular layer of lipids without the polar groups. In other words, the unit membrane is 75 Å thick with a 35 Å thick phospholipid layer between two 20 Å thick protein layers. The plasma membrane surrounding the cell is thicker at the free surfaces of the cell than where it is in contact with other cells.

In unit membrane model the protein layers are assymetrical. On the outer surface it is mucoprotein while on the inner surface it is non-mucoid protein.

Some recent studies have indicated that the arrangement of lipids and proteins in the cell membranes may be different in different cells. Artificial membranes have been prepared consisting of a bimolecular layers of phospholipids with proteins adsorbed on both the surfaces of lipid layer. The electron micrographs of such osmium fixed synthetic membranes revealed two dark bands 20–25 Å thick and a clear space of 20–25 Å between them. The appearance of these artificial membranes is much similar to that of plasma membranes.

Robertson's unit membrane model has been widely accepted as it accounted for a number of properties of cell membranes which are as follows:

1. The densely packed bimolecular lipid layer easily accounted for the presence of 40% lipid by weight in the membrane.
2. The model accounted for the three layered staining pattern of fixed membranes as observed in electron micrographs.
3. Phospholipids spontaneously form a bimolecular system *in vitro* when added to an aqueous environment and there is no requirement for work input to maintain the minimum energy conformation of the synthetic membrane.
4. Hydrocarbons are poor electrical conductors, so the continuous hydrocarbon phase of the natural membrane would explain the known high electrical resistance of membranes.
5. High permeability of natural membranes to non-polar molecules could be explained by their solubility in the nonpolar lipid phase and at the same time accounted for relative impermeability to small ions which do not dissolve readily in this medium.

Some objections to the unit membrane model were advanced during 1960s which are as follows:

(*i*) Unit membrane concept required the uniformity in the feature of membrane but mitichondrial, endoplasmic reticular and plastid membranes have been shown by F.S. Sjostrand (1963) to be different from the plasma membrane.

(*ii*) Mitochondrial and plastid membranes display particulate units in or on them but the plasma membrane does not present such appearance.

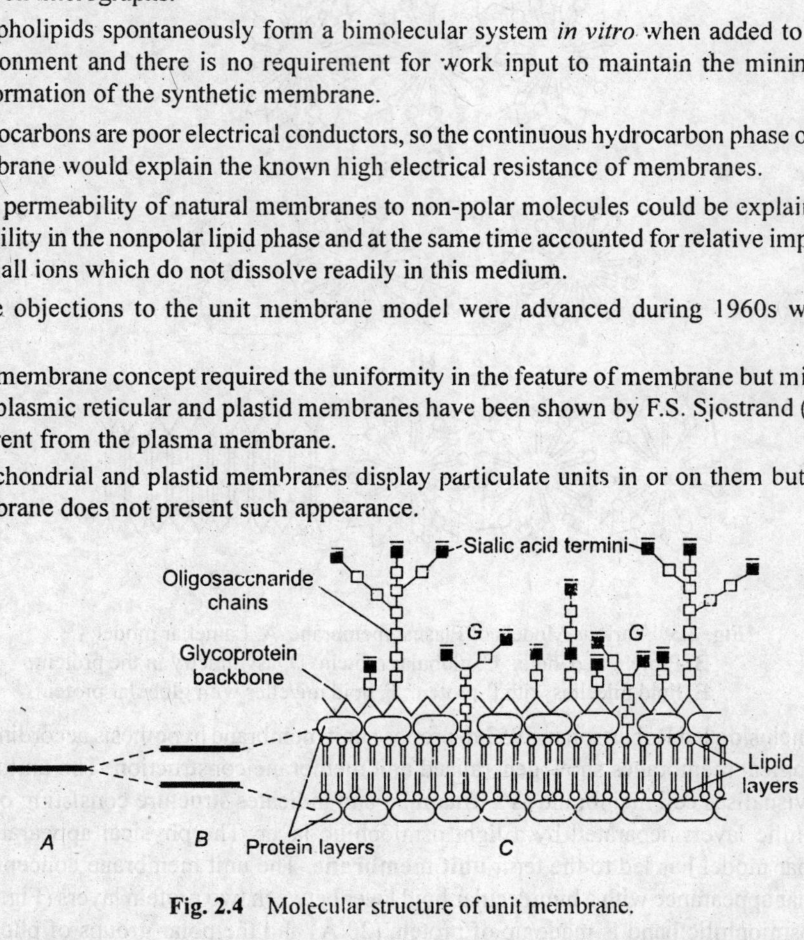

Fig. 2.4 Molecular structure of unit membrane.

(*iii*) The unit membrane appears as a single trilaminar structure in plasma membrane, endoplasmic reticulum, golgi complex and lysosomes but in mitochondria, plastids and nuclei, the boundaries have an outer and an inner membrane, each with unit membrane structural plan.

(*iv*) There can be some variations in the thickness of the unit membranes as a result of varying thickness of osmiophilic and osmiophobic layers.

It is believed that the outer and inner surfaces of membrane are different. The inner surface of membrane is thought to be made up of unconjugated protein and the outer surface layer of conjugated protein and polysaccharides, *mucoproteins*. Attached to the glycoproteins are oligosaccharides-side chains with negatively charged sialic acid terminals. Thus the membrane is asymmetrical in organisation and shows structural polarity (Figs. 2.3 A-F and 2.4). Due to this feature unit membrane model is also called *greater membrane model*.

(III) Kavanau's Lipid Pillar Model

This is a modification of Danielli and Davson trilaminar sandwich model in which lipid layer is visualised in two forms. In one case, it is in the form of pillar while in others it is in the form of flattened discs. The spaces between the pillars act as pores for the passage of ions. The interior of each pillar is formed of non polar faces of phospholipids while the surface is formed of the polar heads. Protein layers are present on both sides of membrane. The main drawback of this model is that it cannot explain active transport or the differential permeability of Na^+ or K^+ ions.

(IV) Models in which proteins are considered to penetrate lipid layers

According to some workers, proteins are considered to penetrate the lipid layer. This model can explain the low surface tension of biological membrane as properly as the trilaminar model.

1. **Benson's Model.** Benson (1966) proposed a cell membrane model on the basis of a study of chloroplast membrane. According to him, the lipids and proteins form a hydrophobic association. The charged polar heads of phospholipid molecules lie on the surface of membrane and are capable of binding ions. The non-polar hydrophobic lipid tails are linked with complementary hydrophobic regions within the interior of proteins.
2. **Lanard and Singers Model.** This model was proposed in 1966. According to this model, one-fourth of the proteins are in helical conformation and the rest form random coils.
3. **Mosaic membrane concept.** According to Baum 1967, "The backbone of membrane is possibly formed by toast shaped cuboidal units of about 80 Å diameter which are covered along the edges as well as the sides by phospholipids". The heads of phospholipids face on the outside while the tails form complexes with the hydrophobic surfaces of the proteins.
4. **Fluid Mosaic Model.** The fluid mosaic model of cell membrane was proposed in 1972 by S.J. Singer and G.L. Nicolson. According to this model, the cell membranes have been visualised as mosaics of lipids and proteins. The lipids are thought to be arranged primarily in a bilayer in which peripheral and integral proteins are embedded to varying degrees (Fig. 2.5). Immunological

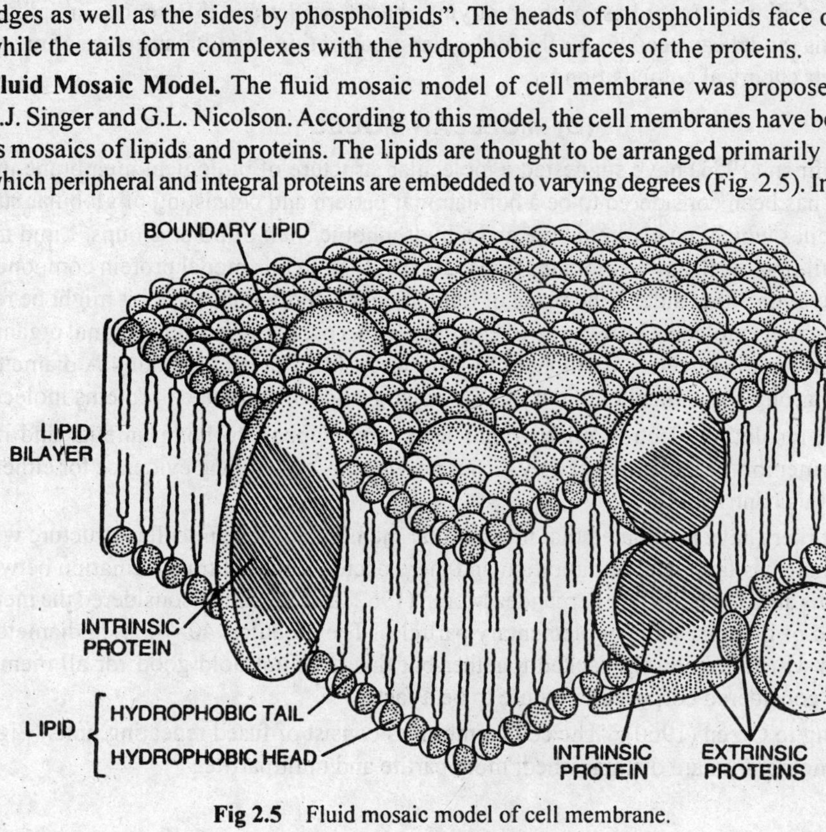

Fig 2.5 Fluid mosaic model of cell membrane.

experiments have indicated the attachment of antibodies to surface-exposed integral proteins such as glycophorine. This suggests that membrane proteins are not fixed within the lipid layer but are free to move laterally like icebergs floating in a sea of lipids. This picture has inspired Singer and Nicolson to coin *fluid mosaic model.* If this analogy of icebergs floating in a sea of lipids is valid then it should be possible to freeze proteins by solidifying the lipid sea in which they float. This is indeed, possible when the phase transition occurs from liquid to solid. At this point the mobility of proteins in the membrane will be checked. Singer and Nicolson considered the lipoprotein association to be hydrophobic and fluidity of the membrane results due to hydrophobic interaction. It should be noted that phospholipids and many intrinsic proteins are amphipatic molecules, *i.e.*, both hydrophilic and hydrophobic groups occur within the same molecule. The globular proteins of the membrane are of two different types: extrinsic (peripheral protein) and intrinsic (integral proteins).

Because of rapid movement of lipid and protein molecules, the fluid mosaic model is different from the static picture of the membrane in Danielli and Davson model. The proteins of the membrane are concerned with the enzymatic activities, transport of molecules and with receptor function. The lipid bilayer acts as the permeability barrier. The fluidity of lipid is supported by many indirect studies based on x-ray diffraction, differentiatial thermal analysis and electron spin resonance (ESR) techniques. These techniques indicate that lipid bilayer has many dynamic properties which are as follows:

1. The rapid motion involving flaxing within each lipid molecule is possible.
2. A rapid lateral diffusion of lipid is possible.
3. Slow motion of lipid molecule from one side of the bilayer to the other is also possible.
4. The lipid molecules might rotate about their axis.

The fluid mosaic model of cell membrane is now widely accepted as it is presumed to apply to membranes of all types regardless of their varying characteristics and differences in lipids protein ratio. In fact this model can account for the molecular organisation and ultrastructure of membranes in-terms of their chemical composition.

(B) MICELLAR MODEL

Hilleir and Hofman (1953) have suggested a molecular structure of biological membrane in which the membrane has been considered to be a non-laminar pattern and consisting of globular sub-units known as micelles which have a lipid core and a hydrophobic shell of polar groups. Lipid micelles are possible building blocks of membranes [Figs. 2.3 (E), (F)]. In this model protein component may form a monolayer on either side of the plane of lipid micelles. Individual sub-unit might be replaced by individual enzyme molecule or by array of enzymes with a precise three dimensional organization. The spaces between the micelles are thought to form water filled pores of about 4 Å diameter lined partly by polar groups of a micelles and partly by polar groups of associated proteins molecule.

The high resolution electron micrographs lend some support to both lamellar and micellar hypotheses for membrane structure yet they do not provide any convincing evidence for either which might resolve this controversy.

Some workers have postulated that intracellular membranes have micellar structure while the plasma membrane has lamellar configuration and there occurs a mutual transformation between the two (De Robertis and others, 1975). Fernandez Moren (1962) and Sjostrand considered the membrane to be composed of globular units or elementary particles. The globules, 40–70 Å in diameter, form repeating units which are closely packed together but this does not hold good for all membranes. Some workers considered corpuscular nature to be a fact.

According to Green (1960), "The cell membranes consist of fused repeating units (Fig. 2.6)". Two types of membranes are distinguished: monopartite and multipartite.

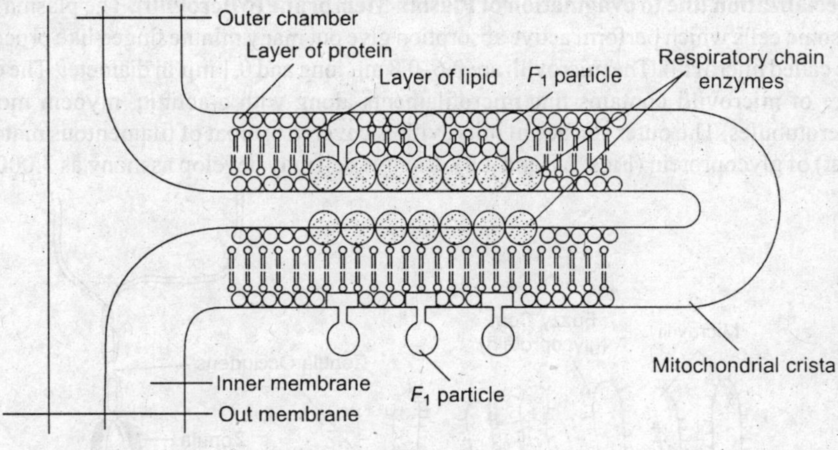

Fig 2.6. Outer Monopartite and inner Multipartite membrane of mitochondria.

In monopartite membrane, the repeating units do not have any projections. Examples of such membranes are the outer membranes of mitochondria and plasma membrane.

In multipartite membranes, the repeating unit is divisible into a base piece, a stalk and head piece. The monopartite repeating units correspond to the base piece of multipartite units. Multipartite membranes are composed of repeating units of two types; macrotripartite and microtripartite units, the former being larger than the latter ones. The macrotripartite units are found in membranes in which electron transfer is coupled with ATP synthesis, *e.g.*, the inner membrane of mitochondria. Membranes with micropartite units have the capacity to carry out ATP energised active transport, *e.g.*, the plasma membrane which transports K^+ and Na^+ and the microvilli which transport sugar and amino acids. In macrotripartite membrane ATPase function is localized in the head pieces while in the micropartite membranes, it is localized in the base pieces.

Protein Crystal Model

David Green and co-workers (1970) have proposed a protein crystal model for cell membrane in which proteins polymerise to form two layers of loosely packed globular protein units of 30–40 Å diameter. The proteins are visualised as having extensive non-polar as well as polar regions on their surfaces. The nonpolar region may exhibit hydrophobic bonding with non-polar groups of phospholipid molecules that fill the cavities between the globular protein units while the polar heads of phospholipid molecules face the membrane surface (Fig. 2.7).

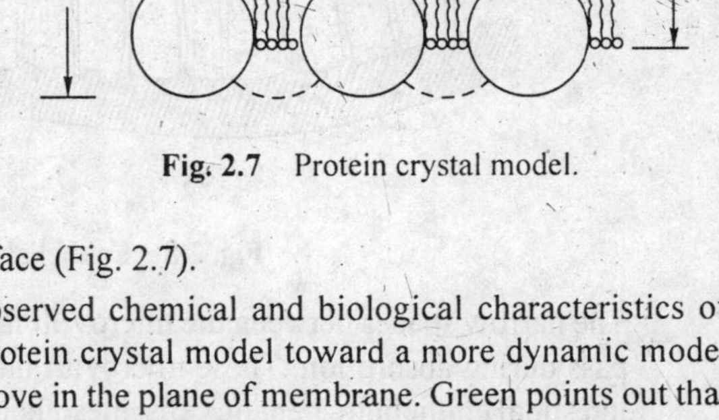

Fig. 2.7 Protein crystal model.

This model accounts for some of the observed chemical and biological characteristics of membranes. Green has recently modified the protein crystal model toward a more dynamic model in which proteins and lipids are considered to move in the plane of membrane. Green points out that a consequence of this modification is that the lipid layer forms essentially a continuum and protein-protein interactions are only temporary.

Specialized Structure of Plasma Membrane

In order to perform specialized functions, plasma membranes are variously modified. The most specialized structures of plasma membrane are as follows:

1. **Specialization due to evagination of Plasma Membrane (Microvilli).** The plasma membranes of some cells which perform active absorption give out many minute finger-like processes. These are called microvilli. The microvilli are 0.6–0.8 mμ long and 0.1 mμ in diameter. The cytoplasmic core of microvilli contains fine microfilaments along with a-actinin, myocin molecule, and microtubules. The outer surface of microvilli is covered by coat of filamentous material (Fuzzy coat) of glycoprotein (Figs. 2.8 and 2.9). A single cell may develop as many as 3,000 microvilli.

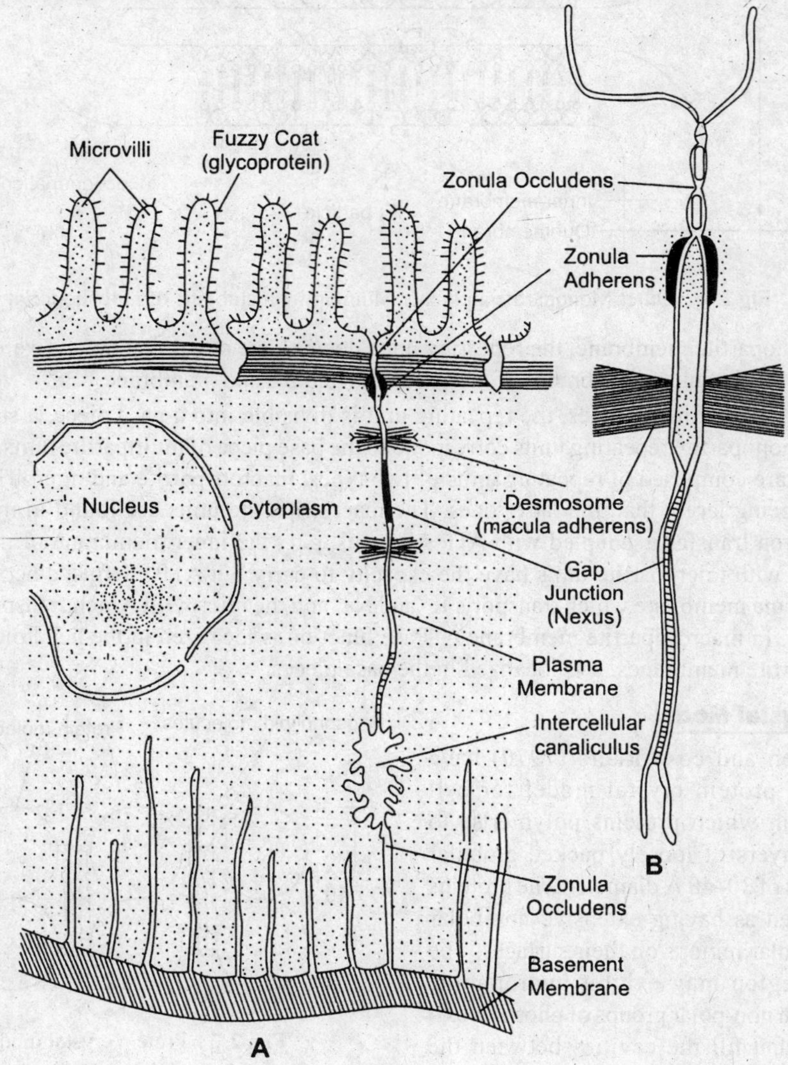

Fig. 2.8 Epithelial cell with microvilli.

The narrow spaces between the microvilli form a kind of sieve through which substances may pass during absorption. These microvilli increase the absorptive surface of the cells, such as intestinal epithelium, hepatic cells, mesothelial cells of convoluted tubules of kidney, epithelial cells of gall bladder, uterous and yolk sac, etc.

2. **Invaginations of Plasma Membrane.** Certain cells which perform active transportation, such as cells of kidney, may show infoldings or invaginations of plasma membrane. The bases of these folds may develop septa and thus narrow compartments are formed.

These invaginations are in contact with many mitochondria which possibly provide energy to the plasma membrane for the active transportation of solutes.

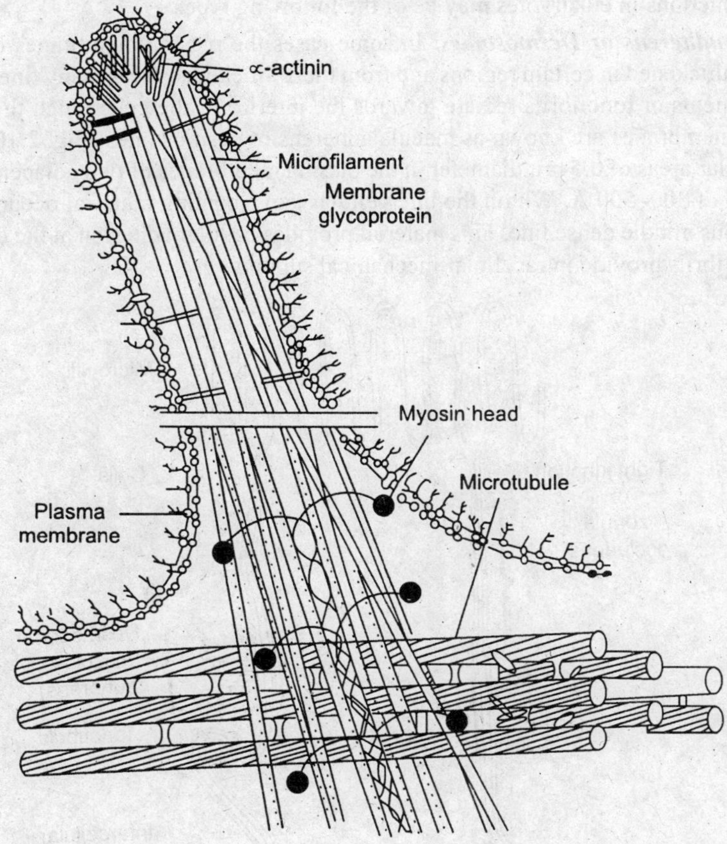

Fig. 2.9 Detailed structure of microvilli.

3. Specialization of Plasma Membrane due to contact (Junctional Complex). Though the adjacent cells remain separated by a uniform intercellular gap of 100-200 Å, their plasma membranes may come too close or may go too far. Sometimes the contact between the cells may result in the modification of plasma membrane. In the specialized regions the membrane proteins of adjacent cells interact in a highly specific way to provide a intercellular junctions. These are called *Junctional complexes*. Some junctions are devices to make the adjacent cells adhere together, some are barriers to prevent fluids from passing between the cells and some junctions serve primarily to establish channels of communications from one cell to other. So the cell junctions are assigned various names that describe their shapes and relative closeness with the adjacent membranes, such as:

(*i*) *Zonule*—A stripe that passes completely around the cell.

(*ii*) *Macule*—It is a contact spot in the plasma membrane.

(*iii*) *Occludens*—It is a zone where the outer surfaces of the two membranes are very close or even fused at several points containing sealing strands which prevent the free passage of substances across the membranes.

(*iv*) *Adherens*—It is a wide separation band between the cell membranes.

Types of Junctions in Eukaryotes

Intercellular junctions in eukaryotes may be of the following types:

1. ***Macula adherens or Desmosomes.*** In some cases the plasma membranes of adjacent cells become thickened in certain regions and from these thickened areas many fine filaments called tonofilaments or tonofibrils radiate towards the interior of the cells. Such thickened areas of plasma membranes are known as macula adherens or desmosomes (Fig. 2.10). Desmosomes are circular areas of 0.5 mμ diameter in the plasma membranes of two adjacent cells separated by a gap of 300–500 Å. Within the intercellular gap a coating material occurs which forms a continuous middle dense line. This material provides cellular adhesion at the desmosomes and the tonofibrils provide intracellular mechanical support.

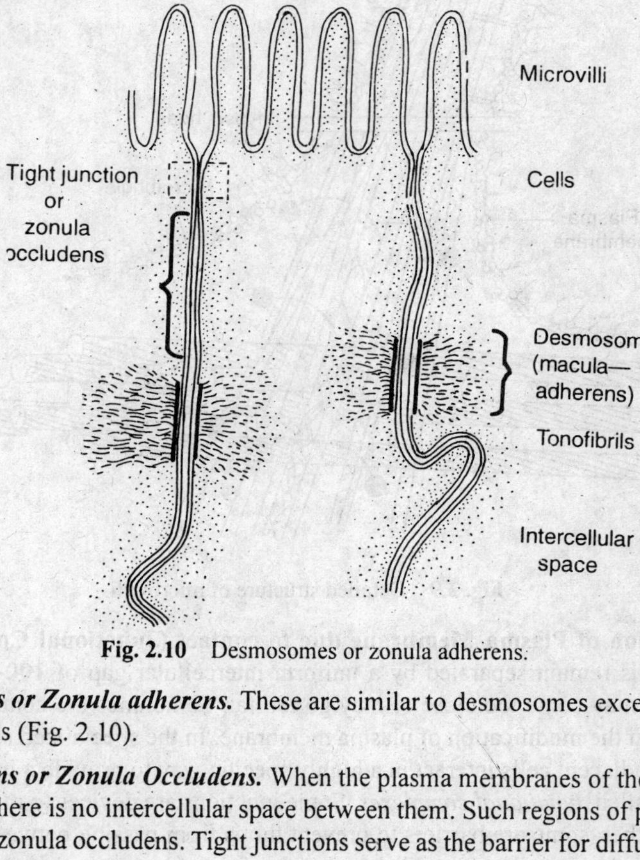

Fig. 2.10 Desmosomes or zonula adherens.

2. ***Terminal bars or Zonula adherens.*** These are similar to desmosomes except that they lack in the tonofribrils (Fig. 2.10).

3. ***Tight junctions or Zonula Occludens.*** When the plasma membranes of the adjacent cells are so close that there is no intercellular space between them. Such regions of plasma membranes are known as zonula occludens. Tight junctions serve as the barrier for diffusion of substances between the adjacent cells.

4. ***Gap junctions or Macula Occludens.*** When there exists a gap of 20 Å between the plasma membranes of the adjacent cells, these gaps are called gap junctions. The gaps are hexagonal and 70–75 Å wide. They provide intercellular communication by allowing the direct passage of molecules and ions between the cells.

5. ***Interdigitation.*** Plasma membranes of adjacent cells may give out interlocked finger-like projections known as interdigitation.

Functions

The plasma membrane is of great physiological importance because of its semi-permeable nature. It controls the cellular semi-permeability, resorption, secretion and excretion.

It regulates the cytoplasmic contents of a cell and acts as a selective barrier between the living cell contents and its external environment because of the fact that it permits only those substances to pass through which are needed by the cytoplasm and does not allow those substances to enter which are not required. Plasma membrane is permeable to water. The diffusion of ions through plasma membrane depends upon the concentration and electrical gradients.

Plasma membrane is not only a passive diffusion barrier but also carries catalytically active regions. The machinery which utilizes energy for carrying out osmotic work is also located in or near plasma membrane.

The important means by which materials can be transported into the cell are "phagocytosis" and "pinocytosis". The process of ingestion of solid substances by the cell is referred to as phagocytosis and the ingestion of fluid material in bulk is known as pinocytosis (Lewis, 1931). These two processes involve invagination of plasma membrane to form vacuoles containing solids or liquids. The membrane bound vesicles or vacuoles are termed phagosomes and pinosomes respectively (Fig. 2.11). The phagosomes and pinosomes along with external materials pass into the cell where they fuse with the secretory granules. In the food vacuoles the digestion of food substance takes place and the digested food is diffused to the ground plasm.

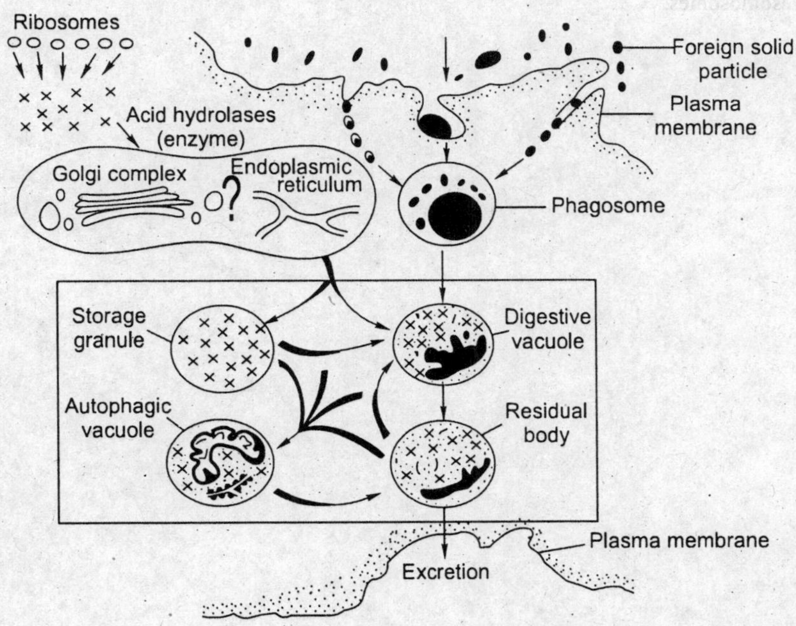

Fig. 2.11 Phagocytosis in the cell.

A reverse process to phagocytosis or pinocytosis may occur for throwing the waste materials out through plasma membrane.

The plasma membrane is the integral organelle of the cell without which protoplasm cannot survive. Like other organelles, the plasma membrane repairs itself if it is injured.

QUESTIONS

1. What do you mean by cell membrane? Discuss the functions of cell membranes.
2. Discuss unit membrane concept.
3. Discuss the various models for the structure of plasma membrane.
4. Write short notes on:
   (*i*) Unit membrane,
   (*ii*) Lipid bilayer hypothesis,
   (*iii*) Fluid Mosaic model, and
   (*iv*) Protein micellar model.
5. Discuss briefly the different types of junction complexes in cell membrane.
6. Write brief notes on:
   (*i*) Phagocytosis,
   (*ii*) Microvilli, and
   (*iii*) Desomosomes.

3 CHAPTER Plastids

Plastids are unique and discrete cytoplasmic organelles found in the cells of most of the plants and certain unicellular organisms particularly protozoa, but not in the cells of higher animals. Although Comparethi (1791) noticed the presence of green granules and Sacks (1865) also observed special bodies in the plant cells yet the term *plastid* was first used by A.F.W. Schimper in 1883. Plastids are coloured or colourless organelles most often occurring as spherical or disc-like bodies in free state in the cytoplasm of the eukaryotic plant cells. In fungi, bacteria and blue-green algae plastids are lacking. However, the electron microscopic studies have revealed that in photosynthetic bacteria and blue-green algae some distinct lamellar structures bearing photosynthetic pigments undoubtedly occur in the protoplasm. These pigmented lamellae are named *chromatophores* by some authors. The term chromatophore was coined by Schmitz (1882).

Classification

The obvious variations found among plastids involve colour or pigment content, size and number. Plastids may contain one or more pigments or no pigment and on the basis of pigment content or colour, plastids have been classified into the following types:

(*i*) Leucoplasts (colourless plastids).

(*ii*) Chloroplasts (green plastids).

(*iii*) Chromoplasts (plastids coloured other than green).

The plastids are generally coloured owing to the several kinds of colouring substances called pigments. Some important pigments and their respective colours are listed below:

Pigment	*Colour*
1. Chlorophylls	Green
2. Carotenes	Yellow or red
3. Xanthophylls	Yellow, Orange or brown
4. Phycoerythrin	Red
5. Haematochrom	Red
6. Phycocyanin	Blue-green

These pigments are found in the plastids. However, there are certain pigments, such as anthocyanins which are not found in the plastids but are dissolved in the ground substance of cytoplasm. The blue, red and pink colours of some flowers and young foliage leaves appear due to the presence of anthocyanins.

The *chloroplast* which contains chlorophyll pigments is the chief site of photosynthesis in plant cells. The plastids containing both chlorophylls and carotenoid pigments are active in photosynthesis but those containing only carotenoid pigments and lacking in chlorophyll are photosynthetically inactive. The coloured plastids generally contain more than one pigment but the colour of the plastid is determined by that pigment which masks the colour of other pigments. The chloroplasts found in the cells of most green plants contain a small amount of carotenoids, such as carotenes and xanthophylls in addition to chlorophylls but carotenoids are not always visible because chlorophylls mask their colours.

Chromoplasts are pigmented plastids bearing colours other than green. They may be yellow, orange, brown or red. Yellow or orange chromoplasts occur in petals, fruits and roots of some higher plants. Generally, these plastids contain carotenoids and a reduced chlorophyll content and are photosynthetically less active. In some cases chromoplasts contain carotenoids only and chlorophyll is lacking. The red colour of ripe tomatoes, for example, is due to the presence of chromoplasts that contain the red pigment *lycopene*, a member of carotenoid group. Whenever the chromoplasts contain a small amount of chlorophylls in addition to a large quantity of carotenoids, green colour of chlorophylls is not manifested because of masking effect of the major carotenoid pigments. Chromoplasts containing various pigments are found in different groups of algae. *Phaeoplasts,* brown coloured chromoplasts having *fucoxanthin* as major or principal pigment occur in brown algae, diatoms, dinoflagellates, etc. and *rhodoplasts,* red coloured chromoplasts having phycoerythrin as major pigment occur in red algae. Certain carotenoids such as fucoxanthin and phycoerythrin absorb light for photosynthetic reaction and transfer the light energy to chlorophylls. In pigmented bacteria, however, the pigments do not show identical reactions.

Leucoplasts are the plastids devoid of pigments. They are involved in the synthesis and storage of the various kinds of carbohydrates (starch), oils and proteins. If they store starch, they are called *amyloplasts*. When leucoplasts store oils, they are termed *elaioplasts.* Leucoplasts concerned with storage of protein are called *proteinoplasts* or *aleuroplasts*. Leucoplasts are rod-like or spheroid in shape [Fig. 3.1 (A)] and occur in embryonic cells, sex cells and the cells of the storage organs. They are of common occurrence in cells that have been protected from light.

According to Schimper, "Transformation of one type of plastid into another type is possible in the following way":

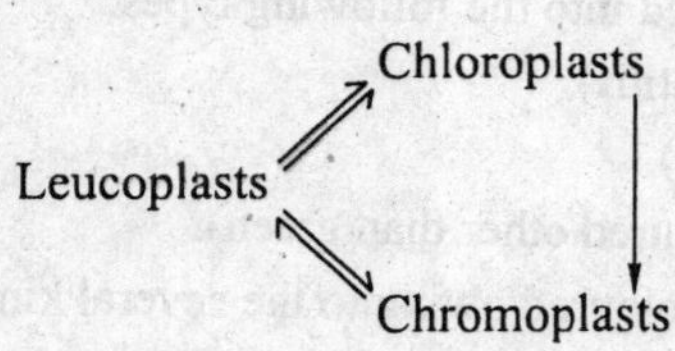

CHLOROPLASTS

Chloroplasts are the most common and familiar plastids of most plant cells. Chloroplasts are discrete membrane found green coloured structures in eukaryotic plant cells. They are of greatest biological importance, since by photosynthesis they produce chemical energy. They can be readily seen in most plant cells at low magnification under light microscope.

Morphology

The chloroplasts remain homogeneously distributed in the groundplasm of plant cells. In some plant cells they are found concentrated around the nucleus and in some they may be found just beneath the plasma membrane. The *shape*, *size*, and *distribution* of chloroplasts may vary in different cells within a plant.

Shape. Plastids in the cells of higher plants show a great deal of variation in shape. The cells of leaves contain many chloroplasts of spheroid, ovoid or discoidal shape. Some may be club shaped having narrow middle zone and bulging ends. They may be vesicular with a colourless centre in some cells. In algae, cells often possess a single large chloroplast which may be star-shaped, band-shaped, cup-shaped, discoidal or reticulate [Fig. 3.1 (B)].

Size. Chloroplasts show considerable variation in size. The average diameter of chloroplasts of higher plant is 4 to 6 μ and thickness 1 to 3 μ. The size is, more or less, constant for a given cell type in a plant but in polyploid cells they are comparatively larger than those in the corresponding diploid cells. Plants growing in shade have chloroplasts that are larger in size and contain more chlorophylls than the chloroplasts found in plants growing in sunlight.

Number. The number of chloroplasts in a cell appears to be relatively constant in different plants and the variation in number whenever seen is related with the physiological state of the cell. When the number of plastids is insufficient it is increased by division of existing plastids and when the number is in excess it is decreased by degeneration.

The cells in most algae possess only a single chloroplast. The cells of the higher plants may have 20 to 40 chloroplasts. Haberlandt (1882) found at an average 36 chloroplasts in each palisade cell and 20 in each spongy parenchyma cell. In moss genus *Mnium*, cells have been found to possess an average of about 106 chloroplasts each.

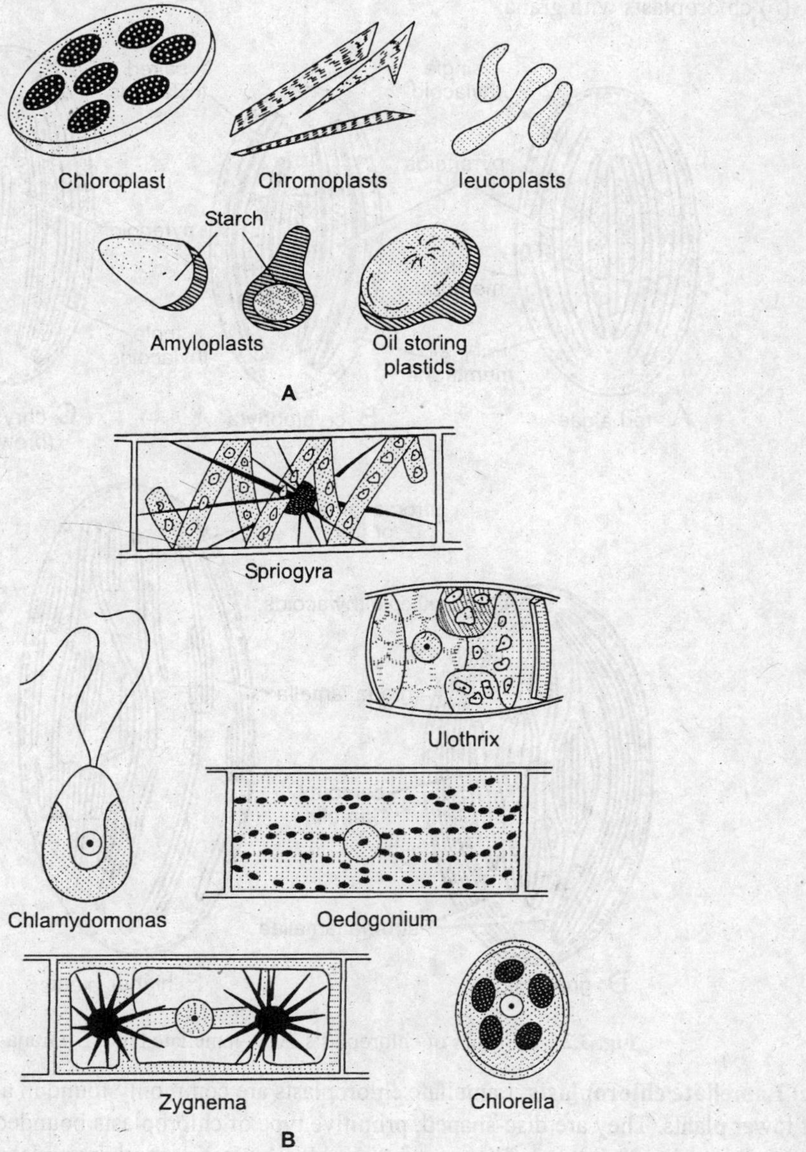

Fig. 3.1 A. Different types of plastids; B. Different shapes of chloroplasts found in algae.

Critical observations of the plastids have shown that they are displaced and deformed by action of cytoplasmic streaming. Besides, some sort of amoeboid movement or contraction may be influenced by illumination. Isolated chloroplasts of spinach show decrease in their volume when they are exposed to light and photophosphorylation is initiated. This effect is reversible. Considerable contraction of chloroplast takes place in dark if ATP is added to the medium. Two different proteins having contractile properties, which may account for this reversible change in the volume of plastids under different sets of conditions, have been extracted from isolated chloroplasts (Ohnishi, 1964). Chloroplasts show greater resistance to osmotic changes and fixatives than other types of plastids as well as mitochondria.

Ultrastructure

There are two types of chloroplasts [Fig. 3.2 (A)]:

(*a*) lamellate chloroplasts.

(*b*) chloroplasts with grana.

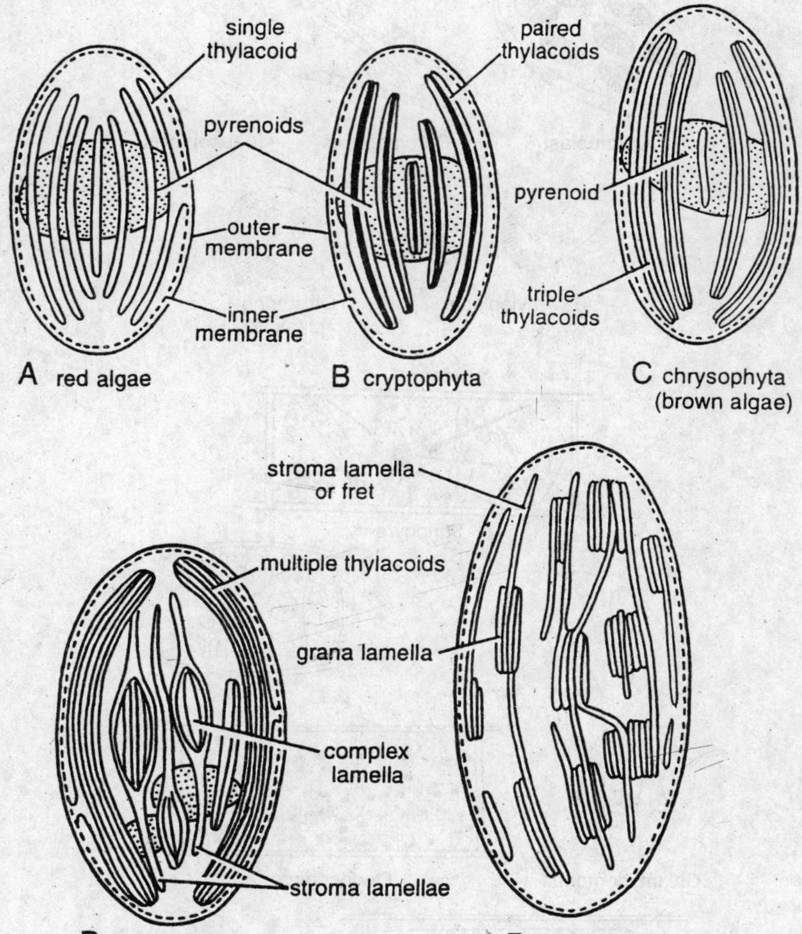

Fig. 3.2 A–Types of chloroplasts. A-D–lamellate type; E–Grana type.

(*a*) Lamellate chloroplasts. Lamellate chloroplasts are commonly found in algae and in some cells of lower plants. They are disc-shaped, primitive type of chloroplasts bounded by two differentially permeable unit membranes of lipoproteins. Inside the plastid membrane, plates containing pigments (chlorophylls, etc.) are orderly placed in a colourless matrix or stroma. Menke (1961) termed the lamellar units as *thylakoids* but Ruthsager and Palade (1957) preferred to call them pigmented *discs*. In section it appears like a pair of parallel membranes joined at each end enclosing narrow space termed *loculus.* The thylakoids in algae are generally long and may run nearly the entire length of the chloroplasts. In many green algae, the thylakoids occur in multiple layers forming grana-like structures called stacks or bands. The number of discs in a stack varies from 2 to 20 or even more. Usually there are four to six thylakoids in a stack. Thylakoids in the stack do not adhere and remain separated by a narrow space called interdisc or interthylakoid space. In a number of other green algae, the thylakoids do not form stacks. They come together and move apart in an irregular manner [Fig. 3.2 (B)]. In prokaryotic photosynthetic cells such as photosynthetic bacteria and blue-green algae chloroplasts do not exist as separate organelles and pigmented lamellae or vesicles are distributed throughout the cytoplasm.

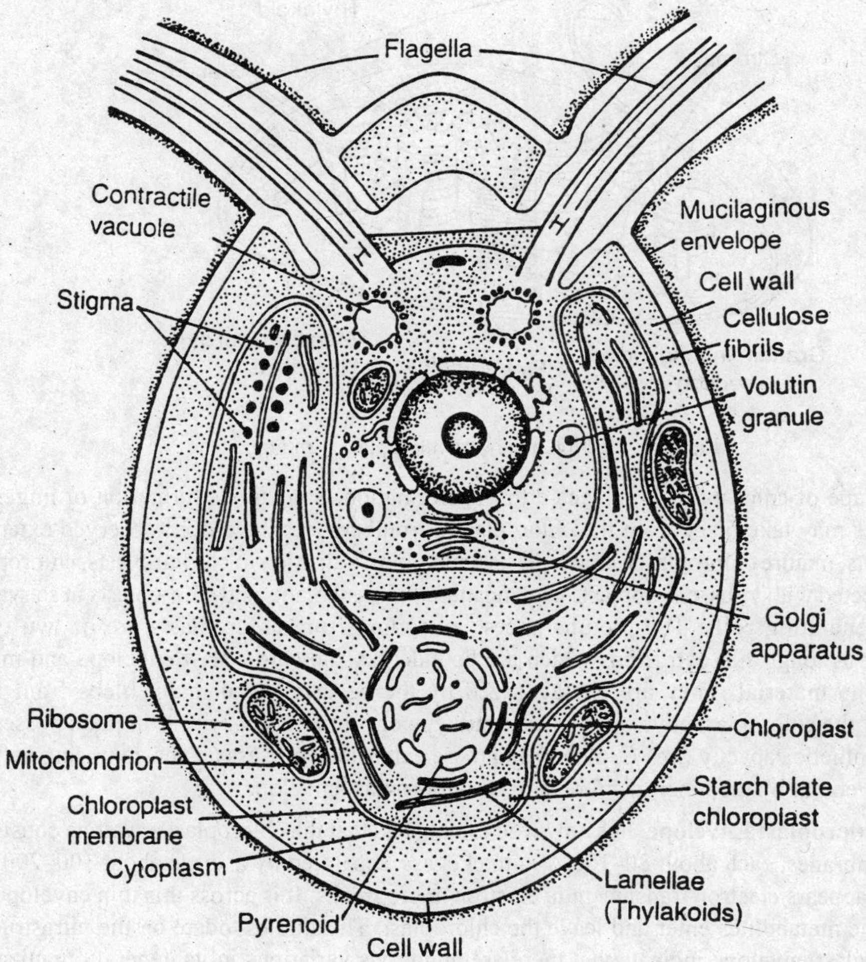

Fig. 3.2 B–Chloroplast of *Chlamydomonas*.

As regards the relation of outer limiting membrane of the plastids with other cytoplasmic membranes, there exist considerable differences. In several cases the plastid envelope is sometimes continuous with elements of endoplasmic reticulum. Besides, connections with other cytoplasmic membranes are quite common in forms with single plastid, *e.g.*, in *Chlamydomonas, Rhodomonas,* etc. In *Ochromonas* there is a double membrane envelope outside the usual outer limiting membrane of the plastid. This outer envelope is continuous with nuclear envelope but does not contain pores. Similar relationships exist in many groups of algae. In chlorophyceae and Rhodophyceae, however, no additional envelope is known to occur outside the usual outer limiting membrane of the plastids.

Chloroplast with grana. This type of plastid occurs generally in the mesophyll cells of the higher plants. The mature chloroplast of the higher plant can be divided into the following *three* parts which are interrelated structurally and differentiated functionally (Figs. 3.3 and 3.4):

(*i*) The chloroplast envelope,

(*ii*) The stroma comprising the internal matrix material, and

(*iii*) The grana fretwork system comprising the internal membrane system.

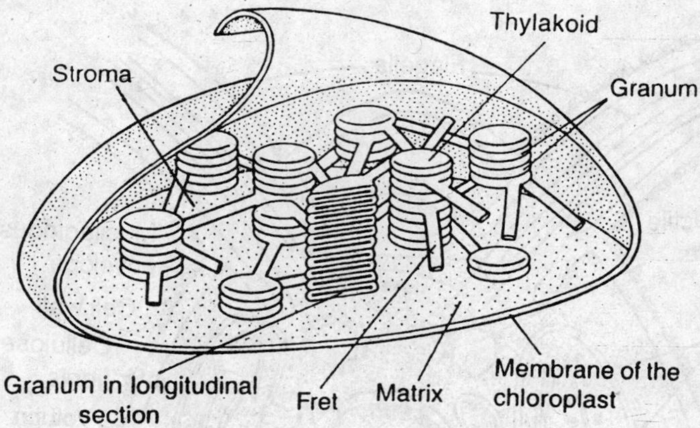

Fig. 3.3 Sectional view of chloroplast with grana.

The shape of chloroplast is subjected to transformation. Sometimes formation of finger like protuberances may take place from the chloroplasts. Protuberances have been observed extending from elioplasts, mature chloroplasts, chloroplasts undergoing transition to chromoplasts, chloroplasts in tissues infected with virus, chloroplasts in manganese deficient plants and chloroplasts in starvation-stressed degenerating cells. The protuberances extending from chloroplasts are of two types. Some appear as long thin extrusions which are bounded by the chloroplast envelope and mainly contain stroma material. They are characterized by the occurrence of many blebs from inner membrane of the chloroplast envelope. The second type of protuberances occur in plant possessing high photosynthetic capacity (*i.e.*, C_4 plants) and these extensions possess an anastomosing network of tubules given out by the inner membrane of the chloroplast envelope.

The Chloroplast Envelope. It is now well established that the chloroplast envelope consists of two unit membranes, each about 80–100 Å in thickeness, separated by a space about 100–200 Å in width which appears electron translucent in electron micrographs. It is across this thin envelope that photosynthetic metabolites enter and leave the chloroplast. The studies todate on the ultrastructure of the chloroplast envelope indicate that there are numerous variations in its general organization. In most of the mature chloroplasts the inner membrane of the envelope invaginates to form small vesicular or flattened or finger-like invaginations. In some cases, much longer invaginations run parallel to the envelope. Schotz and Diers (1967) have reported that the invaginations are plate-like structures which sometimes enclose small enclaves and pockets of stroma.

Very little is known about the intermembrane space of the chloroplast envelope. Recently, Sabricis Gordan and Galston (1970) using *histochemical* method for the fine structural localization of nucleoside phosphates have reported the presence of a Mg^{++} dependent ATPase between the two unit membranes of the envelope. The enzyme appeared to be light activated and these investigators have suggested that it might be involved in the light dependent contraction of chloroplasts.

The stroma. Inside the envelope there is a proteinaceous matrix called *stroma*, containing starch grains, osmiophilic plastoglobulin, phytoferretin granules, protein grains, ribosomes, RNAs and DNA. The stroma also contains enzymes that are involved in dark reactions of photosynthesis, carbohydrate synthesis, protein synthesis and synthesis of chlorophylls, carotenoids and other pigments (Figs. 3.3 and 3.4).

Starch grains are commonly observed in the stroma of chloroplasts. The grains appear nearly oval in most electron micrographs although they may vary from oblong to round.

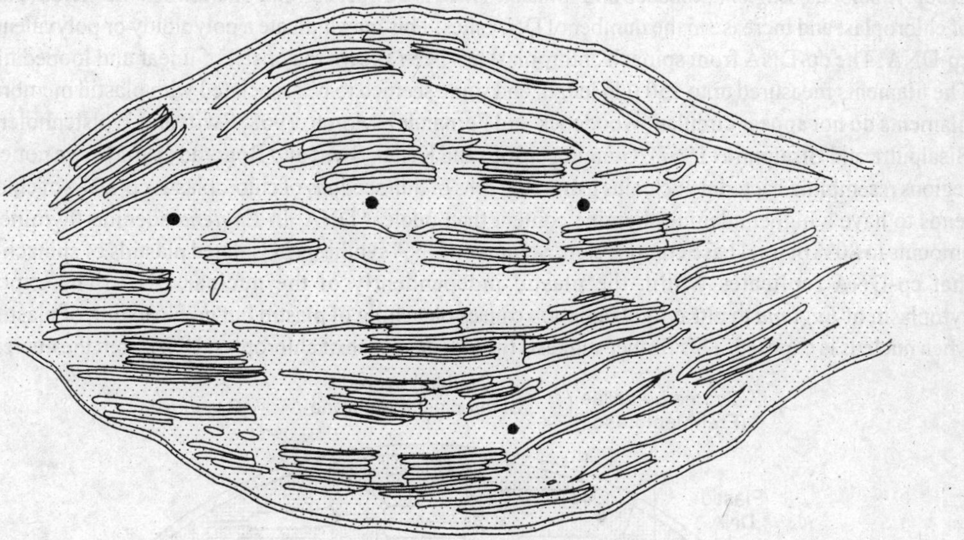

Fig. 3.4 Electron micrograph of chloroplast.

In the chloroplasts fixed with osmium tetraoxide there is characteristic occurrence of round electron opaque bodies in the stroma. These osmiophilic bodies or droplets are called *plastoglobuli*. The plastoglobuli appear free in stroma and may occur singly or in clusters. They may, however, occur associated with grana fretwork system. The size of these plastoglobuli in normal photosynthesizing chloroplasts may be 100 Å or more. The size and the number of these osmiophilic droplets may increase with the advancement of age of plastids (Dodge, 1970) and may also increase during transition from chloroplasts to chromoplasts. It is generally suggested that plastoglobuli represent the reservoirs of excess lipid (Greenwood *et al*, 1963 and Bailey and Whyborn, 1963). Plastoglobulin in chloroplasts of photosynthesizing leaves contains little or no chlorophyll or carotenoid pigments. With the advancement of plastid age or during chloroplast to chromoplast transition, the composition of the globules changes and carotenoid pigments accumulate in them.

Hyde *et al*. (1963) reported the occurrence of an iron protein complex in pea embryos and termed it phytoferretin because of the similarity in structure and composition to ferretin present in animal cell. Phytoferretin particles are electron opaque and measure about 100 Å in diameter. They consist of iron containing cores and the electron translucent regions around the core, the proteinaceous shell (Robards & Robinson 1968). They are of common occurrence in proplastids and differentiating plastids or in senescing chloroplasts and developing chromoplasts. Similar particles have been found in the stroma of chloroplasts in virus infected tissues. Hyde *et al* (1963) have suggested that phytoferretin represented a reservoir of stored iron in the form of a non-toxic iron containing protein.

Another characteristic feature of chloroplast is the presence of ribosomes and RNA in the stroma. The plastid ribosomes are evident particularly after staining with uranyl acetate, but not observed in permanganate fixed cells. The plastid ribosomes in thin sections measure about 170 Å in diameter and are thus smaller than the ribosomes found in groundplasm which measure about 250 Å in diameter. The plastid ribosomes frequently appear as free 70S particles. However, Brown and Gunnina (1966) have observed groups of ribosomes in developing chroloplasts and Bartels and Weier (1967) have also described helically arranged polysomes in the stroma of proplastids of *Triticum*.

Cytological, electron microscopic, autoradiographic and biochemical studies have established beyond doubt the presence of cp-DNA in developing and mature chloroplasts. At ultrastructural level DNA appears as ring-shaped irregular mesh of 25 Å fibril in the stroma (Ris and Plant, 1962; Kislev, Swift and Bogorad, 1965). One or several cp-DNA sites may be located in the stroma of chloroplast of several plants including *Chlamydomonas, Euglena,* tobacco and spinach. There is a direct correlation between the increase in the size of chloroplast and increase in the number of DNA sites. This may indicate a polyploidy or polyvalency of the cp-DNA. The cp-DNA from spinach chloroplast has been shown to consist of linear and looped filaments. The filaments measured upto 150 μ in length and some seemed to be associated with plastid membrane. The filaments do not appear circular (Woodcock and Fernandez-Moran, 1968; Swift and Wolstenholen, 1969). Bisalputra and Bisalputra (1967), Gunning (1965) and Kisler, Swift and Bogorad (1965) have noticed DNA regions resembling bacterial nucleoids in micrographs of thin sections of chloroplasts. cp-DNA of chloroplasts tends to have a higher adenine-thymine content than nuclear DNA. In *Euglena* adenine-thymine content amounts to 80% of total base composition. Studies on DNA synthesis in isolated plastids of spinach indicate that cp-DNA replicates within the plastid independently of the nuclear DNA. The exposure of cytoplasm of *Euglena* to ultraviolet radiation causes bleaching of plastid but no bleaching of plastid occurs when nucleus is exposed. This further suggests the independence of chloroplast DNA from the nuclear DNA.

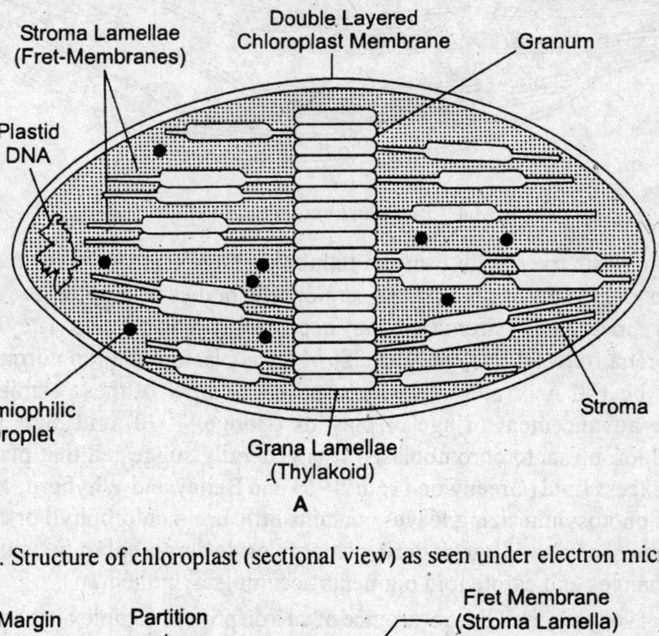

Fig. 3.5 A. Structure of chloroplast (sectional view) as seen under electron microscope.

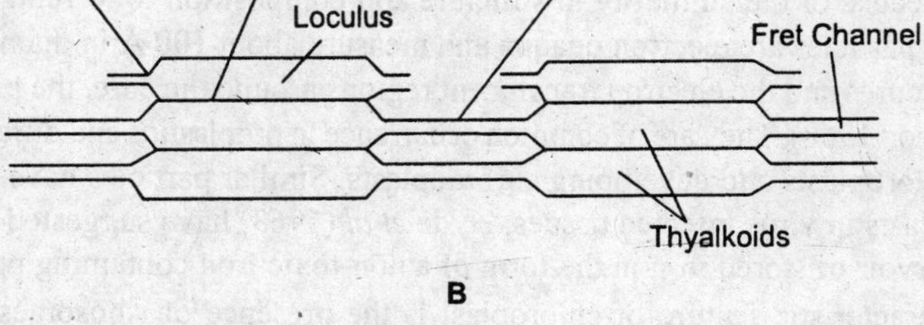

Fig. 3.5 B. Structure of grana.

Grana fretwork system. Meyor (1883) and Schimer (1885) recognised that the internal region of the chloroplast was divided into two phases: the *stroma* and numerous greenish granules, the *grana*. The presence of *grana* was a serious point of controversy until the mid 1930s when Doutreligne (1935), Weier (1936) and Heitz (1936) established beyond doubt that they were not artifacts. The lamellar nature of granum became clear only after the early electron micrographs of chloroplasts were published by Granick and Porter (1947), Steinmann (1952) and some other workers.

Electron micrographs reveal that chloroplast consists of a system or series of fine pigmented plates or lamellae enclosed within the plastid membrane. The small stacks of pigmented lamellae are called **grana** (singular-**granum**). These are flat disc-shaped or cylindrical structures, 0.3 to 1.7 μ in diameter and about 0.2 μ in thickness. There is considerable variation in the number of grana in a single plastid, depending upon the species. The primitive forms tend to have small number of grana than the higher forms. The granum consists of a few to more than fifty disc shaped superimposed parallel membranous compartments which are called thylakoids. These membranous lamellae are frequently fused in a stack. In some cases, the granum extends like a cylinder across the entire width of the plastid. The length of thylakoids may vary in an individual granum and often one or more compartments may extend from one granum to another (Fig. 3.5). Frequently, where thylakoids extend from granum to granum the entire complex may appear branched. The membrane bound channels interconnecting thylakoids of adjacent grana are called the ***frets*** or ***intergranal lamellae*** or ***stroma lamellae.*** The grana are bounded by marginal end membranes which isolate the internal regions of the grana from the stroma. The *marginal end membranes* of grana are continuous with the membranes of the channels interconnecting the thylakoids of the adjacent grana. Recent electron microscopic studies of chloroplasts have provided evidence that the membranes within the chloroplasts form a continuum and that the loculi (cavities of the thylakoids) of the grana communicate *via* fretwork connections (Fig. 3.6). A granum may consist of many compartments but the simplest granum is defined as consisting of two thylakoids and recognized by the presence of a single partition separating two loculi. The entire complex of grana together with membrane bound frets is referred to as *grana-fretwork system.*

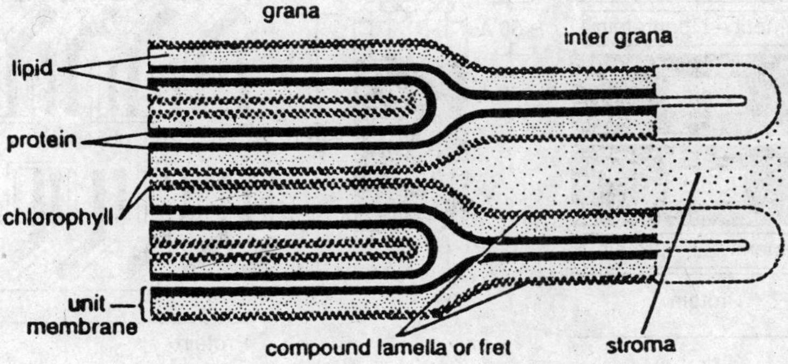

Fig. 3.6 Structure of granum.

The nature and organization of the inter granal membranes have been a focal point of discussion. The evidences accumulated from recent studies of the ultrastructure of chloroplasts indicate that frets are more comparable to flexible channels which are often branched and extend to and connect two or more grana, with the result that the fretwork appears as highly fenestrated sheets. Close observations of the connections of stroma channels with grana revealed that they were often connected with two or more thylakoids of the same granum. The number of frets connected to an individual thylakoid varies and tangential and serial sections indicate that the frets are arranged in a spiral fashion around the grana.

As pointed out earlier, the grana are bounded on all sides and are in contact with the stroma by a membrane which is nearly **85 Å** thick. The fret membranes also measure approximately 85 Å in thickness and are continuous with the marginal end membranes of the thylakoids. Thus, it appears that these membranes are structurally and functionally identical. The partitions of grana are apparently a complex of two membranes which in many instances are separated by a space called *A-space* that measures about 24 Å in width. The partitions usually run across the granum form one margin to the other but sometimes a small gap between the end of the partition and the marginal membrane of the granum is left which results in the confluence of the loculi of adjacent thylakoids. The two membranes of partition may be due to invagination of the marginal membrane. However, the thickness of partition including the A-space is less than the sum of the thicknesses of two marginal membranes. The A-space may appear electron translucent or electron opaque depending upon the fixation, dehydration and staining methods used.

Several views have been put forth to explain the relationship between granum lamellae and stroma lamellae. According to Skinmann and Sjostrand (1955) stroma lamellae develop from granum discs. Hodge *et al.* (1955) expressed that granum discs are locally swollen thylakoids. In some cases, the stroma lamellae may or may not terminate in granum and so they may be continuous or discontinuous (Fig. 3.6). As regards their origin, the stroma lamellae and grana lamellae are similar. Several investigators have published micrographs which apparently indicate that the granal fret membranes are tripartite structures analogous to the unit membrane consisting of central electron translucent bimolecular layer of phospholipids and two layers of proteins, one on each side of lipid layer. The chlorophylls and other pigment molecules are located in the trilamellar limiting membranes of thylakoids and protein matrix binds chlorophylls and lipids (Weier and Banson, 1967), [Fig. 3.7 (A)]. Several investigations have made it clear that chlorophyll is present in partitions and particularly along the A-Space. Since A-Space is generally visible when chlorophyll is retained and diminished or when chlorophyll is removed. These observations support the suggestion that chlorophyll is a major component of A-space in the partition. Several lines of evidence indicate that chlorophyll is also present in the fret membranes.

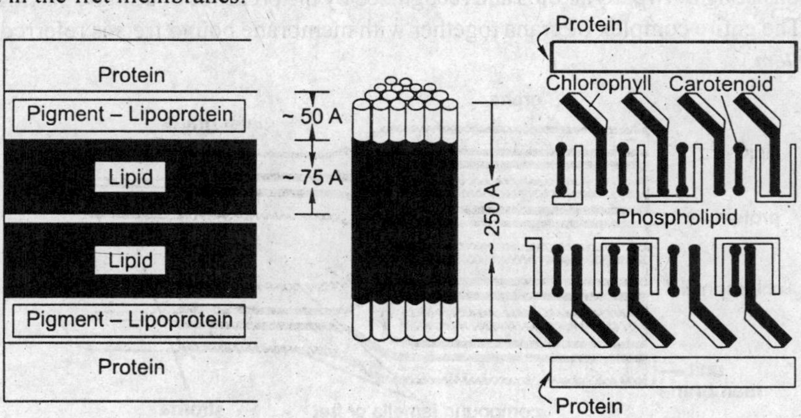

Fig. 3.7 A. Molecular structure of lamella of granum, B. Quantasomes.

The loculi and fret channels are known to swell or contract in response to hypotonic or hypertonic medium which indicates that the loculi and fret channels are hydrophilic and possibly similar in nature. The semi-permeable nature of marginal end and fret membranes indicates that these membranes are, if not identical, probably closely related.

The thickness of the partition varies in different species and it is greatly affected by different chemical reagents. The efficiency of photosynthesis is a function of the wavelength of impinging light. If the chloroplasts are simultaneously activated at wavelengths 6800 Å and 6000 Å, there is a marked stimulation of the photochemical reaction. On this ground existence of two systems of light absorption have been postulated; **photosystems I (PSI)** which is stimulated at wavelength shorter than 700 nm and does not yield oxygen and **photosystem II (PSII)** which is stimulated at wavelength shorter than 680 nm and yields O_2. Photosystem I is made up of units containing about 200 molecules of chlorophyll a, 50 molecules of carotenoids and cyt. f and PC. One molecule of pigment P 700 is related to this unit and that pigment molecule can trap exitons and release electrons. Photosystem II is made up of units containing about 200 molecules of chlorophyll a, some 200 molecules of chlorophyll b, c or d and molecules of xanthophyll, cyt.b_6 and P.Q. The complete set of pigments and lipids containing two photosystems is thought to be clustered in structural and functional unit. Recent methodological advancements, special embedding technique in an aqueous medium, shadow cast technique in electron microscopy have revealed more information about the cytochemical localization of both photosystems. Cytochemical investigations have indicated that the products of photo-oxidation accumulated in the loculi of the thylakoids which suggested that photosystem I was localized at the inner face of thylakoid membrane. The photo-reduction of ferrocyanide which indicates the localization of photosystem II causes dense deposits in the partitions between thylakoids, indicating thereby possible localization of PS II at the outer surface of the membrane.

Steinmann with the help of shadowed preparations of isolated chloroplast lamellae demonstrated the existence of a repeating structure on the inner surface of the membranes of thylakoid (Park and Biggins, 1964). Park and his associates have shown this repeating structure to consist of closely packed arrays of particles or granules (Park, 1963). Each granule is oblate sphere shaped 185 Å long, 155 Å wide and 100 Å thick with a molecular weight of 2×10^5 (Park and Biggins, 1964). This sub-unit has been termed **quantosome** by Park and co-workers and is considered the smallest photosynthetic particle which will perform the light reactions of photosynthesis. The quantosomes seem to be buried in the lamellae of the membranes either evenly distributed as in chloroplasts of lower plants or in packets of grana and stroma lamellae as in the chloroplasts of higher plants [Fig. 3.7 (B)].

The quantaromes are the smallest units capable of carrying out the photochemical reactions and each of them contains about 200 chlorophyll molecules as well as enzymes necessary for the light reactions of photosynthesis. This is not certain at present whether quantosomes represent photosynthetic units within the thylakoid membranes. Souer and Calvin (1962) have suggested that 3 to 6 quantasomes may aggregate together to form a large particle of 500 × 100 Å.

The quantasomes may be present on stroma lamellae also but in relatively much smaller number. Branton and Park (1967) have recognized the following three types of membranes in the plastids on the basis of the size, nature and arrangement of quantosomes in the membranes of thylakoids:

(*i*) membranes with quantosomes,

(*ii*) membranes with smaller particles (110 Å), and

(*iii*) membranes with rough surface and with a few of no particles.

In effect, elementary particles of mitochondria can be homologised with quantosomes. Both of these particles demonstrate an association of electron transport system with the unit membranes and electron transport system is accompanied by phosphorylation. Electron transport system contain electron carriers cytochromes b_6, b_3, f (a hydrophobic protein of 1,00,000 daltons, isolated by non-polar solvents) tightly bound to membrane structures. Chloroplasts also contain ferredoxin, a protein of about 11,600 daltons and containing iron and sulphur but lacking in porphyrin group, plastocyanine, a copper-protein complex and two quinones-vitamin k_1 and plastoquinones, the latter being similar to ubiquinone.

CHROMOPLASTS

During the transformation of chloroplast into chromoplast, yellow coloured droplets termed as globulins appear first and in the course of development chlorophylls and starch gradually decrease, large globuli are formed and arranged along the plastid membrane, lamellar structure breaks down and stroma is disorganized resulting in the empty appearance of the plastid centre. During the transformation of leucoplasts into chromoplasts certain fibrils appear which give rise to crystals filling up the whole plastids. The crystals are found in the form of sheet-like structures containing large quantities of carotenoids.

Origin and Development of Plastids

Many theories have been advanced to explain the origin and development of plastids.

(*a*) **Monotropic development.** According to this theory, the plastids are autonomous structures which can change from one form to another in either reversible or irreversible way.

$$\text{Chloroplast} \xrightarrow{\text{Absence of light}} \text{Yellow coloured plastides (irreversible)}$$

$$\left.\begin{array}{l}\text{Chloroplast}\\ \\ \text{Leucoplasts}\end{array}\right\} \longrightarrow \begin{array}{l}\text{Chromoplasts (irreversible)}\\ \text{(grontic or old age stage)}\end{array}$$

(*b*) **Development of Plastids from Proplastids.** Plastids have received much attention in recent years and much work has been done on their development by Muihlethaler (1957), Gustafsson (1961) and many others. Several workers have suggested that the chloroplasts or any other plastids develop from tiny cytoplasmic bodies called *proplastids* or *plastid precursors*. These are structureless bodies which are bounded by a double-layered membrane and contain a clear matrix. They resemble mitochondria to some extent but they differ from the latter in having flattened infoldings that run parallel rather than perpendicular to the surface. The plastid precursors are self-duplicating and contain DNA and RNA. The proplastids of higher plants are nearly 0.5 μ in diameter. With the help of electron and florescence microscopes Epistin, Hyman and Schiff (1961) have observed the proplastids in dark grown *Euglenae* which were 1 μ in diameter and were similar to the plastid precursors of the higher plants. Proplastids develop into the chloroplasts when they are exposed to light. Various stages in the development of chloroplast are illustrated by a series of diagrams in Fig. 3.8.

The proplastids develop first in the cytoplasm of meristematic cells. On exposure of these cells to light the proplastids appear to lengthen and there develop 'blebs' from the inner side of the double layered membrane. Starch grains now appear inside them. According to Von Wettstein (1959), the steps involved in the development of chloroplasts from proplastids are as follows:

(*i*) Synthesis of vesicles within the plastid primordia,

(*ii*) Aggregation of these vesicles to form long chains of interconnected vesicles,

(*iii*) Rearrangement and fusion of the chains to form parallel double membranes or lamellae,

(*iv*) Multiplication of lamellae by linear splitting, and

(*v*) The growth and differentiation of lamellae into grana and stroma lamellae and in the meantime break down of starch also occurs. In higher plants the lamellae of chloroplasts appear to develop in bundles and form replicas (grana) in parallel geometric pattern. The general pattern of chloroplast formation is similar in all the higher plants. In short, all the pigments are concentrated in grana and the grana are distributed in stroma which forms the body of chloroplast.

(*c*) **Division and Budding.** In cryptogams nev plastids originate by division and budding of the mature chloroplasts. During the division ring-like constriction appears which gradually deepens and finally divides the plastid into fragments (Fig. 3.9). The fragments then differentiate into mature plastids. Budding of plastids takes place only under certain special circumstances, as for example, during the regeneration of a plant from excised leaves.

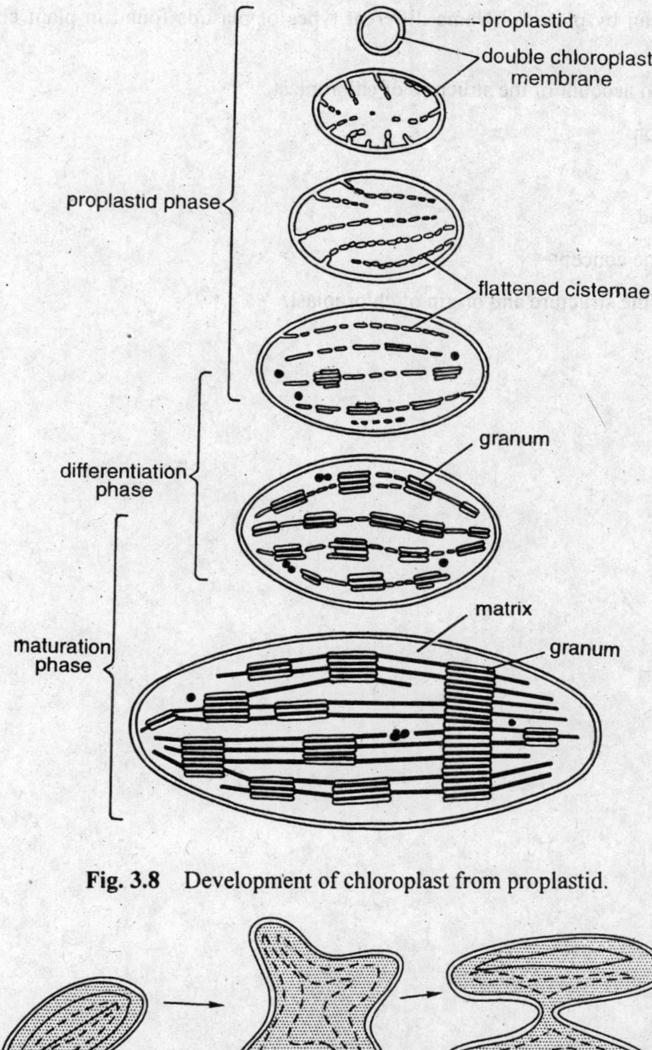

Fig. 3.8 Development of chloroplast from proplastid.

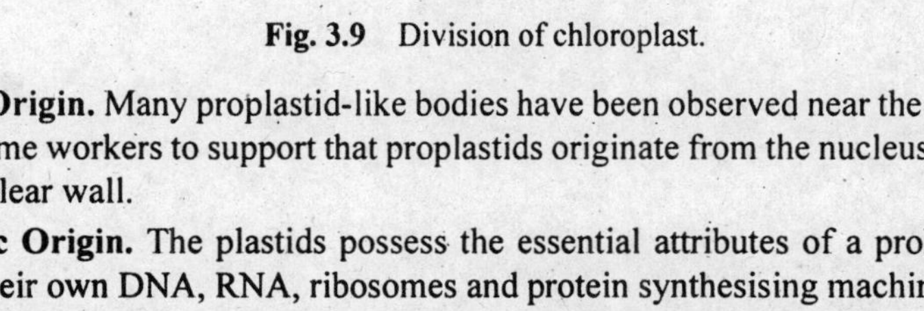

Fig. 3.9 Division of chloroplast.

(*d*) **Nuclear Origin.** Many proplastid-like bodies have been observed near the nucleus. This point has led some workers to support that proplastids originate from the nucleus due to evagination of the nuclear wall.

(*e*) **Symbiotic Origin.** The plastids possess the essential attributes of a prokaryotic cell. They possess their own DNA, RNA, ribosomes and protein synthesising machinery. Mutations can occur in their hereditary material causing alterations in their structures and functions. On the basis of these striking similarities between the plastids and prokaryotes, it has been suggested that plastids have evolved from some prokaryotes like blue-green algae which have permanently settled in the plant cells. The presumed symbiotic cells in the process of becoming plastids underwent several changes during which their autonomy was restricted and part of the genetic control of their development was assumed by the nucleus of the host cell.

QUESTIONS

1. What do you mean by plastids? Name different types of plastids found in plant cells and give their functions.
2. Give an illustrated account of the structure of chloroplast.
3. Write short note on:
   (*i*) Thylakoid,
   (*ii*) Stroma, and
   (*iii*) Quantasome concept.
4. Describe in brief the structure and origin of chloroplast.

4

CHAPTER Mitochondria

There are several synonymous terms for mitochondria such as chondriochonds, chondriomes, mitosomes, chondriosomes and so on. It was Kolliker who observed granule-like structures in the muscle cells of insects in the year 1880. Flemming (1882) named them 'fila' and later on Altmann (1890) named them 'bioplasts'. In 1897, Benda demonstrated similar objects in cells and assigned the name mitochondria to them.

Meves described mitochondria in plant cell in 1904. The advent of electron microscopy took the study of cell structure to a new level and as early as 1947 Buchholz published some electron micrographs of mitochondria teased out of *Tsuga* eggs and maize pollen mother cells but little internal details were available from the crude preparations. The internal structure of mitochondria was described independently by Palade and Sjostrand from ultrathin sections of animal cells in 1953 and the same basic organization has subsequently been shown to hold for all types of mitochondria.

Lewis and Lewis (1914) demonstrated the possibility that mitochondria are concerned with some metabolic activity of the cell and Hogeboom (1948), Kennedy and Lehninger (1948, 1950) and Lehninger (1951) have shown that mitochondria are the chief sites for cellular respiration.

The mitochondria are virtually present in all the aerobic cells. They are absent in bacteria, other prokaryotic cells and mature red blood cells of the multicellular organisms.

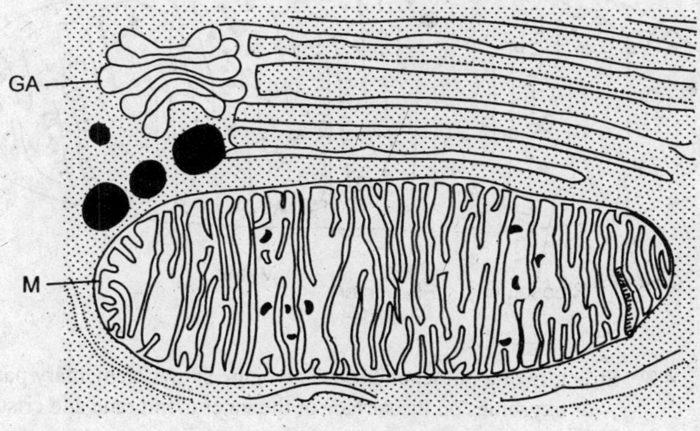

Fig. 4.1 Mitochondria (M.) and Golgi apparatus (GA) as seen in electron micrograph.

The mitochondria are colourless bodies widely distributed in the ground substance of the cytoplasm. These are easily differentiated from other cytoplasmic components by staining process. They are selectively stained by a special stain *Janus green.* The mitochondria are of different shapes. They may be fibrillar, spherical, rod shaped and oval and they may change from one form to another depending upon the physiological conditions of the cells. Mitochondria measure usually from 0.2μ to 2.0 μ in diam. and reach a length up to 40 μ. Generally, the length of mitochondria ranges from 3 to 5 μ and the size of mitochondrion is greatly influenced by the cell environment (pH, osmotic pressure, etc.). They contain numerus enzymes which take part in the oxidative steps of Kreb's cycle in respiratory process. The high energy phosphate compounds such as Adenosine diphosphate (ADP) and Adenosine triphosphate (ATP) are also synthesized and stored in mitochondria. These phosphate compounds after breakdown liberate tremendous amount of energy which is required in the completion of many chemical processes of the cytoplasm. This is why mitochondria are regarded as 'power house' of the cell. They are the principal but not the only sites of oxidation since the oxidation of some compounds also takes place in the ground substance of cytoplasm with the help of enzymes present therein. Mitochondria are found in association with other cytoplasmic organelles in some cells (Fig. 4.1).

Morphology of Mitochondria

Light microscopy can reveal no more than overall shape of mitochondria and for their internal structure one has to depend on electron microscopy. The ultrastructure of cell reveals that mitochondria are special type of containers bounded by two membranes of lipoproteins. The mitochondrial membranes are 50 to 70 Å thick and they can be resolved to display the triple layered unit membrane profile. These membranes are sturdy but flexible allowing contraction and expansion. The *outer membrane* is of smooth outline and shows high degree of stretching power. The *inner membrane* is highly infolded or involuted to form plate or *tube-like* structures, called *cristae mitochondriales* or *mitochondrial cristae*, of variable form and number. Except in the regions where it infolds to form cristae, the inner membrane follows the contours of the outer membrane. The two membranes are separated by an *intermediate space* or *intermembrane space* or outer chamber of fairly constant width which is 20 to 60 Å wide and continuous with the *intracristal space* or *lumen*. The internal space bounded by inner mitochondrial membrane is called *inner chamber*. In a number of cell types connections have been observed between the outer and inner chambers (Figs. 4.2 and 4.3).

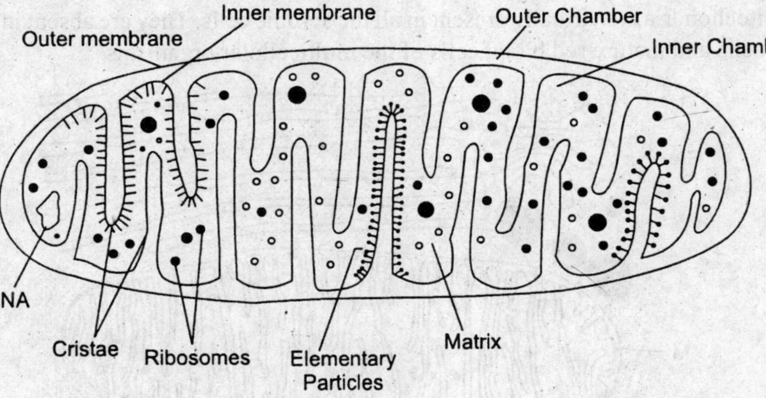

Fig. 4.2 Structure of a mitochondrion. The figure shows elementary particles or oxysomes on the surface of the inner membrane and cristae.

(*i*) **The mitochondrial membranes.** In recent years the two mitochondrial membranes and the chambers they limit have been separated by density gradient contrifugation method. The outer mitochondrial membrane can be separated by causing a swelling and breakage followed by contraction of the inner membrane and matrix. The outer membrane is much lighter requiring stronger centrifugal force or less gradient for separation and is more susceptible to fixation damage, sometimes disintegrating while the inner membrane is still intact. The other difference between these membranes lies in their osmotic behaviour. The inner mitochondrial membrane is an osmotic barrier to substances such as sucrose and contracts and expands in response to changes in sucrose concentraction. The outer mitochondrial membrane is permeable to electrolytes, water, sucrose and molecules as large as 10,000 daltons. The inner membrane is normally impermeable to ions, as well as to sucrose.

The inner mitochondrial membrane and matrix both intact when separated from the outer membrane are collectively referred to as *mitoplast.* The mitoplast possesses pseudopodic processes and is able to carry out oxidative phosphorylation. The outer membrane has about 40 per cent lipid content as compared to 20 per cent in the inner membrane, contains more cholesterol and is rich in phosphotidyl inositol but lower in cardiolipin. The lipid/protein ratio in outer membrane is quite different from that of inner membrane (0.8 in outer membrane and about 0.3 in the inner membrane). The outer mitochondrial membranes have been found to be pitted in six species of higher plants. The pits measure 25-30 Å in diameter (Parsons, Bonner and Verboon, 1965). In animal mitochondria, the outer membranes have not revealed any perforations.

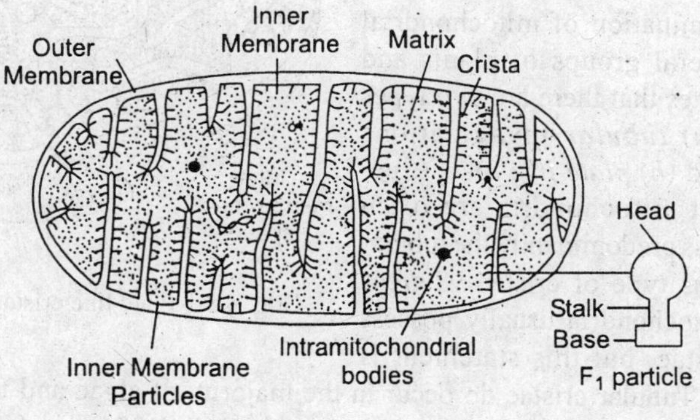

Fig. 4.3 Mitochondrion in longitudinal section.

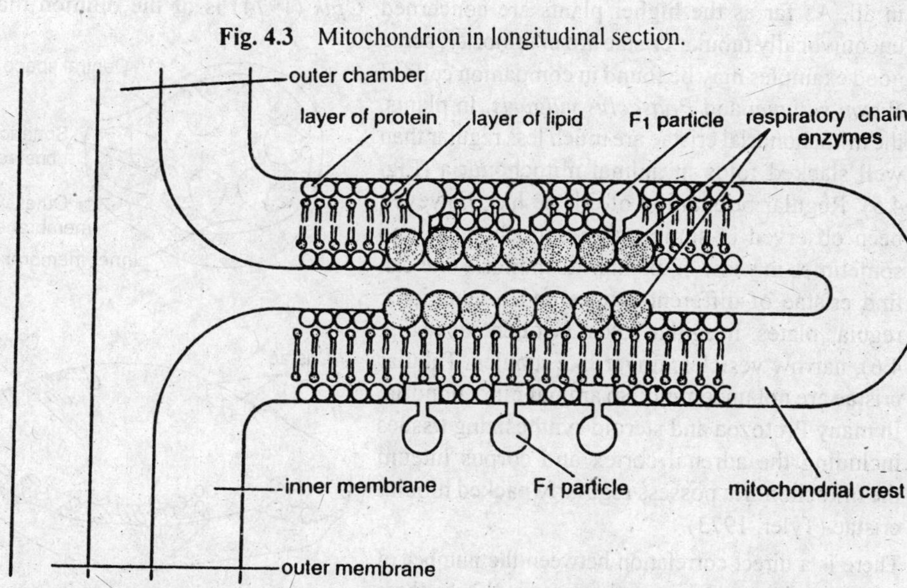

Fig. 4.4 Mitochondrial membranes and the molecular composition of inner membrane.

As regards the true width of intermembrane space, there still exists an uncertainty and, indeed, the very existance of intermembrane space and the intracristal luman in mitochondria is doubtful as, in fact, different fixatives and treatments give different results. Both the common fixatives, osmium tetraoxide and potassium permanganate, usually produce a narrow but distinct inter membrane space. Freeze-etching which is assumed to stabilize organelle morphology unchanged, has shown that the two membranes of yeast mitochondria are closely adpressed leaving no intermembrane space, while the intracristal spaces are narrow, nearly 50 to 100 Å across (Moor and Muhlethaler, 1963). The mitochondrial space appears very narrow in freeze etched onion (*Allium cepa*) root tip cells (Branton and Moor, 1964). Quick freezing of animal tissues have yielded micrographs showing no gap between the outer and inner membranes and no gap between the membranes of the cristae (Malhotra 1965, Steinert, 1969). Thus, there may be much less space between the membranes *in vivo* than that indicated by conventional electron microscopy. The intermembrane and intracristal spaces have so far revealed no visible structure and appear electron translucent. They are generally assumed to contain an aqueous fluid.

(*ii*) **The Cristae.** The inner mitochondrial membrane limiting the mitoplast has two faces; the outer *cytosol* or *C-face* toward the outer chamber of mitochondrion and an inner *matrix* or M-face toward matrix. A careful examination of mitochondrial cristae in several groups of plants and animals indicates that there are two types of cristae: (*a*) *tubular* invaginations (*microvilli*) and (*b*) *plate-like* folds (true-cristae). Plant mitochondria are often stated to possess predominantly the tubular or microvillous type of cristae whereas the animal mitochondria usually possess plate-like cristae, but this statement is too sweeping. Tubular cristae do occur in the majority of algae and fungi but by no means in all. As far as the higher plants are concerned, *Opik* (1974) is of the opinion that really unequivocally tubular cristae are in minority; some good examples may be found in companion cells of *Pisum sativum* and *Phaseolus vulgaris*. In plants, the mitochondrial cristae are much less regular than well stacked folds in animal mitochondria (Fig. 4.5). Regular orientation of cristae has, however, been observed in some pollen mitochondria and sometimes in yeast. In plant mitochondria one can find cristae of different shapes and orientations: regular plates, irregular folds, swollen sacs (Fig. 4.6), narrow vesicles, finger like tubules. Tubular cristae are not uncommon in animal mitochondria. In many Protozoa and steroid synthesizing tissues including the adrenal cortex and corpus luteum the mitochondria possess regularly packed tubular cristae (Tyler, 1973).

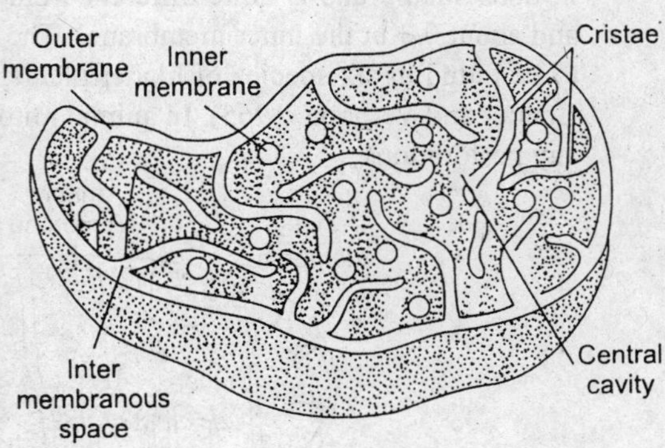

Fig. 4.5 Plate like cristae in mitochondrion.

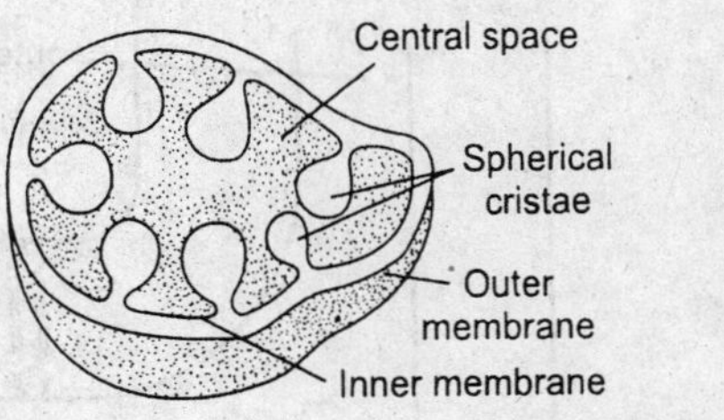

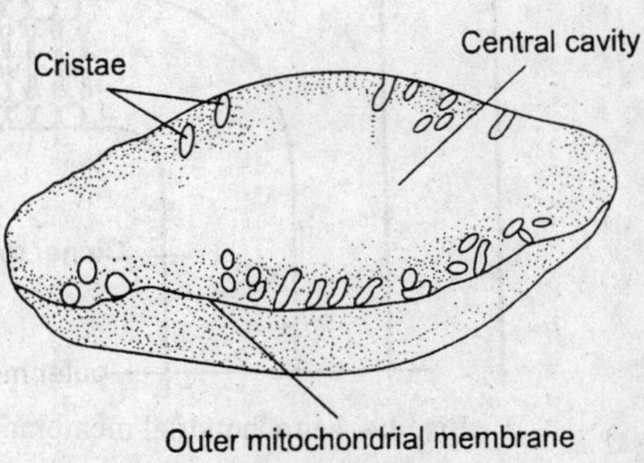

Fig. 4.6 Sac-like Mitochondrial cristae.

There is a direct correlation between the number of cristae and the volume of the matrix. Where there are relatively few cristae, there is much matrix and increase in the number of cristae affects the amount of matrix in the inner compartment of the mitochondrion.

(*iii*) **Mitochondrial particles.** A critical examination of negatively stained preparation, as well as positive contact preparations (Ashhurst, 1965) have revealed the existence of regular array of particles along the surface of cristae or inner mitochondrial membrane facing the matrix. Discovered first by *Fernandez-Moran,* they are called inner membrane *sub-units* or *elementary* *particles* or *Factors of Racker* (F_1) or *oxysomes* (Chance and Parsons). These elementary particles have been observed on the cristae of both plant and animal mitochondria. According to Fernandez-Moran, each of these repeating units is tripartite structure consisting of three parts, a polyhedral or roughly spherical *head,* a narrow stem or *stalk* and a roughly cuboid *base.*

Tripartite unit = basepiece + stalk + head piece.

The stalks (narrow stems) measure nearly 40 to 60 Å in length and 30 to 40 Å in diameter, the heads measure approximately 75 to 110 Å in diameter and bases measure approximately 40 Å. The centre to centre distance between the sub-units measures nearly 110 Å (Parsons, 1963). The base pieces of these particles form a part of the inner membrane surface (Fig. 4.4). The mitochondrial membranes have about 2,700 particles per square micrometer and depending upon the size and type of mitochondrion, there are from 10,000 to 1,00,000 elementary particles per mitochondrion. The oxysomes occur in thickness of the inner mitochondrial membrane but they can be seen out of the membrane surface only under special mounting conditions. These particles play an important role in respiration. Functionally, the elementary particles are associated with the presence of enzymes for oxidative phosphorylation and with mitochondrial adenosine triphosphatase (AT Pase). Green (1959) and Green and Hatefi (1961) have proposed a model for sructure of the inner cristal membrane based on subfractionation and electron microscopy. The membrane has been fractionated into the following four functional complexes each with a total molecular weight of approximately 5×10^5 and containing enzymes, structural proteins, phospholipids and insoluble cofactors.

Complex I — NADH-Co Q reductase, catalysing electron transport from NADH to Co. Q.

Complex II — Succinate-Co Q reductase, catalysing electron transport from succinate to Co-Q.

Complex III — Co. Q-cytochrome c reductase, catalysing electron transport from Co. Q to cytochrome c.

Complex IV — Cytochrome c oxidase.

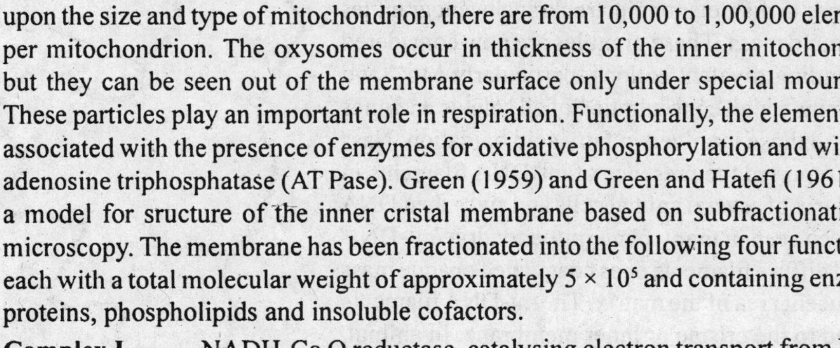

Complexes I, III and IV or II, III and IV in sequence constitute the electron transport chain. CoQ and Cyt. c are considered to act as mobile connecting links or mobile electron carriers between complexes.

The basepieces which are joined laterally through their phospholipid moieties in the membrane are not detachable from the membrane without destroying it, while the stalks and heads can be detached leaving the membrane still intact. The detachable portion is believed to be identical for all. Green and Silman (1967) considered electron transfer chain to be located exclusively in the base piece of elementary particles. The stalk contains a protein of low molecular weight (OSCP, oligomycin sensitivity conferring protein) that binds oligomycin and does not show any enzyme activity. The only enzyme activity detected in the head piece is ATP synthetase, which is probably active in ATP synthesis under normal conditions. The stalk probably acts as a link for couplig electron transport in the base pieces to terminal stages of oxidative phosphorylation (*i.e.*, ATP synthesis) in the head piece. According to Green, the complete electron transfer chain corresponds in size to the elementary particle. Chance, on the other hand, suggests that each elementary particle carries a single electron carrier component and some additional proteins. Chance and Parsons (1963) have suggested the term *oxysome* to represent a factor for coupling phosphorylation to respiration. The elementary particle would then consist of oxysomes or parts thereof.

Matrix. The inner chamber contains a translucent substance, the mitochondrial matrix. The solid component of the matrix is chiefly protein (about 56%). The matrix presents a finely granular texture or may give impression of a meshwork of fibrils. In the matrix there may occur electron-opaque granules, nearly 60 Å or more in diameter or electron translucent granules in amorphous ground mass. These granules are now considered to be sites for binding divalent cations, particularly Mg^{++} and Ca^{++}. Matrix contains a high concentration of all soluble enzymes of Krebs' cycle and those involved in fatty acid oxidation. Nass and Nass (1963) reported the presence of mt-DNA filaments in mitochondrial matrix of several animal cells and since then DNA filaments have been seen in many plant mitochondria. mt-DNA occurs as fine and diffuse filaments or as coagulated opaque mass in electron translucent area of the matrix. The mt-DNA filaments may be connected to the cristae or inner membrane. In animal, mitochondrial DNA is small, about 5 μ long and ring-shaped (Fig. 4.7) like bacterial or phage DNA, but it is linear in plants except yeasts (26 μ long). Generally a mitochondrion contains only one mt-DNA containing region, but sometimes several such regions may be observed, especially in elongated mitochondria.

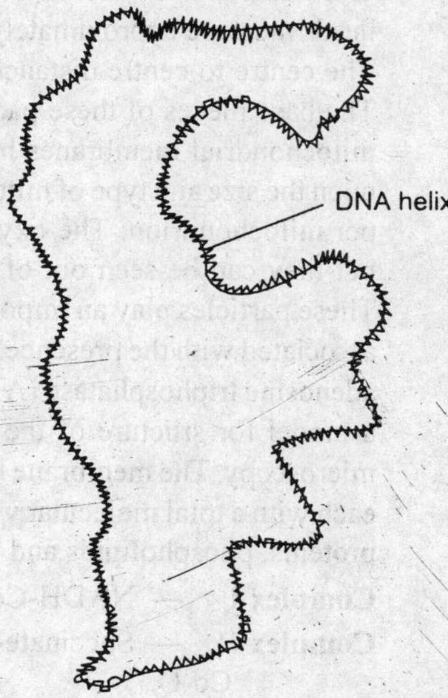

Fig. 4.7 Mitochondrial DNA.

Mitochondrial DNA is different from nuclear DNA in having different purine and pyrimidine bases. Rabinowitch (1968) has shown that mitochondrial DNA contains more Guanine and Cytosine (GC) contents than the nuclear DNA. Mitochondria have a DNA polymerase different from nuclear DNA polymerase.

In mammalian cells mitochondrial DNA constitutes less than 1% of nuclear DNA. It has very low molecular weight (about 10^6) and contains few coded information. Mitochondria can use their DNA to synthesize complementary RNA. Like plastids, mitochondria have their separate protein synthesizing machinery, *i.e.*, they have their own t RNAs (transfer RNAs, small sized 55S ribosomes, aminoacids and enzymes necessary for activating aminoacids. Mitochondrial ribosomes (mitoribosomes) can be seen in the matrix, looking slightly smaller than the cytoplasmic ribosomes (cytoribosomes). They may be found randomly distributed or can form small groups, possibly polysomes. Sometimes mitoribosomes appear associated with cristae or inner membrane. They can synthesize some, but not all of their constituent proteins. They cannot synthesize cytochrome c, the synthesis which takes place on the ribosomes in extramitochondrial cytoplasm. Mitochondria synthesize structural proteins that are thought to act as self-assembly organizers for the lipids, the enzymes and the other proteins which constitute a complete membrane system. Parson and Simpson (1967) have shown that the mitochondria contain the enzyme DNA polymerase which helps in the replication of mitochondrial DNA. Mitochondrial DNA is self-duplicating unit and it has been suggested that the duplication of mitochondrial DNA is controlled by nuclear genes. Mitochondria are, therefore, considered as semi-autonomous structures in the cells.

Structural variations

The structure of mitochondria is not constant in all plants and animals. The mitochondria present several structural variations. These variations are caused due to the differences in number of cristae, size, shape and arrangement of cristae and fusion of mitochondria themselves.

1. **Variations due to number of cristae.** The number of cristae per unit volume varies greatly in the mitochondria of most plant and animal cells and the number of cristae seems to be correlated with oxidative activity in mitochondria.
2. **Variations due to the structure and orientation of cristae.** Normally cristae in rod shaped mitochondria are oriented perpendicular to the long axis of the mitochondria, but this is too a general statement. The cristae may be parallel to long axis of the mitochondrion, as in marrow and striated muscles or they may be arranged as vesicles or occasionally branched to form a network of connecting chambers, as in human leucocytes, protozoa, parathyroid gland and corn mitochondria, or they may be extremely reduced. In some mitochondria the cristae may be arranged in concentric rings and in some cases they may be haphazardly arranged.
3. **Variations due to fusion.** De Robertis (1957) has reported a peculiar state in spermatids of certain insects in which all mitochondria in the cytoplasm aggregate around the nucleus and they all fuse together to form a single mitochondrial body.

Chemical Composition

The gross chemical composition of mitochondria varies in different cells of both plants and animals. Typically, however, by dry weight mitocondria are about 65 to 75% protein and nearly 25 to 30% lipids. Of the lipid component, 90% is phospholipid and 10% carotenoids, cholesterol, vitamin E and other traces. Mitochondria contain sulphur, iron, copper and some vitamins in traces which are mostly related to enzyme activities. They also contain small amount of mt-DNA and about 0.5% RNA. When mitochondria are separated from a cellular environment and ruptured, some of the enzymes associated with matrix are released as soluble proteins while the other enzymes remain firmly bound to the membranes. The mitochondria contain about 70 enzymes and about a dozen co-enzymes and numerous co-factors. Soiuble enzymes from mitochondria include all enzymes of the tricarboxylic acid cycle (TCA) except some dehydrogenases, those that catalyze β oxidation of fatty acids and others that catalyze transamination of aminoacids and synthesis of protein. Membrane bound enzymes of mitochondria are these essential for electron transport chain and oxidative phosphorylation. Mitochondria do not contain the enzymes for anaerobic glycolysis which occurs in the groundplasm. Lehninger (1969) has identified the following enzymes in different regions of mitochondria:

Part of Mitochondria	Enzymes
Outer membrane	Monoamine oxidase, Rotenone-insensitive NADH, cytochrome c reductase, Kynuranine hydroxylase and fatty acid Co. A ligase.
Outer mitochondrial Chamber	Adenylate kinase and Nucleoside diphosphokinase.
Inner membrane	Respiratory chain enzymes, ATP synthatase, Succinate dehydrogenerse, β-Hydroxybutyrate dehydrogease, Canitine fatty acid acyl transferase.
Matrix	Malate and Isocitrate dehydrogenases, Fumerase and aconitase, citrate synthetase, α-keto acid dehydrogenases and β-oxidation enzymes, enzymes concerned with transamination of aminoacids and synthesis of proteins.

Development of Mitochondria

In the meristematic cells of higher plants the mitochondria are not well differentiated and their cristae are irregular, sac-like and a large proportion of matrix is occupied by electron translucent regions with conspicuous DNA filaments. Ribosomes and ribosome like particles can also be seen frequently in them. The undifferentiated mitochondria undergo a series of changes in the course of maturation and the development takes place in the following steps:

(*i*) Mitochondrial size may increase.

(*ii*) Changes in the opacity of matrix and dilation of the cristae take place.

(*iii*) With the advancement of cell age the DNA filament and the surrounding electron translucent areas become less conspicuous or may become completely obliterated (Street, Opik and James, 1968; Opik, 1965). This does not necessarily mean a disappearance of the mitochondrial DNA which can still be isolated from the mature tissues.

(*iv*) The mitochondrial number per cell increases as the cells grow; accurate numerical estimations, however, are not easy and only a few have been made. In maize root cap the number of mitochondria per cell increases from 50 in the cap initial to approximately 175 in the mature cells (Clowes and Juniper, 1964).

Mitochondria do persist in dry dormant seed tissues of higher plants and in spores. The protoplasmic changes which occur during dehydration greatly lower the membrane contrast. Additionally, the number and width of cristae per mitochondria may decrease as seeds dry out, as for examples, in Pea cotyledons (Bain & Mercer, 1966) and radicle (Perner, 1966). DNA and granular mitoribosomes may still be discernible.

Origin

There are at least four different views regarding the origin of mitochondria and these are as under:

(*i*) Autonomous or *de novo* origin.

(*ii*) Origin from promitochondria.

(*iii*) Origin by division or budding of pre-existing mitochondria.

(*iv*) Nuclear origin.

(*v*) Prokaryotic origin.

(*i*) **Autonomous or *de novo* origin of mitochondria.** According to this view, the mitochondria are autonomous organelles with independent hereditary action and leading symbiotic life. The mitochondria are supposed to have originated *de novo* from the simple building blocks such as aminoacids and lipids. Some observations are available where mitochondria were completely damaged and it was found that new chondriomes developed *de novo.* The significance of this observation can be questioned and it is still not clear whether mitochondria are fully autonomous or semi-autonomous bodies.

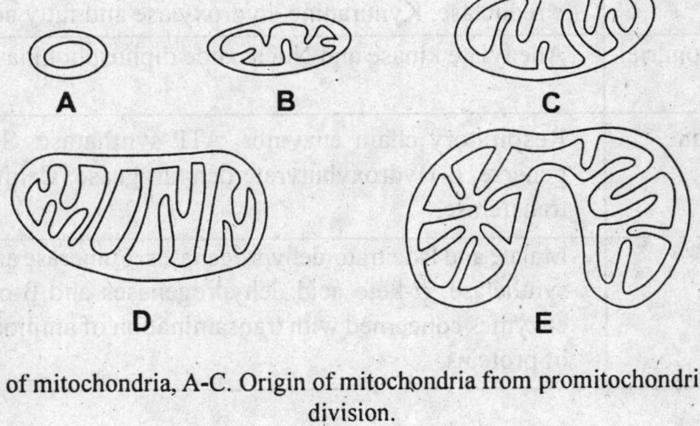

Fig. 4.8 Origin of mitochondria, A-C. Origin of mitochondria from promitochondria, D-E. Origin by division.

(*ii*) **Origin from promitochondria.** Like plastids, the mitochondria also originate from the special type of cytoplasmic precursor bodies which are called the 'promitochondria'. These bodies do not have cristae and cannot use oxygen. Such promitochondria have been observed in anaerobically grown yeast. The precursor bodies after a few hours of exposure to oxygen develop into normal mitochondria with cristae [Figs. 4.8 A, B and C].

According to Morrison (1966), "The promitochondria might originate from endoplasmic reticulum or plasma membrane". The process is illustrated in Fig. 4.9. The hypothesis is not supported by direct evidence, hence not valid at present.

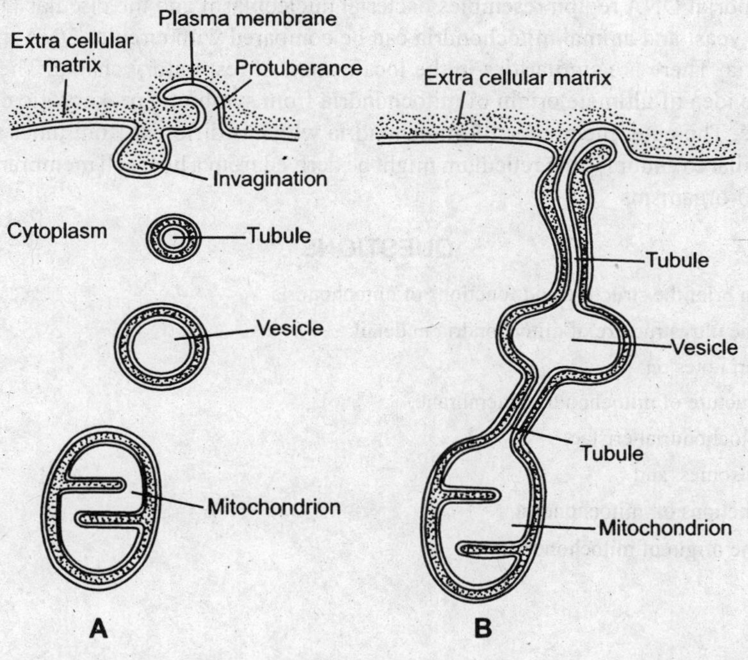

Fig. 4.9 Hypothetical diagram showing origin of mitochondria from plasma membrane.

(*iii*) **Origin by division and budding of pre-existing mitochondria.** Mitochondria have been found to undergo division and budding [Figs. 4.8 D and E]. It has been observed with time-lapse cinomatography that mitochondria gradually elongate and then fragment into smaller mitochondria. This observation has been verified in *Neurospora* by radioactive tracer technique. Budding of mitochondria may occur in differentiated cells giving rise to mitochondrial initials. These can then increase in size and develop cristae. As regards the timing of mitochondrial division, very little informations are available. In shoot meristem of *Epilobium hirsutum* the mitochondrial multiplication occurs during interphase (Anton-Lamprecht, 1967).

(*iv*) **Nuclear origin of mitochondria.** Sometimes the double layered nuclear wall evaginates to form vesicles which when detached from the nuclear wall give rise to promitochondria or protoplastids. In such a development, the mitochondria are independent bodies. A claim for the evagination of mitochondria from the nuclear membrane in developing egg cells of the fern *Pteridium aquilinum* has been made by Bell and Muhlethaler (1964) on electron microscopic evidence, but it has been firmly denied by other authors (Diers, 1964; Tourte, 1968) and recent works from Bell's own laboratory has failed to produce supporting evidence from radioactive labelling experiments (Sigee and Beu, 1971).

Mitochondrial initials may also represent derivation from plasma membrane. Robertson (1964) has suggested a scheme whereby mitochondria may arise from invaginations of plasma membrane (Fig. 4.9) or from endoplasmic membrane.

(*v*) **Prokaryotic origin or symbiotic hypothesis.** When Altman (1890) first described bioplasts, he considered them to be *Element-organismen* evolved from and related to free living forms such as bacteria. According to symbiotic hypothesis mitochondria may be considered as intracellular symbiotic organisms living in association with higher cells. The homologies between mitochondria and bacteria are numerous and when considered from an evolutionary viewpoint, they may be more than circumstantial. The size of a mitochondrion is of the same order or magnitude as that of bacterial cell. Bacteria bear oxidative enzymes in the cell membrane. The protein synthesizing system in mitochondrion is of bacterial type and the mitochondrial DNA region resembles bacterial nucleoplasm and the circular DNA molecules found in yeast and animal mitochondria can be compared with circular DNA molecule found in bacteria. There are similarities in the localization of respiratory chains. These similarities make the idea of ultimate origin of mitochondria from symbiotic prokaryotic organisms very attractive. The outer membrane of mitochondria which is different from inner membrane but quite similar to endoplasmic reticulum might be derived from a host cell membrane surrounding the micro-organisms.

QUESTIONS

1. Discuss in brief the structure and functions of mitochondria.
2. Discuss the ultrastructure of mitochondria in detail.
3. Write short notes on:
   (*i*) Structure of mitochondrial membrane,
   (*ii*) Mitochondrial cristae,
   (*iii*) Oxisomes. and
   (*iv*) Functions of mitochondria.
4. Discuss the origin of mitochondria.

5 CHAPTER

Endoplasmic Reticulum and Ribosomes

ENDOPLASMIC RETICULUM

The cytoplasmic matrix of eukaryotic cell contains a complex network of membrane bound structures called endoplasmic reticulum. The term endoplasmic reticulum was introduced by Porter and Kallman (1952). By the use of ultrathin sections and improved fixation techniques developed by Palade and Porter (1954) it was finally recognised that the endoplasmic reticulum presented cavities of a great varieties of shapes and dimensions surrounded by membranes. The endoplasmic reticulum is found almost universally in eukaryotic cells. It is lacking in the bacterial and myxophycean cells.

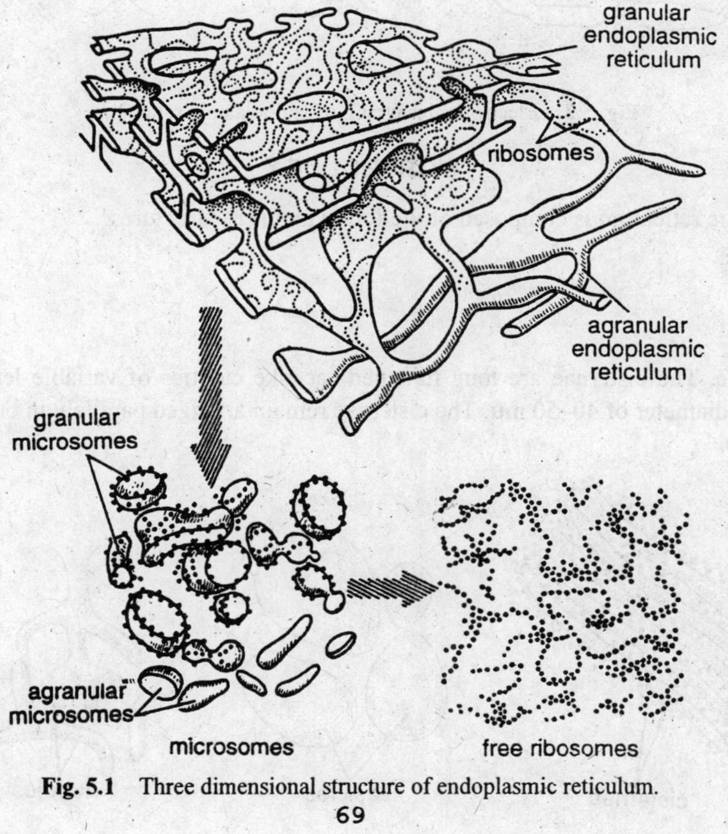

Fig. 5.1 Three dimensional structure of endoplasmic reticulum.

It is a system or network of inter-connected membrane-bound fine tubules called **canaliculae** (Figs. 5.1 and 5.2). Obviously, endoplasmic reticulum (ER) is a hollow system. Sometimes, it appears as a continuous system, connected on one side to plasma membrane and on the other side to the nuclear envelope. This continuity is not recognizable in the thin sections of the cell.

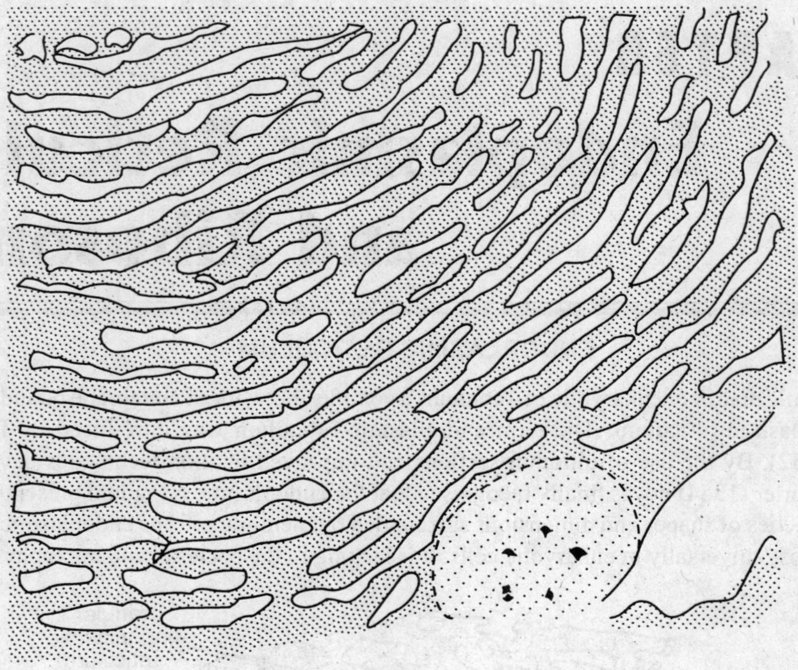

Fig. 5.2 Electron micrograph of endoplasmic reticulum.

MORPHOLOGY

The endoplasmic reticulum is composed of the following three structures:

1. Cisternae.
2. Vesicles.
3. Tubules.

1. Cisternae. The cisternae are long flattened sac like cavities of variable length and width having a diameter of 40–50 mμ. The cisternae remain arranged parallely in bundles or stacks (Fig. 5.3).

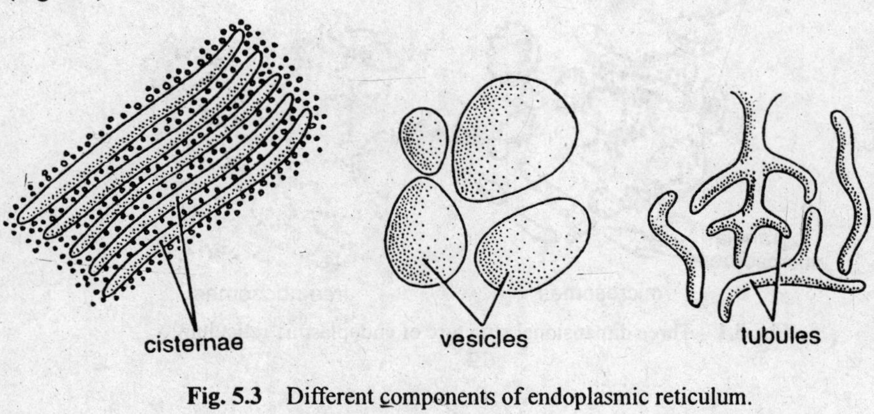

Fig. 5.3 Different components of endoplasmic reticulum.

2. **Vesicles.** The vesicles are oval membrane-bound vacuolar structures having a diameter of 25–50 mμ. They often remain free in the cytoplasm and occur in most of the cells.

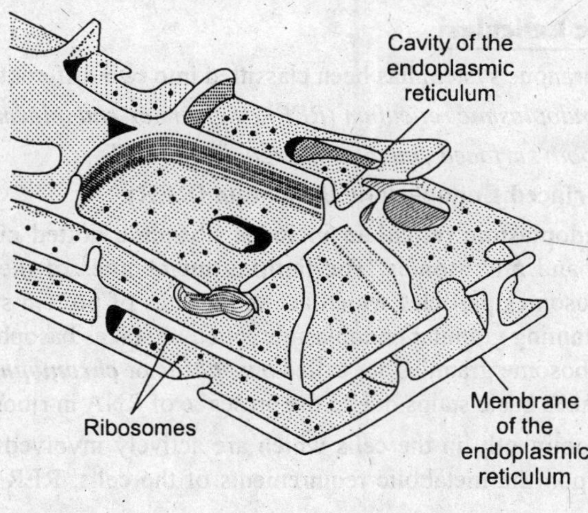

Fig. 5.4 Endoplasmic reticulum in three dimensional view.

3. **Tubules.** Tubules or canaliculae are branched structure which along with cisternae and vesicles form a network. The tubules are about 50–190 mμ in diameter and found almost in all the cells. Ultrastructural studies have revealed that the cavities of cisternae, vesicles and tubules of endoplasmic reticulum are bounded by a thin unit membrane of 50–60 Å thickness (Fig. 5.4). Thus, they are thinner than the plasma membrane. It may show a trilaminar structure in some regions, a micellar or fluid mosaic configurations in other regions. The tubules extend in whole cytoplasm except in the golgi body area.

The membranes of endoplasmic reticulum are studded with single layer of opaque particles of ribonucleoprotein called the ribosomes (Fig. 5.5). The endoplasmic reticulum in certain region of the cell may be devoid of ribosome particles.

The ribosome particles are firmly attached to the membranes and they can be separated from the membranes when drastic forces and techniques are applied on them. In one such technique the cell is treated with sodium deoxycholate; this chemical dissolves the membranes and releases the ribosomes. It has been shown that the ribosomes attached to endoplasmic reticulum are the chief sites for the protein synthesis in the cytoplasm. Usually pores or annuli are lacking in the membranes of canaliculae in the majority of cells but in certain cases the pores or annuli have been noticed, as for example, in the E.R. of invertebrate cells, and oocytes and spermatocytes of the vertebrates. According to Merriam (1959) and Kessel (1963), the annulate lamellae of endoplasmic reticulum arise by the evagination of nuclear envelope.

Several other membranous systems, such as infoldings of plasma membrane, golgi bodies and vacuoles, when properly analysed have been found in direct continuity with the ribosome dotted endoplasmic reticulum. On this

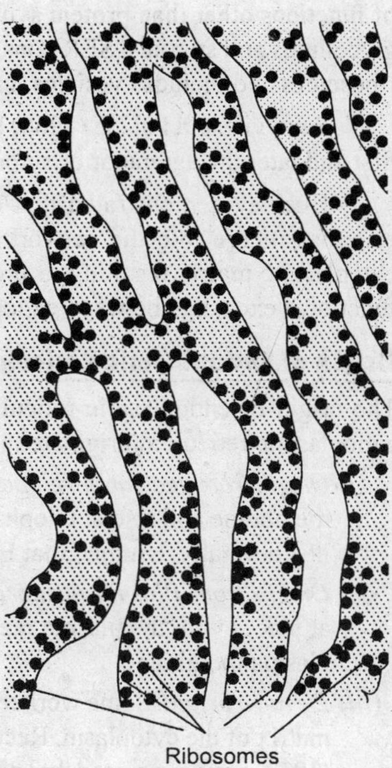

Fig. 5.5 Endoplasmic reticulum with ribosomes.

ground it has been suggested that all the intracellular membranes should be recognised in one single endoplasmic membranous system. This is ***endomembrane concept***.

Types of Endoplasmic Reticulum

The endoplasmic membranous system has been classified into two different sub-systems:

1. *Rough surfaced endoplasmic reticulum (RER) or granular endoplasmic reticulum.*
2. *Agranular or Smooth surfaced endoplasmic reticulum (SER).*

Granular or Rough surfaced Endoplasmic Reticulum (RER)

The rough surfaced endoplasmic reticulum includes ribosome dotted cisternae and tubules of endoplasmic reticulum and microsomes. The limiting membranes of these tubules are studded with single layer of ribosomes particles which act as the sites of protein synthesis. The region of cytoplasmic matrix containing granular endoplasmic reticulum takes basophilic stains and is named as ***ergastoplasm***. The ribosome granules are called ***basophilic*** or ***chromophilic substances*** or ***Nissl bodies***. The reticulum takes basic stains due to the presence of RNA in ribosomes.

RER occurs predominantly in the cells which are actively involved in protein and enzyme synthesis. Depending upon the metabolic requirements of the cells, RER may be converted into SER or *vice-versa.*

The microsomes are special type of organelles in the cytoplasm which are described as small vesicles bounded by thin surface membranes of lipoproteins impregnated with small ribosome particles. These are supposed to have developed when canaliculae of the ergastoplasm are broken into small spheres. The microsomes lie scattered in the cytoplasm. These RNA rich bodies are the chief sites for protein synthesis and their membranes are involved in the steroid synthesis (Fig. 5.10B).

Agranular or Smooth surfaced Endoplasmic Reticulum (SER)

The smooth surfaced endoplasmic reticulum occurs generally in those cells which perform a variety of functions other than protein synthesis. SER is characteristic of cells synthesising phospholipids, glycolipids and steroids. SER consists usually of smooth walled tubules which are not studded with ribosomes and includes vesicles, vacuoles, infoldings of plasma membranes, golgi body etc.

Seikevitz and Palade (1966) have reported that the granular endoplasmic reticulum originated first and later the agranular or smooth surfaced reticulum developed from the former.

Buckley demonstrated that endoplasmic reticulum might be both mobile and elastic. The cisternae merge into the network and endoplasmic reticulum continuously sweeps through the cytoplasmic matrix resulting in proper distribution of the endoplasmic reticulum components and functional exchange between endoplasmic reticulum components and the matrix.

Origin of Endoplasmic Reticulum

The origin of endoplasmic reticulum is not exactly known. But it is normally assumed that the endoplasmic reticulum originated in one of the following ways:

(*i*) *Origin from nuclear envelope.* Since the membranes of endoplasmic reticulum resemble nuclear membrane and also at telophase stage the ER membranes are found to form nuclear envelope, it is normally assumed that ER originated by evagination of the nuclear membranes.

(*ii*) *Origin from Plasma membrane.* Since the tubules of endoplasmic reticulum are connected at places with the plasma membrane it is presumed to have originated by invagination of the plasma membrane.

(*iii*) *De novo origin.* Some workers believe that endoplasmic reticulum has a *de novo* origin from the matrix of the cytoplasm. Recently some workers like Leskes *et al.* (1971) and Eytan and Ohad (1972) have suggested that daughter cells receive full set of membranes from their parental cell and, therefore, the question of *de novo* synthesis of membranes does not arise and endoplasmic reticulum arises by the expansion of pre-existing membranes.

Enzymes of Endoplasmic reticulum

The membranes of the canaliculae of endoplasmic reticulum are reported to contain many kinds of enzymes which perform various important functions. The important enzymes occurring in the membranes are the stearases, NADH-cytochrome c reductase, NADH diaphorase, glucose 6 phosphatase and Mg^{++} activated ATPase. Certain enzymes of the endoplasmic reticulum are involved in the biosynthesis of phospholipids, ascorbic acid, glycerides, steroids and hexose, etc.

Functions of Endoplasmic Reticulum

The endoplasmic reticulum is concerned with several important activities of the cells such as secretion, storage, circulation and so on. The important functions of endoplasmic reticulum are listed under the following heads:

(A) **Functions common for RER and SER**

1. The endoplasmic reticulum acts as ultrastructural skeletal framework in the cell and provides mechanical support.
2. The endoplasmic reticulum acts as transporting system and transports various products to different organelles.

   RER → SER → Cisternae of golgi body → Vacuoles or Secretary granules
3. The exchange of molecules through osmosis, diffusion and active transport occurs through the membranes of endoplasmic reticulum.
4. The endoplasmic membrane contains many kinds of enzyme which perform various synthetic and metabolic activities. Further the membranes of ER provide increased surface for different enzymatic reactions.
5. ER membranes are reported to form new nuclear envelope at telophase stage.
6. ER membranes are found to conduct intra-cellular impulses.

(B) **Specific functions of SER**

1. Synthesis and storage of lipids and lipoproteins.
2. *Glycogenolysis.* Break down of glycogen through the action of glucose 6 phosphatase occurs in SER.
3. *Detoxification.* Harmful materials such as drugs, pollutants as well as metabolic wastes are converted by SER into the substances suitable for excretion by the cell. The detoxification may take place through oxidation, reduction, hydrolysis or covalent linking which make the poisonous substances either inactive or more soluble so that they may be easily eliminated.
4. *Synthetic functions.* SER is involved in the synthesis of a variety of substances such as steroids, pigments and cell wall materials.

(C) **Specific functions of RER**

The major function of RER is the synthesis of proteins. In fact, the array of ribosomes of RER provides the surface for association of active enzymes, amino acids during the synthesis of proteins. The newly synthesized protein instead of passing into the cytoplasm first passes into the cisternae of RER and is thus protected from protease enzymes of cytoplasm. The growing protein chain folds into its secondary and tertiary structures and are thus trapped in the cisternae of RER (Figs. 5.6 and 5.10).

Glycosylation of Secretory Proteins. Nearly all the proteins for secretion are glycoprotein excepting albumin. Addition of carbohydrate to polypeptide chain is called glycosylation. While the polypeptide chain is growing, sugar molecules appear to be added one at a time usually from the nucleotide UDP in the cisternae of RER (Fig. 5.6). However, if the glycoprotein is to contain a terminal glucose, Fucose or Sialic acid, the sugars are added in the golgi complex and canaliculae of RER where appropriate sugar transferases are localized.

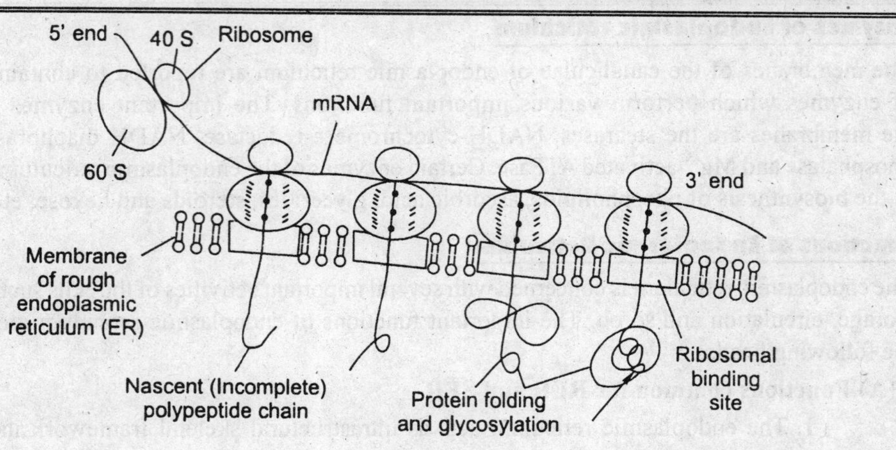

Fig. 5.6 Synthesis of secretory proteins on RER.

RIBOSOMES AND POLYRIBOSOMES OR POLYSOMES

In the cytoplasm of the cell there are found particles composed of RNA (ribonucleic acid) and proteins. These are called *ribosomes*. These particles occur in all the cells that synthesize proteins and they are involved in the synthesis of cellular proteins. The sizes of ribosomes vary somewhat, being approximately 150 Å in bacteria, chloroplasts and mitochondria and 140 to 200 Å in the cytoplasm of eukaryotic plant and animal cells. The ribosome particles are the integral part of rough surfaced endoplasmic reticulum and microsomes. However, it has now been shown that these particles exist in free state in the cytoplasmic matrix. In young meristematic cells, the ribosome granules lie freely in the cytoplasmic matrix and the young endoplasmic reticulum appears to be smooth surfaced. In bacteria and other prokaryotes the ribosomes seem to exist freely in the cell. Ribosomes are often attached in rows or in clusters. These clusters or strings of ribosomes are called ***polyribosomes*** or ***polysomes*** (Fig. 5.7). The polysomes consist of such ribosomes as are actively engaged in protein synthesis and are held together by a string of messenger RNA (mRNA).

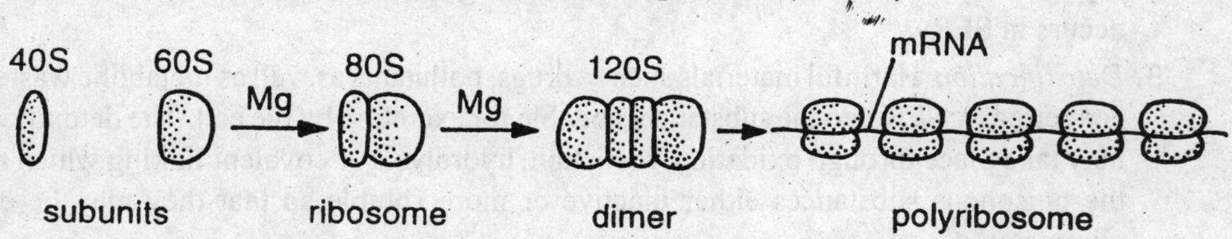

Fig. 5.7 Sub-units of ribosome, whole ribosome, dimer and polyribosome.

Structure of Ribosomes. Ribosomes are generally defined by their sedimentation coefficients or Svedberg units or 'S' values (Svedberg unit is calculated on the basis of several factors). Essentially it represents sedimentation constant (in seconds) per unit field of force and is equal to 1×10^{-13}). On the basis of sedimentation coefficients there are two categories of ribosomes:

(*i*) Ribosomes found in bacteria, blue-green algae and those found in chloroplasts and mitochondria of higher organisms exist as *70 S particles* (varying from 67 S to 73 S).

(*ii*) Ribosomes found in the cytoplasmic matrix of eukaryotic cells exist as 80 *S particles* (varying from 77 S to 81 S).

Each ribosome consists of two sub-units: a smaller sub-unit (molecular weight 8,00,000) and a bigger sub-unit (molecular wt. about 108 million).

70 S ribosome particles can be split into *50 S larger sub-unit* and *30 S smaller sub-units* (Figs. 5.8 and 5.9). Actually the two sub-units of ribosomes found in bacteria occur freely in cytoplasm and they unite only during protein synthesis. 80 S particles can be split into *60 S larger sub-unit* and 40 S *smaller*

sub-unit (Fig. 5.10). The 40 S sub-unit occurs above the 60 S sub-unit forming a caplike structure and the 60 S sub-unit is dome-shaped and it remains attached to the membranes of canaliculae in fixed ribosomes. The structure of ribosomes is still less worked out. According to Nauninga (1967), the 50 S sub-unit of 70 S ribosome is pentagonal campact particle of 160 to 180 Å with a round area of 40-60 Å in diameter. Florends (1968) has noticed a pore-like transparent area in 50 S sub-unit which does not allow entrance of the proteolytic enzymes or ribonuclease enzyme. Such pores also occur in 60 S sub-units of 80 S ribosomes. It is not exactly known how the two sub-units come together to make up the whole ribosomes. The association and dissociation of these two sub-units depend on Mg^{++} ion concentration, low Mg^{++} concentration causes separation of two sub-units while the increased Mg^{++} ion concentration causes association of two sub-units into complete ribosomes. At high Mg^{++} ion concentration in the cytoplasmic matrix two or more ribosomes become associated with one another to form dimer or polymer.

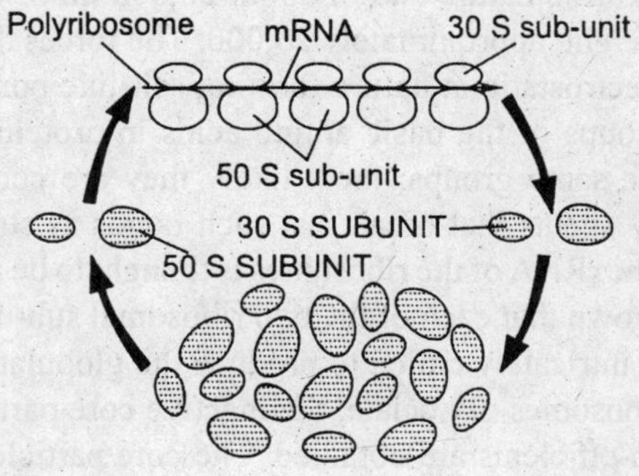

Fig. 5.8 Polysome and subunits of ribosomes of prokaryotic cell.

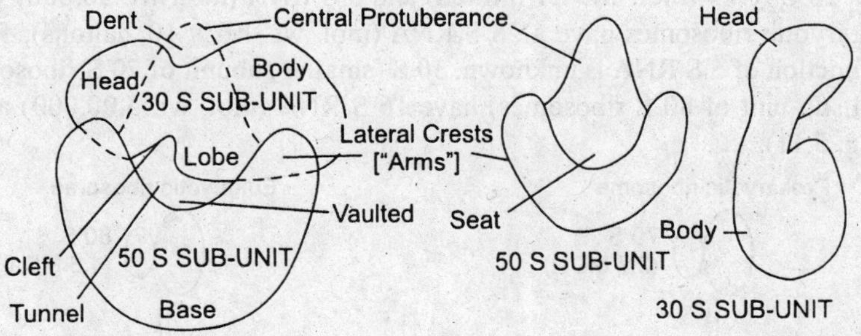

Fig. 5.9 70 S ribosome with larger 50 S and smaller 30 S sub-units.

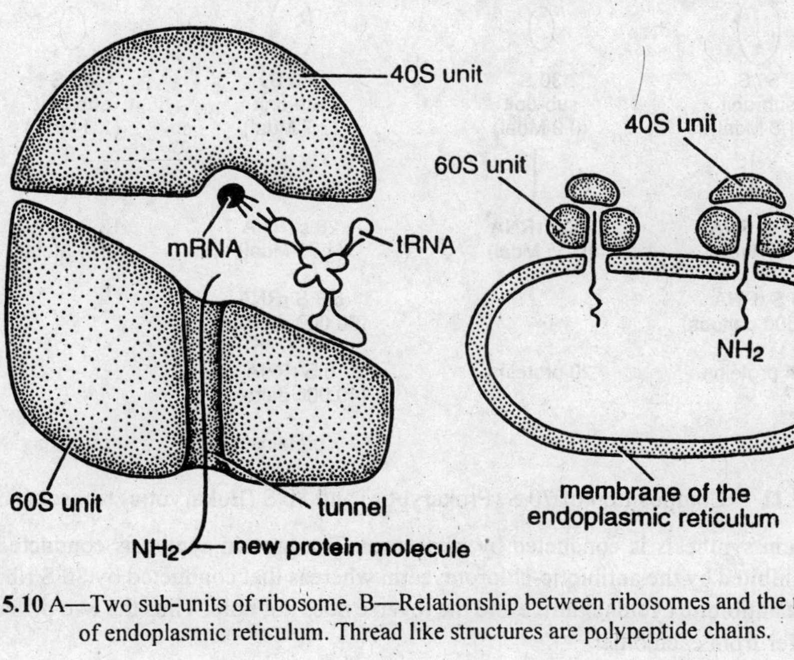

Fig. 5.10 A—Two sub-units of ribosome. B—Relationship between ribosomes and the membrane of endoplasmic reticulum. Thread like structures are polypeptide chains.

The chemical constitution of ribosomes is more or less the same in all the cases and they contain 40 to 60% of rRNA, the rest being all proteins of different kinds. The ribosomal proteins are mostly of basic nature and of about 20,000 mol. wt. Ribosomal RNA (*r* RNA) has a very high molecular weight, approximately 20,000. The forces that hold ribosomal RNA to ribosomal proteins are either electrostatic in nature forming salt-like bonds between phosphate groups of *r* RNA and the amino groups of the basic amino acids in proteins or bonds involving magnesium complexing between the same groups; most likely they are combination of both. RNA can be removed from protein by agents that break just such bonds as strong salt solutions or magnesium complexing reagents. The rRNA of the ribosomes is thought to be almost entirely covered by the proteins. Spirin (1964) has shown that each of the two ribosomal sub-units has the form of a long RNA-protein filament which is intricately coiled to produce the globular form of mature ribosome. When two sub-units of 70 S ribosomes dissociate, the inactive core particle and split proteins (SP) with different sedimentation co-efficients are obtained. The core particles can further be dissociated into RNA and core protein and the split proteins can further be fractionated into acidic proteins (SP-A) and basic split proteins (SP-B).

The larger subunit of ribosome is composed of one large RNA molecule and a small RNA molecule and several protein chains (nearly 30 different proteins) while the smaller sub-unit consists of just one RNA molecule and several (nearly 20) protein chains. 50 S sub-units of prokaryotic ribosome have a 23 S RNA (mol. wt. 1.1 million) and 5 S RNA (mol. wt. 30,000) while the 60 S sub-units of eukaryotic ribosomes have a 28 S RNA (mol. wt. 1.6 × 10^6 daltons), 5.8 *r* RNA and 5 S RNA. The function of 5 S RNA is unknown. 30 S (smaller subunit of 70 S ribosomes) and 40 S sub-units (small sub-unit of 80 S ribosomes) have 16 S RNA (mol. wt. 5,00,000) and 18 S RNA respectively (Fig. 5.11).

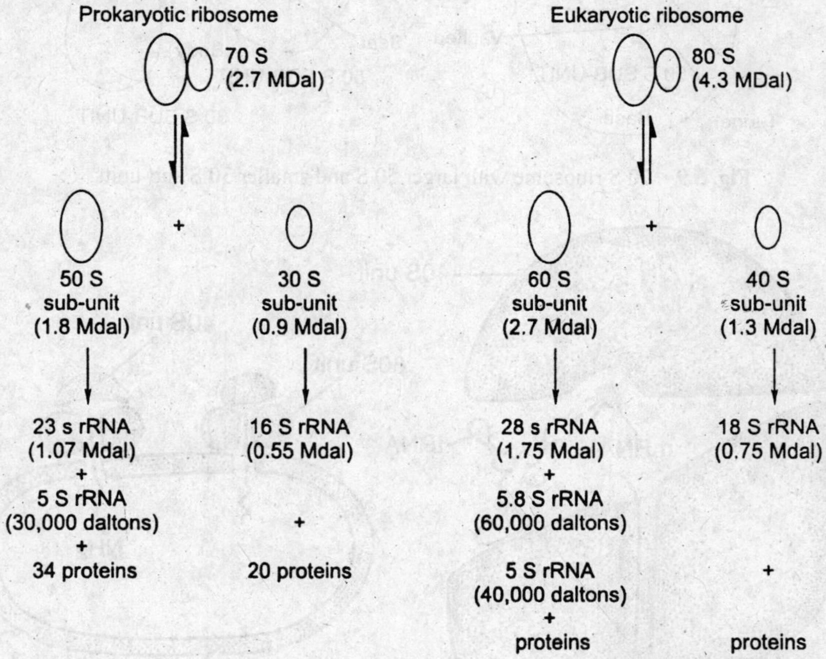

Fig. 5.11 Composition of 70 S (Prokaryotic) and 80 S (Eukaryotic) ribosomes.

The protein synthesis is conducted by ribosomes. The protein synthesis conducted by 70 S ribosomes is inhibited by the antibiotic-chloromycetin whereas that conducted by 80 S ribosomes is inhibited by the antibiotic cycloheximide and the reverse does not hold. Thus the two different types of particles differ from each other.

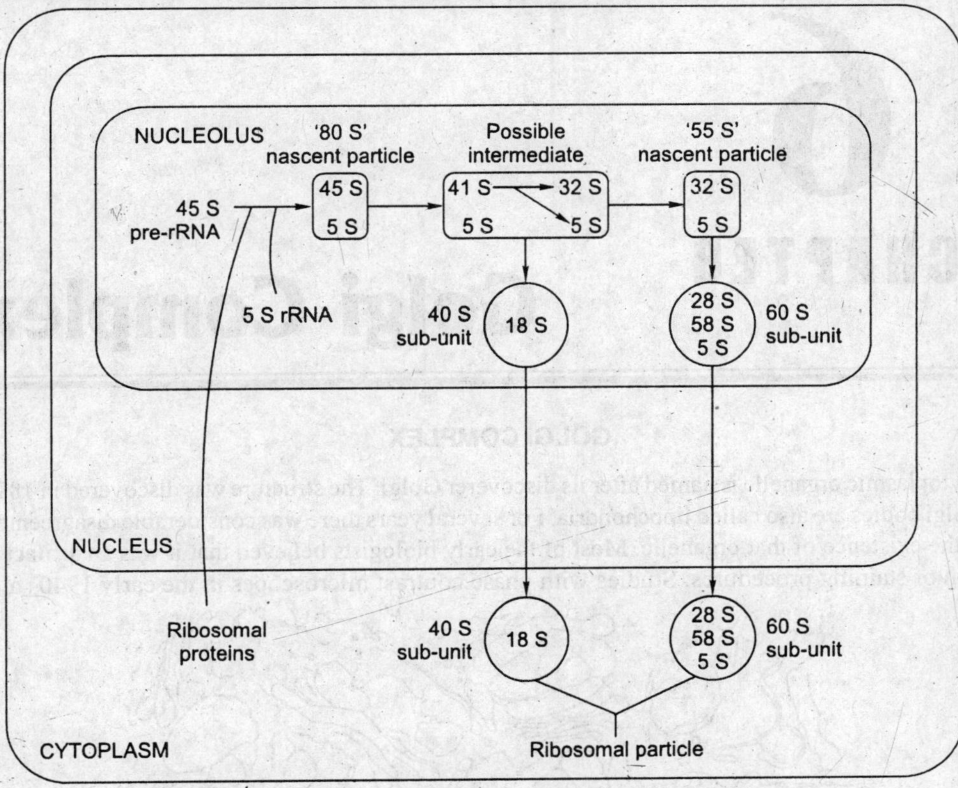

Fig. 5.12 Formation of ribonucleo protein in the nucleolus and their processing into mature ribosomal sub-units.

Ribosomes are not self-replicating particles. Recent studies have established that the cytoplasmic ribosomal sub-units (large and small) are synthesized in the nucleolus from where these sub-units, along with residual accessory proteins are rapidly transported to the cytoplasm where discarding the accessory proteins, the ribosomal sub-units combine with *m* RNA to form polyribosomes (Fig. 5.12).

QUESTIONS

1. What is endoplasmic reticulum ? Discuss different components of endoplasmic reticulum.
2. Describe the structure and functions of endoplasmic reticulum.
3. Discuss briefly the different types of endoplasmic reticulum and their functions.
4. Give an illustrated account of the structure, composition and synthesis of ribosomes of prokaryotes and eukaryotes.

6 CHAPTER

Golgi Complex

GOLGI COMPLEX

This cytoplasmic organelle is named after its discoverer Golgi. The structure was discovered in 1898. The golgi bodies are also called lipochondria. For several years there was considerable disagreement about the existence of that organelle. Most of the early biologists believed that it was an artifact of fixation or staining procedures. Studies with phase contrast microscopes in the early 1940s also

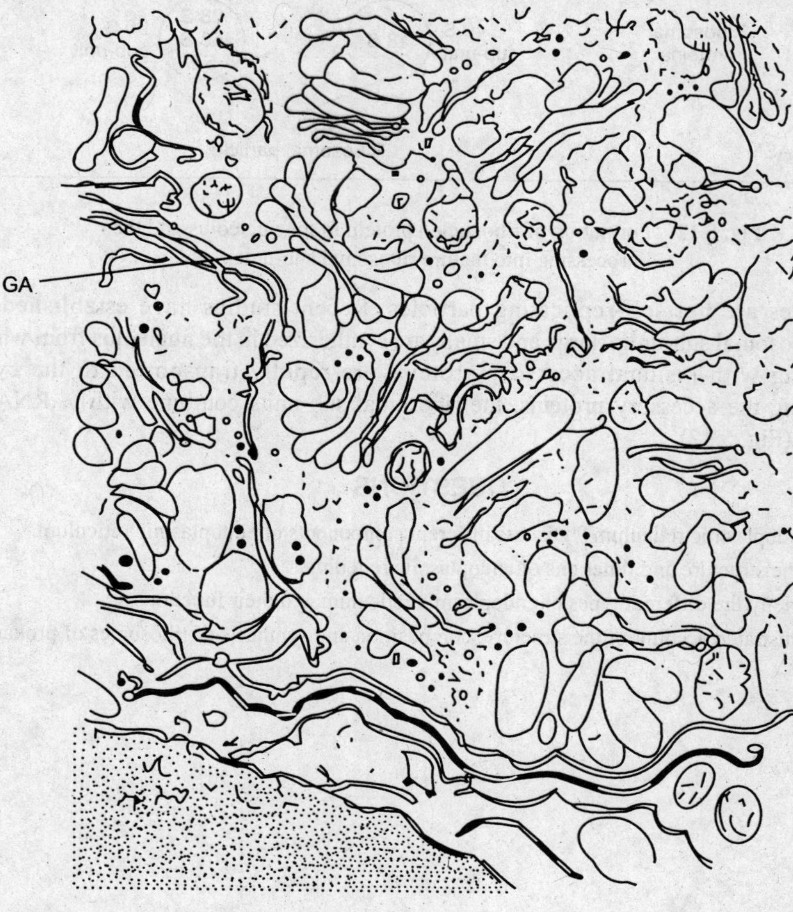

Fig. 6.1 Electron micrography of root cell of maize showing golgi bodies (GA) in ground plasm.

indicated the existence of golgi bodies. The study of electron micrographs of thin sections of cells in 1950s finally proved beyond doubt the existence of golgi bodies in all the cells of eukaryotes (Fig. 6.1). The golgi apparatus does not exist in the prokaryotes.

Structure. The electron microscopic studies have revealed that this organelle consists of series of compactly grouped smooth contoured membrane limited vesicles of variable shapes and dimensions and variable number of small vacuoles (Fig. 6.2). They are selectively stained with Neutral red stain and differ from mitochondria in staining property because they do not take Janus green stain (special stain for mitochondria). When the cytoplasm is centrifuged the mitochondria settle down first and golgi bodies afterwards. This indicates that the golgi bodies are lighter than the mitochondria. The presence of golgi bodies in plant cells has been denied by some early cytologists, but the electron micrographs in recent years have revealed that these bodies are of universal occurrence in both plant and animal cells.

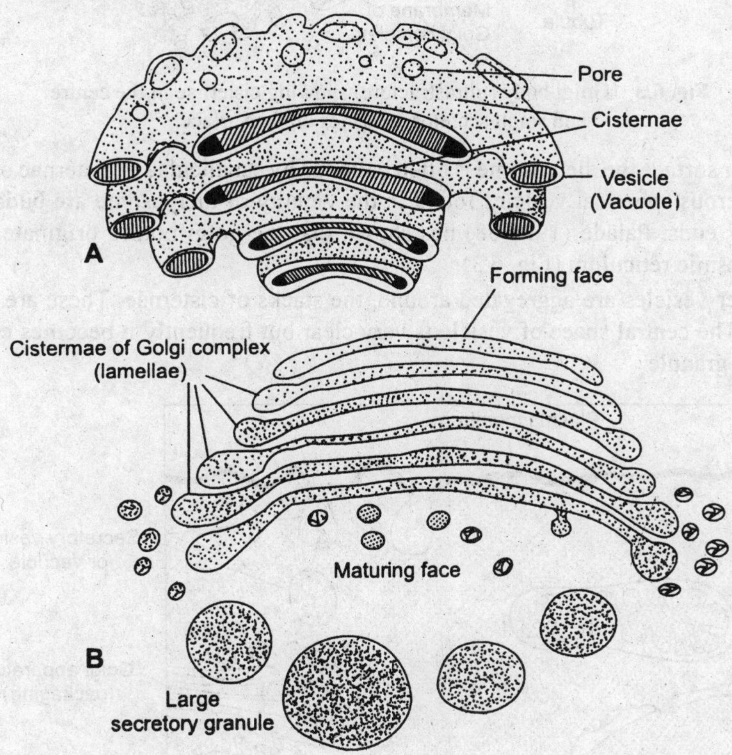

Fig. 6.2 Golgi body; A. Dictyosome, B. Dictyosome and secretory vesicles.

The vesicles of golgi bodies are chiefly of two types:

1. Small and spherical vesicles.
2. Broad flattened vesicles in parallel or often in semi-circular array, the cisternae (Singular—cisterna) (Fig. 6.3).

The cisternae are characterized by their dilated edges. They are compactly arranged in parallel fashion. The stack of flattened cisternae or saccules is known as '***dictyosome***'. The dictyosome has a polarity; its convex side forming the outer face and concave side forming the inner face. The cisternae on the outer face are very flat and thin whereas those on the inner face or concave side are comparatively much dilated and thick. The cisternae on the outer face react only with silver salt and osmic acid while those on the inner face do not react with silver salt and osmic acid. Thus the outer and inner faces are accordingly known as osmic or argentophilic and non-osmic or argentophobic

The number of vesicles per dictyosome varies presumably because of different functional stages of golgi complex. The unit membrane of these cisternae are about 35 Å thick, smooth surfaced, and not associated with ribosome granules.

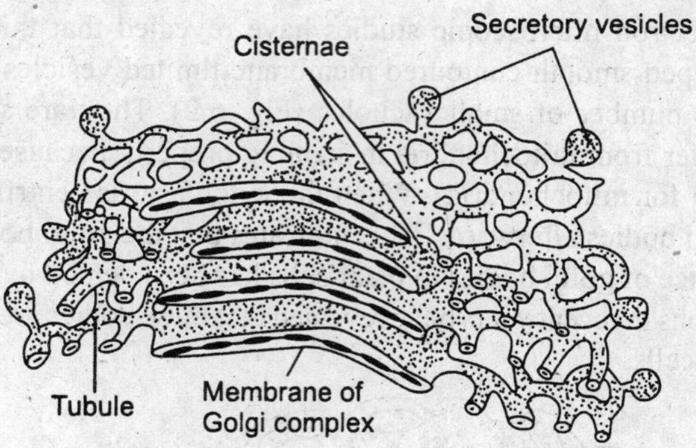

Fig. 6.3 Golgi body (ultrastructure) showing cisternae in the centre and small secretory vesicles on the surface.

On its outer surface the dictyosome is often bounded by canaliculae or cisternae of endoplasmic reticulum. Numerous spherical vesicles found in the vicinity of dictyosome are budded off by the cisternae at their ends. Palade (1956-58) has shown that the golgi bodies originate from smooth surfaced endoplasmic reticulum (Fig. 6.3).

The smaller vesicles are aggregated around, the stacks of cisternae. These are also bounded by membranes. The central space of vesicle is very clear but frequently it becomes condensed and appears as small granule.

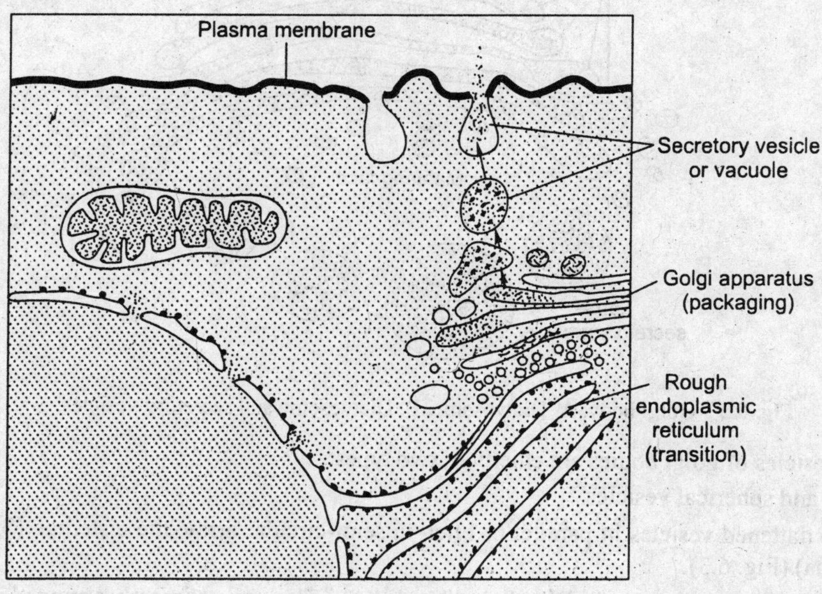

Fig. 6.4 Relation between the endoplasmic reticulum and a dictyosome *via* transition elements.

Functions

The golgi apparatuses are of usual occurrence in the secretory cells where they are involved in secretory process. In plant cells these bodies secrete mucoproteins, slime, mucilage, lipoproteins and other proteinaceous substances. Besides they also synthesize hemicelluloses and pectic group of polysaccharides especially during cell division. In non-secretory cells the golgi apparatus is assigned some other functions. The dictyosomes release substances in the form of tiny vesicles budded from the saccules on the inner face. These vesicles having limiting membrane of their own fuse together to form secretion granules of ever increasing sizes. These granules are generally transported to the cell boundary where their membranes fuse with plasma membrane and finally their contents are discharged out of cell by ***ectocytosis*** (Fig. 6.4). Some proteins synthesized in granular endoplasmic reticulum, particularly those which need to be coupled in varying proportions with polysaccharides to form muco or glucoproteins, pass through the cavities of the reticulum into the dictyosomes and from there they are discharged as secretion granules. This transfer of proteins appears to take place through small vesicles budded from smooth cisternae of endoplasmic reticulum in the vicinity of dictyosomes. The loss of membranes due to the budding of vesicles on the inner face of the dictyosomes appears to be compensated by contribution of vesicles by the smooth surfaced endoplasmic reticulum on the outer face. This indicates that new saccules are formed on the outer face of dictyosomes while the internal saccules break up into secretion vesicles. This hypothesis is supported by the following observations:

(*i*) The number of saccules in dictyosome changes according to the physiological condition.

(*ii*) The number decreases in starving cells and they completely disappear in enucleated cells.

The membranes of cisternae offer surfaces for enzymatic activities. Sometimes, the vesicles act as a system of channels collecting intracellular metabolites and fluids.

QUESTIONS

1. Describe the structure and functions of Golgi body.
2. Describe the different structural components of Golgi body.
3. Discuss the origin and functions of Golgi body.
4. Write short notes on:
   (*i*) Dictyosomes, and
   (*ii*) Secretory vesicles.

7 CHAPTER

Lysosomes and Microbodies

LYSOSOMES

In the cytoplasmic matrix of the cells, there occur variously shaped bodies usually bounded by a single surface membrane and containing hydrolytic enzymes. These are called **lysosomes**. These enzyme containing bodies play important role in the digestion or lysis of intracellular substances, so they are called lysosomes.

Lysosomes were first reported by Christian de Duve and co-workers in Belgium in 1955 following their extensive work on the biochemical identification of certain hydrolytic enzymes in the liver cells of rats. During that period pioneering works were carried out by Novikoff in America when he was able to complement the biochemical work on Lysosome with observations at light microscopic and ultrastructural levels.

Occurrence

Lysosomes appear to be absent in prokaryotes. Cytochemical and electron microscopic studies have revealed the presence of membrane bound enzyme containing bodies in animal tissues. It is not certain whether the structures equivalent to animal lysosomes are present in plant cells. This uncertainty is partly due to strict adherence to the biochemical definition of lysosomes and partly because of technical difficulties in sub-cellular fractionation of plant tissues. Particles containing hydrolases and having characters of lysosomes have been visualized in many plant cells such as tobacco and maize. The presence of lysosomes has been demonstrated in several slime moulds, fungal hyphae and algae. In 1964, P. Matile reported the occurrence of lysosomes in *Neurospora.* Among the algae acid phosphatase has been located in lysosomes of *Euglena* and a few other species.

Structure of Lysosomes

Lysosomes represent a class of morphologically heterogeneous cytoplasmic particles. The polymorphic nature of lysosomes has been attributed to specific functions, substances they contain and stage of digestion of those substances. Their size ranges from 0.25 to 0.8μ in diameter. In mammalian kidney cells they may be as large as 5μ and they may be even larger in phagocytes. The density of lysosomes lies between that of mitochondria and microsomes (fragments of endoplasmic reticulum). Initially they were discovered by differential centrifugation and not by electron microscopes. They were originally obtained by sub-fractionation of the classical mitochondrial fraction. The classical mitochondrial fraction was sub-fractionated into the heavier and lighter components. The light mitochondrial sub-fraction or L-fraction was later named lysosome (lytic bodies). The lighter fraction (lysosomes) lacked cytochrome oxidase but had high concentration of acid phosphatase which became active only after severe mechanical disruption. The absence of cytochrome oxidase in lysosomes distinguishes them from mitochondria.

Lysosomes are bounded by a single limiting membrane of lipoprotein that is homologous with the unit membrane. Chemically lysosome is a bag packed with a variety of hydrolysing enzymes. The surface membrane is impermeable or very little permeable to substrates of the enzymes contained in the lysosome. The internal organization is quite variable; some lysosomes have solid or very dense content, others have a dense outer zone with a less dense core and still others have a cavity or vacuoles.

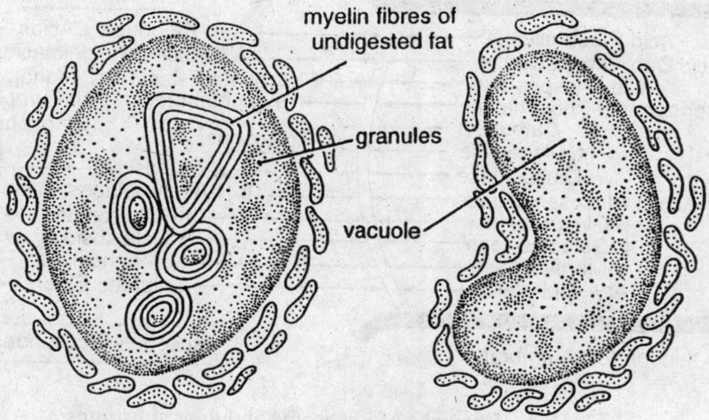

Fig. 7.1 Lysosomes in kidney cell of rat.

When the surface membrane is dissolved the enzymes are set free in the cell sap which greatly alter the whole metabolic system of the cell. The membrane of lysosome is easily dissolved by the action of protolytic enzymes in hypotonic media.

The substances which cause instability of the membrane are called **labilizers**. Other substances, called **stabilizers**, stabilize the membrane. Among the many substances known to labilize lysosome membrane are vitamins A, E and K, progesterone, testosterone, β-*estradiol, proteases, digitonin*, ubiquinone, x-irradiation, ultraviolet irradiation, polyene antibiotics, endotoxins and bile salts. Though the exact mechanism of functioning of labilizers is not known, these substances probably affect the membrane by (*i*) changing the fluidity of the lipid part of the membrane; and (*ii*) increasing the permeability of the lysosome membrane to small solutes and disruption of membrane due to osmotic swelling. Some of the membrane stabilizers are cholesterol, cortisone, chloroquine, heparin. The lysosome membranes have limited permeability which possibly checks the direct access of hydrolytic enzymes to cellular components.

Polymorphism: On the basis of morphological and functional criteria, a variety of lysosomes can be recognised in different cells as well as within a single cell. Two basic forms of lysosomes have been distinguished (Novikoff *et al,* 1964; De Duve and Williams, 1966); (*i*) *the primary lysosomes;* and (*ii*) *the secondary lysosomes.*

Primary lysosomes (storage granules): They originate from the endoplasmic reticulum or are cut off indirectly from the tips of Gogli saccules and have not yet been involved in digestive process.

Secondary Lysosomes: These are digestive vacuoles which are sites of digestive activity. Lysosomes belonging to this group may be classified into two separate types; (*a*) *heterophagic*; and (*b*) *autophagic* on the basis of exo- or endogenous origin of the material undergoing digestion. Each of these types may be further sub-divided as follows:

(*i*)-*Pre-Lysosomes* whose enzymes have never been engaged in hydrolysis.

(*ii*)-*Lysosomes* which are sites of present digestive activity.

(*iii*) *Post-Lysosomes* which have lost their enzymes.

Lysosomes start functioning when the cell takes up substances either by phagocytosis or pinocytosis. Undigested substances or residues within the lysosomes usually remain packed with different inclusions. Under certain circumstances lysosomes start digesting the very cell in which they occur by opening their membrane and setting free their digestive acid hydrolases.

The Lysosomes contain some 40 enzymes. Some of the important acid hydrolases identified from lysosomal fractions are as follows (Fig. 7.2):

1. *Phosphatase*: Acid phosphatase
   Acid phosphodiesterase

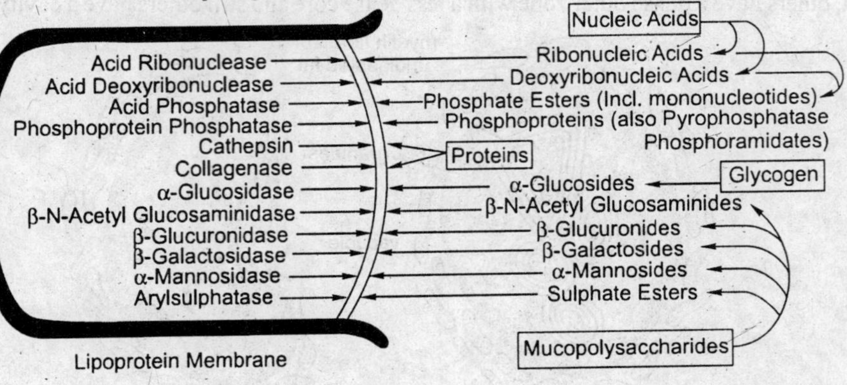

Fig. 7.2 Enzymes of Lysosome and their substrates.

2. *Glycosidases*: β-galactosidase
   β-glucuronidase
   β-N- acetylglucosaminidase
   α-glucosidase
   α-manosidase
3. *Proteases*: Cathepsin A
   Cathepsin B
   Collagenase
   Peptidases
4. *Lipases*: Acid lipase
   Phospholipase A
   Phosphotidic acid phosphatase
5. *Nucleases*: Acid ribonuclease
   Acid deoxyribonuclease

It is not necessary that all the hydrolytic enzymes should occur in a single lysosome. These enzymes have been located in a variety of tissue types. Lysosomes obtained from one source do not present a homogeneous picture under the electron microscope. This suggests that a particular lysosome may contain only a few or even a single species of enzymes. Some evidences do indicate that various hydrolytic enzymes are associated with different lysosomes in a variety of cells. These hydrolases have an acid optimum of about pH5. The enzymes of lysosomes become active only when the surface membrane is ruptured.

Functions of Lysosomes

1. ***Extracellular Digestion.*** Lysosomes are small bags containing digestive enzymes. They behave like tiny time bombs waiting for their explosion in the cytoplasm. When the limiting membrane ruptures, the digestive enzymes are released which take part in the digestion. Sometimes lysosomal enzymes may be released out side the cell where they digest extracellular substances. Saprophytic fungi and other micro-organisms utilize extracellular digestion of complex substrates in the habitat and degrade them into simpler soluble forms which are then absorbed.

2. ***Intracellular Digestion.*** The digestive enzymes released in the cytoplasm may be involved in *autophagy* or *heterophagy.*

   (*a*) ***Autophagy*** refers to digestion of endogenous materials or breakdown of molecules and pieces of cytoplasmic materials within the cell. The simpler substances formed after the digestion are then utilized in the synthesis of some other substances. This recycling of cell component is called turnover. Actually the digestive materials are non-functional part of other organelles like mitochondria, endoplasmic tubules, enclosed in a vesicle or digestive vacuole called *autophagic vesicle* or autophagosome. Primary lysosomes fuse with the autophagic vesicles as a result of which lysosomal enzymes are discharged into them. In this way secondary Lysosomes are formed. The presence of hydrolytic enzymes in the autophagosomes and progressive disintegration of enclosed organelles give indication of digestive activity (Fig. 7.3). Autophagy may bring about cellular digestion after the death of a cell and so it brings about the self-clearance of dead cells. This is why Christian de-Duve called lysosomes as the suicide bags.

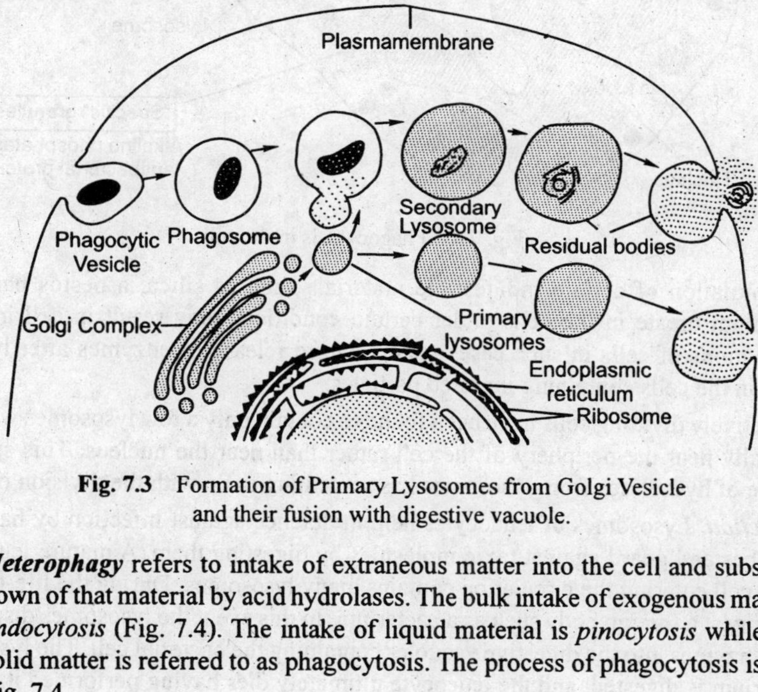

Fig. 7.3 Formation of Primary Lysosomes from Golgi Vesicle and their fusion with digestive vacuole.

   (*b*) ***Heterophagy*** refers to intake of extraneous matter into the cell and subsequent breakdown of that material by acid hydrolases. The bulk intake of exogenous material is called *endocytosis* (Fig. 7.4). The intake of liquid material is *pinocytosis* while the intake of solid matter is referred to as phagocytosis. The process of phagocytosis is illustrated by Fig. 7.4.

3. There is eviuence that lysosomal acid hydrolases are involved in release of certain hormones from secretory cells of certain glands, *e.g.*, thyroid hormones are released by hydrolysis of thyroglobulin.
4. The enzymes released from acrosome vesicle, the giant lysosomes of sperms, dissolve the cortical granules, the structure surrounding the egg nucleus and help in the penetration of sperm nucleus into the egg.
5. ***Role in Metamorphosis.*** During the development of embryo, several tissues become functionless which are digested by the enzymes released from lysosomes and the digested materials are then absorbed by the surrounding cells. Rudolph Weber has reported the absorption tail of tadpole in this way.
6. ***Lysosomal activity in relation to Pathology.*** Several recent pathological studies have indicated that irregularities in lysosomal activity may cause fever, congestive heart failure, hepatitis, Pylonephritis, hypertension, joint injuries, leucocyte granules and tissue injuries. Release

of nucleases (DNAase and RNAase) which attack DNA and RNA respectively may cause chromosomal breaks and rearrangement and in this way it can result in structural abnormalities in chromosomes. The chromosomal abnormality may induce certain harmful mutations in characters. Sometime it may lead to carcinogenesis (cancer formation). Malignant cells are also found to have abnormal chromosomes. Partial deletion of chromosome 21 of man has been associated with chronic myeloid leukemia.

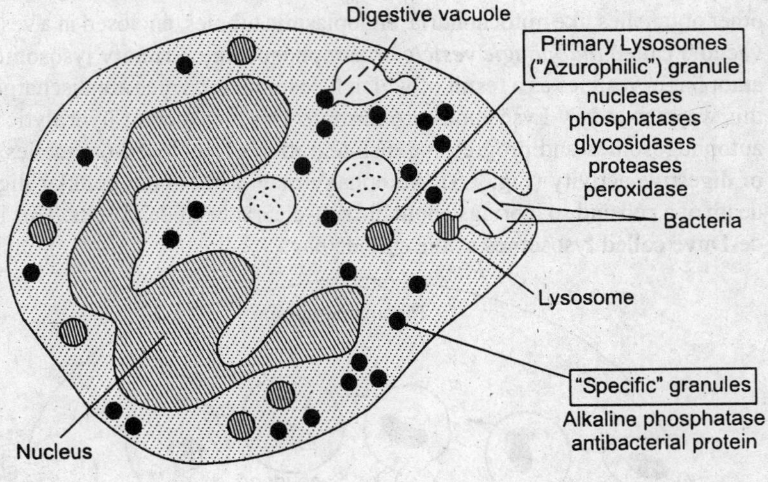

Fig. 7.4 Phagocytosis in the cell.

Accumulation of certain indigestible materials such as silica, asbestos particles, crystals of sodium ureate in the cells under certain conditions may result in cell inflamation. The inflamations of cells in such cases results due to release of enzymes after lysosomal break down in the cells containing ingested particles.

The actively dividing cells have been found to contain only a few lysosomes which are located generally near the periphery of the cell rather than near the nucleus. This suggests that the release of hydrolases from lysosomes has something to do with the division of cell.

7. ***Protection.*** Lysosomes of leucocytes help in defence against infection by bacteria and other microbes and guard against toxic molecules by digesting them. A mature leucocyte or white blood cell entering the circulation contains many lysosomes. During the life-time a leucocyte may ingest a foreign body such as a bacterium. In this event the lysosomes disappear releasing their enzymes into the digestive vacuoles containing the bacterial cell. The lysosomes rupture, bacterium is digested, and the leucocyte ultimately dies having performed its major function in the body.

In animal body several kinds of cells have short lives such as, outer layer of skin and the mucous membrane lining the body. The short lived cells are being continuously replaced and the lysosomes of dead or degenerating cells release their enzymes into the body of the cells so that the whole cell may be digested. The process of tissue degeneration or *necrosis* can be attributed partly to lysosomal function.

MICROBODIES

Microbodies are spherical or ovoid organelles of various sizes, 0.2 to 1.2 mμ in diameter surrounded by a single trilaminar unit membrane and containing dense or sometimes crystelline materials (matrix). Recently, they have been found to contain enzymes, particularly catalases and oxidases.

Some authors suggest that microbodies may be precursors of mitochondria but this suggestion has not been widely accepted. Others are of the opinion that microbodies are related to lysosomes. But present indications are that they are independent organelles with specific functions. Lysosomes and microbodies appear to be the cells way of storing enzymes in large quantities while at the same time keeping the bulk of these enzymes out of circulation.

In general, two types of microbodies are distinguished; *peroxisomes* and *glyoxysomes.* The peroxysomes are so named because of their peroxidative activity. Similarly, glyoxysomes possess all or part of the enzymes involved in glyoxylate cycle in addition to some other enzymes.

Microbodies have been reported in the cells of both plants and animals. They have been isolated from protozoa, yeast, algae, mesophyll cells of plant leaves, and cells of other plant tissues, birds, amphibians, and a variety of mammalian tissues including liver, kidney, digestive tract, brain, lung, muscles, testis.

The membrane of microbody is thinner than that of lysosome and plasma membrane and is more or less similar to the endoplasmic reticulum and the outer mitochondrial membrane. The matrix of microbody is very little as compared to that of mitochondria and plastids. The membrane of peroxysome permits the passage of substrates like amino acids α-hydroxy acids, and uric acid. They have unrestricted permeability as compared to lysosome and mitochondria membrane but they appear to be impermeable to pyridine nucleotides. Peroxysomes appear to arise from canaliculae of endoplasmic reticulum.

Peroxysomes are probably associated with metabolic processes, *e.g.*, photorespiration in the plant cells. They are also known to protect cellular organelles from the toxic effects.

QUESTIONS

1. What are lysosomes? Describe the structure and function of lysosomes.
2. Describe the different types of lysosomes found in cell and discuss their importance.
3. Write brief notes on:
   (*i*) Polymorphism in Lysosome,
   (*ii*) Autophagy,
   (*iii*) Phagocytosis,
   (*iv*) Lysosomal enzymes,
   (*v*) Microbodies, and
   (*vi*) Peroxisome.

8

CHAPTER

Nucleus

STRUCTURE OF NUCLEUS

The protoplast of eukaryotic cell contains prominent spherical or sub-spherical body called the nucleus (Pl.— nuclei) which controls all the life activities. As a rule, there is only one nucleus in one cell, but there are certain exceptions to this general rule. There may be many nuclei in a syncytium which is formed by fusion of many uninucleate cells. Multinucleate cells (coenocytes) are commonly found in plants. Coenocytic condition results due to repeated nuclear divisions without cytokinesis. Sieve cells (phloem) of engiosperms and mammalian red blood cells lose their nuclei during the process of maturation. The enucleated cells are incapable of growth and division. The cells of bacteria and blue-green algae do not possess organised nuclei. The nuclei in some cells may be seen in the centre

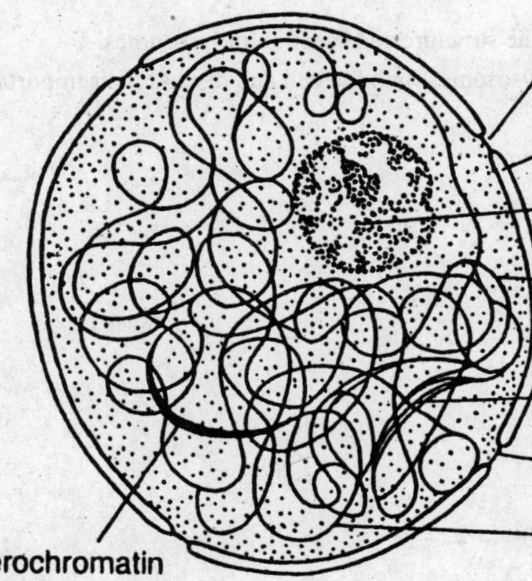

Fig. 8.1 Structure of a nucleus.

but in the majority of cells they may occur at any place in the cytoplasm. There is but little optical differentiation between the nucleus and the cytoplasm. The following are the principal components of the interphase nucleus (Fig. 8.1):

(*a*) Nuclear envelope,

(*b*) Nuclear sap,

(*c*) Nucleolus, and

(*d*) Nuclear reticulum or chromatic reticulum.

Nuclear envelope

The nuclear membrane or karyotheca is the outer boundary of nucleus. It is found in the nuclei of all the eukaryotic cells. Electron microscopy has revealed that it is composed of two concentric unit membranes; an outer membrane and an inner membrane of lipoproteins (Fig. 8.2). According to Du Praw, each membrane is about 80-100 Å thick. The outer and inner membranes are separated by a clear space of 100 to 150 Å (Du Praw, 1970; Robertis, 1970). The space between the two unit membranes is known as **perinuclear space** or **perinuclear cisterna**. Outer nuclear membrane is rough surfaced due to attached ribosomes. Sometimes it becomes outfolded to give rise to endoplasmic reticulum. The inner nuclear membrane is free from ribosomes and sometimes it is associated with chromatin. In many vertebrate cells there is a layer of densely packed fibres just inside the inner nuclear membrane forming a third concentric sheath of uniform thickness (about 300 Å). This is known as fibrous lamina (Fawcett, 1966). It is not known whether these layers have any special function other than mechanical support.

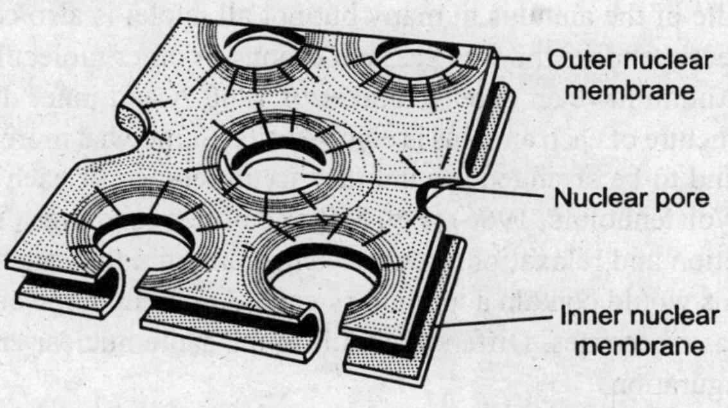

Fig. 8.2 Three dimensional structure of nuclear envelope.

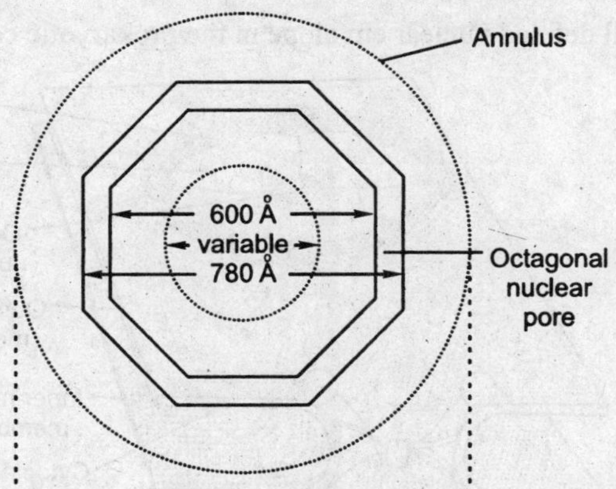

Fig. 8.3 Structure of a nuclear pore.

The nuclear envelope possesses special circular structures, the ***annuli*** (singular–**annulus**) through which certain materials can pass. The presence of annuli is the most characteristic structural feature of the nuclear envelope (Fig. 8.2).

Afzelius (1963) found that the nuclear envelope in luminescent marine dinoflagellate *Noctiluca* has a single membrane only which is 150 Å thick and completely lacking annuli. Annuli can be observed only with electron microscope. They were first discovered by H.G. Callan and Tomlin in 1950. Some authors call the entire annular structure a '**pore**', some others call the entire structure an annulus and still some others use both terms. The annulus is a thickened ring surrounding a central pore (Du Praw, 1970). Since they almost do not represent permanent openings in the nuclear envelope, the term 'pore' and 'pore complex' are misleading (Du Praw, 1970). Thus, it seems preferable to use the term 'annulus', in precise sense, signifying a thickened, electron dense ring of material set into the nuclear envelope and functioning by some unknown mechanism to control the passage of particles in and out of the nucleus. It allows the passage of materials but only through a restricted region. That is not simply a pore but a directional mechanism of transport.

Some authors have suggested that typical nuclear envelopes are penetrated by permanent pores of 500-800 Å diam., but some kinds of protein molecules only 100 Å or less in diameter are unable to enter the nucleus from the cytoplasm. This clearly indicates that there is no existence of permanent openings of about 800 Å diam. (Feldherr and Feldherr, 1960). On the other extreme, the nuclear envelope is not completely closed to passage of large particles, but this maintains a very sensitive control over which particles pass and in which direction. One of the best established features of annulus ultrastructure is that inner and outer nuclear membranes come together in the margin of each annulus. The middle of the annulus in many but not all nuclei is also covered by a thin single membrane, which evidently checks the free passage of ions or other molecules across the envelope (Wiener *et al.,* 1965). Annuli have an outer diameter of 1200 Å and inner diameter between 0 and 500 Å (Fig. 8.3). The structure of each annulus is eight-partite. Somewhat more specifically, 8 spheroid particles have been found to be arranged around the circumference of each annulus in the nuclear envelope of *Amoeba* (Wolstenholme, 1966) (Fig. 8.4) and of pea seedlings (Yoo and Bayley, 1967). Annulus shows contraction and relaxation. In unstretched annuli a thin membrane is present in the centre. A relaxed annulus would contain a temporary central opening large enough to allow passage of resolvable particles as ribosomes. Different annuli in the same nuclear envelope can exist in an expanded or open configuration.

The nuclear membrane can easily be observed in the metabolic nucleus. It disappears at certain stage of nuclear division and reappears on the completion of the division process.

There is no well defined nuclear envelope in the prokaryotic cells.

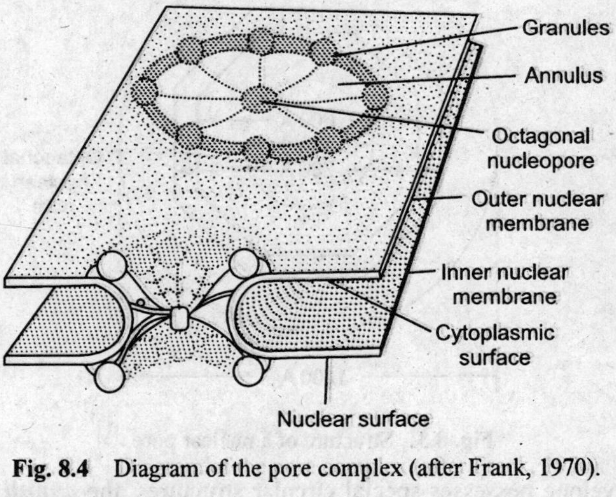

Fig. 8.4 Diagram of the pore complex (after Frank, 1970).

Nuclear sap or Nucleoplasm

The undifferentiated protoplasm or ground substance present inside the nuclear envelope is called nuclear sap or nucleoplasm or karyolymph. The nucleoplasm seems to contain granules of various sizes and densities and is yet of unknown composition and function. During nuclear division, the nucleoplasm is of course continuous with cytoplasmic matrix. The other nuclear components such as the chromatic reticulum and nucleolus remain suspended in the nucleoplasm. Nucleoplasm contains nucleoproteins and many other inorganic and organic substances, such as, nucleic acids, proteins, enzymes, lipids and minerals.

Nucleolus

It is an intranuclear organelle of eukaryotes. The term nucleolus was coined by *Bowman* (1840). It occurs as a spherical, acidophilous structure suspended in the nucleoplasm either in central or eccentric position. The nucleolus is found in close association of a specific chromosome at a fixed point called nucleolus-organising region (Fig. 8.5). The number of nucleoli present in each nucleus is fixed for the cells of a particular species of plant or animal and it depends upon the number of chromosomes or set of chromosomes. In many plants and animals, there is only one nucleolus for each haploid set of chromosomes. In *Vicia faba*, for example, each diploid nucleus contains two nucleoli. In some organisms there may be two or more nucleoli for each haploid set of chromosomes and in some others the number of nucleoli within the nucleus changes with the advancement in the age of the cell. Many lower organisms, such as bacteria, yeasts and some algae, undifferentiated embryonic cells of amphibians, and even certain mammalian cells like erythrocytes, reticulocytes and spermatozoa lack nucleoli. The nuclei of polyploids may have more nucleoli than those of diploids.

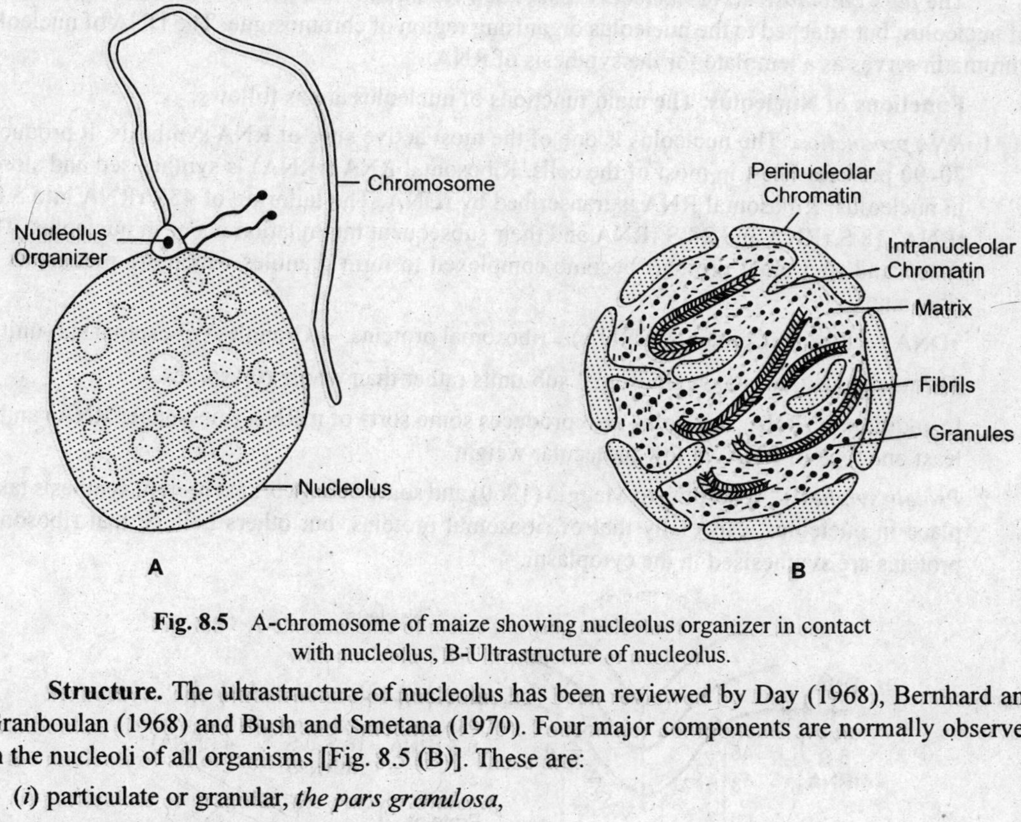

Fig. 8.5 A-chromosome of maize showing nucleolus organizer in contact with nucleolus, B-Ultrastructure of nucleolus.

Structure. The ultrastructure of nucleolus has been reviewed by Day (1968), Bernhard and Granboulan (1968) and Bush and Smetana (1970). Four major components are normally observed in the nucleoli of all organisms [Fig. 8.5 (B)]. These are:

(*i*) particulate or granular, *the pars granulosa*,

(*ii*) fibrillar component, *the pars fibrosa*,

(*iii*) nucleolar associated chromatin, *the pars chromosoma*,

(*iv*) amorphous part, *the pars amorpha.*

The pars granulosa. The particulate component is made up of ribonucleoprotein granules organised as ribosome like particles, 150 to 200 Å in diameter. Although these particles are similar in form, chemical content and staining reaction to cytoplasmic ribosomes, they are slightly smaller than the ribosomes. These granules contain 25 S RNA and proteins in 1 : 2 ratio. Birnstiel *et al.* (1963) have referred these particles as nucleolar ribosomes.

The *pars fibrosa* comprises of the protein containing fibrils, 50 to 150 Å in diameter.

The granules of particulate components are connected together by protein fibrils of *pars fibrosa* and may exhibit tortuously coiled filamentous structure. Thus *pars granulosa* and *pars fibrosa* make up the microscopically identifiable permanent structure called *nucleolonema.* The nucleolonema is considered as one of the important and persistent spongy components of nucleolus. It never disappears completely in course of any metabolic change. Its existence has been noted even during cell division. Thus, there is no question of it being developed *de novo.*

The *pars amorpha* or *amorphous* part is mostly protein and consists of tightly packed fibrils and particles in some organisms.

In animal nucleolus, the nucleolonema generally appears embedded in pars amorpha (*i.e.*, intermingled), whereas in plants the peripheral nucleolonema surrounds the centrally located pass amorpha. In either case, the nucleolonema tends to be associated with chromosomal heterochromatin.

The *pars chromosoma* or *nucleolus associated chromatin* is a network penetrating the body of nucleolus, but attached to the nucleolus organising region of chromosome. The DNA of nucleolar chromatin serves as a template for the synthesis of RNA.

Functions of Nucleolus. The main functions of nucleolus are as follows:

1. *RNA production.* The nucleolus is one of the most active sites of RNA synthesis. It produces 70–90 per cent RNA in most of the cells. Ribosomal RNA (rRNA) is synthesised and stored in nucleolus. Ribosomal RNA is transcribed by rDNA. The tailoring of 45S rRNA into 5.8S rRNA, 18 S rRNA and 28 S rRNA and their subsequent methylation occur in nucleolus. The rRNA and ribosomal proteins become complexed to form granules which are precursors of ribosomes.

   rDNA → Fibrils (Containing rRNA) + ribosomal proteins → Granules (ribosomal sub-units)

   The nucleolus thus makes ribosomal sub-units rather than whole ribosomes.

   In addition to rRNA, nucleolus also produces some sorts of mRNA (messenger RNA) and at least one type of RNA of low molecular weight.

2. *Protein synthesis.* According to Maggio (1960) and some other workers protein synthesis takes place in nucleolus, especially that of ribosomal proteins, but others believe that ribosomal proteins are synthesised in the cytoplasm.

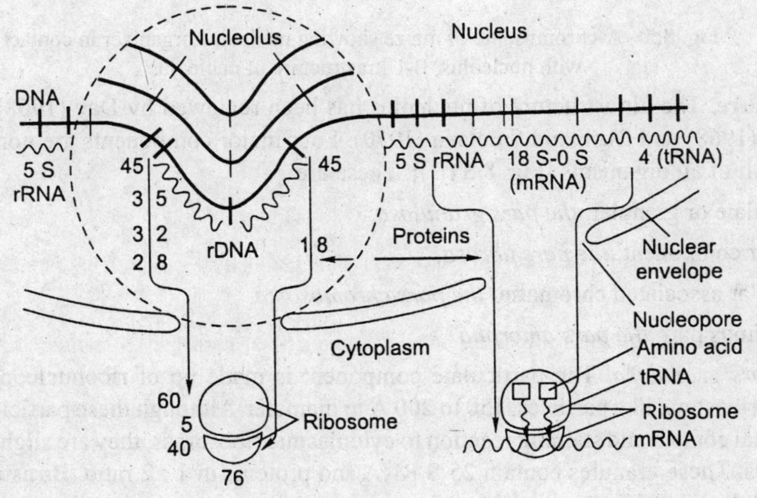

Fig. 8.6 Diagram of the transcription and transport of nuclear RNAs in a eukaryotic cell and participation of tRNA and mRNA in protein synthesis.

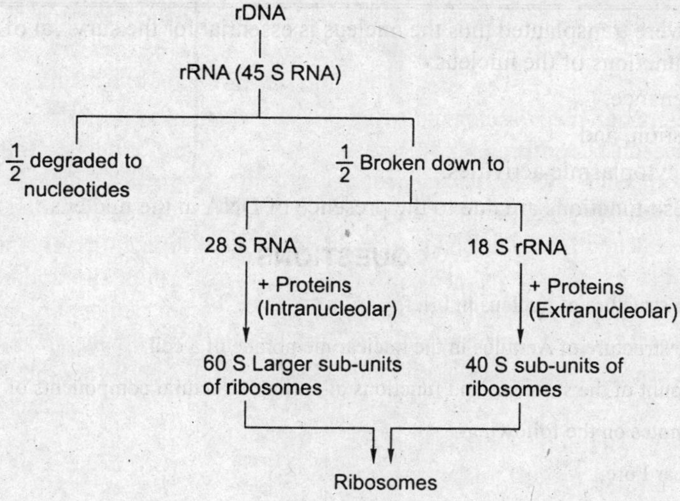

3. *Ribosome formation.* In eukaryotes, rDNA serves as template for 45 S rRNA. Half of the 45 S rRNA is broken down to form 28 S and 18 S rRNAs and the other half is degraded to nucleotides within the nucleolus. 28 S rRNA combines with intranucleolar proteins to form 60 S sub-units of ribosomes and the 18 S rRNA combines with extra nucleolar proteins to form 40 S sub-units of ribosomes. The 60 S and 40 S sub-units of ribosome combine to form ribosome granules (Fig. 8.6).

Chromatic reticulum

In the nucleoplasm a dark staining network is seen which is termed *nuclear reticulum* or chromatic reticulum. The threads of reticulum are formed of *chromatin.* The chromatin network can easily be seen in the interphase nucleus under compound microscope. During the cell divisions chromatin fibres become thick ribbon-like structures and are called *chromosomes.*

The network can readily be stained with basic dyes, such as acetocarmine, haematoxylene, crystal violet and Feulgen stain. The chromatin is a characteristic substance in the nucleus which is of great importance to the cell.

The fibres of chromatin are twisted, uniformly distributed and anastomosed. Two types of chromatin have been recognised. These are (*i*) heterochromatin; and (*ii*) euchromatin. Heterochromatin is darkly stained, condensed and well delimited region in the chromatin thread. The light stained and diffused regions of the chromatin is called **euchromatin**. The heterochromatin and euchromatin both contain DNA which is the basic genetic material, and thus do not differ in this regard. Thus heterochromatin and euchromatin are two different states of chromosomes rather than substances. Spectrophotometric measurements have disclosed that the quantity of DNA per unit area may be two or three times higher in the heterochromatin than in the euchromatin in the same nucleus.

Chromosomes

The term chromosome refers only to the deeply staining DNA containing bodies observed in dividing cells of nucleate organisms. Chromosomes are nuclear components possessing a special organization, individuality and function. They are capable of reproducing themselves without change of morphology and physiologic behaviour in successive generations.

The detailed account on the morphology of chromosomes is given in a separate chapter.

A nucleus of a cell which is not in the process of division is known as *interphase* or *metabolic nucleus.* A cell without nucleus cannot survive for long. Unicellular organisms like *Amoeba* and *Acetabularia* from which the nuclei were removed were unable to survive for long unless new nuclei from other cells were transplanted thus the nucleus is essential for the survival of the cell. There are three important functions of the nucleus:

(*i*) Cell maintenance,

(*ii*) Cell replication, and

(*iii*) Control of cytoplasmic activities.

Most of these functions are due to the presence of DNA in the nucleus.

QUESTIONS

1. Describe the structure of nucleus in brief.
2. Describe the structure of Annulus in the nuclear membrane of a cell.
3. Give an account of the structure and functions of various structural components of Nucleolus.
4. Write short notes on the following:

    (*i*) Nuclear Pore,

    (*ii*) The Pars granulosa, and

    (*iii*) Function of Nucleolus.

9 CHAPTER

Chromosomes

In the nucleoplasm, a dark staining network is seen which is termed *nuclear reticulum* or *chromatic reticulum*. The threads of reticulum are formed of nucleic acid-protein complex called *chromatin* which is now-a-days referred to as ***nucleohistone***. During the interphase or metabolic phase, the condensed portions of chromatin threads (heterochromatin) can be distinguished from the less dense regions (euchromatin) in electronmicrograph of a cell. The chromatin of interphase nucleus represents an aggregation of chromosomes. During cell division the chromatin fibres become greatly condensed and appear as ***chromosomes***. They are invisible in metabolic nucleus due to high water content in nucleoplasm. Chemical analyses consistently show that chromatin is composed of DNA, RNA, histones and non-histones (acidic proteins) (in sperm nuclei of some fishes and birds histones are replaced by more basic proteins, the protamines). The term chromosome refers only to the deeply staining DNA containing filamentous bodies observed in the dividing cells of nucleate organisms. The chromosomes are nuclear components possessing special organic individuality and functions. They bear genes or hereditary units and are capable of reproducing themselves without involving any change in morphology and physiology at successive generations.

The chromosomes can be readily stained with basic dyes such as basic fuchsin (Feulgen's stain), carmine, haematoxylin, crystal violet etc.

Historical Note

It was W. Fleming (1879) who found that the nuclei of cells contain some material that became intensely coloured at certain time. He coined the word ***chromatin*** for the stainable fibrous material of the nucleus. The present name *chromosome* (Gr. *Chroma* = colour, *soma* = body) was coined by W. Waldeyer (1888) for the dark staining thread-like bodies of nucleus. W.S. Sutton and T. Boveri (1902) suggested that chromosomes were the bearers of hereditary potentialities. The fine coiled filamentous component of chromosome was named *chromonema* by Vejdovsky (1912). Morgan (1933) suggested that chromosomes were responsible for the transmission of hereditary traits from one generation to another. By 1953, it was established that the genes of all higher organisms were made of DNA. The chromosome number simply represented the number of packages into which the DNA of a species is divided. De Robertis (1964), Setelo and Wettstein (1965), Good Enough and Levine (1974), Du Praw (1965, 1968) , Solari (1965), Bradbury *et al.* (1981) have made good contributions on the structure and molecular organization of chromosomes.

Number of Chromosomes

Every cell of the body is provided with a packet of genetic material, the nucleus which contains chromosomes. Normally, all the individuals of a species have fixed number of chromosomes in their nuclei. The characteristic group of chromosomes in a cell is spoken of as its complement. The number of chromosomes is not necessarily sufficient for identification of species, since several species may have the same chromosome number, as for example, white squirrel, human and evening bat all have forty-six chromosomes and rice, tomatoes, pines, white oaks and grasshoppers all have same chromosome number *i.e.*, twenty four yet they are separate species. The chromosomes of different species do not have the same quality and quantity of DNA. Here quality means the composition and arrangement of nucleotide sequences and quantity refers to their number.

Generally, the somatic cells of animals and higher plants are *diploid* (*i.e.*, they have two sets of chromosomes-2n) while their gametes, as a rule, are haploid as they contain only one set of chromosomes. The number of chromosomes present in the gamete is referred to as haploid or gametic chromosome number or 'n'. In diploid organisms half of the chromosomes of a nucleus are contributed by male gamete (paternal chromosomes) and the remaining chromosomes are contributed by the female gamete (maternal chromosomes) in the sexual reproduction. The haploid set of chromosomes is also known as ***genome***. In eukaryotes, the minimum visible complement of indespensible chromosomes, each represented only once is referred to as ***basic chromosome set*** and the number of chromosomes in a basic set is called ***basic chromosome number*** or base number. This is symbolised by 'x'. In diploid cells, each chromosome of one haploid set has a partner or ***homologue*** in the other set. Homologous chromosomes are those which are identical in size and carry identical and complementary genes. The chromosome number may change in some organisms and this change may be of two kinds (*i*) Euploidy; and (*ii*) Aneuploidy. Euploidy refers to a condition where the set of chromosomes is kept balanced. If the change in chromosome number is brought about by the increase in the number of set of chromosomes beyond 2n so that the nucleus contains three (3n), four (4n) or more sets, this condition is referred to as ***polyploidy***. When the change in chromosome number involves loss or gain of one or more chromosomes causing the set to be unbalanced, the situation is called ***aneuploidy***. The different types of aneuploids are monosomics (2n – 1), trisomics (2n + 1), nullisomics (2n – 2), tetrasomics (2n + 2), double monosomics (2n – 1 – 1), double trisomics (2n + 1 + 1), etc.

In polyploid individuals which possess more than two haploid sets of chromosomes in their somatic cells, it becomes necessary to establish an ancestral primitive chromosome number or basic chromosome number (x). In common wheat (*Triticum aestivum*), for example, with 2n = 42, n = 21 and x = 7, the somatic chromosome number is, therefore, hexaploid (6x). In some organisms, chromosome number may vary in the cells of different tissues of the body, as for example, in mammals liver cells show chromosome numbers which may be multiple of the normal diploid number.

The number of chromosomes may vary from species to species. The lowest chromosome number in animals has been recorded in round worm (*Ascaris megalocephala*) with 2n = 2. In *Mesotoma* (flat worm) and *Ophryotrocha puerilis* (polychaete), haploid chromosome number (n) is 2. In *Ascaris megalocephala* which has a single chromosome in the gamete, the chromosome is compound which divides into as many as 42 chromosomes in the somatic cells. In plants, the smallest chromosome number has been reported in a member of Asteraceae (compositae), *Haplopappus gracilis* with 2n = 4. In some organisms the chromosome number may be as high as several hundreds in diploid cells. The highest chromosome number in animals is reported in *Aulacantha*, a radiolarian, in which 2n is approximately 1,600 chromosomes. In animals chromosome numbers up to 50 are not uncommon. The higher plants, such as the angiosperms and animals such as primates, the members of the taxonomic groups often have chromosome numbers within a limited range. In angiosperms the most frequent haploid number is 12 and the great majority of the members of this group have at an average haploid numbers 8 to 16. Similarly, fungi also have a definite range of chromosome number, the haploid number ranging from 3 to 8. In primates, the haploid number ranges from 16 to 30.The man has 23 pairs in somatic cells. Some species of plants and animals with their respective chromosome numbers are listed in Table 9.1.

Table 9.1 Chromosome number of some animals and plants

Name of animal	*Diploid chromosome number (2n)*
Paramecium (*Paramecium aurelia*)	30 – 40
Hydra (*Hydra vulgaris*)	32
Round worm (*Ascaris megalocephala*)	2
Silk worm (*Bombax mori*)	56
House fly (*Musca domestica*)	12
Fruit fly (*Drosophila melanogaster*)	8
Honey bee (*Apis mellifica*)	32, 16
Mosquito (*Culex pipiens*)	6
Frog (*Rana tigrina*)	26
Chicken (*Gallus domesticus*)	78
Pigeon (*Columba livia*)	80
Mouse (*Mus musculus*)	40
Rabbit (*Oryctologus cuniculus*)	44
Common rat (*Rattus rattus*)	42
Guinea pig (*Cavia cobaya*)	64
Dog (*Canis familiaris*)	78
Cat (*Felis domestica*)	38
Horse (*Equus caballus*)	64
Pig (*Sus scrofa*)	40
Goat (*Copra hircus*)	60
Chimpanzee (*Pan troglodytes*)	48
Man (*Homo sapiens*)	46
Name of Plant	***Diploid chromosome number (2n)***
Chlamydomonas (*Chlamydomonas reinhardii*)	10, 12, 16, (Haploid sets)
Bread mold (*Mucor heimalis*)	2
Yellow pine (*Pinus ponderosa*)	24
Cabbage (*Brassica oleracea*)	18
Radish (*Raphanus sativus*)	18
Papaya (*Carica papaya*)	18
Cotton (*Gossypium hirsutum*)	52
Coffee (*Coffea arabica*)	44
Sunflower (*Helianthus annuus*)	34
Potato (*Solanum tuberosum*)	48
Tomato (*Lycopersicum solanum*)	24
Tobacco (*Nicotana tabacum*)	48
Banana (*Musa paradisiaca*)	22, 44, 55, 77, 88
Garden pea (*Pisum sativum*)	14
Bean (*Phaseolus vulgaris*)	22
Orange (*Citrus sinensis*)	18, 27, 36
Apple (*Malus silvestris*)	34, 51
Indian corn (*Zea mays*)	20

Name of Plant	*Diploid chromosome number (2n)*
Barley (*Hordeum vulgare*)	14
Bread wheat (*Triticum vulgare*)	42
Rice (*Oryza sativa*)	24
Bajra (*Sorghum vulgare*)	10, 20, 40
Sugarcane (*Saccharum officinarum*)	80
Field bean (*Vicia faba*)	12
Onion (*Allium cepa*)	16

Morphology of Chromosomes

The chromosomes are filamentous bodies found in the nucleus and are visible during cell division. They are linkage structures each consisting of a linear sequence of genetic information or genes. The chromosomes of prokaryotes and eukaryotes are essentially similar in two main activities: (*i*) they are concerned with the transmission of genetic information from cell to cell and generation to generation, (*ii*) they are concerned with the ordered release of genetic information to control cellular functions and development. They are autoduplicating structures of differential complexities in eukaryotes and prokaryotes and their morphology and organization are organism specific. The chromosomes of prokaryotes are generally double stranded circular DNA molecules but in some viruses they may be single stranded. In eukaryotes they present a complex structure. Most of the chromosomes in a cell are called **autosomes** which carry genes for most of the body characters. Besides, there may be one, two or more chromosomes which carry genes for determination of sex as well as for sex-linked characters. These are called **sex chromosomes** (as for example, X-chromosomes and Y-chromosomes). The standard chromosomes forming the chromosome complement of a cell are called **A-chromosomes**. In some species one or more extra chromosomes are present in addition to A chromosomes. They are called *B* ***chromosomes*** or ***supernumerary chromosomes*** or ***accessory chromosomes***.

Size of the Chromosomes

The chromosome size may vary in different closely related genera as well as in different species and varieties of the same genus. The chromosome size is measured normally at mitotic metaphase. They may be as short as 0.2μ in fungi and birds and as long as 30μ in some plants like *Trillium*. Generally, most of the chromosomes at mitotic metaphase fall within a size range of 3.5μ in fruitfly (*Drosophila*) to 6μ in man and 8–10μ in maize. The plants in general have larger chromosomes than the animals. Further a great variation may be seen in the sizes of chromosomes in different tissues within the same organism. The giant chromosomes of salivary gland in Dipteran insects may attain a length of nearly 300μ and a breadth 10μ. In amphibian oocytes, lampbrush chromosomes may be 800μ in length whereas their somatic chromosomes are only a few microns in length. In plant *Mediola* the root tip chromosomes are 50% longer than the shoot tip chromosomes. The organisms with less number of chromosomes have, in general, comparatively larger sized chromosomes than those with many chromosomes. Monoctyledons, in general, have larger chromosomes than dicotyledons which often contain greater number of chromosomes. The plants, in general, have large-sized chromosomes in comparison of animals. Further, all the chromosomes in a cell may not be alike. In single set of chromosomes the largest chromosome may be as much as 50 times larger than the shortest one. On the other hand, in some organisms the size difference among the chromosomes of the set may not be well marked and all the chromosomes are of about the same length.

The Structure of Eukaryotic Chromosomes

The detail structure of chromosome varies from cell to cell and major changes are associated with cell division. Regardless of differences in detailed morphology, both direct and indirect evidences make it clear that each chromosome is an individual entity which, barring accidents, is retained throughout the life of the cell.

The detailed morphology of the chromosome is best studied during cell division, particularly at metaphase and anaphase stages when the chromosomes are in maximum contraction state. The morphology of chromosomes differs from phase to phase in the continuous process of cell growth and division. During the **interphase**, when cells are not dividing, the nucleus has a granular appearance because the chromosome material or chromatin remains dispersed throughout the nucleus. At the time of cell division the chromosome material becomes highly condensed and appears in the form of delicate, contractile, poorly stainable interwoven thread. In the initial stage of cell division they gradually become thick and filamentous and show distinct individuality. When completely differentiated, each chromosome consists of two chromatids held together at a point along their length. The point at which the two chromatids of a chromosome are held together appears as a constriction this is ***primary constriction*** or ***kinetochore*** or ***centromere***. In the metaphase chromosome the centromere stains lighter than other segments. The centromere divides the chromosome into two parts, each called a ***chromosome arm***. In some of the chromosomes there may be additional constriction, ***the secondary constriction*** which indicates the position of ***nucleolar organiser region*** (Fig. 9.1). Thus the structures present in the chromosome include:

(*i*) the chromonema,

(*ii*) primary constriction,

(*iii*) secondary constriction and Nucleolar organiser region, and

(*iv*) telomeres.

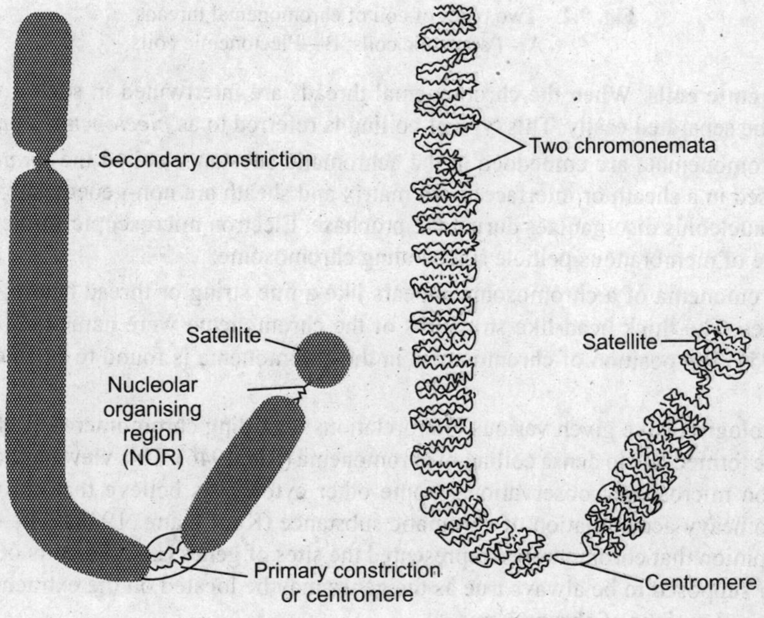

Fig. 9.1 Chromosome structure. A. External structure. B. Helically coiled chromonemata.

The Chromonema

Each chromosome is formed of two identical spirally twisted delicate threads which lie so close to each other that it is difficult to see them separately by ordinary microscope (Figs. 1.43 A and B). These two sub-units are called chromatids. If the chromosomes are treated with trypsin to remove proteins, each chromatid is seen to consist of two or more sub-units. These are termed *chromonemata* (singular-*chromonema*) by Vejdovsky (1912). The question as to how many such chromonemata are there in a chromatid has long been a controversial matter. Some workers have suggested that a chromatid may be composed of two chromonemata while others have suggested a strong possibility for more than two chromonemata. The number of threads may vary in different phases of cell division because at one stage the chromatid may contain one thread and at the other phase it may contain two, four or more chromonemata (Trosko and Wolff, 1964). The chromonemata may remain coiled with each other. The coils may be of the following types:

(*i*) **Paranemic coils.** When the chromonemal fibrils are coiled in such a way that they are easily separable from each other, such coils are called ***paranemic coils*** (Fig. 9.2A).

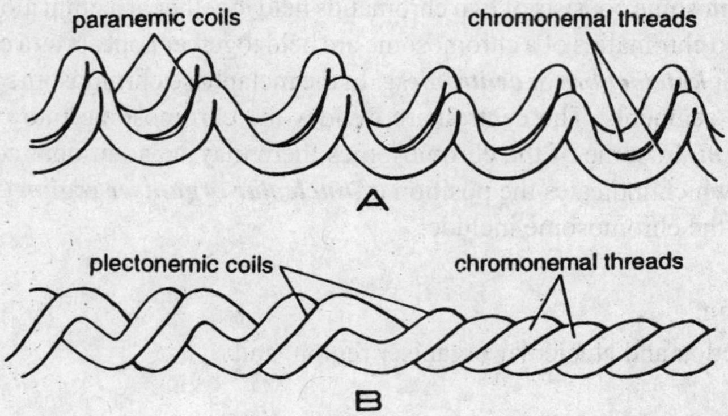

Fig. 9.2 Two types of coil of chromonemal threads. A—Paranemic coils; B—Plectonemic coils.

(*ii*) **Plectonemic coils.** When the chromonemal threads are intertwined in such a way that they cannot be separated easily. This type of coiling is referred to as *plectonemic coil* (Fig. 9.2B).

The chromonemata are embedded in the achromatic substance called the '*matrix*' which, in turn, is enclosed in a sheath or interface. Both matrix and sheath are non-genetic substances which appear when nucleolus disorganizes during the prophase. Electron microscopic studies have denied the occurrence of membranous pellicle surrounding chromosome.

The chromonema of a chromosome appears like a fine string or thread bearing a number of granular bodies. The thick bead-like structures of the chromonema were named chromomeres by Balbiani (1875). The position of chromomeres in the chromonema is found to be fixed for a given chromosome.

The cytologists have given various interpretations regarding chromomeres. A chromomere is supposed to be formed due to dense coiling of chromonema (Ris, 1946). This view has been confirmed by the electron microscopic observations. Some other cytologists believe that chromomeres are formed due to heavy accumulation of chromatic substance (Kauf mann, 1948). The early workers were of the opinion that chromomeres represented the sites of genes (*i.e.*, the units of heredity) but it is no longer supposed to be always true as the genes may be located on the extrachromomeral or interchromomeral regions of chromonema.

Primary Constriction or Kinetochore or Centromere

The principal landmarks which may be seen in the chromosomes at metaphase are (*i*) primary constriction or centromere or kinetochore to which spindle fibres or microtubules are attached, and (*ii*) in some chromosomes one or more secondary constrictions. The position of centromere is fixed in a given chromosome. The primary constriction or centromere stains lighter with Feulgen stain (basic fuchsin) than the other parts of chromosome, especially during metaphase and anaphase. It indicates that DNA extends through this region (Kezer and McGregor, 1971). The presence of DNA has been demonstrated in the kinetochores of several organisms, that forms centromeric heterochromatin.

The ultrastructure of centromere has been difficult to resolve because of certain limitations in the preparation techniques. Recent observations with electron microscope have revealed some structural details of centromere. There may not be a common pattern of structure in plants and animals. In a metaphase chromosome the kinetochore consists of two or more centromeric chromomeres of varying sizes and interchromomeral fibrillae. The fibrillae are strands of chromonemata and their number in a kinetochore varies somewhat, but usually only a few are visible. In some of the kinetochores centromeric chromomeres may not be visible and only thin fibrillae can be seen in them. In most of the kinetochores there is a definite symmetry in the arrangement of fibrillae and chromomeres. There is a central component (the interior zone) consisting usually of one or two chromomeres and the arrangement of chromomeres and fibrillae is same on both the sides of central component . In some of the metaphase chromosomes four granules or chromomeres may be seen within the kinetochore and they are arranged in a square. During anaphase when the two chromatids separate, each of them becomes daughter chromosome. Previously it was believed that the centromere duplicated sometime during the metaphase but it is now clear that the centromere is actually duplicated before metaphase of mitosis and anaphase I of meioses in most of the organisms and the chromatids are held together not by a single kinetochore but by an adhesion of chromatids. The centromere in the chromosome of Chinese hamster consists of a dense core surrounded by a wider less dense zone. The dense core is composed of axial fibrils which are 50–80 Å thick and cohelically coiled. The less dense zone shows several microfibrils which loop out on the outer surface (poleward) of each chromatid at metaphase. The loops disappear after the chromosomes have reached to the spindle poles.

During cell division, kinetochore participates in the organization of microtubule units into spindle fibres and, in fact, its role in the movement of chromosomes during anaphase can be considered its principal function. Some investigators have suggested that the chromosome movement is at least partly autonomous and that the centromere simply provides force for the movement.

The position of centromere is fixed in a particular chromosome. Levan, Fredga and Sandberg (1964) have proposed a nomenclature for centromeric position on chromosomes.

Normally, the chromosomes possess a single centromere *i.e.*, they are **monocentric**. But, in some instances, the chromosome may have two centromeres (*i.e.*, **dicentric chromosome**) or many centromeres (*i.e.*, **polycentric chromosome**). Sometimes the chromosome may be devoid of centromere. Such a chromosome is said to be acentric.

The position of centromere may vary in different chromosomes. The parts of a chromosome on each side of centromere are called **arms**. The centromere plays important role in directing the chromosome movement during cell division. According to the positions of centromere and relative arm lengths, the monocentric chromosome may be designated as follows (Fig. 9.3):

1. **Metacentric.** In this the kinetochore is located at or near the mid-point (*median*) so that the arm ratio is 1 : 1 or nearly so. Such chromosomes are V shaped at anaphase stage.
2. **Submetacentric.** When the centromere is located slightly away from the mid-point so that the two arms are unequal, its position is said to be **submedian** and the chromosome is called **submetacentric**. Such a chromosome appears L shaped during anaphase.
3. **Acrocentric.** When the centromere is located away from the mid-point and is found near one end of the chromosome, its position is said to be **subterminal** and the chromosome is called **acrocentric**. Acrocentric chromosomes appear rod shaped or j-shaped during anaphase and have one long arm and the other very short arm.

   In submetacentric and acrocentric chromosomes one arm is longer than the other. The shorter arm is also designated '*p*' arm (french word petite meaning small or short) and the longer arm is termed '*g*' arm.
4. **Telocentric.** When the centromere is located at one end of the chromosome it is said to be **terminal**. The chromosome with terminal centromere is called **telocentric**. Such a chromosome is characterised by a single arm. Telocentric chromosomes appear rod-shaped during anaphase.

Telocentric chromosomes are found only rarely. They have been reported by Marks (1957) in some species of plants, protozoa and mammals. Truely, telocentric chromosomes are sometimes unstable.

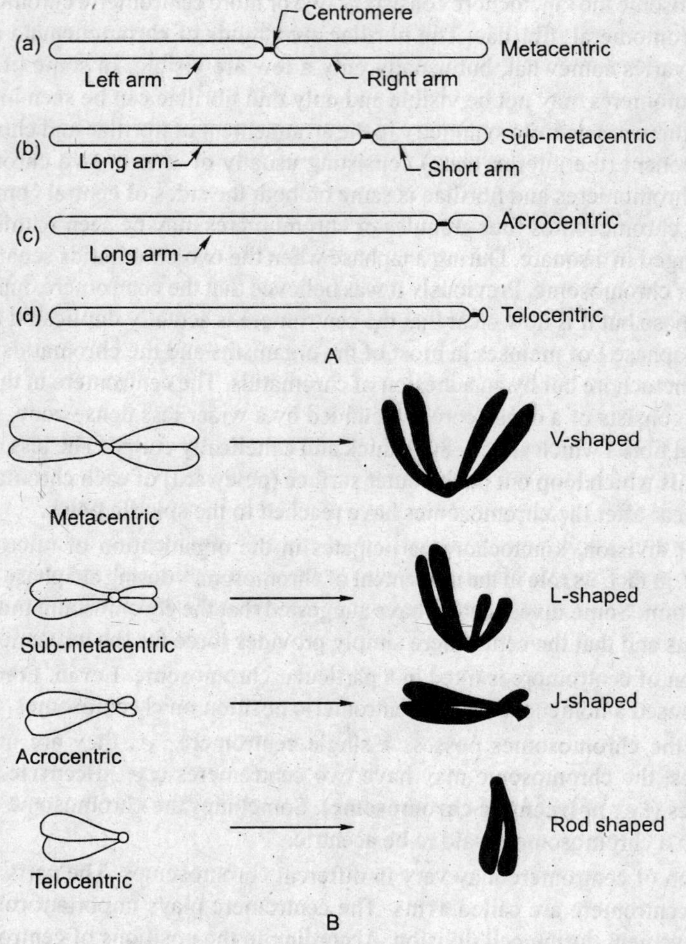

Fig. 9.3 A. Four major types of monocentric chromosomes on the basis of centromere position and consequent relative arm lengths: (*a*) Metacentric, (*b*) Sub-metacentric, (*c*) Acrocentric, (*d*) Telocentric. B. Shape of different chromosomes during anaphase of metosis.

Isochromosome. Isochromosome is a chromosome in which the two arms are identical. It arises by an error in the plane of centromere division. Occasionally one chromatid of the telocentric chromosome after duplication of arm may swing around the centromere to produce isochromosome with two identical arms.

Normally, the centromere divides in a plane parallel to the longitudinal axis of the chromosome but in some cases the division of centromere occurs at right angle to the chromosome axis (mis-division) which may result in the formation of two telocentric chromosomes (Fig. 9.3), each of which is relatively stable and functional because of its kinetochore construction. Such mis-division of kinetochore has been observed in *Pisum*, *Fritillaria*, and *Triticum* and at least in one animal, the grasshopper *Microstethus*. These form-categories are not sharply distinct but grade imperceptibly into each other. Whatever its position, the centromere has a fixed position in each chromosome and maintains its position throughout the life time of chromosome and from generation to generation. Thus a chromosome can be recognized partly by position of its centromere.

Dicentric chromosomes. In some instances, chromosomes with two centromeres may be formed. They are called dicentric chromosomes. Dicentric chromosome may be produced in inversion of heterozygotes as a result of pairing between structurally different homologous chromosomes and subsequent crossing over between the nonsister chromatids. If the two centromeres of a dicentric chromosome move to the opposite poles of the spindle during anaphase, the chromosome breaks. Rarely, a new additional centromere may appear on the monocentric chromosome resulting in an abnormal dicentric chromosome. Such a centromere is called **neocentromere**. The occurrence of neocentromere has been reported in *Zea mays.*

Chromosome with Diffuse kinetochore and Polycentric Chromosome

In some animals for example, coccids (*Homoptera*) and monocotyledons (Juncales and Cyperales), there is no single localized centromere and the chromosome is provided with more than one centromere, *i.e.*, the kinetochore is diffused along the entire length of the chromosome. Such chromosomes are called **holokinetic chromosomes** (Sybenga, 1972) (Fig. 9.4). According to Levan's nomenclature diffuse kinetochore is called lateral centromere. The chromosome with diffuse kinetochore behaves as if the entire chromosome were a kinetochore. During cell division, spindle fibres are associated all along the length of chromosome and each member chromatid moves to its respective pole in a plane perpendicular to the axis of the chromosome. The two chromatids separate from each other from their entire length at once.

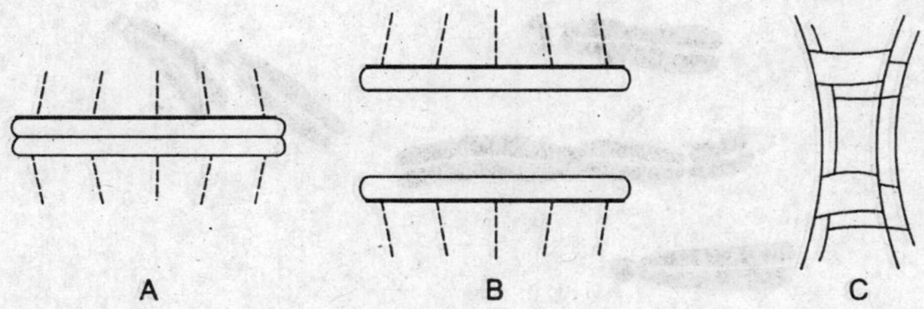

Fig. 9.4 A—Holokinetic chromosomes at mitotic metaphase; B—At anaphase; C—Meiotic metaphase I bivalent of a holokinetic chromosome pair (after Sybenga, 1972).

The polycentric chromosome often resembles chromosome with diffuse kinetochore in behaviour as in coccids. The polycentric chromosome differs from the chromosome with diffuse kinetochore in only one respect that it regularly fragments into smaller, stable functional chromosomal units. The polycentric chromosomes are found in *Parascaris univalens* (*Ascaris megalocephala univalens*) in which gametes contain only a single chromosome. After fertilization 2 chromosomes of somatic cells of embryo fragment and give rise to many small functional chromosome units. Somatic tissues in this worm, frequently have as many as 42 chromosomes per cell.

Acentric chromosome. Sometimes chromosome may break into two pieces so that only one part retains the centromere and the other is devoid of centromere. The chromosome part lacking in centromere is said to be *acentric.* Since acentric chromosomes do not have centromeres, spindle fibres are not attached with them and so they are unable to move to the poles during anaphase. Such a chromosome is usually lost from the cell and may result in deficiency and sometime lethality.

Secondary constriction. Besides the primary constriction or centromere, the arm of one or more chromosomes may show one or more constrictions called **secondary constrictions**. Since such constrictions are constant in position and extent, they are useful in identifying particular chromosomes in a set. Secondary constriction may be short or long. It is often associated with nucleolus during interphase and may take part in the reorganization of nucleolus during telophase stage of cell division. For this reason, such a secondary constriction is called **the nucleolus organizing region or secondary constriction I**. In some cases, nucleolus organizer exists at the terminal position in the chromosome but go undetected cytologically because of its small size.

Under the light microscope the nucleolus organizer appears as a constriction near one end of the chromosome. Normally, in each diploid nucleus there may be two chromosomes with nucleolus organizing region but in some diploids and polyploids, the number of nucleolar organizers may be more than two. The part of chromosome extending beyond nucleolus organiser is very short and appears like a dot or satellite. This extension is called **satellite body**. Chromosomes bearing satellites are called ***SAT-Chromosomes***. Satellite bodies vary in size according to the position of the secondary constriction. The satellite may be minute dot-like in many cases but in some cases large satellite may be seen as for example, in *Vicia faba* which may be connected to the main body of chromosome by very light-staining strands (Fig. 9.5 A). Observations of human chromosomes have revealed that five pairs of chromosomes, numbered 13, 14, 15, 21 and 22 have nucleolus organizing regions.

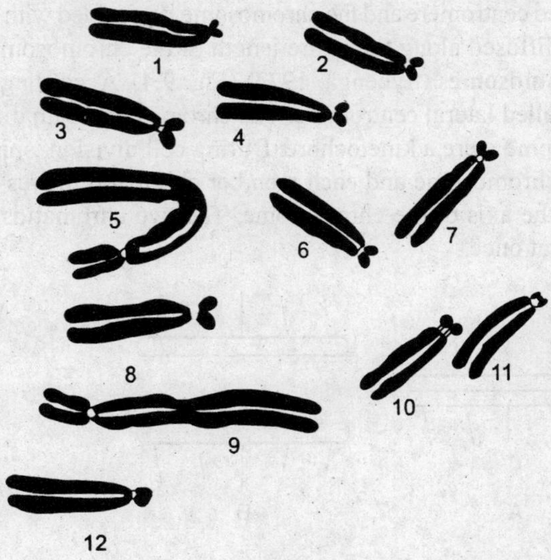

Fig. 9.5 A Chromosomes of *Vicia faba*.

In a few cases a single chromosome has two secondary constrictions. The secondary constriction other than nucleolus organizer is referred to as ***secondary constriction II***. The location of secondary constriction II is fixed for a given chromosome and is a diagnostic characteristic of the particular genome. It has been suggested that secondary constriction represents the site of breakage and reunion in chromosome.

Telomeres. The ends of a chromosome are called telomeres. The telomeres behave differently from the interstitial portions of chromosome. Each telomere has polarity and it prevents the ends of the chromosomes from attaching to each other. A broken or fractured chromosome will not attach to a telomere unless that telomere is also fractured. If the ends or telomeres are broken off either spontaneously or by induction with x-rays or Gamma-rays they will not unite with each other and may disappear from the nucleus in subsequent cell divisions. The cut end of the chromosome is unstable and may unite with broken end of other chromosome resulting in change of morphology. Thus the telomeres maintain the individual integrity of each chromosome. They are specially modified for attachment with nuclear envelope. The ends of the chromosomes are usually found associated with the nuclear wall during interphase and early prophase.

Ring chromosome. When the two cut ends of a chromosome unite, the ring chromosome is formed (Fig. 9.5 B).

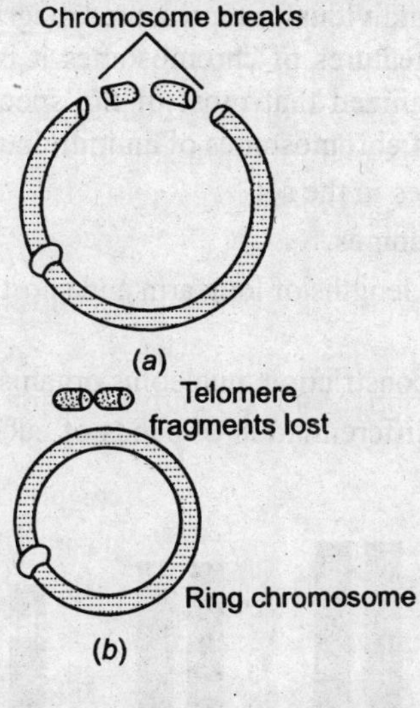

Fig. 9.5 B Origin of ring chromosomes.

Karyotype. The term *karyotype* usually refers to the general morphology of a set of chromosomes at somatic metaphase of an individual. In karyotypic studies pairs of homologous chromosomes of a cell are arranged in a decreasing series. (Fig. 9.6).

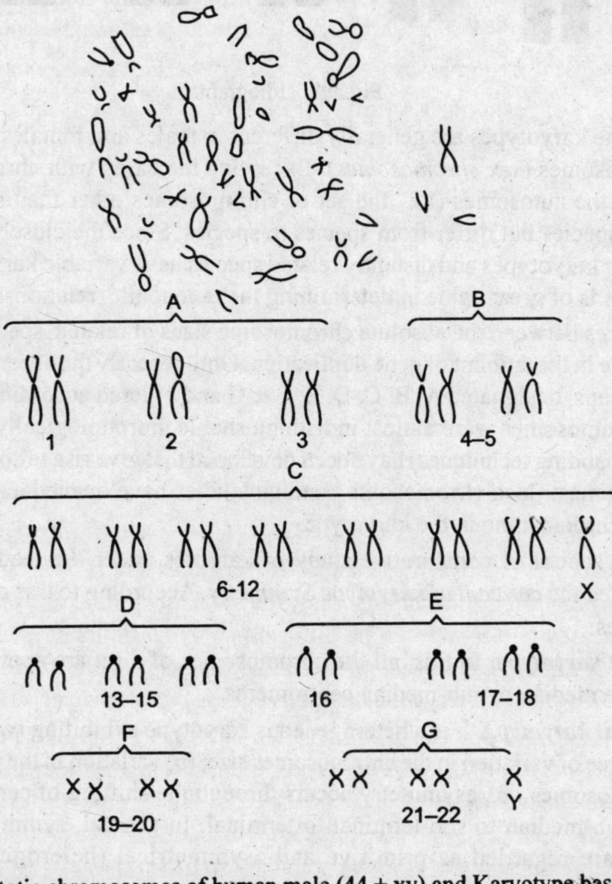

Fig. 9.6 Mitotic chromosomes of human male (44 + xy) and Karyotype based on the same.

When the karyotype of an individual is represented diagrammatically in the form of histogram showing all the morphological features of chromosomes it is called an ***idiogram*** or **karyogram** (Fig. 9.7). It is now well recognized that most of the species possess a definite chromosome individuality. The morphology of chromosomes of an individual includes the following:

(*i*) the number of chromosomes in the set,

(*ii*) relative lengths of chromosomes,

(*iii*) arm ratios (*i.e.*, ratio of the lengths of long arm and short arm as determined by the position of the centromere),

(*iv*) the position of secondary constriction, nucleolus organisers and satellite, and

(*v*) the staining reaction and differentiation of bands of euchromatin and heterochromatin in the chromosomes.

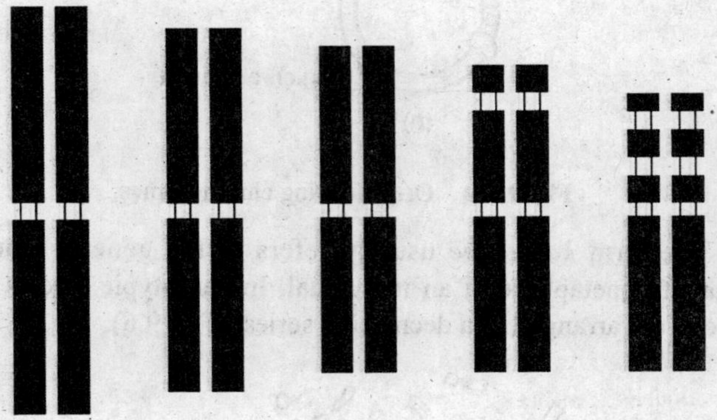

Fig. 9.7 Idiogram.

In animals, the karyotypes are generally different in males and females owing to the presence of X and Y chromosomes (*sex chromosomes*). Excepting the cases with chromosomal aberrations, the karyotypes for the autosomes (*i.e.*, the set of chromosomes other than sex chromosomes) are similar within the species but differ from species to species. Since the closely related species have, more or less, similar karyotypes and distantly related species have variable karyotypes, a comparative study of karyotypes is of great value in determining the taxonomic relationship among the species.

The differences between the absolute chromosome sizes of related species or genera probably reflect the difference in the amount of gene duplication. Until recently the chromosomes were grouped into only seven groups, designated A, B, C, D, E, F & G and ordered according to their sizes. Within each group the chromosomes were almost indistinguishable morphologically. But now the staining techniques (called banding techniques) have been developed that give rise to specific patterns of bands along the chromosomes. Such chromosome banding studies have opened a new way that serves to differentiate each chromosome in the karyotype.

The Russian school of comparative study of karyotes under the leadership of G. Levitzky (1931) has developed the ***concept of karyotype Symmetry***. According to this concept, the karyotypes may be of two types:

(*i*) ***Symmetrical karyotype.*** In this, all the chromosomes of a set are approximately of the same size and have median or sub-median centromeres.

(*ii*) ***Asymmetrical karyotype.*** It is a heterogeneous karyotype exhibiting two main characteristics; (*a*) good degree of variation in the chromosomes size; (*b*) variation in the position of centromeres of the chromosomes *i.e.*, asymmetry occurs through the shifting of centromere position from median to sub-median to sub-terminal to terminal. In general, symmetrical (homogeneous) karyotypes are regarded as primitive and asymmetrical (heterogeneous) karyotypes as specialized and advanced.

Supernumerary or B-Chromosomes

Many species of plants and animals contain, in addition to normal constant complement of A-Chromosomes, a variable number of minute and usually heterochromatic chromosomes. These are called **supernumerary chromosomes** or **extrachromosomes**, or **accessory chromosomes** (McClung, 1900) or **B-chromosomes**. (Randolph, 1928). B-chromosomes are distinguished from the other smaller chromosomes of the normal complement in their staining properties. In some cases, however, the distinction on the basis of staining reaction seems impossible. In *Tradescantia* and *Trillium*, for example, supernumeraries appear to be largely euchromatic and in maize they contain both kinds of chromatin.

B-chromosomes were first discovered by Wilson (1905) in hemipteran insect and by Randolph (1927–28) in maize. Subsequently, their presence in higher plants was reported by Darlington (1937), Hakansson (1945), Fernandes (1946), Melander (1950) and many others.

The occurrence of B-chromosomes has been reported now in more than 375 families. B-chromosomes differ from normal or A-chromosomes in the following respects.

1. **Morphology.** They are usually much smaller in size than the smallest A-chromosome. In gross morphology most of the accessory chromosomes are acrocentric or telocentric and those which are metacentric are usually, if not always, isochromosomes in which the two arms are genetically identical. Accessory chromosomes are frequently and not exclusively heterochromatic. In ***Tradescantia* and *Trillium*,** B-chromosomes appear to be largely euchromatic and in maize they contain both kinds of chromatin (Figs. 9.8 A and B).
2. **Genetic effectiveness.** Accessory chromosomes normally do not strongly influence the phenotype and viability of the organisms carrying them.
3. **Numerical variation.** The number of accessary chromosomes may vary in different cells, tissues, individuals, and populations.
4. **Meiotic behaviour.** The B-chromosomes do not pair with A-chromosomes and there is no chiasma formation. They show lower degree of pairing among themselves, early separation from each other at late prophase I and metaphase I, lagging during anaphase and elimination or preferential non-disjunction during meiosis and post-meiotic divisions (Fig 9.8).

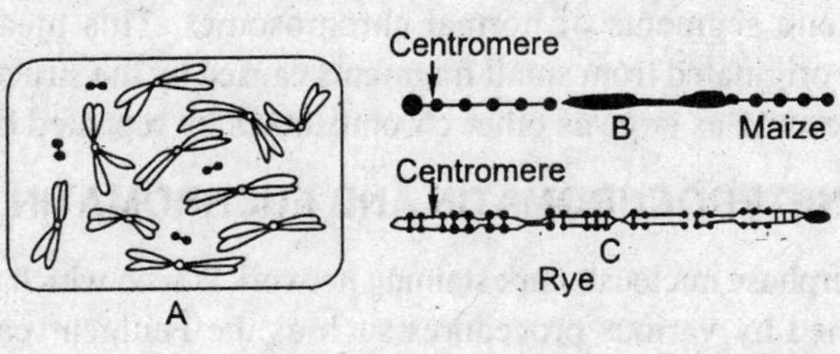

Fig. 9.8 A. Cell with normal or A and B dot like chromosomes, B & C. B chromosomes of maize B and rye C at pachytene (after Sybenga, 1972).

5. **Mitotic behaviour.** During mitotic division of cell, there may be lagging or elimination of B-chromosomes. They may also show polymitosis or preferential distribution.

According to White (1945), there are two types of B-chromosomes:

(*i*) *First types* which are mitotically stable and all the cells of carrier organisms have the same number of B-chromosomes.

(*ii*) *Second type* which are mitotically unstable, giving rise to different numbers of B-chromosomes in different cells of some individual.

Accumulation systems of B-chromosomes are different but normally fall into two categories:

1. Accumulation system in which B-chromosome undergoes distorted segregation at meiosis and is thereby preferentially transmitted to gametes.

2. Accumulation system in which B-chromosome undergoes meiotic non-disjunction in some particular direction or tissues.

Although no example is known of accessory chromosomes affecting the external morphology of carrier plants, they are known to affect the over all vigour of the plants as well as their pollen fertility. Vegetative vigour is usually reduced by high numbers of accessory chromosomes such as 10 or more in *Zea mays*, but in some species such as cultivated rye (*Secale cereale*) a marked decrease in vigour occurs in plants which have one or two B-chromosomes. Generally polyploids are much less likely to contain accessory chromosomes than their diploid relatives. Pollen fertility is more sensitive to the presence of accessories. In *Sorghum purpurea and Seriseum* any increase in the number of accessaries causes reduction in pollen fertility.

Favourable effects of accessory chromosomes may also be seen in some cases. A special type of accumulation system of B-chromosomes in population results in a tendency for the B-chromosomes to increase in frequency in males and to decrease correspondingly in females. In *Festuca pratensis* vigour may be increased by the presence of one or two B-chromosomes. In diploid species of *Achillea*, *A.asplenifolia* and *A.setacea,* plants having two B-chromosomes are vigorous and more fertile than plants lacking them altogether. In cross *A.asplanifolia* X *A. setacea* though not in reciprocal, F_1 plants could be obtained only when the female parent contained accessories. In this case, the presence of accessories in the stylar tissue of *A. asplanifolia* apparently promotes the growth of the pollen tube of *A. setacea.* Plants of these species or their hybrids which contain uneven or high number of B-chromosomes are relatively infertile.

It is generally believed that B-chromosomes have adaptive importance, since their occurrence and frequency within the species vary in the populations of different origin and habitats. In some instances, as in *Festuca pratensis* (Bosemarr, 1936), *Centaurea scabiosa* and *Lilium medeoloides* populations containing B-chromosomes exist in restricted regions of Geographic range of the species which is associated with the specific factors of environment or population structure. In Sweden, populations of *Festuca pratensis* having B-chromosomes occur chiefly in regions of highly clay soils.

A few workers have suggested that B-chromosomes might be parasitic and maintain more or less independently of any adaptive benefit or handicap.

The origin of B-chromosomes is not known, yet it is presumed that they represent centromere containing heterochrometic segments of normal chromosomes. This means that supernumerary chromosome might have originated from small fragments caused by the structural rearrangement and might have eventually become as large as other chromosomes by repeated duplication.

HETEROCHROMATIN AND EUCHROMATIN

In the nucleoplasm of interphase nucleus a dark staining network is seen which is formed of **chromatin**. When chromatin is stained by various procedures such as the Feulgen reaction which is specific for DNA and examined under light microscope, some regions are stained darkly whereas the other regions are stained lightly. The chromatin is Feulgen positive material observed during the interphase in the nucleus. It is a complex of DNA and proteins. During cell division, chromatin fibres become condensed to form thick ribbon-like structures, the chromosomes. Certain segments of chromosomes or the entire chromosomes become more condensed than the rest of the sections of the karyotype during interphase and early prophase. Thus the chromatin occurs in two forms; **heterochromatin and euchromatin.** The condensed regions of the chromosomes are referred to as heterochromatic and the non-condensed segments as **euchromatic** (Fig. 9.9). The heterochromatic regions stain darkly whereas the euchromatic regions stain lightly. During mitosis, the heterochromatic regions are localised at the telomeres, the centromeres and intercalary regions of the chromosomes. The bulk of the chromatin is made up of euchromatin. Within a chromosome there may be small areas of dark staining connected by lightly staining regions. Recently, staining procedures have been developed that result in patterns of darkly and lightly stained regions or bands. The banding patterns are highly specific to each chromosome pair and have been used in unequivocal identification of 23 pairs of chromosomes in man, as well as in a wide range of other organisms.

Although the genetic difference between the heterochromatin and euchromatin is not clear as both contain DNA, the two represent two different states of the same substance. In electron micrographs the two differ in their physical structures. Heterochromatin is composed of 250 Å fibrils whereas euchromatin contains 30 to 80 Å thick fibrils. So, heterochromatin is a condensed coiled state of chromatin and it contains two or three times more DNA than euchromatin. There is good evidence that the heterochromatic regions are genetically inactive as they are not involved in the synthesis of RNA and proteins.

Heterochromatic regions may stain more strongly or weakly than the euchromatic regions (heteropycnosis) and this differential staining reaction is governed by the degree of coiling of the strands of chromosomes. Where the strands are highly condensed there is greater density of chromatin material and hence darker staining than in noncondensed regions. Heteropycnesis may be positive where there is overcondensation or may be negative where there is under-condensation. Chromosomes which remain condensed during interphase are called *heterochromosomes*, as for example, the sex chromosomes of insects. The chromosomes which remain uncondensed during interphase are called *euchromosomes*.

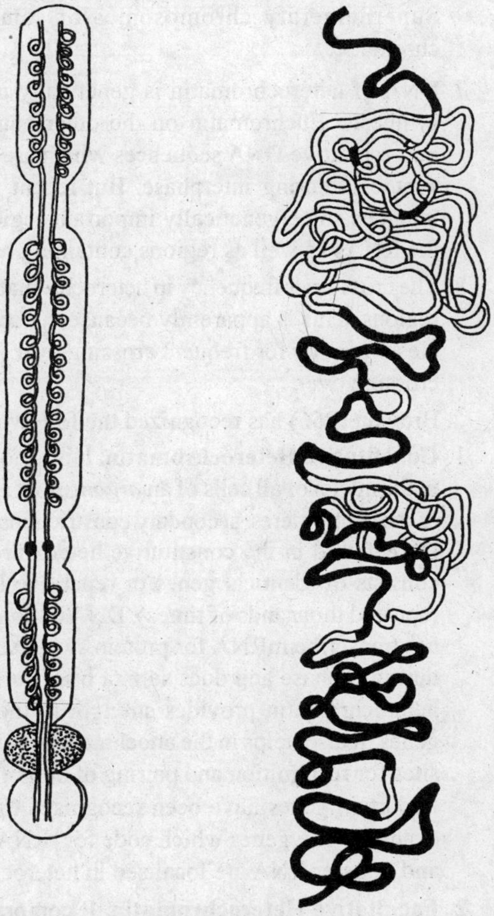

Fig. 9.9 A model of constitutive heterochromatin in a mammalian metaphase chromosome. B. Euchromatin and heterochromatin.

The main differences between the heterochromatin and euchromatin are listed below:

1. Heterochromatin stains deeply while euchromatin stains lightly.
2. Heterochromatin is found in the condensed regions of the chromosomes and represents the densely packed regions of chromatin fibrils while the euchromatin is found in the diffused or loosely coiled regions of chromosome.
3. Heterochromatin replicates at the end of S phase of cell cycle while the euchromatin replicates during early stage of S phase.
4. Heterochromatin does not become acetylated whereas euchromatin contains acetyle group in its histones during interphase.
5. Heterochromatin is more labile than euchromatin and is affected by temperature, sex, age, proximity to the centromere, etc.
6. Heterochromatin is relatively inert metabolically and the heterochromatic segments contain a few genes in relation to their length. In *Drosophila melanogaster* the Y chromosome is totally heterochromatic in nature. Male flies lacking the Y chromosome are phenotypically normal. They are, however, sterile owing to the failure of normal sperm formation during spermatogenesis.

Supernumerary chromosomes of some plants, for example, maize, are also heterochromatic.

7. DNA of heterochromatin is genetically inactive as it does not transcribe mRNA for protein synthesis. Euchromatin on the other hand, is genetically active and is partly composed of non-repetitive DNA sequences which are genetically active and transcribe mRNA for protein synthesis during interphase. But recent findings have suggested that heterochromatin also includes some genetically important regions such as nucleolar organisers, genes for some of the RNAs as well as regions containing highly repeated nucleotides.
8. The cross-over frequency in heterochromatic regions of chromosomes is less than in euchromatic regions. This is apparently because the condensed regions of chromosome fibres do not come close together for frequent crossing over. This situation protects vital genes from the effects of crossing over.

Brown (1966) has recognized the following two main types of heterochromatin:

1. **Constitutive Heterochromatin.** It is a most common type of heterochromatin seen consistently in the nuclei of all cells of an *organism*. It is generally found in blocks, usually around areas such as the centromeres, secondary constrictions, telomeres or as bands in other parts of chromosomes. DNA found in the constitutive heterochromatin is most inactive during protein synthesis. It consists of identical genes or repetitive DNA segments (*i.e.*, their nucleotide sequences are repeated thousands of times). DNA of constitutive heterochromatin replicates readily but does not transcribe mRNA for protein synthesis. This DNA is late replicating as it fails to replicate during S phase and does so in a brief period just before G2 period of cell cycle. Constitutive heterochromatin provides strength to the centromere and also acts as spacer between vital genes. It also helps in the attachment of chromosomes with the nuclear membrane and provides sites for recognition and pairing of homologous chromosomes during meiosis. Recently some important genes have been recognised to be present in the constitutive heterochromatin, for example, polygenes which code for rRNA in nucleolus organizer and those making 5 S rRNA and transfer RNA are localised in heterochromatin regions.
2. **Facultative Heterochromatin.** It comprises about 2.5% of the genome and is metabolically inactive. It reflects the existence of a regulatory device to adjust the doses of certain genes in the nucleus. In animal cells this type of heterochromatin is acquired during embryogenesis. It originates through a process called ***facultative heterochromatisation*** during which a chromosome or a set of chromosomes becomes heterochromatic (turned off) in cells of one sex while the remaining sets of chromosomes become euchromatic (turned on) in the cells of opposite sex. The best known case is that of X chromosomes in mammalian female, one of which is active and remains euchromatic whereas the other X chromosome becomes inactive and forms *sex chromatin* or *Barr body* at interphase. In human, the sex chromatin appears in the embryo between the sixteenth and eighteenth days after fertilization. Before that time both X chromosomes are euchromatic. In some insects, one of the two X chromosomes may be heterochromatic and the other one euchromatic. In plants, Britten and Kohne (1968) have denied the occurrence of facultative heterochromatin.

Nature and Functions of Chromatin in Different Regions of Chromosome

Region of Chromosome	*Chromatin type*	*Function*
1. Telomeres (ends of the chromosome)	Heterochromatin with Repetitive DNA	Maintains the individuality of chromosomes
2. Centromeres (primary constriction)	Heterochromatin (Two types: α & β)	Involved in the separation of chromosomes during cell division.

Region of Chromosome	*Chromatin type*	*Function*
	(*i*) Heterochromatin α is compact, resistance to destaining and present around the centromere (*ii*) Heterochromatin β is diffused, less resistance to destaining and present in the centromere	
3. Nucleolar organizer or Secondary constriction I	Euchromatin	It contains genes for the synthesis of 18 S and 28 S rRNAs
4. Secondary constriction II	Euchromatin	DNA synthesises mRNA throughout cell cycle except during mid mitosis. It also synthesises 5 S rRNA.
5. Satellite	Heterochromatin	It is mostly inactive during protein synthesis.
6. Chromosome arms	(i) Euchromatin forms the bulk	DNA of Euchromatin synthesises mRNA during interphase
	(ii) Heterochromatin intercalates within the euchromatin	(*i*) DNA of heterochromatin is highly repetitive or redundant which appears to compartmentalize the chromosome into several functional parts as for example, nucleolus, centromeres, and multiple initiation points of DNA replication. Some of the repeated sequences act as spacers between genetically active regions. (*ii*) It may also serve as transcription stops, initiation points for RNA polymerase, or sites for attachment with nuclear membrane. (*iii*) The sequences may also be used in pairing of homologous chromosomes during meiosis (Yunis and Yasmineh, 1971)

SPECIAL TYPES OF CHROMOSOMES

In certain organisms, there are special tissues in which the chromosomes are extraordinarily long and they present special morphology. The lampbrush chromosomes of vertebrate oocytes and chromosomes of salivary gland cells of dipteran larvae are the examples of *giant chromosomes*.

Polytene chromosomes or salivary gland chromosomes

The giant chromosomes were first observed in the cells of salivary glands, gut, trachea and other body parts of dipteran insects by E.G. Balbiani in 1881. The name polytene was assigned to these chromosomes by Kollar. The polytene chromosomes are much larger than the normal somatic chromosomes. How this increase in size of chromosomes is brought about is not known. The giant chromosomes consist of a bundle of chromonemal fibrils which arise by a series of about 10 consecutive duplications of the initial chromonemata that increase the DNA content about 1,000 times the DNA content of somatic cells (Fig. 9.10). Because of the multistranded condition, these chromosomes are called polytene chromosomes. The process of reduplication of strands without separation is called endoduplication. The homologous polytene chromosomes always remain closely paired in mitotic prophase (Fig. 9.11). This is called somatic pairing and these chromosomes are thought to be in permanent prophase. So the number of salivary gland chromosomes always appear to be half the number of chromosomes in normal somatic cells. In *Drosophila melanogaster*, the giant chromosomes are found in the form of five long and one short strands. The centromeres of all these chromosomes fuse to form a single more or less amorphous mass called *Chromocentre* (Fig. 9.11).

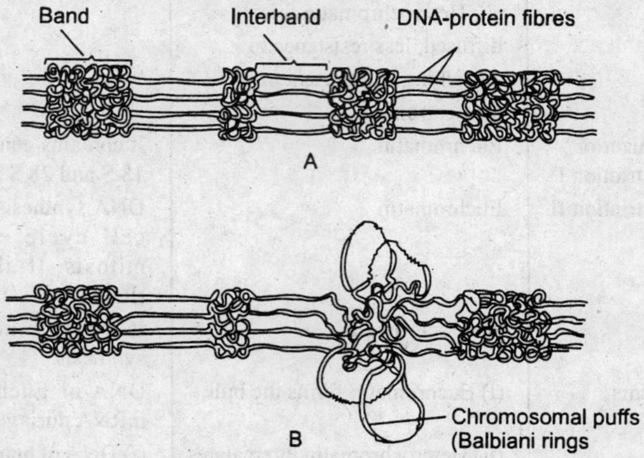

Fig. 9.10 A—Model of Dupraw and Rae showing band and interband structure in giant chromosome.
B—The same model showing a possible method of "puffing" in one of the bands.

The polytene chromosomes bear along their entire length a series of dark bands alternated by light bands or interbands. The dark bands are narrow or broad disc shaped structures. They are euchromatic in nature and contain large amount of DNA, small amount of RNA and certain basic proteins. They are Feulgen positive and absorb ultraviolet (UV) light of 2600 Å. The light bands or interbands are fibrillar, feulgen negative, heterochromatic regions containing small amount of DNA, large amount of RNA and acidic proteins and they absorb little amount of UV light.

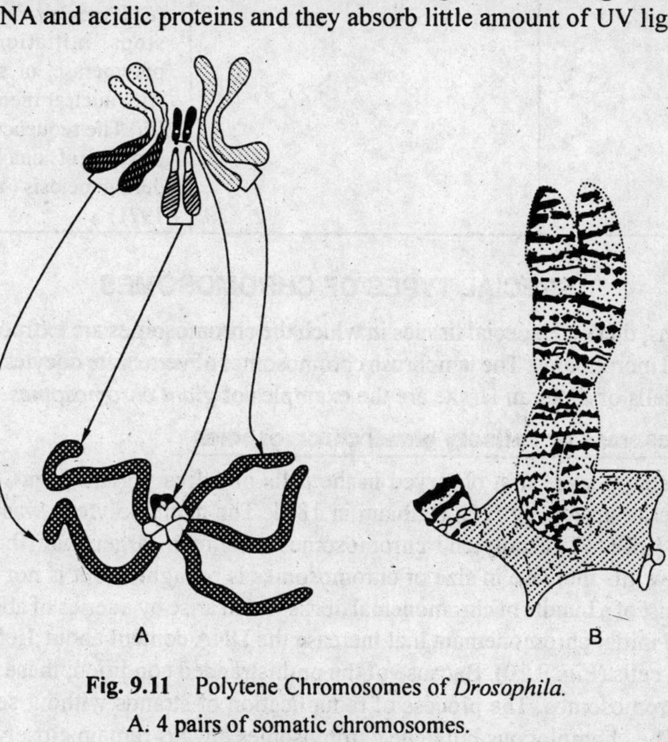

Fig. 9.11 Polytene Chromosomes of *Drosophila*.
A. 4 pairs of somatic chromosomes.
B. One polytene chromosome pair showing dark bands and light bands.

The number, distribution and localization of discs or bands are notably similar in homologous polytene chromosomes of *Drosophila*. During certain developmental stages, the single bands or adjacent bands of polytene chromosomes produce local reversible swellings which are called '*chromosomes puffs*' or *bulbs*. The chromonemata of polytenic chromosomes give out many series of loops laterally. These loops or rings are known as the Balbiani rings and they are rich in DNA and RNA (Fig. 9.12).

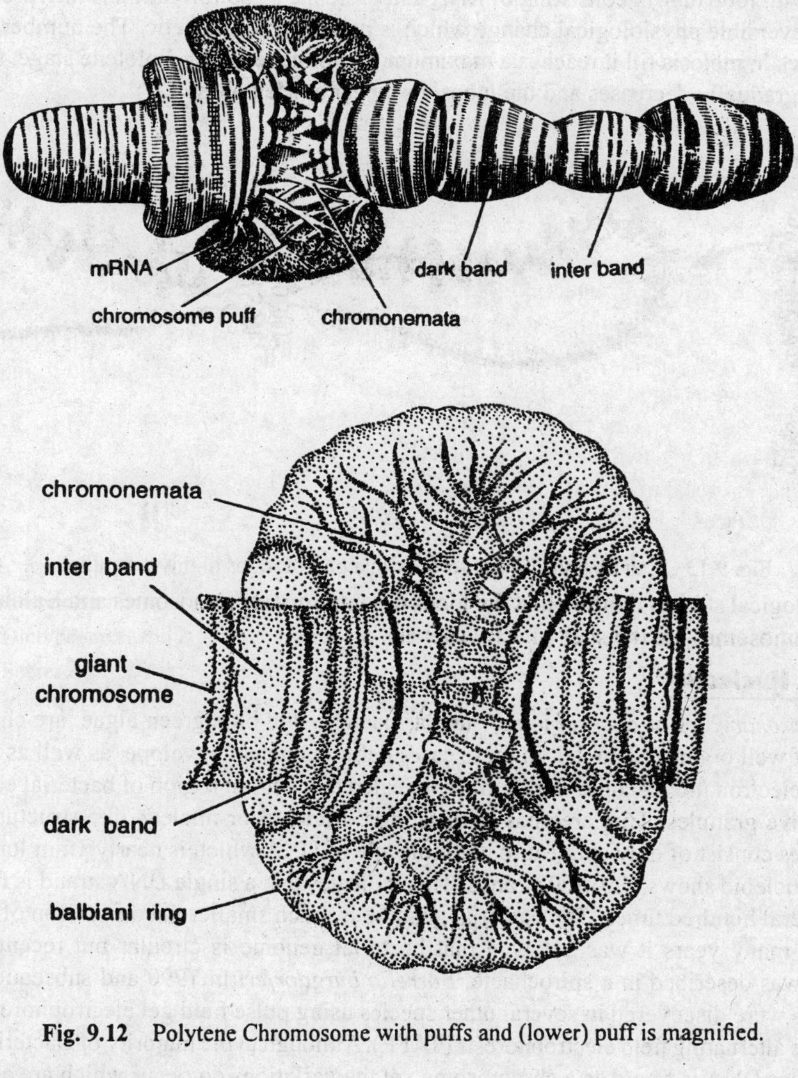

Fig. 9.12 Polytene Chromosome with puffs and (lower) puff is magnified.

Lampbrush Chromosomes

Lampbrush chromosomes are special type of giant chromosomes found in the nuclei of oocytes of many vertebrates, such as fishes, amphibians, reptiles and birds during the prolonged diplotene stage of first meiosis. They are also found in the nucleus of ***Drosophila*** spermatocyte. Lampbrush chromosomes were first observed by Flemming in 1882 and given the name by Ruckert in 1892. These chromosomes may sometime become even larger than the polynemic or polytenic chromosomes of salivary glands of dipterans. The largest chromosomes may sometime be as long as 1 mm in urodele amphibians. These chromosomes consist of main axis and many fine lateral projections or loops which give them the appearance of a test tube brush or ***lampbrush*** (Fig. 9.13). Actually, the main axis consists of four chromatids of two bivalent chromosomes and the chromonemeta of these chromatids give out fine loops at the lateral sides. Only the kinetochore bears no lateral loops. Ris (1957) studied the loops with electron microscope and suggested that the loops were integral parts of chromonemata which are extended in the form of major coils (Fig. 9.13). Generally one to nine loops may arise from a single chromomeral area. The size of loop varies from an average of 9.5 μ in frog to nearly 200 μ in newt. These loops probably consist of one DNA double helix from which fibrils project which are covered with loop matrix consisting of RNA and proteins. Loop formation is interpreted by Gall (1958) as a reversible physiological change which is probably non-genetic. The number of pairs of loops increases in meiosis till it reaches a maximum in diplotene. After diplotene stage, the number of loop pairs gradually decreases and the loops disappear at Metaphase I.

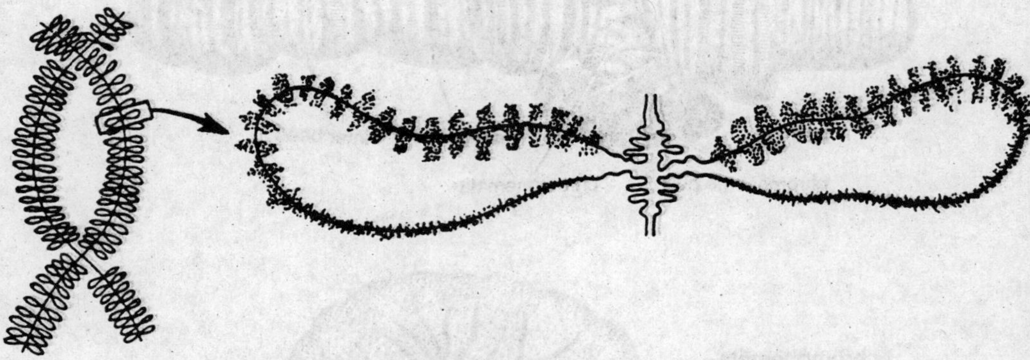

A B

Fig. 9.13 Lampbrush Chromosome with a pair of loops highly magnified.

Physiological studies indicate that the loops of lampbrush chromosomes and Balbiani rings of polytene chromosomes are the sites of active genes.

Prokaryotic Nucleoids

The cells of *prokaryotes* or *akaryobionta,* such as bacteria and blue green algae, are characterized by absence of well organized nuclei. In these cases definite nuclear envelope, as well as nucleoli, is lacking. The electron microscopic studies have revealed that central region of bacterial cell contains feulgen positive granules. That area has been termed **nucleoid** or nucleus like structure. Feulgen positive bodies consist of only single circular molecule of DNA which is nearly 1mm long. In cross section, the nucleoid shows 500–900 strands. This indicates that a single DNA strand is folded forth and back several hundred times. The bacterial genome is much smaller in comparison of eukaryotic genome. For many years it was believed that bacterial genome is circular but recently a linear DNA strand was described in a spirochaete, *Borrelia burgdorferi* in 1990 and subsequently linear chromosomes were discovered in several other species using pulse field gel electrophoresis (PFGE) and transverse alternating field electrophoresis (TAFE). Although in the majority of Bacteria generally a single circular DNA is found as a chromosome yet the variations do occur which are as follows:

(*i*) 2 or 3 circular chromosomes in some

(*ii*) One circular and one linear chromosome in some

(*iii*) Only a single linear chromosome.

The exceptional examples are listed in Table 9.2

Table 9.2 Examples of bacteria with abnormal chromosomes

Bacterium	*Number and shape of chromosome*	*Size of chromosome (in Kilobases)*
Brucellia melitensis	2, circular	one 2100 kb and other 1150 kb.
Leptospira interrogans	2, circular	one 4400–4600 kb and the other 350 kb.
Rhodobacter *spheroides*	2, circular	3046 kb and 914 kb.
Pseudomonas cepacia	3, circular	3400 kb, 2500 kb, 900 kb.
Agrobacterium tumifaciens	2, one circular & one linear	3000 kb 2100 kb.
6 streptomyces species	one, linear	8,000 kb.

Plasmid

Bacterial cell, in addition to single major circular chromosome, contains one or more minor chromosomes, called ***plasmid.*** Plasmid genome is a small circular molecule of DNA. Recently the presence of one or more linear plasmids has also been demonstrated in some bacteria as for example, 2 circular plasmids in *Rhizobium meliloti,* 3 linear plasmids in *Borellia* sp., one linear plasmid in *Bacillus polymyxa*, two linear plasmids in *Rhodococcus*, one linear plasmid in *Streptomyces sp* and one linear in *Thiobacillus versatus*. Further work by a number of workers has demonstrated that the network of thread consists of single chromosome in the form of a ring. It is known that the larger part of the DNA in bacterial nucleoid is naked and is not combined with histones. The nucleoid of *Escherichia coli* has a point of contact with plasma membrane.

In blue-green algae, these chromidia are aggregated in the centre of the cell. This also provides clue that the true nucleus with definite nuclear membrane is probably evolved from the primitive type of nucleus.

Chemical Composition of Chromosomes

The term chromosome refers to deeply stained filamentous body formed of chromatin . The chromatin is composed of DNA, RNA, histones and non-histones. The purified chromatin isolated from the interphase nuclei shows the following composition.

(*i*) *DNA* 30—40%
(*ii*) *RNA* 1—10%
(*iii*) *Proteins* 50—65%

The chemical composition of chromatin varies considerably from species to species and in different tissues of the same species. The chemical composition of chromosomes also differs at different stages of the cell cycle. At metaphase chromosomes contain on an average 15–20% DNA, 10–15% RNA and 65–75% proteins. Out of different chemical components of chromatin, DNA is the only component, the amount of which exhibits a constant relationship with the stage of chromosome replication and the increase or decrease in chromosome number. DNA found in the chromatin may be of two types:

(*i*) Unique DNA
(*ii*) Repetitive DNA

(*i*) **Unique DNA.** It is DNA consisting of segments whose base sequences are present in a single copy per genome (no repetition in base sequences). Unique DNA segments occur in all the structural genes for histones, ribosomal RNA and storage proteins. The proportion of unique DNA in genome of different species ranges from 8% (rye) to 70% (man), the average or common values ranging between 40 and 70%. In bacteria the unique DNA represents about 99.7% of the total genome.

(*ii*) **Repetitive DNA.** The DNA in which the base sequences are repeated several to million times per genome is called repetition or **repetitive DNA (rep-DNA)** by Britten and Kohne (1960). The proportion of repetitive DNA in the genome varies in different species. In man about 30% of genome is represented by repetitive DNA. Repetition DNA is characterised by:

(*a*) the number of nucleotides pairs per sequence,

(*b*) the specific order of base pairs per sequence, and

(*c*) the number of copies of a sequence per genome.

A particular family of repeated sequences can exist either dispersed between unique DNA sequences or as clusters of repeating nucleotides units localised in one or few areas called regional repetition in the genome (Lee Thomas, 1973). The repetitive DNA sequences are grouped into the following classes on the basis of number of copies per genome; highly repetitive, moderately repetitive DNA and middle repetitive or intermediate fraction.

1. **Highly repetitive DNA.** Such DNAs have base sequences repeated more than 10^5 and upto several million copies per genome. These may constitute upto 10% of the genome and consist of very short nucleotide sequences (sometimes more than 100 nucleotides long). Highly repetitive DNA sequences differ in density from the remaining DNA and are separated as minor bands through Cs Cl_2 density gradient centrifugation (Satellite DNA).

   Highly repetitive sequences tend to be associated with constitutive heterochromatic regions of chromosomes particularly in the centromere and chromosome ends. It appears to be transcribed to a very small extent, if at all, within the cell. The probable important functions of such DNAs are participation in chromosomal organisation, involvement in chromosome pairing during meiosis, involvement in recombination and protection of genes for histones and ribosomal RNA.

2. **Moderately repetitive DNA.** Such DNAs consist of base sequences repeated 10^2 to 10^4 times per genome, each base sequence having 300 ± 200 base pairs. They are interspersed in the majority of instances between the regions of unique or non-rep-DNA and may constitute 20–80% of the genome. In man, repetitive DNA fraction is about 13% of the genome. In some eukaryotes this class of DNA may be found in heterogenous state as it may contain many different sequences with different degrees of repetitions.

   Intermediate repetitive DNA. Intermediate fraction transcribed at least in part, includes sequences complimentary to the principal ribosomal RNA (100–500 copies), transfer RNA (16–100 copies) of each species and 5 S rRNA.

In Eukaryotic chromosomes, alternating interspersion of rep-DNA and unique DNA sequences encoding the protein structure shows a high degree of order. The function of most rep-DNA sequences, besides those coding for rRNA, tRNA, 5-S RNA and histones, is presently an open question. Probably rep-DNA represents excess baggages of genome (junks). They do not transcribe but keep the chromosomes in shape (house keeping). They serve as gene spacers. They are the material for creating new genes. Regulation of genetic transcription is an another function of rep DNA.

PHYSICOCHEMICAL ORGANIZATION OF CHROMOSOME

The major chemical components of the chromosomes are DNA, proteins (histones and non-histones) and small amount of RNA. Besides these, calcium and some other minerals are also present. **Histones** are basic proteins that are charged positively at neutral pH. **Non-histone proteins** are acidic. **Histones** are small proteins containing some 100–200 aminoacids, about 20 to 30 per cent of which are lysine and arginine. The positive charges on histones enable them to bind to DNA by electrostatic attraction to the negatively charged phosphate groups in the sugar-phosphate backbone of DNA. Histones also bind tightly to each other. Histones in both DNA-histones and histone to histone bindings play important role in chromatin structure-DNA packaging. They are present in chromatin of all higher eukaryotes in the amount equal to that of DNA weight/weight. The histones present in the nuclei of all higher eukaryotes including man are of five different types. These are called H1, H2a, H2b, H3 and H4. In the sperms of some eukaryotes, however, a special class of small basic proteins called protamines replaces histones. There is a remarkable constancy of histones H1, H2a, H2b, H3 and H4 in all types of organisms and among widely divergent species. This shows that histones have been highly conserved during evolution. They are present in molar ratios of approximately $1H_1$: 2H2a : 2H2b : 2H3 : 2H4. They are specifically complexed with DNA to form basic structural unit of chromatin-the nucleosome. Acidic proteins or non-histone proteins consist of several heterogeneous proteins of different compositions and are thought to be involved in the regulation or expression of some genes or groups of genes.

In the past, electron microscopy of the cell did not provide any clear picture of organization of chromosomes. Then the whole mounts of chromosomes were studied with the help of electron microscopes. Such studies demonstrated the presence of very fine fibrils of 2–4 nm thickness in the chromosomes. Since the thickness of DNA is 2 nm it was presumed that the single fibril present in the chromosomes should possibly correspond to single DNA strand. However, in these fibrils protein content was not clear. Thin sections of metaphase chromosomes and interphase nuclei show numerous fibres of 200–300 Å. Unsectioned chromosomes and ruptured interphase nuclei show nucleoprotein fibres of about 300Å average diameter. These fibres are known as *Chromatin fibres* which are the basic unit of chromosome structure.

The quantitative measurements of DNA have been made in a number of cases. The most convenient unit of measurement of DNA is picogram (1 pg = 10^{-12} gm or 10^{-9} mg) which is equivalent to 31 cms of double helical DNA. The quantity of DNA varies greatly in cells of different kinds of organisms, as for example, Bacterial DNA measures only 1.0–1.5 mm, human diploid cell has 174 cm of DNA (5.6 picograms), *Trillium* cell has 37 metres (120 picograms) and *Drosophila* salivary gland chromosome has 9 metres (293 picograms). All the cells in a human being may have DNA equal to whether it is 12.5 or 2.5 km (100 gm) a length which is equal to 100 times the distance from earth to sun.

It is one thing to determine the chemical composition of the chromosome, but quite another to work out the molecular arrangement of nucleic acids in the chromosomes. Chromosome thickness is usually hundred times that of DNA and length of DNA found in a chromosome is several hundred times that of chromosome. How does this condensation takes place? What are the components that are involved in packaging process? What is the packaging scheme? These are the most challenging and frustrating problems for the biologists today. For any acceptable model of chromosome structure, there are two basic requirements:

(*i*) the packaging of such a long DNA molecule into such a small segment of metaphase chromosome so that linear sequence of genes must be preserved, and

(*ii*) the replication and segregation of DNA should be explained.

MODELS OF CHROMOSOME STRUCTURE

As regards the physicochemical organization of chromosomes, several models have been proposed which may be discussed under the following groups:

(*i*) multiple strand or multineme models, and

(*ii*) single strand or unineme models.

Multiple Strand Models

The multiple strand models have been proposed in light of the observed variation in the thickness of the nucleoprotein fibrils, from 30 Å to 200 Å or even 500 Å. In multineme model, the chromosome is supposed to contain many DNA-protein fibrils. The evidence in favour of this model comes from the studies on the chromosomes of *Tradescantia* (8 fibrils of 200 Å each/chromatid), mosquito (16 strands per chromatid), *Amoeba proteus* (8 strands/chromatid).

Ris Model. From the electron microscopic studies of the chromosomes of plants and animals during meiosis and mitosis, Ris (1957) concludes that chromosome is multistranded (polytene) and it is built like a cable with numerous identical sub-units. Kaufmann and Mcdonald (1957) supported Ris. However, the possibilities that the chromosome is formed of single long DNA cannot be ruled out by these observations because multiple strands observed in electron micrograph might represent a single strand folded forth and back upon itself many times.

Single Strand Models

In bacteria and viruses there is no well organised chromosome. The genetic material in these cases appears to be DNA. DNA molecule in bacteria forms bacterial chromosome and is sometimes termed as *chromoneme* or *genophore*. In bacteriophage, DNA molecule may be 50 μ long. In bacterium it may be 1 to 1.5 mm long and in large chromosomes the DNA molecules may be several centimetres long. At present, most investigators agree that chromosomes have a single DNA-histones system. Whitehouse and Hastings (1965) suggest that a chromatid can be looked upon as one DNA molecule. Taylor working with the chromosome structure of *Vicia* root tips and Gall and Callan (1962) working with the lampbrush chromosomes of *Triturus* oocytes suggest that chromatid is a single stranded structure. They are of the view that many genes may be represented in one DNA molecule. There appears no evidence at present to support the occurrence of more than one DNA molecule in normal chromosome. If it is supposed that single molecule of DNA is extended along the entire length of chromosome the DNA duplication seems to be difficult in view of long distance which it should unwind and fold during the process of division.

Taylor-Freese Model. J.H. Taylor (1957) suggests that chromosome consists of a long two layered protein backbone from which DNA coils branch off just like the legs of a centipede. Taylor (1959) proposed a new model based on the one suggested by Freese (1958). In the model, several linkers have been shown, some of which could open at certain operational sites for the control and order of the nucleotide sequences (Figs. 9.14 and 9.15). Several polynucleotide chains may be involved to stabilize the structure during duplication. The linkers join the terminal phosphates of polynucleotide chains.

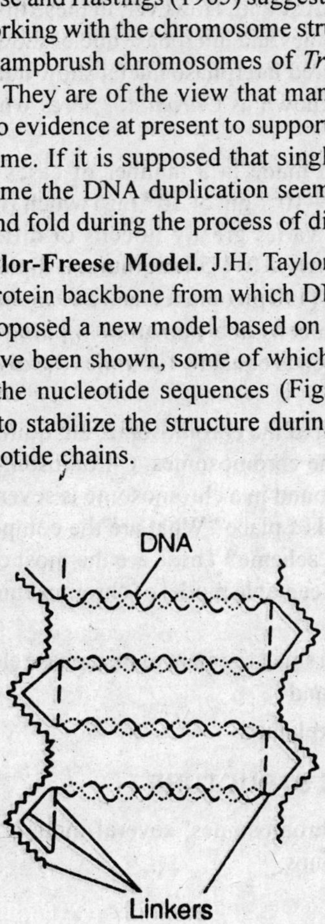

Fig. 9.14 DNA molecules extended in the chromosome. On the sides are shown two strands of histone.

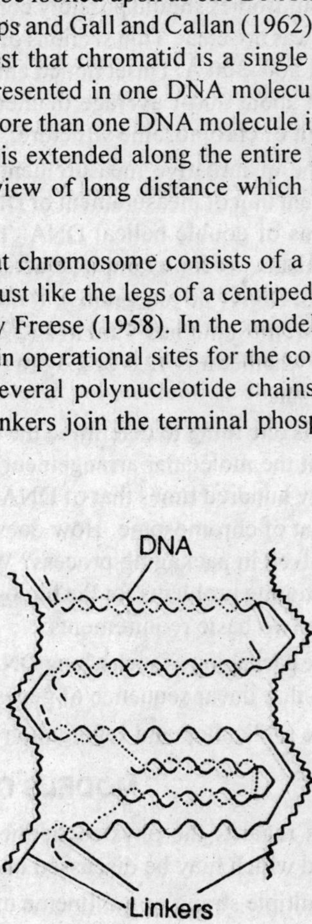

Fig. 9.15 Taylor's model to explain the proposed DNA duplication in a chromosome.

If it is supposed that a chromosome has a single DNA molecule then the question arises as to how the histones fit into chromosome structure. The unequal spacing of two polynucleotide strands of DNA leaves two spiral grooves or furrows. Watson and Crick in 1953 have suggested that the molecular structure of DNA was such that fully extended polypeptide chains (histone strands) could be fit in the spiral grooves along the helix of the polynucleotide chains. Feughelman *et al* (1955) confirmed this. Wilkins (1957) also tentatively suggested that in DNA-histone complex there might be a polypeptide strand in each groove of DNA.

Proteins (histones) of chromosomes probably play the following roles:

(*i*) They provide characteristic morphological shapes and structures to chromosomes.

(*ii*) Histones of chromosomes show species specificity and their main function consists in protecting the DNA strands from breakage and allowing them to fold into condensed configurations.

(*iii*) They also play a part presumably in meiosis at the time of chromosome pairing. Moses and Coleman (1964) have found that pairing cores of the homologous chromosomes, when seen under electron microscope, at first meiosis are composed of proteins and it may be that the close pairing of homologous chromosomes comes about through some sort of interaction between histones and proteins of pairing cores.

(*iv*) They also play important role in the control of gene action and in cell differentiation.

Folded Fibre Model of Du Praw

According to *"folded fibre model"* proposed by E.J. Du Praw in 1965, each eukaryotic chromosome is composed of a single greatly elongated and highly folded nucleoprotein fibril of about 320 Å thickness (the thickness may vary from 100–400 Å). This nucleoprotein fibril is composed of a single, linear, double stranded DNA molecule of 20 Å thickness which remains coated with histones and non-histones, variable amounts of different kinds of RNA and divalent cations (Ca^{++} and Mg^{++}). The nucleoprotein fibril is coiled in packing ratio of 6 : 1 to form a short fibre of 80–100 Å diameter (packing ratio is the ratio of fully extended length to packed length). This fibre is called *Type A Fibre.* Type A fibre, in turn, is supercoiled in packing ratio of 10 : 1 to form a *Type B fibre* of 200–350 Å diameter which is similar to bead seen in electron micrographs of chromatin fibres (Fig. 9.17A).

Fig. 9.16 Chromosome at metaphase. Each of the two chromatids are composed of densely packed mass of fibres.

Du Praw has tried to explain how the nucleoprotein fibril of chromosome replicates during interphase producing two sister fibres and how it is held at the centromere where the DNA segment remains unreplicated until the anaphase (Fig. 9.17B).

The chromatin fibre undergoes replication in centromeric region also and the two sister chromatin fibres undergo extensive coiling to give rise to two sister chromatids. Further folding and supercoiling reduce of chromatin fibres reduce the length and increase thickening and stainability of the chromosome. The available evidences support this model.

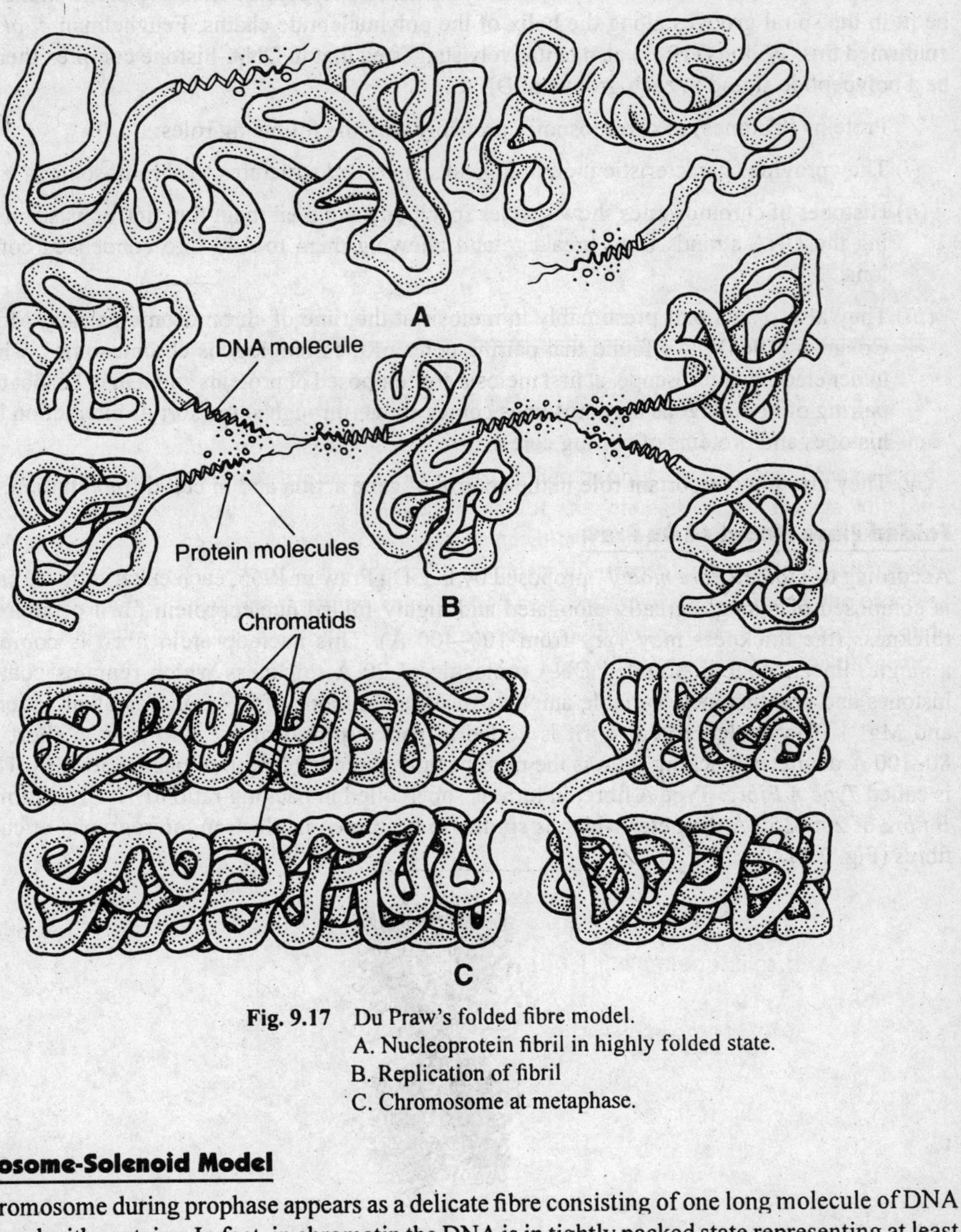

Fig. 9.17 Du Praw's folded fibre model.
A. Nucleoprotein fibril in highly folded state.
B. Replication of fibril
C. Chromosome at metaphase.

Nucleosome-Solenoid Model

The chromosome during prophase appears as a delicate fibre consisting of one long molecule of DNA complexed with proteins. In fact, in chromatin the DNA is in tightly packed state representing at least a hundred fold contraction. This is achieved by several levels of folding. In 1974, AL olans and D.E. olans reported the result of electron microscopy of chromatin fibres obtained from interphase nuclei.

When observed under electron microscope, the chromatin appears to be made up of a series of ellipsoidal beads joined by thread-like structures (Fig. 9.18). They named those bead like bodies as ν (***nu***) ***bodies*** and the connecting thread as linker DNA. Later in 1975, P.M. Outdet and others called those nu bodies as ***nucleosomes.*** When chromatin is digested with **endonuclease**, the enzyme, first attacks linker DNA in the regions between nucleosomes. As a result of this, the nucleosomes are released but the DNA in the nucleosome is protected from digestion by endonuclease. The pieces of DNA associated with nucleosomes are about 200 bps long. This indicates that there are about 200 bps in DNA from linker to linker. Further digestion by endonucleases causes digestion of linker DNA also leaving behind nucleosome cores with a piece of DNA that is about 146 base pairs long. **Nucleosome core** is disc shaped or flat sphere with a diameter of about 110 Å and thickness of about 60 Å.

The nucleosome core contains a histone octamer of two molecules each of histones H2a, H2b, H_3 and H_4 and DNA invariably forming 1 and 3/4 (1¾) coils around the core. The association of core DNA with histone octamer is tight one and the DNA is well protected against enzyme action (*i.e.*, endonuclease enzyme does not digest any DNA associated with histones). The size of linker DNA varies from 8 bp to 114 bp in different species. Linker DNA forms string part between the nucleosomes and is nuclease susceptible (Fig. 9.19). H1 histone is located outside the nucleosome cores. It holds the two strands of DNA together as they enter and leave core. Both linker DNA and nucleosome core are associated with other chromosomal proteins. It is bound to the edges of the nucleaosome discs across the linker DNA and stabilizes about 166 base pairs long DNA. H1 histone molecules seal off two turns of DNA around the histone octamer core (Fig. 9.19). The structure of nucleosome is essentially invariable in all the eukaryotes.

A.

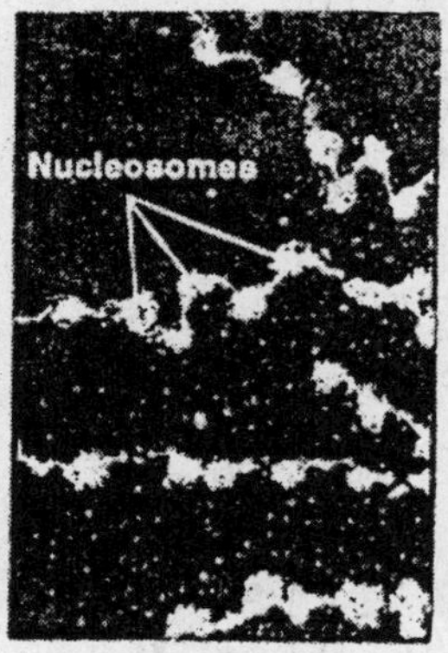

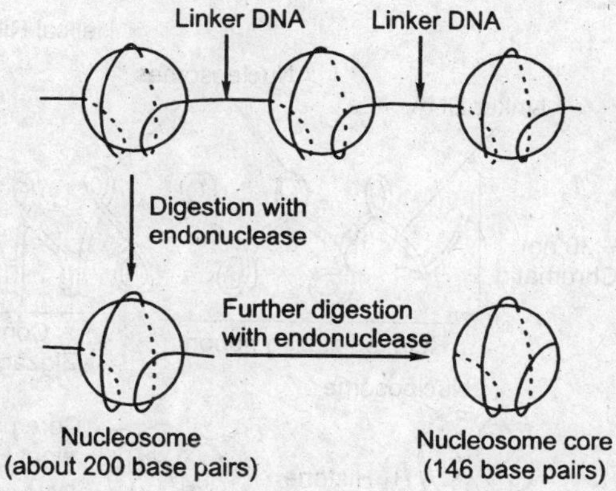

B.

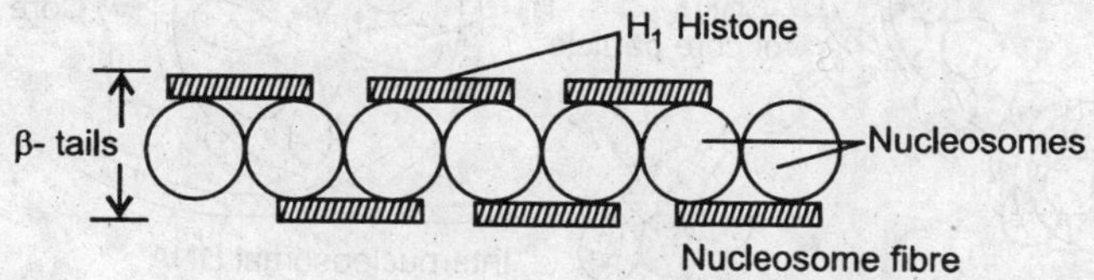

Fig. 9.18 A. Electron micrograph of chromatin showing beadlike structures on a string. Each bead is nucleosome. B. Nucleosome fibre

Fig. 9.19 A nucleofilament digested by endonuclease produces nucleosomes (DNA about 200 base pairs) which on further digestion produces nucleosome core (DNA about 146 base pairs).

The most complex level of packaging involves higher-order coiling of the DNA-histone complexes to produce chromosome morphology as shown in the figure. The supercoiled nucleosome fibre is called **solenoid**. A number of acidic non-histone chromosomal proteins are thought to be involved in further coiling and packing of solenoid. The exact roles of non-histone proteins in chromosome structure and function are unknown. Some of these proteins are believed to have structural roles in the formation of 2,000 to 4,000 Å supersolenoid. When viewed under electron microscope, the DNA-Protein complex in chromosomes appears as a fibre or nucleoprotein filament of about 10 nm diameter (Fig. 9.20). The solenoid structure has a diameter of about 300 Å. Obviously the DNA can be compacted extensively by superimposing coils on coiled structures during cell division.

Under some conditions, nucleosomes are packed together without any linker DNA between them resulting in 100 Å thick chromatin fibre called nucleosome fibre.

The chromatin exists in a more highly coiled state than the 100 Å thick nucleofilament while the packing of DNA in the nucleosomes is less well understood in the higher order of folding of the nucleosome in the chromosomes. Above the level of nucleosome and 100 Å nucleoprotein filament, there is a next level of packing which forms 300 Å thick chromatin fibre. Examination of isolated

300 Å chromatin fibre reveals the complex nature of the chromosome structure. The arrangement of nucleosome in the fibre is difficult to define. The next level of packing beyond 300 Å chromatin fibres are poorly understood. The diameter of interphase chromosome may be 3,000 Å while the diameter of a metaphase chromosome may be 7,000 Å. To achieve these levels of compaction, the simplest concept is that 300 Å fibre becomes looped and coiled like a rope in various ways to give morphologies characteristic of these chromosome.

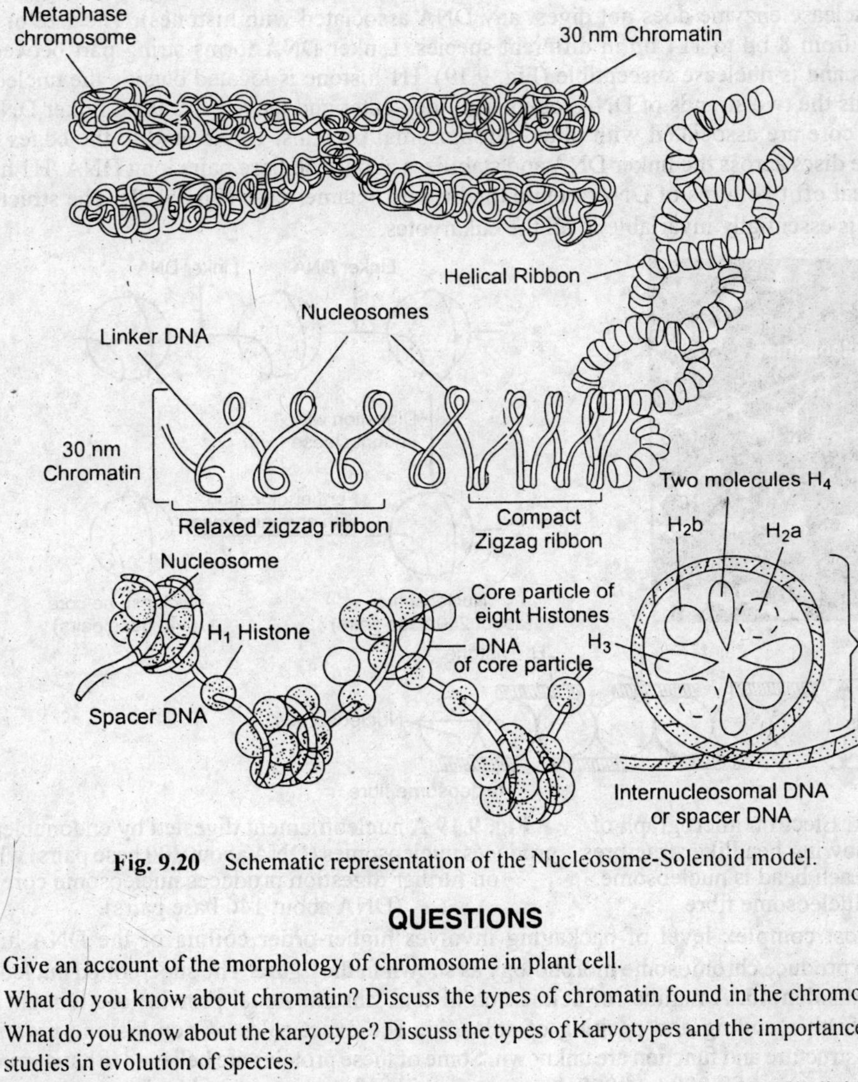

Fig. 9.20 Schematic representation of the Nucleosome-Solenoid model.

QUESTIONS

1. Give an account of the morphology of chromosome in plant cell.
2. What do you know about chromatin? Discuss the types of chromatin found in the chromosome.
3. What do you know about the karyotype? Discuss the types of Karyotypes and the importance of karyotypic studies in evolution of species.
4. What do you know about the basic chromosome number? Discuss the classification of individuals according to their chromosome number.
5. Describe the structure of prokaryotic chromosome. How does it differ from eukaryotic chromosome?
6. Describe the ultrastructure of chromosome and discuss the functions of chromosome.
7. Give an account of the accessory or B-chromosomes and mention their genetic and evolutionary significance.
8. Discuss briefly the models proposed to account for the structure of chromosome.
9. Write brief notes on:

   (*i*) Nucleosome, (*ii*) Polytene chromosome,

   (*iii*) Lamp brush chromosome, and (*iv*) Centromere.

CHAPTER 10

Nucleic Acids

The substance that was eventually proved to be genetic material was discovered in 1869, only four years after the announcement of the Laws of Heredity by Mendel and 84 years before DNA became central issue of modern molecular biology following the publication of famous *double helix* model of DNA by Watson and Crick. In the year 1869, a 22-year-old Swiss physician and chemist named Friederich Meischer isolated a substance that was quite different from carbohydrates, fats, and proteins. He obtained a new and previously unknown biological compound from the nuclei of pus cells and named that as '*nuclein*'. Meischer reported this work in 1871. In 1874, he reported an improved procedure for preparing nuclein from the sperm nuclei of Salmon. In 1881, Zacharias identified nuclein with chromatin. Hertwig (1884) stated "I believe that I have at least made it highly probable that nuclein is the substance responsible not only for fertilization but also for the transmission of the hereditary characters." Later it was discovered that neclein had acid properties. Altmann in 1899 introduced the term nucleic acids to replace nuclein. By 1944, Casperson and Brachel independently suggested that nucleic acids were directly connected with protein synthesis. In 1944, O.T Avery, C. M. Macleod, and M. McCarty conducted a remarkable experiment and suggested that bacteria could be genetically transformed by DNA.

Chemical Composition of Chromatin

Chemical analyses have shown that chromatin is made up of the following classes of chemical compounds.

1. Nucleic acids.
2. Basic proteins or *histones* (*protamines* in sperm nuclei of some fishes and birds).
3. Acidic proteins (*Non-histones*).
4. Lipoproteins.
5. Associated inorganic molecules.

Nucleic Acids

The nucleic acids are of two types:

(*i*) Deoxyribonucleic acid (DNA), and (*ii*) Ribonucleic acid (RNA).

DNA is present in the cells of all plants, animals, prokaryotes and in a number of viruses. DNA molecule appears to be the most stable substance that is present in the chromosome. It has also been reported in some cytoplasmic organelles like chloroplast, mitochondria and centrioles. The amount of DNA per set of chromosomes is constant for a particular species. The DNA content of a diploid somatic cell (2n) is exactly double that of haploid cells (n). There is a close relationship between DNA content and chromosome size. It has been found that about 31cm of double helix DNA equals a pikogram (1pg = 10^{-12}g). In man a diploid cell contains about *5.6 pg of DNA which may correspond to about 174 cm of DNA*.

RNA is found associated with chromosomes, in nucleolus, in the nuclear sap, ribosomes, chloroplast, mitochondria and some other cytoplasmic organelles. According to function, RNAs are of two kinds:

(*a*) **Genetic RNA.** In some viruses which contain RNA, RNA is the sole genetic material which carries all hereditary responsibilities.

(*b*) **Non-genetic RNA.** In cellular organisms DNA is genetic material and RNA molecules do not carry out genetic functions. In these organisms various genetically controlled cellular functions are performed by different kinds of RNA called non-genetic RNAs.

Histones or Basic Proteins

Chromatin of all eukaryotic cells is a nucleoprotein complex in which DNA strand is associated with proteins. The associated proteins are of two kinds; *basic proteins (histones)* and *acidic proteins (non-histones)*. Histones are small proteins that contain between 100 and 200 aminoacids of which 20 to 30% are lysine and arginine. The histones bear positive charges which enable them to bind to DNA primarily by electrostatic attraction to the negatively charged phosphate groups in sugar-phosphate backbone of DNA. Histones also bind tightly to each other and both DNA-histone and histone-histone binding are important for chromatin structure. The histones found in all eukaryotic chromosomes are five distinct types:

(*i*) H1 which is very rich in the basic amino acid lysine (mol. weight–approximately 21,500).

(*ii*) H2a and H2b which are lysine-rich (mol. wts. 14000 and 13775 respectively).

(*iii*) H3 and H4 which are rich in basic amino acid arginine (mol. wts. 15,320 and 11,280 respectively).

The five histones are present in molar ratios of approximately 1H1 : 2H2a : 2H2b : 2H3 : 2H4. They are complexed with DNA to produce the basic structural sub-units of chromatin, **the nucleosome**. The histones have been highly conserved during evolution and four of the five types of histones are similar in all higher eukaryotes. The sperms of a few eukaryotes are exceptions where the histones are replaced another class of small basic proteins called *protamines*.

The chromatin shows about equal amounts of histones and DNA.

Non-histones

The non-histones or acidic proteins are found more or less firmly associated with DNA-histone complexes. In contrast to the histones, the molecules of non-histone proteins are numerous and heterogeneous. Their number in most of the organisms may be more than 100 and they may include the enzymes involved in the synthesis of DNA and RNA, *i.e.*, DNA polymerase, RNA polymerase and protein molecules involved in the control of DNA and RNA synthesis. The content of non-histones varies throughout the cell cycle and from one differentiated cell to another but the histones remain constant in both these instances. The amount of non-histone proteins generally equals the DNA-histones combined.

Inorganic Molecules

Certain small inorganic molecules are also present in the chromosomes. These maintain the integrity of the chromosomes. Phosphorus, potassium, sodium, calcium and magnesium salts are very common in the nucleic material.

Lipoprotein Complexes

Fats may also be found in the chromosomes, generally in association with proteins.

Components of Nucleic Acids

Nucleic acids, both DNA and RNA are linear polymeric macromolecules. The monomeric unit is termed ***nucleotide***. It is ***deoxyribonucleotide*** or ***deoxyribotide*** in the case of DNA and *ribonucleotide* or *ribotide* in RNA. A nucleotide consists of three components: (*i*) *a nitrogenous base,* (*ii*) *a pentose or 5 C-sugar,* and (*iii*) *a phosphate group.* In the nucleotide, the combination of sugar with a nitrogenous base constitute a ***nucleoside***. The hydrolysis of nucleic acids reveals a clear cut picture of their various components.

Nucleic acids
(Polynucleotide macromolecules)
↓ Hydrolysis
Monomer units or Nucleotides
(Nucleotide = Nitrogenous base + pentose sugar + phosphate group)
↓ Hydrolysis
Phosphoric acid | Nucleoside
(Protein base + 5-C sugar)

The various components of nucleic acids are discussed as follows:

Nitrogenous bases or Proteins. The nitrogenous bases present in the nucleic acids are of two types:

1. Pyrimidines, and 2. Purine bases.

1. **Pyrimidine bases.** The pyrimidines are single ring compounds with nitrogen in positions 1' and 3' of a benzene ring. There are three pyrimidine bases in the nucleic acids. These are Cytosine (C), Thymine (T) and Uracil (U).

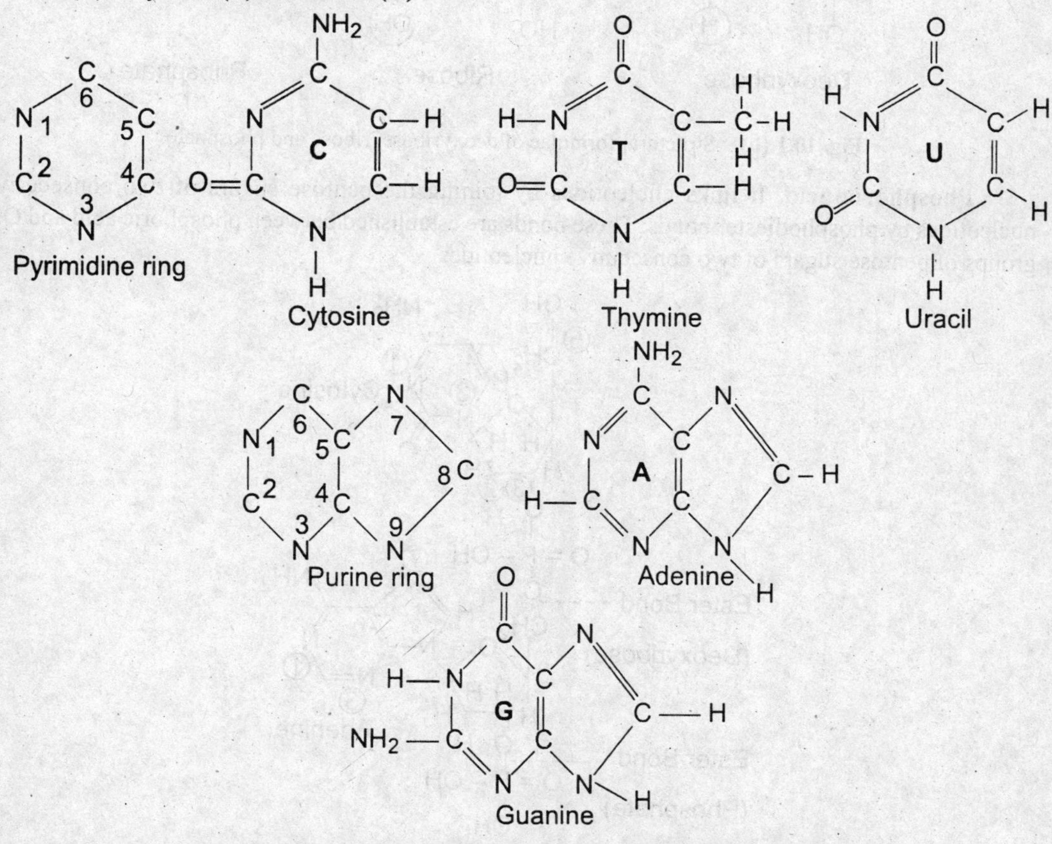

Fig. 10.1 (a) Structural formulae of pyrimidine and purine bases.

Cytosine is found in both DNA and RNA while thymine is characteristic of DNA. Uracil is characteristic of RNA.

2. **Purine bases.** The purines are double ring compounds. A purine molecule has a 5 membered imidazol ring joined to a pyrimidine ring at positions 4' and 5'.

In nucleic acid, there are two purine bases, namely, Adenine (A) and Guanine (G). These are common to both DNA and RNA. The structural formulae of these bases are shown in Figure 10.1 (*a*).

In certain RNA molecules particularly the transfer RNA (found mainly in cytoplasm and nucleolus), a large proportions of bases are methylated, *i.e.*, they are in the form of methyl adenine, methyl guanine and methyl cytosine.

Pentoses. Pentoses are 5 carbon ketosugars. They are of two types; one for each type of nucleic acid; ribose sugar in RNA and deoxyribose in DNA. In deoxyribose sugar, oxygen in the second carbon is lacking. Both ribose and deoxyribose have a pentagonal rings with 5 carbons. Two carbons at 3 and 5 positions are linked to phosphoric acid and carbon 1is linked to the base. Deoxyribose is responsible for positive Fuelgen reaction which is specific for DNA. The structural formulae of the two pentose sugars are given in Fig. 10.1 (*b*).

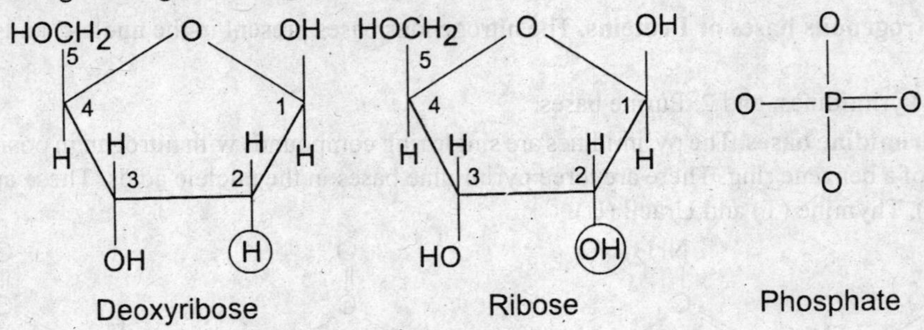

Fig. 10.1 (b) Structural formulae of deoxyribose, ribose and phosphate.

Phosphoric acid. It links nucleotides by joining the pentose sugars of two consecutive nucleotides by phosphodiester bonds. These bonds are established between phosphoric acid and OH groups of pentose sugars of two consecutive nucleotides.

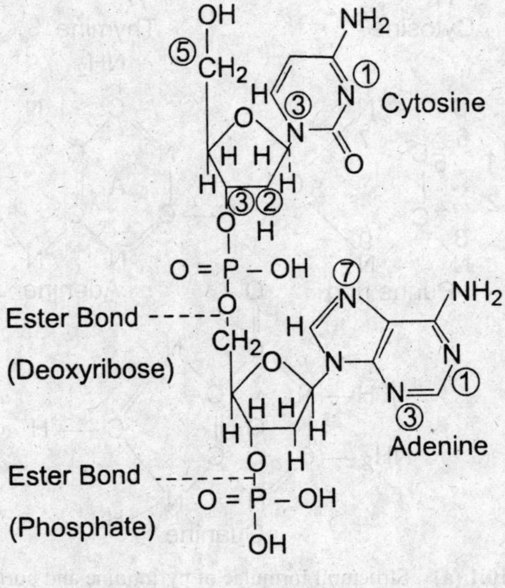

Fig. 10.1 (c) 3' 5' Phosphodiester bond.

On one side of phosphate group, OH group of 3′C of pentose in one nucleoside is involved in the formation of one ester bond and on the other side, OH group of 5′C of pentose sugar of next nucleoside forms a second ester bond [Fig. 10.1 (*c*)].

In this way, two of the three acid groups of a phosphoric acid are used in joining the sugars of two nucleosides. The remaining third acid group enables the molecule to form ionic bonds with basic proteins (*i.e.*, histones and protamines).

The DNA and RNA differ from each other in the kinds of bases and sugars found in them. Bases and sugars of RNA and DNA are listed in the following Table 10.1:

Table 10.1. Protein bases and sugars of DNA and RNA.

	Deoxyribose nucleic acid	*Ribonucleic acid*
1. Bases		
(*i*) Pyrimidine bases	Cytosine	Cytosine
	Thymine	Uracil
(*ii*) Purine bases	Adenine	Adenine
	Guanine	Guanine
2. 5C-sugars (Pentoses)	Deoxyribose sugar	Ribose sugar

The four kinds of deoxyribo-nucleosides and deoxyribonucleotides are as follows (Table 10.2):

Table 10.2. Nucleosides and nucleotides of DNA.

Nitrogen base	*Base+deoxyribose= Deoxyribonucleoside*	*Deoxyribonucleoside+ phosphate=Deoxy-ribonucleotide*	*Abbreviation for nucleotide*
Adenine (A)	Deoxyadenosine	Deoxyadenylic acid or Deoxyadenosine mono-phosphate	d AMP
Guanine (G)	Deoxyguanosine	Deoxyguanylic acid or Deoxyguanosine mono-phosphate	d GMP
Cytosine (C)	Deoxycytidine	Deoxycytidylic acid or Deoxycytidine mono-phosphate	d CMP
Thymine (T)	Thymidine	Thymidylic acid or Thy-midine monophosphate	TMP

The different types of deoxyribonucleotides and ribonucleotides (ribotides) besides occurring in DNA and RNA respectively also occur in nucleoplasm and cytoplasm, but in their triphosphate forms.

Polynucleotides

As stated earlier, the nucleic acids are polymeric molecules consisting of a number of nucleotides in series. The connecting links between consecutive nucleotides are of ester type in which the hydroxyl groups of 3 and 5 carbons of two different sugars form a double ester bond with phosphoric acid. This is known as phosphodiester bond [Fig. 10.1 (*c*)]. In both DNA and RNA, the bases are attached to pentose moiety by a covalent bond between 1′C of sugar and the 9-position nitrogen of the purines or 3-position nitrogen of pyrimidines.

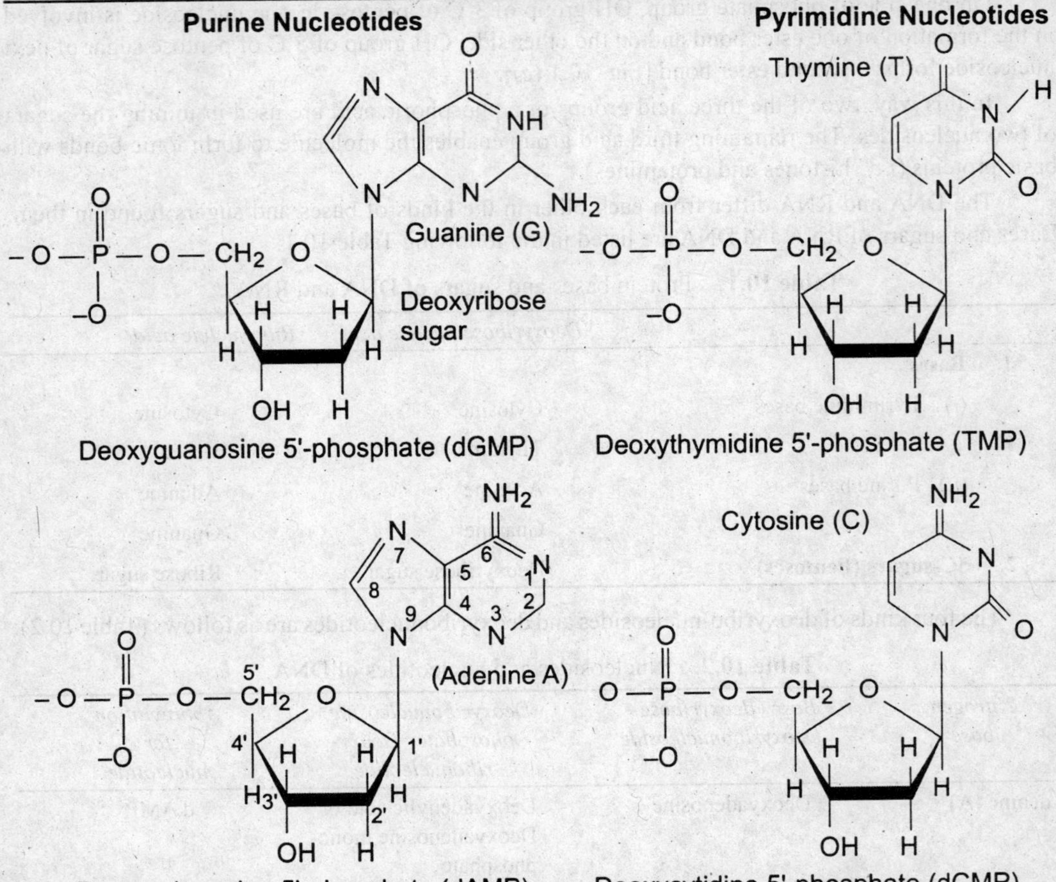

Fig. 10.2 Structural formulae of deoxyribonucleotides.

DEOXYRIBONUCLEIC ACID

Structure of DNA

DNA is a polymeric macromolecule consisting of many deoxyribonucleotides in a linear sequence (Fig. 10.3). Four different deoxyribonucleotides are the major components of DNA which are distinguished by four nitrogenous bases. The four bases characteristic of deoxyribonucleotides are purine derivatives; adenine (A) and guanine (G) and pyrimidine derivatives; thymine (T) and cytosine (C). In working out the structure of DNA, the following three types of studies have been very helpful:

1. **Studies on base pairing and base equivalence.** DNA consists of two unbranched chains of polynucleotides. The backbone of these nucleotide chains is quite regular and consists of alternate sugar and phosphate group joined with 3′ and 5′ phosphodiester linkages. The sugar molecule is attached on one side with a base. In 1950, E. Chargaff together with his colleagues E. Vischer and S. Zamenhop analysed DNA of several different organisms and pointed out that in DNA the amount of adenine equals the amount of thymine (*i.e.*, $A = T$) and the amount of guanine equals that of cytosine (*i.e.*, $G = C$). This important observation has now become a universal law that holds true for every replicative DNA found in nature. Within the limits of experimental error the ratios of $\frac{A}{T}$ and $\frac{G}{C}$ are equal to 1 except for the isolated coliphage $\phi \times 174$.

The ratio $\frac{A+T}{G+C}$ is not equal to one and has been used to characterise the DNA from a particular source.

2. **Studies based on X-ray Diffraction**. While the work on the base composition was proceeding in Chargaff's laboratory, Rosalind Franklin (25.7.1920–16.4.1938-London) and Maurice H. F. Wilkins (Born Dec. 15, 1915-New Zealand) at Kings College London were using x-ray diffraction patterns for the analysis of DNA structure. They observed that DNA from different sources had remarkably similar x-ray diffraction patterns. Astbury had earlier shown "a 3.4 Å repeat, *i.e.*, the distance between any two consecutive nucleotides in the polynucleotide chain was 3.4 Å. To this Franklin and Wilkins added a 34 Å repeat which was intriguing because nothing in the polynucleotide structure appeared to correspond with this dimension. With the x-ray diagram, Franklin and Wilkins also showed that DNA molecule is a very long thin helical strand of about 20 Å diameter. Thus, x-ray diffraction study illucidated the geometry of molecules *i.e.*, their arrangement in space and determined the relative position of atoms making up the molecule (Fig. 10.3). The diffraction pattern obtained by Wilkins and Franklin suggested that the DNA molecule was a double stranded structure. It was in marked contrast to the three stranded structure Pauling had proposed earlier. The spatial geometry of DNA was determined to be in the form of an alpha (α) helix, a spiral configuration maintained by intramolecular hydrogen bonds. It was then important to decide whether bases pointed outward or toward each other in the centre of the molecule.

(*a*) James Waston, born—
6th April, 1928, Chicago.

(*b*) Francis Crick, born—
8th June, 1916, Northampton

Fig. 10.3 (*a*) and (*b*)

Pauling suggested that bases pointed toward outside but Franklin felt that she had evidence that phosphate pointed toward outside and that the bases were in the centre.

3. **Studies based on Titration**. The analytical studies suggested that the two long nucleotide chains were held together by hydrogen bonds between base residues.

This was the situation in 1951 when James D. Watson, [Fig. 10.3 (*a*)] a 22-year-old American postdoctoral research fellow, arrived in Cambridge and met Francis H.C. Crick, [Fig. 10 3 (*b*)], a physicist working for his Ph. D. degree in biophysics. Although they were to work on different problems, they decided to collaborate on the study of DNA. In 1953, Watson and Crick proposed a double helical model of DNA molecule based on the research of Chargaff, Wilkins, Franklin, Pauling and others. To explain their model of DNA Watson and Crick postulated several remarkable features which are as under:

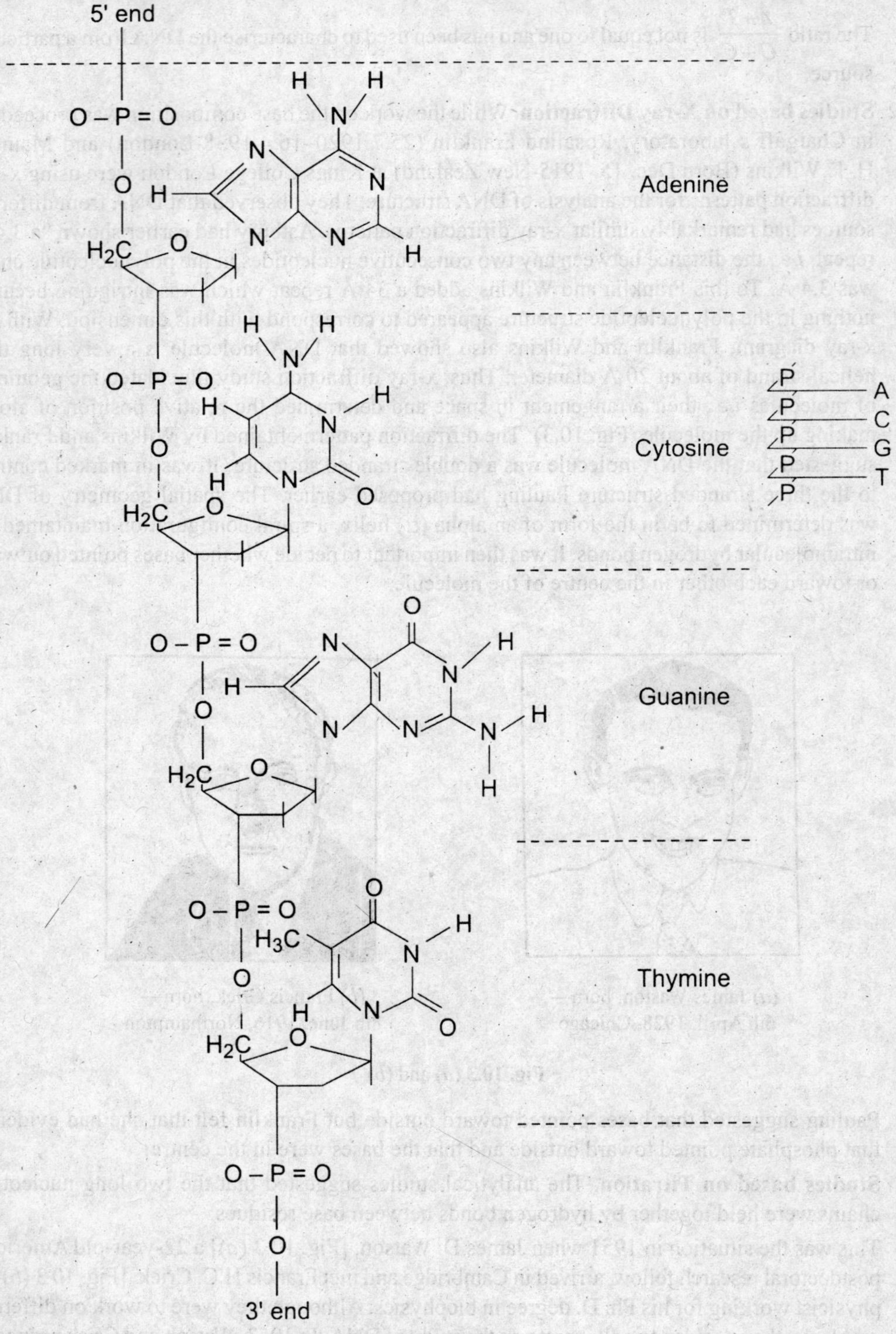

Fig. 10.4 A polynucleotide chain of DNA (Primary structure).

1. The DNA molecule is built up of a long chain of deoxyribonucleotides. The chain of nucleotides is referred to as polydeoxyribonucleotide. The sequence of nucleotides in

chain makes up the primary structure of DNA (Fig. 10.4). The number of polynucleotide chains in a DNA molecule is two and not three.

2. The chains follow right handed helix with 10 bases in one complete turn of the spiral.
3. The two chains are coiled plectonemically, *i.e.,* in an interlocked way about the same axis (*Secondary structure*).
4. Polynucleotide chains have polarity. The pentose sugar at one end of the chain has 5′ hydroxyl or phosphoryl group (5′ end) and the sugar at the other end has a 3′ hydroxyl group (3′ end) as shown in Figure. One of the chains ascends and the other descends (*i.e.*, two chains have reversed polarity or in other words sequences of the atoms in two chains run in opposite directions) (Figs. 10.5 & 10.6). Thus if one strand of polynucleotides has phosphodiester linkages established in 3′→5′ direction, the complementary strand has phosphodiester bonds in just reverse or 5′←3′ direction. In natural polynucleotides, the entire chain has polarity, *i.e.,* all the 5′ carbons point in the same direction so that the chain must end in 5C at one end and 3C at the other end.

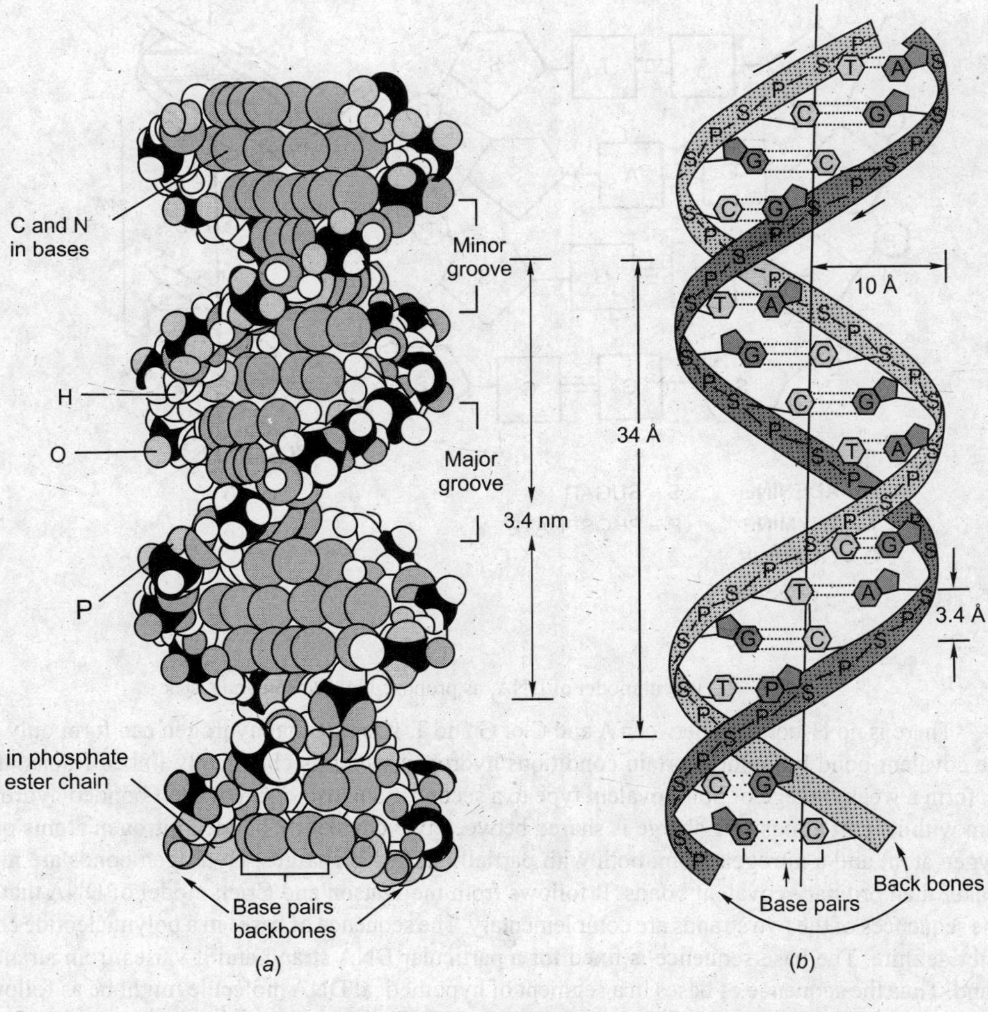

Fig. 10.5 A double helical model of DNA molecule as proposed by Watson and Crick. (*a*) A three dimensional model, (*b*) Double helical model of DNA.

5. In the double helix the phosphates of the nucleotides are on the outside and the bases on the inside. The nucleotides are set in the planes at right angles to the axis of the helix and spaced at intervals of 3.4 Å or 0.34 nm. (1 nm =1 nanometer = 10^{-9} m).
6. The two chains are held by hydrogen bonds which are established between the base pairs, or in other words bases from one polynucleotide chain are hydrogen bonded to bases of the opposite chain.
7. The pairing is highly specific because there is a fixed distance of 10 or 11 Å (1.11 nm) between the two sugar moieties in the opposite nucleotides. Purine base normally pairs with a pyrimidine base (Fig. 10.3). Thus A—T, C—G, T—A and G—C pairs are formed. A and T are joined with 2 hydrogen bonds and C and G are joined with three hydrogen bonds (Figs. 10.6 and 10.7). Adenine cannot pair with cytosine because there would be two hydrogen atoms near one of the bonding positions and none at the other. Similarly guanine cannot pair with thymine.

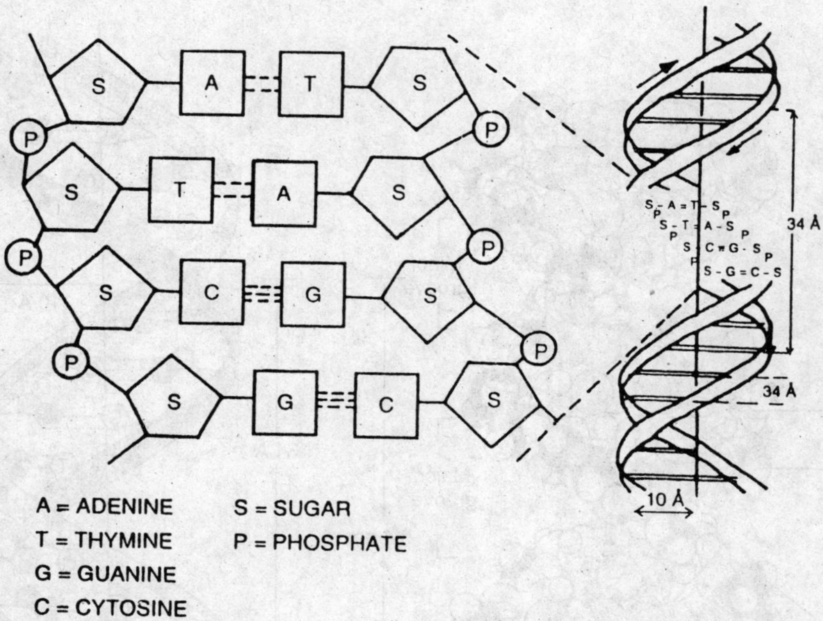

Fig. 10.6 Helical model of DNA, as proposed by Watson and Crick.

There is no H-bonding between A and C or G and T. It is true that hydrogen can form only one true covalent bond but under certain conditions hydrogen which is chemically linked to one atom can form a weak linkage of non-covalent type to a second atom also, *i.e.*, covalent bonded hydrogen atom with its partial positive charge is shared between two covalently bonded nitrogen atoms or an oxygen atom and a nitrogen atom, both with partially negative charges. Hydrogen bonds are much weaker than ordinary covalent bonds. It follows from the Watson and Crick model of DNA that the base sequences of the two strands are complementary. The sequence of bases in a polynucleotide chain is not definite. The base sequence is fixed for a particular DNA strand and it varies from strand to strand. Thus the sequence of bases in a segment of hypothetical DNA molecule might be as follows :

A–T–T–A–C–A–G–A–G–C–T First strand

T–A–A–T–G–T–C–T–C–G–A Second strand

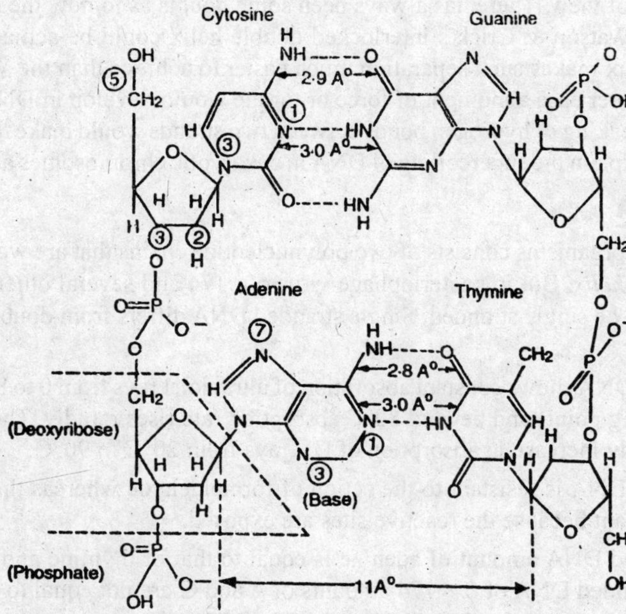

Fig. 10.7 A segment of DNA molecule.

Watson and Crick model found immediate support from the works of Wilkins, Stokes and Wilson (1953), Franklin and Gosling (1953), Feughalman, Langridge, Seeds, Hooper, Hemilton *et al.* (1955). For this brilliant piece of work Watson, Crick and Wilkins were jointly awarded Nobel Prize in 1962. Actually, this work was jointly conducted by many scientists. Dr. Watson said sometime later, "actually it was matter of 5 persons : Maurice Wilkins, Rosalind Franklin, Linus Pauling, F. H. C. Crick and me."

In the cases where DNA double helix does not exist with free ends, the DNA molecule is freely movable and can take over any shape. In some cases the molecule becomes a close ring. Such twisted DNA double helix is said to have a *tertiary structure,* as for example, DNA in bacteria, plastids, mitochondria and some viruses.

Alternative DNA Models

The double helix model for DNA structure proposed by Watson and Crick has been the focal point in biology for more than 20 years. The impact and importance of the model has been so great that scientists did not feel any necessity to re-examine the x-ray diffraction data on DNA structure proposed by Watson & Crick. It is only recently that physical scientists have begun to re-examine x-ray data interpreted by Watson and Crick and suggest some alternative conformations for DNA. It is established beyond doubt that DNA consists of two polynucleotide chains held together by hydrogen bonds between specifically paired bases. But the way in which the polynucleotide chains are associated has been challenged in one particularly alternative structure. In this alternative structure, DNA are viewed as *side by side* double helix rather than being interlocked double helix as suggested by Watson & Crick. G. A. Rodley *et al.* (1976) working in New Zealand and V. Chandrasekharan *et al.* in India have independently proposed a structure of B DNA which is different from the model suggested by Watson and Crick. According to them DNA is formed of alternative right handed and left handed helices arranged side by side. This structure has been called right-left handed helix (RL helix) or *side by side* double helix.

Although interlocked and side by side or RL versions of double helix are compatible with x-ray scattering and other physical data, yet *side-by-side* model does have one interesting feature from the genetic point of view. There has always been some doubts as to how the strand separation and unwinding of the Watson & Crick's interlocked double helix could be achieved during DNA replication. The RL helix makes such separation much easier to achieve than the Watson and Crick model because in the latter case a moment of force or torque would develop in DNA strands during unwinding and mere breaking of hydrogen bonds between two strands would make them free. The RL model can also easily explain the supercoiling of DNA in eukaryotic chromosomes and its replication.

Single Stranded DNA

DNA molecule in most organisms consists of two polynucleotide chains that are wound around each other to form a *double helix*. But in bacteriophage virus $\phi \times 174$ and several other bacteriophages, DNA has been found to be single stranded. Single stranded DNA differs from double stranded DNA in the following respects.

1. Double stranded DNA shows constant absorption of ultraviolet rays from 0 to 80°C (80°C being the critical melting point) and beyond 80°C absorption rate rises rapidly. The single stranded DNA shows steady increase in absorption of UV rays from 20°C to 90°C.
2. Double stranded DNA is resistant to the action of formaldehyde whereas the single stranded DNA is not resistant because the reactive sites are exposed.
3. In double stranded DNA amount of adenine is equal to that of thymine and G is equal to C. But in single stranded DNA of $\phi \times 174$ amounts of A and G are not equal to those of T and C respectively and the proportion is 1 : 1.33 and 0.98 : 0.75.
4. Single stranded DNA is circular while the double stranded is linear.

During replication single stranded DNA becomes double stranded (replicative form).

Localization of DNA in the Cells

DNA can be located in the cells in the following ways:

1. **Feulgen Staining.** Feulgen (1912) found that when DNA is hydrolysed with warm Schiff's reagent it turns reddish purple. During this reaction Schiff's reagent reacts with aldehyde groups in sugar deoxyribose. In 1924 Feulgen developed basic fuchsin staining technique which gives positive indication of presence of DNA in the chromosomes.

2. **Deoxyribonuclease reaction.** DNA can be removed by a specific enzyme deoxyribonuclease. After removal of DNA, the nucleus does not give the Feulgen reaction.

3. DNA absorbs ultraviolet rays at wave-length of 2600 Å. By this method it is possible to locate DNA without staining the chromosomes. Caspersson and others have used this technique to measure nucleic acid contents of the nuclei.

Different Forms of DNA (Conformational Flexibility of DNA)

DNA can exist in five different forms; A, B, C, D and Z forms of which **B *form*** (B DNA) is the most common of the configurations. B DNA is a right handed double helix in which there are 10.0 base pairs per turn (34 Å) of the helix and base pairs are more or less perpendicular to the helix axis.

The structure of DNA molecules changes as a function of their environment. The exact conformation of any DNA molecule or segment of DNA molecule will depend on the interaction between molecule and its environment. When humidity is high, it exists in ***B*-form**. When humidity is relatively low DNA exists in ***A*-form** which has 11 base pairs each turn. Since *A*-DNA occurs under low humidity conditions it has been suggested that it may be formed physiologically as a result of interaction with hydrophobic molecules or changing cellular conditions. Other forms *i.e.*, ***C-DNA***, ***D-DNA***, and ***Z-DNA*** differ from ***B-DNA*** in respect of the direction of helical coiling and/or spacing and inclination of base pairs (Fig. 10.8).

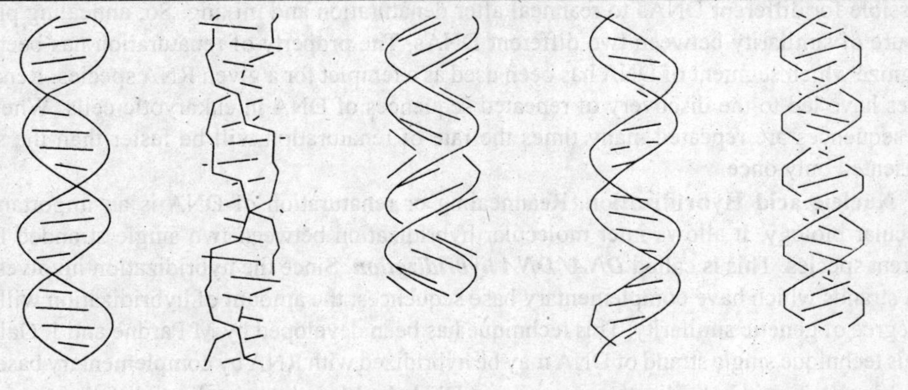

Fig. 10.8 Different forms of DNA.

In 1979, Andrew Wang, Alexander Rich and their associates of MIT, U.S.A. have proposed a new model of DNA called ***Z-DNA*** (named because of zig-zag appearance of the sugar-phosphate backbone). ***Z-DNA*** is a left handed double helix in which the base pairs are arranged peripherally to the helix axis and there are 12.0 base pairs per turn of the helix. B and Z forms of DNA are interconvertible and are the major forms of natural DNA.

The other two forms, ***C-DNA*** and ***D-DNA*** have been described both in natural and synthetic states. They are right handed double helical forms which have 7.9–9.6, and 8.0 pairs of bases per turn respectively. In ***A-DNA*** the base pairs are considerably tilted from the axis of helix at about 19° angle. Because of this displacement, the depth of the deep groove is increased and that of the shallow groove decreased, axial rise being only 2.56 Å.

In ***C-DNA***, the base pairs are arranged towards the middle of the helix and are inclined at an angle of 7.8° (less than that of ***A-DNA***). ***D-DNA***, as viewed along the helix axis, is not circular but hexagonal in cross section. In this form the base pairs are arranged towards the middle of the helix and are tilted at an angle of 16.7° from the axis of the helix.

Denaturation and Renaturation of DNA

Soon after the publication of Watson and Crick's model of DNA, many experiments were carried out by Warner, Rich and others using synthetic polynucleotide strands which showed that the complementary strands have marked tendency to intertwine in pairs and form right handed double helices. Recent investigations of DNA in solution or when packed into native virus particles have shown that the double helix often departs from perfect regularity and may partially untwist to permit local separation of two polynucleotide chains (called "breathing"). The two polynucleotide strands of a DNA molecule remain held by weak hydrogen linkages. Thus they can easily be separated. R.B. Inman (1966) showed that if λ DNA is treated with alkali (high pH) for short periods, DNA strands in the region rich in A—T pairs tend to separate before regions containing more G—C. This is called ***partial denaturation*** or ***partial melting***. The reason of this partial denaturation is that an A—T pair shares two hydrogen bonds, whereas a G—C pair shares three hydrogen bonds. By preventing the complimentary strands from further pairing through the use of formaldehyde, Inman observed a highly reproducible pattern of "bubbles" along the length of DNA molecule. This pattern is called denaturation map. Paul Doty and Julius Marmur (1960) observed that when DNA is heated to 100 °C or slightly below the boiling point of water, all the hydrogen bonds between the complementary bases are destroyed and the DNA becomes single stranded. This is called ***melting*** or ***denaturation***. If the solution is cooled slowly, some double stranded DNA may result which may be normal. This process is called ***renaturation*** or **reannealing**. The renaturation occurs when two single strands run parallel to each other in such a way that the complementary base sequences align. This makes it possible for different DNAs to reanneal after denaturation and mixing. So, annealing provides a measure of similarity between two different DNAs. The property of renaturation has been used to recognize which segment of DNA has been used as a templet for a given RNA species. Renaturation studies have led to the discovery of repeated sequences of DNA in eukaryotic cells. When certain base sequences are repeated many times the rate of renaturation will be faster than the sequence represented only once.

Nucleic acid Hybridization. Reannealing or renaturation of DNA is an important tool in molecular biology. It allows inter molecular hybridization between two single stranded DNAs of different species. This is called ***DNA/ DNA hybridization***. Since the hybridization involves parts of DNA strands which have complementary base sequences, the amount of hybridization will indicate the degree of genetic similarity. This technique has been developed by M.Pardne and J. Gall (1970). By this technique single strand of DNA may be hybridized with RNA by complementary base pairing.

In nucleic acid hybridization experiment DNA double helices are heated so that they denature into single strands. At this stage the denatured DNA strands are mixed with single-strand fragments of DNA or RNA obtained from other source (*i.e.*, another plant or animal species). The mixture is then allowed to cool slowly so that reannealing may take place. If the fragments and the denatured DNA carry homologous base sequences, hybrid DNA/DNA or DNA/RNA duplexes will be formed. Hybridization will fail if homologous base sequences are not available in the fragments involved.

REPLICATION OF DNA

There are three possible methods of replication of double stranded DNA, namely (*i*) conservative, (*ii*) dispersive, and (*iii*) semi-conservative.

(*i*) **Conservative DNA replication**. In this mechanism of DNA replication two parental polynucleotide chains of DNA remain together and the newly formed polynucleotide chains form the daughter molecule of DNA.

(*ii*) **Dispersive DNA replication**. In this replication mechanism, the two helical strands are broken along their length and produce small fragments. Each segment of the broken strands replicates and then all become randomly connected to form two new molecules.

(*iii*) **Semi-conservative replication**. According to Delbruck and Stent (1951), the two strands of a DNA molecule would separate from each other, maintain their integrity and both the strands will synthesize their complementary strands from the pool of nucleotides. The result would be that the newly synthesized molecules would carry or conserve one of the two strands of parent molecule. In this process the synthesis of DNA strands goes hand in hand with replication and is not complete till the synthesis of new polynucleotide chains.

The greatest attraction of the Watson and Crick DNA model for the geneticists is its ability to provide a simple mechanism for gene duplication. These two workers suggested that the two strands of a DNA molecule manufacture the replica of their own and in the process of replication each DNA strand might serve as a template (base) for the synthesis of its complementary strand (semi-conservative method). Watson and Crick hypothesized that:

1. Before duplication, the H-bonds are broken. Then unwinding and separation of the two chains take place.
2. Then follows the formation of new complementary polynucleotide chains along the sides of old DNA strands. So two pairs of polynucleotide chains are formed from the original one pair. In the semi conservative model of replication, each newly formed double helix of DNA contains one old and one new polynucleotide strand.

   According to the present day theory, the duplication of DNA in living cells including bacteria, plants and animals takes place in semi-conservative manner (Fig. 10.9).

   As suggested by Watson and Crick, the two strands uncoil and separate. Once the two strands are free, the base sequences which according to present view carry the genetical codes are exposed to the surrounding nuclear fluid that contains all kinds of chemical building blocks.

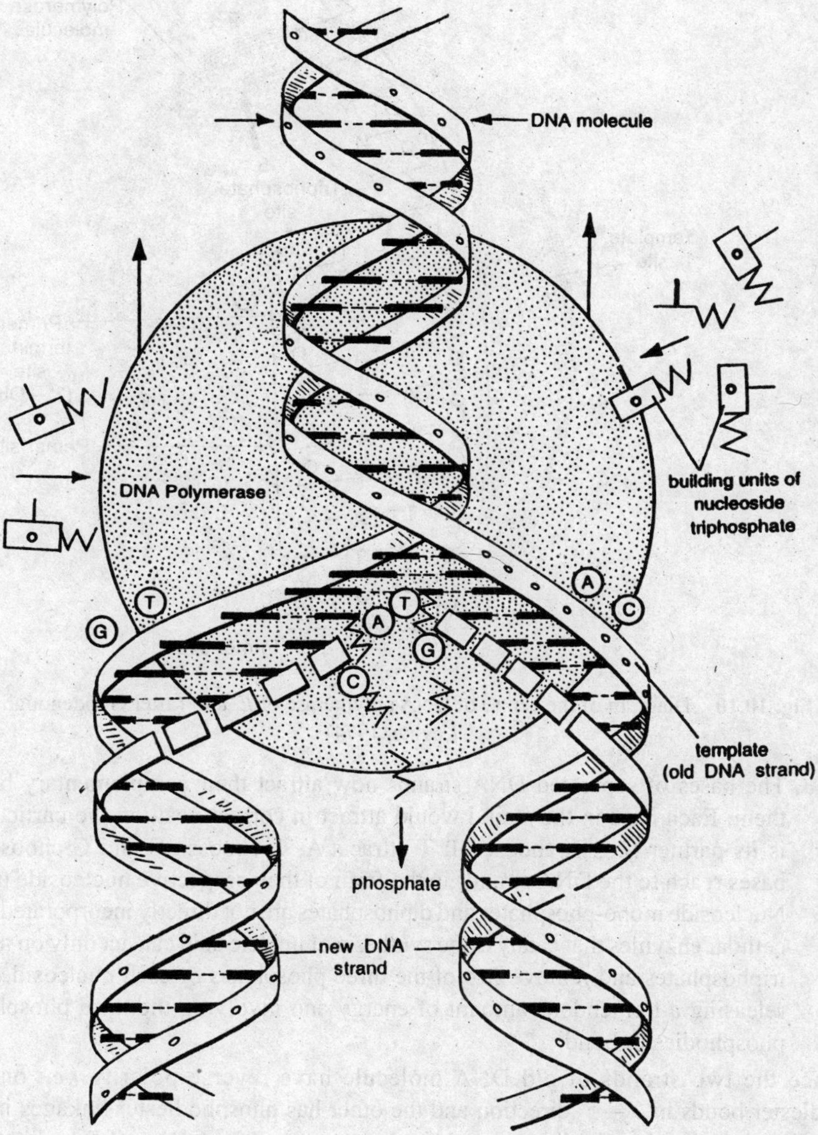

Fig. 10.9 Semi-conservative replication of DNA molecule.

The DNA molecule is assembled from precursor deoxyribonucleotide 5-triphosphates namely deoxyadenosine triphosphate (d-ATP), deoxyguanosine triphosphate (d-GTP), thymine triphosphate (TTP) and deoxycytidine triphosphate (d-CTP) in presence of the enzyme *DNA-polymerase* (Fig. 10.10).

$$\left.\begin{matrix}\text{n d TPPP}\\ \text{n d GPPP}\\ \text{n d APPP}\\ \text{n d CPPP}\end{matrix}\right\} + \text{DNA strand} \xrightarrow{\text{DNA polymerase}} (\text{dTP} - \text{dGP} - \text{dAP} - \text{dCP})$$

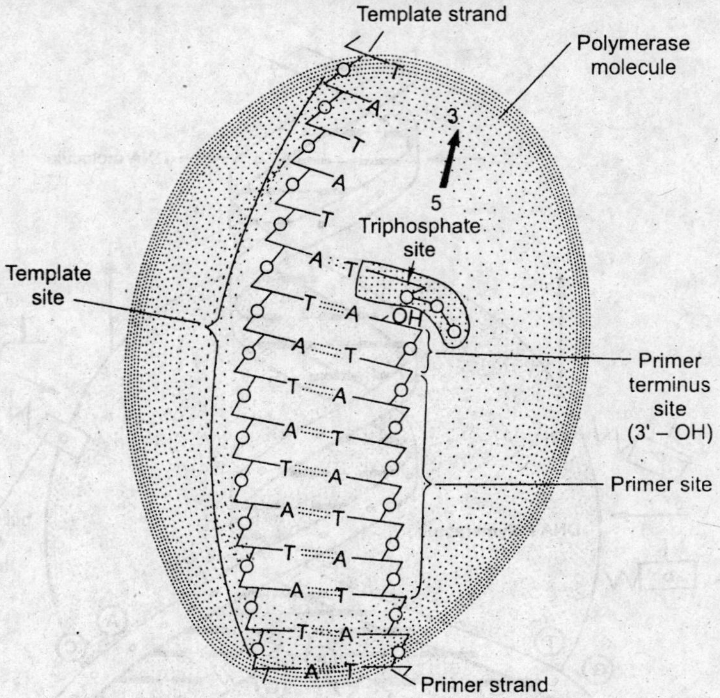

Fig. 10.10 Diagram of the centre of DNA polymerase of *E. coli* (after Goodenough and Levine, 1974).

3. The bases of separated DNA strands now attract their complementary bases towards them. Each base in the strand would attract in correct position the particular base that is its partner (*i.e.*, A chooses T, T attracts A, G chooses C and C chooses G). These bases reach to the DNA strands in the form of their respective nucleoside triphosphates. Nucleoside mono-phosphates and diphosphates are not directly incorporated in DNA. The cellular enzymes that catalyse the synthesis of nucleic acid can act only on the nucleoside triphosphates and remove two of the three phosphates of each nucleoside triphosphate releasing a tremendous amount of energy and involving the third phosphate group in phosphodiester bond.

Since the two strands of old DNA molecule have reverse polarity *i.e.*, one strand has phosphodiester bonds in 3′—5′ direction and the other has phosphodiester linkages in just reverse or 5′—3′ direction, the phosphodiester bonds in the new strands will be established in the directions just reverse to those of old strands (Figs. 10.11, 10.12).

The newly synthesized chain starts with a 5′ phosphate end. Two terminal inorganic phosphates of deoxyribonucleoside triphosphate are split off as pyrophosphates and a phosphodiester bond is established between the 3′ OH group of a nucleotide with 5' phosphate group of the next nucleotide in the newly developing DNA strand in 5′ —3′ direction (Fig. 10.13).

Watson and Crick theory of DNA replication was tested by Messelson and Stahl in 1958. Using the isotopic and centrifugation techniques they confirmed the mechanism of DNA replication and called that type of replication as semi-conservative replication because each of the two resultant daughter DNA molecules retains or conserves one parental polynucleotide strand (Fig. 10.14).

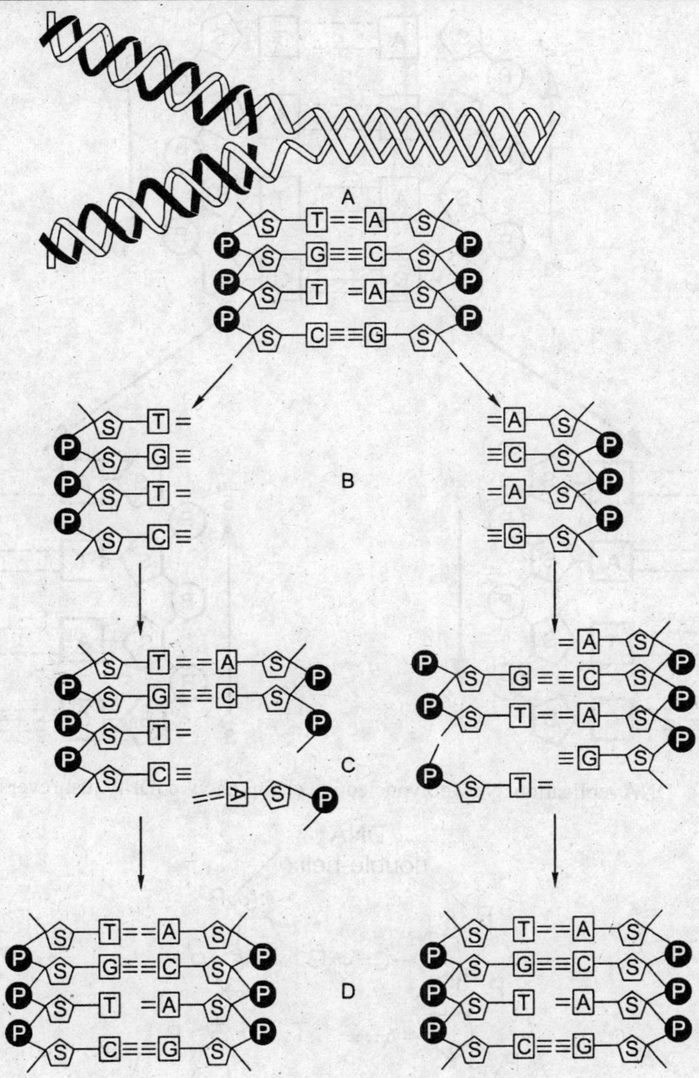

Fig. 10.11 Method of DNA duplication. New polynucleotide chains are developing on the old strands.

A. 2 polynucleotide chains of old DNA.

B. 2 polynucleotide chains separated.

C. Assemblage of complementary nucleotides on the old strands.

D. Synthesis of new strands on old strands complete.

The synthesis of DNA in vitro (*i.e.,* in test tube) has been demonstrated by Kornberg and his associates in 1958. Lehman, Bessman, Sims and Kornberg (1958) purified an enzyme from the bacterium *Escherichia coli.* They took DNA subunits (different types of deoxyribonucleoside triphosphates), specific enzyme-DNA-polymerase and the isolated DNA strands to demonstrate the DNA synthesis in virto. DNA strand was a necessary primer. The DNA strands obtained from different sources can be used as primers in the resynthesis. In the experiment they demonstrated a newly made DNA which was complementary to primer DNA. Bessman, Lehman, Sims and Kornberg (1958) have shown that in the process of polymerisation, the triphosphates retards the rate of reaction and they have also shown that no synthesis of new DNA took place in the absence of primer DNA. The synthesis of DNA in test tube strongly supports template hypothesis of DNA replication as proposed by Watson and Crick.

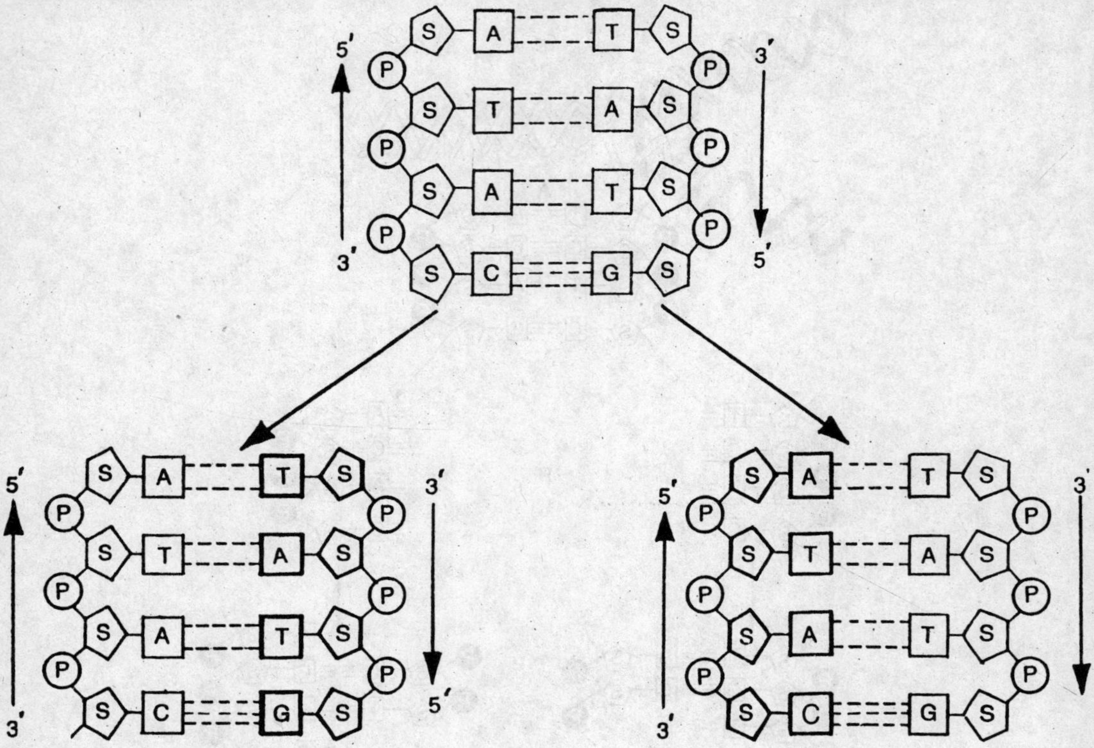

Fig. 10.12 DNA replication. New polynucleotide chains show polarity just reverse of old.

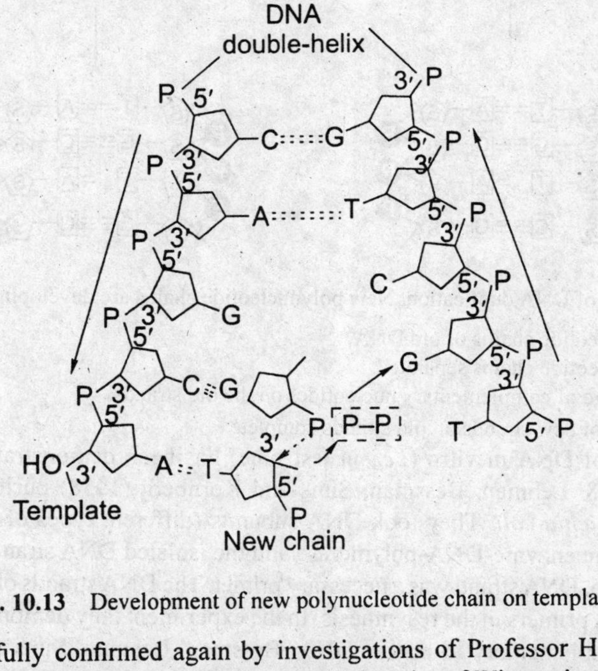

Fig. 10.13 Development of new polynucleotide chain on template.

This has been fully confirmed again by investigations of Professor H.G. Khorana and his associates at the Institute of Enzyme Research in the University of Wisconsin. It is possible to make short polynucleotide chains analogous to DNA by purely chemical processes. Thus d A (Deoxyadenylic acid) and d C (Deoxyribocytidylic acid) were induced to combine to form a dinucleotide, *i.e.,* d AC and then six of these dinucleotides were combined to give the repeating unit d AC AC AC AC AC AC.

In the same way a stretch of nucleotides containing complementary unit TG was put together. On mixing these two products they obtained, a small piece of double stranded DNA containing alternately arranged A and C in one strand and T and G in the other. When that DNA was mixed with the DNA polymerase and the necessary precursors, it was found that extensive DNA synthesis occurred.

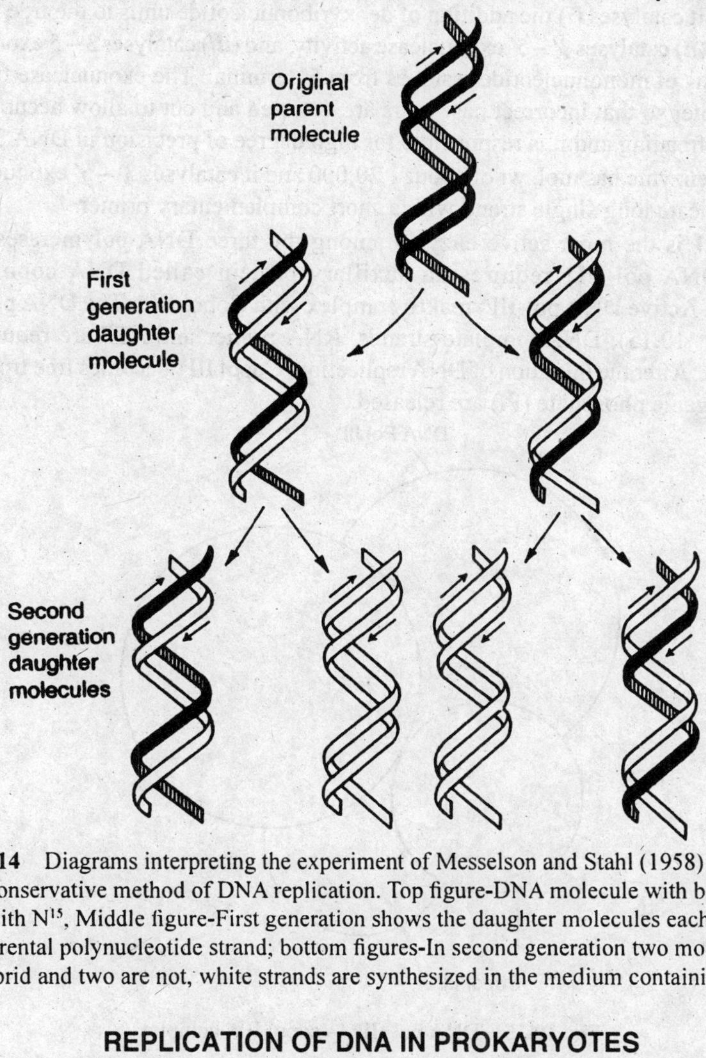

Fig. 10.14 Diagrams interpreting the experiment of Messelson and Stahl (1958) to show the semiconservative method of DNA replication. Top figure-DNA molecule with both strands labelled with N^{15}, Middle figure-First generation shows the daughter molecules each with one parental polynucleotide strand; bottom figures-In second generation two molecules are hybrid and two are not, white strands are synthesized in the medium containing N^{14}.

REPLICATION OF DNA IN PROKARYOTES

Meselson and Stahl experimentally demonstrated that the replication of circular DNA in *E. Coli* is semiconservative type but they did not explain whether the replication is initiated at one point or at several points simultaneously. Cairns (1963) provided conclusive evidence on the basis of visual observation of replicating chromosome in which DNA was labelled with tritiated thymidine (^{3}H TdR) that the bacterium *E. coli* contained single initiation point for the replication of circular DNA which proceeds bidirectionally from that point. The initiation point (ori) is a unique gene region from where the synthesis of new DNA proceeds outwardly in both clockwise and anticlockwise directions until both ends meet on the opposite side of DNA circle.

Now at least three different DNA polymerases with almost similar catalytic properties have been distinguished in *E. coli*. These DNA polymerases are (*i*) DNA polymerase I (DNA pol I), (*ii*) DNA polymerase II (DNA pol II), and (*iii*) DNA polymerase III (DNA pol III). It appears that DNA polymerase III is the main enzyme which is concerned with the DNA replication.

Although DNA pol I also takes part in DNA replication, it is largely used in DNA repair process, *i.e.*, for filling the gaps between small Okazaki fragments that are formed during DNA synthesis.

As regards DNA pol II, no specific function has been attributed to that.

DNA pol I which is a single polypeptide chain with 1,000 aminoacid residues has a mol. wt. of about 1,09,000. It catalyses (*i*) the addition of deoxyribonucleotide units to the free 3′ OH end of a polypeptide chain, (*ii*) catalyses 3′ – 5′ exonuclease activity, and (*iii*) catalyses 3 – 5 exonuclease action involving hydrolysis of mononucleotide residues from 5′ terminal. The exonuclease functions like a correcting type writer so that incorrect base pairs are detected and cut to allow accurate replication. This is called proofreading and it is responsible for high degree of precision in DNA Synthesis.

DNA pol II enzyme has mol. wt of about 1,20,000 and it catalyses 3′ – 5′ exonuclease activity. It is unable to replicate long single strand with a short complementary primer.

DNA pol III is the most active enzyme among the three DNA polymerases. Biologically active form of DNA pol-III* requires an auxiliary protein called DNA copolymerase III* (DNA Copol III*). Active DNA pol-III* makes complex with Copol III*. The DNA pol III* – Copol III* complex (Fig. 10.15), DNA template strands, RNA primer and ATP are required to initiate replication of DNA. After the initiation of DNA replication, Copol III* becomes free from the complex and ADP and inorganic phosphate (Pi) are released.

Fig. 10.15 DNA pol III*— copol III* complex.

Mechanism of DNA Replication

As far as the initiation of replication is concerned the process involves the following steps:

(*i*) Recognition of origin of initiation point.

(*ii*) Unwinding of double helix and separation of two DNA strands.

(*iii*) RNA priming.

(*iv*) Continuous synthesis of a new leading strand on one DNA template in 3′ – 5′ direction and formation of small DNA fragments beyond RNA primers on the second DNA template strand to produce new lagging strand.

Recognition of Origin of Initiation Point

None of the three DNA polymerases known in prokaryotes and eukaryotes can initiate new DNA chains. Instead, initiation of DNA synthesis involves synthesis of a short primer to which deoxyribonucleotides are added by action of DNA polymerases.

Before DNA polymerase starts its action, presumably at several points in duplex DNA, some unique proteins of low molecular weights bind to one of the strands especially in those regions which are rich in A = T base pairs and are susceptible to separation of the two strands of DNA.

RNA Priming

DNA replication is discontinuous and always proceeds in 5′ – 3′ direction on both strands of duplex DNA [Fig. 10.15(*a*)]. The next step, after recognition of initiation point is the generation of short RNA primers, *i.e.*, fragments of RNA which develop from precursors ribonucleoside triphosphate by the action of RNA polymerase or primase. The primer base sequence is directed by the nucleotide sequence of DNA template strand and are 50 – 100 nucleotides long. Initiation of replication process starts by RNA polymerase which recognises the initiation point (Fig. 10.16). After formation of RNA primer, DNA polymerase III begins to add deoxyribonucleotide units that are generated from deoxyribonucleoside triphosphates to 3′ terminal of RNA primers. These R-DNA fragments are called *Okazaki fragments*. They grow in 5′ – 3′ direction (Fig. 10.17).

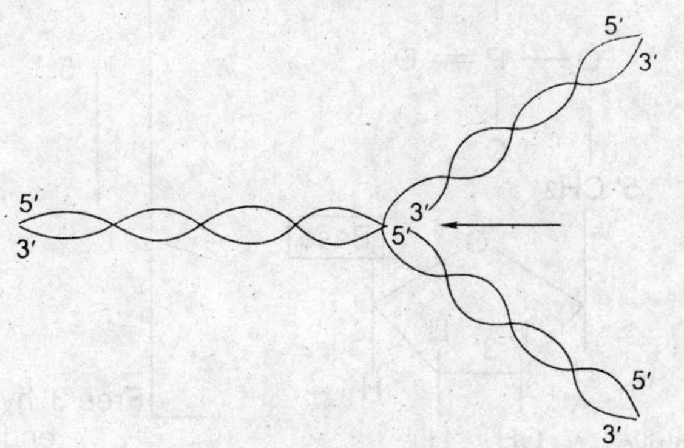

Fig. 10.15 (a) Formation of new DNA strands on both the strands of old DNA in 5′ – 3′ direction.

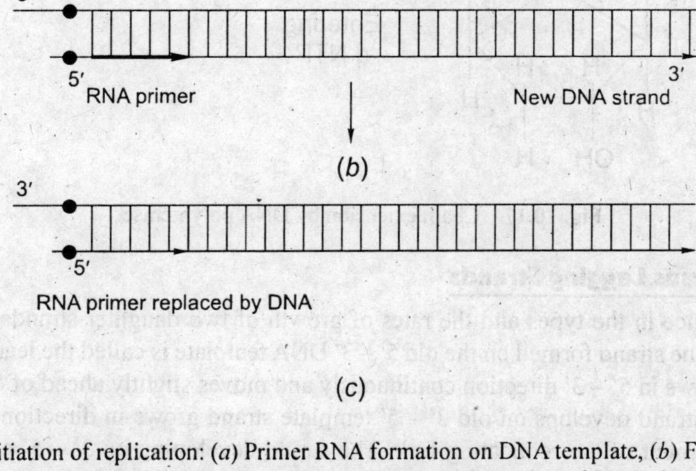

Fig. 10.16 Initiation of replication: (*a*) Primer RNA formation on DNA template, (*b*) Formation of new DNA next to primer RNA, (*c*) Excision of RNA primer and formation of DNA piece in its place.

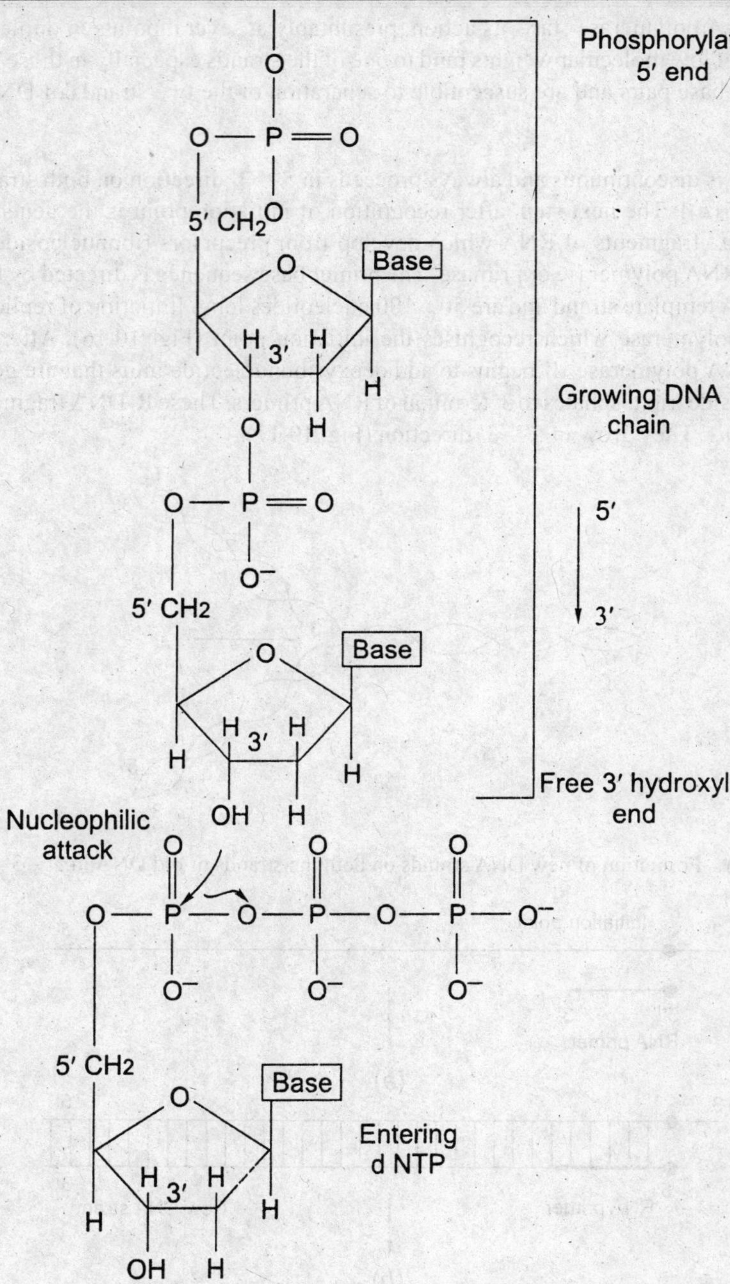

Fig. 10.17 Chain extension by DNA polymerase.

The Leading Versus Lagging Strands

There is a difference in the types and the rates of growth of two daughter strands formed on two DNA templates. The strand formed on the old 5′ – 3′ DNA template is called the leading strand. The leading strand grows in 5′ – 3′ direction continuously and moves slightly ahead of the other strand. The second new strand develops on old 3′ – 5′ template strand grows in direction opposite to the direction of movement of the replicating fork. The strand developing on 3′ – 5′ old strand grows discontinuously and is assembled from a group of DNA fragments made individually in 5′ – 3′ direction. That is called lagging strand (Fig. 10.18). Thus, following the initiation a number of short segments of DNA are formed in presence of DNA polymerase and ATP. During the process ATP is hydrolysed to ADP and Pi. and as soon as the elongation process is completed the following three important events occur (Fig. 10.19).

(*i*) Excision of RNA primer fragments [Fig. 10.19 (*a*)].

(*ii*) Excision of RNA primers is followed by filling up of the gaps by deoxyribonucleotide residues [Fig. 10.19 (*b*)].

(*iii*) Joining together of adjacent ends of DNA fragments by phosphodiester bonds catalysed by DNA Ligases to form a continuous DNA daughter strand [Fig. 10.19 (*c*)].

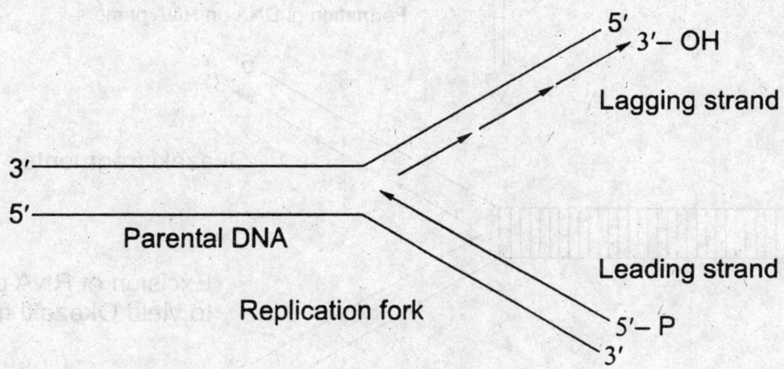

Fig. 10.18 Formation of leading strand in 5′–3′ direction on parental 3′–5′ template DNA and pieces of lagging DNA strands develop in 5′–3′ direction on 5′–3′ strand of template DNA.

Separation of RNA primer takes place in presence of specific enzyme protein. It is possible that the RNA primer is excised residue by residue from 5′ end by exonuclease activity of DNA polymerase I whose primary task is to fill up the gaps left between RNA primed DNA segments (Okazaki segments). Finally adjacent 5′ and 3′ ends are joined by DNA ligase or polynucleotide ligase by formation of phosphodiester bonds.

Replication of Bacterial DNA

DNA thread of bacterial cell is regarded as bacterial chromosome or genophore or chromoneme. In the case of *Esherichia coli* the DNA strand contains about 3 million pairs of nucleotide bases and this is sufficient to form nearly 10,000 genes. The total time taken for replication of DNA in *Escherichia coli* is 40 to 50 minutes at 37°C which means that the synthesis of DNA takes place at the rate of about 1,000 base pairs per second—a very rapid process if done by only one enzyme molecule. Replication of DNA is brought about, as already mentioned, by the enzyme DNA polymerase and the replication starts at the one point of circular chromosome and goes round the circle until a new circle is completed, subsequently, the two daughter chromosomes can be separated. It was shown by Francois Jacob, Sidney Brenner and F. Cuzin (1963) that the replication in fact occurs at the point at which the chromosome is attached to the plasma membrane. Here the replication enzyme DNA polymerase is located and it appears that instead of enzyme moving round the DNA ring, the latter (DNA) moves through the point of attachment called ***swivel point*** and is replicated (J.A.B. Butler, 1970) (Fig. 10.20).

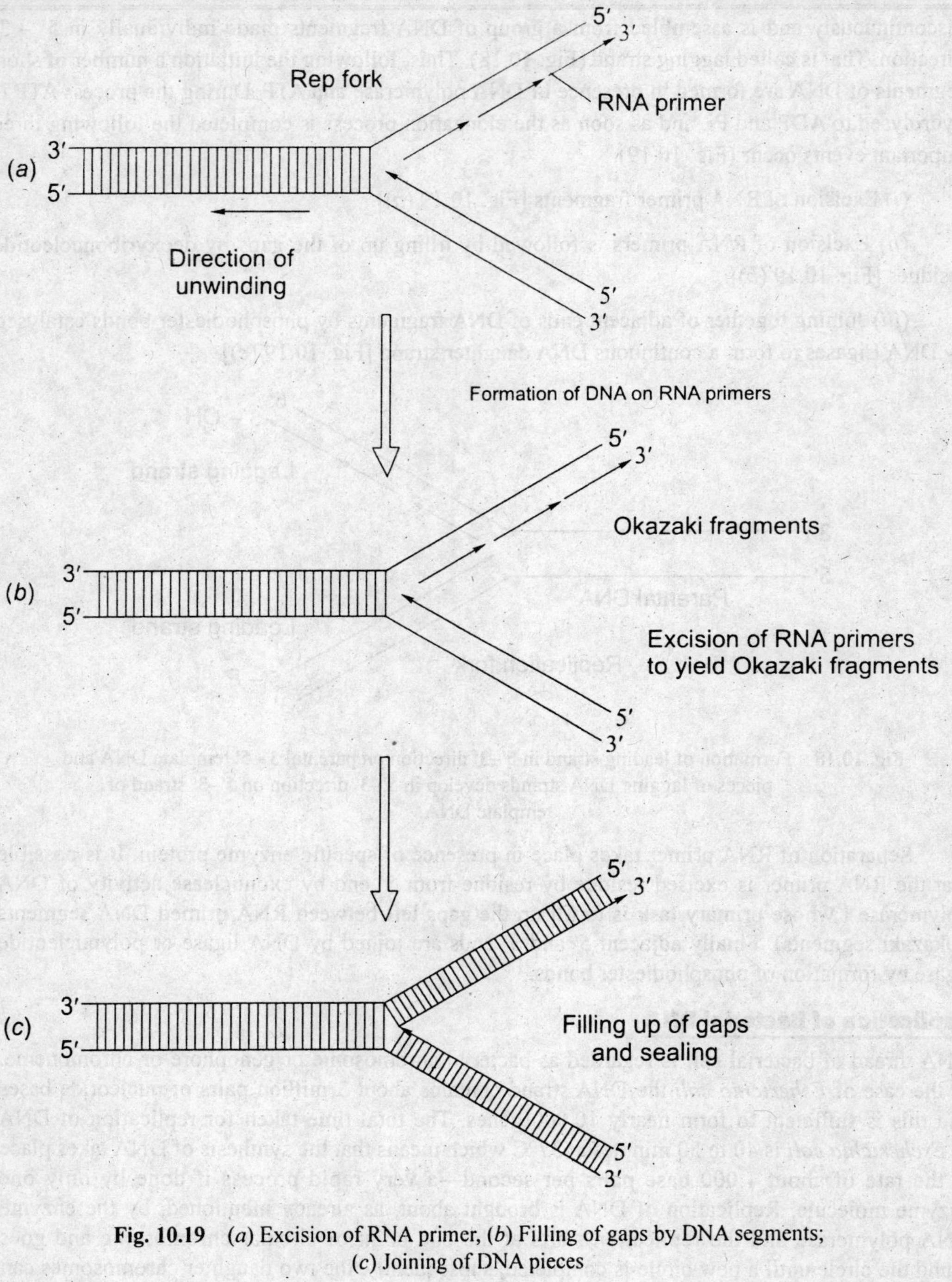

Fig. 10.19 (*a*) Excision of RNA primer, (*b*) Filling of gaps by DNA segments; (*c*) Joining of DNA pieces

Replication of Viral DNA

Viruses contain a nucleic acid and protein sheath. Nucleic acid is DNA in some and RNA in others. Viral nucleic acid contains all the genetic informations necessary for the construction of new viruses. In almost all the viruses genetic material appears to be a single DNA or RNA molecule, either single or double stranded, ranging in length for 1.8 μ in φ × 174 to 56 μ in T_2 virus. DNA of T_4 phage virus

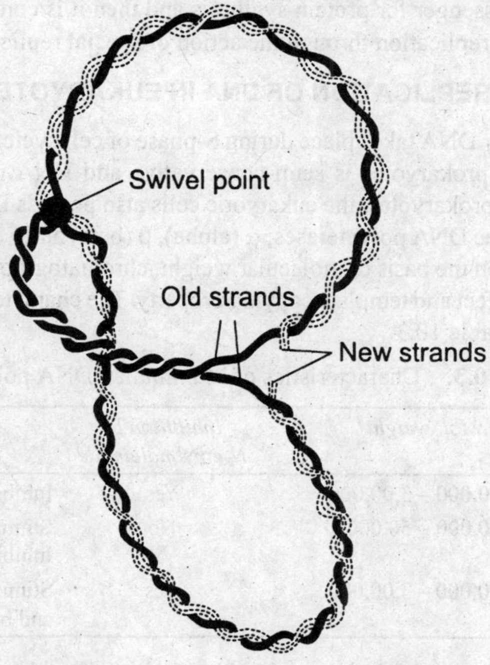

Fig. 10.20 Replication of Circular bacterial DNA.

is circular and has a molecular weight 1.3×10^{10} (*i.e.,* 2,00,000 nucleotide pairs) which is sufficient for 150 to 200 genes. DNA of T_2 phage has about 60,000 base pairs. Coli phage lambda virus has 32 to 50 genes.

Bacterial virus $\phi \times 174$ contains single ring shaped DNA strand containing about 5,500 nucleotides (mol. weight 1.7×10^6). The replication process of $\phi \times 174$ phage virus is very remarkable and extraordinary series of changes takes place when viral nucleic acid is injected into bacterial cell. The normal functions of the cell, *e.g.,* RNA and protein syntheses are disturbed. In the host cell the viral DNA is immediately replicated and as a result double stranded form is obtained. The original strand is called **positive (+ strand)** and the newly synthesized the **negative strand (– strand)**. The two remain associated and form a covalently closed superhelical structure known as parental replicative form (R.F.). The double stranded R.F. is capable of carrying the usual function of DNA, *i.e.,* it produces messenger RNA which become attached to the ribosomes of host cell to make a whole series of new viral proteins and enzyme required for the replication of phage DNA. It has been found that the special protein which is produced by the virus in the early stages of infection checks the synthesis of host DNA. Besides acting as template for forming messenger RNA, the DNA must also arrange for its own replication. This is brought about by a special DNA polymerase enzyme which is formed within the cell at an early stage after infection. The parental R.F. undergoes several rounds of semiconservative replication to produce perhaps some 15 double stranded daughter R.F. molecule. Replication of DNA can occur only at one site within the cell at any time. This site is probably the site of the cell membrane at which replication of bacterial DNA strand normally occurs. Eventually, a shift occurs and positive stranded rings rather that RF duplexes are synthesized. The rings are packed into virus particles and lysis of the host cell occurs.

There are phages which contain RNA instead of the mere usual DNA. The remarkable feature of RNA of these viruses is that it is capable of acting both as a template for replication and also as mRNA for protein synthesis. The RNA in the virus is in single stranded form and when RNA enters the cell it first acts as messenger for protein synthesis and then it is converted into double stranded form which is capable of replication through the action of special replicative enzyme.

REPLICATION OF DNA IN EUKARYOTES

The duplication of nuclear DNA takes place during S-phase of cell cycle. J.H. Tailor in 1957 showed that DNA replication in prokaryotes is semi-conservative and DNA replication in eukaryotes is semidiscontinuous. Like prokaryotes, the eukaryotic cells also possess DNA polymerases and DNA ligases. In eukaryotes three DNA polymerases; α (alpha), β (beta) and γ (gamma) DNA polymerases have been distinguished on the basis of molecular weight, chromatographic properties, sensitivity to N-ethylmaleimide, salt effect and template copying property. The characteristics of these polymerases have been compared in Table 10.3.

Table 10.3. Characteristics of Mammalian DNA polymerases

Type of DNA	*Mol. weight*	*Inhibition by N-ethylmaleimide*	*Salt effect*
α	1,20,000 – 3,00,000	Yes	Inhibited by 725 m M NaCl
β	30,000 – 50,000	No	Stimulated by 120 – 200 M NaCl Inhibited by 50 m M NaCl
γ	1,50,000 – 3,00,000	Yes	Stimulated by 100 – 250 m' M KCl and by 50 m M phosphate

Some eukaryotes have multiple forms of DNA polymerases and some others have single major enzyme. α and β DNA polymerases along with some other proteins usually occur in the replication complex of nuclei and γ DNA polymerases are localised in the mitochondria. Protozoa, fungi and plants do not have β polymerases. The eukaryotic microbes, in general, have α polymerase which is different from α DNA polymerase of mammals. There is too much variability among these DNA polymerases and so it is not possible to make generalization about these enzymes. α DNA polymerase is concerned with the nuclear DNA replication and β polymerase probably functions as a repair enzyme. Eukaryotic DNA ligases require ATP instead of NAD as adenyl donor.

As regards the initiation of DNA replication, Cairns (1963) on the basis of visual observation of replicating chromosome of *E. Coli* in which DNA was labelled with tritiated Thymidine (^{3}H-TdR) pointed out that in bacterium, the replication of circular DNA is initiated at one point. In eukaryotic organisms replication of DNA begins at several precisely defined units along the double stands of DNA. These units are called replication units (RUs) or replicons. Each replication unit has a single specific origin (o) and two termini (t) (Fig. 10.21). In Hela cells (a type of tissue culture cells derived from human cancer cells) the chromosomes may contain as many as 100 such RUs each. This arrangement of RUs in chromosomes is necessary so that enormous length of eukaryotic DNA can be replicated within reasonable time. The spacing between origin points of two adjacent RUs varies from 30,000 to 3,00,000 base pairs.

DNA replication in eukaryotic chromosomes begins by opening of bubbles. The area where double stranded DNA molecule becomes untwisted or denatured for replication is called **replication bubble**. The breakage of H bonds holding the complementary base pairs together is followed by unwinding or untwisting of two DNA strands results in separation of the DNA chains in the replication units. The two separated single strands of DNA act as templates on which new complementary strands of DNA will develop. One or more specific initiation proteins are required to recognise these initiation or origin points. In linear DNA molecules, replication starts on specific initiation points internally and generates new complementary strands which grow bidirectionally towards termini. Though origin and termination signals are little known, perhaps some specific nucleotide sequences act as replication sites which signal the initiation of DNA replication in certain animal viruses such as SV-40 and yeasts.

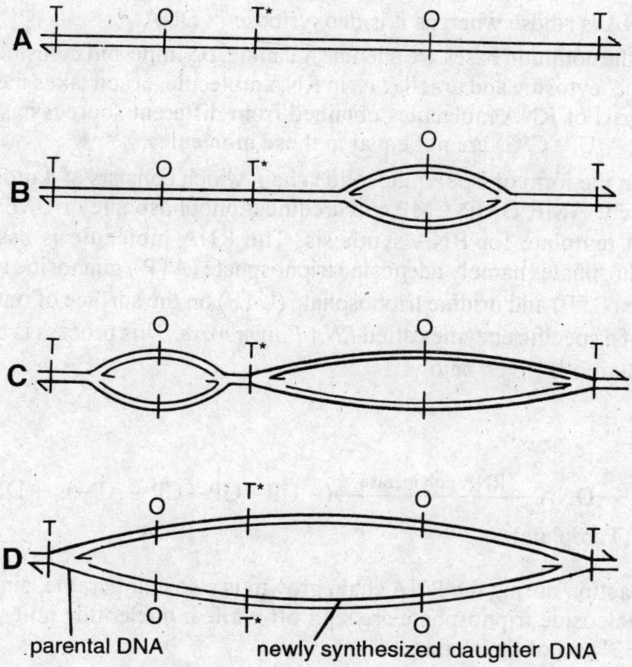

Fig. 10.21 Replication of Eukaryotic DNA: A. A duplex DNA piece with 2 replication units each with 2 termini and one origin point. B. Separation of chains in one RU to form replication bubble and development of new complementary strands in 5′ – 3′ direction on template strands. C. Replication in 2 RUs. D. Joining of adjacent strands of 2 RUs to form continuous daughter DNA.

At replication point, specific initiation protein of replication complex binds to a genetically determined origin sequence to start DNA replication. The rate of movement of replication forks may range from 1,000 to 15,000 nucleotides of DNA per minute at 37°C in E. coli. In eukaryotes the replicating fork moves slowly at about 2,600 bases per minute. In *Drosophila* egg cell about 6,000 replicating units are involved in replication and the replication is completed in about 3 minutes.

In eukaryotic chromosomes initiation of replication of each replication unit begins at origin of replication (o), proceeds bidirectionally producing replication bubbles in DNA and the process terminates after fusion of adjacent replication forks. G–C rich regions of DNA are replicated first and A–T rich regions later during S phase. DNA synthesis also depends on RNA synthesis but the inhibition of RNA synthesis does not have any immediate effect on DNA synthesis. Genetic and biochemical evidences suggest that the start of S phase of cell cycle is determined by a regulatory protein synthesised during G1 phase. That protein acts as positive effector to derepress DNA replication.

The newly formed DNA in eukaryotes becomes complexed with histones to form nucleosomes. The distribution of old and newly synthesised histone octomers in replicated DNA is not clearly understood but there is evidence that there is no mixing of old and new histone octomers in nucleosomes and the old octomers are conserved from generation to generation and the newly synthesised histone octomers are used at the replication forks to form new nucleosomes. Once the nucleosomes are assembled at the replication forks, within a short period, say 15 minutes, the chromatin matures into non-replicating state.

RIBONUCLEIC ACID (RNA)

In broad features, the chemistry of RNA is comparable to the chemistry of DNA. The chemical analysis shows the following two main differences in the polynucleotide chains of DNA and RNA:

(*i*) the sugar in RNA is ribose whereas it is deoxyribose in DNA.

(*ii*) while in DNA, the common bases are adenine, guanine, thymine and cytosine, the RNA contains adenine, guanine, cytosine and uracil (*i.e.*, in RNA molecule, uracil takes the place of thymine). Chemical analysis of RNA molecules obtained from different sources has revealed that base pair ratios (*i.e.*, A/U = C/G) are not equal in those molecules.

RNA is found in the form of a polynucleotide chain which consists of 4 monomeric ribotide or ribonucleotides, namely, AMP, GMP, CMP and uredine monophosphate or UMP.

DNA acts as a template for RNA synthesis. The RNA molecule is assembled from four ribonucleoside 5-triphosphates namely adenosine triphosphate (ATP), guanosine triphosphate (GTP), Cytidine triphosphate (CTP) and uridine triphosphate (UTP) on the surface of one of the two strands of DNA in presence of a specific enzyme called *RNA Polymerase*. This process is called ***transcription*** (Fig. 10.2A). The reaction is given below:

$$\left.\begin{array}{l}\text{n UTP}\\ \text{n GTP}\\ \text{n CTP}\\ \text{n ATP}\end{array}\right\} \underset{\text{(Template)}}{+\text{DNA}} \xrightarrow{\text{RNA polymerase}} \underset{\text{RNA}}{(-\text{UP}-\text{GP}-\text{CP}-\text{AP}-)_n} + \text{DNA} + 4(\text{n})\ \text{PP}.$$

The primary reaction during the RNA chain growth is also comparable, since the two terminal phosphates of ribonucleoside triphosphate are split off while a nucleotide unit with one phosphate group is added to the 3' end of the chain.

Since ribose and deoxyribose differ only in one oxygen atom, while uracil and thymine differ only in a methyl group, interconversion of RNA and DNA precursors takes place very readily in living cells (Du Praw, 1970). Single strands of RNA are polarized 3'–5' polyribotide chains and complementary RNA chains also pair into double helices very readily and hydrogen bonds are established between G—C and A—U base pairs. The three-dimensional structure of RNA is less defined. RNA is commonly found in single stranded state. It does have regions of double helical configuration where the same chain loops upon itself forming hydrogen bonded base pairing structure which may sometime be folded on itself to form localized helical regions. The relationship of RNA polymerase to its template DNA is different from that of DNA polymerase reaction in the following two respects:

(*i*) There is a notable preference for a helical DNA as template and as reported by Hurwitz *et al.* (1963), the two strands of DNA are not dispersed during the reaction.

(*ii*) Only one of the two complementary strands of DNA is copied (or transcribed) to determine the sequence of ribotides in the RNA chain (Hayashi *et al.,* 1964).

The rarity of complementary RNA strands follows from the mechanism of RNA synthesis which produces only one chain at a time.

Transcription

Genome contains complete information for the structure and function of organism, but not all genes are active at any one time. DNA does not act as template for the synthesis of proteins. The genetic information of DNA is transferred to RNA molecules. This process is called ***transcription***. It is actually RNA that acts as a template to produce sequence of aminoacids or protein. This process is called ***translation***.

The common features of transcription in both prokaryotes and eukaryotes are as follows:

1. For transcription two strands of DNA unwind locally and for each gene—only one strand of the double helix DNA acts as template for RNA synthesis. That is called **sense strand** and the opposite strand of DNA is termed **antisense strand**.

2. The product of transcription is single stranded RNA.
3. Ribonucleoside triphosphates–ATP, GTP, CTP and UTP are RNA precursors that are polymerised into RNA chain. Polymerization is catalysed by the enzyme RNA polymerase and Mg^{++}.
4. RNA polymerisation reaction is similar to the of DNA polymerization. RNA polymerase selects ribonucleoside phosphates to form complementary base pairs with the exposed nucleotides of sense strand of DNA.
5. RNA is synthesized in 5′ – 3′ direction *e.g.,* if DNA template strand is 3′– T C C G A T T T –5′; the RNA chain would be 5′– A G G C U A A A – 3′.

The process of transcription on prokaryotic DNA proceeds in the following sequence:

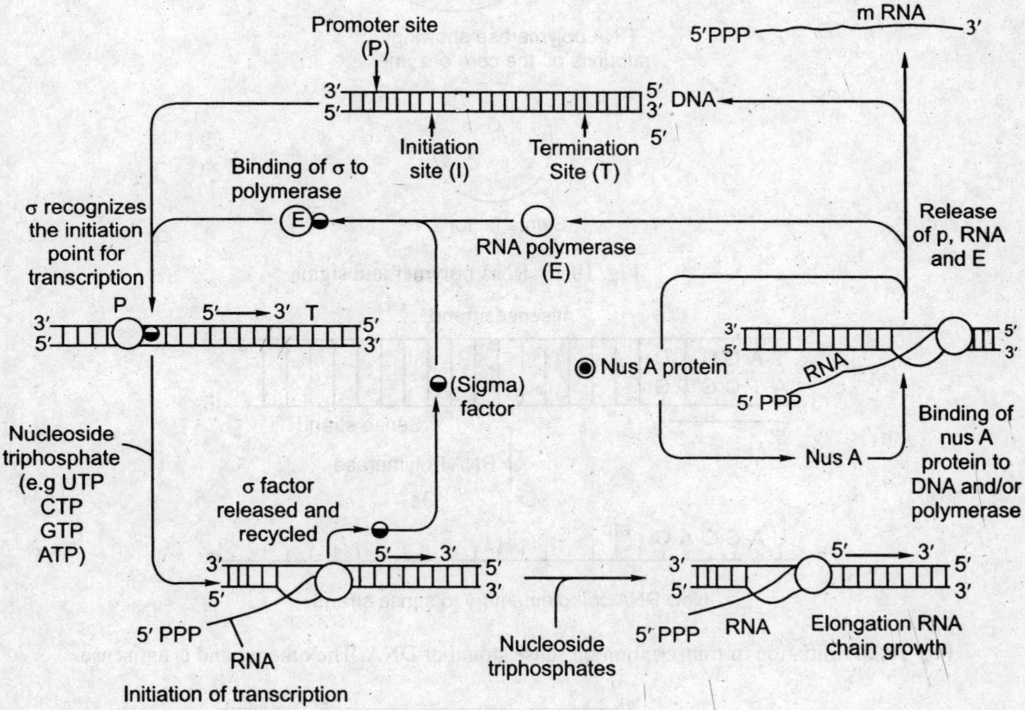

Fig. 10.22 Schematic representation of transcription in prokaryotes

1. Prokaryotic organisms have their own RNA polymerases. Bacteriophages use host's RNA polymerases or code for their own. Transcription requires double stranded DNA, all four ribonucleoside triphosphates, RNA polymerase enzyme with sigma factor, Mg^{++} ions and nus A protein factor. RNA polymerases core enzyme is composed of 4 subunits β', β, and 2 α subunits (Fig. 10.22). β' subunit of core enzyme is DNA binding subunit. It binds to the sense strand of DNA at a specific site called promoter. The other DNA strand which is not transcribed is called anti sense (Fig. 10.23). Each gene (a specific length of DNA on one strand of helix) is divided into *coding* and *non coding regions* (Fig. 10.24). A coding region carries a nucleotide sequence with hereditary information. Coding sequence is flanked on either sides by non-coding sequences. Non-coding sequence upstream is organised into *promoter region* with a set of seven nucleotides in the middle, called **Pribnow Box**. Between promoter and coding sequences is a short stretch of nucleotides called **initiator site** which signals transcription of the coding sequence.

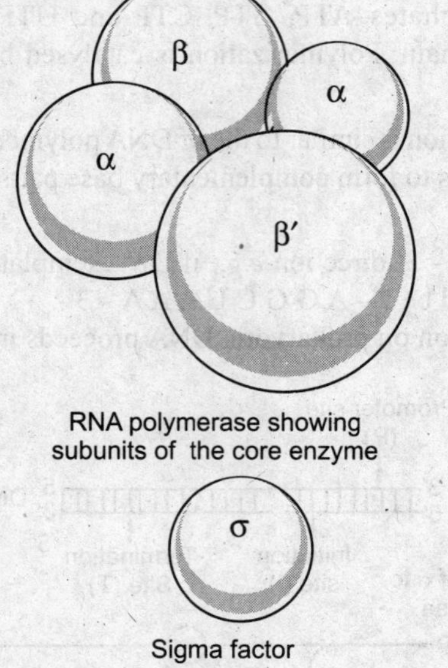

Fig. 10.23 RNA polymer and sigma

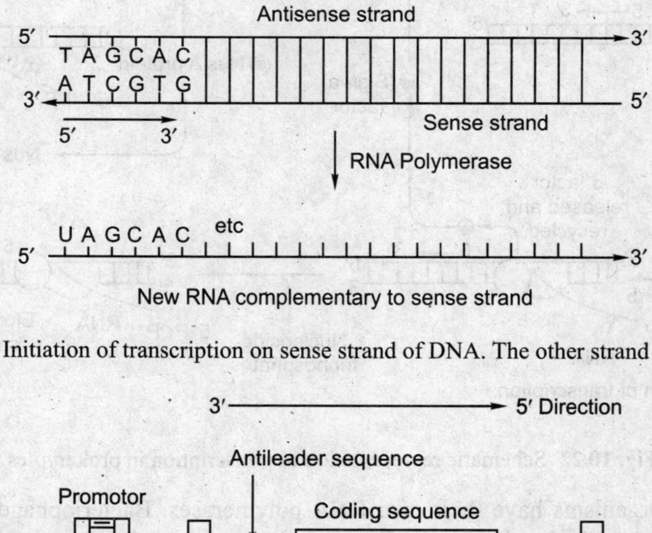

Fig. 10.24 Initiation of transcription on sense strand of DNA. The other strand is antisense.

Fig. 10.25 Schematic representation of the functional organisation of genes in general

The non-coding region downstream has a set of nucleotides called **terminator** which terminates transcription. Thus *Promoter allows* binding of the enzymes for transcription, the initiator gives the start signal for transcription of coding region and the terminator acts as a stop signal for transcription.

2. About 35 base pairs upstream from the start of RNA synthesis is a recognition site. RNA polymerase probably first binds there and then slides to the *Pribnow box* to initiate transcription, β′ subunit of core enzyme binds to the sense strand of DNA at promoter sequence. Before binding to DNA, the core enzyme becomes associated with σ (sigma) factor which plays an important role in transcription. σ factor is a 95,000 Dalton polypeptide. It initiates RNA synthesis at promoter site along the DNA. If σ factor is not linked to RNA polymerase, the enzyme can initiate RNA synthesis, but at random and then both strands are copied instead of one.
3. RNA polymerase then catalyses denaturation or separation of two DNA strands exposing the template strand for initiation of RNA synthesis (Fig. 10.25).
4. RNA synthesis starts, on DNA template in 5′ to 3′ direction. The ribonucleotides come one after another in an array and form complementary base pairs with the bases of exposed template strand of DNA. The consecutive ribonucleotides are linked by formation of phosphodiester bonds in the same way as in DNA synthesis. This process is catalysed by β subunit of RNA polymerase. When the polymerisation has started, σ factor which acted like a catalyst in initiation of RNA synthesis at specific site dissociates from core enzyme and becomes available for initiating on other transcription site.
5. When the gene has been transcribed, a protein factor called ***nus-A*** becomes linked to RNA polymerase and the process of transcription is stopped. Then nus-A, core enzyme or RNA polymerase and RNA strand are separated. The release of RNA strand from DNA template requires the activity of a ρ (rho) protein.
6. When newly synthesised RNA is displaced, DNA strands renaturate to form double-helical DNA.

Transcription in Eukaryotes

Although RNA synthesis in eukaryotes takes place more or less in the same way as in prokaryotes, yet some differences are there–(*i*) in controlling sequences, and (*ii*) involvement of more than one RNA polymerase enzyme in eukaryotic organisms. In eukaryotes, three RNA polymerases; RNA polymerase I, RNA polymerase II and RNA polymerase III are known which play special roles in RNA synthesis. All the three RNA polymerases are complex polypeptides, each consisting of several subunits.

As regards transcription controlling sequences, the base pair sequences upstream from coding sequence indicate the presence of consensus sequence TATAAA called TATA box or the Goldberg–Hogness box about 30 basepairs upstream from the coding sequence. That is considered to be promotor sequence. Further upstream from TATA Box, some other sequences also play role in regulating transcription. Several types of promoter sequences are used in eukaryotes. The terminator sequences downstream of coding sequence are not properly understood in eukaryotes. In yeast, part of termination sequence is TTTTTATA.

There are different types of RNA which perform different functions. These RNAs are categorised as follows:

A. genetic RNA or Viral RNA

B. non-genetic RNA.

Genetic RNA

In most of the plant viruses and in some of bacteriophages and animal viruses, the hereditary material is single stranded RNA. In a few cases such as the poliovirus, influenza virus and reovirus, the RNA is double stranded. The viral RNA is also called **genetic RNA** because it contains all the informations which are normally found in DNA in higher organisms. Molecular weights of RNA in viruses can vary greatly. Larger virus particle has got larger RNA molecule than the smaller ones. Each Tobacco Mosaic Virus (TMV) with particle weight 4×10^7 contains a single strand of RNA of molecular weight

2.2×10^6. TMV RNA is one of the most intensively studied and one of the best understood molecules of RNA. The double stranded RNA of viruses follows the same basic rules of base pairing as in DNA.

The genetic RNAs found in viruses act as informational molecules. However, the information is not transferred from RNA to RNA. Here the viral RNA acts as template for DNA synthesis by reverse transcriptase (**Reverse transcription**). The DNA strand synthesised on viral RNA is involved in the synthesis of complementary viral RNA. Viral RNA can not replicate by itself while present in the host cell. The viral RNA functions directly as a messenger RNA which in association with ribosomal apparatus of the host cell directs on one hand the synthesis of proteins that form the viral sheath and on the other hand RNA polymerase enzyme which is required in RNA replication.

With the help of RNA polymerase and base pairing rules, the viral RNA synthesizes a complementary RNA chain on its surface and thus double stranded structure is produced.

Non-genetic RNA

The hereditary material in cellular forms of organisms is DNA. RNA, though present in good quantity, does not act as genetic material. It plays very important roles in the synthesis of cellular proteins. The different types of non-genetic RNA found in the cell are as follows:

(*i*) *Ribosomal RNA (rRNA)* which is sedimented at high speed in the ultra centrifuge.

(*ii*) Transfer RNA (tRNA) (formerly called soluble RNA or *s* RNA) which remains in the supernatant after the sedimentation of ribosome.

(*iii*) Messenger RNA (*m*RNA) which incorporates labelled uracil before other RNAs.

Ribosomal RNA (*r*RNA). It is the most stable RNA which occurs in ribosomes. About 80% of the total cellular RNA occurs in ribosomes. Ribosomal RNA makes up about 40-60% of the total weight of ribosomes. Each ribosome consists of one large and one small subunit. Both subunits must associate to function in protein synthesis. The association and dissociation of two subunits depend on the Mg^{++} concentration. The ribosomes are usually characterized by their sedimentation coefficients obtained by analytical ultracentrifugation experiments. The ribosomes of eukaryotes sediment at about 80S whereas in bacteria they sediment at 70S. Both these types of ribosomes are structurally similar and are complexes of *r*RNA and proteins. The ribosomes of eukaryotic cells contain; 28S RNA, 5.8S RNA and 5S RNA in the larger (60S) subunit and 18S rRNA in the smaller (40S) subunit. The ribosomes found in prokaryotes and those found in the plastids and mitochondria of eukaryotes also contain three kinds of *r*RNA; 23S RNA and 5S RNA in larger (50S) subunit and 16S RNA in smaller (30S) subunit. Ribosomal RNA contains the four major RNA bases with slight degree of methylation. The $\frac{G}{C}$ and $\frac{A}{U}$ ratios in the ribosomal RNAs of different sources vary considerably. The one regularity that has been observed is that *r*RNA from all sources has a G + C contents more than 50%. Its molecule appears as single polynucleotide strand which is unbranched and flexible (primary structure) but at low ionic strength, it shows random coiling. At

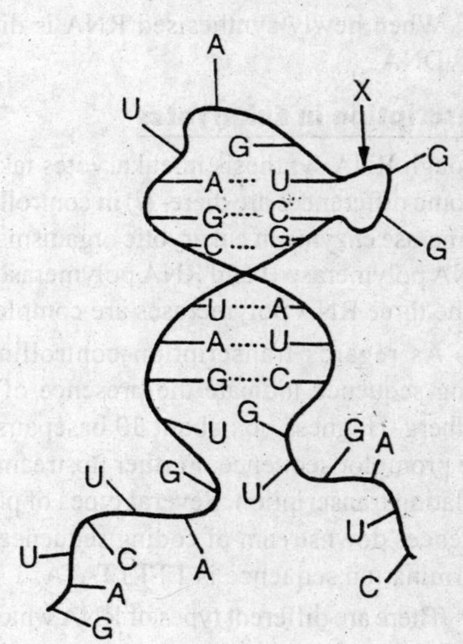

Fig. 10.26 Molecular structure of *r*RNA showing helical regions with complementary base pairs and a looped outer region (Davidson, 1972).

high ionic strength, the molecule shows helical regions with complementary base pairing and looped outer region (Secondary structure) (Fig. 10.26).

Physical studies of *r*RNA, such as hyperchromicity, melting point and viscosity studies, suggest that a great deal of base pairing occurs within the isolated *r*RNA molecules, but this is not clear how regularly periodic it is or whether it has any relevance to the structure of the RNA in the ribosomes.

In eukaryotic cells of plants and animals the different types of *r*RNA molecules are transcribed on *r*DNA in the nucleolar organizer region of a particular chromosome to which nucleolus is attached (Fig. 5.11).

The important steps in the synthesis and maturation of eukaryotic ribosome are outlined below:

(*a*) The ribosomal DNA (*r*DNA) located in the core of the nucleolus synthesizes 45S RNA which is a large initial precursor molecule for 28 S RNA and 18 S RNA.

(*b*) 45 S RNA is combined with proteins to form a complex of S = 80. The proteins of this complex may include ribosomal and other types.

(*c*) 45 S RNA is cleaved into 18 S and 32 S pieces. During this period changes in protein composition may take place and side by side loss of some accessory proteins and alterations in the configuration of the two particles may also take place.

(*d*) 5 S RNA of non-nucleolar origin is incorporated with the 32 S RNA to form a large 65 S subunit which is in due course of time converted into a 60 S or larger subunit of ribosome.

(*e*) 18 S RNA piece produces 40 S or smaller subunit of ribosome.

(*f*) The ribosomal subunits along with residual accessory proteins leave the nucleoplasm and come to cytoplasm through annuli of nuclear envelope. In the cytoplasm the two ribosomal subunits (60 S and 40 S) become associated to form a whole ribosome (Fig. 10.27).

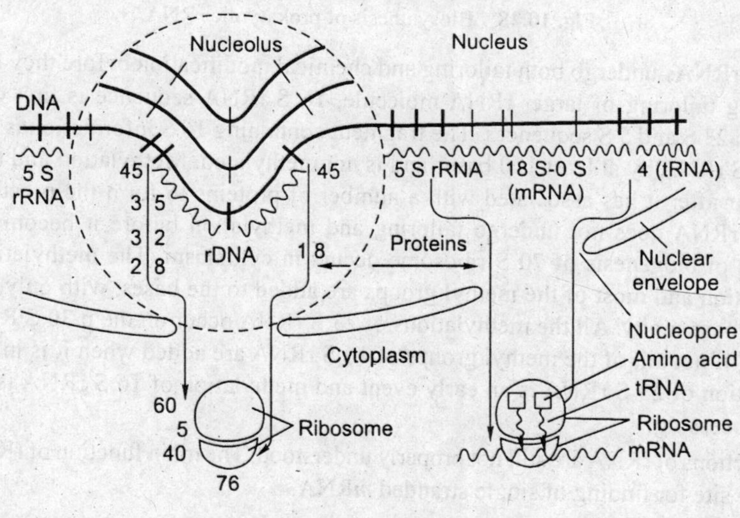

Fig. 10.27 Diagram of transcription of different types of RNA in the nucleus and their transportation to the cytoplasm.

As regards the biosynthesis of prokaryotic ribosomes, enough informations are available about their protein and RNA components. According to Smith *et al.* (1968), genes coding for 5 S, 23 S and 16 S *r*RNAs found in 70S ribosomes of bacteria and according to M Nomura and Associates (1969) 16 S, 23 S and 5 S RNA sequences found in 50 S ribosomes of *Escherichia coli* are tightly clustered in the region of bacterial DNA and are present in several copies. The primary transcript of each repeat unit in a precursor-*r*RNA (Pre *r*RNA) of 30 S (p 30 S) which is cleaved to produce p16 S, p 23 S, and p5 S RNAs, which are immediate precursors of mature 16 S, 23 S and 5S *r*RNAs respectively.

Normally 30 S Pre *r*RNA molecule is not observed in wild type cells because the cleavaging occurs while p 30 S is still in the process of transcription. (Fig. 10.28).

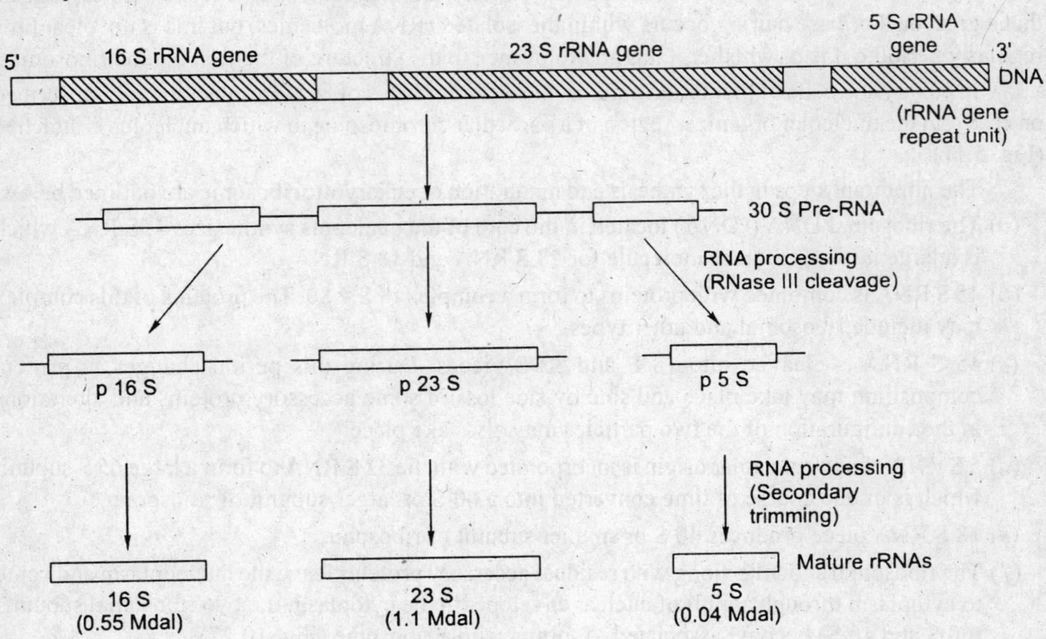

Fig. 10.28 Biosynthesis of prokaryotic *r*RNA.

The pre-rRNAs undergo both tailoring and chemical modification before they become mature rRNAs. During tailoring of larger rRNA molecule, 16 S rRNA sequence is first cleared off and separated from 23 S and 5 S sequences. The fragment containing 16 S information is still larger than the mature 16S rRNA by at least 100 bases and is not methylated. Methylation and tailoring of this molecule occur after it has associated with a number of proteins to form the precursor ribosomal subunits. 5 S rRNA does not undergo tailoring and methylation before it becomes mature. The whole process of biogenesis of 70 S ribosome occurs in cytoplasm. The methylation thus occurs after transcription and most of the methyl groups are added to the bases, with only a few added to the 2-OH of ribose moity. All the methylations of 23 S rRNA occur on the p 30 S RNA component whereas most, if not all, of the methyl groups of 16 S rRNA are added when it is in a mature form. Thus methylation of 23 S rRNA is an early event and methylation of 16 S rRNA is a late event in the rRNAs processing.

The functions of rRNA are still not properly understood. The main function of rRNA is evidently to serve as the site for finding of single stranded *m*RNA.

Transfer RNA

When a cell homogenate containing 10^{-2}M of Mg^{++} is centrifuged at high speed (100,000 Xg for 120 minutes) the RNAs of high molecular weights bound to ribosomes are sedimented and the supernatant liquid so obtained contains a fraction of RNA which is called transfer RNA or tRNA. Earlier, it was called soluble RNA (sRNA). RNAs available in the supernatant liquid have a *"transfer function"* and act as "adaptors" (or connectors) between the amino acids and messenger RNA. Transfer RNAs make up about 15% of the total cellular RNA. Transfer RNAs are the small natural nucleic acids with sedimentation coefficient of 4. The nucleotide chain length is remarkably uniform for all the tRNAs of prokaryotes and eukaryotes, ranging from 76 to 85 nucleotides. The cells probably contain more than 60 different types of tRNA molecules, molecular weight ranging from 25,000 to 30,000 daltons (one dalton = Atomic weight of hydrogen). Each molecule has a specificity for a given amino acid to which it can attach by a covalent bond. Holley *et al.* (1965) showed that tRNA which is specific for alanine in yeast cells consists of 77 nucleotides (mol. wt = 26,600) including 8 A, 12U, 25G, 23C and 9 unsaturated bases (mostly methylated bases). Medison *et al.* (1967) have shown that tyrosine tRNA consists of 78 nucleotides and Zachan *et al.* (1967) have shown that serine tRNA is composed of 85 nucleotides. The A/U and G/C ratios approach equality, so that there is a possibility for considerable base pairing within the molecule. tRNA is actually a single polynucleotide chain which is apparently bent in the middle and the two arms form loops. Thus the molecule becomes clover leaf like. All tRNA molecules have the sequence CCA at the 3′ end and G residue at 5 end.

(5′) G............ tRNA .... CCA(3′)

It is at the 3′ end of the chain (CCA-OH end) that the activated amino acid is attached during the protein synthesis. tRNA molecule appears as a single strand of polynucleotides (primary structure). Holley *et al.* (1965) first proposed the clover leaf pattern of tRNA molecule (Secondary structure) (Fig. 10.29).

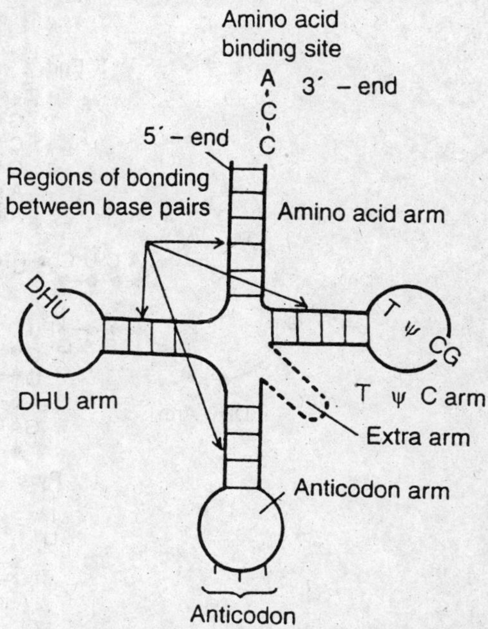

Fig. 10.29 Diagrammatic representation of secondary structure of tRNA. DHU = Dihydrouridine Ψ = Pseudouridine.

Lake and Beeman (1967) have shown that tRNA molecule consists of 3 folds giving it a shape of the clover leaf. According to clover leaf model the single polynucleotide chain of tRNA is folded upon itself to form 5 arms. As a result of folding 5′ and 3′ ends of the polynucleotide chain come near each other. Some of the arms may be differentiated into stem and a loop. In the stem region juxtaposed bases are complementary and paired but in loop there is no pairing. As regards the structure and function of tRNA molecule the following general conclusions have been made (Fig. 10.30):

(*i*) The sequence of bases is such that in each case the molecule can form a hydrogen bonded "clover leaf" configuration in which CCA bases at 3′ end stick out. The last residue, adenylic acid (A), is the amino acid attachment site. At the 3′ terminal, the sequence CCA — OH is added to all tRNAs post transcriptionally.

(*ii*) All tRNAs have one part (right loop of the chain) which is common and which consists of seven unpaired bases including pseudouridine (Ψ). It is known as the TΨC arm. It may be involved in the binding of the tRNA to the ribosome.

(*iii*) All the tRNA molecules contain 7-15 unusual bases, many of which are methylated or demethylated derivatives of A,U, G and C. Unusual bases are nucleotides of methylinosine (MeI), pseudouridines (ΨU or UH_2), methylated purines (methylated A and G), methylated pyrimidines such as methylated thyamine or ribothymine, 5 methyl cytosine and others. Not all these unusual bases are present in one tRNA.

(*iv*) There is one nucleotide triplet in the bottom loop of the chain which is different in all tRNAs examined so far. That is called anticodon. The anticodon is site which probably is complementary to the codon on the messenger RNA and is the site of base pairing between tRNA and mRNA. The nucleotide to the 5′ side of anticodon is always U and that to the 3′ side is always a modified purine.

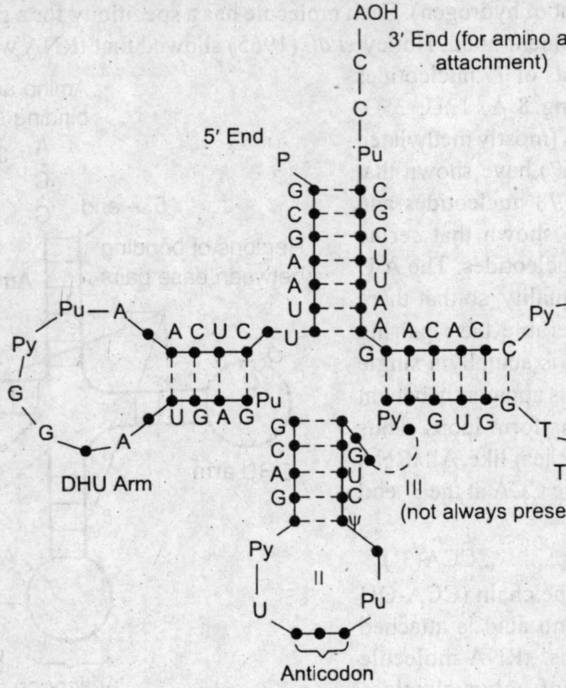

Fig. 10.30 tRNA molecule.

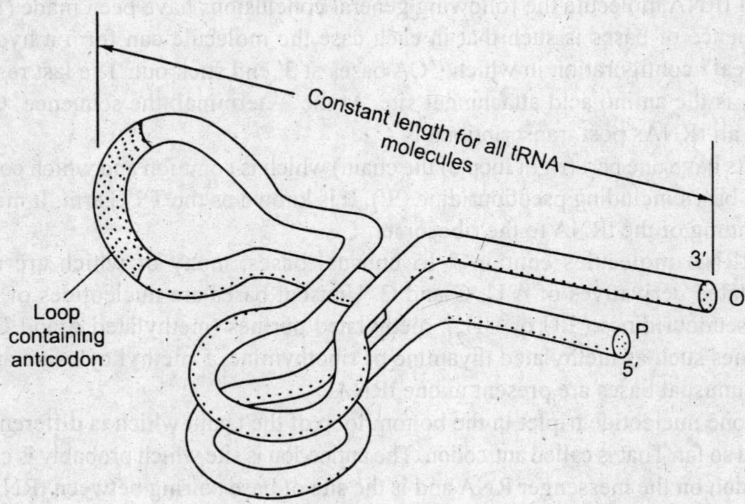

Fig. 10.31 A mode of tRNA.

(*v*) In the left loop of all tRNA there occurs DHU (Dihydrouridine) base. The DHU arm has a site for the recognition of the amino acid activating enzymes.

(*vi*) One of the most important structural similarities among the clover patterns of various tRNA molecules is that the overall distance from CCA at one end to the anticodon at the other end appears to be constant (Fig. 10.31). The difference in nucleotide numbers in different molecules being compensated for by the size of the little loop or "extra arm" located between the right loop and the bottom limb (Figs. 10.29 and 10.30). Unusual bases, such as, dimethyl guanosine, dime A, MeG are located in regions not forming hydrogen bonds. Dimethylguanosine usually occurs in the corner between the left and bottom loops. This suggests that the role of usual bases is to help to determine the three-dimensional structure of tRNA which is of crucial importance.

Tertiary structure of tRNAs. The application of x-ray crystallography to the stable crystals of tRNA has revealed the tertiary structure of tRNA including the location of the major groups. The data revealed that all the double helical stems predicted by the clover leaf model do exist and, in addition, some hydrogen bonds bend the clover leaf model into a stable, rough L-shaped appearance, in which the amino acid binding CCA-OH group at 3' end of the chain is located at the opposite end from the anticodon loop.

Biosynthesis of tRNA

Unlike mRNAs the tRNA molecules are modified extensively. In deed, mature tRNAs not only have modified bases but are considerably shorter than the primary gene transcript. In both prokaryotes and eukaryotes primary gene transcripts called *Precursor-tRNAs* (*Pre-tRNAs*) are produced which are later clipped and trimmed to produce mature tRNA molecule.

In prokaryotes pre-tRNAs contain either single *t*RNA molecule with extra leader and trailor sequences at the 5′ and 3′ ends or contain several species of tRNA molecules each. In the latter case the tRNA molecules are separated by spacer nucleotide sequences. At least two enzymes take part in the processing of pre-tRNAs:

(*i*) *Ribonuclease P* catylysing the removal of leader sequence at 5′ end and

(*ii*) *Ribonuclease Q* catalysing the removal of 3′ trailor sequence. Another enzyme is also involved in the removal of spacer sequences from multi tRNA precursor to liberate tRNA molecules.

In eukaryotes the pre-tRNAs are larger by some 15-35 nucleotides than the mature tRNAs. The processing of pre-tRNAs presumably occurs, although, the enzymes involved and the sites of their action are not yet known.

The transfer RNAs show specificity in attaching to amino acids (Fig. 10.32). So, there must be at least 20 different tRNAs in the cytoplasm of a cell.

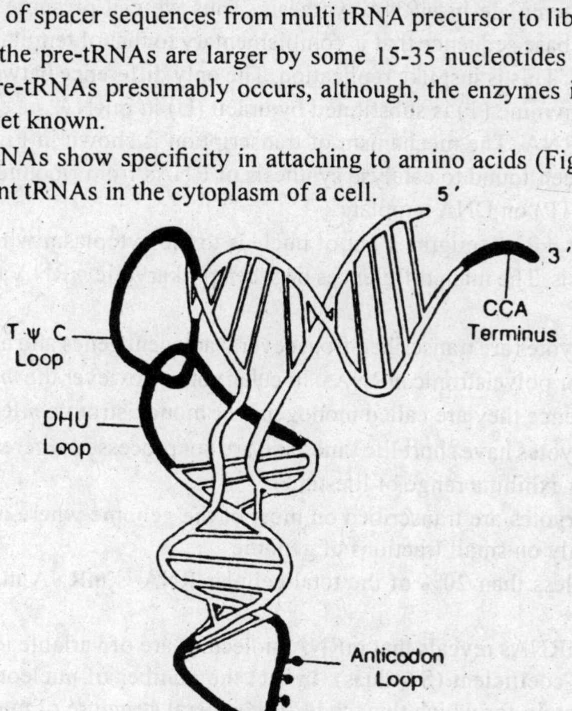

Fig. 10.32 Tertiary structure of tRNA.

The function of tRNA is to carry a particular amino acid at a definite place on the mRNA. Four recognition sites in tRNA to achieve this purpose are as follows (Fig. 10.32):

(*i*) *The amino acid attachment site* which is CCA sequence at the 3′ terminal.

(*ii*) *Anticodon* which is a group of middle three bases on the bottom loop of tRNA molecule. Anticodon of tRNA recognises three complementary bases of a codon in mRNA molecule.

(*iii*) *Ribosome recognition site* which is common to all tRNA molecules and is found in TΨC loop. This consists of GTΨCG sequence.

(*iv*) *Amino acid activating enzyme recognition site* or *synthetase site* found in left arm which activates the enzyme and charges specific amino acid with tRNA.

Messenger RNA

Messenger RNA. Jacques Monod and Francois Jacob (1961) of Pasteur Institute in Paris coined the term "messenger RNA" to describe the template RNA that carried genetic information for protein synthesis from nuclear DNA to ribosomes in the cytoplasm, Messenger RNA makes up less than 5% of the total cellular RNA. Messenger RNA is normally attached to the smaller subunits of Ribosomes and the length of mRNA is such that it connects a number of ribosomes and, in deed, such clusters of ribosomes called polyribosomes or polysomes have been observed with electron microscopes. *m*RNA is assembled on a DNA template. Since DNAs of different organisms differ only in the sequence of their bases, mRNAs which are formed from them must reflect different base sequences. The mRNA is complementary copy of DNA. This was demonstrated experimentally in 1960 by Marmur and Lane. They showed that if DNA molecules are heated to 100°C, the two polynucleotide chains separate because the H bonds holding the base pairs of the opposite strands are disturbed. This process is known as *'DNA denaturation'*. If the temperature is lowered, the complementary strands again unite. This is *renaturation* of DNA. These investigations also showed that isolated DNA strands could make hybrid or the *m*RNA. Several recent experiments indicate that only one strand of double stranded DNA molecule acts as template in *m*RNA synthesis. Thus we can presume that *m*RNA is single stranded. *m*RNA shows base sequence that is complementary to that of template DNA. This process is called "*transcription*". This is just like replication. The only difference between the two processes is that in transcription thymine (T) is substituted by uracil (U) in mRNA, *i.e.*, opposite adenine (A) of DNA comes U in mRNA. The mechanism of transcription is shown in Fig. 10.25. The enzyme RNA-polymerase has been found to catalyse synthesis of RNAs from ribonucleoside triphosphates (ATP, GTP, UTP, and CTP) on DNA template.

In eukaryotes, the *m*RNA migrates out of nucleus to the cytoplasm where the ribosomes are active in protein synthesis. The main differences between prokaryotic *m*RNA and eukaryotic mRNA are as follows:

(*i*) *m*RNAs of prokaryotes are transcribed from several adjacent genes and as such the *m*RNAs are called polygenic or polycistronic *m*RNAs. In eukaryotes, however the *m*RNAs are transcribed on single genes hence they are called monogenic or monocistromic *m*RNA.

(*ii*) *m*RNAs in prokaryotes have short life times and are not processed whereas those of eukaryotes are processed and exhibit a range of life-times.

(*iii*) *m*RNAs of prokaryotes are transcribed on most of the genome where as those of eukaryotes are transcribed only on small fractions of genome.

At any one time less than 20% of the total cellular RNA is mRNA in either prokaryotic or eukaryotic cell.

Examination of mRNAs reveals that mRNA molecules are of variable length with fairly wide range of sedimentation coefficient (S values). In fact the number of nucleotide pairs in a gene is related to the size of protein for which they code. The general structure of mature mRNA molecule is shown in Fig. 10.33. All messenger RNAs are transcribed from the structural genes, *i.e.*, genes that code for proteins. Each mRNA has the following three main parts (Fig. 10.33):

(*i*) *Leader sequence* at 5′ end, the length of which may vary in different mRNAs and within this sequence is present the information for the ribosome to initiate protein synthesis at the correct position. The leader sequence is not involved in the translation process (protein synthesis),

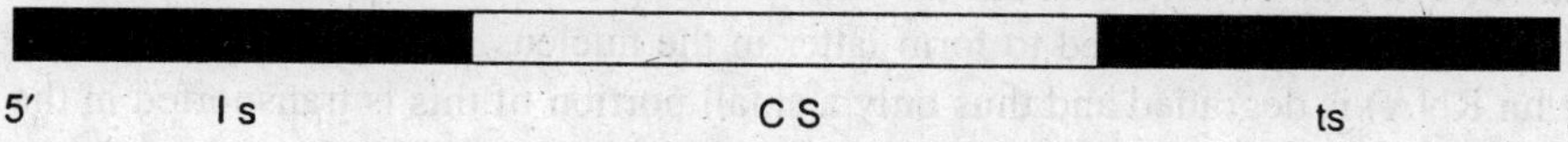

Fig. 10.33 Parts of mRNA: 1s = leader sequence, CS = protein coding sequence, ts = trailer sequence.

(*ii*) *Protein coding sequence* following the leader sequence which is translated into the amino acid sequence of polypeptide, and

(*iii*) *Trailer sequence at 3′ end* which is not translated and is variable in length in different mRNAs. The polycistronic mRNAs found in prokaryotes have leader and trailer sequence on the sides of the coding sequences.

The DNA of eukaryotic chromosome is organized into nucleosome structures. Many studies have shown that active chromatin is more sensitive to deoxyribonuclease attack than the non-transcribing chromatin. This indicates that transcribed segments of chromatin must have a different structural conformation than the inactive chromatin. This is, however, not due to any major change in the nucleosome structure. Unlike, prokaryotic mRNAs, mRNAs, of eukaryotes are modified at both 5′ and 3′ ends. The modifications occur after transcription under the influence of specific enzymes. Messenger RNA has the following structural features:

1. ***Cap.*** At the 5′ end of the mRNA strand there occurs a cap which is blocked by methylated structure (any of the four nucleotides with 2-OH methyl ribose). Messenger RNA without cap shows poor binding of ribosomes and so the cap of mRNA influences the rate of protein synthesis (Fig. 10.34).

Fig. 10.34 The cap at 5' end mRNA molecule.

2. ***Non-coding region 1 (NC1).*** Next to cap is non-coding region of 10-100 nucleotides which is rich in A and U bases and does not translate into protein.
3. ***The Initiation Codon.*** This is AUG codon which initiates the polypeptide chain.
4. ***The Coding region.*** Next to initiation codon comes the coding region which consists of about 1500 nucleotides on the average. This translates a particular protein molecule.
5. ***Terminating codon.*** Next to the coding region there is termination codon which terminates protein synthesis on mRNA. In mRNA of eukaryotes, codons UAA, UAG and UGA act as termination codons.
6. ***Non-coding region (NC2).*** This region lies next to termination codon which consists of 50-150 nucleotides. This region does not translate into protein. This region contains AAU sequence in all the cases.
7. ***Poly A sequence.*** At the 3′ end there is a polyadenylate or poly-A (as AAAAA-) sequence which initially consists of 200-250 nucleotides. Poly-A sequence is added in the nucleus before mRNA is transferred to cytoplasm. Poly-A sequence becomes reduced in extent with the age.

Split Genes and mRNA Production in Eukaryotes

In eukaryotic nucleus a rapidly labelled RNA, called heterogenous nuclear RNA (hn RNA) species can be isolated which is different from rRNA and tRNA or their precursors. The hn RNA is heterodispersed in length and is much longer than the cytoplasmic *m*RNA and has both 5′ and 3′ poly- A modifications characteristic of cytoplasmic *m*RNAs. The hn RNA acts as pre- mRNA or precursor to the functional *m*RNAs, the former being processed to form latter in the nucleus. A large fraction (approximately 90% of the hn RNA) is degraded and thus only a small portion of this is transported in the form of mRNA. In 1977, A.J. Jeffreys and R.A. Flavell discovered that some 600 base pairs (bp) sequence was inserted within the coding sequence for 146 aminoacids β globin chain in rabbit. The insert has been named an *intervening sequence* or *intron* and the coding sequences the *exons*. The introns are transcribed and are not translated. An intron is a sequence in gene that separates coding sequences of genes. A large number of coding genes contain introns in eukaryotes. In higher eukaryotes, they are more extensive than in lower eukaryotes. Not all eukaryotic protein coding genes have introns. Histone genes of invertebrates are uninterrupted. In general, there is no definite pattern to show as to which type of genes are interrupted and which type of genes are not. Simple model for the production of functional mRNA is presented in the Fig. 10.35. At first, both coding sequences and introns are

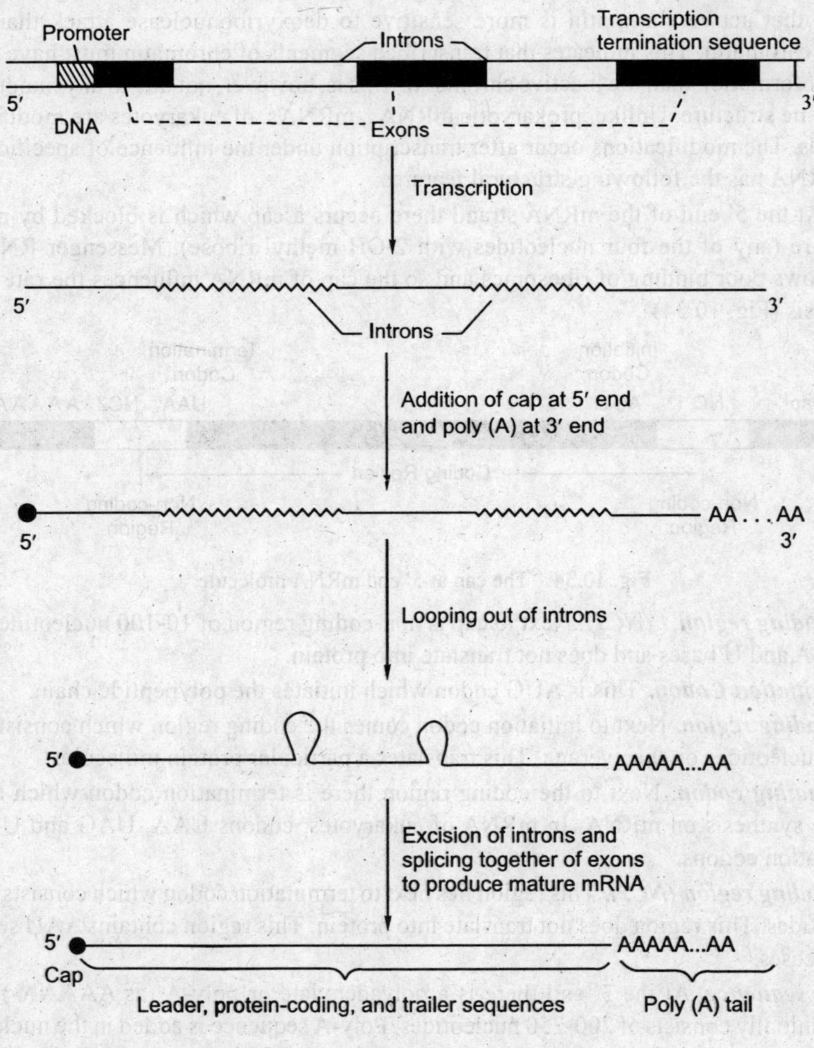

Fig. 10.35 Hypothetical model of mRNA biosynthesis.

transcribed but each intron is subsequently removed in such a way that the adjacent coding genes are spliced together to form a continuous sequence. The process is summarized as under:

(*i*) RNA polymerase binds to the promoter and transcribes on the DNA continuously until a termination is reached, *i.e.*, leader, coding sequences as well as introns and trailer are transcribed.

(*ii*) Transcript becomes capped at the 5′ end. The transcript is then cleaved near the 3′ end and a poly-A tail is added to the new 3′ end to produce a precursor *m*RNA (pre-*m*RNA) which is longer than the mature *m*RNA because of presence of introns. The pre-*m*RNA molecule constitutes hn RNA.

(*iii*) The pre-mRNA molecule is processed to remove introns which are looped out and sub sequently the loops are removed by nucleases. The new adjacent coding sequences are then linked or spliced together. This is called splicing of introns. It is known that all introns start with GU and end with AG and these sequences are presumably crucial for the correct splicing.

Informosome. Many investigators like Spirin, Beltisina and Lerman (1965), Perry and Kelley (1968) and Henshaw (1968) have reported that in certain eukaryotic cells the messenger RNA does not enter the cytoplasm as a naked RNA strand but often remains ensheathed by certain proteins. Spirin has proposed a term informosome for mRNA-protein complex. The proteins of informosome provide stability to the mRNA and protect mRNA from degrading action of ribonuclease enzyme.

Table 10.4. Comparison of Different Types of RNA

	rRNA	*mRNA*	*tRNA*
Percentage of total RNA in a cell	about 80%	3–5%	10–20%
Sedimentation coefficient	28 S, 18 S, 5.8 S 5 S, 23 S, 16 S, 5 S	8 S	3.8 S
No. of nucleotide in each RNA molecule	5 S rRNA–120, 16–18 S RNA: 1,600 to 2,500 nucleotides, 23, 28 S RNA: 3,200 to 5,500 nucleotides	In *E. coli*; 900–1,500 nucleotides	73–93 nucleotides
Mol. weight	23 SRNA: 1.1×10^6 30S RNA: 0.55×10^6	5,00,000	25,000–30,000
Unusual bases	Small number of methylated bases	A few unusual bases	Large number of unusual bases.
Function	Unpaired bases may bind mRNA and tRNA to ribosomes.	Codes the genetic information from nuclear DNA and translates that in the cytoplasm forming specific proteins	Carries aminoacid to mRNA where protein synthesis occurs

QUESTIONS

1. Discuss in brief the structure of DNA and compare it with that of RNA.
2. What is the difference between nucleoside and nucleotide? Draw the four deoxyribonucleosides and their corresponding deoxyribonucleotides.
3. Single Polynucleotide chain contains the base sequence G—U—C—A—G. State whether it is from DNA or RNA and why?

4. If the base sequence in a Polynucleotide chain is G—C—A—A—T—G—C—A, what will be the sequence of bases on the complementary chain and why?
5. Give a brief account of semiconservative duplication (replication) of DNA.
6. What kind of evidence indicates that:
   (*i*) DNA can reproduce itself and
   (*ii*) The nucleotides occur in matched pairs in DNA molecule?

   or

   Discuss at least two experiments which demonstrate the semiconservative mode of DNA replication.
7. Write short notes on:
   (*i*) Pyrimidines,
   (*ii*) Purines,
   (*iii*) Polynucleotide,
   (*iv*) Double helix,
   (*v*) Watson and Crick,
   (*vi*) Kornberg, and
   (*vii*) Khorana.
8. What is the relationship between DNA and chromosome structure? Discuss the different concepts on the arrangement of DNA in the chromosome.
9. Describe the structure and functions of different types of RNA in the chromosome.
10. Discuss some examples where:
    (*i*) DNA is single stranded.
    (*ii*) RNA is double stranded.
11. How are the duplication of genetic RNA of viruses and DNA duplication in bacteria accomplished?

11 CHAPTER Division of Cell

One of the important properties of the protoplasm is reproduction or "self-perpetuation device". In reproduction new protoplasmic units are created which resemble the old protoplasm. Reproduction at any level requires addition of sufficient amount of raw materials to the protoplasm. The addition of the specific raw materials to the protoplasm is possible only when the metabolic activities are regulated properly. The controlled metabolism results in the increase in the protoplasm of a cell. This, in turn, may be followed by increase in the number of cells, *i.e.*, reproduction of cells.

Cell Cycle

In both unicellular and multicellular eukaryotes, the cell reproduction is a cyclic process of growth, nuclear division and usually cytoplasmic division. This is called ***cell cycle***. In somatic cells, the cell cycle consists of the following four phases (Fig. 11.1):

1. G_1 (gap1) phase,
2. S (synthesis) phase,
3. G_2 (gap 2) phase, and
4. M (mitosis) phase.

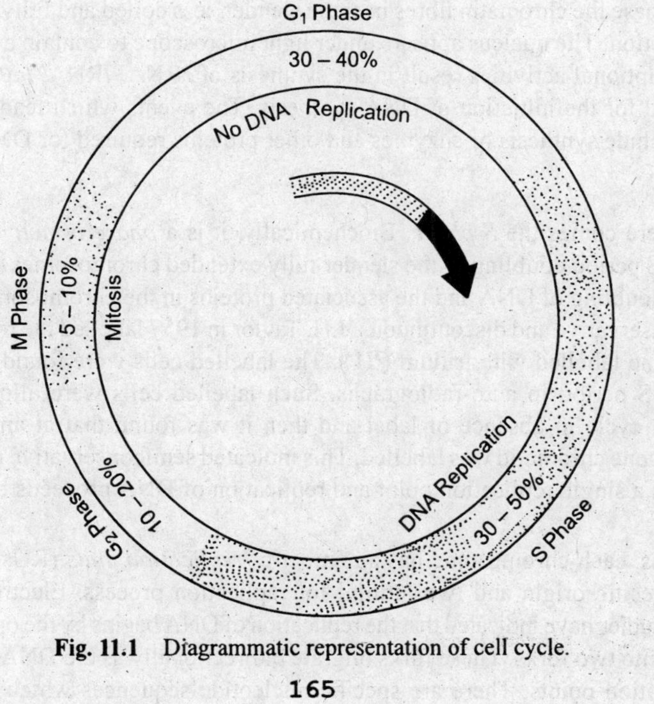

Fig. 11.1 Diagrammatic representation of cell cycle.

The first three phases, *i.e.*, G_1, S and G_2 are described in the ***interphase*** or ***metobolic phase*** of the cell.

The G_2 phase is followed by M phase during which the nucleus and cytoplasm divide to produce two daughter cells. The daughter cells then begin a new cell cycle. The division of cells is always followed by subsequent enlargement of daughter cells. By making direct or indirect measurement of growth characteristic of organism, it is possible to calculate a "***generation time***" which is the average time that has elapsed between successive cell doublings during the exponential phase of growth. The generation time for the bacterium *Escherichia coli* under favourable growth conditions can be as low as 20 minutes while for mammalian cells in tissue cultures it may be as many hours. Much of the information about the molecular biology of the cell cycle has come from the studies based on cell cultures. For eukaryotic cells, in general, the relative time spent in each of the four phases of cell cycle varies, as the generation time itself does.

G_1 Phase

The G_1 phase is set in immediately after the cell division. It is characterised by a change in the chromosome from the condensed mitotic state to the more extended interphase state and by a series of metabolic events leading to initiation of DNA replication. G_1 phase of cell cycle varies in length from cell to cell within the same cell population. The length of this phase is more than the other three phases. This period represents in general 25 to 40% of the generation time of a cell. The cause of variability in G_1 is not known, although there is an evidence to suggest, that the protein content of the cell possibly determines the length of G_1.

The G_1 is a very significant phase of cell cycle in the sense that the cells which stop dividing normally remain arrested in this phase. Within G_1 phase, there is a definite check point at which DNA synthesis is initiated and once the biochemical events associated with that point have occurred the cell proceeds towards division. There are specific molecular signals at that point which decide whether a cell has to continue its proliferation or it has to specialize into a permanent or non-dividing cell.

During G_1 phase the chromatin fibres become slender, less coiled and fully extended and more active for transcription. The nucleus appears under light microscope to contain a network of delicate fibres. The transcriptional activities result in the synthesis of *r*RNA, *t*RNA, *m*RNA and a series of molecules required for the initiation of DNA synthesis. The events which lead to the initiation of DNA synthesis include synthesis of enzymes and other proteins required for DNA synthesis.

S Phase

After G_1 phase there comes the *S phase*. Biochemically, it is a *phase of active DNA* and *histone synthesis*. During S period doubling of the slender fully extended chromosomes takes place which is accomplished by doubling of DNA and the associated proteins in the chromosomes. The replication of DNA is semiconservative and discontinuous. J.H. Taylor in 1957 labelled the root tip cells of broad bean with thymidine labelled with tritium (^3H). The labelled cells were found to contain labelled chromatin during S period in auto-radiographs. Such labelled cells were allowed to go through one more division cycle in absence of label and then it was found that at any point along each chromosome, only one chromatid was labelled. This indicated semiconservative replication of DNA. In bacteria there is a single replication point and replication of DNA proceeds bidirectionally from that point.

In eukaryotes, each chromosome has a number of *replication units* (RUs) or replicons, each of which has a specific origin and two termini for replication process. Electron microscopy and autoradiographic studies have indicated that the replication of DNA begins by the opening of a "bubble" in DNA resulting into two forks. These forks migrate bidirectionally as the DNA is synthesised until they reach termination points. There are specific nucleotide sequences which mark the initiation and termination points. Careful examination of electron micrographs of replicating chromosomes shows the spacing between the replication units which, in a wide range of eukaryotes, ranges from 7 to 100 mm (30,000 to 300,000 base pairs). After the termination process, the adjacent replication forks are fused. Since DNA replication is dependent on protein and RNA synthesis for the overall replication of chromosome, it is necessary that new proteins must be synthesised. Although some of the histone protein is synthesised during G_1 phase, most of it is synthesised during S phase. Further, synthesis of ribosomal RNA must continue from G_1 to S phase if DNA synthesis is to start.

The eukaryotic chromosomes consist of DNA-histones complexes, called nucleosomes. When the DNA is replicated and new histones are synthesised the two are complexed to form nucleosomes. The distribution of old and newly synthesised histones in the nucleosomes of newly replicated DNA is not clearly understood but it is clear that the old octamers of histones are conserved from generation to generation and the octamers of newly synthesised histones are used at the replication forks to form new nucleosomes.

G_2 phase

G_2 phase follows the S phase. The metabolic significance of this phase is not fully understood as yet. This phase represents to some 10 to 25% of the generation time of the cell. In G_2, the chromosome consists of two strands or chromatids. The cells that are arrested at the transition between S and G_2 will have twice the usual amount of DNA. Tetraploid nuclei are occasionally found in various tissues, as for example, in liver cells, heart muscles etc.

The critical event at G_2 check point is concerned with the condensation of chromatin. The condensation process involves high order folding of the chromatin fibre but the mechanism for this is poorly understood. The chromatin condensation is associated with accumulation of a cytoplasmic substance. The existence of chromatin condensing factor has been demonstrated by R.T. Johnson and P.N. Rao (1970, 74) by using Sendai virus to fuse He La cells (a type of cell derived from human cancer cell) at different stages of cell cycle. When a mitotic cell was fused with a cell at G_2 phase, chromatin condensation took place at once in G_2 nucleus, producing normal looking chromosomes. When a mitotic cell was fused with a cell at G_1 phase chromatin condensation was demonstrated in G_1 nucleus also. Similarly, condensation of chromatin in S nucleus was also observed when mitotic cell was fused with a cell at S phase, although multiple fragments of condensed chromatin instead of complete chromosomes were seen. The chromatin condensation is thought to be associated with, and perhaps due to, phosphorylation of histone H_1 (Marks, Paik and Borun, 1973; Ord *et al.* 1975; E.M. Bradbury, 1975).

The syntheses of RNAs and proteins occur during most of the cell cycle, but the two processes may be inhibited during S phase and almost completely suppressed during late prophase. Another important event that takes place during G_2 period is the synthesis of the tubulin proteins which is later polymerized to produce microtubules that make up spindle apparatus during M phase.

M phase

M phase follows G_2 phase. During this phase the cell divides into two daughter cells. The chromosomes are duplicated during interphase and they are distributed tc the progeny cells by division process. After M phase, the resulting daughter cells then enter the G_1 phase of next cell cycle. The cells which after completion of mitosis do not enter G_1 are referred to as G_0 cells.

During M phase, RNA synthesis stops at late prophase and resumes at telophase. The protein synthesis drops drastically and during that period RNA synthesis also stops. These changes in the synthetic activity may be attributed to the non-availability of transcription sites owing to highly condensed state of the chromosomes. If no mRNA is available to ribosomes, protein synthesis cannot take place. The nucleolus and nuclear wall break down goes side by side with the inhibition of RNA and protein synthesis. These structures are reformed in daughter nuclei in late telophase when the synthesis of RNA and proteins resume. The molecular signals for these changes are not known. The behaviour of chromosomes during different phases of cell cycle, as illustrated by De Robertis *et al.* (1975) are shown in (Fig. 11.1).

Types of Cell Division

The cell divisions are of two types.

1. **Direct cell division or Amitosis.** In this process, the cytoplasm and the nucleus of a cell divide directly by constriction and form two nearly equal daughter cells (Fig. 11.2). Amitotic division is commonly seen in the cells which lack distinct nuclei and chromosomes, as for example, the bacterial cells (Fig. 11.2).

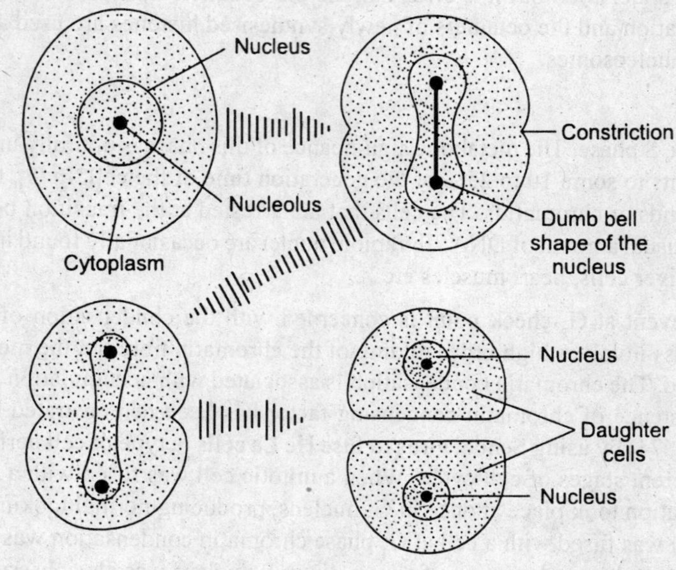

Fig. 11.2 Diagrammatic representation of amitosis.

2. **Indirect cell division**. This type of division completes in two distinct successive processes:
   (*i*) Division of nucleus (karyokinesis).
   (*ii*) Division of cytoplasm (cytokinesis).

   Indirect division are of two types:
   (*a*) *Mitosis or equational division.* It does not involve reduction in the chromosome numbers of nuclei.
   (*b*) *Meiosis*. This reduces the chromosome number of the nucleus to half.

MITOSIS

Walter Fleming (1878), a German biologist, was the first to study the cell division. He stated that the nucleus of a dividing cell undergoes a series of changes. To this process he called mitosis. Thus mitosis refers only to karyokinesis. In this process all the chromosomes undergo longitudinal doubling, the halves of the chromosomes then separate into two similar groups and the latter finally give rise to two daughter nuclei. In other words, the daughter nuclei have qualitatively and quantitatively the same types of chromosomes which the original nucleus possessed.

'Mitosis' in its loose sense has been used by the biologists to denote the division of the nucleus and cytoplasm both. Some cytologists define mitosis as a process involving equal division of the nucleus followed by equal division of cytoplasm. It is also called equational division. Mitosis commonly occurs in all the somatic cells of body.

The suitable materials for the study of this process in plants are the root tips of onion and marginal tissue of the young leaf of *Tradescantia* (Fig. 11.3).

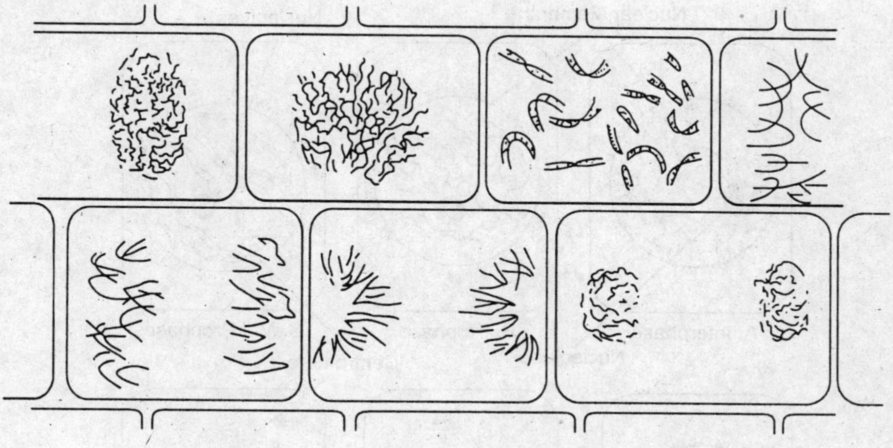

Fig. 11.3 Mitosis in cells of root tip.

The process of mitosis is described below:

1. Karyokinesis or Division of Nucleus

The process is continuous one, but for the convenience in description it is divided into several successive steps. These stages are as follows:

(*a*) Interphase or metabolic state or resting state,

(*b*) Prophase,

(*c*) Metaphase,

(*d*) Anaphase, and

(*e*) Telophase.

(*a*) **Interphase.** When the cell is not in division stage its nucleus is said to be in interphase or metabolic state (Figs. 11.4 A and 11.4 B). In this stage the nucleus has a definite nuclear wall. The karyolymph (nuclear sap) is very dense with one or more nucleoli and inconspicuous chromatic reticulum in it.

(*b*) **Prophase**. This is the first visible step in the nuclear division (Figs. 11.4 A and 11.4 B). During this stage, several changes take place in the nucleus. The nuclear reticulum and nucleoli become clear due to dehydration of the nuclear sap at this stage. Stainability of the reticulum increases. The chromatic network dissociates and the chromosomes become distinct. The chromosomes show relic coiling and appear as if they are longitudinally split into two parts. The two components of the chromosome are called chromatids. The two chromatids lie closely parallel and are still connected with each other at a single point, called the centromere or kinetochore. In the beginning, the chromatids of the chromosomes are spirally coiled. This type of coiling is known as relational or plectonemic coiling (Fig. 11.5) but uncoiling takes place at the end of prophase. Each chromatid consists of many deeply staining bodies, called the chromomeres. The chromomeres are held together by much finer thread named the chromonema. Probably, this duplication takes place during interphase, sometimes before prophase. The nucleolus remains attached to a chromosome near the centromere, but in the late stage it begins to disappear and simultaneously the matrix sheaths appear around the chromosomes. The chromatids now become shortened and thickened. The centrosome, which is found only in the cells of some lower plants and all the animals, divides in the early stage into two star shaped centrosomes

(Figs. 11.6 B and D). The centrioles move and come to lie on the two opposite sides of the nucleus. The disappearance of nuclear membrane finally terminates prophase.

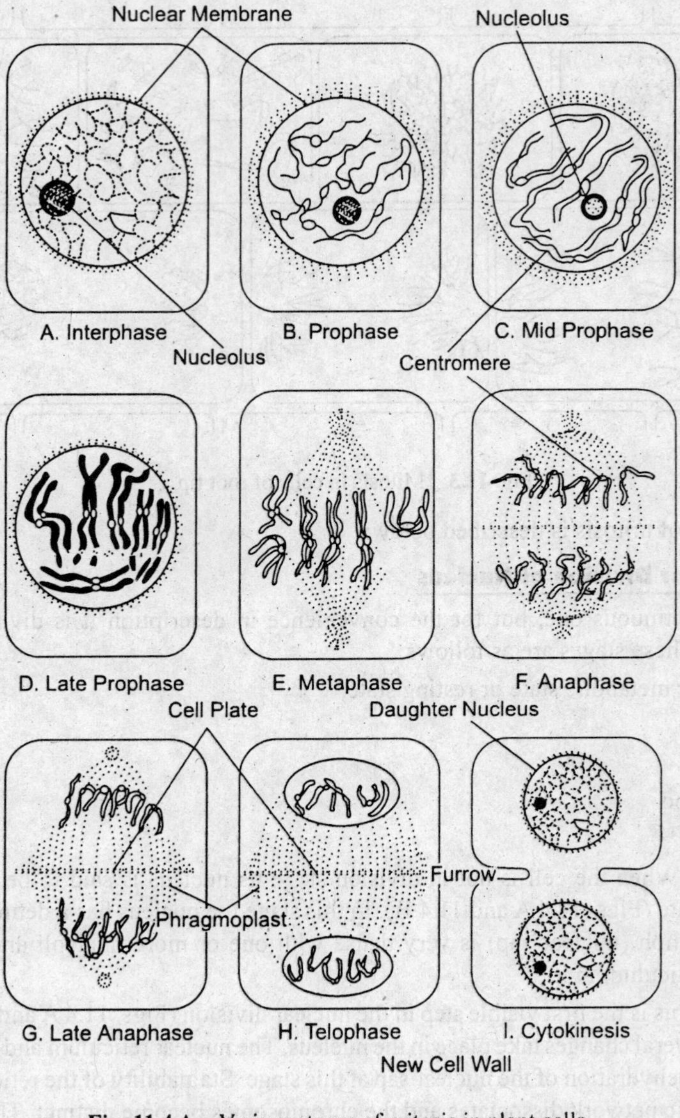

Fig. 11.4 A Various stages of mitosis in plant cells.

(*c*) **Metaphase.** The beginning of the metaphase is marked by the dissolution of nuclear membrane and simultaneous appearance of spindle fibres (Fig. 11.4 B and 11.6 D). The spindle is a structure that is formed during nuclear division or mitosis and is the part of the machinery responsible for equidistribution of the hereditary material into two daughter cells. It was long a subject of dispute whether the spindle fibres exist in the living cell and whether they are responsible for the anaphasic movements of the chromosomes. The reality of fibres in the living cells has now been proved. By electron microscopy, the spindle fibres appear to be microtubules. At metaphase stage, the chromosomes can be easily counted and their size and shape can be determined. The chromosomes at this stage become arranged in the equational plane of the cell and their kinetochores (centromeres) lie in the centre of the spindle. Some of the spindle fibres now become connected with the centromeres of the chromosomes. They are called chromosomal

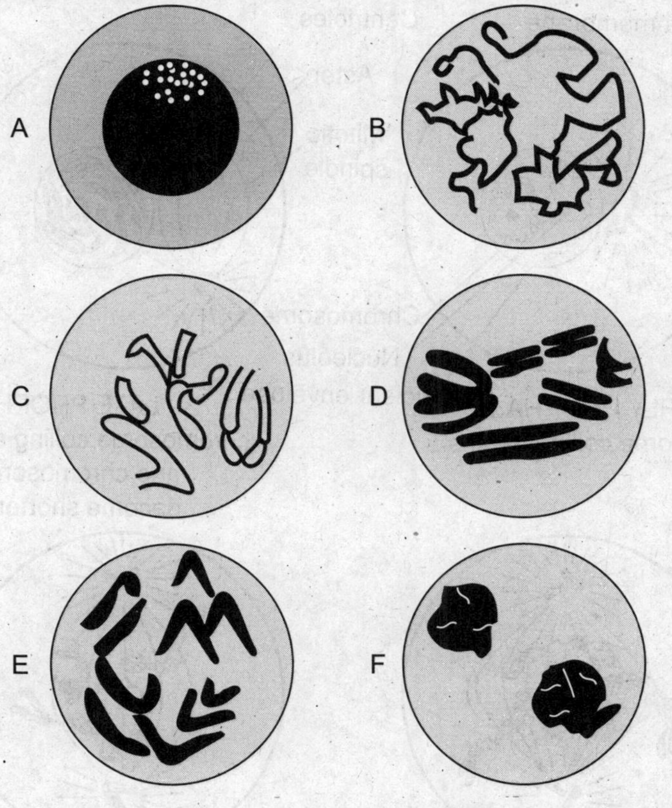

Fig. 11.4 B Stages of Mitotic Division in *Trillium* sp. (*a*) Metabolic phase, (*b*) Early prophase, (*c*) Late prophase, (*d*) Metaphases, (*e*) Anaphase, (*f*) Telophase.

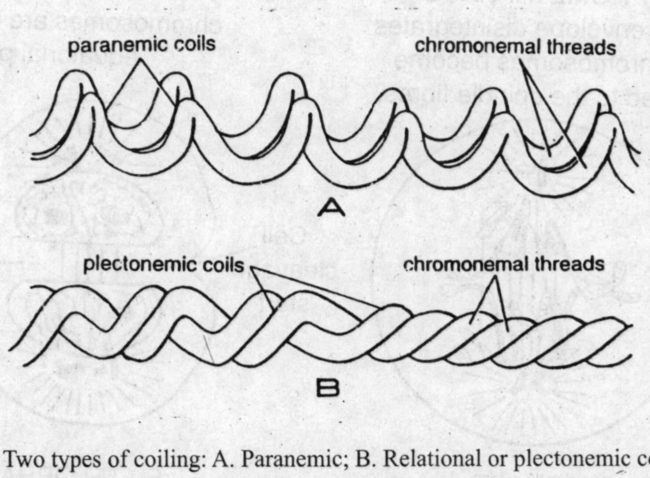

Fig. 11.5 Two types of coiling: A. Paranemic; B. Relational or plectonemic coiling.

fibres or half fibres. The other fibres run uninterruptedly between the two poles of the spindle. They are termed continuous fibres. The spindle attachment region determines the future shape of chromosome. If the centromere is located terminally the chromosomes will be rod shaped, if it is sub-terminal then the chromosomes will be 'j' shaped, and if it is median then the chromosome will be 'V' shaped at the anaphase. At the late metaphase, the centromere of each chromosome divides in the longitudinal plane forming two daughter centromeres which are attached to the

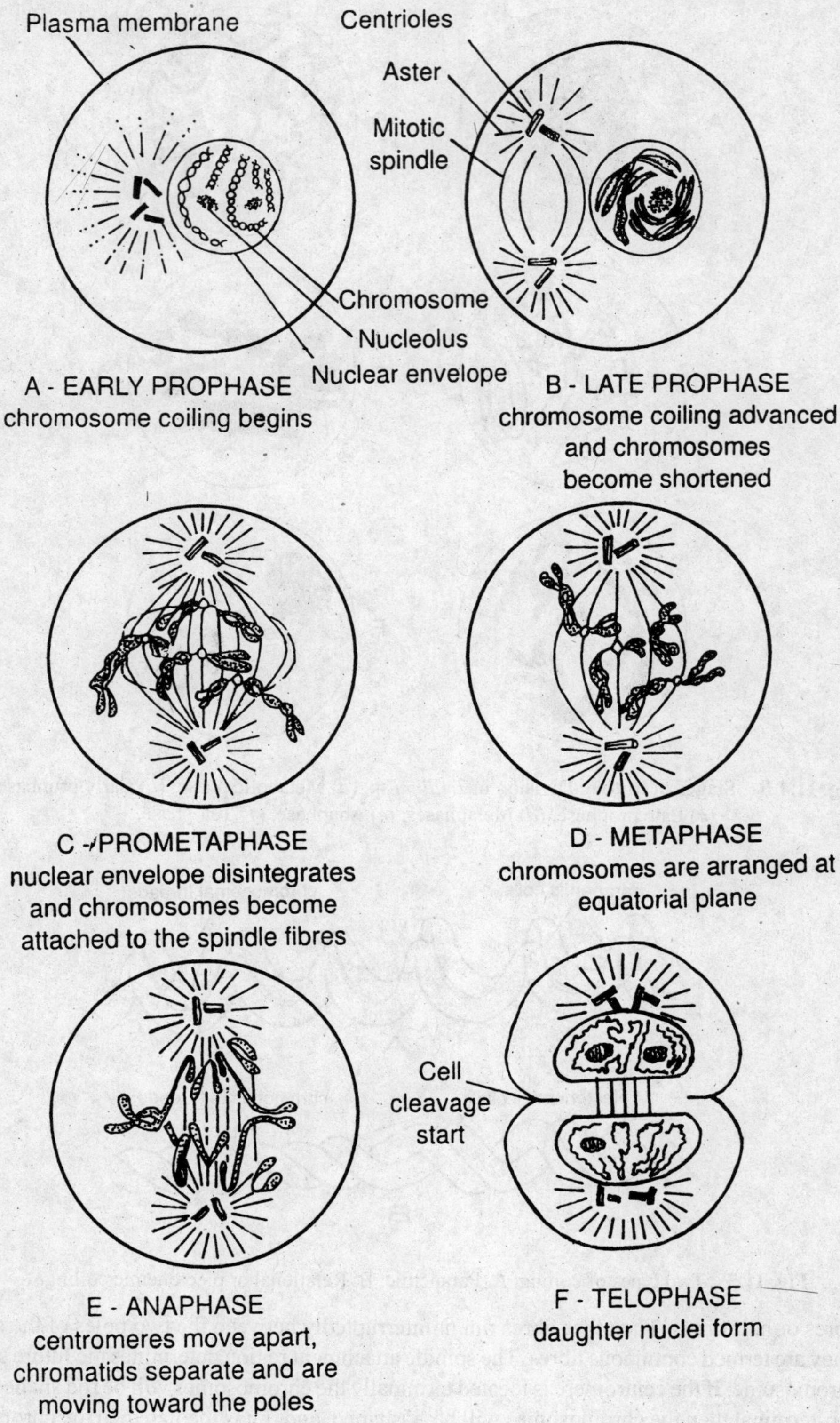

Fig. 11.6 Stages of mitosis in animal cells.

chromatids of their respective sides. Thus, the two chromatids become completely free from each other. One chromatid of each chromosome is attached with spindle fibre of one pole and the other chromatid is connected with the fibre of the opposite pole.

(*d*) **Anaphase.** The beginning of anaphase is marked by doubling of the centromeres and appearance of fibrils between the daughter centromeres. Now the sister chromatids of the chromosomes begin to move apart towards opposite poles of the spindle (Figs. 11.4 A and 11.4 B, 11.6 E). The movement of the chromatids appears to result from the interaction of special region on the chromosomes (centromeres) with the microtubules. It seems as if some sort of repulsive force develops between two sister centromeres. The nature of repulsive force is not clearly understood, but it seems possible that the movement is due to changes in the configuration of the protein chains which make up the spindle fibres. The spindle fibres are contractile in nature. This exerts a contractile force on the centromeres. The contractile force, probably, causes the anaphasic movement of daughter chromosomes (chromatids).

At the late anaphase stage, one of the two chromatids of each chromosome reaches to one spindle pole and the other moves in the opposite direction to the other spindle pole. During the anaphasic movement of the chromatids, the gel fibrils (interzonal fibres) between the two centromeres lengthen and the chromosomal fibrils contract and finally disappear. When the two sets of chromosomes reach to the opposite poles, the interzonal fibres of the spindle between the two poles form the 'stem body'.

(*e*) **Telophase.** The last visible stage of nuclear division is telophase. This follows the anaphase (Fig. 11.4 B and 11.6 F). At this stage of mitosis, the chromatids at each pole behave like daughter chromosomes. The daughter chromosomes begin to lose their matrix sheaths. This disappearance of matrix coincides with the reappearance of nucleolus. Spindle fibres disappear and nuclear membranes appear around the two opposite chromosomal sets. The telophase ends when the two daughter nuclei are completely organized from the two sets of the daughter chromosomes. The karyokinesis is now followed by cytokinesis.

2. Cytokinesis

The division of cytoplasm is known as cytokinesis. It is accomplished either through the formation of a cell plate in between two newly formed daughter nuclei or by means of peripheral furrowing.

(*a*) **Cell plate formation**. Earlier it was thought that the spindle fibres in the equatorial region of the cell after karyokinesis thickened to form an intermediate cell plate dividing the cytoplasm into two.

This cell plate later on splits longitudinally to form two cytoplasmic membranes for daughter protoplasts. The daughter protoplasts in plant cells then secrete calcium pectate which is deposited in between two membranes in the form of middle lamella.

In recent years it has been suggested that a transverse cleft or fluid film is formed as a result of disappearance of spindle fibres in the region of equational plate (Fig. 11.4). This fluid film is believed to be formed by fusion of many small vacuoles or droplets. The daughter protoplasts secrete calcium pectate in the cleft to form the middle lamella. This type of cell wall formation is exemplified by Bryophyta and Tracheophyta.

(*b*) **Furrowing.** In this method, a peripheral cleavage furrow appears gradually between the two daughter nuclei (Fig. 11.4 I). The furrow deepens and when the edges of the furrow meet in the centre of the cell, the protoplast is divided into two daughter cells. The daughter protoplasts then secrete intermediate cell wall. This type of cross wall formation is seen in bacteria, fungi and animal cells.

Significance of Mitosis

Mitotic division is a common method of multiplication of cells in the body. During mitosis the chromosomes split longitudinally and the chromatids of these chromosomes separate into two equal groups which finally form two daughter nuclei. In this way the same genetic constitution is maintained qualitatively as well as quantitatively in the chromosomes of the two daughter nuclei formed by this process.

MEIOSIS

In the sexually reproducing organism two sex cells or gametes containing haploid number of chromosomes (n) fuse and form a zygote. The zygote possesses a nucleus which contains double number of chromosomes. Such cells are commonly symbolised as diploid or 2n cells. If mitosis is the only method of nuclear division the number of chromosomes would be doubled in succeeding generations. But, the chromosome number for a particular animal or plant species is fixed. It is, therefore, necessary that the gametes must be provided with haploid chromosome number in order to keep the number of chromosomes of a given species constant in each generation. The number of chromosomes of a gamete is symbolised by *'n'* (haploid or gametic chromosome number). The mechanism which reduces the double (diploid or $2n$) complements of chromosomes to haploid set (n) in each sexual generation is called meiosis.

Weismann (1887) was the first to point out the reduction of chromosome number in the reproductive cells. Later Farmer and Moor (1905) proposed the term meiosis for the special type of cell division that reduces the chromosome number by half from diploid to haploid condition. The products of meiosis have half as many chromosomes as the parental cell possesses.

During meiosis, homologous chromosomes of a diploid nucleus pair, replicate only once producing $4n$ and undergo assortment so that each of the four daughter cells resulting from two successive divisions of the meiotic cell or meiocyte receives one representative of each chromosome set. Thus the diploid chromosome number is meiotically reduced to haploid number (n), characteristic for the gametes or haploid phase of the life cycle.

Types of Meiosis

The reduction of chromosome number takes place either before or after reproduction. In animals the reduction division occurs during germ cell formation, but in higher plants it occurs generally in the spore mother cells during sporogenesis (spore formation). On the basis of time at which meiosis takes place, three different types of meiosis have been recognised:

1. **Terminal or gametic meiosis.** It is found in animals and a few lower plants. This type of meiosis occurs just before the formation of gametes or gametogenesis.
2. **Initial or zygotic meiosis.** This type of meiosis is seen in plants with haplontic life cycle and it occurs in zygote just after fertilization.
3. **Intermediary meiosis or sporic meiosis.** It is a characteristic of species with diplo-haplontic life cycle. This type of meiosis takes place at sometime intermediate between fertilization and formation of gametes. It is involved in the production of spores or sporogenesis.

Meiocytes. The cells which divide by meiosis are called *meiocytes.* In animals the primary oocytes and primary spermocytes are meiocytes. In most of the higher plants the meiocytes are sporocytes (spore mother cells) giving rise to microspores or pollen grains (male sex) and megaspores (female sex). In haploid plants, the zygotes act as meiocytes.

Important materials for the study of this process in plants are young anthers of *Tradescantia, Datura,* maize, *Vicia* and onion.

Process of Meiosis

Meiosis completes in two successive divisions:

1. Meiosis I or reduction division or heterotypic division.
2. Meiosis II or homotypic division.

Both these divisions involve two steps:

A. Division of the nucleus or karyokinesis

B. Division of cytoplasm or cytokinesis

1. Meiosis First

It is very important division in which the chromosome number of a cell is reduced to half in daughter cells. It is also called heterotypic division or reduction division.

A. Karyokinesis

The nuclear division of first meiosis includes the following four successive stages:

(*i*) Prophase I,

(*ii*) Metaphase I,

(*iii*) Anaphase I, and

(*iv*) Telophase I.

(*i*) **Prophase I**. The prophase of heterotypic division is a long process and is further divided into five successive sub-stages for convenience in description.

These sub-stages are:

(*a*) Leptotene or Leptonema.

(*b*) Zygotene or Zygonema.

(*c*) Pachytene or Pachynema.

(*d*) Diplotene or Diplonema.

(*e*) Diakinesis.

Prior to onset of prophase I, the nucleus is said to be in metabolic stage. The metabolic nucleus is bounded with a definite nuclear membrane. The nuclear substance is very dense, and contains one or more nucleoli and inconspicuous chromatic reticulum. DNA synthesis takes place during the S-phase of premeiotic interphase, but the linkers remain in single state (Uhl, 1965).

Leptotene

It is a short process. At this stage, the nucleus size increases considerably. The chromosomes appear as long slender threads bearing numerous chromomeres in a bead like fashion (Fig. 11.7 A and 11.10A). The chromosomes are evenly dispersed in the nuclear field, but sometimes they are present only in one part of the nucleus leaving the other side completely empty. The latter condition is called "synizesis". Sharp termed it, "bouquet formation".

The chromosomes are widely despiralized showing only relic or minor coils. They appear as optically single chromonemal threads and the differentiation into chromatids is not yet recognizable although DNA is doubled during S phase. They are so fine that microscopic observations are difficult and even if they are double stranded, the two strands are entirely beyond the resolution limits of the light microscopes. In some cases, the nucleolus produces by budding process subsidiary nucleoli. There is a marked increase in RNA and protein synthesis during leptotene which accounts for the increase in size and budding of nucleolus.

Zygotene

At the beginning of this stage, the homologous chromosomes start pairing (Figs. 11.7B and 11.10B). Here morphologically and genetically similar chromosomes are generally referred to as homologues, but some cytologists hold the view that only those chromosomes should be called homologues which pair at zygotene stage. One chromosome in each pair of homologues is maternal, *i.e.*, it comes from the mother and the other is paternal, *i.e.*, from the father side. The phenomenon of chromosome pairing is known as **synapsis**. The pairing of homologous chromosomes is initiated by some sort of attraction force. The early suggestion was that the electrostatic forces were involved in the synapsis. The second idea based on physical observations holds that hydrodynamic force which develops in the nucleoplasm causes vibration in the chromosomes. Due to this force the homologous chromosomes move slowly but surely to form the synaptic pairs.

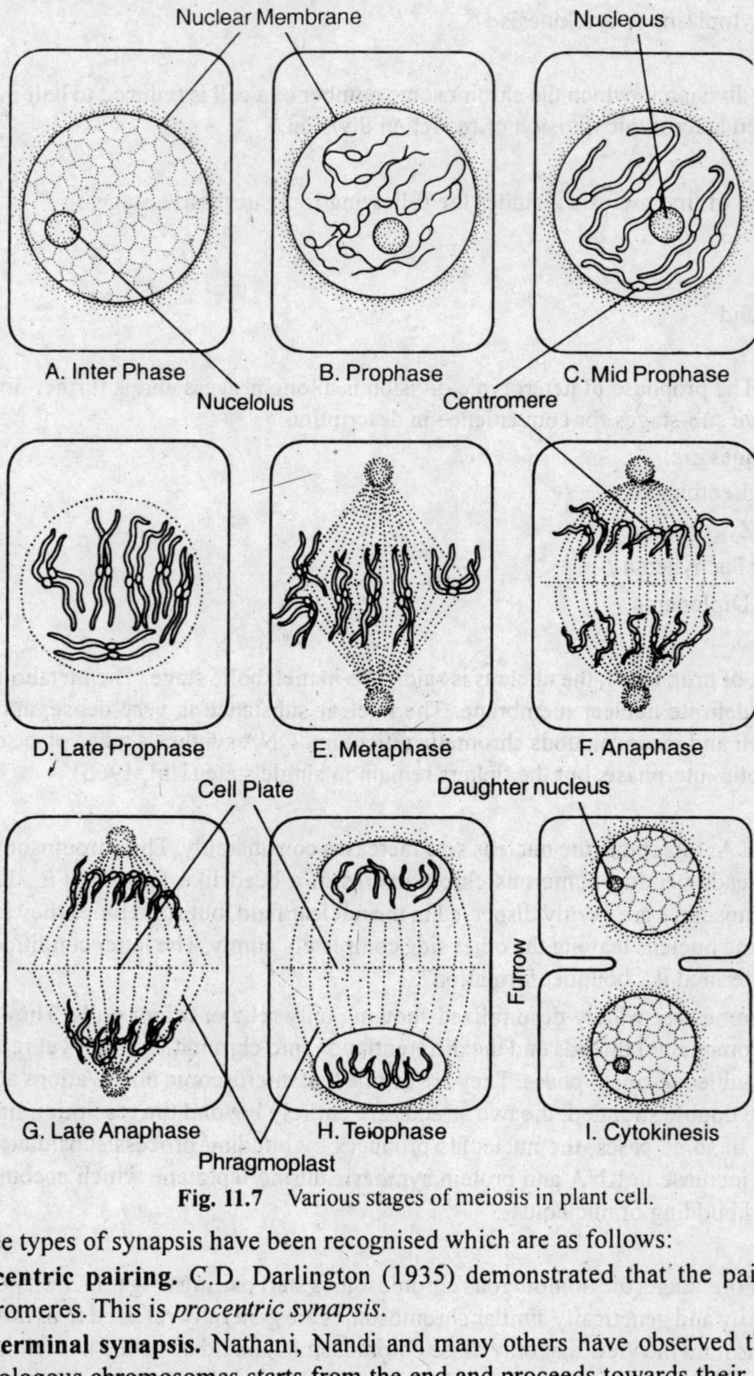

Fig. 11.7 Various stages of meiosis in plant cell.

Three types of synapsis have been recognised which are as follows:

(*i*) **Procentric pairing.** C.D. Darlington (1935) demonstrated that the pairing starts at the centromeres. This is *procentric synapsis*.

(*ii*) **Proterminal synapsis.** Nathani, Nandi and many others have observed that the pairing of homologous chromosomes starts from the end and proceeds towards their centromeres. This is *proterminal synapsis*.

(*iii*) **Localised or Random pairing.** In this type of synapsis the pairing of homologues occurs at several points at random. Synapsis continues along the length of chromosomes until it is complete and no unpaired region is left. Why do the homologous chromosomes pair in meiosis only? There are two theories to explain the phenomenon of pairing:

(*a*) **Precocity Theory.** According to Darlington, the chromosomes in early meiotic division are in single state while those in the mitosis are double from the very beginning. The argument then is that 'pairing force' is satisfied by similar chromatids of the mitosis chromosomes and thus no residual force is left which can bring the homologous chromosomes together. In meiosis during the leptotene stage the chromosomes are in singlet state, thus the pairing force would exist between the homologues. In light of the fact that DNA synthesis and chromosome replication occur during S phase of pre-meiotic interphase this theory is not substantiated.

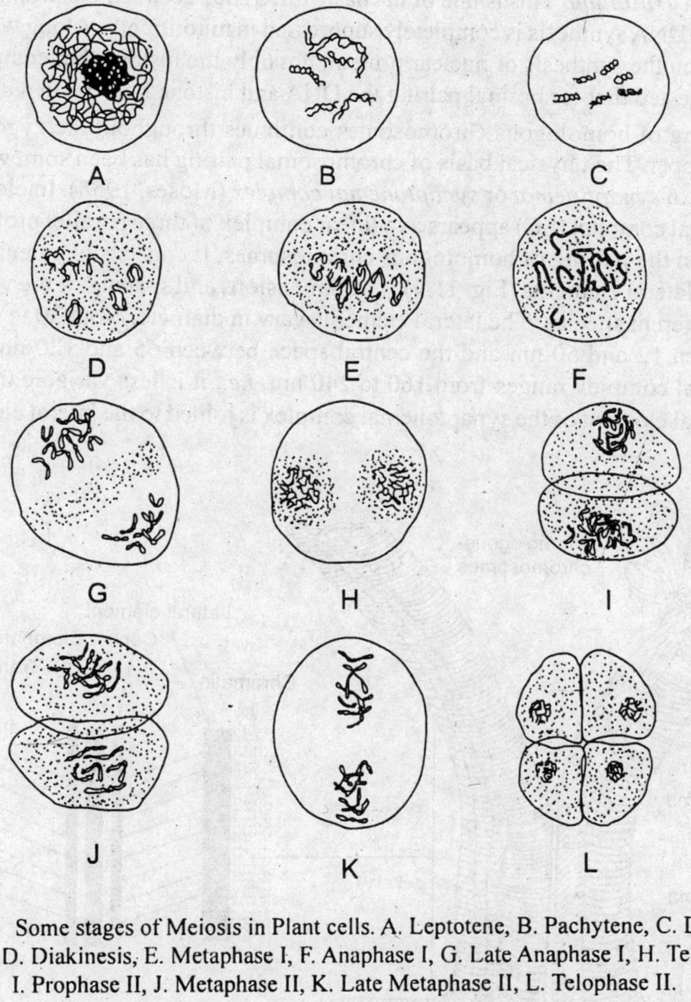

Fig. 11.8 Some stages of Meiosis in Plant cells. A. Leptotene, B. Pachytene, C. Diplotene, D. Diakinesis, E. Metaphase I, F. Anaphase I, G. Late Anaphase I, H. Telophase I, I. Prophase II, J. Metaphase II, K. Late Metaphase II, L. Telophase II.

(*b*) **Retardation Theory.** According to Sax and others, the pairing of homologues occurs on account of retardation of cellular metabolism during early prophase I. Since the prophase I is of long duration, the reduced metabolism results in uncoiling of relic spirals of chromosomes and subsequent point to point pairing of homologous chromosomes in Zygotene. This theory is also not very satisfactory but may have some support.

Sometimes, pairing may also take place between two partially homologous or dissimilar chromosomes, although synapsis may involve considerable distortion of the longer chromosome. Partially homologous chromosomes are referred to as *homoelogous chromosomes* (Huskins, 1932). *Homoeology* designates the residual homology of originally completely homologous chromosomes. In case of diploid organisms, the number of pairing configurations, the *bivalents*, is equal to half their somatic chromosome number. If there are more than two sets of homologous chromosomes in a cell as for example, in autotetraploids, the pairing takes place between two chromosomes only in one region. In autotetraploids in which there are four homologues, two chromosomes pair in one region and the other two will pair with unpaired parts of the partly paired homologues. The pairing configurations consisting of more than two homologues are called multivalents like trivalents, quadrivalents and so on.

Stern has recently provided evidence for the synthesis of a small amount of DNA during zygotene and pachytene in *Trillillium.* This is one of the basic difference between the prophase of mitosis and meiosis because DNA synthesis is completely suppressed in mitotic cells. Along with DNA synthesis there also goes on the synthesis of nuclear protein, possibly the histone that complexes with DNA. It has been suggested that for normal pairing the DNA and histone must be in a definite proportion.

The pairing of homologous chromosomes continues throughout the zygotene in a manner resembling a zipper. The physical basis of chromosomal pairing has been somewhat elucidated by the observation of *synaptinemal* or *synaptonemal complex* (Moses, 1958). In electron micrograph the synaptonemal complex (SC) appears as a linear complex of three parallel proteinaceous strands situated between the pairing of homologous chromosomes. It consists of a dense *central element* flanked by two lateral elements (Fig. 11.9). The dimensions and spacing of the various component may vary in different species. The lateral elements vary in diameter from 30 to 65 nm, the central element between 12 and 50 nm and the central space between 65 and 120 nm. The total width of synaptonemal complex ranges from 160 to 240 nm, *i.e.*, it is less variable than its component parts. The central element of the synaptonemal complex is joined to the lateral elements by delicate

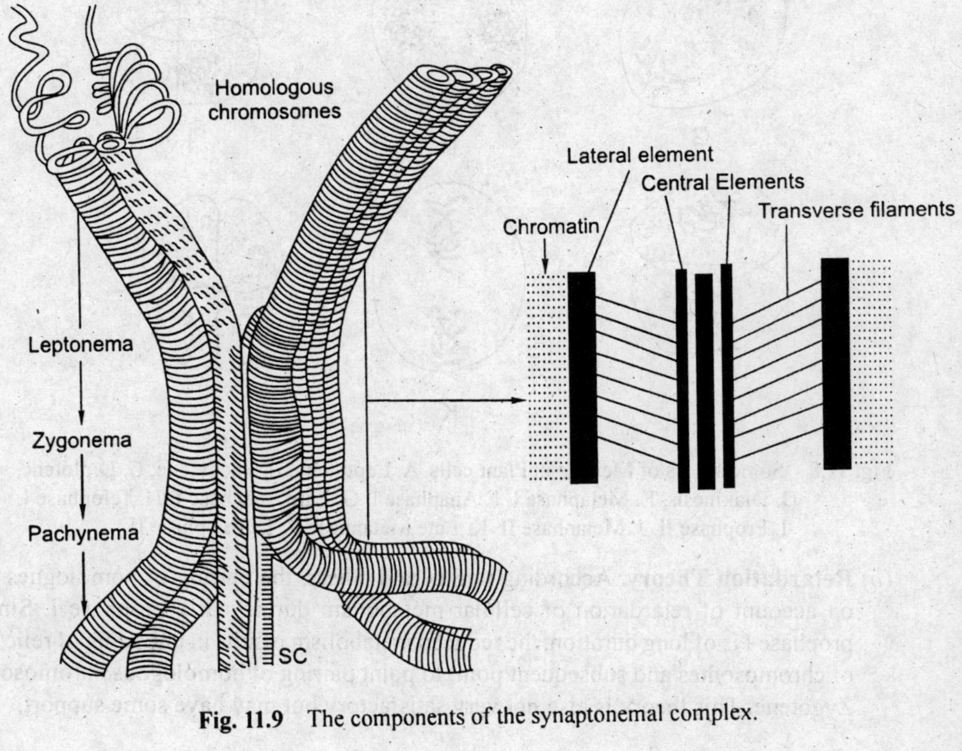

Fig. 11.9 The components of the synaptonemal complex.

transverse filaments or *L.C. fibres* that may serve to align and join two homologous chromosomes via lateral elements and to stabilize the pairing homologues at a fixed distance. The transverse filaments are fine (1.5 to 2.0 nm), straight, non-looping protein fibres that originate in the lateral elements, have a remarkable fixed length and in central region, together with additional protein material, they form central element (Socare and Moses, 1973). The synaptonemal complexes of most organisms terminate at the nuclear envelope. The attachment point may be polarized or may show no definite pattern of arrangement.

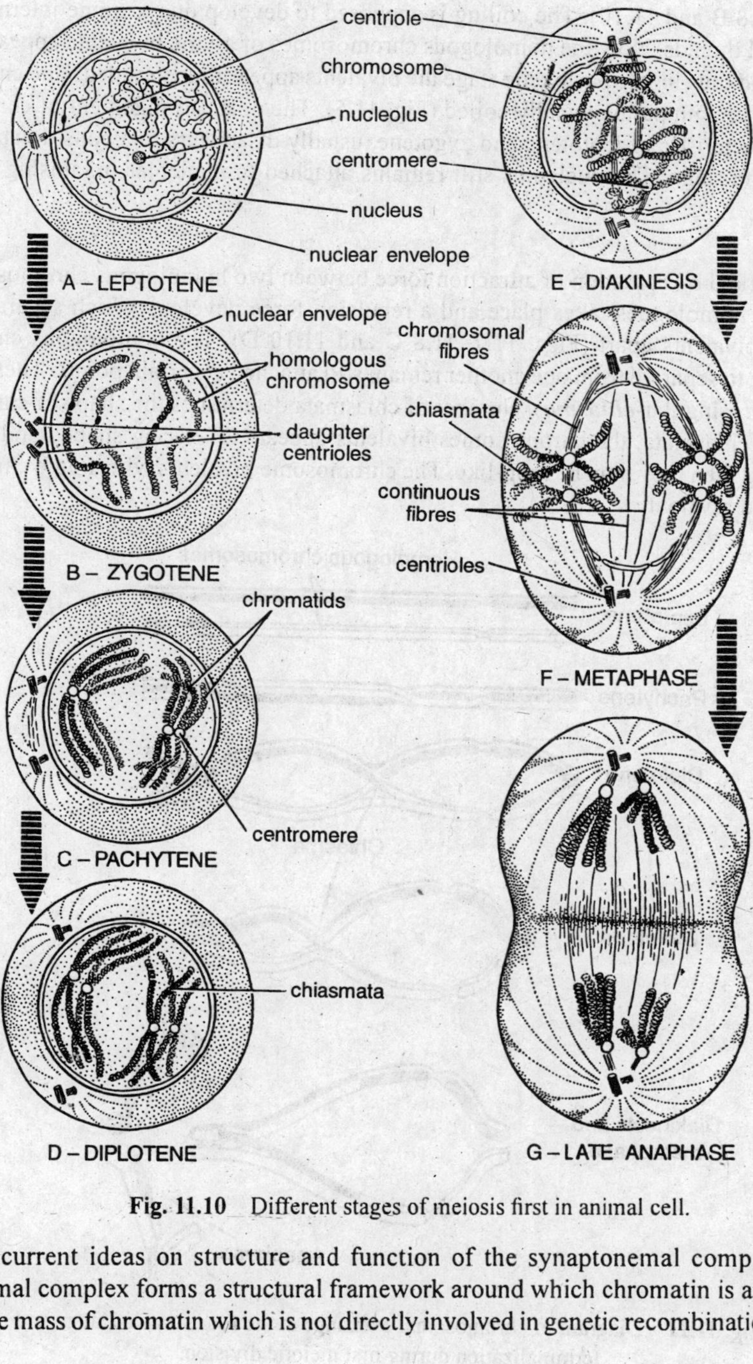

Fig. 11.10 Different stages of meiosis first in animal cell.

The current ideas on structure and function of the synaptonemal complex propose that synaptonemal complex forms a structural framework around which chromatin is arranged in such a way that the mass of chromatin which is not directly involved in genetic recombination, is sequestered on the outer surface of the lateral elements and only small stretches of chromatin penetrate the lateral element and enter the central space where they engage in molecular pairing and genetic exchange. Usually the synaptonemal complex is dissociated from the chromosomes at diplotene stage and then dispersed or aggregated into *polycomplexes.*

Pachytene

The pachytene starts when the pairing is complete. It is usually of long duration. The chromosomes show thickening and shortening. Two components of each bivalent coil around each other by relational coiling (Figs. 11.8 B and 11.10). The coiling is supposed to develop due to some internal forces in the chromosomal threads. Now the homologous chromosomes of all the bivalents appear to be split longitudinally and thus in late pachytene stage the bivalents appear four stranded. The chromatids of each homologue remain paranemically coiled (Fig. 11.5). The polar arrangement of chromosomes, which is so characteristic of leptotene and zygotene, usually disappears to a considerable extent in the late pachytene stage. The nucleolus still remains attached to nucleolar organising region of a particular tetrad.

Diplotene

This stage is marked by cessation of attraction force between two homologous chromosomes. Now uncoiling of the homologues takes place and a repulsion force develops which separates the two partners of the bivalents apart (Figs. 11.7, 11.8 C and 11.10 D). The homologous chromosomes although tending to separate from one another remain still attached at certain points. These points are called ***chiasmata*** (singular-***chiasma***). Number of chiasmata determines the shape of chromosomes. If there are two chiasmata, the chromosomes bivalents appear 'O' or ring shaped and if there are many such points then they may be loop-like. The chromosome bivalents with single chiasma at one end is termed rod bivalent.

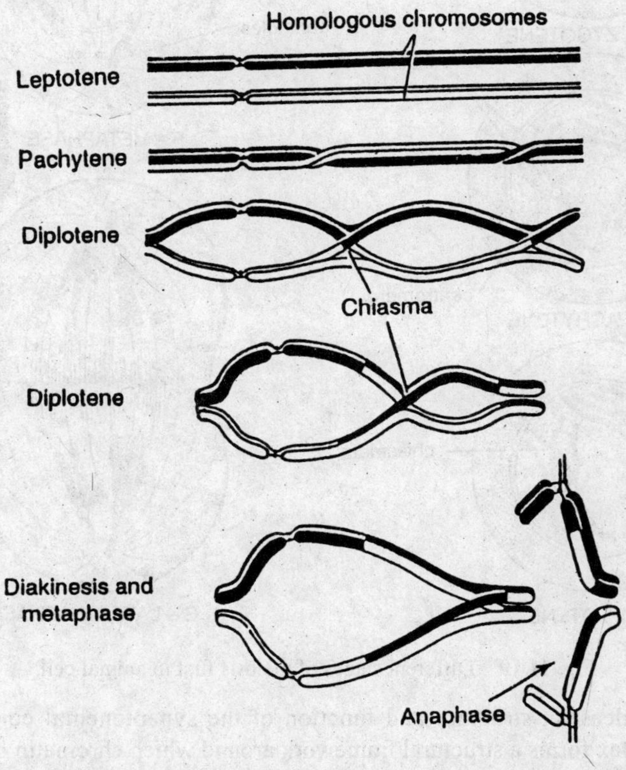

Fig. 11.11 Diagram of chromosomal exchange at the four stranded stage and terminalization during first meiotic division.

At chiasmata, the sister chromatids of bivalent are broken and rejoined crosswise, *i.e.*, reunion takes place between segments of non-sister chromatids as is shown in Fig. 11.11. This phenomenon is known as ***'crossing-over'***. Crossing-over may take place in pachytene stage after the duplication of chromosomes. Crossing-over involves the exchange of the chromatid segments and in this way it results in new types of chromatids. If there is no exchange of segments in the reunion, then the original type of chromatids will be formed. Stern and Hotta (1969) have reported that the breakage of chromatid is facilitated by an enzyme *the endonuclease* which is reported to increase in the nucleus during the pachytene stage. After the division of chromatids, the segments of non-sister chromatids are united and an enzyme *ligase* is known to help in this process (Sterm and Hotta, 1969).

In the late stage, the loops between the chiasmata become wider and tend to move away from centromere towards one or both the ends of the chromosomes. This process is known as **terminalisation**. Nucleolus becomes disorganised and simultaneously the matrix sheaths appear around the chromosomes. The chromosomes become very much coiled and shortened. The nuclear membrane remains visible.

Diakinesis

The last substage of prophase I is diakinesis. In this stage, chromosomes appear highly contracted due to formation of major coils. The separation of homologous chromosomes is complete. The nuclear membrane begins to disappear gradually (Figs. 11.7 E, 11.8 D and 11.10 E). Exchange of parts between the chromatids of homologous chromosomes may also take place at this stage. The matrix sheath obscures the internal structure of the chromosomes. The surfaces of the chromosomes appear rough. The sheath is visible only under the light microscopes. De Robertis (1954), on the basis of observations made with electron microscope, denied the existence of the matrix sheaths round the chromosomes.

Metaphase I

The metaphase stage of first meiosis is marked by complete disappearance of the nuclear membrane. Spindle fibres develop from the nuclear wall material and ground substance. Four stranded chromosome bivalents are arranged at the equator of the spindle in such a way that their individual ***undivided centromeres*** lie in the middle of spindle equidistant from the poles (Figs. 11.7F, 11.8E and 11.10F). Parts of both the homologues of a bivalent on one side of the centromeres show coiling in one direction, but those on the other side of the centromeres show coiling in reverse direction. The spindle fibres become attached with the centromeres of bivalents; one centromere of each bivalent becomes attached with the fibre of one pole and the other centromere attaches with the fibre coming from the opposite pole.

Anaphase I

The metaphase I is terminated by anaphase I. In the anaphase I, the centromeres of homologous chromosomes of bivalents repel each other (Figs. 11.7G, 11.8F, G and 11.10G). After the separation of centromeres, the homologous chromosomes begin to move apart towards the two opposite poles of spindle. The chromatids of chromosome are not separated at the spindle attachment region as the centromere remains undivided in the heterotypic division. At late anaphase, one homologue of each bivalent reaches to one pole and the other to the opposite pole. Though there are chances that the maternal and paternal chromosomes may go to different poles yet their movements is independent and at random. The future haploid cells would, therefore, contain a mixture of paternal and maternal chromosomes. If there are two bivalents in nucleus four different combinations are possible from these two pairs. If there are 'n' pairs of chromosomes, the possible number of chromosome combinations is calculated by a formula $(2)^n$. In human, for example, there are 23 pairs of chromosomes, so the possible number of chromosome combinations is $(2)^{23}$ or 83, 88, 608 even after neglecting the additional possibilities provided by crossing over.

Telophase I

When the chromosomes reach to the poles, they become shortened in length as the spirals are brought closer. Each chromosome is surrounded by thick matrix sheath. The major coils disappear, and minor coils then become very clear. The chromosomes assume regular zig zag shapes. Two daughter nuclei with nuclear membranes and nucleoli are formed from two groups of chromosomes (Fig. 11.7 H).

B. Cytokinesis

The two newly formed nuclei with reduced chromosome number are sometimes separated by a cross wall. Thus a dyad is formed. In some cases, however, cytokinesis is postponed till the end of the second division.

Interphase

This stage may be very short or long or in some cases absent. In the cases, where interphase is absent, the doubled chromosomes directly pass to the prophase of homotypic division or second meiosis without undergoing any remarkable change. However, in the cases where interphase exists, the daughter nuclei increase in volume and become ovoid. The major spirals are further despiralized and minor spirals also uncoil to a certain extent. The matrix seems to be lost and the threads appear more clear and delicate.

2. Homotypic division or Second Meiosis

A. Karyokinesis

It is a simple mitotic division. The nuclear division includes the following four successive stages :

(*i*) Prophase II,

(*ii*) Metaphase II,

(*iii*) Anaphase II, and

(*iv*) Telophase II.

Prophase II

This stage is of very short duration. The chromosomes do not undergo any appreciable change. They are already split into two sister chromatids during pachytene of first prophase. The two chromatids of each chromosome remain loose except at the centromere. At late prophase, the chromosomes are much thicker and shorter than at early stage (Figs. 11.7 I, 11.8 I and 11.12 A_1 and A_2). The centromeres of the chromosomes remain still undivided. Disappearance of the nuclear membrane and simultaneous formation of the spindle terminate the prophase II. In some cases the prophase II is absent and the chromosomes after anaphase I directly form metaphase II plate.

Metaphase II

The doubled chromosomes in each daughter cell move in middle region of the cell and are finally arranged on the equator of the newly formed spindle (Fig. 11.7 I and 11.8 J). The centromere of each chromosome now divides in the longitudinal plane to form two daughter centromeres; each daughter centromere thus becomes the centomere of the chromatid. Finally, the centromeres of chromatids become connected with attachment fibres from their respective spindle poles.

Anaphase II

The anaphase II starts with the repulsion of centromeres of sister chromatids (Figs. 11.7, 11.8 K and 11.12B_1, B_2). Now the daughter chromatids of chromosomes start moving apart towards the opposite poles of spindle. At the end of this process one chromatid of each chromosome goes to one pole and the other to opposite pole.

Telophase II

When the daughter chromosomes have reached to the spindle poles, the chromonemata anastomose to form the nuclear reticulum. The nucleolus begins to organise on one chromosome and the nuclear membranes appear around the chromosomal groups (Fig. 11.7 L and 11.8 L).

B. Cytokinesis

Intermediate walls develop between two daughter nuclei of each cell after the completion of telophase

II. Division of the cytoplasm is accomplished either by furrowing or by cell plate formation. Finally, four cells, each with single haploid nucleus, are formed in meiosis (Figs. 11.12 C_1, C_2, D_1 and D_2).

From the above analysis of meiosis, it is concluded that the chromosomes in the leptotene stage are found in singlet state. At zygotene stage homologous singlet chromosomes pair and then the reproduction (duplication) of each chromosome occurs during pachytene. After duplication, the homologous chromosomes show a tendency to repel each other during diplotene and diakinesis stages. The duplication of chromosomes is followed by two divisions. First division is heterotypic or reductional which separates the homologous chromosomes into two equal sets. Thus the original number of chromosomes is reduced by half. The second division is a simple mitosis which takes place in the two nuclei resulted from heterotypic division and separates the daughter chromatids of the chromosomes. The two successive divisions of diploid nucleus thus result into four nuclei having haploid set of chromosomes.

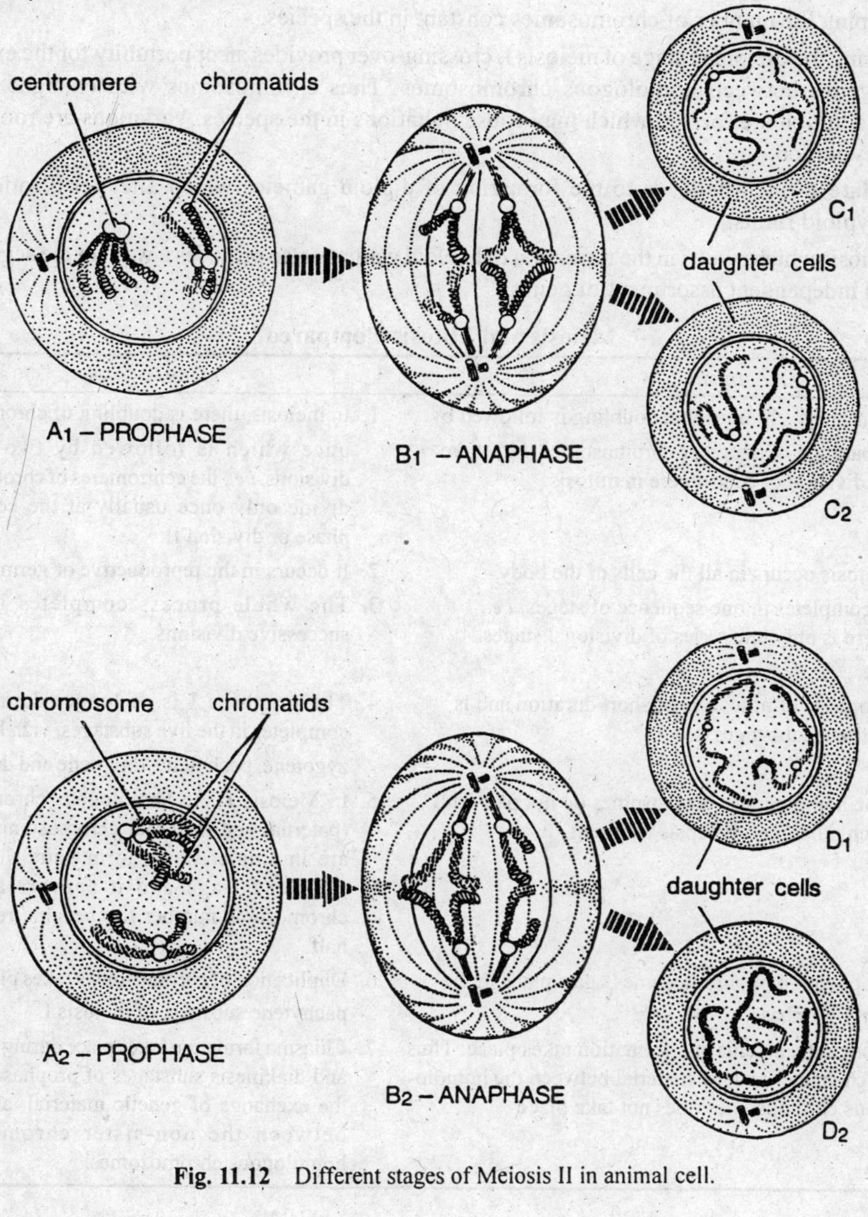

Fig. 11.12 Different stages of Meiosis II in animal cell.

The discussion ends with one question—what is the need of second meiosis or homotypic division if the reduction of the chromosome number has already taken place in the first meiosis? No proper explanation is given for this question, but during the gametogenesis, second division is of much importance in the organisms which have 2n chromosomes.

Significance of Meiosis

Meiosis is a complex process and is of great significance in the life cycle of plants and animals which reproduce sexually. Significance of meiosis is as follows:

1. If the mitosis is the only method of cell division, after every act of fertilization the chromosome number of individual will become doubled in subsequent generations. The increase in number of chromosomes is harmful and creates imbalance between cytoplasm and the nucleus which may ultimately cause several variations and malformations in the organisms. Meiosis helps in keeping the number of chromosomes constant in the species.
2. During the diplotene stage of meiosis I, crossing-over provides an opportunity for the exchange of genes between homologous chromosomes. Thus chromosomes with changed genetic constitution are formed which may cause mutations in the species. Variations are root causes of evolution.
3. Failure of meiosis leads to the formation of diploid gametes which after fertilization form polyploid forms.
4. Meiosis which results in the formation of haploid gametes also provides a way for the segregation and independent assortment of genes.

Meiosis and Mitosis Compared

Mitosis	*Meiosis*
1. In mitosis, chromosome doubling is followed by separation of daughter chromosomes, *i.e.*, there is division of centromere in mitosis.	1. In meiosis, there is doubling of chromosomes once which is followed by two nuclear divisions, *i.e.*, the centromeres of chromosomes divide only once usually at the equational phase or division II.
2. Mitosis occurs in all the cells of the body.	2. It occurs in the reproductive or germ cells.
3. It completes in one sequence of stages, *i.e.*, there is only one series of divisional stages.	3. The whole process completes into two successive divisions.
Prophase	
4. Prophase of mitosis is of short duration and is without sub-stages.	4. The prophase I is of longer duration and completes in the five substages, viz., leptotene, zygotene, pachytene, diplotene and diakinesis.
5. The homologous chromosomes do not pair with each other, *i.e.*, synapsis is absent.	5. In Meiosis I, the homologous chromosomes (paternal and maternal chromosomes) which are in singlet states form pairs during the zygotene stage and it appears as if the chromosome number has become reduced to half.
6. Duplication of chromosomes takes place in the early prophase.	6. Duplication of chromosomes takes place in the pachytene substage of meiosis I.
7. Generally, no chiasma formation takes place. Thus, exchange of genetic material between the homologous chromosomes does not take place.	7. Chiasma formation takes place during diplotene and diakinesis substages of prophase I. Thus, the exchange of genetic material takes place between the non-sister chromatids of homologous chromosomes.

Mitosis	*Meiosis*
Metaphase	
8. Chromosomes appear two stranded.	8. Since homologous chromosomes do not separate in prophase I, the chromosome pairs appear in tetrad state, *i.e.*, four stranded in metaphase I.
9. The centromere of each chromosome divides into two and thus the two chromatids of the chromosome become free from each other.	9. Centromeres of the homologous chromosomes do not divide, thus their chromatids do not become free in the metaphase I.
Anaphase	
10. The two chromatids of each chromosomes move towards the opposite poles of spindle.	10. In meiosis I, the two homologous chromosomes of each bivalent separate and move towards the opposite poles of spindle during anaphase stage.
11. The daughter chromosomes appear in singlet state.	11. The separated chromosomes are two stranded but their centromeres remain still in undivided state. However, the centromere is spli into two during the Metaphase II.
12. The chromosomes are long and thin.	12. The chromosomes are short and thick.
Telophase	
13. The telophase always comes after anaphase.	13. The first telophase is sometimes omitted.
Significance	
14. The chromosome number in each daughter nucleus remains the same as in parent nucleus.	14. In meiosis, the chromosome number of parent is reduced to half in the daughter nucleus.
15. A diploid cell will produce two diploid cells by a mitosis.	15. A diploid cell produces four haploid cells by a meiosis.

QUESTIONS

1. Describe in detail the process of Mitosis. What is the significance of mitosis in unicellular and multicellular organisms?
2. Explain and illustrate the statement that mitosis is essential in the asexual reproduction of unicellular organism and the growth of multicellular ones.
3. Describe the details of first meiosis with particular emphasis on prophase I. Compare and contrast the mitosis with meiosis.
4. Which of the two processes-mitosis and meiosis has greater significance in genetic studies and why?
5. When do the chromosomal strands become double in:
   (*i*) a somatic cell, and (*ii*) a pollen mother cell?
6. What do you understand by
   (*i*) Reduction or disjunctional division, and
   (*ii*) Equational division?
7. Write short notes on:
   (*a*) Spindle,
   (*b*) Homologous chromosomes,
   (*c*) Synapsis,
   (*d*) Chiasmata, and
   (*e*) Terminalization.
8. Explain, why do you expect differences between the cells arising from the meiosis and not from mitosis?

12 CHAPTER Development of Genetics as a Branch of Biology

All living things have a tendency to produce young ones. The offsprings of all the organisms (plants and animals) resemble their parents in several respects. Why is it so? It is a question related to the phenomenon of heredity. Heredity literally means *"tendency of like to beget like"*, *i.e.*, all living things tend to produce young ones like themselves, for example, cow can produce only a cow; an elephant can produce a little elephant but never a tiger. Similarly bean, gram and mango plants can produce only bean, gram and mango respectively. Sometimes new types of plants and animals develop when unlike parents are crossed. Is it not heredity then? Yes, it is a pure and simple case related to heredity. Thus heredity makes new types of organisms come out in some matings between two unlike parents. Very often the offsprings are unlike the parents, as for example, sometimes deaf children are born from the parents with normal hearing. This is also due to heredity.

The offspring may be similar to their parents but they are never exact repetitions. They differ not only from one another but from their parents also in many characters. These differences are called ***variations*** which may be of two types:

1. Variations which are inherited from one generation to other, and
2. Variations which are non-heritable and appear in the organisms when they grow in the changed environmental conditions, such as temperature, food, moisture, light and so on.

The branch of biological science which deals with heredity and variations is termed ***Genetics.*** The word 'genetics' (a Greek word meaning to generate) was proposed by William Bateson (1902). This branch of science seeks to understand the laws governing the transmission of hereditary potentialities from parents to offsprings.

History and Development of Genetical Science

It is difficult to imagine what ideas prehistoric men might have had about the phenomenon of heredity. Hippocrates (400 B.C.), Aristotle (350 B.C.) and other Greek philosophers believed that reproductive material came from all parts of individual's body and characters were directly transmitted to the progeny, and thus according to them, the inheritance of the characters was direct. The ideas of the other hypotheses were baseless and fantastic. Those philosophers initiated fantastic stories to explain reproduction and sex determination, such as, banana plants were supposed to have resulted from a cross between *Acacia* tree (Babool) and palm; the giraffe was thought to be a hybrid between camel and leopard, ostrich (Suturmurga) was supposed to have developed when camel mated with sparrow. Many such stories were repeated before seventeenth century.

The science of heredity and variation found its way to development when details of sexual reproduction were discovered. The reproductive parts of the plants were reported for the first time by N. Grew in 1682 and sexual reproduction in plants was described for the first time by Camerarious in 1694. A.D. Leeuwenhoek observed sperms of several animals in 1677 and suggested their association with eggs.

Credit also goes to Camerarious for developing for the first time a plant hybrid from a cross between hop plants and hemp. Thomas Fairchild (1717) obtained an artificial hybrid by crosspollinating light red flowered Sweet Williams with pink flowered related species of Sweet Williams and noted that the characters of both parents (mother and father) developed in the offsprings. The hybrid was called "Fairchild Sweet Williams" or "Fairchild mule". His experiments were designed carefully and the results were actual and not accidental. Later on many other workers performed similar artificial pollinations between different related species and found, more or less, similar results.

J. Swammerdam (1679) suggested that the development of an organism is a simple enlargement of a minute preformed individual called homunculus which could be present in the sperm or in ovum (Fig. 12.1). This is called ***"preformation theory"***. Ovists thought that homunculus was present in the ovum while the animaculists insisted that preformed homunculus existed in the sperms. K.F. Wolff (1738–1794), a German scientist rejected the theory of preformation and suggested that the gametes contained undifferentiated living substances capable of forming organized body after fertilization. This idea was called the "**theory of epigenesis**". According to this theory many tissues and organs absent in the gametes develop subsequently *denovo* due to mysterious vital forces.

Linnaeus (1707–1778) believed that new plant species arose by natural hybridization. Joseph Koelreuter put it to the test of experiment during the years from 1760 to 1770. He made 500 different crosses and maintained the records up to second filial generation. He observed in his experiments that the hybrids between plant varieties appeared either intermediate or resembled one parent or the other. He also noticed uniformity of all the hybrids in first filial generation of all the crosses. In the second filial generation, the progeny showed tendency to reversion to original parental characters (segregation). These observations led Kohlreuter to conclude:

1. that both male and female parents contribute equally to the formation of their offsprings, and
2. that since the hybrids tend to revert sooner or later to parental species they cannot form new species. Carl Gaertner in the years 1830s and 1840s repeated and expanded Kohlreuter's experiments. In 1849, he published an account of his 10,000 breeding experiments together with copious references to the works of others in his book in German. Such crosses were made by Thomas Andrew Knight in 1780s, John Goss and Alexander Seton in 1820s, French botanist Charles Naudin. Those workers noticed the same mode of appearance of characters in the hybrid as had already been observed by Koelreuter.

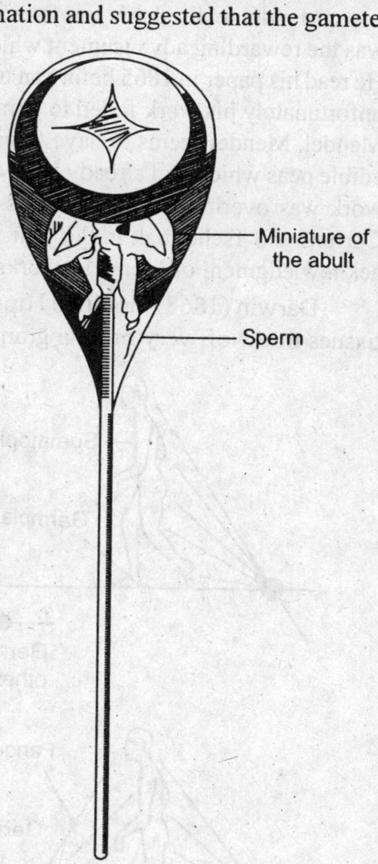

Fig. 12.1 Drawing of homunculus by Harsoeker.

A careful perusal of early literatures reveals that numerous hybridization experiments were performed during 18th and 19th centuries in search of the laws governing the mechanism of heredity but the results did not lead to any appreciable hypothesis.

Greek thinkers of eighteenth century were of the view that inherited characters of the individuals were acquired through direct contact with environment.

The French scientist Jean Baptiste Lamarck (1744–1829) formulated the theory of inheritance of acquired characters (**Lamarckian theory** or **Lamarckism**). In the theory, he stated *that only those organs of the body developed well which were in great use over long period of time and the organs which were in less or no use became reduced and finally disappeared.* The theory was criticised by several workers because it failed to demonstrate the mechanism experimentally.

Gregor Johann Mendel (1822–1884) began his experiments of cross breeding of different varieties of garden peas in 1856 with the intention to discover the relation between hybridization and history of evolution of organic forms. He got the results similar to those obtained by his predecessors. The serious mistakes which Mendel's predecessors committed were:

1. that they considered several characters at one time. Due to that they got confused and could not draw any conclusion from the results they obtained.
2. that they neither grew all the hybrids they got from various crosses nor they could maintain the record of number of progeny of each kind.

Mendel realised this and found a statistical law for the hybrids of second filial generation. This was the rewarding advancement which enabled him to detect the underlying mechanism of heredity. He read his paper in 1865 before an unappreciative audience of Natural History Society in Brunn, but unfortunately his work failed to attract their attention and remained unrecognised in the life time of Mendel. Mendel seems to have found the details of all early experiments in crossing of varieties of edible peas which had already been carried out by Knight (1789), Goss (1820) and others. Mendel's work was overlooked for 34 years. It was rediscovered by three independent workers De Vries, Correns and Tschermak in the year 1900. The classical works of these workers ensured worldwide acknowledgment of Mendel's works.

Darwin (1868) modified Hippocrate's classical concept and suggested that all the cells and tissues threw off very minute granules called "gemmules" or "pangenes" and these multiplied,

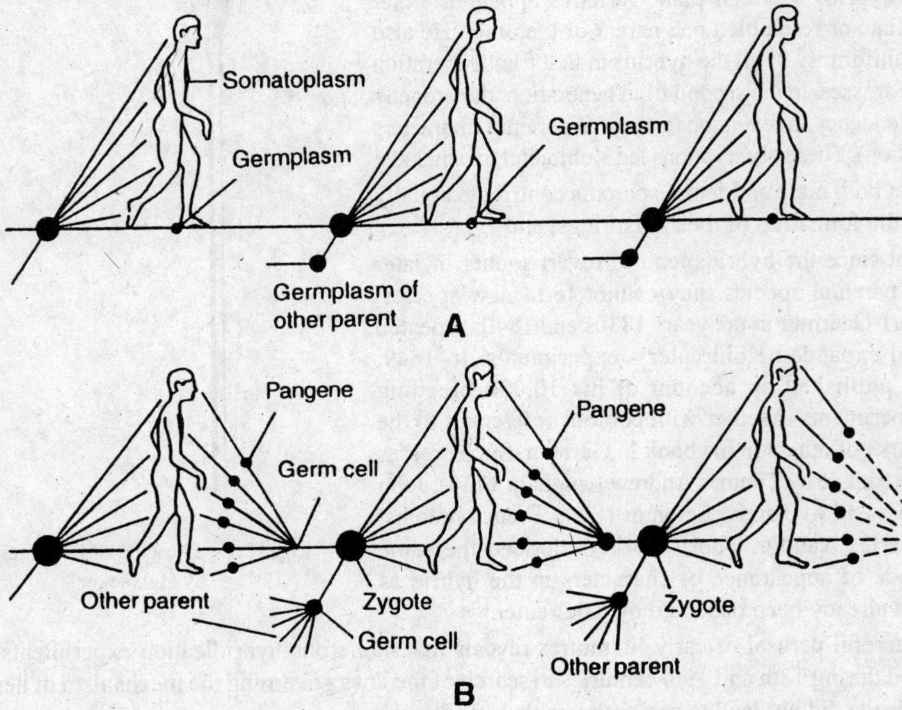

Fig. 12.2 Diagram showing inheritance of germplasm (A) and pangenes (B).

circulated throughout the body of organism and were transmitted to the germ cells (gametes), which thus contained a multitude of minute granules from each individual part of body. This is called *"**theory of pangenesis**"* (Fig. 12.2). Darwin regarded these particles as bearers of hereditary characters and thought that they were responsible for the appearance of ancestral characters in the offsprings (Origin of Species, Ch.27).

The above idea of Darwin could attract little support by the time Mendel's work was rediscovered. Much more knowledge accumulated about the chromosomes, and the biologists began to realise the important roles of chromosomes in explanation of Mendelian laws.

The theory of pangenesis was disproved by Galton (1823–1911) and A. Weismann (1835–1934). Weismann demonstrated that pangenesis could not be verified. He cut the tail of mice for 22 generations and got tailed mice in all the generations.

The experiment of mutilation may appear crude, but results definitely contradicted the theory of pangenesis and Lamarckism. Weismann proposed the *"**theory of germplasm**"* according to which (*i*) the body of an organism contains two types of cells; somatic cells having somatoplasm and reproductive cells or germ cells with germplasm, (*ii*) the somatoplasm cannot form germplasm but germplasm is meant for reproduction and can form somatoplasm as well and (*iii*) changes in the structure of somatoplasm caused by environment (acquired characters) cannot influence the reproductive cells or germplasm whereas the changes occurring in germplasm influence the progeny.

The field of modern genetics started progressing fast when Thomas Hunt Morgan (1910) put forward his *"**chromosome theory of heredity**"*. The theory states that

(*i*) the chromosomes are the bearers of hereditary characters,

(*ii*) the genetic determinants (genes) for many characters are arranged along the chromosomes in definite linear fashion, and

(*iii*) the gene is a factor for determining one or more hereditary characters in all organisms.

T. H. Morgan made investigations on the nature of genes and received Nobel Prize in 1934.

After Morgan, De Vries and many other popular biologists opened new chapters in the history of genetics. In recent years, biochemistry provided new dimensions to both genetics and evolution. H.J. Muller got Nobel Prize in 1946 for the discovery of induction of mutations by x-rays.

Biochemistry has contributed several very important ideas concerning heredity which is named "Biochemical Genetics". This branch of science is concerned with the chemical and physical nature of hereditary material and the mechanism which governs the action of genes. G. W. Beadle and J. Lederberg have made outstanding contributions to biochemical *genetics* and got Nobel Prize in 1958. A. Kornberg and Ochoa got Nobel Prize in 1959 for their work on the chemistry of DNA and RNA.

Watson, Crick and Wilkins got Nobel Prize in 1962 for the discovery of double helix model of DNA. Jacob, Monad and Leowulf discovered regulator genes and operator genes which regulate the activity of structural genes. Nurenberg, Hargovind Khorana and Holley worked independently on the genetic code and for their outstanding work they shared the Nobel Prize in 1968.

QUESTIONS

1. Give a brief account of history of development of genetics.
2. Write short notes on:
   (*i*) Theory of germplasm.
   (*ii*) Lamarckism.
   (*iii*) Theory of pangenesis.

13 CHAPTER

Mendel and his Laws of Heredity

The credit goes to Gregor Johann Mendel for making first effort in the field of heredity and formulating the basic laws of inheritance. He is now appropriately famous as 'the father of science of heredity' (Fig.13.1). He was, of course, not the first man to discover any general law of inheritance as Kohlreuter and several other predecessors had already obtained the results similar to those of Mendel but had failed to explain the problem. Mendel had two very important qualities with him which enabled him to achieve a good scientific record. These were:

1. his ability to ask the right questions from nature, and
2. his ability to interpret nature's answers patiently and correctly.

Fig. 13.1 Gregor Johann Mendel (1822–1884).

Mendel was born in the family of an ordinary farmer in a Sicilian village of Heinzendorf in 1822. His father created deep interest in him in gardening. Mendel received his early education at his home and then he attended a preparatory schooling. After completing preparatory course, he completed two-year course in philosophy. During the latter part of preparatory course he had to meet financial burdens of the family. He entered Augustinian religious establishment of St. Thomas at Brunn in Maravia, Austria (now Czechoslovakia) on the advice and help of his teacher, Professor Franz. At the age of 25 he became priest (Monk) in 1847. Later Mendel was appointed as substitute teacher of Physics and natural history in a secondary school of Brunn. There he served for some fourteen years. In his spare time Mendel grew several varieties of pea in his monastery garden. He conducted his famous garden pea experiments in his garden during the years 1856 to 1864. He also studied science and mathematics at the university of Vienna. There he realised the value of statistics. He became abbot of the monastery in 1868 and died in 1884.

Mendel's Experiments

Mendel's predecessors working on several varieties of plants differing from one another in many complex characters produced fertile hybrids by artificial pollination, but they failed to explain the mechanism of heredity.

The following were the two reasons behind their failure:

1. They considered simultaneously many characters in which parents differed. This made them confused. Due to confusion they could neither trace the individual characters through successive generations nor they could maintain the complete numerical records of results.

2. Secondly, they believed that the hereditary characters of two parents (mother and father) become thoroughly mixed in the offspring.

Mendel did realise the above two causes of his predecessors' failure. In order to overcome those difficulties he carefully planned experiments exactly on the same pattern as his predecessors had already followed. He began his breeding experiments with different varieties of garden peas (*Pisum*) in 1856. He performed those experiments not with the intention of testing direct action of foreign pollen or to see if two kinds of pollen acted identically, but to discover the relationship between hybridization and the history of evolution of organic forms. He spent first two years in selecting varieties with distinctive and contrasting characters and in making sure that the parental stocks under study were pure, *i.e.*, their characters were permanent and true breeding. In other six years he set up experiments for breeding the varieties with contrasting characters. As indicated in his field diary, Mendel studied several sets of such characters. Out of some twenty or thirty characters, he finally selected the following seven sets of contrasting characters.

Characters			*Dominant*	*Recessive*
Stem Characters:	(*i*)	Length:	Tall	Dwarf
	(*ii*)	Flower position	Axillary	Terminal
Pod Characters:	(*iii*)	Pod colour	Yellow	Green
	(*iv*)	Shape	Smooth or Inflated	Constricted
Seed Characters:	(*v*)	Shape	Round	Wrinkled
	(*vi*)	Seed coat colour	White	Grey
	(*vii*)	Colour of cotyledons	Yellow	Green

Mendel studied all the above seven pairs of differing characters individually; first in original stocks and then in the hybrids which presented combination of such characters in different generations. He classified the progenies according to their characters and also maintained complete counts of individuals belonging to particular classes. It is only on account of quantitative method of investigation Mendel could achieve success in his work. One of the most important experiments which Mendel performed is briefly described below.

In an artificial cross two pure breeding plants, one with tall stem (6ft. in height) and the other with small or dwarf stem (about 1½ ft. in height), were cross pollinated. Mendel accomplished this cross pollination by opening immature flower bud before its anthers had matured. Firstly, he removed the stamens from flower bud and then pollinated its stigma with pollens from the desired stock. The cross-pollinated flowers were covered in order to prevent foreign pollens of undesired stocks from reaching the stigma. The seeds obtained from that cross on germination developed plants which represented the first filial generation (F_1). They were not intermediate between the tall and dwarf as might be expected, but were all tall plants. Similar results were obtained regardless of whether pollen grains were taken from dwarf plants (stamenate) and tall was used as the female (pistillate) parent or *vice versa.* Mendel made crosses to study the inheritance of six other pairs of characters and noticed that of the two contrasting characters of two different parents only one appeared in the F_1 hybrids and the other remained latent or unexpressed. The particular parental character which appeared unchanged in the hybrid plants of F_1 generation was, therefore, called "**dominant**" and the unexpressed character as "recessive". Thus by this definition, the tall character is dominant and dwarf is recessive in the above cross.

Mendel allowed the hybrid plants of first filial generation to self-pollinate (selfing). The seeds obtained from F_1 plants were planted which developed into plants of second filial generation (F_2). In F_2 generation both tall and dwarf characters appeared in the ratio of approximately 3 tall to one dwarf (3 : 1). Mendel took 1064 F_2 hybrid seeds and grew them into plants. Among F_2 plants, 787 were tall and 277 short (dwarf), *i.e.*, in the ratio of 2.84 tall to 1 dwarf or on allowing for experimental error due to chance, it comes to 3 : 1. Plants of F_2 generation were allowed selfing and the F_3 generation was raised. Mendel then noticed that F_2 dwarf plants produced seeds which developed into dwarf plants on germination or in other words F_2 plants with recessive characters bred true on selfing and of the F_2 tall (3/4 of the total F_2 population), 1/3 produced only talls whereas the remaining 2/3 produced both tall and dwarf plants in 3:1 ratio in F_3 generation. The experiment can be summarized in the graphic form as follows:

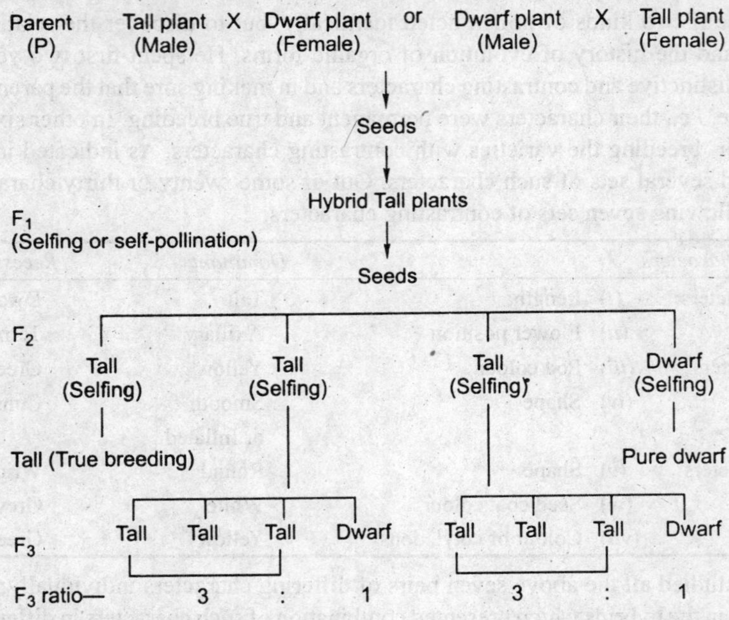

Mendel was very much impressed when he found that all the seven pairs of differentiating characters which he combined by crossing gave similar results. He found in each of his crosses that

1. one of the two contrasting characters was dominant and the other was recessive;
2. the recessive characters did not disappear permanently but only their appearance was checked by dominant ones in F_1 generation. The parental characters combined in the hybrids reappeared in second generation in the ratio of 3 dominant to one recessive;
3. the F_2 recessives always bred true to their characters but of the F_2 dominant types one-third were pure whereas 2/3 were hybrid types which when allowed selfing segregated into dominants and recessives in the ratio 3 : 1 in F_3 generation.

Mendel continued his valuable investigation for eight years and examined in great detail about 10,000 plants. The data and inferences drawn from experiments were set forth in a paper entitled **"Experiments in Plant hybridization."** Mendel read his paper before the members of the *Brunn Natural History Society* on February 8, 1865 and got that published in the proceedings of that society in 1866. Unfortunately, his work failed to attract the attention of the people and remained unrecognised in the life of Mendel. Importance of his work was realised after his death when three biologists, De Vries in Holland, Correns in Germany and Von Tschermak in Austria worked independently on the same problem and arrived at the similar conclusion in 1900.

Terminology used in Mendelian Laws

The knowledge of the following technical terms is essential for understanding Mendel's Laws of inheritance.

1. **Hybrid.** The product of a cross between two stocks belonging to the same genus or species or between two different genera is called hybrid.
2. **Hybridization.** It is a phenomenon which speaks about the mixing of two unlike genetic constitutions.
3. **Alleles (Allelomorphs).** This term was coined by Bateson and Saunders (1902) for characters which are contrasting or differing from one another.
4. **Monohybrid Cross.** It is a cross between two parents differing in only one pair of alternating characters.
5. **Dihybrid Cross.** It is a cross between two parents differing in two pairs of contrasting characters.
6. **Polyhybrid Cross.** This is a cross between two stocks differing in more than two sets of alternating characters.
7. **Reciprocal Crosses.** A set of two reciprocal crosses means that the same two parents are used in two experiments in such a way that in one cross one individual is used as female parent and the other as male parent and in the second cross of the same genotype sexes of the parents are reversed, as for example, cross between a tall male and a dwarf female is reciprocal of a cross between a tall female and dwarf male.
8. **Dominant character.** Mendel introduced the term for a character which manifests itself in all the members of first filial generation from a cross between two pure breeding parents differing in respect of this character, *i.e.*, in a cross involving two different alleles one which suppresses the other character is called dominant.
9. **Recessive.** The suppressed character which does not appear in first filial generation is called recessive.
10. **Back Cross.** It is a cross between a pure parental strain and its hybrid progeny from previous cross.
11. **Homozygous (Homozygote).** This term was coined by Bateson and Saunders (1902). When the diploid individuals carry sets of similar traits or characters (example, TT or tt) they are said to be homozygotes. They develop when both the gametes (male and female), each carrying the same factor, fuse. Such plants breed true to their characters (Gk. *Homo*, alike).
12. **Heterozygous (Heterozygote).** (Gk. *Hetero*, different). This term was proposed by Bateson and Saunders (1902) for a zygote or a diploid individual developed from it (zygote) which carries both factors of a pair of alleles. It develops when two gametes with different genetic constitutions fuse. Such individuals do not breed true.
13. **Phenotype.** The term was coined by Johannson (1909) for the visible characters or external appearance of an organism with respect to particular character or a group of characters.
14. **Genotype.** The term was proposed by Johannson (1909) for hereditary or genetic constitution of an individual.

MENDEL'S LAWS OF INHERITANCE OR LAWS OF HEREDITY

Mendel crossed different varieties of garden pea (*Pisum sativum*) with the intension to discover relationship between hybridization and history of evolution of organic forms. He classified and recorded the number of individuals in progenies of each cross and compared the proportion on mathematical pattern. This led Mendel to discover a precise pattern and so to formulate basic laws of heredity. These laws are:

1. Mendel's first laws or law of segregation or law of purity of gametes.
2. Mendel's second law or law of independent assortment.

1. Mendel's First law or Law of Segregation or Law of Purity of Gametes

Mendel formulated his first law from the conclusions drawn from his monohybrid experiments (*i.e.,* from crosses in which only one set of differentiating characters was taken into consideration in the parental stocks). This law states that if a pair of contrasting characters or allelomorphs originally found in two parental stocks (male and female) are brought together in the hybrid, they remain unchanged without contaminating each other. In the first filial generation (F_1) only one of the two characters appears in the hybrids (dominance) and the other character (allele) remains latent or unexpressed. These two coexisting dominant and recessive characters of F_1 hybrid segregate or separate from each other during formation of gametes. *This is law of segregation.* Since in F_2 generation recessive types reappeared, it was apparent to Mendel that heredity constitution of an individual was not necessarily revealed by its external appearance.

The results of monohybrid experiments led Mendel to conclude that the germ cells or gametes from each parent, though at that time little was known of the cytology, carried "**some thing**" which was responsible for the appearance of particular character in adult plant. To that "some thing" he called ***germinal unit*** or ***factor***. These are now known as **genes**. He maintained that those factors remained unchanged even when the characters did not appear in the progeny, as for example, the unit for recessive character in F_1 generation. The théory also suggested that the factors separated when the germ cells were formed in the next generation. Professor Charlotte Auerbach in his book "Science of Genetics" writes, "This theory were so far ahead of what was then known about the cell structure and fertilization that modern geneticists stand amazed at the brilliance of mind which starting from very simple observations could push so far, so confidently and so daringly into completely unexplored territories."

When the segregation takes place in F_1 hybrids, two types of gametes are formed and a gamete contains only one unit character of each allelomorphic pair. Therefore, half of the gametes will have dominant factor and the remaining half will carry recessive factor. This rule applies to both male and female gametes. Thus the gametes formed from F_1 hybrids are always pure for a particular character and one may carry either dominant or recessive factor but not both. This is why law of segregation is known as "law of purity of gametes". If self-pollination is allowed, two kinds of eggs will form all possible combinations with two kinds of male gametes. Now there appear two probabilities:

(*i*) the union may occur between male and female gametes carrying similar factor or alleles. In this case the zygote will receive either both dominants or both recessives.

(*ii*) the union may take place between female and male gametes, one carrying dominant unit and the other with recessive unit.

Zygotes will be of different behaviour depending upon the ways in which the gametes unite. The zygotes which possess set of identical alleles will always develop into pure breeding individuals. These are called **homozygotes**. They may be either pure recessives or pure dominants. The second type of zygotes which contain both the factors of a set of differentiating alleles are termed **heterozygotes** and individuals developing from them are called **heterozygous**. They always develop into individuals which will never breed true to their characters. Thus, in the segregation process hybrid types produce parental types as well as hybrids. In other words, the law of segregation may be defined as the nonmixing of alleles in the hybrid. What it means is that if good allele and bad allele come together in the hybrid neither they will fuse nor will they influence each other, but will continue to be together in the hybrid generation after generation. A good allele is as good when it leaves the hybrid as it was when it entered the hybrid and the bad allele is just as bad as it was when it entered the hybrid.

Mendel found that in the second filial generation both dominant and recessive characters appeared in 3 to 1 ratio. Further analysis of F_2 individuals has shown that they are in the ratio of one homozygous dominant: two heterozygous dominants: one homozygous recessive (1 : 2 : 1). What was the significance of these simple numerical ratios ? These ratios enabled Mendel to formulate his first law of heredity (law of segregation).

Mendel explained his results making use of letter symbols. He introduced symbol "A" for dominant and 'a' for recessive factors. Capital letter is used for dominant character and small letter for recessive one. This use of letter symbol is most significant as it implies for the existence of "some thing" or factor or character (gene).

Now returning to a cross between tall and dwarf peas, if we assume that the tallness is governed by factor 'T' and dwarfness by factor 't' then the genetic composition of pure tall and pure dwarf parents will be TT and tt respectively. Both these parents are homozygous and if selfing is allowed in them, the tall will always produce only tall plants and similarly the dwarf plant will produce dwarf plants.

During development of reproductive cells or gametes the segregation or separation of factors or genes occurs. Here each gamete produced by homozygous tall parent will carry only one 'T' gene and similarly each gamete from homozygous dwarf parent will carry only one 't' gene. Therefore, fusion of gametes from tall parent with gametes from dwarf parent will produce zygotes of only one kind, all with genetic composition 'Tt'. Because the dominant factor T is present in all the F_1 offsprings, they all are tall (dominance). The F_1 generation now possesses a pair of genes, one for tallness and the other for dwarfness, so it should be called heterozygous (Tt). According to law of segregation, two genes (T and t) of heterozygous tall will segregate during the development of gametes and as a result, two kinds of gametes will be produced, half of the gametes will have 'T' gene and the remaining half will carry 't' gene. Here egg cells carrying T factor may be fertilized by male gametes carrying T or by those carrying t, and in the same way eggs carrying t factor may fuse with male gametes carrying either T or t factor. In other words, the fertilization may be at random. As a result of chance combination, an approximate ratio 3 tall plants to 1 dwarf plant would be expected.

Parents (P): Pure breeding Tall X Pure breeding dwarf
Genotypes: TT ↓ tt

Gametes (T) (Cross) (t)

↓

F_1 Tt (Impure breeding or heterozygous tall)

T t

(Segregation):
Gametes

F₂ — ♂ gametes / ♀ gametes

♂ \ ♀	T	t
T	TT	Tt
t	Tt	tt

Summary : F_2 Genotype: 1TT: 2 Tt: 1tt
F_2 phenotype: 1 pure tall : 2 heterozygous tall : 1 pure dwarf

It is evident from the above graphic representation that among F_2 tall plants there are two categories, some carry T T and some Tt. Though they are similar in their external appearance yet their genetic constitutions are different. The entire genetic constitution of an individual, both expressed and latent, makes up its **genotype**, and the appearance or assemblage of characters which are expressed in the individual, as for example, tallness or dwarfness, constitutes its **phenotype**.

Back Cross. When heterozygous individuals of F_1 generation are crossed with one of the parental types (P) individuals it is known as **back cross**. The back cross may be of the following two types:

(*a*) **Out cross.** When the F_1 hybrid individuals are crossed with dominant homozygous parent it is called out cross. In this type of back cross all the offsprings will show dominant trait but genotypes will be in the ratio 1 homozygous dominant. 1 heterozygous dominant, as is shown in the following table.

(*b*) **Test cross.** When F_1 hybrid individuals are crossed with homozygous recessive parent it is called ***test cross***. When heterozygous talls of F_1 are crossed with the recessive dwarf variety, half of progeny will be tall and half will be dwarf in F_2 generation as illustrated in the following table. This is a back cross. Because it is used in detecting whether a given plant is homozygous or heterozygous, it is also known as *test cross*.

		Test cross			Out cross		
Phenotype	:	F_1 Tall	×	Dwarf	F_1 Tall	×	Pure tall
Genotype	:	Tt		tt	Tt		TT
Gametes	:	T & t		t	T & t		T
Back-cross	:	Tt		tt	1 TT		1 Tt
offspring	:	Tall		dwarf	Tall		Tall
Summary	:	1Tall		:1 Dwarf	1 Pure tall	:	1Heterozygous
Offspring ratio	:						or hybrid tall

Other Example of Law of Segregation

The law of segregation holds good not only in plants but also in animals. Here one original experiment conducted by a well known geneticist T.H. Morgan on *Drosophila* is cited to illustrate the law of segregation in animals (Fig. 13.2).

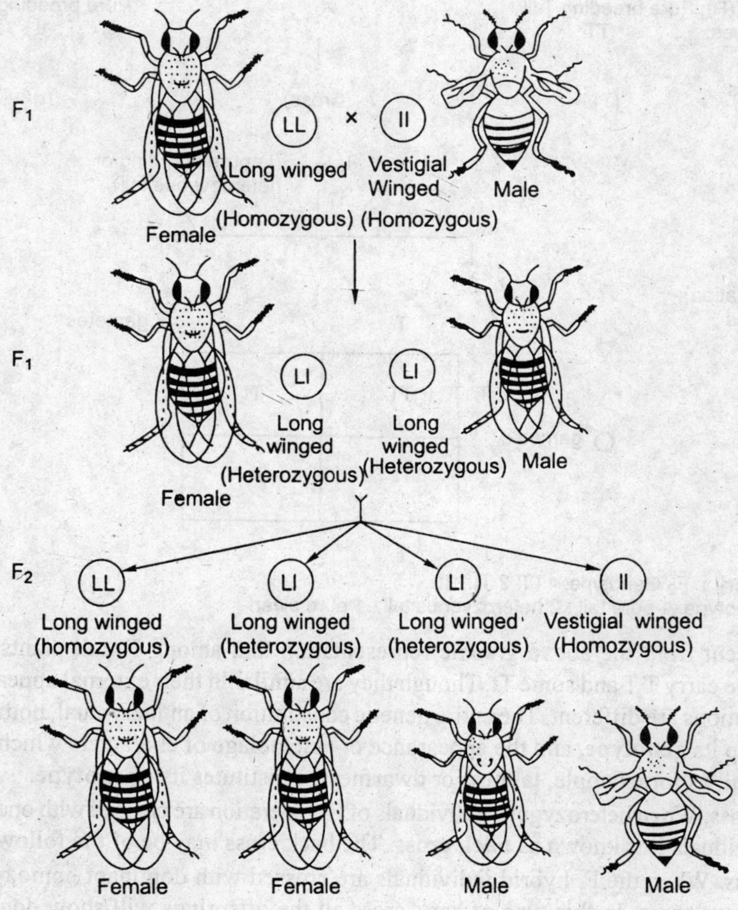

Fig. 13.2 A monohybrid cross between long-winged and vestigial-winged in *Drosophila*

Morgan crossed a homozygous longwinged (wild type) *Drosophila* with a homozygous vestigial-winged fly and found that all the F_1 hybrids were longwinged. When F_1 hybrids were allowed mating among themselves, they produced longwinged hybrids and vestigial winged hybrids in three to one (3 : 1) ratio in F_2 generation. The mechanism of segregation is illustrated in Fig. 13.2.

2. Mendel's second law or Law of Independent assortment (Dihybrid and Tri-hybrid experiments).

Mendel conducted breeding experiments to study not only the monohybrid crosses but also the dihybrid and tri-hybrid ones in which the parental stocks differed in two and three characters respectively. He planned his experiments on dihybrid and tri-hybrid crosses with the intention to discover the relationship amongst the different pairs of contrasting characters. Mendel also wanted to know whether the different pairs of alleles tended to remain together in parental arrangement in the progeny or they were independent.

Dihybrid Cross. In a dihybrid experiment, a pure breeding variety of pea with round (smooth) seeds and yellow cotyledons was crossed with a pure breeding variety with wrinkled seeds and green cotyledons. F_1 seeds were found to be all round and yellow (round dominated over wrinkle and yellow cotyledon dominated over green one). It was immaterial which of the parents provided the pollen and which acted as pistillate parent. It also made no difference whether both dominant traits had originally been in one parent or each of the parents had contributed one of the dominant alleles. When F_1 hybrids were allowed self-pollination, in the F_2 generation four different combinations were noted. Of the total 556 F_2 seeds, 315 produced plants with round- yellow seed; 108 round-green; 101 wrinkled-yellow and 32 wrinkled green. These different combinations, *i.e.*, round yellow, round green, wrinkled yellow and wrinkled green, were found to fit very nearly into a ratio 9 : 3 : 3 : 1 which Mendel recognised as multiple of 3 : 1, *i.e.*, $(3 + 1)^2$. Mendel saw that the ratio of 9 : 3 : 3 : 1 would be obtained if the two sets of traits inherit independently of each other, since two independent 3:1 ratio would be combined to simplify mathematical probability. The results of dihybrid cross can be shown graphically as follows:

Parents (P) Round, yellow X Wrinkled, green
(Cross pollination)
↓
F_1 Round, yellow (Hybrids)
↓
Selfing (Self pollination)

F_2	Round, yellow		Round, green		Wrinkled, yellow		Wrinkled, green
	9/16	:	3/16	:	3/16	:	1/16
	:Ratio 9	:	3	:	3	:	1

From the results of dihybrid experiment, Mendel concluded the following points:

1. the members of two sets of alleles segregated in F_2 generation.
2. the alleles of one set behaved independently with respect to those of the other set at the time of combination, *i.e.*, they are independently assorted, as for example, round character appeared in combination with green and wrinkled appeared with yellow in the above cross.

The above two points are read in Mendel's second principle called *Law of Independent Assortment.*

The independent assortment and random combination can be more easily illustrated if parent (P) are represented by letter symbols for 2 sets of alleles or genes (factors) under consideration. Here 'R' denotes for roundness and 'r' for wrinkleness and Y and y are used for the factors controlling yellow and green character of cotyledons respectively. Now the mode of inheritance in the dihybrid cross can be explained in the following way:

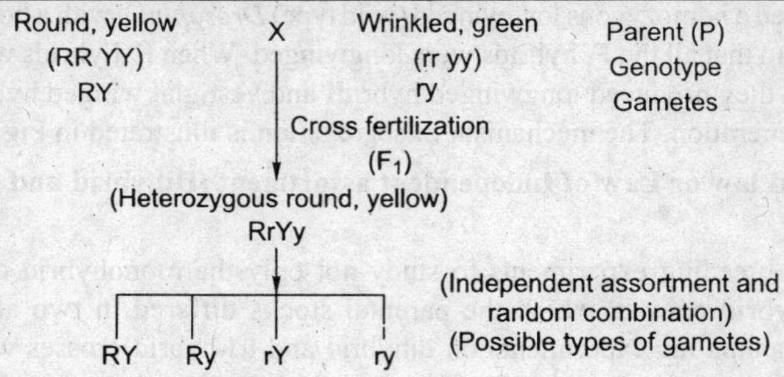

After fertilization, the possible combinations will be 4 × 4 = 16. This is shown in the following checker board (Fig. 13.3).

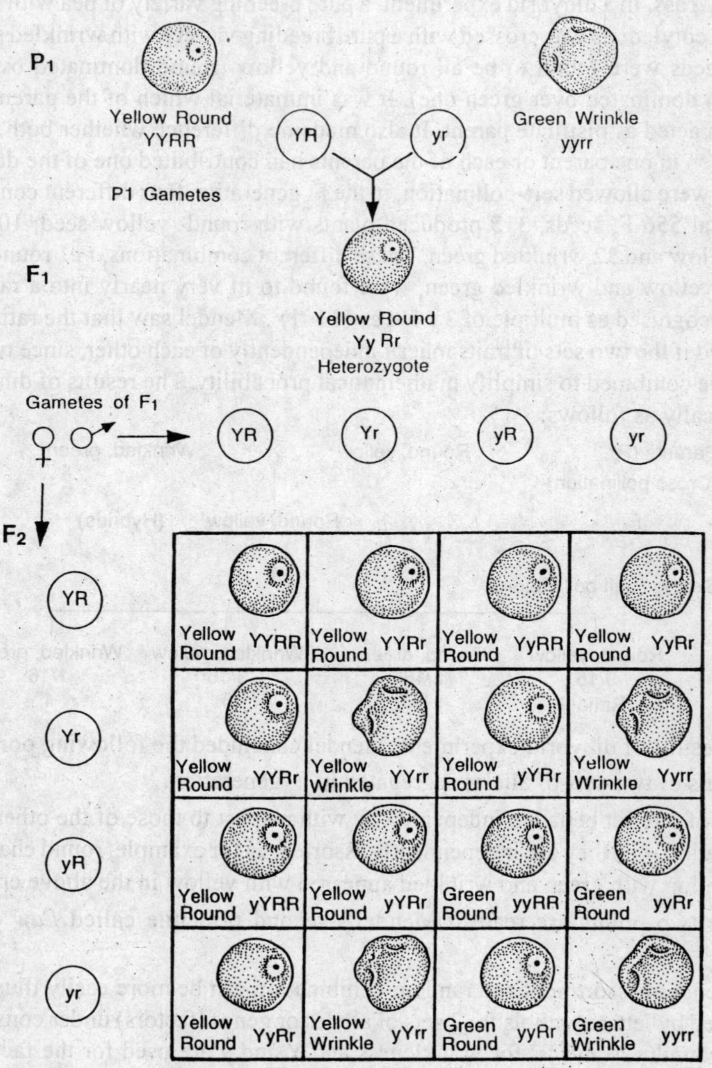

Fig. 13.3 A dihybrid cross between yellow, round and green wrinkled pea.

Summary Chart:

S.No.	*Phenotypes*	*No.*	*Genotypes*	*Genotypic frequency*
1	Round, yellow	9	RRYY	1
			RRYy	2
			RrYY	2
			RrYy	4
2	Round, green	3	RRyy	1
			Rryy	2
3	Wrinkled, yellow	3	rrYY	1
			rrYy	2
4	Wrinkled, green	1	rryy	1

This method of representation is called Punnett square and is named after an English geneticist, R.C. Punnett who first used it to illustrate possible combinations resulting from the crosses. The checker board method is time consuming so in its place the following easy and quick device is now used. This is known as forked line process. It is explained below:

Round, yellow × Wrinkled, green

Here dihybrid cross is resolved into two monohybrid crosses. Thus round and wrinkle characters form one set and yellow and green form the other set. These two sets of contrasting alleles operating simultaneously will segregate in two monohybrid ratios 3 : 1. Thus 3 round : 1 wrinkled and 3 yellow : 1 green will be produced in F_2. Since these characters are combined at random, the 3 : 1 ratio of round and wrinkled may be combined with 3 : 1 ratio of yellow and green, Thus:

3 Round < 3 Yellow = 9 Round, yellow
3 Round < 1 green = 3 Round, green

1 Wrinkled < 3 Yellow = 3 Wrinkled, yellow
1 Wrinkled < 1 green = 1 Wrinkled, green

In this way, we get the same dihybrid ratio 9 : 3 : 3 : 1. The same ratio can be obtained by checker board method:

	Monohybrid ratio of cross between Round and Yellow	
Monohybrid ratio of cross between yellow and green	3/4 Round	1/4 Wrinkled
3/4 Yellow	3/4 × 3/4 = 9/16 Round yellow	3/4 × 1/4 = 3/16 Wrinkled yellow
1/4 green	1/4 × 3/4 = 3/16 Round green	1/4 × 1/4 = 1/16 Wrinkled green

The genotypic expression may be illustrated as follows:

I set of alleles	II set of alleles	Combination	Phenotypes
1 RR	1 YY —	1 RRYY	Round yellow
	2 Yy —	2 RRYy	Round yellow
	1 yy —	1 RRyy	Round green
2 Rr	1 YY —	2 RrYY	Round yellow
	2 Yy —	4 RrYy	Round yellow
	1 yy —	2 Rryy	Round green
1 rr	1 YY —	1 rrYY	Wrinkled yellow
	2 Yy —	2 rrYy	Wrinkled yellow
	1 yy —	1 rryy	Wrinkled green

Summary

Round yellow	Round green	Wrinkled yellow	Wrinkled green
9	3	3	1

Dihybrid Test Cross. In the F_2 progenies, only one of the four phenotypes, *i.e.,* wrinkled and green will be homozygous for both the characters and the other three phenotypes can be homozygous or heterozygous for one or both the characters. In order to find out whether any individual is homozygous or heterozygous for both characters, test crosses are made. If the individual is heterozygous for both the characters (as F_1) and it is crossed with a double recessive, a phenotypic ratio 1 : 1 : 1 : 1 will be obtained. This is clear from the following example:

RrYy (Round and yellow green) × rryy (Wrinkled and green seeds)

Gametes: RY, Ry, rY, ry, — ry

gametes→♀ ↓ ♂	RY,	Ry,	rY,	ry,
gametes ry	RrYy	Rryy	rrYy	rryy
Phenotypic ratio	Round yellow 1	Round green 1	Wrinkled yellow 1	Wrinkled green 1

Tri-hybrid experiment. Mendel also crossed two parents which were different from each other in three sets of contrasting characters. In one trihybrid experiment, pure breeding plants with tall stem, round and yellow seeds were crossed with pure plants with dwarf stem, wrinkled and green seeds. In this example 3 monohybrid sets operate simultaneously. Supposing that tall dominates over dwarf, round over wrinkled and yellow over green, the F_1 progeny would be cent percent tall, round and yellow. Here expected F_2 results are shown mathematically by $(3 + 1)^3$ which may be expanded to 27 : 9 : 9 : 9 : 3 : 3 : 3 : 1. The result may be explained graphically by resolving the trihybrid cross into 3 monohybrid ratios and multiplying them in the following manner:

Phenotypic ratios

3/4 Tall	3/4 Round	3/4 Yellow	= 27/64 Tall, Round, Yellow
		1/4 Green	= 9/64 Tall, Round, Green
	1/4 Wrinkled	3/4 Yellow	= 9/64 Tall, Wrinkled, Yellow
		1/4 Green	= 3/64 Tall, Wrinkled, Green
1/4 Dwarf	3/4 Round	3/4 Yellow	= 9/64 Dwarf, Round, Yellow
		1/4 Green	= 3/64 Dwarf, Round, Green
	1/4 Wrinkled	3/4 Yellow	= 3/64 Dwarf, Wrinkled, Yellow
		1/4 Green	= 1/64 Dwarf, Wrinkled, Green

F_2 phenotypic ratio = 27 : 9 : 9 : 9 : 3 : 3 : 3 : 1

Genotypic expression of trihybrid cross can be presented in the following way:

Allelic sets I set	II set	III set	F_2 genotypes	F_2 phenotypes
1TT	IRR	1YY	— 1 TTRRYY	— Tall, Round, Yellow
		2Yy	— 2 TTRRYy	— Tall, Round, Yellow
		1yy	— 1 TTRRyy	— Tall, Round, green
	2Rr	1YY	— 2 TTRrYY	— Tall, Round, Yellow
		2Yy	— 4 TTRrYy	— Tall, Round, Yellow
		1yy	— 2 TTRryy	— Tall, Round, green
	1rr	1YY	— 1 TTrrYY	— Tall, Wrinkled, Yellow
		2Yy	— 2 TTrrYy	— Tall, Wrinkled, Yellow
		1yy	— 1 TTrryy	— Tall, Wrinkled, green
2Tt	IRR	1YY	— 2 TtRRYY	— Tall, Round, Yellow
		2Yy	— 4 TtRRYy	— Tall, Round, Yellow
		1yy	— 2 TtRRyy	— Tall, Round, green
	2Rr	1YY	— 4 TtRrYY	— Tall, Round, Yellow
		2Yy	— 8 TtRrYy	— Tall, Round, Yellow
		1yy	— 4 TtRryy	— Tall, Round, green
	1rr	1YY	— 2 TtrrYY	— Tall, Wrinkled, Yellow
		2Yy	— 4 TtrrYy	— Tall, Wrinkled, Yellow
		1yy	— 2 Ttrryy	— Tall, Wrinkled, green
1tt	1RR	1YY	— 1 ttRRYY	— Dwarf, Round, Yellow
		2Yy	— 2 ttRRYy	— Dwarf, Round, Yellow
		1yy	— 1 ttRRyy	— Dwarf, Round, green
	2Rr	1YY	— 2 ttRrYY	— Dwarf, Round, Yellow
		2Yy	— 4 ttRrYy	— Dwarf, Round, Yellow
		1yy	— 2 ttRryy	— Dwarf, Round, green
	1rr	1YY	— 1 ttrrYY	— Dwarf, Wrinkled, Yellow
		2Yy	— 2 ttrrYy	— Dwarf, Wrinkled, Yellow
		1yy	— 1 ttrryy	— Dwarf, Wrinkled, Yellow

It is evident from the above phenotypic expressions that some phenotypes have old combination of characters as are in parent but most of the combinations are new which are generally spoken as recombinations. These combinations are good proofs to suggest that the characters are assorted independently.

Trihybrid test cross. This type of test cross involves crossing of a trihybrid (e.g., TtRrYy) with triple recessive. In this, eight different types will be obtained in 1 : 1 : 1 : 1 : 1 : 1 : 1 : 1 ratio in next generation.

P_1 Hybrid Tall, Round, Yellow TtrRYy X Dwarf, wrinkled, green ttrryy

Gametes

TRY TrY Try TRy tRY trY tRY try try gametes

Progeny→ TtRrYy : TtrrYy : Ttrryy : TtRryy : ttRrYy : ttrrYy : ttRrYy : ttrryy

Causes of Mendel's Success

The common garden pea plants were selected by Mendel as suitable material for his experiments for the following reasons:

1. The plants were annual with well defined characters.
2. The pea plants could be grown easily and crossed readily.
3. Flowers were self-pollinated and the hybrids were fertile.

Mendel succeeded in his experiment due to following points which he always kept in mind.

1. He was careful to work with pure varieties of peas which bred true to their characters.
2. He considered only one set of traits or characters at a time in the beginning.
3. Mendel recorded and classified offsprings of each individual in each generation and compared the results on mathematical pattern.
4. He maintained records of each cross at least up to third generation.
5. Mendel was the first man who decided the question of heredity and had this perception that characters of organisms were governed by a large number of separate units or factors. He also considered that single factor controlled a single character.

Biological Importance of Mendelism

Mendel's experiments have practical applications in plant and animal breeding. It is possible to produce by hybridization desired types which do not occur in nature. The desired traits carried in different varieties can be combined and maintained in a single individual variety. Many new disease resistant and high-yielding crop plants and ornamental varieties are being developed by cross hybridization, for which Mendel's law of segregation and law of independent assortment have provided basis.

QUESTIONS

1. Why did Mendel use pea as experimental material in his experiments? Give a brief account of Mendel's experiments on pea.
2. Define the following terms:
   (*i*) Hybrid,
   (*ii*) Monohybrid cross,
   (*iii*) First filial generation,
   (*iv*) Allelomorphs,
   (*v*) Reciprocal crosses,
   (*vi*) Back cross, and
   (*vii*) Genotypes and Phenotypes.

3. Discuss Mendel's laws of inheritance. Point out the importance of his work.
4. In maize F_1 plants obtained from a cross tall (TT) X dwarf (tt) was back crossed with (*i*) pure tall parent, (*ii*) dwarf parent. Give the gametes of F_1 hybrid and those of the two parents and also show the genotypic and phenotypic ratios of the progenies obtained due to back crossings.
5. One F_1 hybrid obtained from a cross between red flowered (RR) pea variety and white flowered (rr) pea variety was selfed. What will be genotypic and phenotypic ratios of progenies obtained due to selfing ?
6. In *Mirabilis*, plant hybrid from red (RR) and white (rr) flowers is pink (Rr). A plant with pink flower is selfed and is also back crossed with both the parents. Give the genotypic and phenotypic ratios expected of the progenies obtained after selfing and back crossing.
7. In pea tall (T) is dominant over dwarf (t); green pod (G) is dominant over yellow (g) and Round seeds (R) is dominant over wrinkled (r). A homozygous dwarf green wrinkled pea is crossed with a homozygous tall, yellow and round. Utilizing forked line method give the genotypes of parents, and F_1 and F_2 progenies and show how many types of gametes will be formed in the individuals of F_1 generation.
8. In the following crosses give:

   (*i*) the phenotypes of the parents,

   (*ii*) gametes produced by each parent, and

   (*iii*) expected genotypic and phenotypic ratios in the progenies.

   If dominant tall is represented by T, recessive dwarf by t, dominant green pod represented by G and recessive yellow by g and dominant round seeds by R and recessive wrinkled seeds by r gene.

   (*i*) TTGgRr × ttGgrr

   (*ii*) ttggRr × TtGgrr

   (*iii*) TtGGRr × TtGgRr

   (*iv*) TtGgrr × ttGgRr.

14 CHAPTER

Correlation between Mendel's Theory and Chromosomes' Behaviour

Mendel, while explaining the laws of inheritance, suggested that the genes were hereditary units which were transmitted from cell to cell and generation to generation. Actually, Mendel was not at all familiar about physical explanation of his results. In order to explain as to how the genes could behave in such a way that each daughter cell at the time of gamete formation received all that was in the parent cell, and then that became a complete cell and not half a cell or only part of the parent cell several early workers have speculated to a great extent. Wilhelm Roux for the first time pointed out the importance of chromosomes in the heredity and suggested that the chromosomes were strings of beads like structures which were present in the nuclei of the cells. He also suggested the possibility to explain the mechanism of gene transmission from cell to cell.

Sutton and Boveri, in 1904 followed the points of Wilhelm Roux and formulated a theory called *the chromosome theory of heredity*. They in order to explain the law of independent assortment, suggested that if different genes were in different chromosomes completely independent of one another, the results of Mendel's dihybrid crosses could be explained. The similarity between the proposed behaviour of Mendelian factors prior to cell formation and observed behaviour of chromosomes during meiosis will now be apparent. The important points of similarity between the two are as follows:

1. The chromosomes occur in pairs in the cell, so do the Mendelian factors.
2. Homologous chromosomes pair during meiosis and only one member of each pair finds its way into given germ cell. Mendelian factors segregate from one cell to another in similar fashion.
3. The spatial arrangement of any given chromosome pair on the spindle is independent of the arrangement of all chromosome pairs. Mendelian factors are also supposed to assort themselves independently.
4. By the process of fertilization the original number of chromosomes is restored in the zygote, similarly Mendelian factors (pairs of contrasting characters) reduced to one half their original number during germ cells formation are restored by fertilization.

So provisionally, the Mendelian factors may be identified by the physical entities, called chromosomes.

Cytological Basis of Segregation

Mendel knew nothing of the chromosomes and meiosis. Sutton and Boveri in order to explain the law of independent assortment suggested that different genes were in different chromosomes completely independent of one another. Thus, similarity between the Mendelian factors and chromosomes became apparent. The cytological basis of segregation of alleles can be explained if we take the example of an individual which possesses two allelomorphs 'A' and 'a' on two different homologous chromosomes (*i.e.*, A on one chromosome and *a* on the second chromosome. In the prophase stage the two chromosomes will pair and duplicate and in first meiosis A and *a* chromosome wills segregate or separate from each other. In the second meiosis the two chromatids of A chromosomes and those of *a* chromosome separate. Thus it becomes clear that both alleles of heterozygote separate during meiosis (gametogenesis) to produce gametes with only one chromosome with either dominant allele or recessive allele. No gamete will have both the genes or chromosomes of an allelic pair (Fig. 14.1).

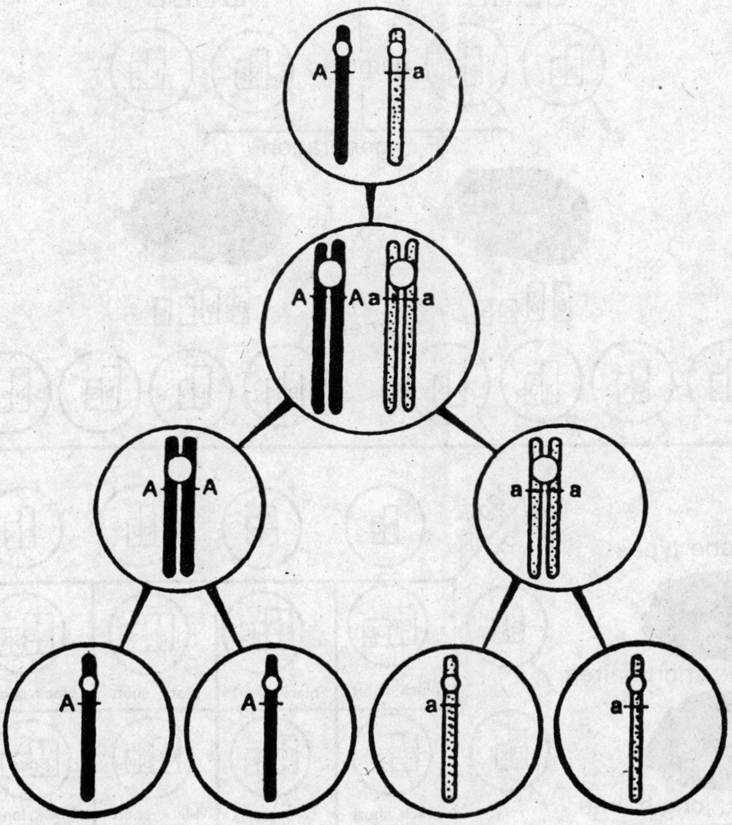

Fig. 14.1 Segregation of a pair of genes localized in a pair of chromosomes. Gametes with only one gene are formed.

Explanation of Law of Independent Assortment

To explain this, take the example of a cross between black short haired guinea pig and a brown long haired guinea pig in which allele B (Black) is in one chromosome and allele b (brown) is in other chromosome. Similarly allele S representing for short hair is in one chromosome and allele s representing long hair is in other chromosome. The mechanism of inheritance is shown in Fig. 14.2.

In 1910 Thomas Morgan pointed out that genes in chromosomes could be separated from each other by an exchange of parts of chromosomes. After the discovery of DNA structure, scientists were certain that the genes of all higher organisms were made of DNA. The chromosome number simply represented the number of packages into which DNA of a species is divided. It also became clear that even if the different species possessed the same chromosome number, they do not have the same quality and quantity of DNA.

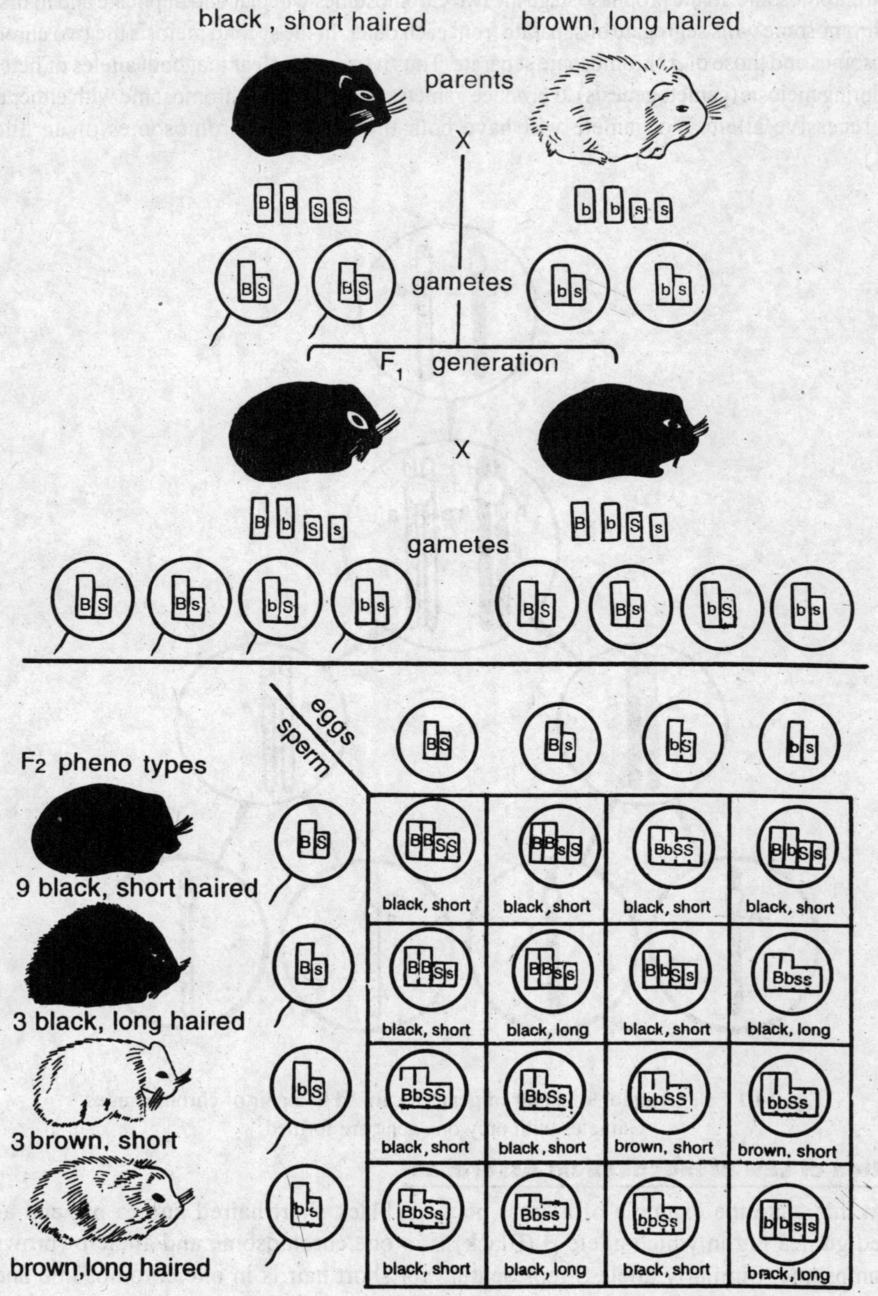

Fig. 14.2 Dihybrid cross between black short haired guinea pig and a brown long haired one showing independent assortment.

QUESTION

1. If gene L is represented by one chromosome and 1 gene is represented by another chromosome, shown by diagrams the segregation of genes in the gametes of an organism which is heterozygous in nature.

15 CHAPTER

The Expansions of Mendelian Laws

Mendel's two laws of heredity—the law of segregation and the law of independent assortment, form a base for predicting the outcome of simple crosses in the eukaryotic organisms. These laws have been explained presuming that the characters of the organisms were determined by *"factors"*. Those factors were later called ***genes*** by W. Johannsen in 1909. According to Mendel, there existed 1 : 1 ratio between the traits and the factors. But subsequent investigations have revealed that the real world of genes is more complex than is revealed by Mendelian principles of heredity and the exceptions and modifications abound. In this chapter we shall see that (*i*) more than two allelic forms of a gene can exist, (*ii*) there are variations in dominance relations, and (*iii*) many phenotypes are determined not by one gene pair but by several gene pairs or by non-allelic gene pairs interacting with one another or by genes interacting with environment.

Multiple Alleles and Complex Loci

In the early history of genetics, Bateson (1906) stated that Mendelian factors were free entities and the presence or absence of a particular factor or gene determined the dominance or recessiveness respectively but when it became clear that the genes were linked to chromosomes and indeed that many genes were linked to chromosomes in linear fashion, the *presence and absence* hypothesis ran into its severe criticism. In the meantime, *multiple allelism* was also discovered. Normally, a given diploid individual can possess only two different alleles for a set of characters. L.Cuenot (1904) demonstrated that the number of alternative forms of a particular gene was not necessarily limited to two, *i.e.,* more than two allelic forms of a gene can map at the same locus. This is called ***multiple allelism***. Multiple alleles obey the same rule of transmission as alleles of which there are only two kinds. Dominance relationships among multiple alleles vary in that for some groups of alleles homozygous and heterozygous genotypes produce different phenotypes, whereas in others the alleles may be arranged in a descending series in which every allele is dominant over all alleles below it.

The human ABO blood group alleles afford a modest example of multiple allelism. There are four blood groups (phenotypes) in the ABO system in which three alleles, I^A, I^B and i, determine the blood groups and only two of the three alleles can be present in any one individual as shown in Table 15.1.

Table 15.1. ABO Blood Group System in Humans

Blood phenotypes	*Genotype*	*Antigen*
A	I^AI^A or I^Ai	A
B	I^BI^B or I^Bi	B
AB	I^AI^B	A and B
O	ii	O

From this Table it is evident that in this allelic series, the alleles I^A and I^B determine a unique form of antigen each, and the allele i determines a failure to produce either forms of antigen. Marriages between blood groups AB ($I^AI^B \times I^AI^B$) will produce A Blood group: AB Blood groups: B Blood groups in 1 : 2 : 1 ratio or genotype ratio $1I^AI^A : 2I^AI^B : 1I^BI^B$.

Incompatibility alleles in plants form good examples of multiple allelism. In many dicots and monocots, *e.g.,* tobacco, petunias, sweet cherries, evening primroses, Brassica, etc., one gene determines compatibility/incompatibility relations with many different allelic forms of the gene present in different plants of any one species. In these cases incompatibility is determined by a single self incompatibility gene S. Generally, multiple alleles (such as S_1, S_2, S_3, S_4...) are common at S-locus. In some cases two genes, each with multiple allelic series are also known. The incompatibility reaction was first explained by East and Mangelsdorf in 1925 in *Nicotiana sanderae.* The incompatibility reaction of the pollen is determined by its own genotype and not by the genotype of the plant on which it is produced. If the pollen grain bears a S allele that is also present in the style, then it will not grow and so fertilization will not be effected. However, if S allele is absent in the style tissue, the pollen grain produces pollen tube containing male nuclei and this tube effects fertilization. The number of S alleles in a series in a species can be very large–over 50 in the evening primrose and *Trifolium* and cases with more than 100 alleles have been reported in some species. The following crosses indicate as to how multiple alleles control compatibility in certain plants:

Compatibility in Certain Plants

Self-fertilization or cross between the parents of same genotype	cross-fertilization	
Parents: $S_1S_2 \times S_1S_2$ Fully incompatible Progeny: none	$S_1S_2 \times S_2S_3$ ↓ Partially compatible Genotype of progeny: S_1S_2 & S_2S_3	$S_1S_2 \times S_3S_4$ ↓ Fully compatible Genotype of Progeny: S_1S_3, S_2S_3, S_1S_4, S_2S_4

Variation in Dominance Relation

The phenotypic ratios obtained by Mendel with garden peas have been observed in a wide range of organisms. Such ratios require that one member of an allelic pair be dominant over the other (***complete dominance***). But, in fact, this kind of relationship between the alleles is quite uncommon. As far as the phenotype of heterozygote is concerned there are three more dominance relations.

Incomplete Dominance. In *Mirabilis Jalapa* plants, when a pureline with red flowers is crossed to a pure line with white flowers, the results obtained are shown in Fig. 15.1.

It is evident from this cross that F_1 have not red but pink flowers. The possible explanation in this case may be that red phenotype and its determining allele R is *incompletely dominant* over the white phenotype and its allele r. **Incomplete dominance** is quite common and it is identified by the intermediate appearance or phenotype of the hybrid (heterozygote) between those of the parents (homozygotes). The phenomenon of incomplete dominance was first clearly described by Kolreuter.

Several examples of incomplete dominance may be found in animals also. One such example is found in Andalusian fowls where black feather colour has been found to be incompletely dominant over the splashed white feather colour. When black feathered fowl is crossed with a white feathered hen, the F_1 individuals are blue feathered. When inbreeding is allowed in F_1 hybrids, F_2 generation is represented by black feathered, blue feathered and splashed white feathered individuals in 1 : 2 : 1 ratio as illustrated by Fig. 15.2.

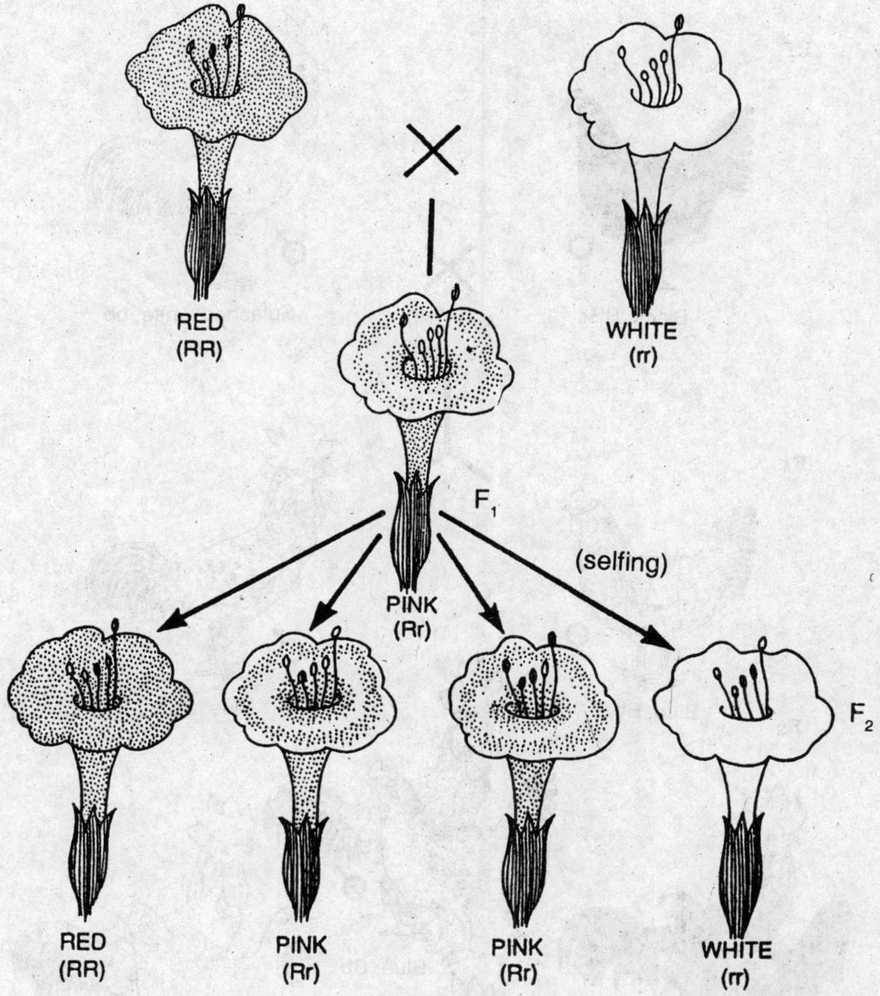

Fig. 15.1 Cross between red and white flowers of *Mirabilis jalapa* showing incomplete dominance.

Overdominance. Another type of dominance relation is ***overdominance*** in which the phenotype of heterozyote is more extreme than that of either homozygous parent. The amount of florescent eye pigment in heterozygous white eyed *Drosophila* exceeds that found in either parent. Over dominance is consequence of the heterozygous state of concerned gene, unlike partial dominance, complete dominance and codominance.

Co-dominance. There is yet another type of dominance relation, the *co-dominance*. In the case of **co-dominance** the heterozygote shows the phenotype of both homozygotes or parents rather than expressing an intermediate phenotype as in *incomplete dominance*. In a sense, then, co-dominance is no dominance at all. ABO blood groups in man is an example of co-dominance in which both the alleles I^A and I^B are equally expressed. These blood groups are governed by presence of an antigen on outer surface of red blood cells. A gene i produces the antigens on the surface of red blood cells. Dominant gene I^A produces antigen A which gives rise to blood group A. Another dominant allele I^B produces antigen B which gives rise to blood group B. In heterozygous condition I^A I^B, both alleles are codominant and they produce their own A and B antigens and so heterozygotes are classified in blood group AB.

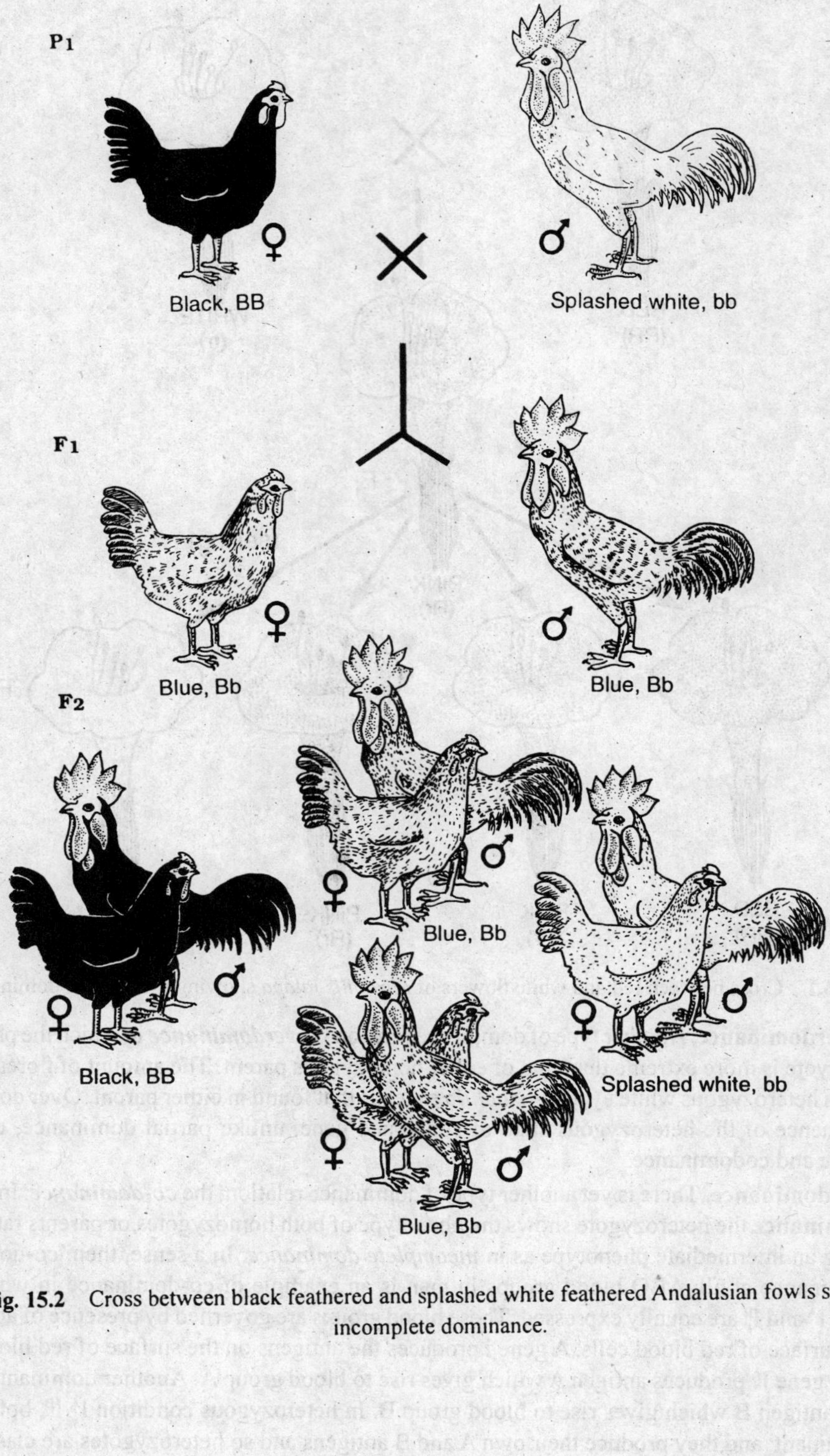

Fig. 15.2 Cross between black feathered and splashed white feathered Andalusian fowls showing incomplete dominance.

Marriages between heterozygotes I^AI^B blood groups result in three types of phenotypes.

Parent I^AI^B ♀ × I^AI^B ♂

gametes I^A and I^B I^A and I^B

♂ \ ♀	I^A	I^B
I^A	I^AI^A	I^AI^B
I^B	I^AI^B	I^BI^B

Genotype		Phenotype
1 I^AI^A	—	1 Blood group A
2 I^AI^B	—	2 Blood group AB
1 I^BI^B	—	1 Blood group B.

M-N blood group system also present a good example of co-dominance. The three blood groups and their genotypes are given in Table 15.2.

Table 15.2. M-N Blood Groups in Human

Blood phenotypes	*Genotypes*
M	L^ML^M
N	L^NL^N
MN	L^ML^N

Blood groups are characterised by the presence of specific immunological antigens on the surface of red blood cells. MN individuals with genotype $L^M L^N$ have both antigens.

Penetrance and Expressivity

Some genes are little influenced by the other genes or the environment and are generally governed by one or more genes with large easily detectable effects. These genes are called **oligogenes**. The traits determined by oligogenes show distinct classes and are known as ***qualitative characters***. On the other hand many characters are affected by the genetic background and by the environment. These characters are governed by **polygenes**. The characters which are affected by polygenes are referred to as ***quantitative characters***. Since the quantitative characters do not show distinct classes, they have to be studied by measurements. A gene does not determine a phenotype by acting alone, it does so in conjunction with other genes and with environment. Generally, the oligogenes express themselves in all the individuals that carry them and their expression is fairly uniform. The ability of a gene to express itself in an individual that carries it is called **penetrance**. An organism may be of genotype aa or A^- (AA or Aa) but may not express the phenotype associated with its genotype because of the modifying effect of the environment or because of modifiers. Such genes are said to have ***incomplete penetrance***. Penetrance is defined as the percentage of individuals with a given genotype which exhibit the phenotype specified by that genotype. For example, if only 10 per cent of the seedlings of lima bean carrying a gene for chlorophyll deficiency exhibit chlorophyll deficiency, this gene has penetrance of only 10 per cent. The gene which expresses itself in every individual that carries it is said to have *complete penetrance.*

Expressivity, on the other hand, refers to the degree or extent to which a given genotype is expressed phenotypically in an individual. Some genes show variable degrees of expression in different individuals. Such genes are said to have variable expressivity which may be slight, intermediate or severe. A gene which expresses itself uniformly in all the individuals carrying it has **uniform** or **complete expressivity** while those genes not expressed phenotypically in all the individuals uniformly are said to have **variable or incomplete expressivity**. The gene causing chlorophyll deficiency in lima bean (*Phaseolus lunatus*) seedlings exhibits variable expressivity in addition to incomplete penetrance. In some seedlings the cotyledonary leaves have no chlorophyll, in some the chlorophyll is absent only in the tips, while in some others the chlorophyll deficiency may be observed in the margins of the leaves. Thus a single gene with variable expressivity may exhibit a number of phenotypes as if more than one gene were involved.

Some genes express themselves in a specific environment such as at particular temperature. The characters whose development depends upon a specific environment are referred to as ***threshold characters***, as for example, a mutant gene in barely (*Hordeum vulgare*) which produces albino seedlings at temperature below 8°C and normal green ones at the temperature above 19°C. The incomplete penetrance of some genes may be due to a threshold requirement which may be helpful in identification of desirable types. For example, disease resistant and disease susceptible plants cannot be distinguished unless they are exposed to disease.

Modifying Genes

There are many genes which seem to have little or no effect of their own but increase or decrease the expression of other major genes. These are called modifiers or modifying genes since they modify the effects of other genes.

Pleotropic or Many-fold Effects of a Single Gene

Mendel for convenience of explanation of his results said that single independent factor controlled a single trait only, but actually it is not always so. Now several such genes are known which govern many traits at a time. In *Drosophila*, the recessive gene in homozygous state produces not only the vestigial wings but also it affects several other traits such as tiny wing-like balancer behind the wings, certain bristles, structure of reproductive organs, less production of eggs and decrease in the longevity. In Jowar (***Sorghum***), the gene that controls the length of internodes also determines breadth and length of leaves and length of inflorescence (penicle). In cotton, the height, boll size, number of ovules per locule and viability of seeds are governed by a single gene. This type of gene controlling several characters at a time is termed as pleotropic gene and the effect is referred to as pleotropic effect. The pleotropic genes sometimes show one major effect and several secondary effects.

Gene Interactions

Most of the qualitative characters are determined by oligogenes or major genes. Nonallelic genes may not function independently in determining phenotypic characters of an organism. The effect of a gene depends not only on its own function but also on the functions of other genes and also on the environment. Interaction between gene products may occur to produce new phenotypes without modifying typical Mendelian ratios, or it may cause modification of Mendelian ratios by one gene product interfering with phenotypic expression of another non-allelic gene or genes. The genetic interactions are discussed under the following heads:

(*i*) Gene-Gene Interactions.

(*ii*) Gene-Environment Interactions.

GENE-GENE INTERACTIONS

Mendel selected seven pairs of contrasting characters in pea each of which was governed by a single gene and all those seven sets of alleles were found on separate chromosome pairs. According to Mendel, there existed 1:1 ratio between trait and factor (gene). In fact, many of the characters in almost all individuals are governed by two or more genes. Such genes affect the concerned traits in various ways and consequently the typical F_2 dihybrid ratio (9:3:3:1), trihybrid ratio (27:9:9:9:3:3:3:1) are modified.

The genes show various relationships with each other in producing a character. These relationships are referred to as gene-gene interactions. Gene-gene interactions are classified on the basis of manner in which they influence each other. The relationships between the interacting genes are of several types, some of the common interactions are as follows:

1. Interaction between dominant factors, 2. Complementary gene action, 3. Supplementary gene action, 4. Epistatic gene action, 5. Inhibitory gene action, 6. Duplicate gene action 7. Multiple gene action or polygenic inheritance, 8. Lethal gene effect.

Various gene-gene interactions are discussed here as follows:

I. Interaction between Dominant Factors (Typical dihybrid Ratio for a single trait)

Some of the heritable characters in plants and animals depend upon the combined effect of two or more of dominant genes. This type of gene interaction produces typical dihybrid ratio 9 : 3 : 3 : 1 in F_2 for a single character.

Bateson and Punnett have explained it by taking example of inheritance of comb types in fowls.

The following four main types of combs (Fig. 15.3) are common among fowls:

(*i*) Rose comb, (*ii*) Pea comb, (*iii*) Walnut comb, and (*iv*) Single comb.

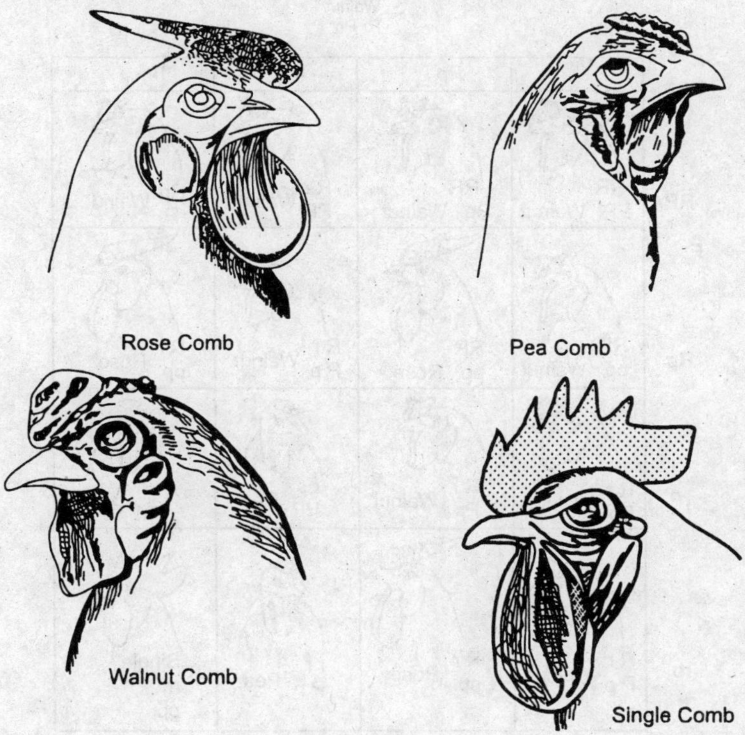

Fig. 15.3 Comb types in fowls.

Among these, pea and rose types are governed by two separate dominant genes which are found on separate chromosomes and single comb is recessive to all of them. If the dominant gene for pea comb is represented by 'P' and that of Rose comb by 'R', the genotypes of the above three homozygous forms will be as follows:

Pea = PP rr

Rose = ppRR

Single = pp rr

When Rose and Pea combed fowls were mated, progeny with a new type of comb, called the Walnut comb (genotype Pp Rr) appeared in the F_2. In the production of Walnut comb, therefore, two independent dominant factors 'P' and 'R' are jointly responsible. When these F_1 hybrids were inbred, they produced all the four types of combs in the typical dihybrid ratio 9 Walnut :3Rose :3Pea :1 Single. This is shown in Fig. 15.4.

From checker board the following ratio of different phenotypes is obtained:

Phenotype	Walnut	Rose	Pear	Single
Number	9	3	3	1

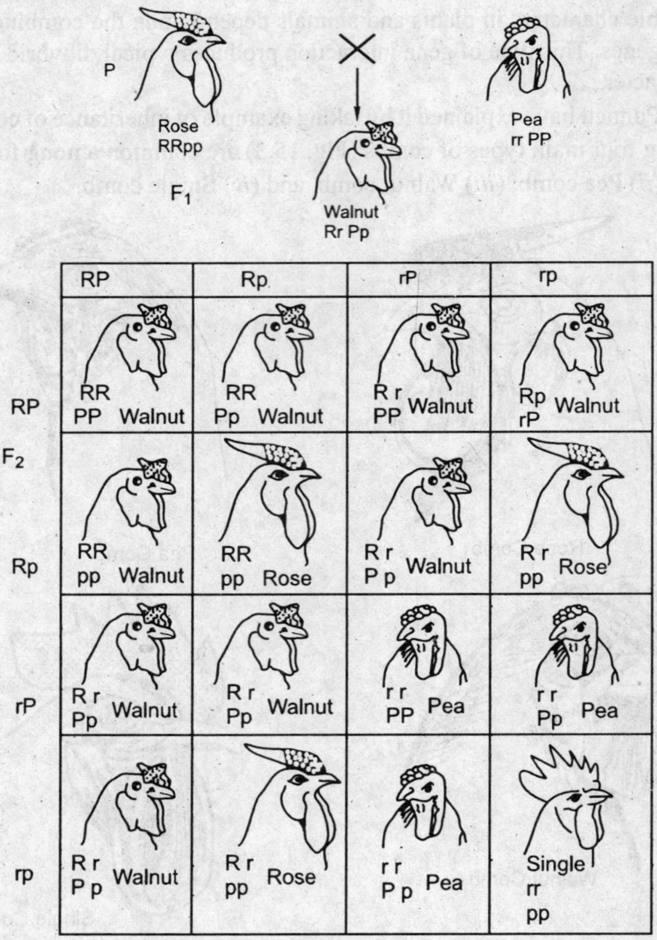

Fig. 15.4 Inheritance of comb types in fowls.

II. Complementary gene action (9 : 7 Ratio)

Certain characters are produced by the interaction between two or more dominant genes inherited from different parents. These genes are complementary to one another, *i.e.*, if any of the two dominant genes is present alone it remains unexpressed, but are expressed only when they are combined together in F_1 through suitable crossing, they complement each other to produce the concerned phenotype.

W. Bateson and R.C. Punnett crossed two white flowered varieties of Sweet pea (*Lathyrus odoratus*). In the experiment all the plants of F_1 generation developed red flowers instead of expected white flowers. These F_1 plants, on self-pollination, produced red and white flowered plants in the ratio 9 : 7 in F_2 generation. In sweet pea, purple or red coloured flowers develop only when two genes C and R are present. Plants with genotypes cc RR or rr cc or cc rr are all white flowered.

Bateson explained his Sweetpea experiment in the following way:

Parents : White flowered Sweet pea × White flowered Sweet pea or CCRR × ccrr
Genotype : CCrr ↓ RRcc (Red) ↓ (White)
F_1 Red flowered plants Cc Rr
Genotype of F_1 CcRr (Red flowered)

Gametes of F_1, CR, Cr, cR, cr

♂ gametes / ♀ gametes	CR	Cr	cR	cr	
CR	CR CR Red	Cr CR Red	cR CR Red	cr CR Red	
Cr	CR Cr Red	Cr Cr White	cR Cr Red	cr Cr White	F_2 generation
cR	CR cR Red	Cr cR Red	cR cR White	cr cR White	
cr	CR cr Red	Cr cr White	cR cr White	cr cr White	

Summary:

Red flowered plants with both complementary genes C and R = 9, White flowered plants having either 'C' or 'R' = 7.

From the checker board it is clear that the red colouration of the flower is due to interaction of two complementary factors C and R. The parents developed white flowers because they lacked either gene 'C' or gene 'R' in their genotypic expressions. F_1 hybrids received both the complementary genes C and R, hence all the flowers in that generation were red. The ratio 9 : 7 obtained in F_2 generation is actually a modification of normal dihybrid ratio (9 : 3 : 3 : 1).

Similar cases are reported in many other plants. In *Sorghum* (Jowar), for example, there are two white grained races which, when crossed, produced F_1 hybrids with brown grains. When brown grained F_1 plants were self-pollinated, they produced brown grained and white grained plants in the ratio 9 : 7.

In paddy, the plants of two varieties with colourless stigmas, when crossed, produced F_1 hybrids with purple stigma. By selfing the F_1 hybrids, plants with purple and white stigmas were obtained in the 9 : 7 ratio in F_2.

III. Supplementary gene action (9 : 3 : 4 Ratio)

Sometimes interaction involves two different types of genes in which one type can express itself independently but other has no expression of its own. When two such genes interact, they produce a new character. In *Sorghum* (Jowar) the black purple colour (P) is dominant over brown (p). Blackish pigmentation is found to change to red in certain crosses. The modifying factor Q borne on another chromosome (non-allelic) has no expression of its own in presence of other recessive factor 'p'. Similarly, the recessive factor 'q' is also without any phenotypic effect but when both dominants P and Q are brought together, the blackish purple colour is changed to red. The theoretical results to be expected from a cross between blackish purple and brown glumed variety can be explained in the following way:

Parents Blackish purple PP qq × Brown pp QQ

↓

F_1 Pp Qq (reddish purple)

Gametes of F_1: PQ, Pq, pQ, and pq

♀ / ♂	PQ	Pq	pQ	pq	
PQ	PQ PQ Red purple	Pq PQ Red purple	pQ PQ Red purple	pq PQ Red purple	F_2
Pq	PQ Pq Red purple	Pq Pq Black purple	pQ Pq Red purple	pq Pq Black purple	
pQ	PQ pQ Red purple	Pq pQ Red purple	pQ pQ Brown	pq pQ Brown	
pq	PQ pq Red purple	Pq pq Black purple	pQ pq Brown	pq pq Brown	

Summary:

F_2 Phenotypic classes	No. of individuals in each class	Ratio
1. Reddish purple	9	
2. Blackish purple	3	9: 3 : 4
3. Brown	4	

An example of similar type is met within rats and rodents. In rats, there are three common skin or coat colours. These are Black, Albino, and Wild type Agouti (yellow hair with black bases). The black coat colour is governed by dominant gene 'C' and albino colour by dominant gene 'A'. The agouti colour is governed by interaction of two dominant genes A and C. 'A' does not show any visible effect if it is alone, but the gene C is able to express itself independently. When a true breeding black rat is crossed with an albino carrying the dominant gene A, all the F_1 hybrids are of the agouti type. When F_1 individuals are inbred, the F_2 generation is represented by agouti, black and albino in 9 : 3 : 4 ratio. These results can be explained in the following way:

Parent : Black × Albino
Genotype : CC aa ↓ cc AA
F_1 Hybrids- Agouti
F_1 genotypes- Cc Aa
Gametes of F_1 CA Ca cA ca

♀ \ ♂	CA	Ca	cA	ca
CA	CA CA Agouti	Ca CA Agouti	cA CA Agouti	ca CA Agouti
Ca	CA Ca Agouti	Ca Ca Black	cA Ca Agouti	ca Ca Black
cA	CA cA Agouti	Ca cA Agouti	cA cA Albino	ca cA Albino
ca	CA ca Agouti	Ca ca Black	cA ca Albino	ca ca Albino

F_2

Summary: F_2 Generation

Agouti = 9
Black = 3 } Ratio – 9 : 3 : 4
Albino = 4

IV. Epistasis (12 : 3 : 1 Ratio)

According to Mendel's laws of inheritance, when two contrasting characters (allelomorphs) are brought together, one of them dominates over the other in F_1. In the phenomenon of epistasis, two independent nonallelic genes affecting the same trait of an individual interact in such a way that one overmasks the expression of the other. Here the gene which prevents the expression of another nonallelic gene is said to be ***epistatic*** and the suppressed gene is said to be ***hypostatic***.

In cholam (*Sorghum caudatum*), the pearly grain colour is dominant over chalky. Pearly colour develops when dominant gene 'Z' is present, and chalky colour develops in absence of it. Another dominant gene 'W' which is responsible for the red colour of seeds, is dominant over the gene Z. When 'W' is present, the grain will always be red irrespective of the presence of other gene Z. In the absence of W, the grain colour will be pearly or chalky depending upon the presence or absence of gene Z. Thus the genotypes of all the coloured grained plants will be as under:

Red grained = WWzz or WWZZ
Pearly grained = wwZZ
Chalky grained = wwzz

When a red grained type with the genotype WWzz is crossed with pearly variety (wwZZ), F_1 offsprings are red grained (Ww Zz). In F_2 generation three kinds of plants; red, pearly and chalky, appear in the ratio 12 : 3 : 1 which is a modified dihybrid ratio. The mode of inheritance can be represented in the following way:

Parents	Red WWzz	× ↓	Pearly wwZZ	
F_1 Hybrid		Red WwZz		
Gametes of F_1	WZ	Wz	wZ	wz

♀ gametes \ ♂	WZ	Wz	wZ	wz
WZ	WZ WZ Red	Wz WZ Red	wZ WZ Red	wz WZ Red
Wz	WZ Wz Red	Wz Wz Red	wZ Wz Red	wz Wz Red
wZ	WZ wZ Red	Wz wZ Red	wZ wZ Pearly	wz wZ Pearly
wz	WZ wz Red	Wz wz Red	wZ wz Pearly	wz wz Chalky

F_2

Summary of F_2 generation:

Phenotypes –	Red	:	Pearly	:	Chalky
Ratio–	12	:	3	:	1

In the squashes (*Cucurbita pepo*), Sinnot has shown that there are three common fruit colours, white, yellow and green. Sinnot has shown that white was dominant over both yellow and green and yellow was dominant over green only. In a cross between white variety (genotype WW YY) and green variety (genotype wwyy), the F_1 progeny appeared white fruited (genotype Ww Yy) which on selfing gave a 12 white : 3 Yellow : 1 green ratio. Here 'W' acts as an epistatic factor and 'Y' is active only with 'w';

Parents	White	× ↓	Green
Genotype	(WWYY)		(wwyy)
F_1 Hybrid		WwYy (white)	
F_1 gametes		WY, Wy, wY, wy	

Gametes ♀ \ ♂	WY	Wy	wY	wy
WY	WY WY white	Wy WY white	wY WY white	wy WY white
Wy	WY Wy white	Wy Wy white	wY Wy white	wy Wy white
wY	WY wY white	Wy wY white	wY wY yellow	wy wY yellow
wy	WY wy white	Wy wy white	wY wy yellow	wy wy green

F_2

Summary of F_2 Generation:

Phenotype-	White	:	Yellow	:	green
Ratio-	12	:	3	:	1

From the above experiment, it is clear that each fruit colour is being governed by two non-allelic genes W and Y. If gene W is in dominant condition, dominant Y gene will not show its effect. Gene Y is active only when it is combined with recessive 'w'.

V. Inhibitory gene action (13 : 3 Ratio)

In this modification or gene interaction, one dominant factor inhibits or suppresses the phenotypic expression of another dominant gene found on different chromosome. This may be explained by taking example of pigmentation of leaf in rice plants. Here purple colour of the leaf is due to factor Lp and green due to lp. The purple colour is dominant over the green. There is another factor 'I'. If dominant factor 'I' is present with the factor Lp, the purple colouration (due to factor Lp) is inhibited and the leaf becomes normal green. The factor 'I' has thus no visible effect of its own but it inhibits the colour expression of gene Lp.

In a cross between purple pigmented (ii Lp Lp) and a green (I I lp lp) rice plant, the F_1 plants were all green (I i Lplp) and in F_2 there appeared green and purple pigmented plants in 13 : 3 ratio. The results are explained below:

Parents Green × Purple
Genotype II lp lp ↓ ii Lp Lp
F_1 Ii Lplp (Green)
Gametes of F_1 – ILP, Ilp, iLp, ilp.

♀ \ ♂	ILP	Ilp	iLp	ilp
ILP	ILp ILp green	Ilp ILp green	iLp ILp green	ilp ILp green
Ilp	ILP Ilp green	Ilp Ilp green	iLp Ilp green	ilp Ilp green
iLp	ILp iLp green	Ilp iLp green	iLp iLp purple	ilp iLp purple
ilp	ILp ilp green	Ilp ilp green	iLp ilp purple	ilp ilp green

F_2

Summary:	**F_2 Phenotype-**	Green	:	Purple
	F_2ratio :	13	:	3

A similar effect is seen in the inheritance of plume colours in fowl. White Wyandotte and white leghorn are two breeds of fowls with white plumes. The white plumage of the white wyandottes is recessive to the coloured plumage. The white leghorns possess a colour factor which is prevented from being expressed by an inhibiting gene. If the inhibiting factor is symbolised by 'I' and the colour factor by 'C', the genotype of white leghorn can be expressed by CCII and that of white Wyandotte by ccii.

When white Leghorns (CCII) and white Wyandotte (ccii) are crossed, the F_1 hybrids are all white (Cc Ii). When the F_1 hybrids are inbred, the offsprings in F_2 appear in the ratio 13 White :3 coloured. The results can be explained in the following way–

Parents White leghorn × White Wyandotte
Genotype CCII ccii (white)
F_1 hybrid CcIi (white)
Gametes of F_1 CI, Ci, cI and ci

Gametes ♂ / ♀	CI	Ci	cI	ci
CI	CI CI white	Ci CI white	cI CI white	ci CI white
Ci	CI Ci white	Ci Ci coloured	cI Ci white	ci Ci coloured
cI	CI cI white	Ci cI white	cI cI white	ci cI white
ci	CI ci white	Ci ci coloured	cI ci white	ci ci white

F_2

F_2 generation — White : Coloured
13 : 3

VI. Duplicate gene action (15 : 1 Ratio)

In this type of interaction two or more genes either individually or in combination produce the same phenotypic effect. An example of this type of interaction is the inheritance of awn in rice grain. In a cross between 'awned' and 'awnless' types, F_1 rice grains were all awned. Awned character is thus dominant over awnless. The F_1 plants on self-pollination give awned and awnless types in 15 : 1 ratio in F_2 generation. This result can be explained if it is supposed that awned character is governed by two dominant genes, say A_1 and A_2. The presence of one or both dominant genes results in awned grains and in absence of both these dominant genes awnless grains develop.

Parents— Awned × Awnless
Genotype $A_1 A_1 A_2 A_2$ $a_1 a_1 a_2 a_2$
F_1 hybrid Awned
$A_1a_1A_2a_2$
F_1 Gametes — A_1A_2, A_1a_2, a_1A_2 and a_1a_2

Gametes ♂ / ♀	A_1A_2	A_1a_2	a_1A_2	a_1a_2
A_1A_2	A_1A_2 A_1A_2 awned	A_1a_2 A_1A_2 awned	a_1A_2 A_1A_2 awned	a_1a_2 A_1A_2 awned
A_1a_2	A_1A_2 A_1a_2 awned	A_1a_2 A_1a_2 awned	a_1A_2 A_1a_2 awned	a_1a_2 A_1a_2 awned
a_1A_2	A_1A_2 a_1A_2 awned	A_1a_2 a_1A_2 awned	a_1A_2 a_1A_2 awned	a_1a_2 a_1A_2 awned
a_1a_2	A_1A_2 a_1a_2 awned	A_1a_2 a_1a_2 awned	a_1A_2 a_1a_2 awned	a_1a_2 a_1a_2 awnless

F_2

Summary:

F_2 phenotypes— Awned : Awnless
Ratio— 15 : 1

From the above cross it is clear that out of sixteen offsprings in the F_2 generation, fifteen have one or both the dominant genes, so they are all awned type and only one is such as is double recessive, hence awnless.

The inheritance of chlorophyll in ragi can be cited as second example. Two factors C_1 and C_2 produced chlorophyll. When a cross was made between chlorophyllous and achlorophyllous parents, F_1 seedlings with factors C_1 C_2 were chlorophyllous (green) and F_2 individuals were green and non-green in ratio 15 : 1.

VII. Multiple gene action or Polygenic Inheritance

The term polygenic inheritance is applied when two or more independent pairs of genes affect the same character in the same way and in additive fashion. In such cases, the net effect on the trait depends upon the combined action of several genes, each of which has a small effect on the same trait. Such genes are called the ***'cumulative genes'*** or ***'polygenes'*** and the trait influenced by them is known as ***'polygenic trait'***. Polygenic traits do not show clear cut difference between individuals and all gradations occur between the two extremes. The characters are thus quantitative instead of being qualitative, as for example, the height, weight, pigmentation, blood pressure, finger prints, ridge counts and intelligence in man, the amount of fruits and seeds produced by plants and milk or meat produced by animals.

The inheritance of pericarp colour in wheat is polygenic according to Herman, a Swedish Geneticist. Nilsson-Ehle (1909) crossed pure red seeded wheat with pure breeding white seeded wheat. The colour of seed coat is dependent in this case on the genotype of the plant and not the seeds because the pigment is derived from maternal tissue. The F_1 plants were found to have red seed. When selfing was allowed in F_1 plants, all the 78 F_2 plants were found to have red seeds instead of red and white in 3 : 1 ratio as expected. He then self-pollinated F_2 plants individually and found that 78 plants could be grouped into the following 4 categories on the basis of the ratio obtained in F_3 generation.

(*i*) Group 1 — F_2 plants producing red seeded and white seeded plants in a ratio 3 : 1. Such plants were eight in number. From this he reasoned that they were heterozygous for one gene pair.

(*ii*) Group 2 — F_2 plants produced 15 red : 1 white; so he reasoned that they were heterozygous for two gene pairs. Such plants were 15 in number.

(*iii*) Group 3 — F_2 plants produced red and white in 63 : 1 ratio (three homozygous recessive $= (1/4)^3 = 1/64$) and they were 5 in number.

(*iv*) Group 4 — F_2 plants produced only red seeded plants and so he reasoned that they were homozygous for dominant alleles for redness of at least one gene pair.

From these ratios one thing became clear that he was dealing with a triplicate gene situation in which a dominant allele of any one of three separate genes (A,B or C) could cause redness and the original cross must have been AA BB CC (red) × aa bb cc (white) and F_1 Aa Bb Cc. The following Table shows the F_2 genotypes and their theoretical probabilities and observed frequencies:

Ratio on selfing F_2 plants	Genotypes of F_2 plants	Probability (Genotype)		Observed ratio
1. 3 : 1	Aa bbcc	1/2 × 1/4 × 1/4	= 1/32	
	or aaBbcc	1/4 × 1/2 × 1/4	= 1/32	
	or aabbCc	1/4 × 1/4 × 1/2	= 1/32	
			= 3/32 = 6/64	8/78
2. 15 : 1	AaBbcc	1/2 × 1/2 × 1/4	= 1/16	
	or AabbCc	1/2 × 1/4 × 1/2	= 1/16	
	or aaBbCc	1/4 × 1/2 × 1/2	= 1/16	
			= 3/16 12/64	15/78
3. 63 : 1	AaBbCc	1/2 × 1/2 × 1/2	= 1/8 = 8/64	5/78

Ratio on selfing F_2 plants	Genotypes of F_2 plants	Probability (Genotype)		Observed ratio
4. 1 : 0	AAbbcc or AABBCC or aa BBcc etc.	1 – 6/64 – 12/64 – 8/64 – 1/64	= 37/64	50/78
5. 0 : 1 (not found)	aabbcc	1/4 × 1/4 × 1/4	= 1/64	0/78

Although Nilsson-Ehle was unlucky not to have picked up the pure white F_2 aa bb cc yet his results statistically fit into the triplicate gene model very well.

Nilsson-Ehle actually observed a lot of variations in the amount of redness in red coloured seeds, some being very dark, some being light and some were of intermediate shades and it was difficult for him to make accurate groupings. However, in a dihybrid selfing (say Aa Bb cc × Aa Bb cc), he was able to classify the progeny plants into 5 groups with the following relative frequencies :

Dark red	1	15
Medium dark red	4	
Medium red	6	
Light red	4	
White	1	

To explain the variation in the degree of redness, he hypothesized that the *total number* of dominant alleles of both gene pairs was determining the degree of colouration. In other words, each dominant gene could be considered to be contributing one dose of redness and the effect of each dose was *equal* and *additive*, the more dominant genes a plant inherited darker the colour of pericarp will be.

This idea fits with the observed ratio well forgetting about the third gene pair, although irrelevant here, we can calculate the probability of getting doses of 0, 1, 2, 3 and 4 dominant alleles from the cross Aa Bb × Aa Bb as follows:

For A and a	For B and b	For Aa and Bb	Total number of dominant doses
AA = 1	BB = 1	1 AA BB	4
	Bb = 2	2 AABb	3
	bb = 1	1 AAbb	2
Aa = 2	BB = 1	2 AaBB	3
	Bb = 2	4 AaBb	2
	bb = 1	2 Aabb	1
aa = 1	BB = 1	1 aaBB	2
	Bb = 2	2 aaBb	1
	bb = 1	1 aabb	0

Relative frequencies of different dominant doses:

4 dominants —1
3 dominants — 4
2 dominants — 6
1 dominant — 4
0 dominant — 0

The inheritance of fruit shapes in pumpkin (*Cucurbita pepo*) is another example of cumulative effect. A cross between two different races with spherical fruits gives disk shaped F_1 offsprings which on self-pollination give 9 disk :6 sphere :1 cylindrical in F_2 generation.

Another interesting example of the interaction of multiple factors is the inheritance of the skin colour in man. C.B. Devenport (1913) considered that two pairs of genes account for the difference of pigments between negro and concasian (white people). He conducted his most significant studies in Bermuda and Jamaica where interracial marriages between coloured and white people were relatively frequent and legitimate births were common.

Let us assume that negroes have two pairs of dominant colour genes $P_1\ P_1\ P_2\ P_2$ and whites have only recessive genes $p_1\ p_1\ p_2\ p_2$. A marriage between a negro and a white would give F_1 with genotype $P_1\ p_1\ P_2\ p_2$. The individuals of this generation are called mulattoes. Their colour will be intermediate between two parents. Marriage between mulattoes produces offsprings (F_2) ranging in colour from "Pure Negro" through all the intermediate shades to pure white. The pattern of inheritance can be shown in the following way:

Parents Negro White

Genotype $P_1P_1P_2P_2$ $p_1p_1p_2p_2$

F_1 ↓ mulattoes

F_1 Genotype $P_1p_1P_2p_2$

F_1 gametes — P_1P_2, P_1p_2, p_1P_2, p_1p_2

♂ Gametes ♀ Gametes	P_1P_2	P_1p_2	p_1P_2	p_1p_2
P_1P_2	P_1P_2 Black P_1P_2	P_1p_2 Dark P_1P_2	p_1P_2 Dark P_1P_2	p_1p_2 Dark p_1p_2
P_1p_2	P_1P_2 Dark P_1p_2	P_1p_2 Dark P_1p_2	p_1P_2 Dark P_1p_2	p_1p_2 Dark P_1p_2
p_1P_2	P_1P_2 Dark p_1P_2	P_1p_2 Dark p_1P_2	p_1P_2 Dark P_1p_2	p_1p_2 Dark p_1P_2
p_1p_2	P_1P_2 Dark p_1p_2	P_1p_2 Dark P_1p_2	p_1P_2 Dark P_1p_2	p_1p_2 White p_1p_2

F_2

Other investigators have suggested that the range in the skin colours between negro and white is due to more than two pairs of pigment forming genes. Gates (1953) has suggested three pairs of genes. Some workers have suggested the number of colour genes from 2 to 20 pairs.

In *Drosophila,* eye colour inheritance is governed by multiple alleles. The wild type eye colour is red in *Drosophila* but the gene for red eye has mutated to white, eosine, cherry wory, apricot and many other colours.

The principle of mulitiple gene inheritance or quantitative inheritance is now considered as most important principle of genetics. It has been strengthened greatly by the use of statistical methods devised by Fisher (1947). Practical applications are being made in plants and animal breeding for the improvement of field quality and colours.

VIII. Lethal Factor (2 : 1 Ratio)

Lethal genes are Mendelian factors whose substitution for their normal allelomorphs results in the developmental or physiological changes and finally causes the death of an organism. The lethal genes may cause death of organism at any time in the life cycle of organism from gamete to adult. Lethal genes may be dominant or recessive, for example, yellow coat colour in mouse is a heterozygous (Yy) character. Yellow mouse with homozygous dominants (YY) does not survive. Yellow coated mice never breed true. When inbred, these yellow mice always give a progeny of 2 yellow : 1 black instead of monohybrid 3 : 1 ratio. The result of mating between two heterozygous mice may be represented as follows:

Yellow (Yy) × Yellow (Yy)

1YY	2Yy	1yy
Pure yellow (homozygous lethal) (dies)	yellow	Black

The homozygous offsprings (YY) die, because the gene Y in the homozygous condition has lethal effect on the embryo. Thus monohybrid F_2 ratio 1 : 2 : 1 is converted into 2 : 1 ratio by the death of homozygous yellow (Fig. 15.5). This was verified when spontaneously aborted fetuses were observed in uteri of females.

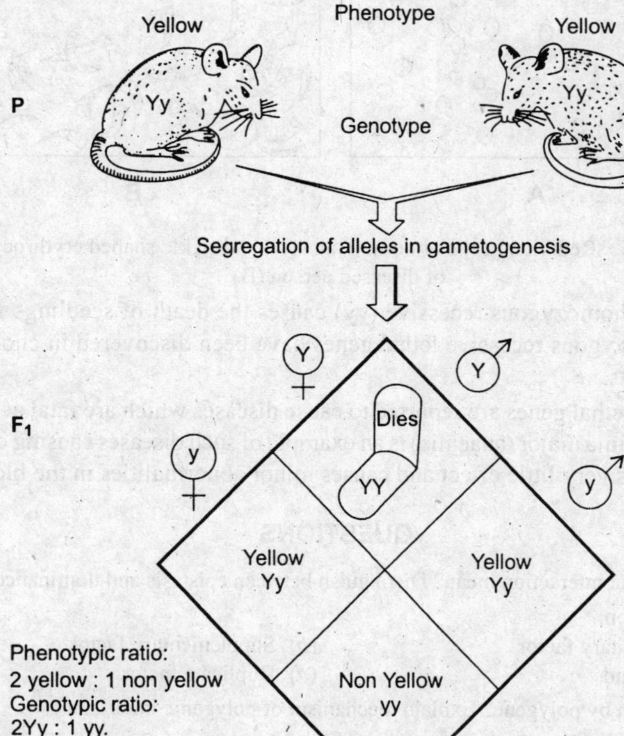

Fig. 15.5 2 : 1 ratio due to homozygous dominant lethal genes in mice.

In fowl, the creeper gene'C' is lethal in homozygous dominant condition. Creeper fowls with heterozygous genotype (Cc) have short crooked legs. Normals have two recessive genes. When two creepers were mated, a ratio of 2 creepers to 1 normal appeared as shown below:

Creeper (Cc) × Creeper (Cc)

ICC	2Cc	1cc
Dies	Creeper	Normal

In human beings a few lethal genes are known which are associated with fatal diseases. *Infantile amaurotic idiocy* is caused by a recessive allele in homozygous condition. The bearer of this genotype loses its eye sight and becomes mentally degenerated in the early childhood and finally dies. *Colley's anemia* or *sickle cell anemia* which is common in certain tribes of Africa is yet another example of fatal disease associated with lethal allele. There are two dominant genes Hb_1^A and Hb_1^S. Individuals with genotype $Hb_1^a\ Hb_1^s$ (heterozygous) do not show symptoms of this disease and are phenotypically similar to those with genotype $Hb_1^a\ Hb_1^a$ (homozygous). Individual with genotype $Hb_1^S Hb_1^S$ (homozygous) only suffers from the syndrome (disease) sickle cell anemia which leads to death at the adolescence stage. In this disease, the erythrocytes or red blood cells become distorted and become sickle shaped. These distorted cells block the blood capillaries and impede circulation. Besides, these cells do not help in the exchange of respiratory gases (Fig. 15.6).

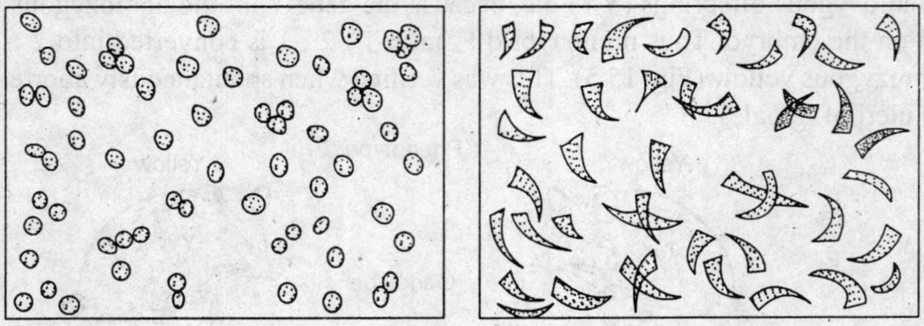

A B

Fig. 15.6 Red blood cells of normal man (A) and sickle shaped erythrocytes of diseased person (B).

In rice, a lethal homozygous recessive (yy) causes the death of seedlings within ten days of growth. Similar homozygous recessive lethal genes have been discovered in cholam, maize and in other Indian crop plants.

In man, several lethal genes are reported to cause diseases which are fatal at different stages of development. Thalassemia major (anaemia) is an example of such diseases causing death in childhood. Thalassemia minor has very little effect and causes minor abnormalities in the blood.

QUESTIONS

1. What does the gene interaction mean? Distinguish between epistasis and dominance.
2. Write short notes on:
   (*a*) Complementary factor (*b*) Supplementary factor
   (*c*) Epistasis, and (*d*) Duplicate factor.
3. What do you mean by polygene? Explain mechanism of polygenic inheritance.
4. In sweet pea, factor C or P alone produces white flowers and C and P if went together produce purple flowers. In the following crosses the genotypes of the parents are given, note the phenotypes of these parents and work out genotypic and phenotypic ratios of the progeny. (*a*) CCPp × ccPp. (*b*) CcPp × CcPp. (*c*) ccPp × CCpp and (*d*) Ccpp × ccPp.
5. Nelson-Ehle made crosses between two types of oat, one with white hulled seeds and the other with black hulled seed, the F_1 was black hulled which on self-pollination gave 428 black hulled, 106 grey hulled and 36 white hulled in F_2 generation. Give the genotypes of the parents, F_1 and F_2.
6. Discuss different variations in dominant relationship.
7. Write short notes on:
   (*i*) Multiple allelism,
   (*ii*) Co-dominance, and
   (*iii*) Lethal factor.

16 CHAPTER

Linkage and Crossing Over

LINKAGE (NON-INDEPENDENT ASSORTMENT)

When Mendel's work was rediscovered, thousands of experiments were conducted by the early biologists on the diverse groups of plants and animals. The results obtained in those cases were quite contrary to the principles of independent assortment, but none of the interpretations was good enough to justify for an alternative hypothesis to Mendel's principle of independent assortment. Sutton and Boveri noted that chromosomal behaviour in meiosis at the time of gametogenesis paralleled the genetic events described by Mendel. They concluded that the genes (hereditary factors or units) must be located in or on the chromosomes. Mendel's choice of the seven pairs of characters was indeed fortunate one, for the garden pea has seven pairs of chromosomes and each of the seven pairs of characteristics was associated with a different pair of chromosomes. William Bateson and R.C. Punnett (1906) reported a dihybrid experiment in sweet pea (*Lathyrus odoratus*) where segregation ratio could not be explained by Mendel's law of independent assortment. They crossed purple flowered and long pollen grained sweet pea with another variety of sweet pea which had red flowers and round pollen grains. The purple colour of the flower (r^+) and long pollen grains (l^+) had been proved by previous experiments to be governed by separate genes and were dominant over red colour of the flower (r) and roundness (l) of pollen grains respectively. So, as expected, in F_1 they found all hybrid plants with purple flowers and long pollen grains. When F_1 plants were self-pollinated a wide divergence was noted in F_2 generation from the expected normal dihybrid ratio, *i.e.*, 9 : 3 : 3 : 1.

Phenotypic class	*Observed numbers*	*Expected numbers on the basis of 9:3:3:1 ratio*
Purple-long	296	242
Purple-round	19	80
Red-long	27	80
Red-round	87	27

W. Bateson and R.C. Punnett were the first to observe that two pairs of genes did not necessarily assort independently of each other. But they believed that the genes were not located on chromosomes and so rejected the chromosome theory. To explain their results they proposed that independent assortment had occurred but was followed by differential sex reproduction. That is to say, during gamete formation, the cells containing both dominant (r^+l^+) and both recessive (rl) genes underwent mitotic reproduction to a greater extent than the others thereby giving specific ratio. This is ***reduplication hypothesis***.

Bateson and Punnett performed another cross in which they took a sweet pea plant with purple flower and round pollen (r^+r^+ll) and another plant with red flower and long pollen grain (rrl^+l^+). The

F_1, as expected, were with purple flowers and long pollen grains (r^+rl^+l) which on selfing produced F_2 plants.

The segregation in F_2 generation was again not in way as could be expected from the normal dihybrid ratio *i.e.*, 9 : 3 : 3 : 1.

P. Purple, round (r^+r^+ll) X Red, Long (rrl^+l^+)

F_1 Purple, Long (r^+rl^+l)

Selfing

F_2 observed Numbers	Purple, long	Purple, round	Red, Long	Red, round
	(226)	(95)	(97)	(1)

Coupling and Repulsion

Bateson and Punnett pointed out that when two or more dominants entered from same parents they tended to remain together and did not assort independently so that the recombinants were fewer than the parental types. Similar was the case with recessive alleles also. Bateson and Punnett called this tendency of both dominants or both recessives introduced in the cross by the same parents to remain together in subsequent generation more often as ***"coupling"***. Conversely, it was found that two dominant genes introduced in the cross by different parents, as shown in the second experiment tended to remain apart in subsequent generation. This is called ***repulsion***. Thus if a chromosome carries two genes A and B and its homologue carries recessive alleles a and b, A and B are said to be in coupling. Possibilities are also there that in a chromosomal pair, one contains genes A and b and its homologue contains a and B. In such a condition A and B genes are in repulsion. Two other terms *cis* and *trans* have also been used to denote coupling and repulsion respectively. The theory of coupling and repulsion is now discarded. Actually, Bateson and Punnett did not relate the mechanism with the chromosomes. Thus they missed true explanation by the narrow margin.

This discrepancy or abnormality in the earlier mentioned dihybrid result was explained in 1910 by Thomas Hunt Morgan who found similar gene behaviour in *Drosophila melanogaster* while studying the behaviour of two different gene pairs. One gene pair affects eye colour (pr responsible for purple eye and pr^+ for red eye) and the other pair affects wing length (gene vg responsible for vestigial and vg^+ for normal wing). He crossed flies with prpr vgvg and $pr^+pr^+vg^+vg^+$ and then test-crossed F_1 females $prpr^+vgvg^+$ with prpr vgvg male and obtained the following results:

Red eye, Normal wing	1339
Purple eye, Vestigial wing	1195
Red eye, Vestigial wing	151
Purple eye, Normal wing	154
	2,839

Obviously, this is a drastic deviation from the expected 1 : 1 : 1 : 1 ratio of a dihybrid test cross and again indicates coupling of genes. The most common classes are the two gene combinations pr^+vg^+ and pr vg originally introduced by parental flies.

A cross between two parental types with genotypes pr^+pr^+vgvg and pr pr vg^+vg^+ (each parent being homozygous for one of the recessive genes) produced F_1 individuals with genotype $pr^+pr\ vg^+vg$ and the test cross pr^+ pr vg^+ vg ♀ × pr pr vgvg ♂ gave a different result.

Red eye, normal wing	757
Purple eye, vestigial wing	146
Red eye, vestigial wing	965
Purple eye, Normal wing	1067
	2935

Again the results of that test cross were not close to 1 : 1 :-1 : 1. The test cross is very important cross for determining whether the two genes are associated or independent of each other. When there is significant deviation from 1:1:1:1 ratio so that too many parental genotypes and too few recombinant genotypes are found, the two genes in question are considered to be associated.

In order to explain his results T.H. Morgan established the theory of genes and very clearly explained the role of chromosomes in inheritance of characters. According to **gene theory** of Morgan the chromosomes are the bearers of genes and the genes are arranged on the chromosomes in linear order. Morgan suggested that the *two pairs of genes under study were present on the same pair of homologous chromosomes.* The tendency of genes to be inherited in group is known as ***linkage***. To explain this phenomenon Morgan and Castle advanced a **linkage hypothesis** which states that

(*i*) Chromosomes are the bearers of hereditary units, the genes.

(*ii*) Genes are arranged on the chromosomes in linear order.

(*iii*) Genes located on the same chromosome are said to be linked and the chromosome as such is a **linkage group**. Thus, there are as many linkage groups in a cell as are the pairs of chromosomes.

(*iv*) Each gene has a definite position (locus) in the chromosomes.

(*v*) The distance between linked genes determines the degree of linkage strength. The closely located genes show strong linkage while the widely located ones show weak linkage.

(*vi*) Closely linked genes more often tend to remain in original combination in the course of inheritance (complete linkage) but some times genes change linkage groups in some meiotic cells because of exchange of chromosome segments).

Morgan showed that coupling and repulsion were the two distinct aspects of the single phenomenon to which he called linkage. Thus when pr and vg genes were introduced from one parent they were physically located on the same chromosome and similarly pr^+ and vg^+ on the homologous chromosome from the other parent. This hypothesis would apply to repulsion as well where one chromosome of a parent carries pr and vg^+ and the other homologous chromosome has pr^+ and vg. This explains why parental gene combinations remain together. In order to explain the nonparental gene combinations, Morgan suggested that when homologous chromosomes pair during meiosis, there is occasionally a physical exchange of chromosome segments by a process called *crossing over* which results in new combination. The mechanism will be clear in foregoing discussion. The original parental gene arrangement on the homologous chromosomes is called *parental combination* and the new gene combinations arising out of crossing over are called ***cross over types*** or ***exchange products*** or ***recombinants***. This hypothesis was later confirmed by F. Janssen who reported his observations on meiotic chromosomes in amphibians and orthopterans and suggested a cytological basis for crossing over. Janssens often noticed that during meiosis when duplicated homologous chromosomes were paired two non-sister chromatids crossed each other while the other two did not. He termed such cross (X) shaped configuration as *chiasma* (plural *chiasmata*). Chiasmata are just what one might expect cross over events to look like. It was not until 1921 however that Bateson abandoned his hypothesis in view of the sufficient counter evidences.

Linkage Group

It is an established fact that the hereditary genes which determine the characters of the individuals are carried in the chromosomes of the germ cells. When organisms were studied for their chromosome number and also for number of Mendelian traits, it was found that the gene number far exceeded the number of chromosomes. As in the organism, number of genes is greater than chromosome number, it can be expected that each chromosome contains many genes. The genes that are located in the same chromosomes are linked to it and would be expected to be inherited as a single group. Thus each chromosome is linkage group. There are as many linkage groups as are the pairs of chromosomes, for example 4 for *Drosophila melanogaster*, 23 for man, 10 in maize and so on. Since the chromosomes ordinarily duplicate at the time of gamete formation, all the genes present on a particular chromosome will be transmitted to the gamete as single block or unit. The random assortment will not take place between genes borne in the chromosomes. This phenomenon is known as linkage. The linkage is not due to any relation between two genes but is simply because they happen to be located in the same chromosome. Experimentally, it has been found that linked genes are not always inherited as single unit, but nearly they tend to remain together. It has also been observed experimentally that larger the chromosomes the larger will be the number of genes on them. Small chromosomes have got small number of genes.

CROSSING OVER AND RECOMBINATION

The linkage of genes on the chromosomes is complicated and it is sometimes not stable. When the chromosomes remain intact, all the genes located in them remain together and linkage is said to be complete. But sometimes it is not so and the linked genes separate during reduction division when the exchange of chromosomal parts takes place between the chromatids of the homologous chromosomes. Thus a limited amount of recombination occurs during first meiosis and consequently genes on a chromosome in coupling state may cross over and become transferred on the opposite member of a homologous chromosome pair and *vice versa.* Morgan and Castle (1912) were the first to introduce the term '**crossing over**' for this exchange of genetic material. Crossing over thus may be defined as the exchange of segments (blocks of homologous genes) between components of a pair of homologous chromosomes. In simple language, the crossing over can be defined as the tendency for genes to enter the gamete in new combinations which are quite different from those of the parents (recombination).

Actually during the zygotene of the first meiotic prophase there occurs pairing or synapsis between homologous chromosomes in such a way that two similar chromosomes lie side by side. In pachytene, the homologous chromosomes split longitudinally into chromatids so that the synapsed chromosomal pair appears four stranded. After this stage synaptic mates repel each other. In this process, the chromatids remain attached at certain points crossing each other. These points are called chiasmata (singular, chiasma). In these points synaptic association is so close that it would seem likely that homologous are held by bonds of some sort. It is presumably at these points that the crossing over and chiasma formation will occur and genic exchange will take place. As a result of crossing over, chromosomes with new gene sequences are formed.

Mechanism of Crossing Over

The important theories which have been proposed to explain the mechanism of crossing over are as follows:

1. **Janssen's Classical Theory.** Janssen (1909) believed that prior to the formation of chiasmata the homologous maternal and paternal chromosomes come in pair and in pachytene stage they become coiled round each other and become doubled (Fig. 16.1). He further suggested that in order to derive chiasmata from such a coil, the paternal and maternal chromatids made contacts at intervals and then one chromatid of a chromosome penetrated that of the other until they were broken, where upon they rejoined in new ways; paternal to maternal and *vice versa*, forming the typical chiasmata between them. The theory did not suggest a satisfactory mechanism by which two chromatids break at precisely equivalent points.

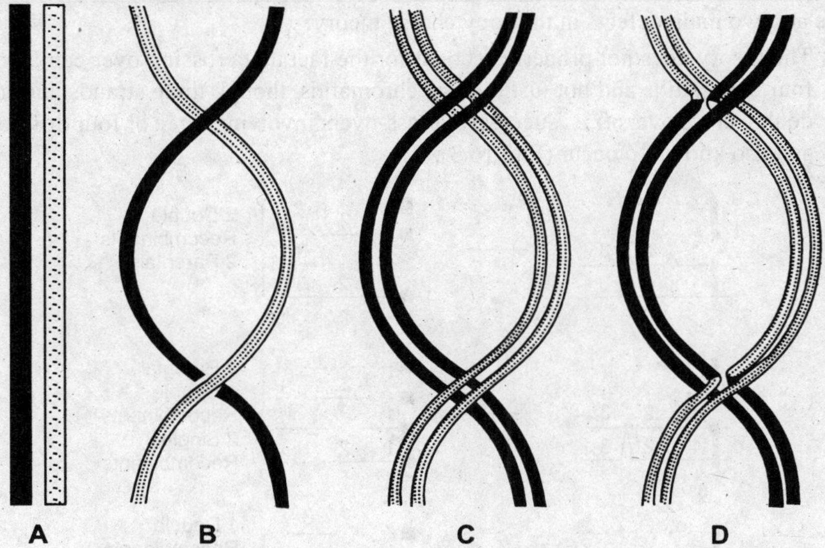

Fig. 16.1 Diagram explaining Janssen's hypothesis of origin of chiasmata.
A. and B. homologous chromosomes; White paternal and Black maternal.
C. Late Pachytene stage showing doubling and breakage.
D. Diplotene stage showing reunion of chromosomal fragments in new pattern.

2. Belling's Copy Choice Theory. In 1931, a cytologist named J. Belling proposed "the copy choice theory". According to this theory, the paired chromosome in first meiotic prophase duplicate their genes before the fibres that join them in tandem are developed. During the process, if the chromosomes are twisted around each other, the connecting fibres may connect genes of one chromosome at some points and adjacent genes produced by other chromosome at the other. In brief, the theory assumes that the crossing over is the direct result of the new chromatids copying partly from one strand and partly from other homologous strand (Fig. 16.2).

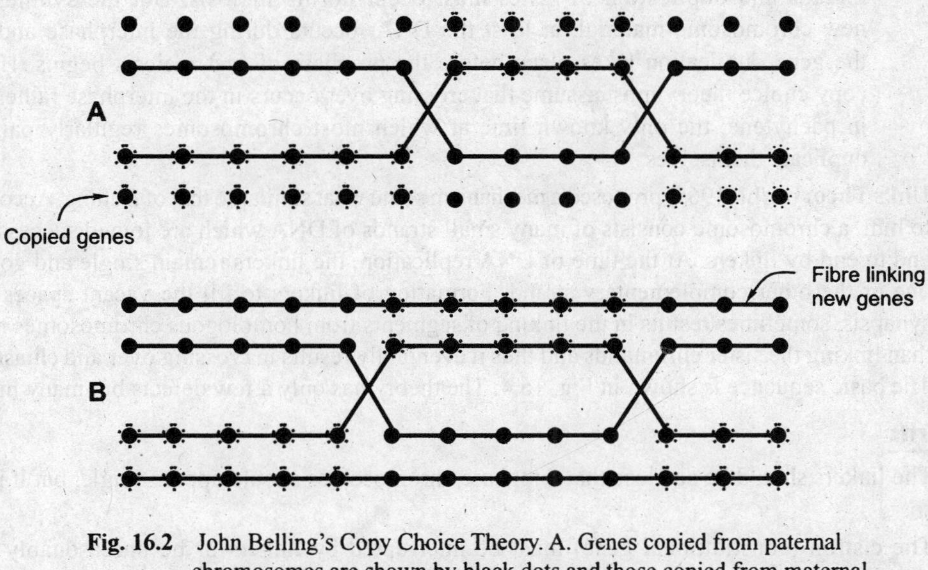

Fig. 16.2 John Belling's Copy Choice Theory. A. Genes copied from paternal chromosomes are shown by black dots and those copied from maternal chromosome are indicated by dotted circles. B. Fibres interlinking new genes of different chromosomes form chiasma type chromatids.

There are two main defects in the copy choice theory:

(*i*) The theory does not properly account for the fact that crossing over can involve all the four chromatids and not just the two chromatids, though three strands and four strands double cross overs (*i.e.*, successive cross-overs involving three or four different strands) are also known to occur (Fig. 16.3).

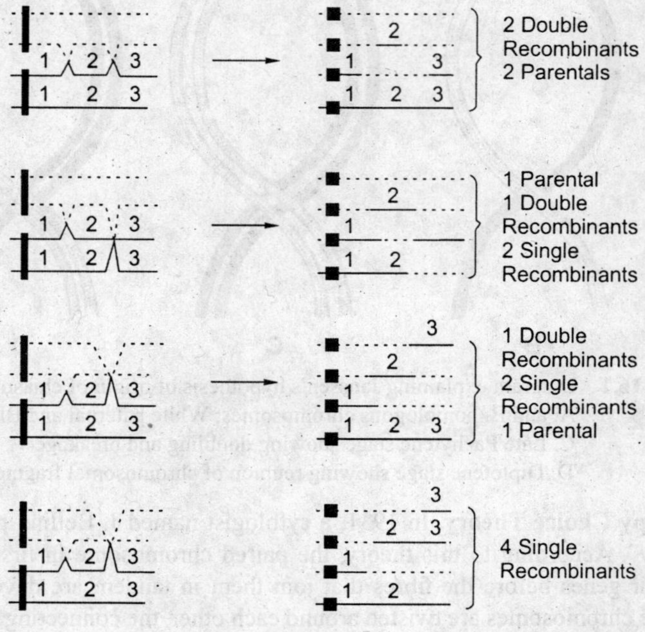

Fig. 16.3 Different types of three strands and four strands double cross over.

(*ii*) The second shortcoming of this hypothesis is that it requires singleness of leptotene threads and duplication of genes must occur during meiosis. But the synthesis of new chromosome material, at least the DNA, occurs during the interphase and thus the gene duplication takes place before the prophase of first meiosis begins. Hence, copy choice theory must assume that crossing over occurs in the interphase rather than in pachytene, the only known time at which most chromosomes regularly pair and duplicate themselves.

3. Uhl's Theory. Uhl (1965) proposed a mechanism some what similar to that of Belling. According to him, a chromosome consists of many small strands of DNA which are joined successively end to end by linkers. At the time of DNA replication, the linkers remain single and go with one or the other complementary strand. Formation of linkers to fill the vacant spaces after synapsis, sometimes results in the linking of segments from homologous chromosomes rather than linking the sister chromatids and thus it eventually results in crossing over and chiasmata. The basic sequence is shown in Fig. 16.4. The theory has only a few defects but many merits.

Demerits

(*a*) The linkers should be so close that telophase chromosomes should appear single, but it is not so.

(*b*) The cistron (constitutional gene) must be interrupted by linkers or be unreasonably long since intragenic crossing over does take place, unless this kind of crossing over has different explanation, which seems impossible.

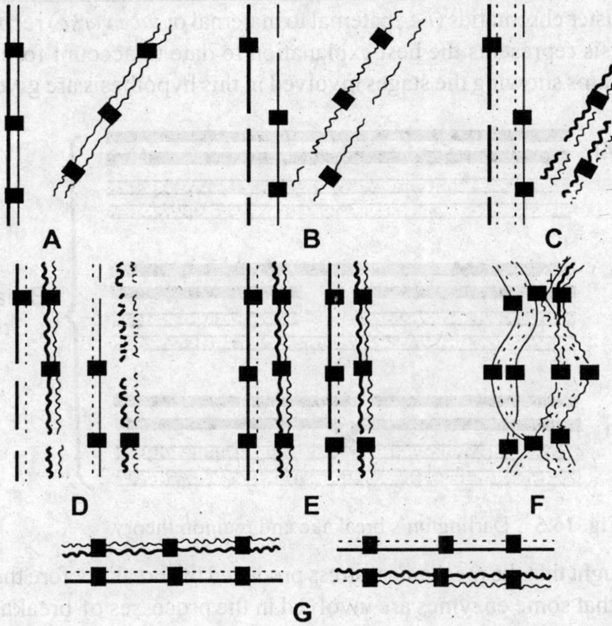

Fig. 16.4 A—Homologous chromosomes consisting of small strands of DNA which are joined end to end by single stranded linkers. B—DNA strands. C, D, E—Formation of linkers to fill the vacant spaces. F—Chiasma formation. G—Cross-over type of chromosomes.

Merits

(*a*) The theory goes far toward reconciling the multistranded nature of the chromosomes with unitary behaviour of chromatids.

(*b*) In this theory, stress has been given on the bond exchange between the adjacent DNA strands of the two homologues and no stress is laid on the physical breakage and reunion.

(*c*) The theory restricts rejoining to successive segments and does not allow for lateral exchange. Presumably, very specific bonds are involved in reunion.

(*d*) Broken chromosome and chromatid ends resulting from the action of radiation or chemicals have apparently free choice and they can rejoin with any other broken end regardless of the direction. Other important points in the theory are:

I. Exchange takes place between chromatids without any breakage.

II. Precision in the process.

III. Positive interference, *i.e.*, successive chiasmata will be minimal distances apart.

IV. Three and four strands double cross-overs and chiasmata can be explained.

V. The synthesis of almost all DNA is supposed to occur prior to synapsis.

4. Darlington's Theory of Crossing Over. A more probable and easy hypothesis of mechanism of crossing over has been advanced by C. D. Darlington in 1935. This is called ***"breakage and reunion theory."*** Darlington postulated that the homologous chromosomes are intertwined during the four stranded stage of first meiosis. The twisting exerts strain on the chromatids. If the stresses are great enough one or more chromatids of the homologues will be broken at one or more points. If more than one chromatid break, there is possibility that the broken chromatids of one chromosome will unite with broken ends of different chromatid forming a chiasma. If the union takes place between sister chromatid parts (*i.e.*, paternal to paternal and maternal to maternal) no genetic consequence is anticipated. However, if the break and reunion occur between non-sister chromatids (*i.e.*, paternal to maternal or *vice versa*) recombinants would result. This hypothesis represents the best explanation to date to account for the formation of recombinants. Diagrams showing the stages involved in this hypothesis are given in (Fig. 16.5).

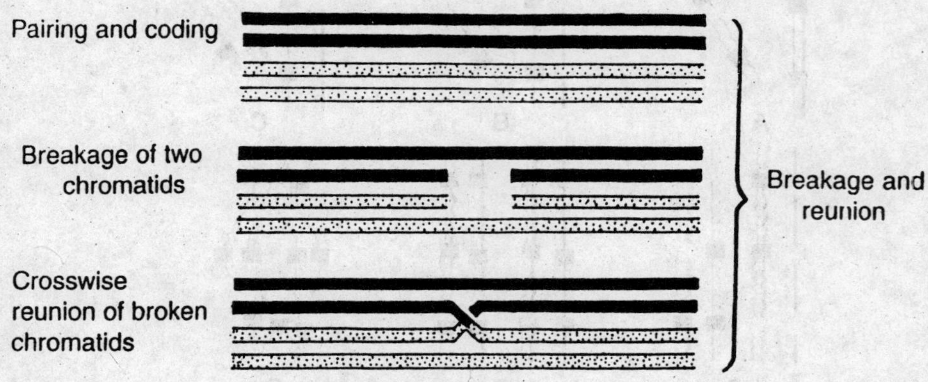

Fig. 16.5 Darlington's breakage and reunion theory.

Previously it was thought that the mechanical stress produced the breaks before the recombination but now scientists believe that some enzymes are involved in the processes of breakage and reunion. The idea that some enzymatic processes bring about breakage and rejoining of DNA strands was first putforth by Howard Flanders and Boyce in 1964. Reproduction of bacterial chromosome, for instance, during conjugation indicates that the ring structure of DNA may be broken without any known mechanical stress. There is also no known twisting of bacterial chromosomes in the recombination process. While considering the reunion of broken DNA strands, regardless of breaks occurred, one must hypothesise an enzymatic process in the formation of covalent bonds necessary to repair the sugar phosphate backbone of the DNA molecule.

Stern and Hotta (1969) are of the opinion that two nuclear enzymes namely *endonuclease* and *ligase* bring about crossing over. The enzyme endonuclease helps in the breaking of the chromatids while the ligase helps in the reunion of the chromatid segments. They have also reported the synthesis of a small amount of DNA during the period of crossing over which is used in the repairing of the broken chromatids.

Somatic Crossing Over

The pairing and crossing over take place not only in gonad cells but also in somatic cells. The process of somatic crossing over is known to occur in a number of organisms including *Drosophila* and maize. In the Diptera insect, the homologous chromosomes have been observed to pair in the early mitotic prophase. The crossing over takes place when the chromosomal threads are plectonemically arranged. Although the somatic crossing over is difficult to detect, there is no doubt that it occurs. The exact mechanism is not yet clear (Fig. 16.6).

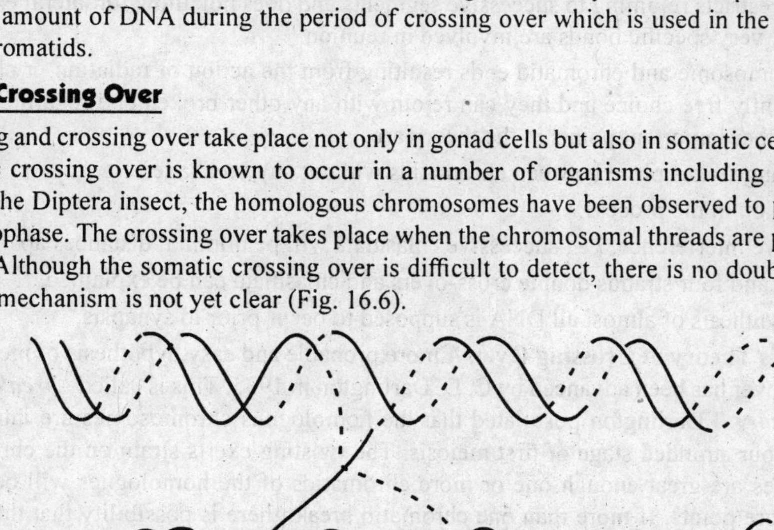

Fig. 16.6 Chiasma formation in somatic chromosomes.

The somatic crossing over may bring several changes in the structure and physiology of the organisms. Majority of individuals are heterozygous for many genes. Somatic crossing over may result in the malfunctioning of parts or organs on account of presence of recessive genes in homozygous condition. The influence of this process may sometimes be very serious and fatal.

Factors Affecting Crossing Over

It has been proved experimentally that the frequency of crossing over between two genes is not completely dependent on their distance but several other physiological and environmental factors can also influence it. Temperature, x-ray and chemical composition of food may change the frequency of crossing over. In *Drosophila melanogaster* it may be reduced with the increase of the age of flies.

Significance of Crossing Over

Crossing over is a widespread phenomenon which is known to occur in all the higher organisms, as well as in most bacteria and viruses. The exchange of genes between nonsister chromatids of homologous chromosomes in this process leads to the production of recombinants. Hence, it is of great importance in the evolution. The phenomenon of crossing over is of paramount importance in plant breeding because new varieties with valuable characters can be evoloved by eliminating undesirable genes through this process. Study of crossing over is very much helpful in mapping of genes on the chromosomes.

QUESTIONS

1. What do you understand by recombination? Discuss the various views available to explain the mechanism of recombination.
2. Briefly discuss the different theories of chiasma formation and describe whether chiasmata are the cause or consequence of crossing over.
3. At what stage of meiosis does crossing over occur?
4. What is the significance of crossing over?
5. Write short notes on:
   (*i*) Linkage,
   (*ii*) Copy choice theory, and
   (*iii*) Breakage and reunion theory of crossing over.

17 CHAPTER

Sex Determination and Sex Linkage

Sex refers to contrasting and complementary characteristics exhibited by male and female individuals within the same species as well as capacity for performing amphimixis and/or genetic recombination by such individuals.

In most of the higher plants and animals male and female parents are separate individuals (Fig. 17.1). But what makes these individuals develop into male and female sexes? This basic question of genetics is related to sex determination. Sex determination and sex differentiation are two different processes.

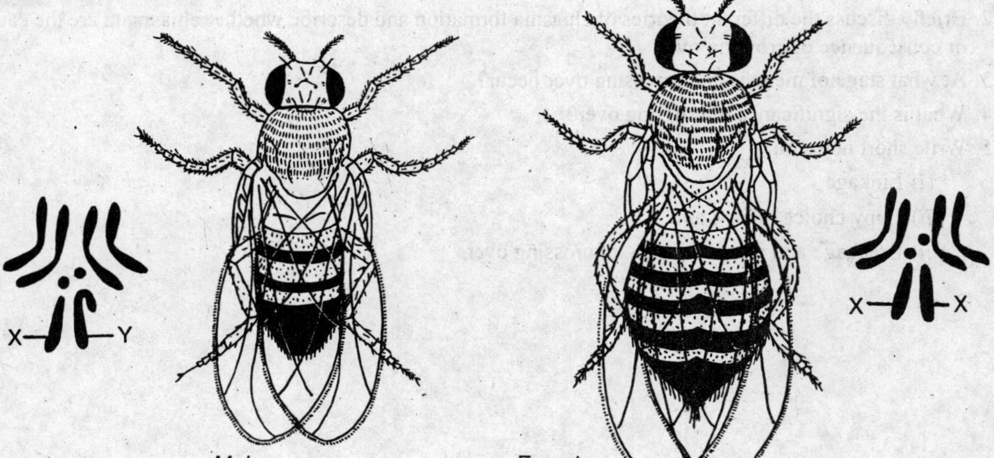

Fig. 17.1 *Drosophila melanogaster.*

The basis for the biological divergence of sexes may be either inherent in the organism or impressed on it by its environment. This process is referred to as sex determination.

According to Darlington and Mather (1949), sex determination is the process by which the male and female gametes after fusion produce zygotes which develop the property of one sex or the other sex and sex differentiation involves chain of developmental processes by which the genetically determined sex becomes translated into characteristic structures, functions and behaviour patterns. The sex determination is complete after fertilization, but sex differentiation starts after fertilization and continues throughout embryonic development, childhood and adolescence (sexual maturity).

Sex Chromosomes

In the sexual reproduction male and female gametes come and fuse to form a zygote. These gametes are specialized cells. In the gamete nucleus, chromosomes may be found in definite number and they may be of two types:

1. **Sex chromosomes.** These are concerned with the determination of sex.
2. **Autosomes.** These are concerned not with sex determination but with the determination of other body characters. These are the chromosomes whose number and morphology do not differ between the males and females of the species.

Henking (1891) while making study of meiosis in Hemipterous insect *Pyrrhocoris apterus* observed deeply staining chromatin element. He named that darkly stained body as "X body" as he was in doubt whether that body was a chromosome. Later on workers who worked on insect cytology identified that body as chromosome. McClung (1902) suggested that the chromosome was concerned with sex determination. This fact was also established by an American scientist E.B. Wilson (1905) who contributed valuable knowledge on the sex determination in the early part of the 20th century. Miss Stevens (1905) noted similar observations in other insects. The X-body, previously reported by Henking, was found to be a sex chromosome which later became popular by the name X chromosome. In 1908, she showed that in *Drosophila melanogaster* there were four pairs of chromosomes in the somatic cells. In the male, one pair of chromosomes was peculiar in the sense that the two components were of unequal sizes and one of the two components of that chromosome pair resembled the X chromosome of female. Wilson (1909) proposed the term 'Y chromosome' for the odd chromosome. So *Drosophila* could be described as XX female and XY male. In the insect genus *Protenor,* the females have XX and males have besides autosomes, only one X chromosome and no Y chromosome (hence XO). In human, there are 23 pairs of chromosomes which may be analysed in the following way:

Female—22 pairs of autosomes (AA) + XX chromosomes.

Male—22 pairs of autosomes (AA) + XY chromosomes.

In *Drosophila,* the chromosomes may be formulated in the following way:

Female—6A + XX

Male—6A + XY.......... (Fig. 17.2).

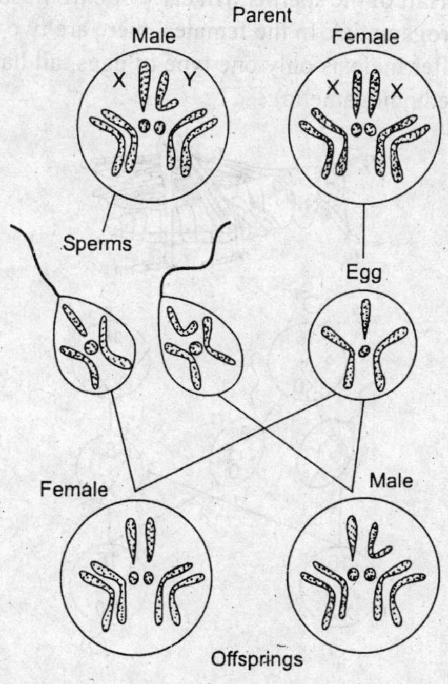

Fig. 17.2 XX female and XY male type of sex determination in *Drosophila melanogaster.*

The X and Y chromosomes differ from each other in their length. In *Drosophila*, the Y chromosome is somewhat larger than the X chromosome and in man reverse is the case. The Y chromosome is largely heterochromatic and has only a few loci which are apparently homologous with those of X chromosome.

MECHANISM OF SEX DETERMINATION

Many theories have been proposed to account for the mechanism of sex determination in plants and animals. Some of them are as follows:

Chromosome Theory of Sex Determination

It has been discovered that sex is inherited as a pair of Mendelian characters and is dependent on the behaviour of particular sex chromosome or a pair of such chromosomes and antosomes donot have any role in it. However, it may be true only in case of mammals and some plants but not in many other animals and plants. The chromosome mechanism of sex determination varies in different organisms. Three well known mechanisms are described below:

(*i*) **XX Female—XO Male Type.** The situation is common in grass hoppers and some of the bugs, where the female parent possesses normal pairs of autosomes and a pair of X chromosomes and the males have normal pairs of autosomes and single X chromosome. The odd chromosome

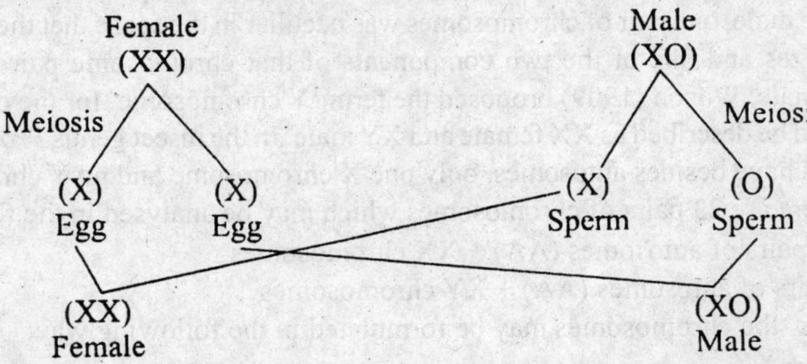

remains unpaired in meiosis at the time of spermatogenesis. After meiosis two types of sperms will be formed. Half of the sperms will carry X chromosomes and remaining half will not (*i.e.*, males are heterogametic). In the females, there are two X chromosomes besides the paired autosomes. So after meiosis only one type of eggs, all having X chromosome will be formed (*i.e.*, females are homogametic).

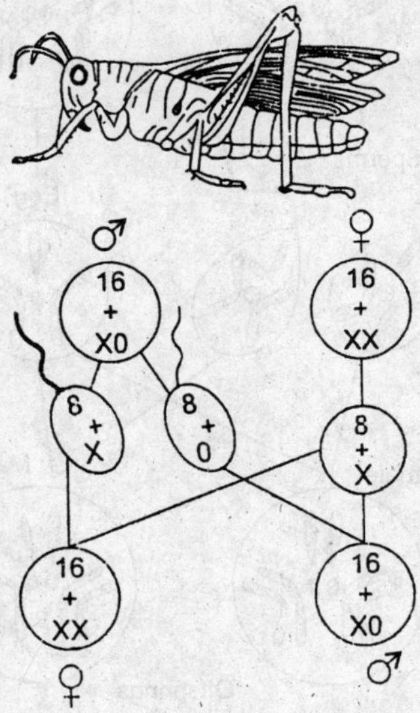

Fig. 17.3 The XX Female-XO Male system of sex determination in grasshopper.

If an egg is fertilized by a sperm containing X chromosome, the resultant zygote will have XX chromosomes and will develop into female individual. But if the egg is fertilized by sperm lacking a X chromosome, the result will be of XO constitution and the offspring will be male. The mechanism is shown in Fig. 17.3.

The theory gained wide acceptance as a common method of sex determination for several years in the early part of twentieth century. The theory holds good in those cases where males have only single X chromosome. It cannot explain those cases where both sexes are characterised by same number of chromosomes.

(*ii*) **XX Female—XY Male Type.** In this type of sex determination, both the sexes have got even number of chromosomes; female sex having two X chromosomes (XX) and the male having one X and one Y chromosome (XY). Females are homogametic and produce eggs each of which carries a X chromosome. The males on the other hand are heterogametic and they produce two different types of sperms; 50% containing X chromosome and remaining 50% with Y chromosome. X egg fertilized by X sperm will give birth to XX female and the egg with X chromosome if fertilized by sperm carrying Y chromosome will produce XY male.

This mechanism is very common in *Drosophila* insects (Fig. 17.2) and many other invertebrates, some fishes and mammals including man (Fig. 17.4).

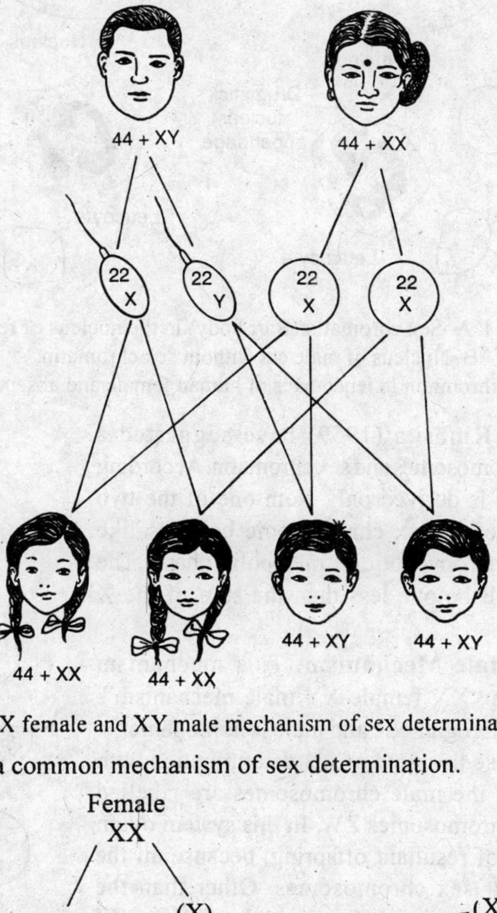

Fig. 17.4 XX female and XY male mechanism of sex determination in human.

In plants also, it is a common mechanism of sex determination.

The evidence that Y chromosome is determiner of fertility and sex of male individuals in man and other mammals comes from the following source:

BARR body. Barr and Bertram (1949) reported the presence of a deeply stained planoconvex, spherical or pyramidal chromatin body of about 0.8 × 1.1 μ size in the nerve cells of female cat which was absent in the male. That body is called *Barr body*. Such a Barr body has also been observed in most of the body cells, *e.g.,* skin, oral epithelium and blood cells of females of human and other mammals (Fig. 17.5). Human females are sex chromatin positive and the males are sex chromatin negative. The Barr body or sex chromatin can be seen in interphase nucleus as a small chromocentre which stains deeply with basic dyes. It can be found in different positions in different cells. It may be seen (*i*) attached to nucleolus in some cells, as in nerve cells, or (*ii*) in free state in the nucleoplasm as in neurons after electric stimulation, or (*iii*) attached to nuclear membrane as in the epithelial cells of oral mucosa, or (*iv*) as a nuclear expansion as in neutrophil leucocytes of human female in which the Barr body appears as small rod, called the *drum stick.*

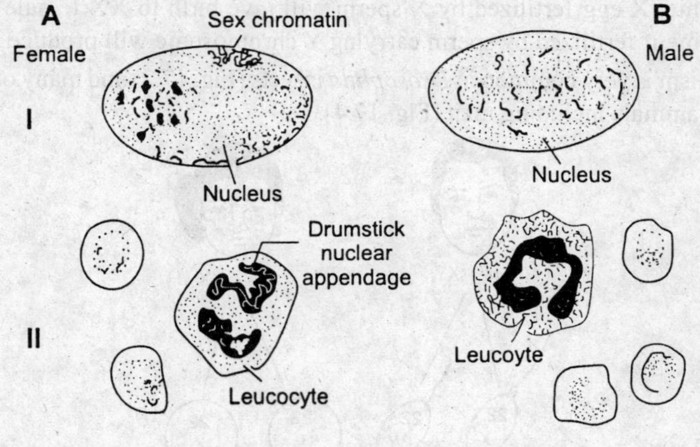

Fig. 17.5 I. A–Sex chromatin (Barr body) in the nucleus of female cat; B–Nucleus of male cat without sex chromatin.
II. B–Sex chromatin in leucocytes of human female and absent in male.

Ohno, Kaplan and Kinosita (1959) have suggested a relationship between sex chromosomes and sex chromatin. According to them the sex chromatin is derived only from one of the two X chromosomes and the other sex chromosome behaves like an autosome and is not heteropycnotic in metabolic phase. The number of Barr bodies will be one less than the sum of the X chromosomes in a cell.

(*iii*) **ZW Female-ZZ Male Mechanism.** This mechanism is just reverse of the "XX female-XY male mechanism". Here the female is heterogametic and male is homogametic. In order not to confuse this type of sex determination with XX–XY mechanism the male chromosomes are labelled ZZ and the female chromosomes ZW. In this system ovum determimes the sex of resultant offspring, because all the sperms, have similar sex chromosomes. Other than the reversal of homo and heterogametic sexes, the ZW♀ – ZZ ♂ system functions similar to XX ♀–XY♂ system. This mode of sex determination has been observed in birds includng fowls (Fig. 17.6), amphibians, reptiles, some fishes, butterflies, moths and in a plant species *Frageria orientalis*.

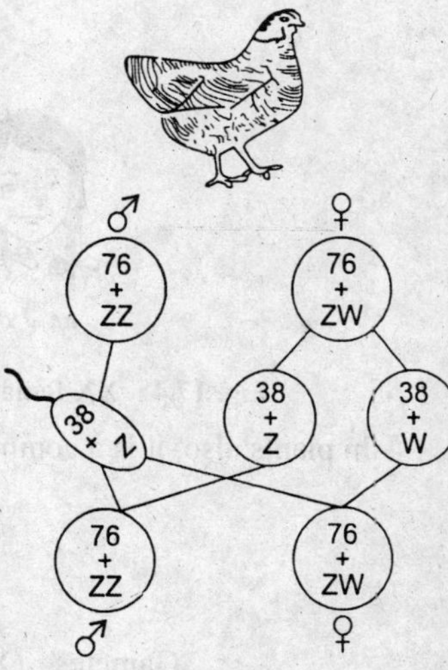

Fig. 17.6 The ZWfemale -ZZ male system of sex determination in fowl.

In the XX female-XO male mechanism, quantity of X-chromosome material probably determines the nature of sex. In XX female-XY male system also, the difference in the quantities of X chromosome in male and female can be regarded as basis for sex determination.

There are two hypotheses to explain XY mechanism of sex determination.

(*a*) According to one hypothesis the X chromosome only determines the nature of sex and Y has no role to play.

(*b*) According to second hypothesis Y chromosome also plays important role in the sex determination and it carries genes that are dominant over those factors which determine femaleness.

1. Meiotic Mix-ups (Non-disjunctions)

The pairing and separation of chromosomes determine at an equal separation of pairs of homologous chromosomes. Occasionally the separation of X chromosomes fails (*i.e.,* they do not disjoin) in meiosis. This leads to the production of two types of exceptional eggs; some carrying two X chromosomes and some carrying no X chromosomes. This is called non-disjunction. The phenomenon has been found to be the main cause in the evolution of individuals with abnormal chromosomal constitution. *Drosophila* has been identified with a wide range of number of X and Y chromosomes. A close study of sexual structure of these individuals has demonstrated that the amount of X chromosome material is the deciding factor in sex determination and the Y chromosome plays minor role in *Drosophila.* Production of various individuals with abnormal chromosomal constitution can be explained in the following manner:

Gametes		Offspring	Sex
Female	Male		
(*i*) XX abnormal egg	X sperm	XXX	Super female or Metafemale (Usually dies)
(*ii*) egg without X (nullo-X egg)	X sperm	XO	Nonfertile male
(*iii*) XX egg	Y sperm	XXY	Normal female
(*iv*) nullo-X egg	Y sperm	OY	Zygote dies

The term superfemale is not intended to indicate greater sexual effectiveness. Such individuals are actually sterile. They have exaggerated female features and the term "Super" refers not to their effectiveness as female and unfortunately, it is most inappropriate term. The sexual structure of individuals with abnormal number of X and Y chromosomes shows that the Y chromosome does not carry male determining factors. The difference between male and females appears to be caused by difference in number of X chromosomes. Both XX and XXY are normal female showing that Y does not affect determination of femaleness by X chromosomes. Both XY and XO individuals are males showing that Y chromosome does not affect the determination of maleness by a single X chromosome. Although the Y chromosome has no male determining factors, it does contain male determining fertility factors. Individuals with only a single X chromosome are male in both external and internal structure but sometimes they are sterile. Recent researches indicate that the sterility in such males is caused by the absence of Y chromosome. The males lacking Y chromosomes produce sperms with short tails. This presumably reduces the effectiveness of sperms in fertilization.

In human, different possible combinations of X and Y chromosomes owing to non- disjunction of sex chromosomes may result in the following types of individuals:

Chromosome combination	*Sex*
XXX	Female with mental abnormality
XX	Normal female

Chromosome combination	*Sex*
XXY	Abnormal male
XXYY	Abnormal male
XXXY	Abnormal male
XXXXY	Abnormal male
XY	Normal male

This Table clearly indicates that in man Y chromosome is concerned with maleness. So the individuals carrying Y chromosomes will be male regardless of number of X chromosomes. The development will be abnormal except in the individual with normal chromosome constitution, *i.e.*, XY and XX. The study of non-disjunction process thus helps in ascertaining the specific role of X and Y chromosomes in sex determination.

2. Genic Balance Theory of Sex Determination

C.B. Bridges in 1925 reported his genic balance theory of sex determination in *Drosophila.* In this theory, he stated the following facts:

(*i*) that genes for femaleness were located in the X chromosome and those for maleness in the autosomes,

(*ii*) that all the individuals carried genes for both sexes—male and female, and

(*iii*) that the ratio between X chromosome and autosomes sets (X chromosome number/No. of autosome sets) in the zygote, regardless of the number of Y chromosomes, indicated the nature of sex. This ratio is called sex index. Sex index value (X/A) 0.5 or less gives males, and if X/A equals 1 or more than one, then it gives females, and if value of X/A is between 0.5 and 1 then intersexes (*i.e.,* individuals intermediate between male and female) will result.

Bridges successfully produced various combinations of X chromosomes and sets of autosomes in *Drosophila*, which are as follows:

Phenotypes		*No. of X Chromosomes*	*No. of sets of autosomes ('A'sets)*	*Sex index = No of X Chromosomes / No. of sets of autosomes*
Super female		3	2	1.5
Normal female	tetraploid	4	4	1.0
	triploid	3	3	1.0
	diploid	2	2	1.0
	haploid	1	1	1.0
Intersex		2	3	0.67
Normal male		1	2	0.50
Super male		1	3	0.33

This balance explanation does not hold true for mammals and also in those individuals where Y chromosomes are known to play some roles in sex determination.

3. Haplo-Diplo Mechanism of Sex Determination

In the haplo-diplo type of sex determination, females are diploid and arise from fertilized eggs and males are haploid and arise from the parthenogenetic development of unfertilized eggs. It is, in fact, very rare, but is common in some species of insect orders, such as Hymenoptera including ants, bees, (Fig. 17.7) wasps, etc., Homoptera, some Arachnoides and Rotifers. The most elaborate studies on this type of sex determination have been made by Whitings (1945) on the wasp, *Habrobracon.* In this wasp, diploid chromosome number is 20 and haploid number is 10. A small number of the male wasps came from fertilized eggs and thus they were diploid, but the majority came from unfertilized eggs and were haploids. All females were found to be diploid. The diploid males which developed from fertilized eggs were poorly viable. Whitings postulated that sex in *Habrobracon* was determined by multiple allelic series of 'S' gene, say, S – 1, S – 2, S – 3, and so on. Femaleness presumably develops only in the state of heterozygosity (say, S – 1/S – 2, S – 1/S – 3) for a number of sex factors. Males are of many kinds with regard to these factors. Homozygosity for most of them, say S – 1/S – 1, S – 2/S – 2, and so on, apparently gives rise to poorly viable males. The surprising fact is that intersexes are rare and some if developed by chance are caused by factor other than the sex mechanism.

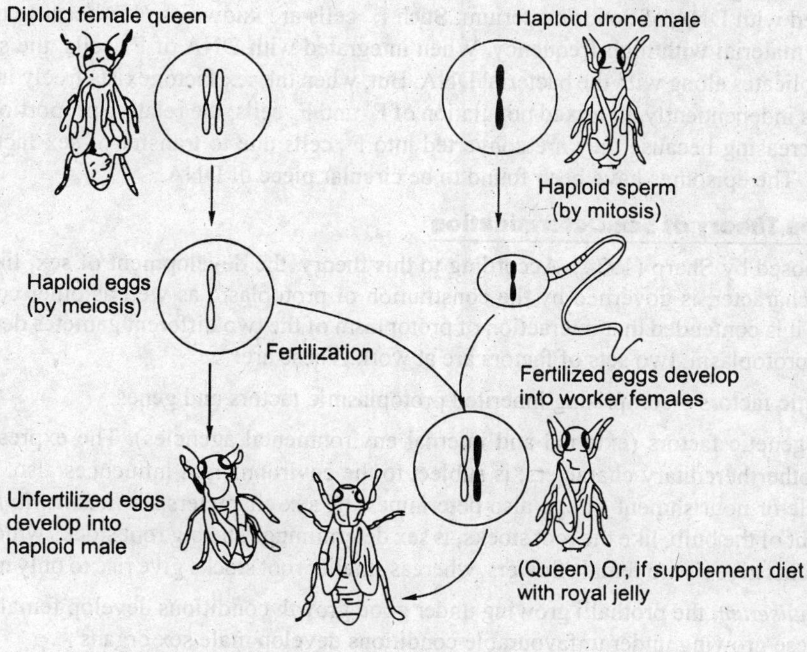

Fig. 17.7 Sex determination by ploidy level in honey bee.

4. Single Gene Sex Determination

In some plants and animals, sex is determined probably by a single gene only. In some micro-organisms, as for example *Chlamydomonas, Neurospora* and yeasts, the sex is determined by a single gene. These micro-organisms have two separate sexual strains or mating types (+ and – strains) which are determined by a pair of alleles. The sexual union takes place only when the gametes from two different strains meet each other. The gametes from the same strain do not fuse in such organisms.

Male and female of the mosquito *Culex molestus* differ from each other only in a single gene. A very interesting situation of single gene sex determination has been observed in maize. In this case, if the plant is monoecious, microspores are produced by staminate flowers in the tassel while the megaspores are produced by female flowers of the ear. In this plant there are two important genes—

(*i*) Recessive gene 'ba' for barren plant, and (*ii*) the recessive gene 'ts' for tassel seeds. Plant with genotype ba/ba, ts/ts is only pistillate, while those with ba/ba, ts^+/ts^+ and ba/ba, ts/ts^+ are only staminate. Thus dioecious sporophyte with single sex can be evolved from hermaphrodite sporophyte by the establishment of following two genotypes:

(*i*) ts^+/ts, ba/ba—normal tassel, no ear → male

(*ii*) ts/ts, ba/ba—pistillate tassel, no ear → female

From this, it is clear that the maleness and femaleness are determined by either heterozygosity or homozygosity for one allele 'ts'.

5. Cytoplasmic Sex Determination

In bacterium *Escherichia coli,* Tatum and Lederberg (1947) have discovered the phenomenon of sexuality. In this case, the sexual process involves unidirectional transfer of genetic material from male (donor) to female (recipient) cell and the type of sex is controlled by cytoplasmic factor, called sex factor (F). The male cells contain sex factor and so they are F^+ cells while the female cells lack it and are F^- cells. The sex factor behaves like an episome which may exist either in free state in the cytoplasm or integrated with DNA of donor cell. For genetic transfer it is necessary that sex factor be integrated with DNA of donor bacterium. Such F^+ cells are known as *HFr* because they transfer the genetic material with high frequency. When integrated with DNA of F^+ cells, the sex factor or episome replicates along with the bacterial DNA. But, when the sex factor exists freely in cytoplasm, it multiplies independently. In mixed population of F^+ and F^- cells, the relative proportion of F^- cells goes on decreasing because they are converted into F^+ cells due to transfer of sex factor from the donor cells. The episomes have been found to be circular piece of DNA.

6. Nutrition Theory of Sex Determination

It was proposed by Sharp (1934). According to this theory, the development of sex, like any other hereditary character, is governed by the constitution of protoplasm as well as other conditions. In this theory, it is contended that interaction of protoplasm of the two different gametes determines the sex. In the protoplasm, two sets of factors are at work. These are:

(*i*) Genetic factors—comprising inherited protoplasmic factors and genes.

(*ii*) Non-genetic factors (external and internal environmental agencies). The expression of sex, like other hereditary characters, is subject to the environmental influences also. Nutrition of female or nourishment of egg also determines the sex characters. In *Arisaemia japanica* the weight of the bulb, like the root stocks, is sex determining. The big root stocks which are rich in nutrients give rise to female flowers, whereas smaller root stocks give rise to only male flowers.

In *Equisetum,* the prothalli growing under good growth conditions develop female sex organs whereas these growing under unfavourable conditions develop male sex organs.

In *Dinophilus,* a marine worm, the size of egg determines the nature of sex. The small eggs produce males and the larger ones, the females.

7. Environmental Sex Determination

Some forms, in order to prevent wastage of males or in the sense of failing to find mate, are adapted to purely environmental sex determination. Under such conditions, the genome of every zygote must contain all the genes necessary for the development of both types of reproductive systems. There must be correspondingly some mechanism which may be unknown to us, for inhibiting the action of some sex genes while permitting others to function. Which particular genes are inhibited and which are permitted to function, all these are controlled either by internal or by external environmental conditions. The marine sessile worm *Bonellia* is a popular example showing environmental mechanism of sex determination. The female animal is about two inches long shows complex anatomical differentiation. Males are of the size of large protozoa and have rudimentary organs (Fig. 17.8). They live in the uterus of female. The larvae that are released in water can meet one of the following two fates:

(*i*) The larvae which come to the surface of mature female and become attached to the female probosis develop into males and then these males migrate to female reproductive canal and live there as parasite.

(*ii*) The remaining larvae which do not reach to the mature female individuals mature into females. It has been demonstrated experimentally that the probosis of mature female to which the larvae become attached scretes a hormone like substance. That substance probably influences the attached larvae to develop into males. It is thought that the hormone like substance suppresses the action of female determining genes of the larvae, thereby allowing only the male determining genes to function.

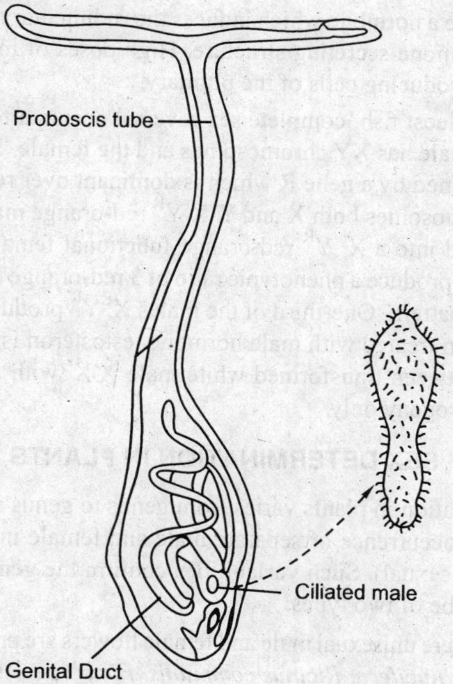

Fig. 17.8 *Bonellia*: female and male.

In the cattle, it has long been known that in dizygotic twins (two-egg, or fraternal twins) in which one is female and the other is male, the female twin in over 90% of the births is phenotypically male-like. This genetic female with male features is called a freemartin or hormonal intersex. The freemartins have external female genitalia but internal sex organs are like those of male. The male twin is usually normal in such cases. According to F. R. Lillie (1917), free martins occur only in the cases where the foetal membranes of the two embryos fuse allowing thereby cross circulation between the two foetuses. Some hormones released by developing male foetus find their way through placenta into the female where they suppress the development of secondary sexual characters and at the same time induce the male sexual characteristics. If the foetal membranes of the two embryos do not fuse, the freemartins do not occur. R.D. Owen and Stormont have supported Lillie's hypothesis on the basis of serum analysis in cattle.

Hormonal Control on Sex Differentiation and Sex Reversion

In some cases sexual differentiation is controlled by hormonal balance of the body. Many cases are known in which the organisms are females for some part of their life spans and then they turn to males or *vice versa*. This phenomenon is known as sex reversion. In these cases the genome is capable of developing either type of reproductive system. Here the determination of sex is a sort of switch over mechanism controlled chiefly by hormone balance. Example of sex reversion is seen in birds, oysters and marine annelids. In birds, only one gonad of female normally develops into functional ovary and the other gonad remains rudimentary. If the functional ovary of a hen is removed the second rudimentary gonad develops into testis and hen is sexually changed to male (cock).

In birds the pituitary gland of female produces female hormone that causes the development of ovary in the beginning but after the development of ovary is complete, the pituitary does not produce female hormones because the hormone secreted by ovary checks it from doing so. The pituitary female hormone suppresses the action of male hormone producing cells of the body. When ovary is destroyed, the cells of the adrenal gland, for the reasons unknown, are able to function before the hormone of gland and secrete a hormone which induces the rudimentary gonad to develop into testis. The testis itself is male hormone secreting structure. High doses of male hormone suppresses the action of female hormone producing cells of the pituitary.

In *Medaka,* a small teleost fish, complete sex reversal occurs after administration of gonadal hormones. In this fish, the male has XY chromosomes and the female, XX chromosomes. The body colour red orange is determined by a gene R which is dominant over recessive gene r. The genes R and r are linked to sex chromosomes both X and Y. X^rY^R red-orange male when treated with female hormone estron is converted into a $X^r\ Y^R$ red-orange functional female. The transformed females when crossed to X^rY^R males produce a phenotypic ratio of 3 red-orange to 1 white and male to female ratio of 3 : 1 on Mendelian pattern. One-third of the males X^RY^R produce only male offspring in F_2. The white female X^rX^r when treated with male hormone testosteron is converted into a X^rX^r white functional male. A cross between transformed white male X^rX^r with a normal white female X^rX^r gives rise to white female progeny only.

SEX DETERMINATION IN PLANTS

Mechanism of sex determination in plants varies from genus to genus and species to species. Some plants show dioecism, *i.e.,* occurrence of separate male and female individuals while others show monoecism (*i.e.,* they are bisexual). Such variabilities confirm the genetic difference among them. The monoecious plant may be of two types:

(*i*) Monoecious plants where unisexual male and female flowers are present in the same sporophyte, for example, in *Cocos nucifera, Ricinus communis, Ficus* species and many others.

(*ii*) In some plants, male and female parts develop in the same flowers. Such monoecious plants are called hermaphrodite.

In the organisms that have the sexes combined in a single individual, there are no special sex chromosomes in the karyotype and the formation of germ cells (microspores and megaspores) in such plants is accomplished by process of histological differentiation.

Lindsay (1930) has classified known cases of sex determination in dioecious plants into many types which are as follows:

(*a*) **Normal XY male-XX female type.** Mechanism is already described in the beginning of this chapter. This is very common mechanism of sex determination in the majority of plants.

In *Sphaerocarpus* (a bryophyte) the sex determination is typically of X – Y type. Allen (1917) studied sex determination in this genus and noted a dot like Y chromosome in male cells and a long rod like X chromosome in female cell. In this bryophyte chromosome formula of male plant is 7A + Y and that of female is 7A + X (Fig. 17.9). The zygote has 14A + X + Y chromosomes. During meiosis, 7 normal pairs of autosomes and one pair of XY are seen. After meiosis half of the spores will, therefore, have 7A + X (female) and other half 7A + Y (male).

(*b*) **XY male and XX female where male is heterozygous and digametic.** In *Melandrium album,* male plant is heterogametic (XY) and the female plant homogametic (XX). In this plant Y chromosome determines maleness irrespective of number of X chromosomes and its absence is responsible for femaleness. Y chromosome can be readily distinguished from X chromosome.

There are two groups in this mechanism of sex determination:

(*i*) Tripartite.

(*ii*) Tetrapartite.

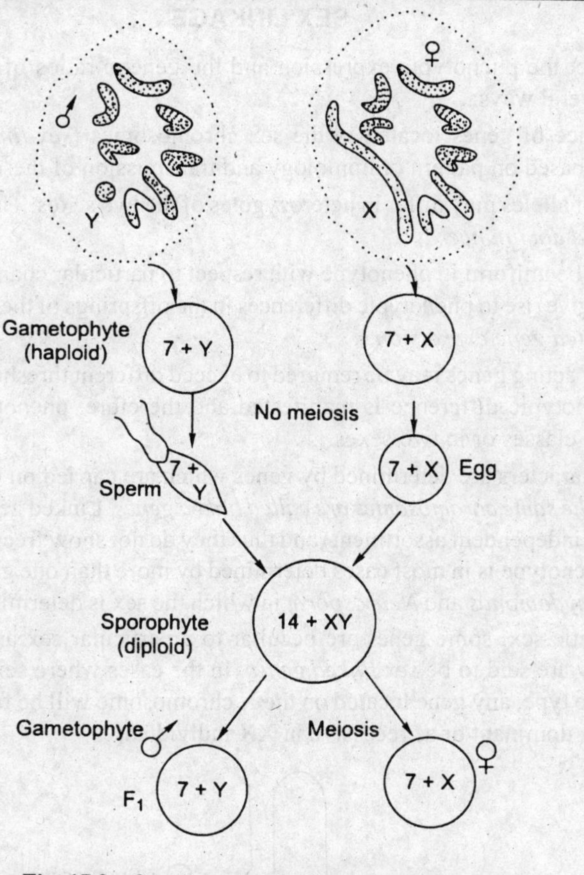

Fig. 17.9 Sex determination in *Sphaerocarpus.*

(*i*) **Tripartite.** The chromosomal constitution of staminate plants is AA (autosomes) + Y_1XY_2 and male gametes are A + X and A + Y_1Y_2. The female plant is 2A + XXX. During meiosis, the tripartite Y_1XY_2 or heterotrivalent chromosomes in male separate in such a way that chromosomes Y_1 and Y_2 go to one pole and the middle X chromosome goes to opposite pole. *Humulus japanicus* (staminate) and *Reumex* species are common example for this.

(*ii*) **Tetrapartite.** In *Humulus lubulus* Sinnot (1929) found the male plant with chromosome constitution AA + X_1 Y_1 X_2 Y_2. The gametes produced by the male plants have chromosome constitution A + X_1 X_2 and A + Y_1 Y_2. Here the tetrapartite chromosomes X_1 Y_1 X_2 Y_2 separate during meiosis in such a way that X_1 X_2 go to one pole and Y_1 Y_2 to other pole. The female plants have AA + X_1 X_2 X_2 X_1 and the gametes are of the A + X_1 X_2 constitution.

(*c*) **XX female-XO male type**. In some plants, the males are of XO types with chromosome constitution AA + XO. They are digametic and hemizygous with gametes of A + X and A + O types. The female plants are of the type 2A + XX with gametes of A + X type.

Hofmeyer (1958) noted that the female plants of *Carica papaya* were heterozygous and digametic. Thus XY female and XX male type of sex determination is observed in this plant.

(*d*) In *Rumex acetosa,* genic balance determines sex and the autosomes play a role in the process of sex determination.

SEX LINKAGE

Sex is able to affect the phenotypic expression and the genetic ratios of chromosomal hereditary determinants in several ways:

(*i*) The inheritance of genes located in the sex chromosomes (*sex-linked inheritance*) follows special rules based on pattern of homology and transmission of these chromosomes.

(*ii*) Dominance of alleles may differ in heterozygotes of the two sexes. This phenomenon is termed *sex-influenced dominance.*

(*iii*) One sex may be uniform in phenotype with respect to particular characters, yet it may transfer genes which give rise to phenotypic differences in the offsprings of the other sex. This is referred to as *sex-limited gene expression.*

(*iv*) Cumulatively acting genes may be required to exceed different thresholds in males and females before a phenotypic difference is manifested and therefore, phenotypic segregation simply results in two classes or in two sexes.

Hereditary characters are determined by genes which are carried on the chromosomes. *Genes that are carried on the same chromosomes are called linked genes.* Linked genes do not obey Mendel's second law (Law of independent assortment) and thus they do not show free recombination. The 'sex' being a complex phenotype is in most cases determined by more than one gene. There are only a few cases, such as *chlamydomonas* and *Neurospora,* in which the sex is determined by a single gene pair.

In homogametic sex, some genes are peculiar to a particular sex and are located in the sex chromosomes. They are said to be *sex linked genes*. In the cases where sex determination is of XX female and XO male type, any gene located on the X chromosome will be fully expressed regardless of whether it acts as dominant or as recessive in XX individual.

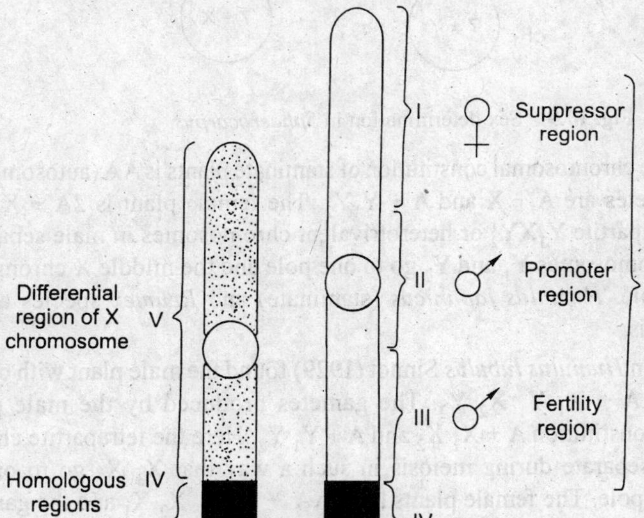

Fig. 17.10 Comparison of X and Y chromosomes of the plant, *Lychnis (Melandrium)*. Regions I, II and III bear holandric genes, and region V. X-linked or sex-linked genes. Genes in region IV are termed incompletely X-linked or sex-linked. *Lychnis* is unusual in that its Y-chromosome is larger than the X-chromosome (after Burns, 1969).

The sex chromosomes of XY males have two regions; (*i*) the pairing segment, and (*ii*) the differential region. During meiosis, the X and Y chromosomes pair only partly. The pairing segments of X and Y chromosomes are said to be homologous and crossing over may take place in that region. The non-pairing segment of each sex chromosome, which appears to hold genes that have no counterparts on the other kind of sex chromosome is called **differential region** (Fig. 17.10). The genes present in the differential region of sex chromosome, whether dominant or recessive show their effect in the phenotype of male. These genes are called *hemizygous* (*half zygous*). Hemizygous genes found in the differential region of X chromosome demonstrate an inheritance pattern and constitute the X-*linkage group.* The genes found in differential segment of Y chromosome show *Y linkage.* The genes present in the pairing regions of both X and Y chromosomes also show inheritance pattern (Fig. 17.11). That is called *X-Y-linkage.* Thus in XY situation, sex-linked genes may be classified into the following three types: (1) X-Linked, (2) Y-Linked or Holandric, and (3) X-Y-Linked.

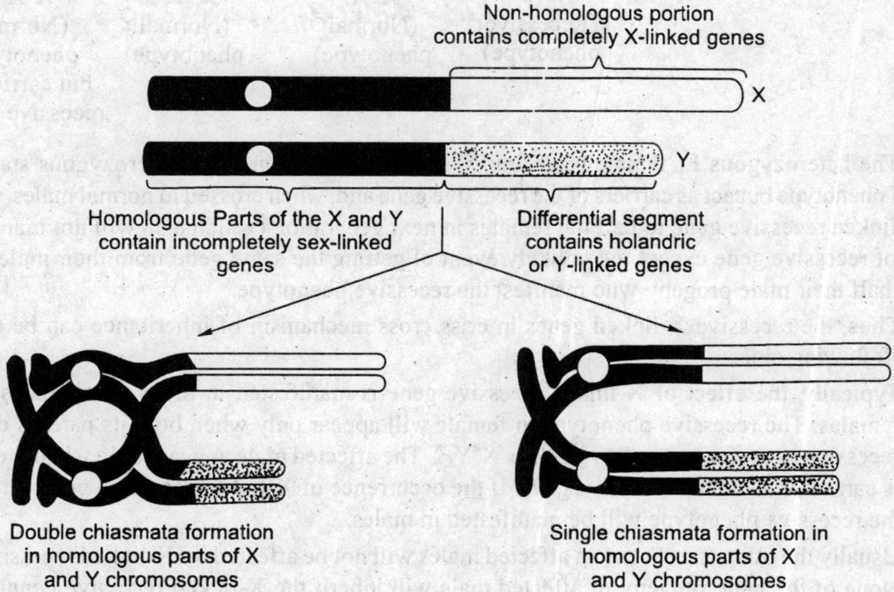

Fig. 17.11 Pairing and differential segments of X and Y chromosomes.

Inheritance of X-linked Genes

The X-linked genes may be recessive or dominant and so in different states they show characteristic patterns of inheritance.

In man, the mutant gene for red-green colour blindness, and the gene for Haemophilia are sex linked and are, linked with X chromosome. Occasionally, both the mutant genes occur in the same family. Such families are of two types:

(*i*) In one type of families, two sex linked genes are brought together through marriage between individuals carrying one or other gene on X chromosome. The mutant genes are carried on different X-chromosomes. In genetical terminology, they are said to be linked in repulsion.

(*ii*) In other families both the sex linked genes are carried on one X chromosome. Actually, single sex linked gene was present on an X chromosome but the second gene appeared on it due to mutation. Such genes are said to be linked in "coupling".

Inheritance of X-linked Recessive Gene

The recessive gene linked to X chromosome shows a characteristic pattern of inheritance. In this, a X-linked recessive gene is transmitted by male parent (father) to heterozygous female in F_1 generation (*i.e.*, to daughter) and then from heterozygous female to one half F_2 males (grand sons). This type of inheritance pattern *from father to daughter to grand sons* is called *criss cross mechanism.* The schematic representation of criss cross mechanism is as follows:

Parents X^AX^A ♀ (Normal phenotype) X X^aY ♂ (Recessive phenotype)

F$_1$ X^AY ♂ (Normal phenotype) X^AX^a ♀ (Normal phenotype but carrier of recessive gene) X X^AY ♂ (Normal phenotype)

F$_2$ X^aY (Recessive ♂ phenotype) X^AY (Normal ♂ phenotype) X^AX^A (Normal ♀ phenotype) X^aX^A (Normal ♀ phenotype but carrier of recessive gene)

The heterozygous F$_1$ females carrying X-linked recessive gene in heterozygous state show normal phenotype but act as carriers of the recessive gene and, when crossed to normal males, transmit the X-linked recessive gene to half the females in next generation (which also will not manifest the effect of recessive gene except the unlikely event of getting the same gene from their male parent) and to half their male progeny who manifest the recessive phenotype.

Thus, the recessive X-linked genes in criss cross mechanism of inheritance can be detected by the following clues:

1. Typically, the effect of X-linked recessive gene is manifested in many more males than in females. The recessive phenotype in female will appear only when both its parents carry the recessive gene, for example, X^AX^a♀ x X^aY♂. The affected male appears only when the mother is carrier of X-linked recessive gene. If the occurrence of X-linked recessive gene is rare then the recessive phenotype will be manifested in males.
2. Usually the offspring (F$_1$) of an affected males will not be affected by X-linked recessive gene. None of the male progeny of affected male will inherit the X-linked recessive gene because the male progeny receives only Y chromosome from its male parent. Thus, the F$_1$ males will not only be free of the recessive phenotype, but also they will not transmit X-linked recessive gene to their offspring.
3. All F$_1$ females of an affected male parent show normal phenotype because recessive gene in heterozygous condition exists in masked state, and acts as carrier of X-linked recessive gene. If such a recessive gene carrying F$_1$ female is crossed to normal male one half the males in next generation will manifest the recessive phenotype.

   In man about fifty-six X-linked genes have been reported. Some of the congenital diseases of human such as haemophilia, colour blindness, anhydrotic ectoderma (non-functioning sweatglands), myopia (short sightedness), juvenile glaucoma (hardening of eye ball) and so on are sex-linked. These diseases are associated with X-linked recessive genes and are common in men.

   (*a*) **Inheritance of X-linked recessive gene for colour blindness in man.** An example of criss cross mechanism is sex-linked colour blindness or red-green blindness in man. In the retina of eyes of man are found colour sensitive cells which differentiate red and green colours. The colour sensitivity is controlled by certain genes located in X chromosome. Individuals having recessive genes for the above colours are unable to differentiate red and green colour and so they are called colour blind or red-green blind. X-linked recessive gene is able to express itself in male because of absence of any allelic gene on Y chromosome. The inheritance of colour blindness is explained in Fig. 17.12.

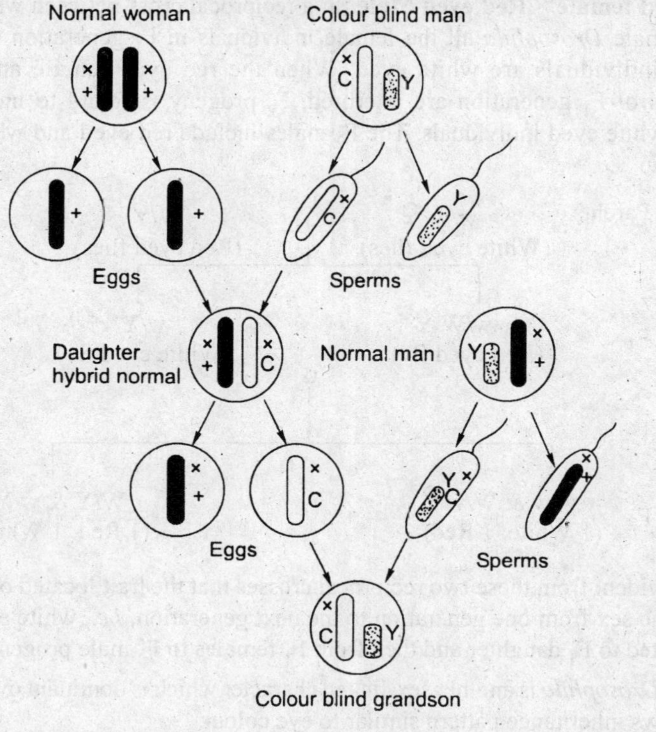

Fig. 17.12 Sex linked inheritance in man from father to daughter to grandson (Criss-cross mechanism).

(*b*) **Inheritance of X-linked gene for eye colour in Drosophila.** One of the classical examples of inheritance of recessive X-linked genes is *eye colour in Drosophila.* According to Morgan (1910), in this fly, the gene for white eye colour is X-linked and recessive to another X-linked dominant gene for red eye colour. The criss-cross pattern of inheritance of sex-linked eye colour in *Drosophila* is clear from the comparison of the following reciprocal crosses (*i.e.,* the crosses in which the genotypes of the parents are reversed).

In the following reciprocal crosses, sex linked red eye character (W) is dominant over sex-linked white eye recessive (w).

(*i*) **Red eyed ♀× White eyed ♂.** When a wild red eyed female is crossed with a mutant white eyed male *Drosophila*, all the F_1 individuals, both males and females, are red eyed. When the red eyed male and female individuals of F_1 are intercrossed, F_2 progeny includes red eyed females and male population consisting of 50% red eyed individuals and 50% white eyed individuals. This explained here as under:

P — W/W ♀ (Red eyed) X w/Y ♂ (White eyed)

F_1 — W/w ♀ (Red eyed) X W/Y ♂ (Red eyed)

F_2 — W/W & W/w♀ (All red) ; W/Y & w/Y ♂ (1 Red eyed: 1 white eyed) (*ii*) **White eyed female × Red eyed Male.** In a reciprocal cross between white eyed female and red eyed male *Drosophila* all the female individuals in F_1 generation are red eyed and all the male individuals are white eyed. When the red eyed female and white eyed male individuals of F_1 generation are interbred, F_2 progeny is found to include 50% red eyed and 50% white eyed individuals. The F_2 males include red eyed and white eyed individuals in 1 : 1 ratio.

Parents w/w ♀ (White eyed flies) u W/Y ♂ (Red eyed flies)

F_1 w/W ♀ (Red eyed fly) u w/Y ♂ (White eyed fly)

F_2 w/w & W/w ♀ (1 White: 1 Red) W/Y & w/Y ♂ (1 Red: 1 White)

It is thus evident from these two reciprocal crosses that the trait located on a sex chromosome alternate the sex from one generation to the next generation, *i.e.*, white eyed trait from father is transmitted to F_1 daughter and then from F_1 females to F_2 male progeny.

Bar eye in *Drosophila* is another sex-linked character which is dominant over normal eye shape. It also shows inheritance pattern similar to eye colour.

(*iii*) **Inheritance of X-linked Dominant Genes.** X-linked dominant genes may be found either on the X chromosome of male or on one of the two X chromosomes of female. The patterns of inheritance of X-linked dominant genes in the two situations are different.

If the dominant gene is linked to X chromosome of male parent, it is transferred to all the females but to none of the males in F_1 (*i.e.*, from father to daughters) as shown below:

Parents X^aX^a ♀ (Recessive) X X^AY ♂ (Dominant)

F_1 X^AX^a ♀ (Dominant) X^AX^a ♀ (Dominant) X^aY ♂ (Recessive) X^aY ♂ (Recessive)

If the dominant gene is linked to one X chromosome of female parent and the male parent is with recessive gene, the X-linked dominant gene from female will be transmitted to one half of the males and females in the next generation as explained below:

Parents X^AX^a ♂ (Dominant) X X^aY ♂ (Recessive)

X^AX^a ♀ (Dominant) X^aX^a ♀ (Recessive) X^AY ♂ (Dominant) X^aY ♂ (Recessive)

In human, some congenital diseases may be caused by X-linked dominant genes, as for example, defective tooth enamel, early wearing down of teeth.

Inheritance of Holandric or Y-linked Genes

The genes on the differential segment of Y chromosome are inherited only by males (holandric genes) *i.e.,* they are transmitted from male to male (from father to son) in human. Such genes as can be passed on to only one sex are referred to as sex-limited. In human, as well as in some other organisms with similar mechanism of sex determination, the presence of Y chromosome and its unique segments determine maleness. Therefore, it is suggested that the genes for maleness (holandric genes) exist on the differential segment of Y chromosome. Hypertrichosis or hairy ear pinna (Fig. 17.13) has been suggested to be possibly controlled by one Y linked gene but no example of such inheritance has been confirmed.

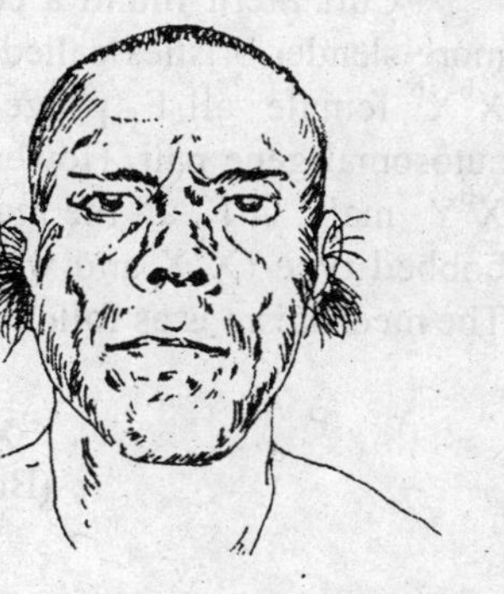

Fig. 17.13 Hairy ear or hypertrichosis in man due to a Y-linked gene.

There is one clear example of Y-linkage in the fish *Lebistes* in which Y-linked 'maculatus gene' determines a pigmented spot at the base of dorsal fin of male. This phenotype is inherited only by males, with transmission from father to son. The females never carry or express this gene.

Inheritance of X-Y-linked Genes

The genes which occur in the pairing segments of X and Y chromosomes are inherited like the autosomal genes. X-Y-linked genes are partially or incompletely linked because crossing over may occur in the homologous sections of X and Y chromosomes. Some of the diseases in human, as for example *Retinites pigmentosa, Xeroderma pigmentation, Epidermolysis bullosa* are X-Y-linked (Fig. 17.14).

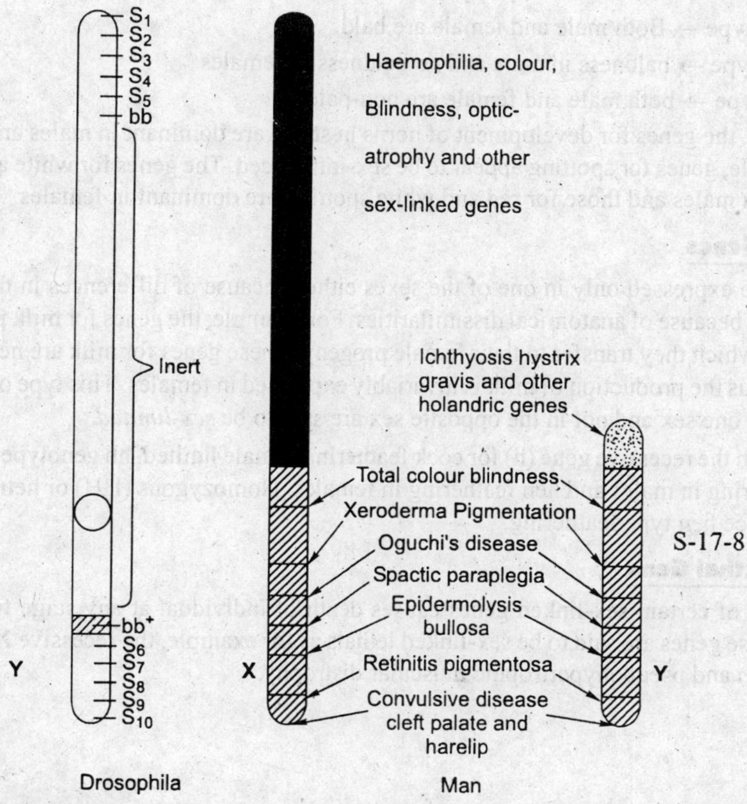

Fig. 17.14 Chromosome map of Sex Chromosome of *Drosophila* and human male.

Curt Stern found a certain non-wild recessive allele causing a phenotype of shorter and more slender bristles called bobbed (b). When a wild type X^+Y^+ male is crossed with a bobbed X^bX^b female, all F_1 progeny are wild type X^+X^b and X^bY^+ as could be expected from the autosomal gene pair. However, if the cross is made between F_1 X^+X^b (bobbed) female and wild X^bY^+ male, in F_2 all the males are wild type (X^+Y^+ and X^b Y^+) and the female progeny includes bobbed type ($X^bX^b$and wild type (X^+X^b) in 1 : 1 ratio. Thus X-Y linkage is revealed in F_2. The mechanism is as follows:

P X^bX^b ♀ (Bobbed) X X^+Y^+ ♂ (Wild)

F_1 X^+X^b ♀ (Wild) X X^bY^+ ♂ (Wild)

F_2 X^+X^b ♀ (Wild) | X^bX^b ♀ (Bobbed) | X^+Y^+ (Wild) | X^bY^+ ♂ (Wild)

Sex influenced Inheritance of Genes

The expression of certain genes located either on autosomes or on the pairing segments of sex chromosomes may be influenced by the sex of the bearer. Baldness in certain families, for example, is hereditary and the gene B for baldness is found to be dominant in males and recessive in females. The genotype Bb expresses itself only in presence of male hormones.

BB genotype → Both male and female are bald

Bb genotype → baldness in male and no baldness in females

bb genotype → both male and female are non-bald.

Similarly, the genes for development of horns in sheep are dominant in males and recessive in females. In cattle, genes for spotting appear to be sex-influenced. The genes for white and mahogany are dominant in males and those for red and white spotting are dominant in females.

Sex Limited Genes

Some genes are expressed only in one of the sexes either because of differences in their hormonal environment or because of anatomical dissimilarities. For example, the genes for milk production are found in bulls which they transfer to their female progeny. These genes for milk are never expressed in males and thus the production of milk is invariably expressed in females. This type of genes which express only in one sex and not in the opposite sex are said to be *sex-limited.*

In chicken the recessive gene (h) for cock feathering is male limited. hh genotype is responsible for cock feathering in males and hen feathering in females. Homozygous (HH) or hetrozygous (Hh) genotypes induce hen type feathering.

Sex-linked Lethal Gene

If the presence of certain sex-linked genes causes death of individual at any stage from zygote to adult stage, those genes are said to be sex-linked lethals as for example, the recessive X-linked genes for *haemophilia* and pseudohypertrophic muscular distrophy.

QUESTIONS

1. Compare and contrast the chromosome theory and genic balance theory of sex determination.
2. Discuss in brief the chromosome theory of sex determination and point out the significance of the study of cases of meiotic non-disjunction in *Drosophila* and human beings.
3. Write short note on:
   (*i*) Single gene sex determination,
   (*ii*) Haplodiplo mechanism of sex determination,
   (*iii*) Sex determination in *Bonellia*,
   (*iv*) Sex reversion,
   (*v*) Sex differentiation, and
   (*vi*) Sex linked inheritance.
4. What do you mean by sex-linked inheritance? Discuss the mechanism of inheritance of genes linked to sex chromosomes.
5. Write notes on:
   (*i*) Sex influenced genes,
   (*ii*) Inheritance of gene for colour blindness in man,
   (*iii*) Sex limited characters.

18 CHAPTER

Cytoplasmic Inheritance or Extranuclear Inheritance

The cytogenetical experiments show that overwhelming majority of inherited characters in higher organisms are governed by the genes that are carried by chromosomes in the nuclei of the cells. Nevertheless, the major part of any individual cell is extranuclear material and so the cytoplasm cannot be regarded as barren genetic ground. It is perhaps, not surprising that some of the characters may occasionally be transmitted from one generation to another through cytoplasm. The transmission of certain genetic information from parents to offsprings through cytoplasm is termed as ***cytoplasmic inheritance*** or ***extrachromosomal*** or ***extranuclear inheritance***. Extrachromosomal inheritance is governed by cytoplasmic factors that exhibit replication and independent transmission.

Evidences for cytoplasmic inheritance have come mainly but not exclusively from the cytogenetical studies on eukaryotic organisms. Some chlorophyll deficiencies or variegations in the higher plants are known to be inherited in this way. Enzyme controlled characteristics in fungi, yeasts (***Saccharomyces***) and ***Neurospora***, killer particles in *Paramecium*, antibiotic resistance in alga *Chlamydomonas,* reduction in the number of setae on the legs of mites of genus *Tetranychus* and breast cancer caused by milk virus in mice are some of the important examples of cytoplasmic inheritance.

Nucleus-Cytoplasm Interaction

The cytoplasm is the centre of metabolic activities in the cell and the nucleus is the ultimate controlling centre. The cytoplasm provides the suitable environment in which the cytoplasmic as well as nuclear genes operate. Many enzymes operate through the cytoplasm and particularly through certain living cytoplasmic structures. Hence it can be expected that the cytoplasm exerts an influence on genic action. In fact, the cytoplasm contains several hereditary units which are termed ***plasmogenes*** or ***cytogenes***. These units which contain their own genetic material are believed to be self-duplicating and are capable of determining hereditary characters. The totality of hereditary factors transmitted through cytoplasm is called ***plasmone***. The cytoplasmic inheritance is not subject to the same general rules as nuclear inheritance (*i.e.,* it does not follow Mendelian rules), but in some cases it seems to be of importance.

The control a nucleus exercises over the cytoplasm is easily demonstrable. If the nucleus of a cell is removed, the cytoplasm may survive for some time but normal function soon ceases. The mammalian red blood cell (RBC) is enucleated in its terminal stage of differentiation and as such it is extremely short-lived.

The Hammerling's experiments involving interspecific nuclear transplant in the marine green alga *Acetabularia* can be cited as an excellent example to show functional dependence of the cytoplasm upon the nucleus. The genus *Acetabularia* is unicellular alga which consists of basal rhizoids, a long stalk and a cap. The nucleus is confined to the rhizoidal region. The cap characteristics are of great significance in the species identification. Hammerling took two species of *Acetabularia*, namely (1) *A. mediterranea* (2) *A. cranulata*.

In *A. mediterranea*, the cap has an average of 81 rays with rounded tips (Fig. 18.1).

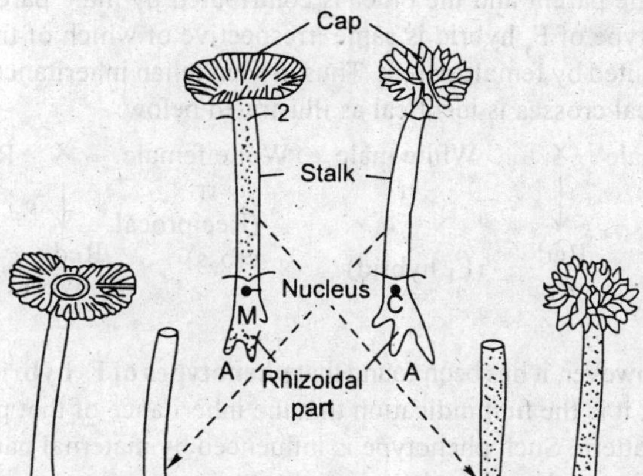

Fig. 18.1 Nuclear transplant experiment in *Acetabularia mediterranea (M)* and *Acetabularia cranulata (C)*.

In *A. cranulata,* average number of rays per cap is 30 and the tips of the rays are pointed.

Hammerling conducted the following interspecific grafting experiment:

1. He took immature plants (in which the cap has not developed) of each species and grafted them together by cutting the tops of the stalks and bringing the cut surfaces into contact. A new stalk soon developed at the point of grafting and this gave rise to a cap which was intermediate in appearance between the caps of the parents. A number of such caps had an average of 42 rays each and hence approached *A. cranulata* but the tips of the rays were usually rounded like those in *A. mediterranea.* The intermediate character of this graft hybrid is to be expected because it contains two nuclei, one from each species. In control experiments, two plants of *A. mediterranea* were grafted together which produced caps resembling the normal caps of species. Similar results were obtained with *A. cranulata.*
2. If rhizoids of *A. mediterranea* (nucleated part) were grafted with the decapitated rhizoidless stalk (enucleated part) of *A. cranulata* and *vice versa*, the caps showed characteristics of the species from which the nucleus was derived. Such grafts contained only one nucleus. The type of cap was found to depend on the length of the stalk grafted. Sometimes, the caps were found to be intermediate in appearance if the enucleate stalk was 1 to 2 cm in length, whereas if the piece of the grafted stalk was about .5 cm. long or less the caps usually resembled that of rhizoid parent. Since the rhizoids represented only a very small proportion of the total cytoplasm, the major changes in the function of the cytoplasm had to be based upon the mode of control exercised by the transplanted nucleus. Thus for the question at hand the answer must be that neither the nucleus nor the cytoplasm exercises exclusive control over the other.

Characteristics and Detection of Cytoplasmic Inheritance

Information concerning cytoplasmic inheritance is meagre but whatever is known about the occurrence and action that is highly confusing. Some of the cytoplasm characteristics appear to be under the ultimate control of chromosomes (Darlington, 1958) but in other cases there is no evidence of such a relationship. Only a few cases of cytoplasmic inheritance, completely independent of nuclear genes, are known. The cytoplasmic inheritance can be detected by special methods. Two rules are used for their detection; one is negative and the other positive. Genes in diploid organisms exist in pair and two members or alternative forms of a single gene are called alleles. The alleles are located on homologous chromosomes at the same position. In the process of hybridization, one allele is contributed by the female parent and the other is contributed by male parent. The outcome of the cross in terms of phenotype of F_1 hybrid is same irrespective of which of the two alleles, recessive or dominant was contributed by female parent. Thus in Mendelian inheritance phenotypic expression of F_1 hybrids in reciprocal crosses is identical as illustrated below:

Red female RR	× ↓	White male rr	White female rr	× ↓	Red male RR
	Red Rr	(F_1 hybrid)	(Reciprocal cross)	Red F_1 hybrid Rr	

In certain cases, however, it has been found that **phenotypes of F_1 hybrids of reciprocal crosses are different**. If it is so, it is the first indication that the inheritance of that particular character may not follow Mendelian pattern. Such phenotype is influenced by maternal parent (mother). This type of inheritance is referred to as *Non-Mendelian* inheritance.

The negative rule states that hereditary traits which are transmitted by cytoplasm must not show Mendelian segregation in crosses. The reciprocal crosses with respect to a particular set of characters governed by a set of extrachromosomal genes yield dissimilar hybrids. The positive rule is based on the fact that most of the cytoplasmically inherited characters would follow the maternal (mother) lines, *i.e.*, ***uniparental mode of transmission***. In higher plants and animals, large ova (macrogametes) contain large amount of cytoplasm but male gametes or sperms (microgametes) have essentially negligible amount of cytoplasm, so under such conditions most of cytoplasmic factors are transmitted to the offsprings through the ova of mother. It is known as ***maternal inheritance*** or ***trans-ovarian transmission***. In this mode of transmission, all the progenies of the parents have maternal features and only female parent can transmit the cytoplasmic characters to the succeeding generation. Maternal inheritance may be, of two kinds:

I. On one hand, some treatments if applied to the female parent may affect its egg's cytoplasm in such a way that her subsequent offsprings are modified in some way. Effects of this kind are called ***dauer-modifications*** or persisting modifications. These characteristically persist for only a small number of generations unless the same treatment is repeated on the progeny. The dauer-modification was first discovered experimentally by V. Jollos. He treated protozoa with chemical poisons and heat shocks and induced several morphological abnormalities in them. These could have been simple environmental modifications but they persisted in the progenies for several generations after the environmental stimulus was removed. Such abnormalities go on decreasing generation after generation and eventually disappear completely, as the original treated cytoplasm decreases gradually through cell division. Fruitflies subjected to heat treatment and bacteria treated with chemicals have provided further evidences.

II. On the other hand, some maternal inheritance are also known which do not depend upon the repeated application of an external stimulus to the cytoplasm. True cytoplasmic inheritance is controlled by independent cytoplasmic genes. Maternal effects reflect the influence of mother's genes on developing tissues and dauer modifications reflect the temporary influence of cytoplasmic structures induced temporarily by environmental conditions.

Some Examples of Cytoplasmic Inheritance

The following are the two major groups of cytoplasmic factors which constitute extra-nuclear genetic systems:

1. The first group of cytoplasmic factors involves intracellular organelles. In eukaryotic cells, two cytoplasmic organelles, mitochondria and the plastids of green plants contain their own genetic materials (*extranuclear genes* or *extra chromosomal genes*). The term extrachromosomal gene is potentially confusing because the genes found in the mitochondria and chloroplasts do in fact comprise extranuclear DNA.
2. The second group of cytoplasmic factors involves intracellular symbionts such as bacteria and viruses that live symbiotically within cells of host organisms. They replicate along with host's genetic material and their presence causes a special modification of phenotype of host organisms.

Genetics of Plastids

In the cytoplasm of the plant cells are found many small cytoplasmic bodies, called plastids. These plastids are of several types, such as, chloroplast, leucoplast, chromoplast and so on. Plastids arise from smaller particles, the proplastids, found in the egg cytoplasm. They reproduce themselves independently of other cell parts by process of division or self-duplication. Plastid characteristics are inherited in the majority of the cases from the female parent but in some cases a few characteristics are transmitted from the male parent also. One of the common examples of plastid inheritance is the variegation found in plants. In variegated condition, some organs of the plant, say the leaves, may show green and white patches or green, yellow, white or pink spots. These spots of varyings colours, show differences in the kinds of plastid. Although in many plants variegations are known to depend on ordinary Mendelian genes (*i.e.,* they are governed by nuclear genes), yet there are a few known cases which show a largely maternal type of inheritance. Inheritance in these cases depends on the nature of the plastids.

Carl Correns (1909) described a case of maternal transmission in *Mirabilis jalapa* (four-O'clock plant). The blotchy or spotted leaves of these variegated plants show patches of green and white tissue but some branches carry only green leaves and others only white leaves. Flowers develop on all types of branches and may be cross-pollinated in different combinations. The following crosses were studied in *Mirabilis jalapa*:

	Parent phenotype		*Progeny*
	Pistillate parent	*Pollen parent*	*Phenotype*
1.	White leaved	Green leaved	White leaved
2.	White leaved	White leaved	White leaved
3.	White leaved	Variegated leaved	White leaved
4.	Green leaved	Green leaved	Green leaved
5.	Green leaved	White leaved	Green leaved
6.	Green leaved	Variegated leaved	Green leaved
7.	Variegated leaved	Green leaved	Variegated, green or white leaved
8.	Variegated leaved	White leaved	Variegated, green or white leaved
9.	Variegated leaved	Variegated leaved	Variegated, green or white leaved

From these experiments two surprising features came to light:

(*i*) *There was a difference in the results of reciprocal crosses;* as for example, Green leaved ♀× white leaved ♂ gave green leaved progeny and white leaved ♀× green leaved ♂ gave white leaved progeny. The differences observed in the results of reciprocal crosses were quite contrary to Mendel's findings. In fact, in conventional genetics, differences in the results of reciprocal crosses are expected and encountered in the case of sex-linked genes. However, the results of the crosses of *Mirabilis jalapa* cannot be explained by sex linkage.

(*ii*) The phenotype of female parent determined the phenotype of all progeny (***uniparental transmission***) and the phenotype of male parent did not contribute anything to the progeny. This phenomenon is referred to as ***maternal inheritance.*** In subsequent generations, the maternal inheritance has been found to be effective in the same pattern as those observed in original crosses. The different leaf colours arise due to the presence of green or colourless plastids in the cells. The inheritance of different leaf colours in *Mirabilis jalapa* might be explained if the plastids are somehow genetically autonomous and are never transmitted through male parent. For an organelle to be genetically autonomous, it must be provided with its own genetic determinants that are responsible for its phenotype. In this case, the plastids carry their own genetic determinants that are responsible for their colour. Since the bulk of cytoplasm of zygote is contributed by the egg and the male gamete contributes negligible amount of cytoplasm, plastids present in the cytoplasm of egg are responsible for the appearance of maternal colour in the progeny and the failure of pollen parent to transmit its colour to progeny is reasonable.

The variegated branches produce three kinds of eggs; some with colourless plastids (leucoplasts), some contain only chloroplasts, and some are with both chloroplasts and leucoplasts. The egg containing green and colourless plastids produces a zygote that also contains both kinds of plastids. In the course of development, some form of cytoplasmic division occurs during mitosis that segregates chloroplasts and leucoplasts into pure cell lines which produce variegation in the leaves of the progeny (Fig. 18.2). This process of segregation of plastids during cytokinesis of mitosis is described as *mitotic segregation.* This process represents the segregation and recombination of organelle genotype. Therefore, it is also called **cytoplasmic segregation and recombination** (CSAR).

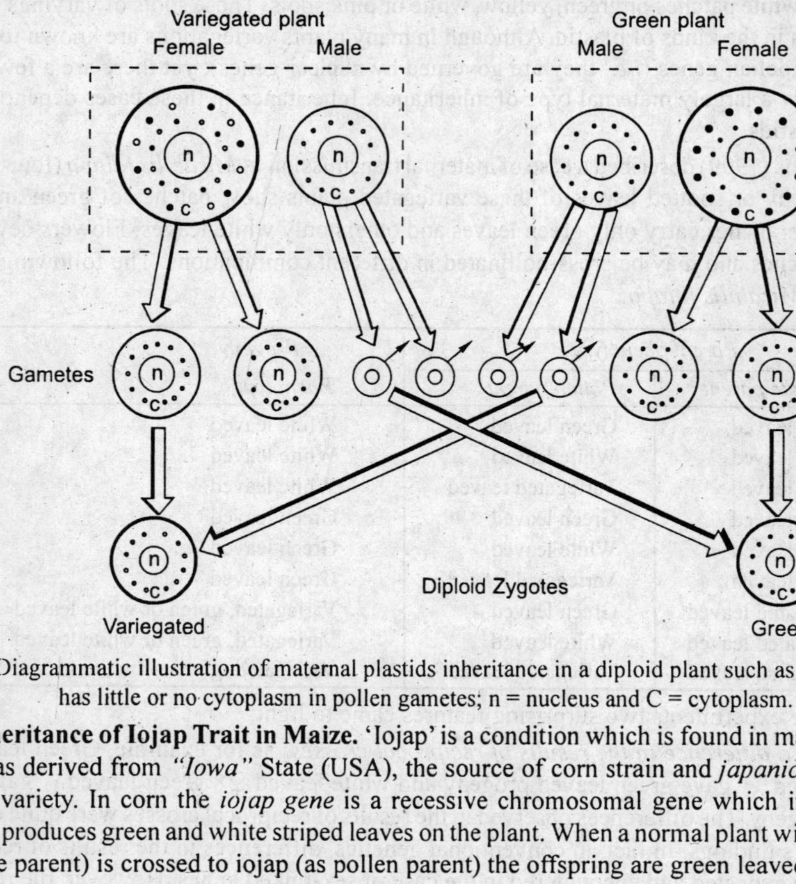

Fig. 18.2 Diagrammatic illustration of maternal plastids inheritance in a diploid plant such as *Mirabilis* which has little or no cytoplasm in pollen gametes; n = nucleus and C = cytoplasm.

Inheritance of Iojap Trait in Maize. 'Iojap' is a condition which is found in maize. The name 'iojap' was derived from *"Iowa"* State (USA), the source of corn strain and *japanica*, the name of a striped variety. In corn the *iojap gene* is a recessive chromosomal gene which in homozygous condition produces green and white striped leaves on the plant. When a normal plant with green leaves (as female parent) is crossed to iojap (as pollen parent) the offspring are green leaved.

Ij/Ij (Green) ♀ X ij/ij (iojap) ♂

F$_1$ Ij/ij (Green)

In reciprocal cross between normal green (♂ parent) and iojap (♀) plant there appear offsprings of three different types as shown below:

Ij/Ij Green ♂ X ij/ij iojap (♀)

F₁ Ij/ij Green Ij/ij white Ij/ij Green and white striped

In iojap plants, green and white striped condition is inherited from female parent (maternal inheritance). It seems that in iojap plants, there occur two types of plastids; normal green and abnormal iojap plastids. During differentiation of female gametes plastids are randomly distributed in the egg cells. If the egg receives normal green plastids it will give rise to green leaved progeny irrespective of which plant acted as pollen parent. If the egg receives abnormal colourless plastids, it after fertilization will produce white leaved offspring. If the egg receives both green and white plastids then after fertilization it will produce offspring with green and white striped leaves. Probably a nuclear gene controls the development of abnormal plastids in the cytoplasm.

If striped leaved F_1 iojap (Ij/ij) female parent is crossed to normal green leaved (Ij/Ij) male parent, the following types of offspring are obtained:

Ij/ij (striped) ♀ X Ij/Ij (green) ♂

Ij/Ij	Ij/ij	Ij/Ij	Ij/ij	Ij/ij	Ij/ij
green	striped	white	green	striped	white

This back-cross experiment shows that green males have no effect upon progeny.

If plastids are self-duplicating structures possessing their own hereditary factors, it should be possible to isolate them and demonstrate that they contain the complete genetic system. Researches in number of different laboratories have demonstrated that the chloroplasts contain DNA, RNA and ribosomes. DNA found in the chloroplast is different from the nuclear DNA, as the **plastid DNA** has low molecular weight in comparison to nuclear DNA. Their base composition is also different from that of nuclear DNA. The lighter DNA has been called ***'satellite DNA'***. DNAs of different types of plastids are self-perpetuating and arise from pre-existing ones. They are not produced by nuclear DNA which would otherwise be capable of producing only one kind of plastid DNA. Evidence for this has been obtained by Gibor and Granic (1962) from the green alga *Euglena.* They irradiated the cytoplasm with ultraviolet (UV) light. When the ultraviolet light was directed at the cytoplasm, the irradiation caused permanent disappearance of the plastid colour and thus the irradiated Euglena in successive generations never developed chloroplast. This indicates that the chloroplasts are highly mutable. Only irradiation of cytoplasm caused irreversible mutation of the plastids which then grew and divided as tiny colourless proplastids in different generations. Here, if it is reasoned that ultraviolet sensitive DNA units of the cytoplasm are formed by the nuclear DNA then the new cytoplasmic DNA units formed by the nuclear DNA should replace the damaged DNA of the cytoplasm and prevent bleaching. Since this did not happen, it is concluded that DNA of plastid is formed from pre-existing DNA. Replication of plastid DNA has been suggested by auto-radiographic studies also. In all cases, there are multiple copies of plastid DNA in each chloroplast. The transcription of RNA from plastid DNA has also been experimentally proved in the isolated chloroplasts. Smillie (1967) demonstrated that the chloroplast ribosomal RNA of *Euglena* engages in complementary base pairing with Euglena chloroplast DNA (Hybridization technique) indicating that information necessary to make chloro-ribosomes resides in the chloroplast's own chromosome. This conclusion has been confirmed for tobacco chloroplasts by Tewari and Wildmann. Their estimation indicated that nearly four sets of ribosomal genes were present in the chloroplast. Their data also indicated that the nuclear DNA contains genes for chlororibosomes as well as for cytoplasmic ribosomes. RNA-protein translation (protein synthesis) system has been demonstrated in isolated chloroplasts. Although protein synthesis in isolated plastids is suggested by incorporation of aminoacids, no net synthesis of a specific protein of the plastids has been demonstrated as yet.

Streptomycin Resistance in Chlamydomonas

Ruthsagar (1965) has reported some cases of extrachromosomal inheritance in green alga *Chlamydomonas reinhardi.* The alga reproduces by asexual as well as sexual means. It does not have different sexes but has positive and negative strains or mating types (mt^+ and mt^-). The sexual reproduction involves fusion between two morphologically similar but physiologically dissimilar gametes of two different mating types (+ strain or mt^+ and – strain or mt^-) and the gametic fusion results in zygote. The sex is determined by a single chromosomal gene. When meiosis occurs in zygote, four haploid daughter protoplasts are formed which give rise to new plants. Out of 4 new plants resulted from a zygote two are of + strain and the other two are of negative (–) strain. Although both the sexes contribute equally to the zygote, there is maternal transmission of certain cytoplasmic traits.

Ruthsagar (1965) isolated two strains of *Chlamydomonas*: one strain was resistant (Sm^r) to 500 µg of streptomycin per ml. of culture solution and the other was sensitive (Sm^s). When the reciprocal crosses were made between the streptomycin resistant (Sm^r) and streptomycin sensitive (Sm^s) strains, the following results were obtained:

$Sm^r\ mt^+$ ♀	X ↓	$Sm^s\ mt^-$ ♂		$Sm^s\ mt^+$ ♀	X ↓	$Sm^r\ mt^-$ ♂
All	Sm^r (F_1 Hybrids)			All	Sm^s (F_1 Hybrids)	

The diploid cells undergo meiosis and give rise to four haploid cells (tetrads) as shown in Fig. 18.3. From these crosses the following two inferences with respect to Sm resistance can be drawn:

1. The F_1 reciprocal crosses differ from each other.
2. The phenotype of F_1 is governed by mt^+ strain *i.e.,* it is maternal inheritance.

Using the analogy of higher organisms, mt^+ is referred to as female and mt^- as male. The mating type genes mt^+ and mt^- segregate in 1 : 1 ratio as expected for true Mendelian inheritance. In the higher organisms formation of zygote involves fusion between an egg and a sperm and the contribution of cytoplasm to the zygote by sperm is negligible. Under such condition it is easy to comprehend the mechanism of maternal inheritance. But in *Chlamydomonas*, male (mt^-) and female (mt^+) gametes being indentical in size contribute equal amount of cytoplasm to the zygote, even then the cytoplasmic features of only mt^+ strain is expressed in F_1, *i.e.,* it is uniparental inheritance. Now the question arises, what happens to the cytoplasmic determinants of mt^- gametes. This problem was solved by Ruthsagar who discovered that the chloroplast DNA of mt^- strain becomes degraded in zygote and the mt^+ gene or a gene closely associated to it specifies a *restriction–modification* system. Here restriction implies degradation and the modification means protection. The system encoding the DNA modifying enzyme, modifies its own DNA which cannot be degraded by the restriction system. The mt^- chloroplast DNA which is not modified or protected is degraded by restriction system of mt^+gamete. It is suggested that mt^+ linked gene encodes an endonuclease enzyme (enzyme which degrades DNA) which differentiates plastid DNA of its own cell from that of mt^- cell or digested due to modification. Since plastid DNA of mt^- strain is degraded after sexual union, no expression of mt^- chloroplast DNA is possible and hence uniparental pattern of inheritance is observed.

In rare cases (one in a thousand), however, the plastid DNA of mt^- cell escapes degradation by restriction system of mt^+ cell and the zygote contains the plastid DNAs from both the cells. Such a zygote is referred to as **Cytohet** or **cytoplasmic heterozygote**. Cytohets are important from the view point of studying the recombination of cytoplasmic genes. Ruthsagar has constructed the genetic map of *Chlamydomonas* chloroplast DNA using the genic analysis of cytohets. This suggests that the inheritance of streptomycin resistance is uniparental and the factor for streptomycin resistance resides in the cytoplasm of + strain or mt$^+$ (Fig 18.3).

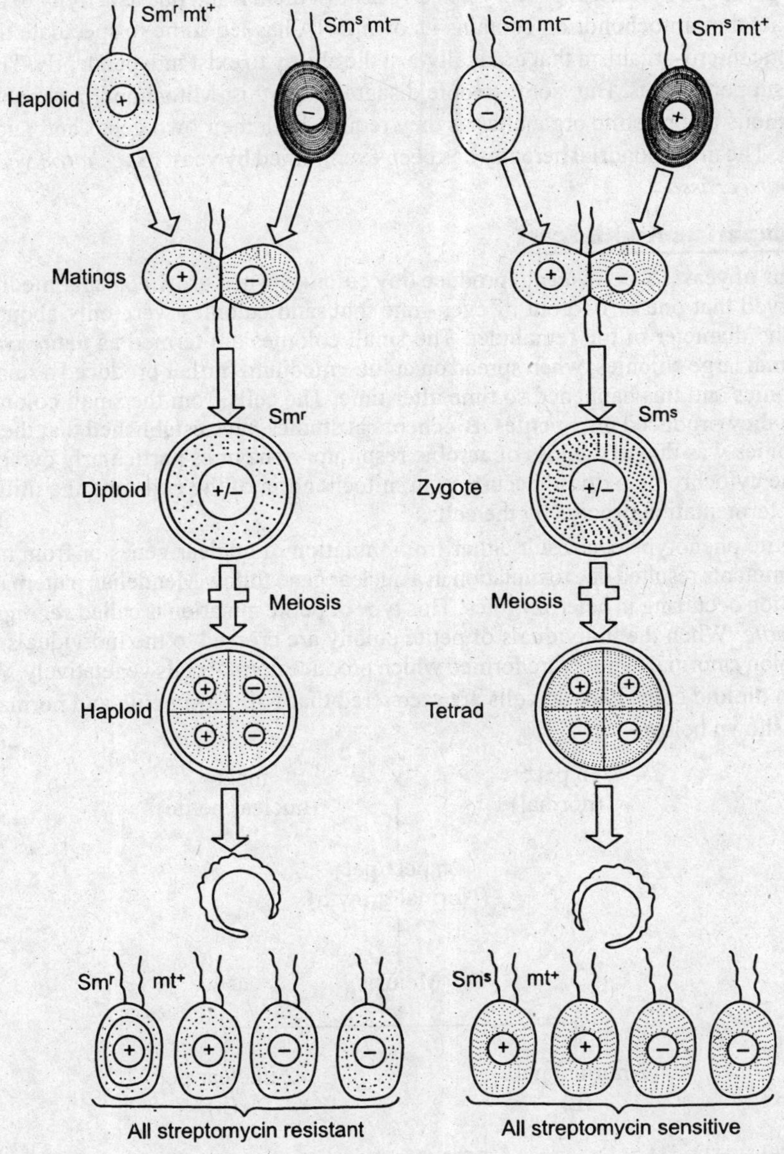

Fig. 18.3 Inheritance of resistance to streptomycin in Chlamydomonas. The plus and minus signs refer to mating types (mt) which is inherited as a single gene difference. The progeny is with rare exceptions always like the plus parent in its reaction to streptomycin, but in these crosses the mating type difference segregates in every tetrad.

Inheritance of Mitochondria

Mitochondria are the cytoplasmic organelles which contain various respiratory enzymes and are main source of energy for the cell. They are found in cell in variable numbers. The complexity of structure of the mitochondria and their similarity in some ways to plastids suggest the possibility that they may be inherited in the same way as the plastids. The mitochondria have a double membrane and contain besides respiratory enzymes of electron transport chain (the cytochromes), their own genetic determinants (DNA). There is certain amount of electron micrographic evidence for the continuity of this cytoplasmic organelle through cell division. Although most of the mitochondrial proteins and enzymes are produced by nuclear genes, yet nearly 20% of them result due to activity of mitochondrial genes. The fact that mitochondrion contains its own DNA has led some to speculate that it evolved from symbiotic micro-organism that gradually lost the ability to exist independently. There is erough evidence in support of this. But, some people disagree with this. Mitochondria cannot be regarded truly autonomous cytoplasmic organelles as they require both their own genes and nuclear genes in order to exist. The mitochondrial heredity has been exemplified by yeast (*Saccharomyces cereviceae*) and *Neurospora crassa*.

Ephrussi's Experiment with Yeast

Certain strains of yeast (*S. cereviceae*) produce tiny colonies when grown on agar medium. Ephrussi (1953) observed that one or two out of every one thousand colonies were only about one-third or one-half of the diameter of the remainder. The small colonies are termed as ***petite colonies***. Cells from the normal large colonies, when spread on culture medium, further produced a small proportion of petite colonies and this happened so time after time. The cells from the small colonies were true breeding and they produced only petites. Biochemical studies have established that the slow growth of petite colonies was due to the loss of aerobic respiratory enzymes particularly cytochrome *a* and *b* and enzyme cytochrome oxidase occurring in mitochondria of the cells and the utilization of the less efficient fermentation process by the cells.

The petite phenotype can result either from mutation of nuclear genes or from mitochondrial genes. Petite mutants resulted due to mutation in a nuclear gene follow Mendelian pattern of inheritance with segregation occurring in heterozygotes. This type of petite mutation is called *segregational petite* or *nuclear petite*. When the individuals of petite colony are crossed to the individuals from normal large sized colony, normal zygotes are formed which produce normal cells vegetatively. When meiosis takes place in diploid cells, haploid cells are recovered that will form petite and normal colonies in 1 : 1 ratio as shown below:

n pet^+ (normal) × n pet^- (nuclear petite)

↓

2n pet^+/pet^-
(Normal growth)

↓

Meiosis

↓

2 normal pet^+ (n) | 2 nuclear pet^- (n)

Bores Ephrussi and his associates (1953) also found that in presence of small amount of acridine dyes such as acriflavine and euflavine many cytochrome deficient petite colonies developed which showed extrachromosomal (non Mendelian) inheritance for petite characters. They were called ***vegetative petites***. The rate of mutation was much higher at lower concentrations than that normally expected for chromosomal mutation. The vegetative petites may arise directly from mutations in mitochondrial genes leading to defective mitochondria.

There are two classes of *vegetative* or *extrachromosomal petites*: (*i*) ***neutral petites***; *and* (*ii*) ***suppressive petites***, which show different patterns of inheritance.

Neutral Extrachromosomal or Vegetative Petites

When a cross is made between wild type haploid yeast and neutral petite haploid yeast, normal diploid offsprings are obtained. The diploid individuals by budding process produce several normal diploids. When meiosis occurs in normal diploids, haploid ascospores are formed which produce normal haploid colonies. If the determinants of this trait were chromosomal, one would expect normal and petite traits in 1 : 1 ratio in the population of haploid spore cells. This suggests that the inheritance is non-chromosomal. The genetic basis of this type of inheritance can be explained assuming the presence of an extrachromosomal or cytoplasmic factor [rho^+] in normal strain and missing or [$rho^-_{[N]}$] in neutral petite mutants. The neutral petites [$rho^-_{[N]}$] usually lack in mitochondrial DNA. Now, if the haploid neutral petite is crossed to haploid normal strain, the diploid would be normal. The normal condition in diploid cells appears because of normal mitochondria with [rho^+] factor contributed by normal haploid strain. These normal mitochondria replicate and are passed on to haploid spores after meiosis in diploid cells. The mitochondria contributed by neutral petite mutant possibly do not replicate and gradually degenerate. So all the haploid spores and their descendants would be normal. The pattern of inheritance is as follows:

n [rho^+] (normal) X n [$rho^-_{[N]}$] (neutral petite)

↓

2n
[rho^+]/[$rho^-_{[N]}$]
Normal growth

↓ Meiosis

No petite — 4 normal haploids

Inheritance Pattern of Suppressive Petites

The suppressive petite mutant shows different behaviour than the neutral petite. When a cross is made between haploid cells of suppressive petite and haploid cells of normal strain, diploid cells are obtained which are in part normal and in part petite and as their name indicates, they can suppress normal aerobic respiration in presence of normal cytoplasm. The normal diploids after meiosis produce normal haploid spores while the diploid petites after budding produce diploids which may be all petites or some normal and some petites. Normal diploids after sporulation produce only normal and no petite. It is thus obvious that suppressive petites follow non-Mendelian pattern of inheritance. The genetic basis of this type of inheritance can be explained by assuming the presence of an extrachromosomal factor [rho^+] in normal strain and [$rho^-_{[S]}$] in suppressive petites.

n[$rho^-_{(s)}$] (suppressive petite) X n[rho^+] (Normal)

↓

2n[rho^+]/[$rho^-_{(s)}$]
(Partly normal and partly petite)

↙ (Immediate meiosis): 4 Petite (n) — No Normal

↘ (Diploids by budding) 1-99% petites + some Normal

Meiosis: 0 petite — 4 Normal (n)

The genetic cause for suppressive petite is mitochondrial mutation. Unlike neutral petites, the mitochondria of suppressive petites contain mutant DNA. The mutant mitochondria can replicate and can be passed on to the progeny cells which can, in turn, express mutant phenotype. In the cross in question it is the relative proportion of normal and mutant mitochondria that determines the phenotype of the particular cell. The diploid cells and haploid spores would be normal if normal mitochondria predominated and they would show mutant phenotype if mutant mitochondria predominated. The lack of normal segregation and also the high mutability of normal colony cells provide good evidence that vegetative petite phenotype is due to extrachromosomal or cytoplasmic genes.

The Poky Strain in Neurospora

There are several examples for mitochondrial enzyme deficiency which are cases of extrachromosomal inheritance. One of the classical examples of extrachromosomal inheritance of plasma genes came from studies of *Neurospora.* In this fungus, there is a slow growing mutant strain called *poky.* The mitochondria contain cytochromes *a*, *b* and *c* which are electron transport proteins necessary for oxidative phosporylation. In poky strains, either cytochrome *a* or cytochrome *b* is absent but cytochrome *c* is present in excess. Poky differs from petite in that the two mutants are not deficient for the same enzymes. When poky as female parent was crossed with a normal strain as a male parent, the progeny were found to be poky. In reciprocal cross (Normal ♀× poky ♂), the progeny were normal. This non-Mendelian uniparental inheritance suggested that the cytoplasm of female parent was important because the only difference between reciprocal crosses was in contribution of cytoplasm. The male gametes in *Neurospora* contribute negligible amount of cytoplasm just as in animals or higher plants. So, it is probable that the factor for pokyness resided somewhere in the cytoplasm. The segregation of poky from normal is never observed and the progeny of poky ♀× normal ♂ will always be poky. Thus nuclear genotype has no effect on this particular phenotype.

Mitochondrial Genome

The genetic material of mitochondria exists in the form of super coiled circular DNA. No histones are associated with this DNA. The mitochondrial DNAs or mt DNAs of animals are much smaller than those of higher plants and fungi and plant mt DNA is much larger than in man. Within mitochondria there are nucleoides similar to those found in bacteria and several copies of mt DNA are found within each nucleoid region. In yeast there may be found 1 to 45 mitochondria in each cell, 10 to 30 nucleoids in each mitochondrion and 4 to 5 mt DNA in each nucleoid. The replication of mt DNA takes place in presence of a specific mitochondrial DNA polymerase.

The origin of mitochondria from the pre-existing ones (hence, its DNA replication) is less clear than of plastids. A clear example of origin of mitochondria from the pre-existing one was reported for a marine alga *Micromonas.* The alga consists of one nucleus, one chloroplast, and one mitochondrion. Electron micrographs show that the three organelles divide independently at the time of cell division. This observation clearly indicates that all mitochondria originate only from pre-existing ones and that somewhere in the cell cycle the mitochondrial DNA must replicate itself. Mitochondrial DNA codes for a number of mitochondrial components. Mitochondria contain ribosomes that carry out protein synthesis. The mitochondrial DNA codes at least for two r RNAs found in these ribosomes, the other r RNAs are encoded by nuclear DNA. The mt DNA also codes for 22 or 23 t RNA and about 10 to 12 proteins found in the inner mitochondrial membrane. The mRNA is transcribed from mt DNA. The mitochondria, thus, have their own protein synthesizing system. The protein synthesis in mitochondria has been found to be similar in some ways to Bacterial protein synthesis. Mitochondrial protein synthesizing systems are quite autonomous and independent of the nucleus. It has been demonstrated that if mitochondrial DNA from *Neurospora* is used *in vitro* transcription and translation system of *Escherichia coli* two proteins, P_1 and P_2 are synthesized. These P_1 and P_2 proteins are demonstrated to be present in the *Neurospora* mitochondria *in vivo*. It is not known exactly what functions these proteins perform. Possibly, they participate in ensuring the precise special arrangement that is a necessary feature of mitochondrial enzyme batteries. Thus they may be part of a scaffold on which Krebs' cycle and electron transport systems found in mitochondria are located. The electron transport enzymes are known to be coded by nuclear genes. In yeast, one nuclear genome is about 10^{10} daltons DNA whereas the mitochondrial DNA genome is only $^1/200$ as big or 5×10^7 daltons; 5×10^7 daltons is about 75,000 nucleotide pairs and of this about $^1/5$ is not transcribed and about $^1/5$ is probably taken up by ribosomal RNA and transfer RNA. The remaining 45,000 nucleotide pairs are equivalent to about 15,000 aminoacid codons which are enough to code for about 75 proteins each having about 200 aminoacids.

Inheritance of Shell Coiling in Snail

One of the classical examples of intricate relationship between maternal genotype and egg cytoplasm "phenotype" was studied in snails by Sturtevant. He showed that there are two strains of water snails (*Limnaea peregra*) that differ each other in the direction of coiling of shell. Looking into the opening of the shell it can be seen that in one strain the shell always coils to the left (sinistral) whereas in the other strain the shell always coils to the right (dextral). In the cross dextral ♀× sinistral ♂ all the F_1 progeny have dextral coils implying that dextral is dominant over sinistral. However, in the $F_1 \times F_1$ cross (*i.e.,* inbreeding), all the F_2 snails are also dextral. The reciprocal cross (dextral ♂× sinistral ♀) produces F_1 progeny that are all left coiler. In this case $F_1 \times F_1$ cross also yields only dextral coils. From these experiments it becomes clear that coiling of snails is not determined by individuals' own genes but by those of mother. The offsprings whose mothers are either homozygous or heterozygous for right coiling are right coilers even if they are homozygous for sinistrality (left coiling). In the same way offspring of left coiling mother are left coilers even if they carry dominant genes for right coiling. The F_2 females of either cross (right coiler) when mated with males of any genotype produce at an average right coilers and left coilers in 3 : 1 ratio. But 3 : 1 ratio appears in F_3 and not in F_2. If F_2

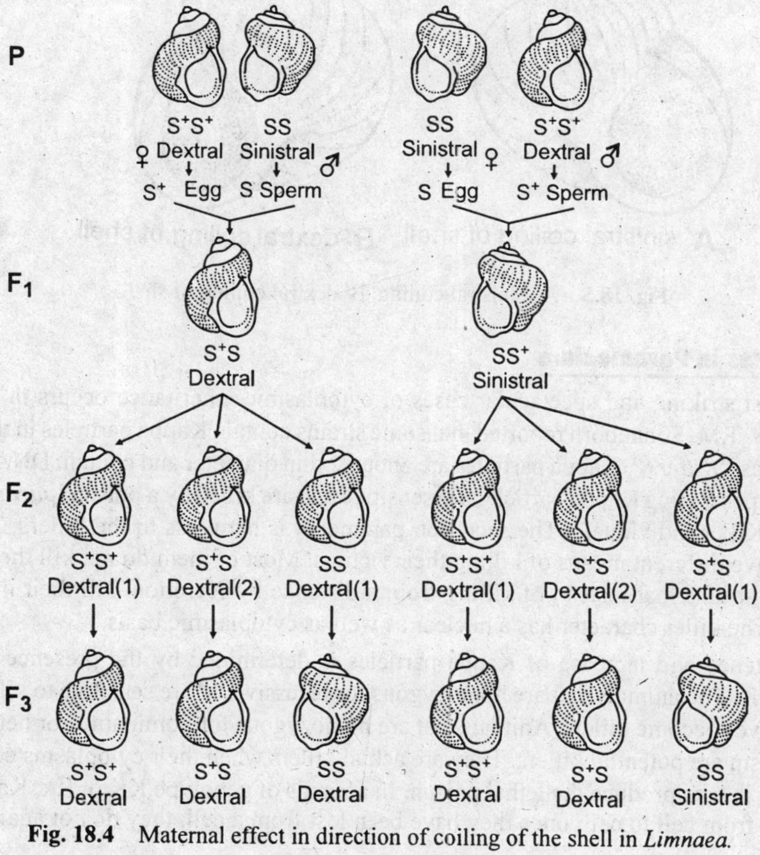

Fig. 18.4 Maternal effect in direction of coiling of the shell in *Limnaea.*

males are mated with homozygous right coiling females, there is no segregation and all their progenies are right coilers, but if they are mated with homozygous left coiling females only left coilers are produced. It is clear then that the paternal (male) genotype is not crucial in determining the phenotype of the offspring. It becomes clear when it is assumed that the genes for left coiling pattern (S^+) and that the phenotypic expression of individuals is determined by the genotype of their mothers. The crosses illustrating the maternal determination of shell coiling in snails are shown in the Fig. 18.4.

Further investigations suggest that coiling depends upon the early cleavages in the zygote. If the spindle is tilted to left of the median line of zygote, the successive cleavages will produce a spiral to left and if the orientation of spindle is tilted to the right of the median line of zygote a dextral pattern will follow. The spindle orientation is governed by the genotype of oocyte from which the egg develops. However, the exact mechanism of spindle orientation is not known (Fig. 18.5).

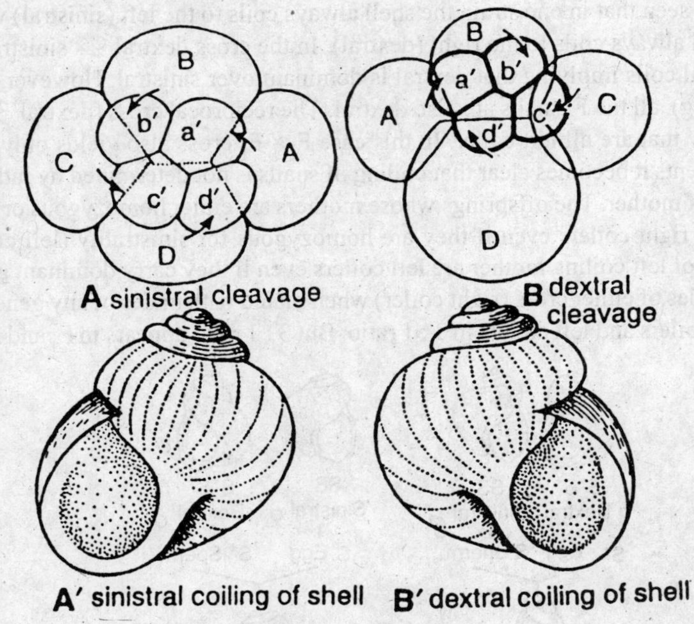

Fig. 18.5 A′ Sinistral coiling, B′ dextral coiling of shell.

Kappa Particles in Paramecium

One of the most striking and spectacular cases of cytoplasmic inheritance occurs in *Paramecium aurelia.* In 1938, T.M. Sonneborn reported that some strains contain Kappa particles in the cytoplasm and are known as *"Killers"*. Kappa particles are about .2 μ in diameter and contain DNA and protein. Individuals not possessing Kappa particles are sensitive and are killed by a poison '***paramecin***' which is secreted by Killer individuals. The secretion paramecin is harmless to the killers. The different killer strains have different means of killing their victims. Most of them do not kill their mates. But there are some strains that instead of killing from a distance by secretion, kill their mates through close contact. The killer character has a nuclear as well as cytoplasmic basis.

The existence and increase of Kappa particles is determined by the presence of a nuclear dominant gene K. The animals that are homozygous for recessive 'k' are sensitive to killing and they cannot themselves become killers. Animals that are homozygous for dominant K or heterozygous in normal cytoplasm are potential killers. They are actual killers when their cytoplasms contain Kappa particles which in turn produce the lethal poison. In animals of genotype KK or Kk, Kappa particles are transmitted from cell to cell; once they have been lost from a cell, they do not again develop by themselves. The individuals with genotype kk may also contain Kappa particles of some sort in the cytoplasm, although this state is unstable and eventually the particles disappear.

Paramecium generally reproduces by *conjugation method*, a system of *parasexual reproduction* (Fig. 18.6) and *autogamy* (Fig. 18.6). When a killer strain of *Paramecium aurelia* with genotype KK conjugates with sensitive strain having genotype kk, the ex-conjugants are all heterozygous (Kk). The genotype Kk of hetrozygous ex-conjugants suggests that they should be identical and killers. But, this is not always the case and Kk hetrozygotes are equally divided into killers and sensitives (Fig. 18.6). If the conjugation is accomplished in a short duration normally no exchange of cytoplasm takes place between the killers and sensitive individuals. When autogamy takes place in hetrozygotes 8 nuclei are formed after meiosis and mitosis divisions and then 7 of the 8 haploid nuclei degenerate and the remaining nucleus undergoes mitosis and the two identical nuclei so formed fuse to form a homozygous diploid. Thus all the sensitive hetrozygotes produce only sensitive offspring and the killer hetrozygotes produce only killer offspring.

Since a heterozygote Kk after autogamy does not produce sensitive and killer types, this pattern of inheritance is non-Mendelian which confirms cytoplasmic basis of killer trait. The inheritance of Kappa was at first considered a good example of cytoplasmic inheritance. These particles are not true cell organelles like the plastids or mitochondria. Closer studies, however, have shown that Kappa particles are infectious and resemble bacterium *Caedobacter taeniospiralis*. Their transmission in cytoplasm from cell to cell is, therefore, more correctly compared with the transmission of parasitic micro-organisms. The toxic substance produced by killer paramecia is diffusible in liquid medium. This is evident from the fact that when the killers were allowed to remain in a fluid medium for a time and were then replaced by sensitive individuals, the latter were killed. The toxic substance had no effect on the killer strain. The fact that kappa particles can be maintained only in the animals with gene K is no rigid argument against the interpretations. Individuals with genotype KK will be sensitive if there are no Kappa particles in the cytoplasm. When such sensitive cells are placed in concentrated suspension of disintegrated killer animals, some of them acquire Kappa from the suspension and are changed to killers. Kappa is subject to mutation and if killer Paramecia are exposed to high temperature, X-rays and chloromycetin they become sensitive which can be crossed to killers in different ways as shown in Fig. 18.6. Sometimes one animal may carry two or more different forms of Kappa.

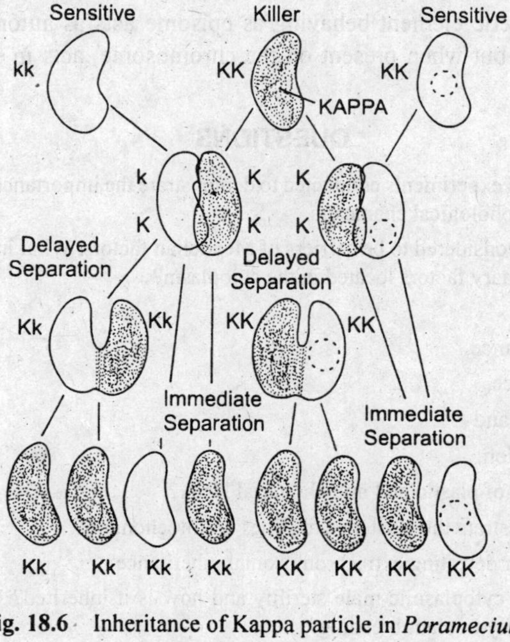

Fig. 18.6 Inheritance of Kappa particle in *Paramecium.*

The toxic substance has been shown to be associated with a particular kind of kappa, occurring in nearly 20% of kappa population. These kappa bacteria possess a refractile protein containing body called *R body.* The bacteria having R bodies are *brights* and they are infected with a virus that dictates the synthesis of viral protein as well as R protein body in kappa bacterium. The virus may act as the toxin in the killing response and *R body* facilitates the penetration of the toxin. The *non-bright* kappa bacteria may also contain virus but the virus may be in provirus state in them.

Episomes

The extrachromosomal genetic system can be influenced by chromosomal genetic system, yet the two systems at this point would appear to be basically distinct from one another. A virus like particle has been described in bacteria. The particle can participate in a cell either as chromosomal or as extra chromosomal element. This is called "episome". The episomes are defined as genetic elements which can exist in two alternative states, either autonomous in the cytoplasm or incorporated in the chromosomes. When present in cytoplasm the episome is free to multiply independently of the cell division, but when incorporated in chromosome the episome is replicated only if the chromosome is replicated. The concept of episome was put forward by Jacob and Wollman in 1958 to explain some of their findings on virus infected bacteria.

In virus infected bacteria, the viral DNA acts as episome. The episome, when it reaches to the chromosomal genetic systems of bacterial cell, causes a break in the bacterial DNA and effects its recombination potentially. Finally the viral DNA becomes attached with the bacterial DNA acting as though it were a part of the bacterial DNA and persists there apparently as a gene. The viral DNA is replicated like bacterial DNA and is transmitted to the daughter cells when cell fission occurs. The addition of viral genetic material to the bacterial DNA brings about many remarkable changes in the phenotypic appearance of bacterium so much so that the bacterium appears like a new species. The episome can, however, be separated from the bacterial chromosome by exposing virus infected bacterial cell to ultraviolet light, and by treating it with vitamin C.

When the viral genetic system, which acts as episome, is located in the bacterial cytoplasm, it replicates there independently and forms a number of genetic elements like itself and causes immediate death of host cell. From the above example of virus infected bacterial cell, it can be inferred that the same genetic element behaving as episome acts as autonomous self-duplicating structure in the cytoplasm but when present on the chromosome, acts in some different way and exhibits other features.

QUESTIONS

1. Describe Hammerling's experiments conducted to demonstrate the importance of nucleus and cytoplasm in determining the morphological characters.
2. The chromosomes are considered to be carriers of Mendelian factors. What in your opinion could be the physical basis of hereditary factors located in the cytoplasm?
3. Write short notes on:
   (*i*) Maternal inheritance,
   (*ii*) Plastid inheritance,
   (*iii*) Kappa particles, and
   (*iv*) Dauer-modification.
4. What is the importance of plastid and mitochondrial DNA?
5. How would you demonstrate the existence of genes in mitochondria?
6. What are the criteria for detecting extrachromosomal inheritance?
7. What do you mean by cytoplasmic male sterility and how is it inherited? What is the importance of cytoplasmic male sterility in plant breeding?

8. In Japanese maples "copper" colour variant is very common. If a copper branch arises spontaneously on a tree, how would you test to see if it is a nuclear or a cytoplasmic mutation?
9. Suppose that a diploid plant A has a different cytoplasm from plant B. You wish to obtain a plant with cytoplasm of plant A and the nuclear genome predominantly of plant B to examine nuclear-cytoplasmic relations. How would you go about producing such a plant?
10. In *Paramecium aurelia*, killers mated with sensitives but the conjugation ended before exchange of cytoplasms between the two conjugants and the ex-conjugants remained killers and sensitives in 1 : 1 ratio, but the sensitive ex-conjugants produced only sensitive individuals after autogamy. Explain these results with the help of suitable line diagrams.
11. Inbreeding in dextral coiling strain of snail produced offspring which were all left coiler (sinistral). What explanation would you offer to explain this mode of inheritance?

19

CHAPTER

Modern Concept of Gene

Mendel while explaining the result of his breeding experiment pointed out that the hereditary characters were governed by some particulate genetic determiners present in the germ cells. The genetic determiners of heredity characters are now called '**genes**'. The term gene was coined by Johannsen (1909). De Vries used the term '**pangen**' and perhaps, from this word the term gene was derived. In the old books of genetics the term 'gene' is treated as merely something which affects the phenotype and behaves as a particle. It should be noted here that nothing was said about the size, shape and chemical constitution of gene in old definitions.

After the rediscovery of Mendelian laws of heredity, the cytology advanced rapidly and reached to the stage where Mendel's hypothetical hereditary factors could soon be correlated with the chromosomes on following grounds:

1. Like the chromosomes, Mendelian factors are carried singly in mature germ cells.
2. Like the chromosomes, the hereditary factors are brought together in pairs by fertilization.
3. Like the chromosomes they separate or segregate at the time of germ cell formation in different generations. The only objection in treating the chromosomes equivalent to Mendelian factors is that the number of Mendelian factors in the organisms far exceeds the number of chromosomes. In 1902, Sutton and Boveri independently suggested the way out for this objection and considered the chromosomes as containers of Mendelian hereditary units. They did not treat chromosome as single heredity unit. In 1914, Thomas Morgan was working with the fruitfly (*Drosophila melanogaster*). He noted that the fruitfly in question had 4 pairs of chromosomes and the hereditary characters were too many. This fact led Morgan to propose **gene theory**. The theory states that—
   (*i*) chromosomes are bearers of hereditary units or genes and each chromosome carries hundreds or thousands of genes.
   (*ii*) the genes are arranged on the chromosomes in the linear order and on the specific regions or loci.

If it is considered that gene is specific particle, then it follows that the chromosome would be a linear array of such particles bound together by non-genetic linkers of some type. Such a morphological differentiation can be observed in the chromosomes under some special circumstances. When a chromosome is observed in the early stages of first meiotic prophase, it appears as a 'string of beads'. The thickened parts or beads of chromosomes are called chromomeres and the string or linker joining the chromomeres is known as chromonema (Fig. 19.1). At one time, chromomeres were treated as genes. But, this is an oversimplified view and is no longer held. The banding pattern of salivary gland chromosome of *Drosophila* (Fig. 19.2) and chromosomal loops in the Lampbrush chromosome (Fig. 19.3) of amphibian eggs are also suggestive of functional discontinuity. Actually, the genic material is not different from other non-genic material in the chromosome. It is, perhaps, suggested by the fact that genetic recombination can and does take place within genic as well as non-genic areas (Pantecarvo, 1958).

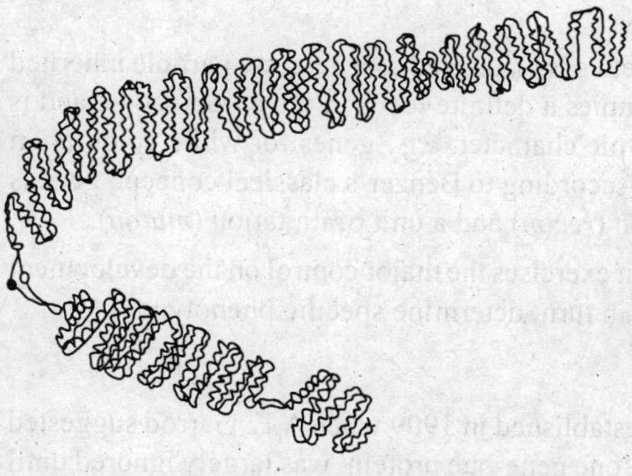

Fig. 19.1 Two spirally twisted chromatids of chromosome (magnified view).

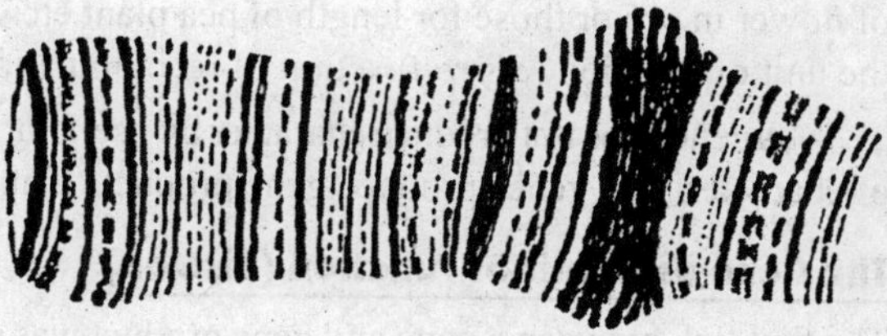

Fig. 19.2 Giant salivary gland chromosome of *Drosophila*.

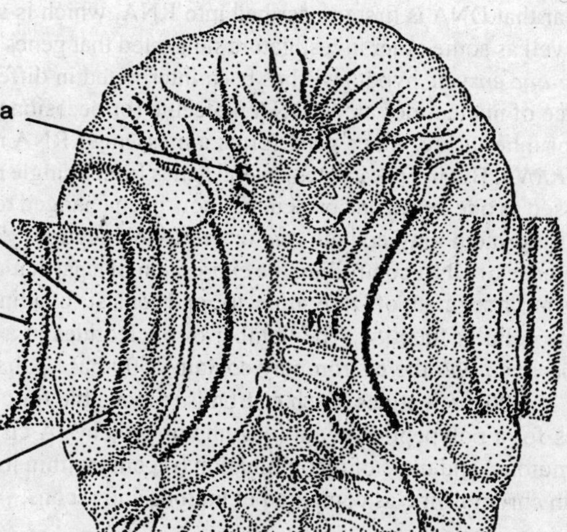

Fig. 19.3 Lampbrush chromosome with a portion highly magnified.

Definition of Genes

As both the operation and the physicochemical nature of gene are not known clearly, the gene can be defined best in terms of its effects. Efforts to formulate theoretical models or hypothesis of gene action have resulted in highly divergent opinions.

There are two schools of thought which define genes in different ways:

(*i*) According to one school of thought, the specific molecule of genic material controls each character.

The exponents of this school are Dobzhansky (1955) and Beadle (1955).

(*ii*) The other school under the leadership of Goldschmidt holds the view that the interactions between all the components of genome are responsible for the manifestation of characters.

Stadler (1955) points out that the differences of opinions arise because one school defines genes as "operational units" whose actions can be demonstrated experimentally and the other school defines the gene in a hypothetical physicochemical sense.

In hypothetical sense the gene of classical genetics is visualised as a descrete particle inherited in Mendelian fashion or as unit of function that occupies a definite locus in the chromosome and is responsible for the expression of a specific phenotypic character, *e.g.*, genes for white colouration of flower in pea or those for length of pea plant etc. According to Benzer's classical concept, gene is the unit of function (*cistron*), a unit of recombination (*recon*) and a unit of mutation (*muton*).

A gene does not produce character by itself but it exercises the major control on the development of character. The genes produce gene products that, in turn, determine specific phenotype.

The Gene as a Unit of Function (Cistron)

The first link between a gene and gene product was established in 1909 when A.E. Garrod suggested that the gene product was a protein. This hypothesis 'one gene-one protein' was largely ignored until around 1940 when it was conclusively demonstrated by Beadle and Tatum (1941) that the genes directed the protein synthesis. Encouraged by this link between gene and protein, investigators worked out the mechanisms by which information in DNA was translated into the structure of protein. Soon it became clear that DNA is first transcribed into RNA, which is subsequently translated into protein. Garrod as well as some other investigators suggested that genes produce proteins that act as enzymes. *One gene-one enzyme* hypothesis has been interpreted in different ways. Gene represents a particular sequence of nucleotides along DNA molecule (or occasionally RNA in certain viruses) that acts as a unit of inheritance. Since gene forms a messenger RNA molecule it can be said *one gene-one messenger RNA*. But in prokaryotes several genes form a single mRNA (polycistronic gene). The concept expressed as one genes one protein has now been changed to *one gene-one polypeptide*, since an enzyme or a protein may contain one or more polypeptides. Stahl had defined gene as a polynucleotide sequence of DNA controlling the expression of a particular trait. The gene as a unit of function thus represents a segment of DNA molecule and consists of a linear sequence of nucleotides which controls some cellular function. The number of nucleotides in a gene may vary in different organisms. In *Escherichia coli* the cistron may contain 1500 base pairs but in some others it may contain as many as 30,000 nucleotides. Each cistron is responsible for coding one mRNA molecule which in turn codes for a polypeptide chain (enzyme or protein). A cistron may have hundreds of units of mutation (mutons) and units of recombination (recons) within it. Therefore, cistrons occupy much greater area in chromosome as compared to mutons and recons.

Gene as the Unit of Recombination (Recon)

The classical studies on the genetics of *Drosophila* indicated that gene was the shortest segment of chromosome controlling phenotypic characters which could be separated from its adjacent segments during crossing over, *i.e.*, the genes were those parts of chromosomes between which crossing over was possible and the crossing over was not supposed to take place within the gene. But recent studies based on the tests for recombination in viruses have shown that in viral DNA strand crossing over could occur not only between the genes but also within the genes. One of the sub-units of gene has been called ***recon***. It is the smallest unit capable of genetic recombination. Recombination studies on microbes indicate that structurally recon may have one or two pairs of nucleotides. Benzer (1955) demonstrated crossing over within the gene in T_4 bacteriophage. T_4 Phage contains one linear chromosome. There are two strains of T_4 phage: wild strain producing smooth edged plaques and the r II mutant strain producing rough edged plaques. r-II mutants are of several types. Benzer found that two adjacent genes rII A and rII B were responsible for rough edged plaque character. Each gene forms a polypeptide and the two polypeptides form a protein. A change in the polypeptide chain would result in a change in the phenotypic expression.

Mutants rII A and rII B may cross in *E. coli.* The progeny produced from the cross were analysed for the frequency of recombination and it was found that some phage particles were of normal type. This was possible only if crossing over occurred within the genes by which segments of A and B genes united to form normal wild genes.

The Gene as a Unit of Mutation (Muton)

Gene has also been defined as an unit of mutation. The smallest chromosomal unit capable of undergoing mutation has been called the ***Muton***. At molecular level a muton consists of one or many pairs of nucleotides within the DNA molecule. Mutation thus may be caused by a change in one or more nucleotides in the DNA molecule.

Muntzing (1961) defines a gene as small segment of chromosome having an unitary biochemical function and specific effect on the properties of individual. According to him the genes may also occur in the cytoplasmic bodies that are sometimes associated with chromosomes or even sometimes occurring free in the cytoplasm.

IDENTIFICATION OF GENETIC MATERIAL

For understanding the nature of genes it is necessary to know the chemical and physical nature of genetic material. At the moment most of our knowledge about this comes from the studies on fungi, bacteria and viruses. Here everybody seems generally agreed on the fact that the genetic materials of lower organisms are similar to those of higher ones. Before deciding the nature of hereditary material one should keep in mind the following three principal characteristics of the material responsible for inheritable traits:

1. It must contain all the informations regarding cell structure, function, development and reproduction.
2. It must be able to replicate accurately so that the progeny cells inherit the same genetic features as the parent cells.
3. It must be capable of undergoing variation through recombination and mutations and exist in infinite forms, otherwise organisms would not be capable of change, and adaptations and evolution would not be possible.

Much of the chemistry of the genetic material was known before its significance in the genetics was recognised. In mid to late nineteenth century and early twentieth century scientists believed that the genetic instructions from the nucleus were carried to the cytoplasm by **protein molecules** that were folded into specific configuration. But several spectacular discoveries during the past four decades have changed the old ideas and now **Deoxyribonucleic acid (DNA)** is considered as the genetic material. Following publication of Mendel's work Johann Friedrich Miescher in 1869 isolated a novel phosphorus bearing compound and named that as nuclein. Miescher and other scientists subsequently demonstrated that the nuclei of all cellular organisms contain nuclein and nuclein was an important constituent of chromosomes. Chromosomes consist of 60% protein and 40% DNA. The following seven direct and indirect evidences from different sources have established that DNA is the essential genetic material (Lima de Faria and Moses, 1966 and others).

1. In prokaryotic organisms (organisms lacking well organized nucleus and chromosomes) to which operationally viruses can also be included, the genetic material is DNA.
2. Deoxyribonucleic acid is confined almost to the chromosomes in the nuclei of eukaryotes. DNA is a stable macromolecule, the content of which is directly related to the chromosome number.
3. The amount of DNA per nucleus is species characteristic and the amount of DNA found in the gamete is only half of the amount of DNA present in the somatic cells of all the members of a species. The diploid condition is restored only after fertilization. Careful measurements of the amount of DNA and proteins in somatic cells and sperm cells indicated that DNA only played

hereditary role and not protein. The measurements indicated that the amounts of protein in sperms and somatic cells bore no definite relationship with each other whereas the amount of DNA in sperms was exactly one-half that in somatic cells.

4. Genes are characterised as self-perpetuating units, DNA molecules too replicate themselves and in the process of replication the old DNA acts as template for new one.
5. The highest efficiency of mutagenesis by ultraviolet light is at the wavelength of 2600 Å which is also wavelength of peak absorption by DNA.
6. The genetic transformation in bacteria is mediated by DNA. The most convincing evidence in support of the fact that genic material is DNA could appear in 1944 when Avery, MacLeod, and McCarty reported their work on transformation in *Pneumococcus*.
7. Hershey and Chase (1952) confirmed that in bacterial viruses (bacteriophages) the genetic material is DNA. Certain RNA viruses are exceptions in which the transfer of genetic information has been taken over by ribonucleic acid. Fraenkel Conrat (1955) working on the reproduction of Tobacco mosaic virus (TMV) proved that the hereditary material in that virus was definitely RNA.

The major experiments which established that DNA is genetic material are discussed here in to brief:

1. Transformation Experiment in Bacteria

Frederick Griffith (1928) for the first time demonstrated *genetic transformation* in bacterium *Diplococcus pneumoniae*, now named *Streptococcus pneumoniae* and Avery, MacLeod and McCarty (1944) later showed that DNA and not the protein, was the carrier of hereditary characters in *Diplococcus pneumoniae* or *Pneumococcus*. This bacterium causes pneumonia in mice and men. There are two strains of *Pneumococcus*. In one strain, capsule layer (slime coat) is formed of a polysaccharide material and colonies are shining and "smooth" ('S' strain). In other strain, cells lack polysaccharide slime layer and colonies formed by such cells are irregular or "rough" ('R' strain). Smooth (S) cells are virulent and cause Pneumonia but rough (R) cells are non-virulent. 'S' *Pneumococci* are classified into types, such as, type I-S, type II-S, type III-S and so on. The cell of 'S' colony may change occasionally into 'R' bacterium but the reverse change (*i.e.,* from R to S) is almost never seen. R cells upon division always give rise to R cells. In the course of his experiment, Griffith injected mice with living II-R *Pneumococci* and found that the mice did not suffer. But when the mice were injected with live III-S type they suffered by pneumonia and died and when heat killed III-S bacteria were injected the mice did not suffer. However, when the mixture of living cells of non-virulent II-R and heat killed III-S cells were injected into the mice, the mice unexpectedly developed pneumonia and died. Post-mortem examination of dead mice showed the presence of both II-R and III-S type of Pneumococci in the heart blood and this led Griffith to conclude that something released from the heat killed III-S cells was taken up by the avirulent R type cells which might have caused genetic transformation of living II-R bacteria into virulent cells (Fig. 19.4).

Avery and his associates at the Rockfeller Institute earnestly pursued to identify the *transforming principle*. Between 1930 and 1933 they demonstrated transformation *in vitro* rather than in the mice. They disintegrated encapsulated cells of type III-S and separated various chemical components (Carbohydrates, proteins, fats, DNA, RNA etc.). Then they took R-II cells derived from type II-S and mixed them separately with different chemical components of III-S cells. Avery and his associates observed that the DNA fraction was able to change some of the R cells (noncapsulated) to encapsulated S cells. The transformed R-II cells were like III-S and were similar in all respects to the cells from which the DNA fraction was taken (Fig. 19.5).

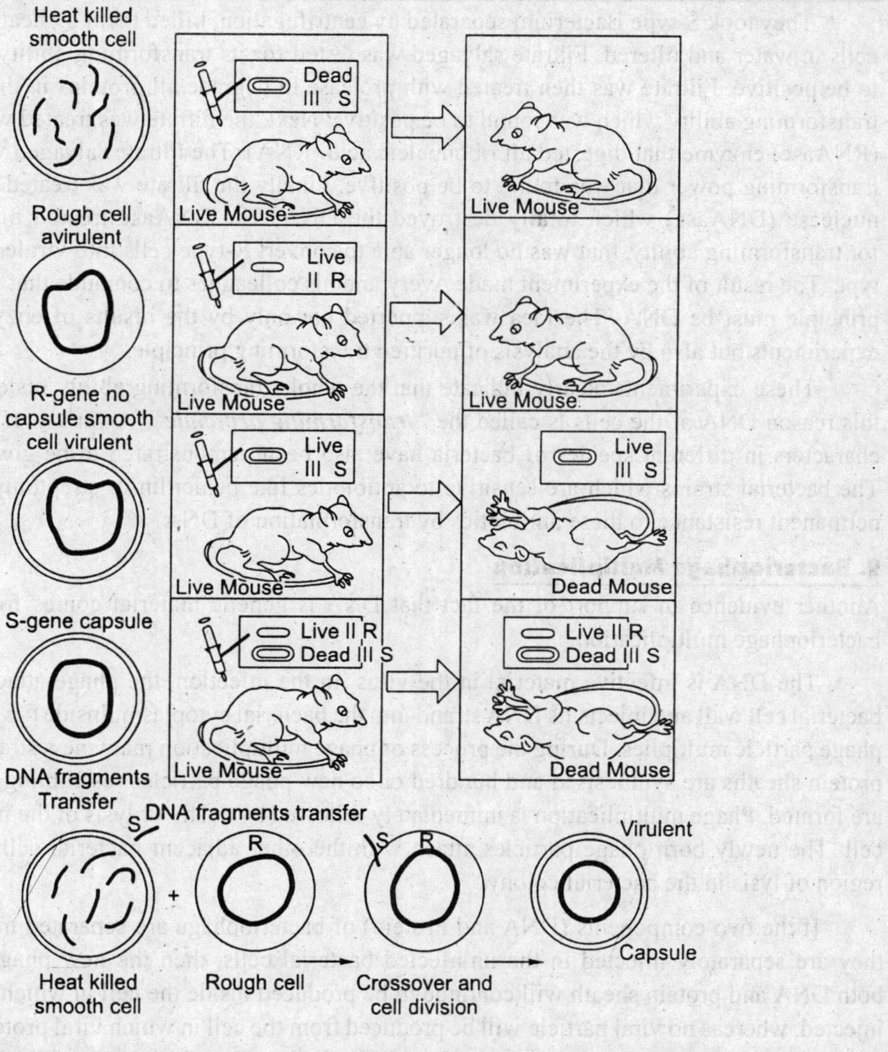

Fig. 19.4 Griffith's experiment that demonstrated the principle of transformation.

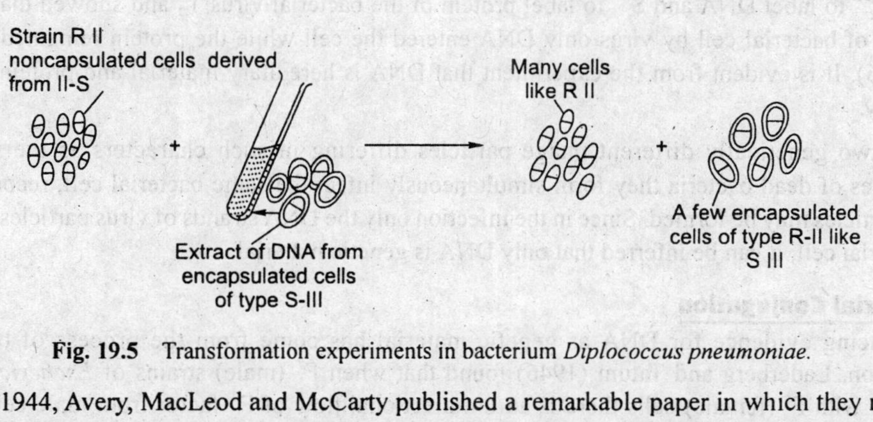

Fig. 19.5 Transformation experiments in bacterium *Diplococcus pneumoniae.*

In 1944, Avery, MacLeod and McCarty published a remarkable paper in which they reported that they had purified the *transforming principle*. Analysis of the molecular composition and weight indicated that transforming factor was DNA.

They took S-type Bacterium separated by centrifugation, killed them by heating, ruptured the cells in water and filtered. Filtrate salvaged was tested for its transforming ability, that was found to be positive. Filtrate was then treated with protease to degrade all proteins in that and tested for transforming ability which was found to be positive. Next, the filtrate was treated with ribonuclease (RNAase) enzyme that digested all ribonucleic acid (RNA). The filtrate salvaged was tested for its transforming power that was found to be positive. Finally the filtrate was treated with deoxyribonuclease (DNAase) which totally destroyed the DNA. The DNAase treated filtrate was tested for transforming ability, that was no longer able to convert R-type cells into virulent cells like III-S type. The result of the experiment made Avery and his colleagues to conclude that the transforming principle must be DNA. The idea was supported not only by the results of enzymatic digestion experiments but also by the analysis of purified transforming principle.

These experiments clearly indicate that the whole transforming ability resides in DNA. For this reason DNA of the cells is called the *"**transforming principle**"*. A variety of other hereditary characters in different species of bacteria have also been demonstrated to be governed by DNA. The bacterial strains which are sensitive to antibiotics like penicillin or streptomycin can acquire permanent resistance to these antibiotics by transformation of DNA.

2. Bacteriophage Multiplication

Another evidence in support of the fact that DNA is genetic material comes from the study of bacteriophage multiplication.

The DNA is infective material in the virus. In the infection, the phage attaches itself to the bacterial cell wall and injects its DNA strand into the bacterial cytoplasm. Inside the bacterial cell the phage particle multiplies. During the process of phage multiplication many new strands of DNA and protein sheaths are synthesised and hundred or so new phage particles resembling the original one are formed. Phage multiplication is immediately followed by death or lysis of the infected bacterial cell. The newly born phage particles attach with the other adjacent bacterial cells and produce a region of lysis in the bacterial colony.

If the two components (DNA and Protein) of bacteriophage are separated mechanically and they are separately injected in the uninfected bacterial cells, then the new phage particles with both DNA and protein sheath will continue to be produced inside the cell in which virus DNA was injected, whereas no viral particle will be produced from the cell in which viral protein was injected. The "Waring Blender Experiment" by Hershey and Chase in which they utilized the radioactive isotope P^{32} to label DNA and S^{35} to label protein of the bacterial virus T_2 and showed that during infection of bacterial cell by virus only DNA entered the cell while the protein remained outside (Fig. 19.6). It is evident from the experiment that DNA is hereditary material and protein is non-hereditary.

If two genetically different phage particles differing in such characters as the sizes of the plaques of dead bacteria they form simultaneously infect the same bacterial cell, recombinant phage particles may be formed. Since in the infection only the DNA strands of virus particles entered the bacterial cell, it can be inferred that only DNA is genetic material.

3. Bacterial Conjugation

A convincing evidence for DNA as genetic material has come from the process of bacterial conjugation. Laderberg and Tatum (1946) found that when F^+ (male) strains of *Escherichia coli* conjugate with F^- (female) cells there is an unidirectional flow of F^+ factor from male cells to F^- female cells so that the latter is converted into F^+ from male strain. The F^+ factor is found to be a fragment of DNA molecule which occurs in the cytoplasm of bacterium.

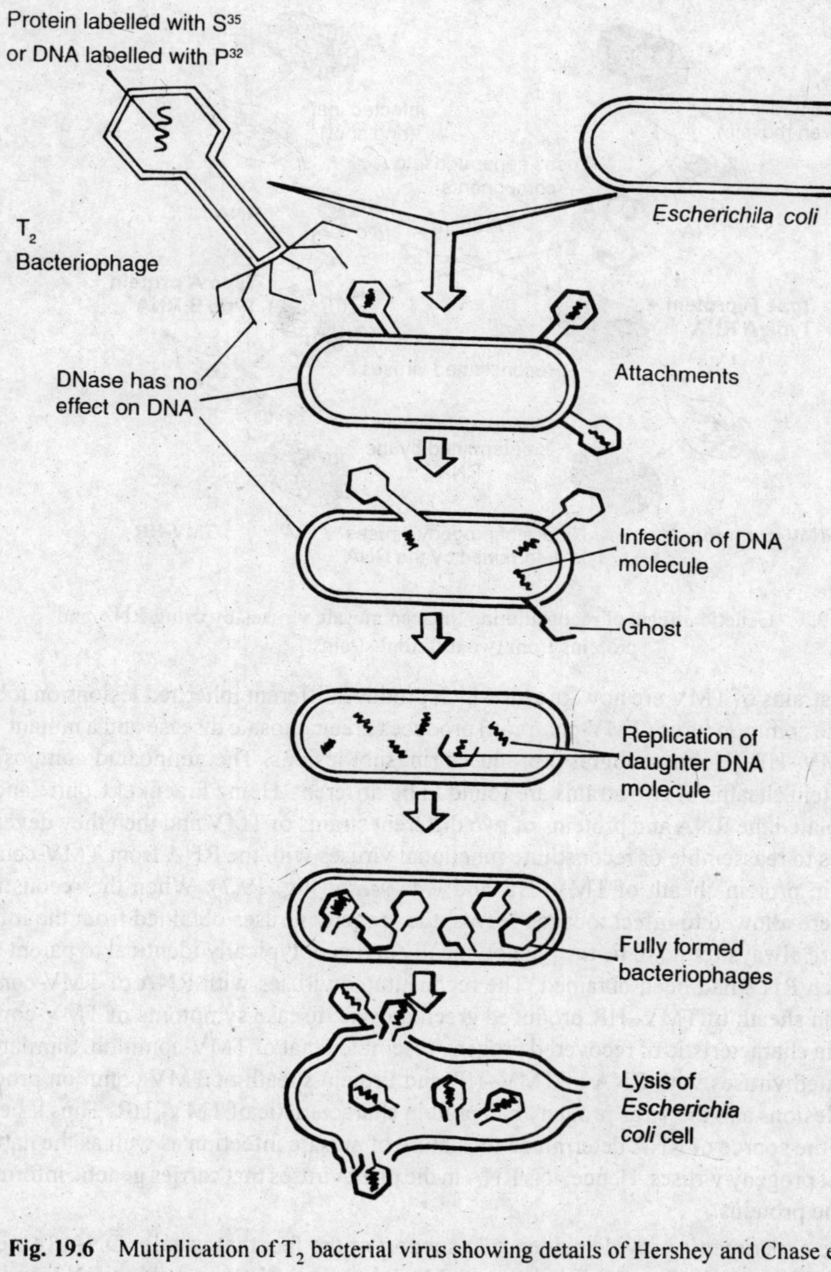

Fig. 19.6 Mutiplication of T_2 bacterial virus showing details of Hershey and Chase experiment.

4. RNA as the Genetic Material in Some Viruses

Tobacco mosaic virus (TMV) which causes mosaic disease in tobacco, consists of an outer protein coat and an inner core of nucleic acid RNA (Fig. 19.7). In 1955, Fraenkel Conrat fractionated RNA and protein of TMV and injected them separately into two healthy tobacco plants. In this experiment they observed that viral RNA, completely free from its protein, could produce new virus particles which also caused mosaic disease in tobacco plant. The inoculation of viral protein did not produce the disease symptoms at all. This experiment showed that viral RNA alone can direct the formation of normal virus (Fig. 19.7).

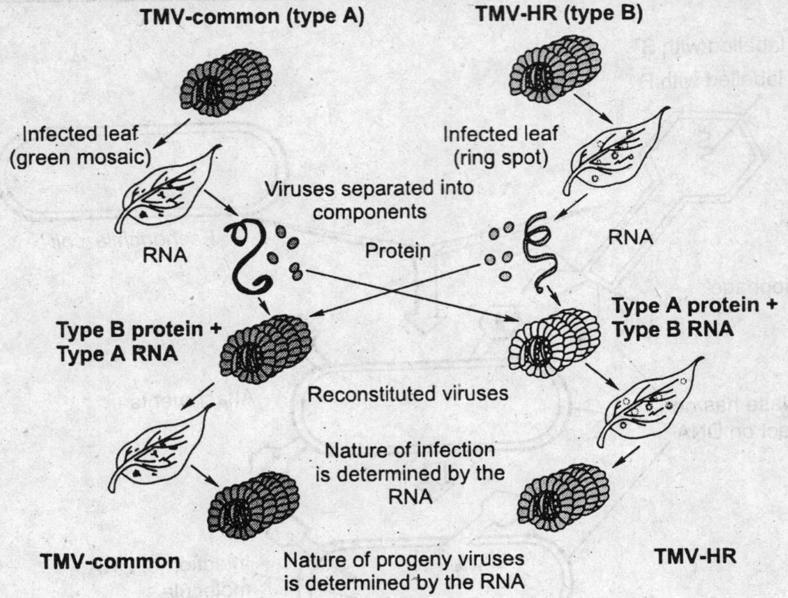

Fig. 19.7 Genetic effects of reconstituting tobacco mosaic viruses by using RNA and proteins from two different strains.

Different strains of TMV are now known which produce different inherited lesions on tobacco leaves. The common virus (TMV-common) produces green, mosaic disease and a mutant TMV strain, TMV–HR (Holmes ribgrass) produces ringspot lesions. The aminoacid compositions of the protein sheaths of two strains are found to be different. Heinz Fraenkel Conrat and Bea Singer isolated the RNA and proteins of two different strains of TMV and then they developed techniques to reassemble or reconstitute functional viruses with the RNA from TMV-common enclosed in protein sheath of TMV–HR and *vice-versa* (Fig. 19.7). When the reconstituted viruses were allowed to infect tobacco leaves, the progeny viruses obtained from the infected leaves were always found to be the phenotypically and genotypically identical to parent strain from which RNA had been obtained. The reconstituted viruses with RNA of TMV-common and protein sheath of TMV–HR produced green mosaic disease symptoms of TMV-common and protein characteristic of recovered progeny resembled that of TMV-common. Similarly the reconstituted viruses with RNA of TMV–HR and Protein sheath of TMV-common produced ring spot lesions and the virus progeny had protein characteristic of TMV–HR. Thus it become clear that the source of RNA determined the nature of mosaic infection as well as the nature of proteins of progeny viruses. Hence, it is RNA in the plant viruses that carries genetic information and not the proteins.

The above experiments provide evidence in support of the fact that usually DNA, (excepting a few RNA viruses) is the genetic substance. Much has been discussed about DNA and RNA in the chapter on chromosome chemistry.

QUESTIONS

1. Define gene. What are different schools of thought to define gene?
2. How will you say that gene is a unit of function, a unit of recombination and a unit of mutation.
3. Describe two experiments which initially demonstrated that DNA is genetic material.
4. In certain viruses no DNA in found. In such cases RNA is the genetic material. Discuss an experiment which shows that RNA carries the genetic information in such cases.

20 CHAPTER

Gene Action

ONE GENE—ONE POLYPEPTIDE CONCEPT

It is a fact that hereditary characters are maintained and transmitted from one generation to another through DNA molecules, because DNA can duplicate itself and duplicated molecules can be passed on to the offsprings. The general activity of genes brings about expression of hereditary traits in the organism. Now the main questions are as to how the genes (DNA molecules) govern the biosynthetic processes of the cells and how these genes control the phenotypic properties of the organisms. The answers to these basic questions were sought in the relationship between genes and specific biochemical reactions.

The heritable changes that geneticists first studied were necessarily those which could most easily be observed. An English physician, Sir Archibald Garrod, made penetrating study of some rare hereditary diseases in human and recognised that certain biochemical deficiencies were caused by enzymatic abnormalities. On the basis of his studies on congenital (existing from the birth) diseases of human, Sir Garrod safely suggested relationship between genes and enzymes. The idea that the action of a gene is concerned with the formation of particular enzyme was ignored by most geneticists for some thirty years.

James B. Summer of Cornel University and John H. North Rop of Rockefeller Institute between 1926 and 1930 showed that enzymes are proteins. The idea about gene and enzyme relationship was revived by George W. Beadle and Edward L. Tatum (1941). From the studies on heritable metabolic abnormalities of the fungus, *Neurospora crassa,* they concluded that all the intermediary biosynthetic steps of a metabolic process were governed by distinct genes. Beadle and Tatum formulated the *one gene -one enzyme concept* in 1944. The theory states that a gene exerts its influence on the phenotype through its role in the production of an enzyme.

Beadle and Tatum studied the genic action in *Neurospora crassa.* Normally the fungus can grow on minimal culture medium containing agar, sucrose, nitrate, inorganic minerals and the only vitamin biotin. This means that this organism can synthesize all other vitamins and aminoacids which are required in its metabolism. When the conidia of this fungus are treated with mutagenic agents (say, X-ray), some of them become unable to grow on minimal medium. These mutant spores are then tested systematically by adding particular vitamins, aminoacids, etc. to the minimal medium to determine what substance or substances they are unable to synthesize. The mutants can be crossed with normal or wild type and their products of meiosis, the 8 ascospores, may be individually tested for their nutritional requirements. If an arginine requiring strain (*a* –) is crossed with normal strain (*a* +), all the 8 ascospores can grow in a medium containing arginine, but only 4 ascospores can grow in a medium lacking arginine. This indicates that a single gene has mutated. The arginine requiring mutant may be more complex because, as shown in the following chain, arginine synthesis involves a chain of intermediate steps, each reaction is controlled by one gene. Three steps have been noted in the conversion of glutamic acid to arginine and each of them has been found to be controlled by one gene.

Glutamic acid
↓ Gene I
Ornithine..........Step I
↓ Gene II
Citrulline..........Step II
↓ Gene III
Arginine ....Step III

The mutation of a single gene leads to the suppression of one step. This can be demonstrated by growing mutant on minimal medium with that substance which cannot be synthesized. In several biochemical studies, the extracts of *Neurospora* have shown that tryptophan like arginine is synthesized in a sequence of chemical reactions. Mutations in tryptophan require strains map at several genetic locations. Each mutant is defective in one of the steps in biosynthetic sequence. The mutation in a specific chromosomal region is reflected by the loss of activity of one enzyme. Thus basic gene-enzyme relationship is clear.

Recent researches have verified the basic conclusion about gene-enzyme relationship. Currently the Beadle and Tatum's concept of one gene-one enzyme has been revised to one gene-one polypeptide chain (protein) in view of the complexity in the structures and the functions of enzymes. The modern researches have proved that gene is DNA which is directly concerned with the synthesis of particular protein.

Gene Action

The expression of genes by genetic transcription into complimentary RNA sequences and subsequent translation of hereditary information contained in mRNA into polypeptide chain which forms the ultimate product of gene action is called primary gene action. The analysis beyond primary action of gene is greatly complicated by the integrated state of cellular and developmental metabolism, by the remoteness of the phenotype from the primary gene action and the number of intervening steps influenced by other genes (gene interaction) and by environmental factors (gene activation). In prokaryotes the transcription and translation of genetic information occur in one cell compartment whereas in eukaryotes the two processes are accomplished into two separate compartments of a cell, *i.e.,* nucleus and cytoplasm. In addition, some genetic information (organelle DNA) is also present and utilized within certain cytoplasmic organelles particularly plastids and mitochondria. The operation of nuclear and extranuclear genomes is coordinated by some mechanism which is not yet fully understood. The genetic regulation of primary gene action in strict sense of the term occurs only at the level of transcription. The whole series of biochemical processes which lead from a gene to the phenotypic expression by which it is recognised is referred to as **gene action system** (Waddington, 1962). Thus gene has two essential functions:

(*i*) replication or self-reproduction; and
(*ii*) intervention in mechanism by which the phenotype of organism is produced in a given environment (*phenogenesis*).

Primary Gene Action (Gene and Protein Synthesis)

The genotype (total genetic material) of the cell determines the potential type of proteins and also determines their relative amounts in the cell. Proteins serve as structural components of the cells which make up the framework of living body. Special types of proteins which act as catalysts in bringing about numerous chemical reactions and control them precisely are called enzymes. In fact, all functions of the living system are carried out by proteins. Thus from the structural as well as functional points of view proteins are important constituents of the cells, or in other words, proteins constitute the fundamental molecular machinery of the cell.

Genes act by controlling the structure and the rate of production of specific proteins (enzymes). Genes are segments of DNA molecule. The DNA of each gene forms a complimentary mRNA strand which attaches to the ribosomes where it serves for coding of a protein (enzyme). The sequence of aminoacids in protein (*i.e.,* the structure of protein or enzyme) is determined by the sequence of nucleotides in mRNA which in turn is determined by the sequence of nucleotides of DNA (gene). A series of enzyme controlled reactions determine traits in an organism. Since the structure of these enzymes is controlled by genes it follows that genes determine traits. The whole events may be summarised as follows:

DNA $\xrightarrow{\text{Transcription}}$ mRNA $\xrightarrow{\text{Translation}}$ Sequence of amino acids (Protein or enzyme) $\rightarrow$ Traits

The basic building blocks of proteins are amino acids. Excepting proline, all other amino acids have a common structure consisting of a central carbon atom (the α-carbon) to which is linked α-amino group ($-NH_2$), an α-carboxyl ($-COOH$) and a hydrogen atom (proton). The other part of aminoacid is called the R-group which varies from one aminoacid to another. It is R-group that gives the aminoacid its chemical properties. The general formula of an aminoacid is illustrated below:

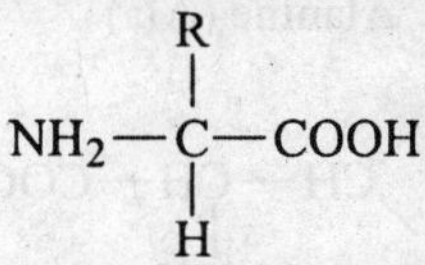

(Aminoacid)

Proteins are polymeric molecules and are formed by combination of many amino acid molecules in linear sequence. The amino acids are joined with one another through special type of bond, called peptide linkage. This peptide bond is established by elimination of a molecule of water between carboxyl ($-COOH$) and amino ($-NH_2$) groups of adjacent amino acids. The peptide bond is of amide ($-CONH-$) type. An example of peptide bond formation is given below:

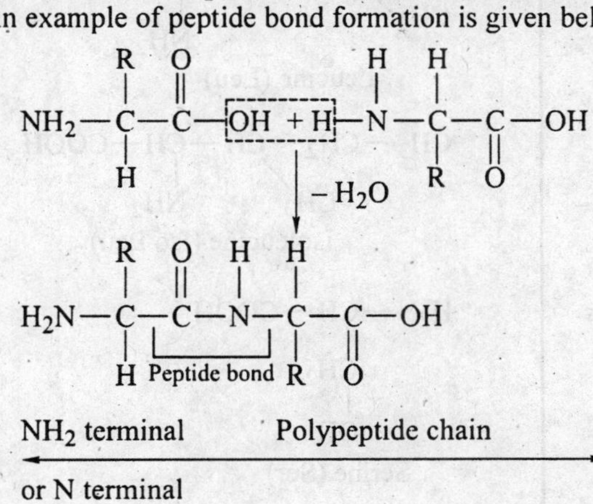

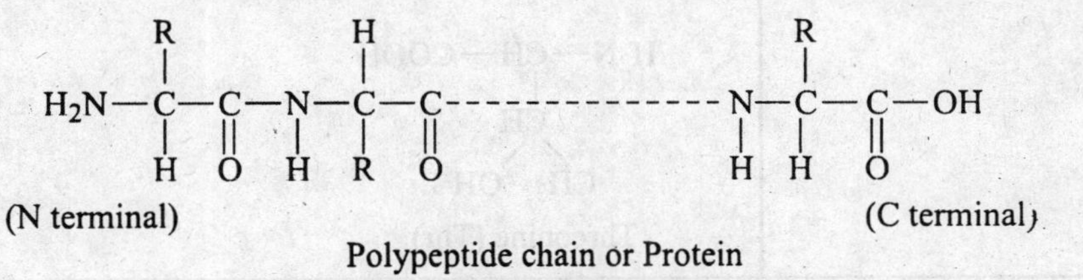

Polypeptide chain or Protein

A general structure of polypeptide chain or protein.

In this way, many amino acids are joined end to end by peptide bonds making a long chain. One end of protein chain contains amino (NH_2) group (amino end) and the other contains carboxylic (– COOH) group (carboxylic end). The linear chain of amino acids formed in this manner is termed as **polypeptide chain** or **protein**. The formation of polypeptide linkage requires the action of enzyme peptide polymerase and energy rich compound guanosin triphosphate (GTP).

In living system about 20 out of 22 different aminoacids are known to take part in the protein synthesis. These are mentioned in the Table 20.1.

Table 20.1 Structure of naturally occurring aminoacids

Group of aminoacids	Name of the aminoacid, symbol and chemical formula
A. **Aliphatic amino-acids.**	
I. *Monoamino mono-carboxylic aminoacids or simple aminoacids.*	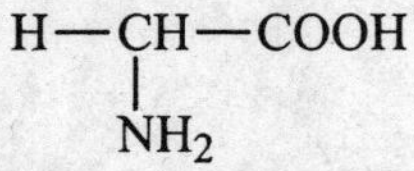 Glycine (Gly)
	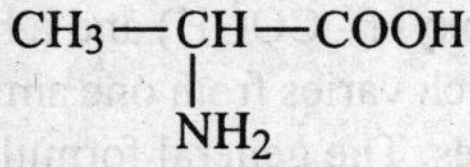 Alanine (Ala)
	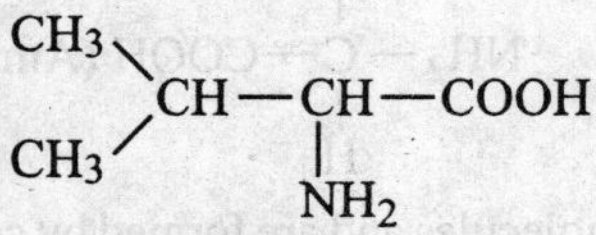 Valine (Val)
	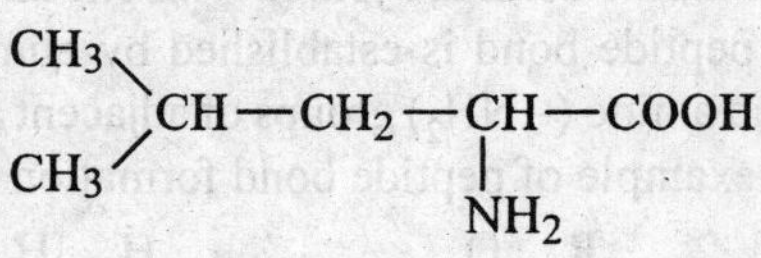 Leucine (Leu)
	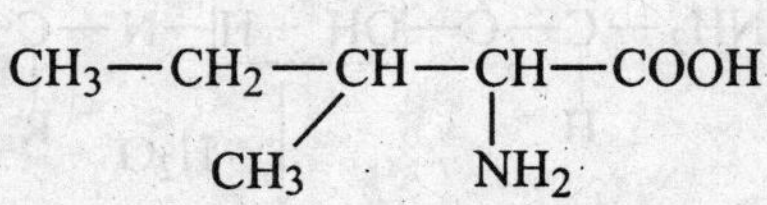 Isoleucine (Iso Leu)
II. *Hydroxyamino acids*	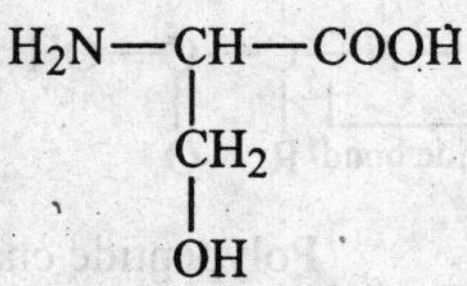 Serine (Ser)
	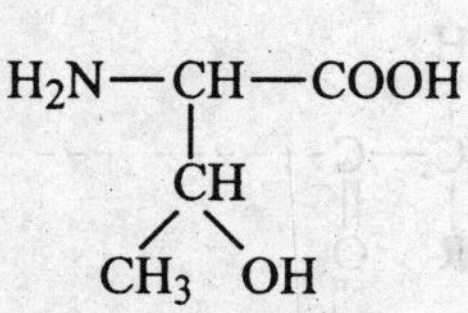 Threonine (Thr)

Group of aminoacids	Name of the aminoacid, symbol and chemical formula
III. *Sulphur contaning amino acids*	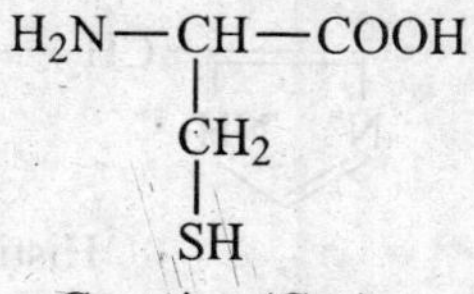 Cysteine (Cys) 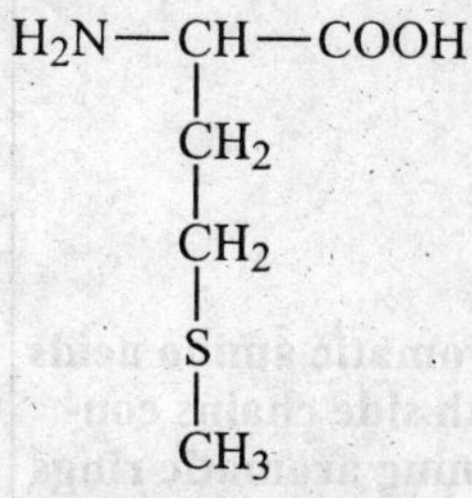 Methionine (Met) 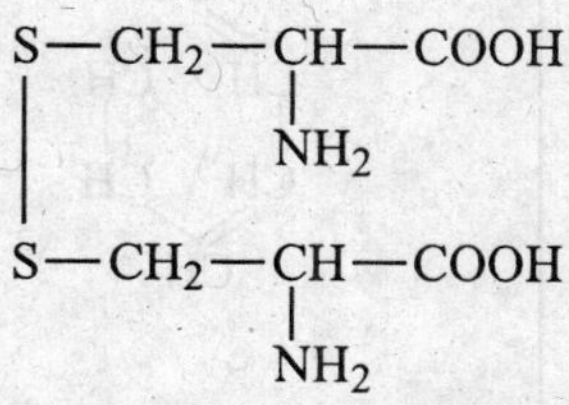 (Cystine, a dimer of cysteine)
IV. *Dicarboxylic amino acids and their amides*	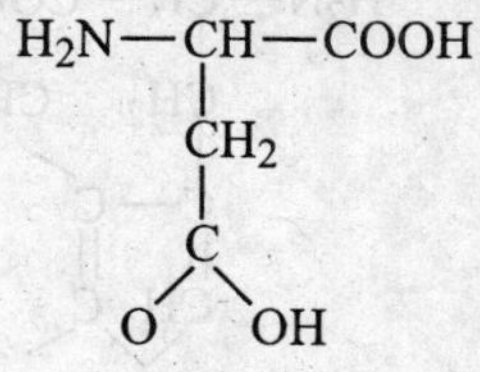 Aspartic Acid (Asp) 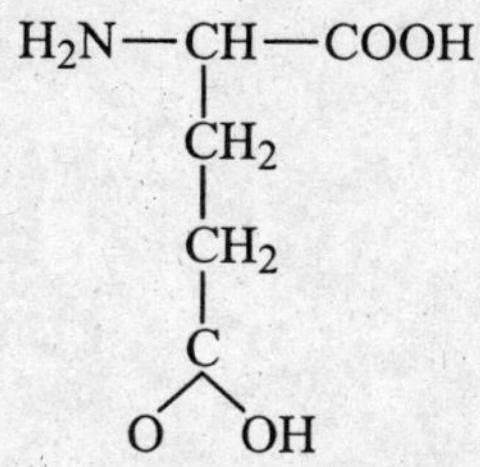 Glutamic Acid (Glu) 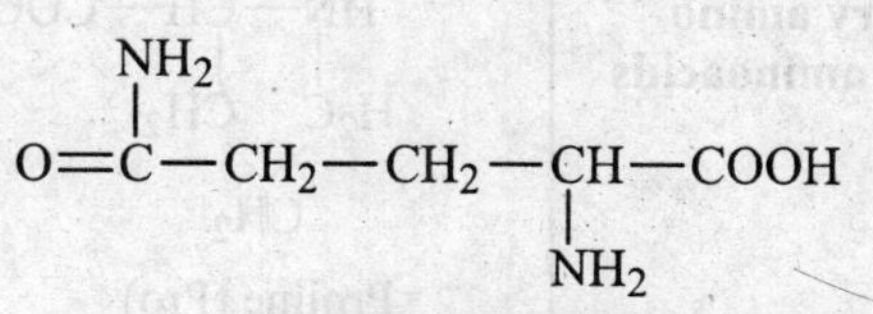 Glutamine (Gln) 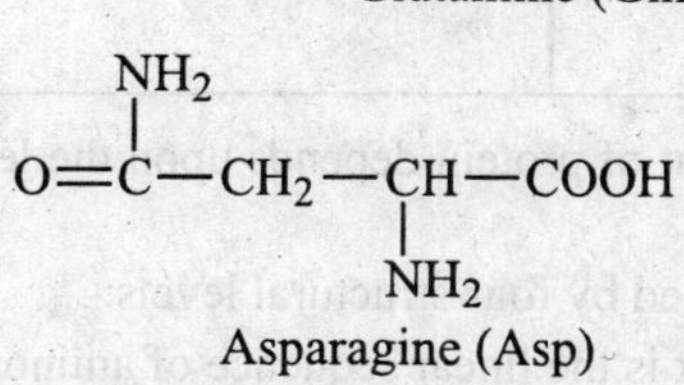 Asparagine (Asp)
V. *Aminoacids with basic groups*	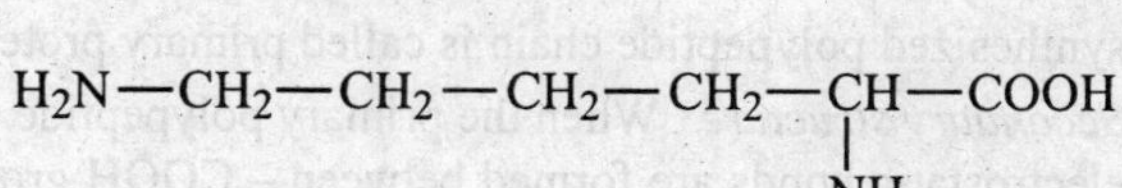 Lysine (Lys) 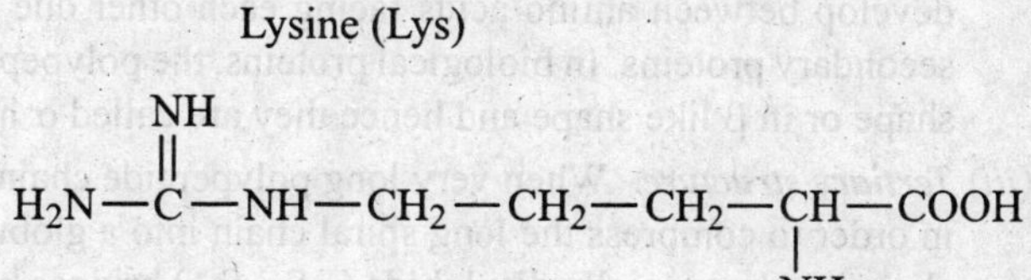 Arginine (Arg)

Group of aminoacids	Name of the aminoacid, symbol and chemical formula
	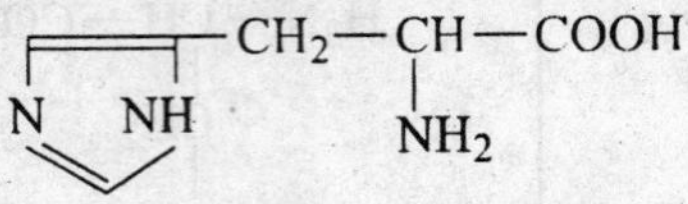 Histidine (His)
B. **Aromatic amino acids with side chains containing aromatic rings**	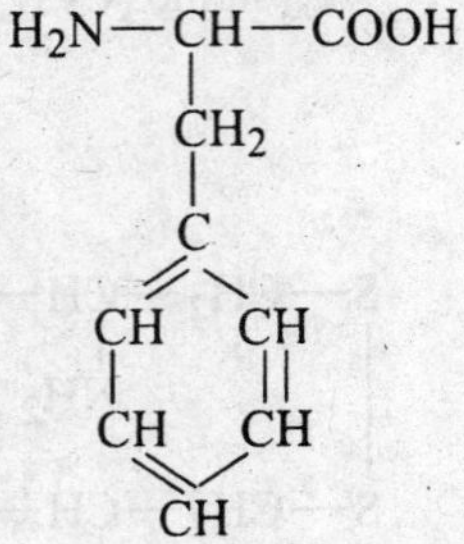 Phenylalanine (Phe) 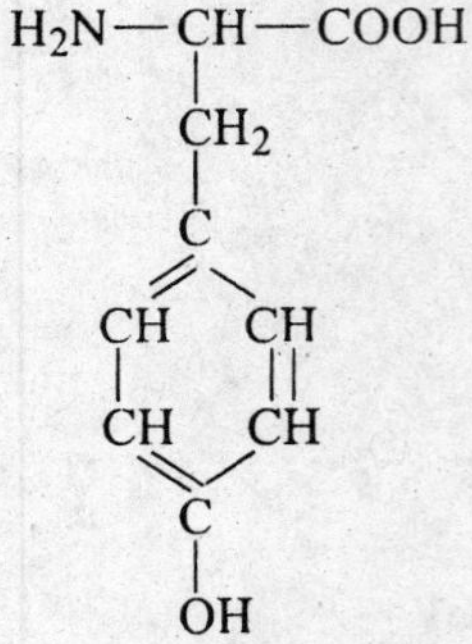 Tyrosine (Tyr) 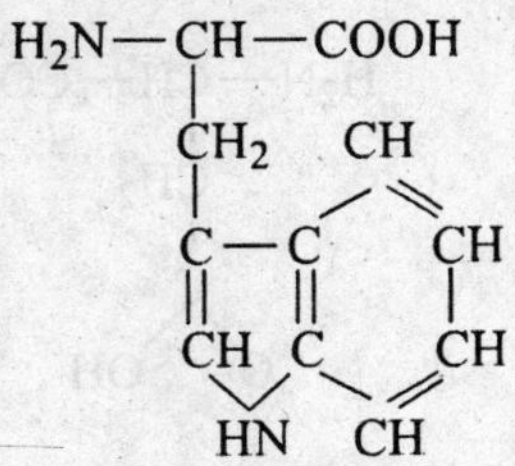 Tryptophan (Try)
C. **Secondary amino group of aminoacids**	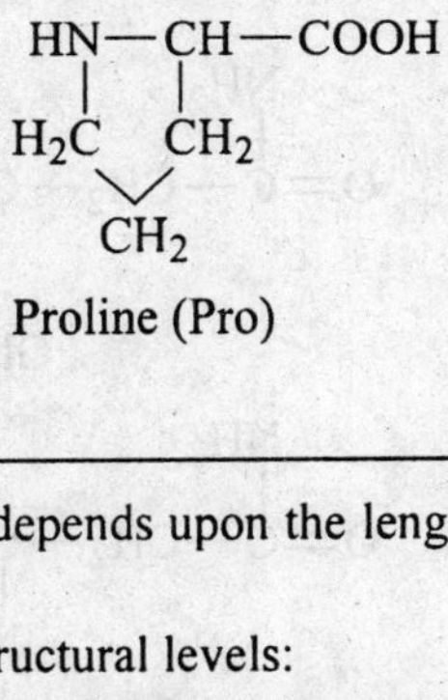 Proline (Pro) 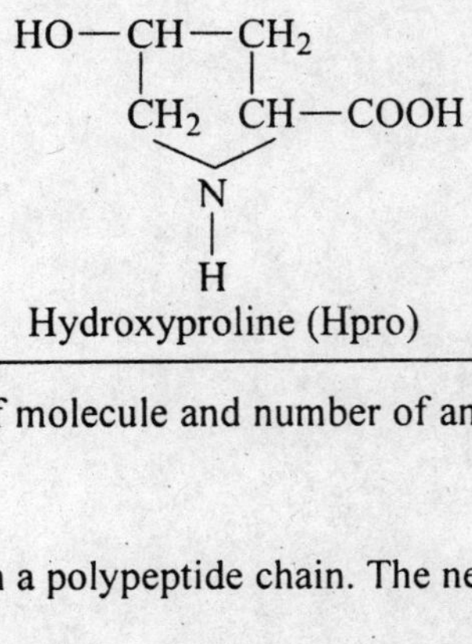 Hydroxyproline (Hpro)

The molecular weight of protein depends upon the length of molecule and number of amino acids in it.

Proteins are recognised by four structural levels:

(*i*) *Primary structure.* It is the linear sequence of aminoacids in a polypeptide chain. The newly synthesized polypeptide chain is called primary protein.

(*ii*) *Secondary structure.* When the primary polypeptide chains are twisted or coiled into a helix, electrostatic bonds are formed between – COOH group and NH_2 group and hydrogen bonds develop between amino acids facing each other due to coiling. Such proteins are said to be secondary proteins. In biological proteins, the polypeptide chains remain coiled either in α like shape or in β like shape and hence they are called α helix and β helix respectively.

(*iii*) *Tertiary structure.* When very long polypeptide chain becomes extensively folded and coiled in order to compress the long spiral chain into a globular form and subsequently certain intra chain bonds, specially disulphide (– S – S –) bridges between cysteine residues of polypeptide chain over a vast surface to create interstices between the polypeptide chains, a tertiary structure of protein is resulted. The tertiary structure of a protein often places hydrophobic (water hating) groups on the outside and hydrophilic (water loving) groups on the inside. Examples are insulin enzyme, ribonuclease enzyme etc. The most tertiary proteins act like catalysts or enzymes.

(*iv*) *Quaternary structure.* The association of more than one polypeptide chain to form stable unit corresponds to the quaternary structure. Quaternary proteins result when inter chain bonds or bridges are established so as to link two or more otherwise independent polypeptide chains. Most proteins with molecular weight higher than 20,000 possess quaternary structure and are composed of more than one polypeptide chain. The best example of quaternary protein is haemoglobin (molecular weight 1,00,000) which is formed of two α-and two β-chains.

The chemical and biological individuality of a protein depend upon the order in which amino acids are linked in it. So in a particular protein chain aminoacids are arranged in correct order. But can these aminoacids be so precisely arranged? Yes, it is so, and this precision in the sequence of aminoacids during the synthesis of protein is controlled by DNA molecule which itself is not directly involved in the synthesis. Actually DNA molecules send their message to the sites of protein synthesis through special type of RNA called messenger RNA or mRNA. The information for the structure of polypeptide (protein) is stored in a polynucleotide chain. The sequence of bases in a polynucleotide chain determines the sequence of amino acids in a particular polypeptide. The transfer of information from DNA to *m*RNA and then from *m*RNA to protein (aminoacid sequence) is unidirectional according to Francis Crick (1956) and it does not flow in reverse direction *i.e.,* from protein to RNA to DNA. The DNA molecule is provided with the information for its own replication. Francis Crick termed this flow of information from DNA to RNA to protein molecule as **central dogma**. This is shown in Fig. 20.1.

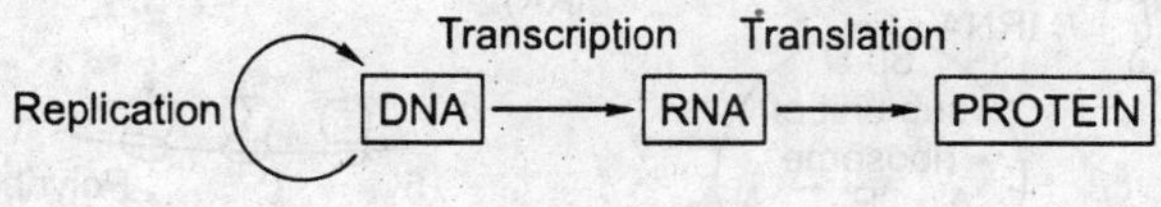

Fig. 20.1 The central dogma of molecular biology.

The central dogma of molecular biology, therefore, involves the following three major processes for preservation and transmission of genetic information:

1. **Replication.** A process which is indicated by the arrow encircling DNA signifying that DNA is template for self replication.
2. **Transcription.** The arrow between DNA and RNA indicates that all cellular RNAs are synthesiged on DNA templates.
3. **Translation.** This is process by which all proteins are determined by RNA templates on the ribosomes.

In certain cells infected with RNA viruses, *e.g.,* TMV, ϕMS_2, ϕR_{17} etc., the viral RNA produces new copies of itself with the help of RNA replicase. Genetic RNA of some viruses, *e.g.,* RSV, some times acts as a template for the production of complementary strand of DNA (**reverse transcription**). On this ground Barry Commoner (1968), however, suggested that flow of information should be cyclic rather than in one way, but such reversals of normal flow of information are rare events.

MECHANISM OF PROTEIN SYNTHESIS (TRANSLATION)

Protein synthesis in the cell is conducted by ribosomes that are found attached to the membrane of endoplasmic reticulum and microsomes, as well as in free state in the groundplasm.

The main components that take part in protein synthesis at cellular level are: 20 different amino acids, different types of RNAs, enzymes, aminoacid activating enzymes, polypeptide-polymerase and energy liberating molecules, such as ATP and GTP. DNA which contains genetic information synthesizes three kinds of RNA:

(*i*) Messenger RNA (mRNA)

(*ii*) Ribosomal RNA (rRNA)

(*iii*) Transfer RNA (tRNA) or soluble RNA (sRNA)

mRNA is copied from DNA molecule. The specific locus of DNA molecule where mRNA is formed is referred to as a structural gene. tRNAs come probably from special genes called determinants for tRNAs.

A Protein Synthesis in Prokaryotes

The mechanism of protein synthesis has been thoroughly investigated in *Escherichia coli.* In bacterial cell, the protein synthesis takes place on 70s ribosomes. The process of protein synthesis in *E. coli* involves the following steps:

1. **Transcription.** The partial uncoiling of two DNA strands occurs. This is followed by the production of single stranded mRNA on one of the two DNA strands. The messenger

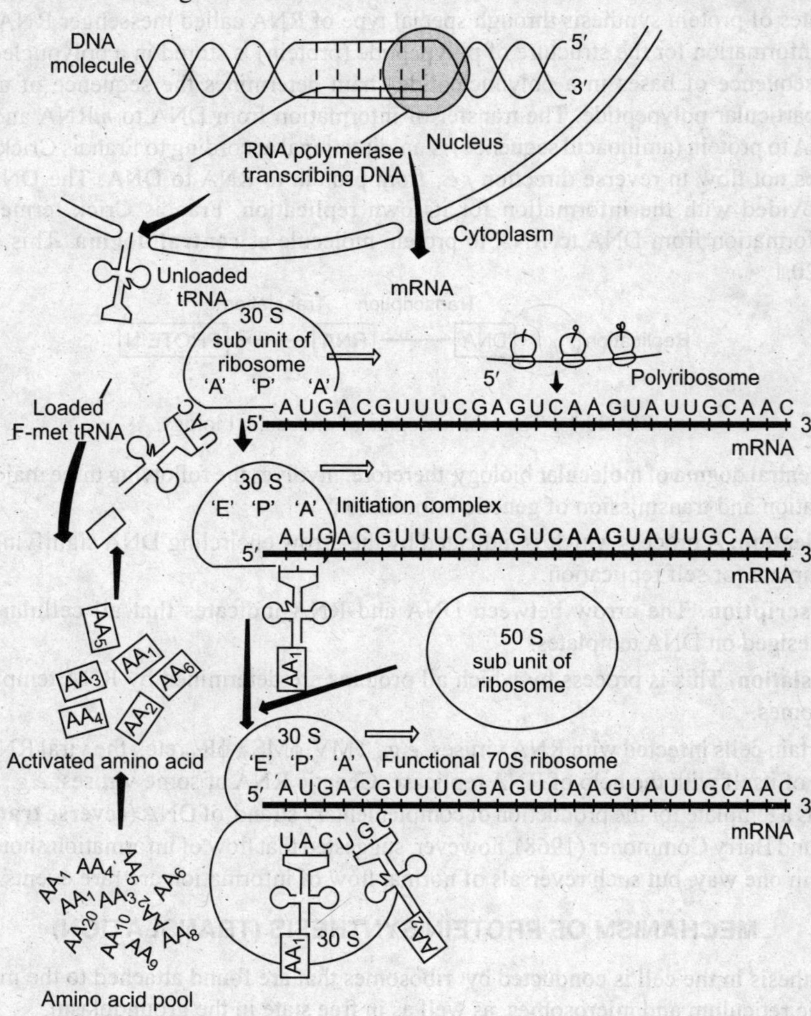

Fig. 20.2 Transcription and translation in *E. coli*; A = aminoacyl site or decoding site, 'P' = Peptidyl site or condensing site, E = Exit site, AA_1 = N-formyl methionine, AA_2 = Threonine.

RNA complement is made in accordance with base pairing rules. This is transcription. The transcription of genetic code of DNA into mRNA is catalysed by the enzyme *RNA polymerase*. *m*RNA carries the information in the form of base triplets for the synthesis of a particular protein (Fig. 20.2).

2. The second step involves the separation of *m*RNA from DNA and then its transfer from nucleus to cytoplasm and final attachment of 5' end of *m*RNA with 30s (smaller) sub-unit of ribosome in presence of protein initiation factor. Before the *m*RNA migrates from nucleus to ribosome in cytoplasm it undergoes process of maturation. In eukaryotes the newly formed RNA is called heterogenous nuclear RNA (hn RNA). Many of the functional RNA molecules including *t*RNA and *m*RNA in eukaryotes are processed from much longer precursor RNAs which are 5,000 to 50,000 nucleotides long and may be 10 to 100 times longer than the mature functional RNA molecules which are derived from them. Precursor RNAs are transcripts of split genes which contain both sequences coding for aminoacids *(exons)* and those not coding for aminoacids (*introns*) interspersed. The non-coding sequences from the pre-RNAs are cleaved out and coding sequences are spliced together to produce functional mature RNA molecules. A few of the eukaryotic genes are not split genes. In prokaryotes some of the RNA molecules are cleavage products of longer pre-RNA.

3. **Translation.** As has been pointed out, *m*RNA determines the sequence of amino acids in the polypeptide chain which in turn is determined by sequence of nucleotides in DNA segment (gene). This process is called translation. Translation involves the following steps which are shown in Figs. 20.2, 20.3, 20.4.

   (*a*) **Activation of aminoacid.** Elsewhere in the cytoplasm, aminoacids are selected for activation. Professor Fritz Libmann and others discovered in 1956 that before amino acids can combine to form proteins, they must be activated and this is achieved by combining with phosphate. The activation involves the reaction between aminoacid and ATP. The reaction is catalysed by specific enzyme *aminoacyl RNA synthetase*. So, for the activation of 20 amino-acids there should be at least a set of 20 such enzymes in the cytoplasm. Amino acid activating enzymes were first discovered by M. Hoagland.

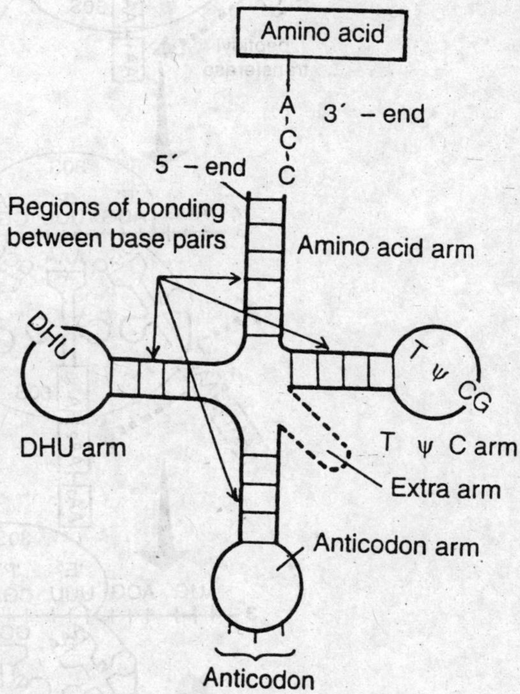

Fig. 20.3 Aminoacyl-*t*RNA (Amino acid is attached to CCA-OH end).

The product formed after activation is *aminoacyl-adenylate* enzyme complex which is energy rich compound.

$$\underset{\text{aminoacid}}{\text{AA}} + \text{ATP} + \underset{\substack{\text{(aminoacyl}\\ \text{synthetase)}}}{\text{enzyme}} \longrightarrow \underset{\substack{\text{(aminoacyl adenylate}\\ \text{enzyme complex)}}}{\text{AA} \sim \text{AMP}} \text{— enzyme} + \text{PP (Pyrophosphate)}$$

(*b*) **Attachment of activated aminoacid to tRNA.** The CCA end of *t*RNA molecule now attaches with specific aminoacid adenylate-enzyme complex. The aminoacyl-adenylate remains attached to the enzyme in the form of monocovalent complex until it is transferred to *t*RNA. The carboxyl group of aminoacid residue of aminoacyl adenylate is transferred to 3' OH group of ribose sugar of terminal adenosine at CCA end of *t*RNA. As a result, aminoacyl-*t*RNA, AMP and enzyme are formed.

AA ~ AMP – enzyme + *t*RNA ⟶ AA – *t*RNA + AMP + Enzyme aminoacyl synthetase.

The transfer of aminoacids to *t*RNA is catalysed by the previous aminoacyl RNA synthetase enzyme itself (Fig. 20.3).

(*c*) The aminoacids-*t*RNA complex then comes to *m*RNA where adapter nucleotide triplet or anticodon of *t*RNA becomes attached with the complementary base triplet (codon) of

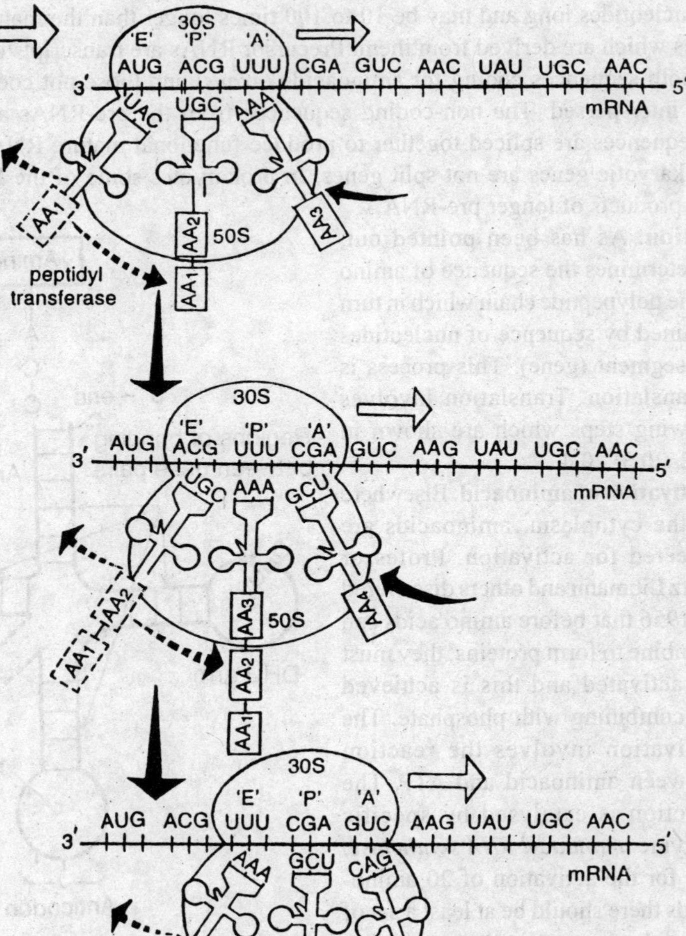

Fig. 20.4 Diagrammatic representation of protein synthesis in *E. coli*. AA_3 = phenyl alanine, AA_4 = Arginine, AA_5 = Valine, AA_6 : Lysine.

*m*RNA. The fate of amino acid is determined at the very moment, it becomes attached with the corresponding *t*RNA. The messenger RNA and *t*RNA-amino acid complex attachment is temporary. Like this, many *t*RNA-amino acid complexes are arranged one after another at proper places on messenger RNA strand in linear fashion. In the attachment, the adapter trinucleotides of *t*RNAs act as anticodons (Figs. 20.4 and 20.5). The *m*RNA molecules have **translation initiation site at 5' end** and the **chain termination site** close to trailor end. The initiation site consists of a codon AUG and unknown secondary structure of *m*RNA. The chain termination site has one of the three codons UAA, UAG and UGA. Mature *m*RNA binds with smaller ribosomal sub-unit in presence of initiation factor IF_2. Soon *t*RNA-N-Formyl methionine complex (F met-*t*RNA) comes from the cytoplasmic amino acid pool and binds with the first triplet codon of *m*RNA to initiate the process of protein synthesis and to form initiation complex. Initiation complex is formed in presence of guanosine triphosphate (GTP) and three protein factors F_1, F_2 and F_3. This is followed by union of bigger sub-unit with smaller ribosomal sub-unit in presence of Mg^{++} and initiation factors F_1, F_2 to form the ribosome. The codes of *m*RNA are not read by a single ribosome but by many ribosomes interlinked by *m*RNA (Polysomes).

(*d*) **Initiation of Protein Synthesis**. The messenger RNA always has first triplet as AUG or GUG at its 5-end and these triplets code for aminoacids N-formyl methionine (F. met) which usually initiate a protein chain. Thus, in all proteins formyl methionine occupies the first place, *i.e.*, ***at***, amino end and when the protein molecules are completely synthesised formyl methionine may be detached from the protein molecules by activity of hydrolytic enzyme deformylase. In formyl methionine-*t*RNA complex the amino group is blocked by formyl group leaving only COOH group free to react with NH_2 group of the second amino acid (AA_2). In this way, polypeptide chain always grows from amino end toward – COOH end.

When one *t*RNA–aminoacid complex attaches to *m*RNA at starting end, then the second *t*RNA-aminoacid complex also comes just after the first and finally the two adjacent amino acids form peptide linkage. Like this several molecules of amino acids will join in a definite order through peptide bonds to form specific protein molecule (Fig. 20.5).

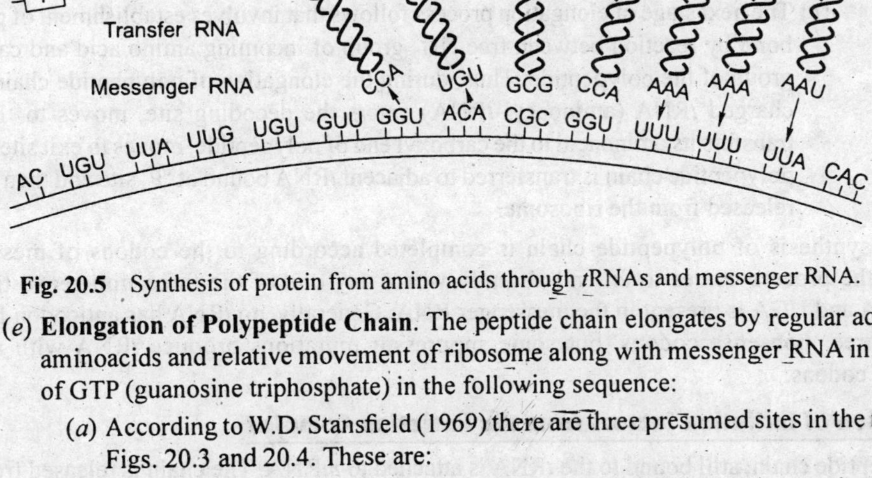

Fig. 20.5 Synthesis of protein from amino acids through *t*RNAs and messenger RNA.

(*e*) **Elongation of Polypeptide Chain**. The peptide chain elongates by regular addition of aminoacids and relative movement of ribosome along with messenger RNA in presence of GTP (guanosine triphosphate) in the following sequence:

(*a*) According to W.D. Stansfield (1969) there are three presumed sites in the ribosome Figs. 20.3 and 20.4. These are:

(*i*) Decoding site or 'A' site which binds the loaded AA~*t*RNA complex with the *m*RNA by base pairing.

(ii) A condensing site or 'P' site or peptidyl site which joins the aminoacid to the growing polypeptide chain.

(iii) An exit site or 'E' site at which *t*RNA detaches from the polypeptide, messenger RNA and ribosome.

*t*RNA with their associated aminoacids will enter the ribosomal site 'A' and will be checked by a 'checking factor' to see if there is a correct fit between the codon on the messengers RNA and the anticodon of *t*RNA. If the fit is incorrect, the *t*RNA is rejected and presumably other *t*RNAs will continue to be tried until the correct one is found.

(b) As the initiating codon AUG or GUG has entered the ribosome and is in position facing the site 'A', the correct *t*RNA, *i.e.*, f-met-*t*RNA is checked against codon. This reaction is facilitated by the presence of initiation factor F_1

$$30S - mRNA + f\text{-met-}tRNA \xrightarrow{F_1} 30S - mRNA - f\text{-met-}tRNA$$

(c) The 30S f-met-*t*RNA along with messenger RNA, then moves from decoding site ('A' site) to peptidyl site ('P' site) and with this the next codon of *m*RNA enters 'A' site where it finds the second correct aminoacyl-*t*RNA. Aminoacyl-*t*RNA (AA_2 – *t*RNA) binds with the codon of 'A' site in presence of GTP and two proteins called transfer factor Tu and Ts which remain associated with ribosomes. In this binding process, a complex is formed from GTP, the transfer factors and the incoming aminoacyl-*t*RNA which ultimately fixes aminoacyl-*t*RNA (AA_2 – *t*RNA) at the 'A' site of ribosome and at the same time releases transfer factors – GTP complex and inorganic phosphate.

(d) Due to the relative movement of ribosome and *m*RNA in presence of single GTP molecule the next codon enters the 'A' site. The A site is now occupied by another aminoacyl-*t*RNA (AA_3 – *t*RNA) corresponding to the next codon of *m*RNA and f-met-*t*RNA reaches at the exit site (E-site) and AA_2-*t*RNA occurs at the P site. Now an enzyme known as transferase I kicks off *t*RNA from formyl methionine and flips the formyl methione (AA_1) to AA_2-*t*RNA bound at the 'P' site. According to Monro (1967) an enzyme known as peptidyl synthetase found in 50S sub-unit helps in the formation of peptide bond. The 'G' *factor* is supposed to release the discharged or deacetylated *t*RNA from the site 'E' of ribosome.

(e) The next stage of elongation process follows that involves establishment of peptide bond by reaction between free NH_2 group of incoming amino acid and carboxyl group of the polypeptide. Thus, during the elongation of polypeptide chain, each charged *t*RNA (aminoacyl-*t*RNA) enters the decoding site, moves to 'P' site, transfers its aminoacid to the carboxyl end of polypeptide, moves to exit site where polypeptide chain is transferred to adjacent *t*RNA bound at 'P' site and then finally released from the ribosome.

The synthesis of polypeptide chain is completed according to the codons of messenger RNA and the process comes to an end abruptly where any one of the three **non-sense triplets** UAG, UAA and UGA is present in the messenger RNA. Generally, no *t*RNA has anticodon for any of these three 'nonsense codons' but some suppressor mutations produce *t*RNA with any of these three codons.

Dissociation of Initiation factors from the Initiation Complex

The polypeptide chain, still bound to the *t*RNA is attached to *m*RNA. The chain is released from the ribosome under the direction of three distinct proteins which are called released factors R_1, R_2 and S. These factors are bound to the ribosome and control the hydrolysis of ester linkage between *t*RNA and the polypeptide chain. Reproduction of a primary polypeptide chain according to specification of *m*RNA is called **translation**.

After the completion of chain the two sub-units of ribosomes separate.

Protein Synthesis on 80S Ribosomes of Eukaryotes

The process of protein synthesis on 80S ribosomes of eukaryotes is found to be more or less similar to that on 70S ribosomes described above (Allende, 1961; Boulter, Elliis and Yarwood, 1972; Hall *et al.*, 1974). However, the process of initiation of polypeptide chain on 80S ribosomes of eukaryotes differs from that of prokaryotes in the following aspects:

1. In eukaryotes in place of formyl methionine methionine acts as chain initiation aminoacid.
2. In eukaryotes smaller sub-unit (40S) associates with methionine-*t*RNA without the help of *m*RNA.
3. For the formation of initiation complex involvement of GTP is not necessary (Zasloff and Ochoa, 1972).
4. The initiation process involves the following steps:
   (*i*) met-*t*RNA + 40S sub-unit ⟶ 40S-met *t*RNA
   (*ii*) 40S-met *t*RNA + *m*RNA ⟶ 40S-*m*RNA-met *t*RNA
   (*iii*) 40S-*m*RNA-met *t*RNA + 60S sub-unit ⟶ 80S-*m*RNA met-*t*RNA initiation-complex.

QUESTIONS

1. What is one gene-one enzyme concept? Explain.
2. Give a brief account of different kinds of RNA and discuss their structure and functions.
3. What do you understand by one gene-one polypeptide concept? Give an account of mechanism of protein synthesis.
4. Give an account of role of transfer RNA in protein synthesis.
5. Write short notes on:
   (*a*) RNA polymerase,
   (*b*) Peptide transferase,
   (*c*) Release factors, and
   (*d*) Initiation of protein synthesis.
6. Explain how an amino acid is activated and then attached to specific *t*RNA. What is the evidence that *m*RNA is long enough to code for several proteins?

21

CHAPTER

The Genetic Code and Regulation of Protein Synthesis

From the study of protein synthesis, it is evident that amino acid sequence in the protein molecule is determined by DNA which carries information for this purpose. But where does the information lie? If we look at DNA chain the Deoxy-ribonucleotides backbone seems to be of no help as it is a structure with repeating nucleotides, AG CT AT GC and so on. Here the DNA bases adenine, guaninee, cytosine and thymine are denoted by standard letters A, G, C, and T respectively. The code lies only in the four bases (A,T, G and C) which appear many times in the chain providing different base sequences to different DNA strands. Actually, the 4 bases in DNA molecules keep in a coded form the message for the structure of protein molecules. Through transcription process the information coded in DNA is transferred to *m*RNA tape. Thus the 'four letter' (four bases—A, T, G and C) language of DNA is transcribed into another complementary four letter language in messenger RNA. The next important step is the actual translation of four letter language

Table 21.1. The different types of mRNA code words.

Singlet Code (4 Words)
A
G
C
U

Doublet Code (16 words)	A	G	C	U
A	AA	AG	AC	AU
G	GA	GG	GC	GU
C	CA	CG	CC	CU
U	UA	UG	UC	UU

Triplet Code (64 Words)					
First base	Second base				Third base
	U	C	A	G	
U	UUU	UCU	UAU	UGU	U
	UUC	UCC	UAC	UGC	C
	UUA	UCA	UAA	UGA	A
	UUG	UCG	UAG	UGG	G
C	CUU	CCU	CAU	CGU	U
	CUC	CCC	CAC	CGC	C
	CUA	CCA	CAA	CGA	A
	CUG	CCG	CAG	CGG	G
A	AUU	ACU	AAU	AGU	U
	AUC	ACC	AAC	AGC	C
	AUA	ACA	AAA	AGA	A
	AUG	ACG	AAG	AGG	G
G	GUU	GCU	GAU	GGU	U
	GUC	GCC	GAC	GGC	C
	GUA	GCA	GAA	GGA	A
	GUG	GCG	GAG	GGG	G

of messenger RNA into specific protein with 20 letter-(amino acids) language. Since the gene is involved in the synthesis of protein and since protein represents in its primary structure linear combinations of 20 amino acids, it is logical to conclude that the coded message of the gene must be in the form of words which determine the sequence of particular amino acids. This is what we mean by genetic code.

But how is it that just four bases in DNA codify the different amino acids in the proteins? If each base specifies to one aminoacid, then only 4 aminoacids can be codified. If two adjacent bases (*i.e.,* two letter words) are utilised, the maximum number of base pairs (words) will be $4 \times 4 = 16$ which can account for only 16 different amino acids. But, the number of different amino acids is 20 for which at least 20 different words will be needed. So two letter codes do not solve the purpose. Now, if we consider three letter words or triplet codes (which include three adjacent nucleotides), then $4 \times 4 \times 4 = 64$ different types of three letter code words will be possible. They can account for 64 different amino acids and that is much more than 20 required. The different types of possible *m*RNA code words in each case have been shown in Table 21.1.

Characteristics of the Genetic Code

1. In *m*RNA, the group of 3 nucleotides that specifies one aminoacid is known as a code word or "**codon**". The message is read as group of three nucleotides at a time. Code is comma-free.
2. Since in triplet coding, the number of codons is more than the number of amino acids, it is hypothesised that the genetic code would be degenerate. **Degenerate coding** means that for some amino acids there would be more than one codon. This means that a messenger RNA, with two exceptions (AUG and UGG), may provide for the incorporation of a particular amino acid more than one triplet. In addition to this, one would expect to find more than one transfer RNA for an amino acid, that is degenerately coded. Recent researches have proved that these expectations are correct. Let us consider here a segment of single chain of DNA molecule containing nine bases (ATAGTCGAT). If it is supposed that the genetic code is commaless, *i.e.,* without *spacer words* (non-sense words) between sense words, then the coded message would be read in the following two ways:
   1. As overlapping words (Fig. 21.1).
   2. As non-overlapping words (Fig. 21.1).

   If the words are read in the overlapping fashion, then the third nucleotide of DNA (*i.e.,* A) appears repeatedly in three triplets ATA, TAG and AGT (Fig. 21.1) but if the code is read in non-overlapping manner then third nucleotide (A) appears only in one code ATA.

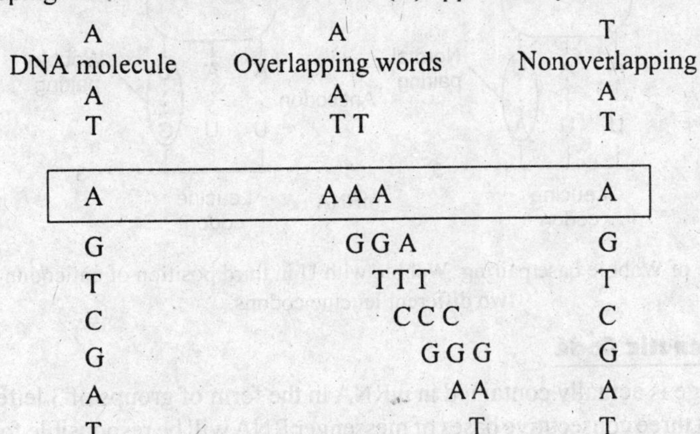

Fig. 21.1 DNA coding.

Now the question arises whether the genetic code is of overlapping type or non-overlapping type? The experimental evidence is in the favour of non-overlapping code, that is, the adjacent codons do not overlap. The argument against the concept of an overlapping code system is based on the fact that a change in third nucleotide of DNA would bring about changes in the three amino acids in the overlapping condition, because all the first three trinucleotides ATA, TAG and AGT would be changed by substitution of a nucleotide other than A in third place. The result would be that protein molecule formed in the changed condition will differ from the normally formed protein in three amino acids. In the non-overlapping code, since only one word will be changed by the substitution of a nucleotide other than A in the DNA molecule at third place, the protein formed in the changed condition will differ from that synthesized in normal condition in only single amino acid. The latter condition has been found really practical. This suggests that a non-overlapping system is involved in genetic coding.

3. The genetic code is universal, that is, all living organisms have the same genetic language. It means that a message from an animal cell will produce the same protein whether it is translated by protein synthesis machinery of a bacterial cell or plant cell.
4. The genetic code uses specific initiation codon and stop codons. AUG triplet codes for methionine and is the initiation signal and if AUG is absent from the 5′ end of *m*RNA, it would not be in a position to carry out translation or protein synthesis. Of the 64 triplet codes three are stop signals or termination signals or 'non-sense' codons, UAG, UAA and UGA, which stop the synthesis of polypeptide.
5. **Wobble Base Pairing.** It has been found from sequence analysis that the pairing of base at the 5′ end of anticodon (that is complementary to third base of codon) is not as strict or precise as are the first and the second bases and it pairs with more than one base at the 3′ end of the codon *i.e.*, U in third position of anticodon may pair with A or G in codon, G may pair with C or U and I (inosinic acid having the base hypoxanthine) may pair with U, C, A. This is called wobble base pairing by Crick (1996) and only certain pairings are possible in this regard. Wobble does not allow any single *t*RNA to recognise four different codons (Fig. 21.2).

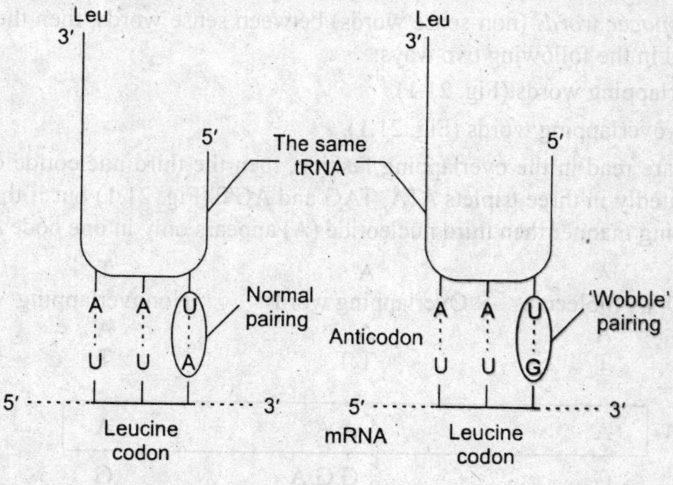

Fig. 21.2 An example of Wobble base pairing. Wobble with U in third position of anticodon *t*RNA recognises two different leucine codons.

Deciphering of Genetic Code

If the genetic message is actually contained in *m*RNA in the form of groups of 3 letter code words, it may be assumed that three consecutive bases of messenger RNA will be responsible for the attachment of one amino acid. Experimental evidence supporting the concept of triplet code was provided by F.H.C.Crick and co-workers (1961). Marshal Nirenberg and his colleague John Matthaei were first to break the genetic code in vitro study of protein synthesis. Usually radioactive amino acids are used. Nirenberg and Matthaei took in a test tube the ribosomes, enzymes and energy liberating compounds and saturated them with the *t*RNAs obtained from the cell extract of a bacterium *Escherichia coli.* Thus they stimulated the protein synthesizing mechanism of a cell in the test tube. In the experiment they observed that the incorporation of the amino acid phenylalanine into protein was considerably stimulated by the addition of an artificial RNA composed of uracil base only (*i.e., m*RNA containing UUUUUUUUUUUU or poly U). This artificial RNA polymer and similar other polymers are synthesized with the help of an enzyme *polynucleotide phosphorylase* discovered by Ochoa, Nobel laureate. Since poly U RNA chain contains only one base U, it is clear that UUU is the code for amino acid phenylalanine, if triplet coding is correct. This opened up new possibility for breaking the genetic code.

After this significant discovery Nirenberg, Ochoa and their co-workers undertook the job of finding codons for various other amino acids with RNA polymers. These RNA polymers consisted of either one type of base like poly U or two bases, as poly AU(-AU-AU-AU and so on) or three types of bases, *e.g.,* poly AUC (AUC-AUC-AUC..........and so on). In some cases the codons for amino acids were identified easily.

Binding technique for breaking genetic code. In many cases, however, there was ambiguity as the base sequence in the artificially synthesized messenger RNA could not be predetermined. Further, it was difficult to determine the sequence of bases in a codon, *e.g.,* in making RNA polymer of three bases–A, U and C, there are six possible ways for the arrangement of bases in triplets AUC, ACU, CAU, UCA and UAC. This was not easy to determine which one codon was valid for what

FIRST LETTER	Second Letter: U	C	A	G	THIRD LETTER
U	UUU, UUC } phe UUA, UUG } leu	UCU, UCC, UCA, UCG } Ser	UAU, UAC } tyr UAA C.T.* UAG C.T.*	UGU, UGC } cys UGA* C.T. UGG trp	U C A G
C	CUU, CUC, CUA, CUG } leu	CCU, CCC, CCA, CCG } pro	CAU, CAC } his CAA, CAG } gln	CGU, CGC, CGA, CGG } arg	U C A G
A	AUU, AUC, AUA } ile AUG met C.I.	ACU, ACC, ACA, ACG } thr	AAU, AAC } asn AAA, AAG } lys	AGU, AGC } ser AGA, AGG } arg	U C A G
G	GUU, GUC, GUA, GUG } val	GCU, GCC, GCA, GCG } ala	GAU, GAC } asp GAA, GAG } glu	GGU, GGC, GGA, GGG } gly	U C A G

*C.T. = chain termination codon

The genetic code as established for the *bacterium Escherichia coli* (*C.I.* = chain initiation; C.T. = chain termination or NONS = Non-sense

A = Adenine, G = Guanine, C = Cytosine, U = Uracil

ala = Alanine, arg = Arginine, asa = Aspartic acid

asp = Asparagine, cys = Cysteine, glue = Glutamic acid

gl = Glutamine, gly = Glycine, his = Histidine

ile = Isoleucine, leu = Leucine, lys = Lysine, met = Methionine

phe = Phenylalanine, pro = Proline, ser = Serine, tyr = Tyrosine

thr = Threonine, trp = Tryptophan, val = Valine.

Fig. 21.3

amino acid. Answers were partly provided by the discovery made by Nirenberg and Leder (1964). They discovered that specific *t*RNA molecules could bind to ribosomal *m*RNA even if the *m*RNA was composed of a single trinucleotide. The binding of *t*RNA to *m*RNA is not possible if the *m*RNA consists of only mono or dinucleotides. This provides further proof for the triplet nature of the codes. Simple trinucleotides of known linear arrangement are easy to synthesize. The trinucleotide group with known base sequence were experimentally tested for their specificity to particular type of amino acid. Thus by sequence analysis of such a system the exact codes could be determined. This is known as *binding technique.* Actually this technique helped a long way in working out the actual code. By this process all the 64 possible triplets were separately tested for 20 amino acids and condons for all the amino acids have been assigned. Figure 21.3 shows different amino acids and their respective codons.

In the Fig. 21.3 triplets UAG, UAA and UGA in *m*RNA are called *non-sense* codons or chain terminators (C.T.) as they do not code for any aminoacid and their presence in *m*RNA results in the stoppage of polypeptide formation. The triplets AUG and GUG appear to have two roles; firstly, they act as chain initiators (C.I.) and secondly, as specifiers of amino acid methionine and valine respectively. The genetic codes are universal, that is, the codes that are specific to particular amino acids in bacteria are also specific to those very amino acids in man, mouse, birds and all other living organism.

Dr. Khorana's method. Dr. Khorana was also working independently on the genetic codes. He was impressed by Dr. Nirenberg's technique but he evolved his own method of cracking the code. He synthesized DNA of known sequence and added that to the cell free system in order to get messenger RNA of known sequence. He began by linking just two DNA bases together and then polymerising the doublats give a long DNA strand containing an alternative sequence of the two bases (say, for example, AC—AC—AC—AC—AC—AC). By repeating trinucleotides sequences such as TTC/AAG, and even tetranucleotide sequences such as TTAC/AATG he synthesized in laboratory eight kinds of artificial DNA with known base sequences which are as under:

Repeating dinucleotide sequences	*Repeating trinucleotide sequences*	*Repeating tetranucleotide sequences*
Poly d—TC : AG	Poly d—TTC : AAG	Poly d—TTAC : AATG
Poly d—TG : AC	Poly d—TTG : AAC	Poly d—TATC : ATAG
	Poly d—TAC : ATG	
	Poly d—ATC : TAG	

The synthetic *m*RNA synthesized on the surface of synthetic DNA with known base sequence then directed the protein synthesis. From the sequence of amino acids in the protein molecules so obtained, the specific codes could be directly established for them. Using this procedure Khorana established all the 64 codons. Some of the salient features of this experiment are summarised here.

1. Repeating two nucleotide sequences in RNA, polypeptides with two amino acids in alternating arrangement were obtained *e.g.,*

   —U-C-U—C-U-C—U-C-U—C

   Serine Leucine Serine

2. Repeating trinucleotide sequences yielded a mixture of three proteins, each with a number of particular amino acid, for example, the repeating sequence—CUU yielded protein with polyphenylalanine, protein with polyleucine and protein with serine. The scheme is as follows:

   —CUU—CUU—CUU—CUU—CUU—CUU—

   CUU = Leucine   Leu Leu Leu Leu Leu Leu

   UUC = Phenylalanine   Phe Phe Phe Phe Phe

   UCU = Serine   Ser Ser Ser Ser

3. Repeating tetranucleotide sequences he got proteins with repeating 4 amino acids sequence as for example:

—UCUA — UCUA—UCUA —UCUA—UCUA—

-U-C-U	-A-U-C	-U-A-U	-C-U-A	-U-C-U-	A-U-C
Ser	Ileu	tyr	leu	Ser	Ileu

These experiments established beyond doubt 64 different codes. These are shown in Fig. 21.3.

In 1959, when Nirenberg was thinking of using RNA of known sequence to work out code words, Dr. Robert Holley and his team at Cornell University took the extremely bold decision to work out the structure of naturally occurring RNA. They chose the smallest form of RNA (*t*RNA) and set out to isolate it from yeast. By 1962 they had isolated and purified 3 yeast *t*RNAs. The purified *t*RNAs was then split into segments by special techniques and then they started to work out piece by piece, the order of bases along the backbone of one of them. In the base sequence of *t*RNA, which carries the amino acid alanine, there was a region containing the triplet CGI (I=Inosine a chemical compound similar to G) or practically CGG. That triplet acted as adapter nucleotide triplet or anticodon for alanine. By March 1965, they had succeeded in their efforts and anticodons for all amino acids had been identified. By determining the structure of *t*RNA molecule Dr. Holley and his colleagues have shown that it was possible to analyse with same technique other nucleic acids also.

For their brilliant analytical work Dr. Holley, Dr. Hargovind Khorana and Dr. Nirenberg were named co-winners of the 1968 Nobel Prize.

REGULATION OF PROTEIN SYNTHESIS (OPERON CONCEPT)

The DNA present in a cell contains genetic information for all the proteins required by an organism and may contain from few to thousands of genes. All the functional genes do not make polypeptides all the times. Though the different types of cells in the body of an organism differ in their structure and function, their genes are identical. As development proceeds the activity of some genes increases while those of others decreases or stops, *i.e.,* the genes are switched *on* and *off* at different times. This process is referred to as **differential gene action.** When the genes are active they direct the synthesis of enzymes and proteins which affect certain phenotypic traits. The metabolic products may influence the synthesis of enzymes (*feed back* or *inhibition of end product*). Although the cell may contain many genes which produce as many enzymes as their number, only the enzymes required at a particular time are produced. Therefore, the flow of information from DNA (gene) to protein is regulated in some way or the other. The mechanism of regulation of gene action has been studied extensively in certain micro-organisms such as *Escherichia coli* and some higher organisms. A hypothesis to explain induction and repression of enzyme synthesis was first put forward by Francois Jacob and Jacques Monod (1961). For this and other brilliant contributions they were awarded Nobel Prize in Medicine in 1965. The scheme proposed by Jacob and Monod is called *Operon Model* which consists of the following components (Fig. 20.3). Operon is a group of **structural genes** whose transcription is regulated by action of **regulator gene, promotor gene, operator gene** and **terminator gene**.

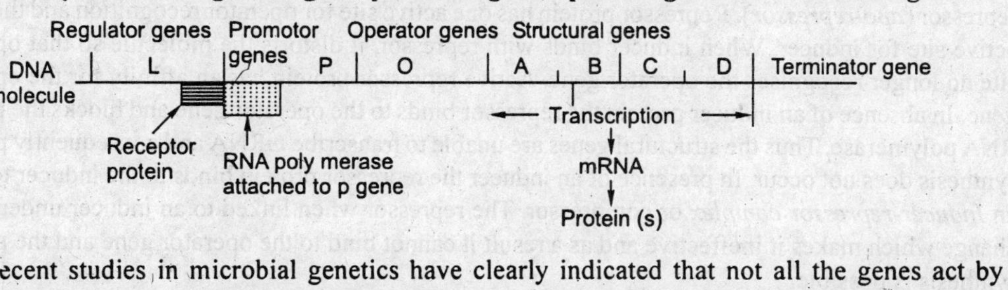

Recent studies in microbial genetics have clearly indicated that not all the genes act by determining the structure of given proteins or enzymes. Some of them regulate the action of other genes and are called **regulator genes**. Regulation may be brought about either by suppression of protein synthesis or by induction of the process.

Structural Gene

These are directly concerned with the synthesis of cellular proteins. The structural genes produce *m*RNAs through transcription and determine the sequence of amino acids in the synthesized proteins. As there are many proteins in the cells, there are as many structural genes under common control. Each structural gene may be controlled independently and that transcribes a separate *m*RNA molecule or all the structural genes under an operon may form one long polycistronic or polygenic *m*RNA molecule.

Eukaryotes have only a single functional site for producing monogenic *m*RNA, whereas, in prokaryotes polygenic *m*RNAs are common. A small segment of DNA molecule in bacterium may have several structural genes which may transcribe one long polycistronic *m*RNA molecules. This controls the synthesis of many proteins.

Operator Gene

Operator gene is located adjacent to first structural gene. It determines whether the structural gene is to be repressed by the repressor protein, a product of regulator gene. The operator gene is recognised by repressor protein which binds to the operator forming an *operator-repressor complex*. The operator genes as well as the group of genes that are under it are known as *operons*. The operator genes are carried in the chromosome and show recombination also.

Promoter Gene

It is continuous with the operator gene and is located just before operator. The complete nucleotide sequence of control region in some bacteria extends from the termination codon (ACT on DNA or UGA on *m*RNA) of the structural gene and induces the promoter and operator genes. The promoter region according to Pribnow (1975) has three essential elements which are constant in position with respect to each other. These elements are:

(*i*) a recognition sequence,
(*ii*) a RNA-polymerase binding sequence, and
(*iii*) an *m*RNA initiation site.

The recognition site is located outside the polymerase binding site, *i.e.*, the region of DNA protected from the action of DNAse. RNA polymerase first binds to DNA by forming a complex with the recognition sequence. Then it binds with the binding sequence to produce the pre-initiation complex. The binding site consists of a group of 7 bases present in a constant position in the protected fragments. RNA initiation site starts the *m*RNA transcription on one or two bases near the binding sequence. The starting sequence is located in the DNA fragment protected from DNAse action.

Regulator Gene

These genes direct the synthesis of protein which may be either an active repressor or an inactive repressor (*apo-repressor*). Repressor protein has one active site for operator recognition and the other active site for inducer. When inducer binds with repressor, it distorts the molecule so that operator site no longer recognises the operator gene. Active repressor protein has an affinity for the operator gene. In absence of an inducer protein the repressor binds to the operator gene and blocks the path of RNA polymerase. Thus the structural genes are unable to transcribe *m*RNA and consequently protein synthesis does not occur. In presence of an inducer the repressor protein binds to the inducer to form an *Inducer-repressor complex* or corepressor. The repressor when linked to an inducer undergoes a change which makes it ineffective and as a result it cannot bind to the operator gene and the protein synthesis is possible.

Effector or Inducer

An inducer or effector is a small molecule—a sugar, an amino acid or a nucleotide that can be linked to a regulator or repressor protein to change its ability to interact with an operator or promoter gene. Jacob and Monod have suggested that a regulator protein will bind to an operator and inhibit operon expression as long as the effector or inducer is absent, but if the effector molecule appears in the intracellular environment it will bind to the regulator protein and thereby pull the regulator away from the operator.

According to Jacob and Monod concept, *operon* is a group of contiguous structural genes showing coordinate expression and closely associated sites. This definition does not presume any particular mechanism of regulation, negative or positive. In the case of negative control of operon expression of the controlling elements are so coded that transcription of the operon occurs in the absence of any cytoplasmic regulatory component (repressor) specific for the particular system. In the case of positive control system the coding at the initiator controlling site prevents the expression of operon in the absence of any specific regulatory product. The operon represents a unit of polarised coordinated genetic transcription in prokaryotes that is initiated and regulated at the operator site. It consists of the promoter, the operator, one or more structural genes coding for enzymes performing related metabolic functions, and the terminator (Fig. 21.4). Transcription of the operon which eventually results in a polycistronic *m*RNA starts at the promoter where the RNA-Polymerase binds to DNA. Synthesis of *m*RNA starts if the operator is free of repressor protein (negative control).

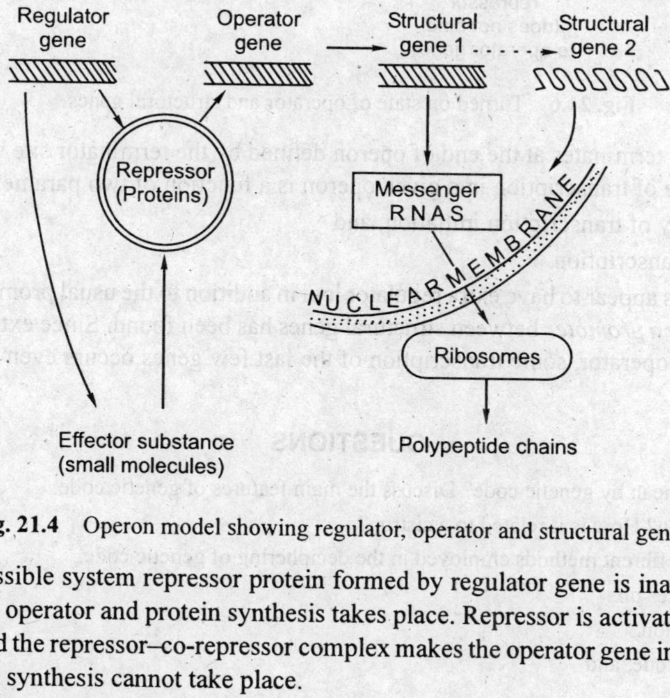

Fig. 21.4 Operon model showing regulator, operator and structural genes.

In the repressible system repressor protein formed by regulator gene is inactive, therefore, it does not block the operator and protein synthesis takes place. Repressor is activated in the presence of co-repressor and the repressor–co-repressor complex makes the operator gene inactive. As a result of this, the protein synthesis cannot take place.

Regulator genes produce macromolecules that act upon the operator genes to turn it on or off. A single regulator gene can affect the synthesis of different proteins whereas the structural gene obeys the one gene-one enzyme rule. If the repressor substances produced by regulator genes act upon operator genes, they depress their activity and then the latter do not permit the structural genes to produce their *m*RNAs. This is said to be the off condition (Fig. 21.5).

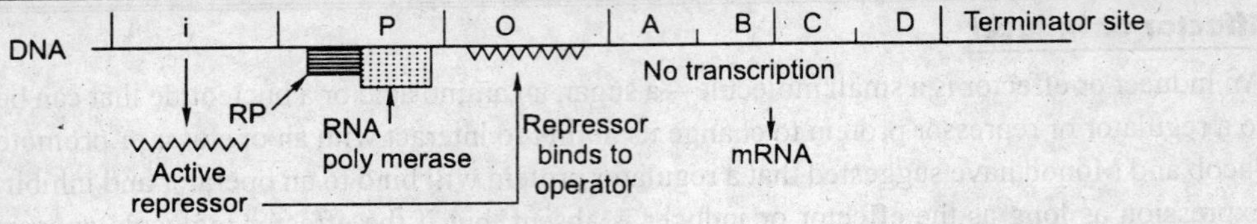

Fig. 21.5 Turned off state of operator and structural genes.

Regulator genes sometime produce inducer substances. If this substance is present in the nucleoplasm, it combines with the repressor substance, these two repressor and inducer substances do not act upon operator genes (O), they activate the operator genes and allow operator genes to turn on the structural genes as shown below in Fig. 21.6.

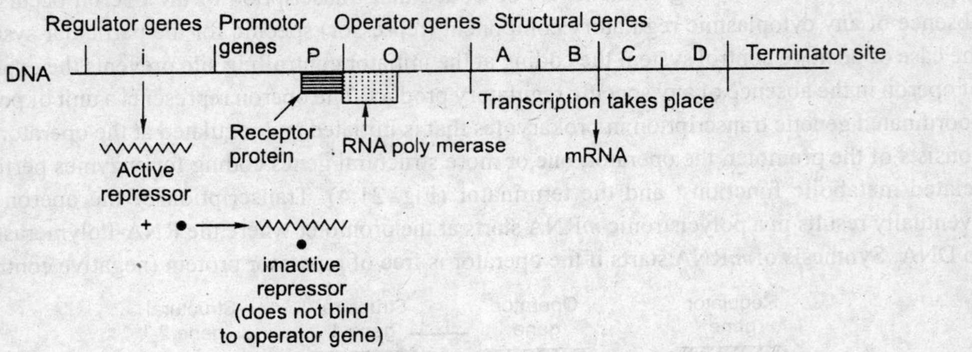

Fig. 21.6 Turned on state of operator and structural genes.

Transcription terminates at the end of operon defined by the terminator site where the *m*RNA is released. The rate of transcription in a given operon is a function of two parameters.

(*i*) The freqyency of transcription initiation, and

(*ii*) The rate of transcription.

Some operons appear to have extra regulator loci in addition to the usual promoter site adjacent to operator, and *extra promoter* between structural genes has been found. Since extra promoter gene is not linked to the operator, some transcription of the last few genes occurs even when the system is switched off.

QUESTIONS

1. What do you mean by genetic code? Discuss the main features of genetic code.
2. What is a codon? How is it related to a cistron?
3. Describe the different methods employed in the deciphering of genetic code.
4. Write short notes on.
   (*i*) Anti codon,
   (*ii*) Triplet code, and
   (*iii*) Operon.
5. Discuss briefly the operon concept and the mechanism of operon function.
6. What do you mean by structural genes, operator genes, regulator genes, promoter genes and inducers? How do they affect one another ?

22 CHAPTER Mutation

The offsprings resemble their parents in one or several respects, yet there are differences between the two. These differences, whether large or small, are called **variations**. The individuals exhibiting changed characters or altered phenotypes are called variants. Some of the variations may be induced by environment, while others may be hereditary. Generally, the variations caused by environment are not permanent, hence non-heritable. But the variations which appear due to changes in the hereditary mechanism are permanent and heritable. *Sudden appearance of marked heritable variation in the nature of any organism in ordinary sense is known as **mutation** and the offsprings with unusual heritable variability in characters are called **mutants***. A *mutant* individual or cell is one in which the changed phenotype is attributable to mutated gene or genes.

Many other definitions for mutation have been proposed from time to time by different biologists. Some are as under:

1. Sudden appearance of new hereditary character in the progenies of plants and animals was referred to as 'sport' or mutation by Darwin.
2. In the broad sense, it covers "any heritable change in the genotype" (Macromutation). In the narrower sense, "mutation is a change of gene" (micromutation) (Sinnot and others).
3. According to Bateson, "mutation is discontinuous variation".
4. According to De Vries, "mutations are sudden and drastic heritable changes not traceable or ascribable to segregation or recombination".
5. Stebbins describes mutation as discontinuous chromosomal change with genetic effect. He further states that the chromosomal change refers to chemical change in a small part of its chromomere, as well as to alteration of its physical structure.
6. Amatto and Otto (1956) have defined mutation as a change in the heredity constitution of a given species.

Fig. 22.1 Hugo DeVries.

Historical

Little was known about mutation before 19th century. It was Darwin who first noticed several sudden changes in the organisms in nature. He called those changes as 'sports'. Several years after Darwin, Bateson marked that some variations were not continuous. They were called *discontinuous variations*. Hugo DeVries (Fig. 22.1), one of the three persons who rediscovered Mendel's laws of inheritance, observed in 1901 sudden changes in *Oenothera lamarkiana.* In *Oenothera lamarkiana,* gigas (large size), nanella (dwarf) and many other unusual changes, such as, changes in the colour and shape of flowers were marked by him. The new forms, he observed, differed in appearance from the normal forms and apparently they arose in small numbers in each generation. On the basis of those observations he proposed a general theory of species formation by means of sudden discontinuous changes. DeVries called such sudden changes as **mutations**. His observations were published in a book entitled *The Mutation Theory.* Simultaneously several unique cases of mutation had been reported in nature. Ancon sheep (Fig. 22.2), one of the most interesting cases of mutation was first noticed in England. The ancon sheep had a short leg (an unusual feature). That type appeared suddenly in the flock of sheep, existed for several years and then disappeared. Again after eighty years, they reappeared in Norway and since then they are existing in some European countries. Besides these, many other remarkable cases of mutation have been noticed in both plants and animals. In 1904, Morgan reported white eyed *Drosophila melanogaster* in the population of red eye flies.

Fig. 22.2 Ancon sheep.

The sudden occurrence of Haemophilia disease in the royal family (Queen Victoria's family) of England deserves special mention here. Queen Victoria came from a family in which Haemophilia was not known, but suddenly the disease appeared in some of her sons and subsequently in several sons of her daughters. Why did it appear ? Perhaps, because something went wrong with the genetic machinery of one of the parents or apparently mutation to Haemophilia took place in germ cells of one of her parents so that she herself was already heterozygous for newly mutated character.

Characteristics of Mutation

Originally the geneticists were of the opinion that mutations were spontaneous and random in effect, *i.e.,* they were occasionally noticed but this statement seems to be confusive. Mutations may occur in any cell at any stage of the life of an organism. Mutations occurring in reproductive cells (sperms and ova) are referred to as *germinal mutations.* In sexually reproducing species only the mutations arising in germ cells are transmitted to the future generation. Mutations arising in body cells other than germ cells are called *somatic mutations*. Somatic mutations, as for example, the fatal cancer of blood or the leukaemia, chronic myeloid, etc. arising in the body cells are not inherited as they affect only the mutant individuals and are not transmitted to future generations. Asexually propagated species in which somatic mutations occur may become established as mutant strains. Many new varieties of fruit trees have originated as *"bud mutations"*. In vegetatively propagated mutant plants, the gamete will be either normal or mutant depending on which cell forms germinal tissue.

A mutation may be *dominant* or *recessive, viable* or *lethal, sex-linked* or *autosomal.* In diploid species, recessive mutation can produce bud mutation or mosaics only when the individual is heterozygous for the gene in question. Mosaic in snapdragon flower, which is half purple and half lavender, sometimes develops on a branch which, in addition to one or more such mosaic flowers, also bears several wholly purple or wholly lavender colour flowers. The plant itself was heterozygous for the recessive lavender genes. In some of the cells from which mosaic bud developed the normal allele has mutated to lavender. Dominant mutations may give rise to immediately observable bud, spores or mosaics.

Mutations are detected when some heritable changes occur in the characters of an organism. Since the characters are governed by genes, any change in them directly reflects the change in the genes (***gene mutation***). So the relationship between a gene and a character is apparent only due to mutation. A gene mutation is also called ***point mutation***. Sometimes, the mutations in genes do not result in any perceptible change in characters. Nevertheless, such changes are important from evolution point of view because they go on accumulating and have a cumulative effect on the phenotype. Normally, the original activity of a gene is lost due to mutation and this is the reason why original or wild type genes are dominant and mutant genes are recessive. In diploid organisms, for each character there are two genes located on two homologous chromosomes. As the mutation is a random process, the chance of simultaneous mutations in both the genes controlling a particular trait are very remote. If the mutation takes place in one of the two genes governing a particular trait, it results in heterozygosity because the other gene remains unaffected. In such a case the mutation may not be immediately expressed because of the presence of the wild or unmutated genes. It may be expressed in later generations in a small number of individuals which are homozygous for mutant gene.

Recessive mutations perpetuate and are maintained in heterozygous state. In haploids, each character is usually governed by a single gene. So every mutation is expressed in them. Undesirable or unfit mutations are eliminated from the natural populations.

Genetic change is important process that generates hereditary variants in natural populations of microbes, plants and animals. In any consideration of ***Change,*** a fixed reference point or ***Standard*** is necessary. In genetics, that fixed reference point or standard is provided by the so called ***wild type*** (*i.e.*, the normal phenotype in wild population of a species). The non-wild type gene is called mutant gene. The mutated gene differs from original one in chemical composition or structure. The mutated gene like the original one must be capable of accurate self-replication in mutated form. Any change away from the standard form is called *forward mutation.* Sometimes mutant genes change towards the standard form. This is called ***reverse mutation, reversion*** or ***back mutation.***

For example:

a^+ (wild type recessive gene) → a (mutant gene)
D^+ (wild type dominant gene) → D (mutant gene)
} Forward mutation

a (mutant gene) → a^+ (wild type gene)
D (mutant gene) → D^+ (wild type gene)
} Reverse or Back mutation

These gene states are arbitrary, *i.e.*, wild type of today may have been mutant gene in the past and *vice-versa*. Mutant is an adjective term and so it should precede a noun such as 'mutant individual' or mutant cell). Even if sometime the noun is left unstated, in such a case mutant always means an individual or cell whose phenotype shows that it bears mutation.

The truly mutated genes do not show reversion to the original gene. However, there are some rare exceptional cases in which reversion may be observed. Each individual gene mutates only very rarely, perhaps once in 1,00,000 or once in a million cells; but as the number of genes in the majority of the organisms is very high, the overall mutation frequency per generation may be considerable.

CLASSIFICATION OF MUTATIONS

Mutations can be classified on several basis. Some of the basis of classification are as follows :

1. Dominance Relationship

Mutations may be classified on the basis of dominance relationship of mutant alleles in to the following types.

(*i*) ***Dominant mutation*** which results into dominant alleles.

(*ii*) ***Recessive mutation*** which results often in recessive alleles.

(*iii*) ***Codominant mutation*** which results occasionally in codominant alleles (e.g. blood group antigens).

(*iv*) Incomplete dominant mutation which results in wild type alleles.

2. The cause of Mutation

Depending on the causes of their origin, mutations may be classified into the following two types :

(*i*) ***Spontaneous mutations.*** These mutations arise spontaneously without any known cause. Spontaneous mutations occur in individuals at quite low but constant rate which ranges between 10^{-7} and 10^{-4}. It may vary from gene to gene. Some genes show exceptionally high rates of spontaneous mutations and are called mutable genes.

(*ii*) ***Induced mutations.*** Induced mutations are produced when an organism is treated with some physical or chemical agents called mutagens. Such mutations are recovered at much higher frequencies than spontaneous mutations.

3. Tissues of origin

Mutations may be classified into (*i*) Somatic mutations and (*ii*) Germinal mutations on the basis of tissues in which they occur as follows :

(*i*) ***Somatic mutations***: The mutations occurring in somatic tissue, *i.e.*, body cells that are not meant for producing gametes are called *Somatic mutations.* Somatic mutations may lead to a sector of identical cells that can be recognised in the background of normal cells. Such mutations can perpetuate in vegetatively propagated plants. (Fig...) when somatic mutation occurs in the vegetative bud, it is termed *bud mutation.*

(*ii*) ***Germinal mutation.*** Germinal mutations occur in the reproductive cells that are destined to produce gametes. These mutations are transmitted to the next generation only through reproductive units. These mutations can be detected only if they appear in the zygote. Such a situation is thought to have occurred in Queen Victoria who carried a germinal mutation in her one X chromosome (sex chromosome) for haemophilia disease. This mutation was transmitted in some of her male descendents. At operational level the detection of germinal mutations depends on the ability to rule out meiotic segregation and recombination as possible causes. True mutations both germinal and somatic, never arise because of segregation and recombination of genes.

4. Effects on Survival

Mutations are classified into the following groups on the basis of their drastic effects on survival of organisms :

(*i*) Lethal mutations, (*ii*) Sublethal mutations or subvital mutations, (*iii*) Vital mutations, and (*iv*) Super vital mutations. Out of these only vital and supervital mutations are useful for study purpose. **Lethal mutation** is recognised through its lethal effect on the organisms. Sometimes the primary cause of death is easy to identity, *i.e.*, certain blood abnormalities, but often the lethal gene is recognisable only by its mortality effect.

Subvital mutations reduce the chances of survival of organisms in which they occur. ***Supervital mutations*** may result in improvement of biological fitness and vigour under certain conditions. Besides, some mutations are neither harmful nor beneficial to the organisms.

5. Environmental influence

The phenotypic consequences of mutation may be so subtle as to require refined biochemical techniques to differentiate mutation from wild type or sometimes so severe as to produce gross morphological defects or death. Based on the method of detection, the following rough classification can be followed :

(*i*) ***Morphological Mutations.*** In this class of mutations, the changes occur in visible properties of an organism, such as changes in shape, colour or size of individual Albino ascospores in *Neurospora*, curly wings in *Drosophila* and *Dwarf* peas are all morphological mutations.

(*ii*) ***Conditional Lethal Mutations.*** In this class of mutations, a mutant allele expresses mutant phenotype under certain specific conditions (called restrictive condition) but expresses a normal phenotype under other conditions (called permissive condition). For example, a certain class of mutations in *Drosophila* is known as dominant heat sensitive lethal. Heterozygotes in this class (say H^+ / H) are normal at 20°C (permissive condition) but die if the temperature raised to 30°C (restrictive condition). Many mutant organisms are less vigorous than normal forms. For this reason, conditional mutants are handy in that many of them can be grown in permissive conditions and then shifted to restrictive conditions for study.

(*iii*) ***Biochemical Mutations***. This class of mutations is identified by the loss of or change in some biochemical functions of the cell and consequently the mutants are unable to grow and proliferate. However the growth of mutant can be restored by supplementing the growth medium with specific nutrients. Biochemical mutants have been detected in microbes. Microorganisms by and large are prototrophic as they are nutritionally self sufficient and can grow on minimal culture medium containing simple in organic salts and soluble organic nutrients. They are called ***Biochemical mutants*** or ***Resistant Mutations.*** When mutant cells or organisms acquire the ability to grow in presence of some specific inhibitor such as cycloheximide or a pathogens to which wild types are susceptible, these are called ***resistant mutations***. Such mutants have been extensively used in production programmes because they are relatively easy to select and maintain.

6. Quantum of Morphological Effects

Depending on the quantum of phenotypic change caused by them the mutations may be classified into:

(*i*) ***Macromutations.*** Macromutations produce large enough changes in the phenotypes which may be detected without any confusion due to environmental effect e.g. short legged mutant in sheep.

(*ii*) ***Micromutations.*** These mutations produce small morphological effect which can be easily confused with the effects produced by environmental factors. They can be detected and isolated by special technique, e.g., mutations in quantitative characters governed by polygenes.

7. Cytogenetical Basis

Mutations have been classified on the basis of cytogenetics into the following types :

(*i*) Gene mutations or point mutations

(*ii*) Chromosomal mutation

(*iii*) Cytoplasmic mutations.

Darlington and Mather have classified mutations into the following five types :

1. Gene mutations
2. Structural mutations or mutations due to change in chromosome structure (chromosomal aberration)
3. Numerical mutations or mutations due to change in genome or basic chromosome number (Phenomenon of Ploidy)
4. Cytoplasmic mutation (mutations in cytoplasm due to change in cell organelles like plastids and mitochondria.

Out of the above mutational classes, numerical or genomic mutations, cytoplasmic mutations are described in separate chapters. Here only structural mutations and gene mutations will be discussed in detail.

CHROMOSOMAL MUTATION OR CHROMOSOMAL ABERRATIONS

Mutations involving either entire chromosome or large portions of chromosomes are referred to as *gross* or ***chromosomal* mutations**. Generally, there are two categories of chromosomal mutations:

(*i*) mutations involving changes in the chromosomal architecture or gross morphology of chromosomes, ***the chromosomal aberrations***.

(*ii*) mutations which result from the changes in the number of chromosomes, the **phenomenon** of ***ploidy*** or ***numerical aberration*** of *chromosomes*.

The phenomenon of ploidy is discussed separately. Though gene mutations are related directly with chromosomal mutations as the genes are parts of chromosomes yet it is possible that chromosomes may become reorganised without involving change in the nature of genes. This type of reorganisation is called chromosomal aberrations or structural abnormality of chromosomes.

The chromosomes generally remain unchanged but under certain natural or artificial adverse circumstances, certain changes may occur in the chromosome structure which may result in loss or gain of genes or may change the positions of genes in the chromosomes.

Structural changes in the chromosomes take place during meiosis. During the pachytene stage of first meiosis, homologous chromosomes replicate and in diplotene stage chiasmata develop. In the latter stage, non-sister chromatids may become broken into pieces. These chromosomal fragments then rejoin in a sequence which may be quite different from the original one. The breaking and reunion of chromatid segments may thus result in a number of abnormalities in the chromosome's structure. These are known as ***chromosomal aberrations***. Chromosomal aberrations refer to changes in the parts of individual chromosomes rather than whole chromosomes or set of chromosomes in a genome. In this process of breaking and reunion, sometimes acentric chromosomes (*i.e.,* chromosomes without centromeres) are formed due to union of two or more chromosomal parts lacking in centromere. Acentric chromosomes are incapable of proper mitosis and they ultimately disappear. Sometimes dicentric chromosomes (*i.e.,* chromosomes with two centromeres) may also develop, but like acentric chromosomes they also do not persist for long. There are a few exceptional cases in which dicentric chromosomes may persist. Chromosomes with single centromere behave normally in mitosis and persist. Any change in the structure of chromosomes may result change in the sequences of genes. This consequently brings about variation or mutation in the characters. There are four types of chromosomal aberrations:

(*a*) Deficiency or Deletion,

(*b*) Duplication,

(*c*) Translocation, and

(*d*) Inversion.

Deficiency or Deletion

Chromosomes contain a number of genes on them. The genes are arranged in linear fashion. The number and also the positions of genes on a chromosome are fixed. When a chromosome is broken into several pieces, the healing (reunion) of the segments takes place and it is possible that the two ends of the fragments unite together leaving one or more acentric parts free (Fig. 22.3). The acentric pieces of chromosomes as mentioned earlier disappear. The healed segments will form a chromosome that will be deficient for some of the genes, particularly those found on the lost part. Depending upon the length of chromosome segment lost in this way, the loss involves one gene or a block of genes. The loss of a section of genetic material and genetic information from a chromosome or linkage structure is termed *deficiency* or *deletion* (Painter and Muller, 1929). Deletions may be produced in several ways, such as by loss of *terminal acentric fragments* or *interstitial segment* of chromosomes (Fig. 22.3).

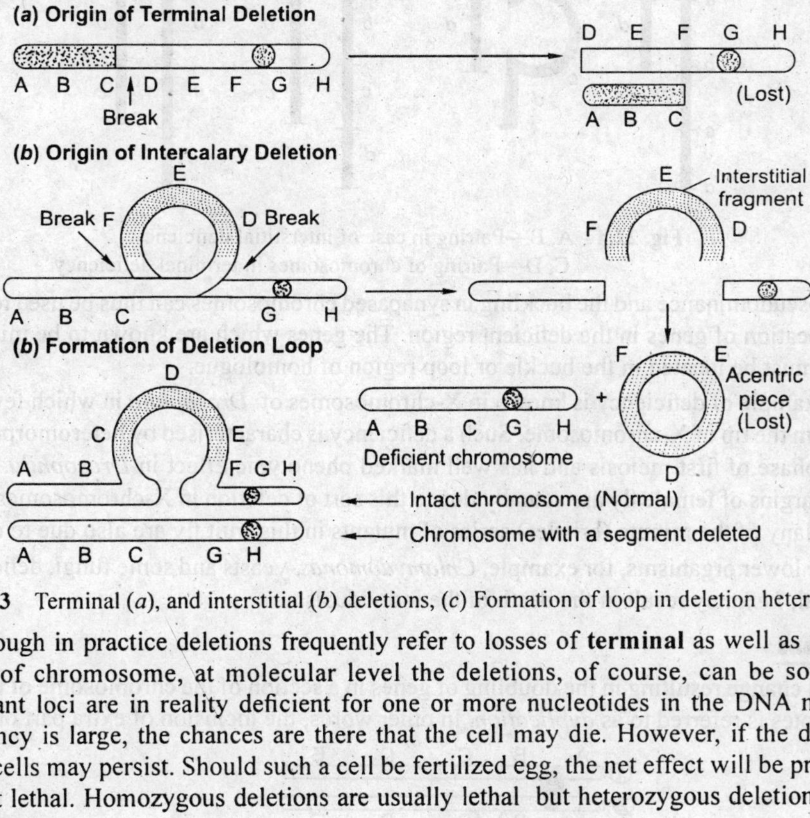

Fig. 22.3 Terminal (*a*), and interstitial (*b*) deletions, (*c*) Formation of loop in deletion heterozygote.

Although in practice deletions frequently refer to losses of **terminal** as well as **intercalary segments** of chromosome, at molecular level the deletions, of course, can be so small that many mutant loci are in reality deficient for one or more nucleotides in the DNA molecule. If the deficiency is large, the chances are there that the cell may die. However, if the deficiency is small, the cells may persist. Should such a cell be fertilized egg, the net effect will be production of a dominant lethal. Homozygous deletions are usually lethal but heterozygous deletions appear as normal mutations.

Deficiency can be detected by its two characteristics, namely, *genetic effects* and *cytological effects.*

Genetic effects of deficiency. These are primarily due to the loss of genetic information and secondarily due to quantitative changes in the genotype as well as the change of genic balance. Deficiencies have been useful in determining the exact locations of genes on the chromosomes. One of the genetic effects of deficiency is **pseudo-dominance.** Pseudodominance occurs when the chromosome that carries the deletion of dominant allele pairs with a normal homologous chromosome carrying the recessive allele. In absence of dominant alleles the recessive alleles would be expressed in phenotype as if they were dominant. This is called **pseudo-dominance.** If some of the corresponding genes have tendency of lethal effect in double dose, they will be lethal in single dose (recessive lethality). The chromosomes with deletion never returns to a normal state.

Cytological effects. The existence of relatively large deletions in the chromosome complements of eukaryotes may be proved cytologically (*i*) by the occurrence of centric and acentric chromosome fragments in mitosis, and (*ii*) by the absence of regional pairing during first meiotic prophase. If the cell is heterozygous for deletion, *i.e.,* it has one normal chromosome and a deficient homologue, then during synapsis, the chromosomes pair precisely gene by gene all along the homologous region and in deficient region, however, the normal chromosome will not pair. Therefore, a **buckling** (*loop*) will develop in the normal homologue at the point of intercalary deficiency (Fig. 22.4). It is, therefore, known as **buckling effect**. Such a configuration is called **deficiency loop** or **compensation loop** and can be observed under microscope.

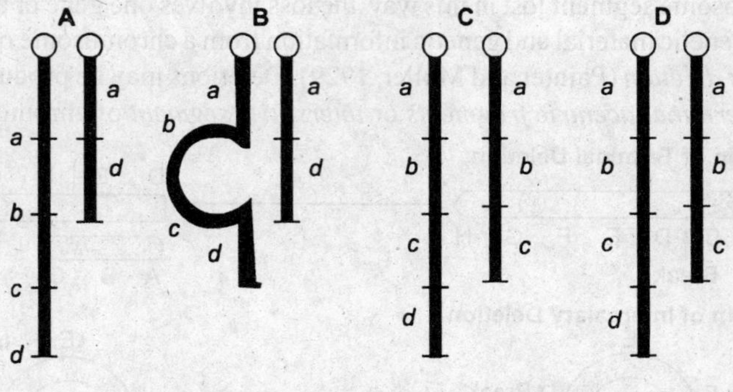

Fig. 22.4 A, B—Pairing in case of interstitial deficiency.
C, D—Pairing of chromosomes in terminal deficiency.

The pseudominance and the buckling in synapased chromosomes can thus be used to determine the exact location of genes in the deficient region. The genes which are known to be missing in the deficiency must be located in the buckle or loop region of homologue.

An example of deficiency is known in X-chromosomes of *Drosophila* in which few bands are missing from the tip of X-chromosome. Such a deficiency is characterised by heteromorphic bivalent during prophase of first meiosis and has well marked phenotypic effect in *Drosophila.* Notches in the wing margins of female fly are actually due to this sort of deletion in X-chromosomes. It is lethal in males. Many of the minute (bristles) series of mutants in this fruit fly are also due to deficiency.

In the lower organisms, for example, *Chlamydomonas,* yeasts and some fungi, deficiencies are totally lethal, *i.e.,* they result in the death of the individuals.

Duplications

A structural change resulting in the doubling of genes in a section of the chromosome of prokaryotes and eukaryotes is referred to as *duplication.* In other words, the inclusion of extra part or duplicated

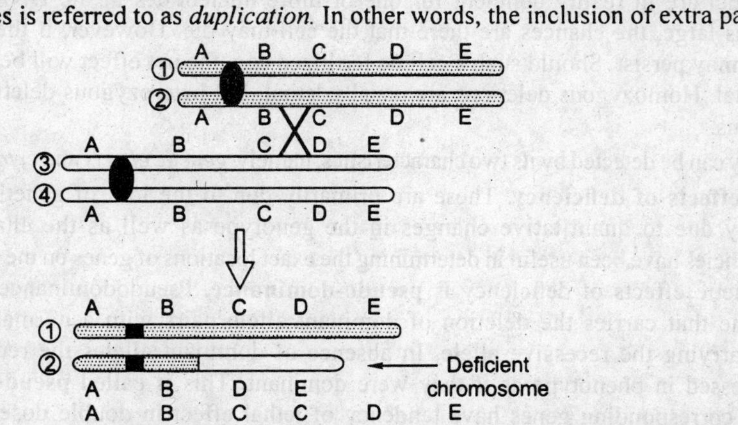

Fig. 22.5 Production of deletion and duplication in chromosomes due to unequal crossing over.

gene sequence of a chromosome beyond the normal complement is called *duplication.* If a segment of one chromosome is incorporated in another homologous chromosome, it is called *intrachromosomal* duplication (Fig. 22.5), but if the duplicated chromosome segment is either incorporated into a non-homologous chromosome or occurs as a fragment in the chromosomal set it is called **interchromosomal** duplication (Fig. 22.5). The duplicated segments contained within a single chromosome could exist in one of the following configurations depending upon the position and sequence of the duplicated genes.

(*i*) *Direct tandem duplication* in which the duplicated gene sequence lies just next to normal corresponding section and the sequence of genes with respect to centromere is the same in duplicated segment as in the normal section of the chromosome. It will be clear from the following gene sequences in the normal and duplicated chromosomes:

a b c d . e f g h i j — Normal
a b c d . e f e f g h i j — Direct tandem duplication
(→ →)

(*ii*) *Reverse tandem duplicaton* in which the duplicated section with reverse gene sequence lies adjacent to normal sequence as shown below:

a b c d . e f g h i j — Normal
a b c d . e f f e g h i j — Reversed tandem duplication
(→ ←)

(*iii*) *Displaced direct duplication* in which the duplicated section is not adjacent or contiguous with the normal section (*i.e.,* separated by other segment). The duplicated and normal gene sequences may be in the same arm (*intra arm*) or in two different arms of the chromosome (*inter arm duplication*):

a b c d . e f g h i j — Normal
a b c d . e f g h e f i j — displaced direct duplication
(→ →)

(*iv*) *Displaced Reverse duplication* in which the duplicated section with reverse gene sequence is separated from normal segment by other segment as shown in the following chromosomes :

a b c d e f g h i j — Normal
a b c d e f g h f e i j — displaced reverse duplication
(→ ←)

(*v*) *Transposed duplication* in which the duplicated gene sequence is attached to another position owing to interchromosomal duplication.

The size of duplicated segment may vary considerably. The smallest duplications that can be investigated cytologically are those of single band of polytene chromosomes. However, the whole arm of the chromosomes may be duplicated, thus giving rise to isochromosome. Where all linkage groups or chromosomes in the haploid chromosome set are doubled, this is referred to as genome mutation.

Duplication may arise in several ways. One of the most common methods is unequal crossing over, a process which produces one chromosome with duplication and another with deficiency. This is primary structural change of chromosome. The schematic representation of gene duplication in chromosome by unequal crossing over is shown in Fig. 22.6.

Duplications may also occur due to crossing over in inverted or translocated segments. (*Secondary structural mutation of chromosome*)

Duplications in general, are much more viable than deficiency. A heterozygous duplication has an appearance similar to deletion. Duplications sometimes appear as dominant mutations. By duplication an allele may be duplicated, triplicated or multiplicated, hence duplications may be utilized in studying the effect of various quantitative doses of members of an allelic set.

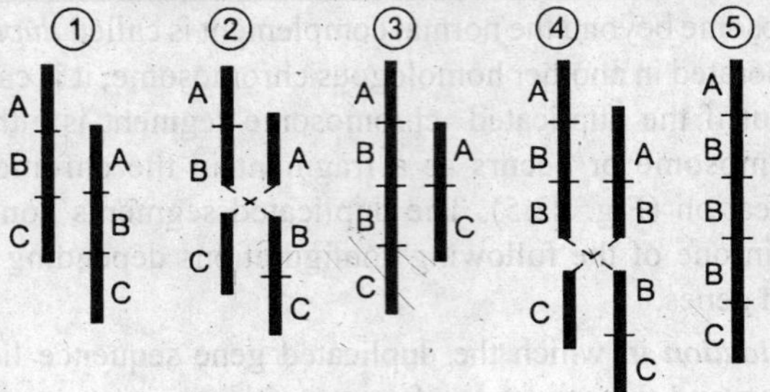

Fig. 22.6 Unequal crossing over in the homologous chromosomes results in the duplication of certain genes. In 3, gene B is duplicate and in 5, gene B is repeated three times.

Well known example of duplication which had a significant impact on genic theory is Bar-eye mutation in *Drosophila.* With the discovery of chromosomal nature of this case, it was found that extra pieces of chromosome were associated with a normal X-chromosome duplicating and triplicating a section of it (Fig. 22.7). The Bar-eye, a sex-linked incompletely dominant mutation responsible for the development of rod-shaped eye with reduced eye facets, appeared spontaneously in a wild type stock of *Drosophila melanogaster* with round eyes. Homozygous stock of Bar-eyed mutants produced flies with normal eye and flies with even more reduced eye (Double Bar) in approximately equal frequency. These observations suggested that Bar locus was very unstable but the appearance of wild type and double bar flies in equal numbers could not easily be explained. Moreover, the two mutant phenotypes appeared in the progeny when Bar-females were crossed with normal males, and not in the progeny of Bar-males crossed with normal females. *Drosophila* is unusual in the sense that meiotic crossing over occurs only in females and not in the males. This observation suggests that mutation was not responsible for the wild type and double Bar progeny and that Bar might somehow be related to meiotic crossing over.

In 1925, Sturtevant using females homozygous for Bar (B) but heterozygous for forked bristles (gene f) and for fused wing veins (gene fu) demonstrated that the normal and double Bar flies were products of crossing over within the Bar locus. Sturtevant found that some normal and some double Bar flies were recombinants, each with one or the other flanking marker ('f' or 'fu').

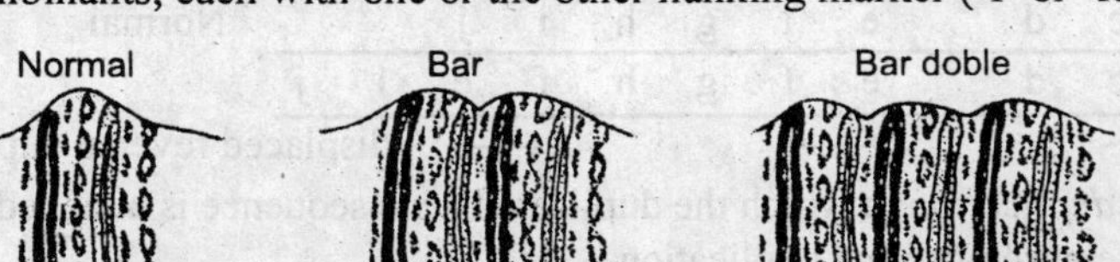

Fig. 22.7 Chromosomes of *Drosophila* showing duplication of gene.

Sturtevant hypothesis of unequal crossing over was supported by Bridges in 1936. Bridges studied the chromosomes of wild type Bar eyed and Double Bar eyed flies and noticed the presence of single 16Å segment in normal eyed, two 16Å segments in tandem in Bar eyed and three continuous 16Å segments in double Bar eyed (Fig. 22.7). By unequal crossing over in homozygous Bar females one 16Å segment is transferred to a chromatid with duplication and consequently a chromosome with only a single 16Å segment (normal type) and another with three 16Å segments (Double Bar) are recovered. Duplication and reduplications of Bar region can be easily seen in the salivary gland chromosomes (Fig. 22.7). The chromosomes of *Drosophila* salivary gland are large and easily distinguishable. They replicate several times but do not divide. So, each chromosome appears like a 'rope' of about 1,000 tightly coiled chromatids—a *polytene chromosome.* The genes for Bar eye and several other characters have been, in fact, cytologically pinpointed to specific bands or chromomeres.

In corn and peas, a number of duplicating factors are known. Duplications like deletions may be so small as to be molecular. Since they change the genic balance, they may produce abnormalities in body characters. Duplications are considered to play a role in origin of new genes through functional diversification of duplicated members.

Like deficiencies, duplications have diagnostic cytological and genetical effects.

Cytological Effects. The existence of relatively small duplications in the chromosome complements of eukaryotes may be proved by the appearance of regional distortions of duplicated chromosome during pairing in first meiotic prophase or somatic pairing in specialized tissue like salivary glands of Diptera.

Crossing over in reverse tandem duplication may result into a dicentric chromosome. This can be frequently observed in maize. As the two centromeres move to opposite poles, a chromosome bridge is formed which later on breaks at any point along the bridge. The broken ends are sticky and the replication of the broken pieces may result in two sister chromatids which may be joined together due to their sticky ends. Thus, whole chromosome arm may be duplicated giving rise to *isochromosomes*. In sporophytic tissues of plants, the isochromosomes are uncommon.

Genetic Effects. The genetic effects of duplications depend on the genetic information the duplicated segments contain and the change in genic balance effected by them. In homo and heterozygous state they may cause an increase or decrease in the viability of their carriers and in extreme cases may act as lethals.

Since the duplications supply the additional genetic material and change the genic balance, they play important role in evolution at individual and population levels. Under evolutionary conditions, small duplications may provide a basis for the mutational differentiation of genetic material and the different copies of the same gene may change in different directions without disturbing the normal functions of an organism. The cases in which different gene pairs affect the same character (as for example, multiple factors, complementary factors) possibly arose initially after duplication of single genes. The repetition of DNA sequences frequently seen in highly evolved organisms is a direct indication to this.

Translocation

Chromosomes may break into two or more fragments, each with a raw end. These segments may reunite and during reunion either the pieces of the same chromosome or the pieces of the non-homologous chromosomes may be fused. Ionising radiations such as X-rays and gamma-rays are frequently used to break chromosomes for producing structural changes. In this process, a part of the chromosome is transferred to another non-homologous chromosome within the chromosome complement. In other words, a **translocation** is a chromosomal rearrangement which involves (*i*) the unidirectional transfer of a chromosome segment and its gene sequence to a different chromosome within the chromosome complement, and (*ii*) the exchange of segments between non-homologous chromosomes. In this way, translocation only results into a change in the sequence and position of genes, their quantity being unaffected. There are two main types of translocations:

1. Simple Non-reciprocal Translocations or Transpositions

In this process, a piece of one chromosome is transferred to a non-homologous chromosome. It can be of two types:

(*i*) **Simple translocation** or **Terminal transposition.**

(*ii*) **Shift translocation** or **Interstitial transposition.**

(*i*) **Simple Translocation.** When a chromosome breaks into two parts due to external or internal stress, one of the two segments of the broken chromosome may become attached to the natural end of the nearest chromosome which may not be its homologue. This is simple translocation. Suppose, there are two non-homologous chromosomes A B C D E F G and T U V W X Y Z. If F G part of the first chromosome is transferred to second chromosome, a new chromosome T U V W X Y Z F G would result as shown in Fig. 22.8. In this non-standard or translocation chromosome, F G is said to be in simple translocation state.

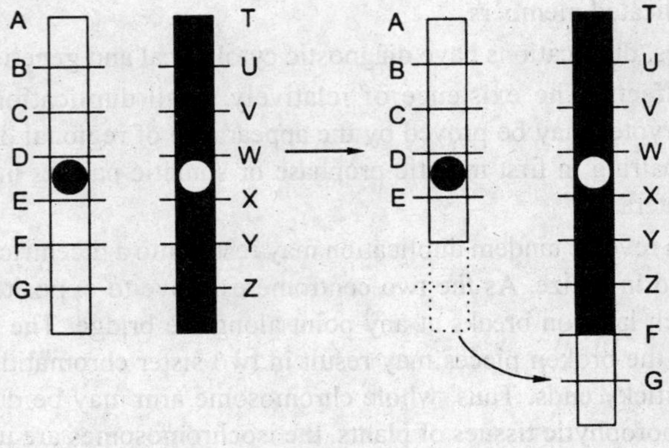

Fig. 22.8 Simple translocation.

In the normal course, the terminal transpositions are not common because the raw ends or the telomeres of unbroken chromosomes are not sticky.

(*ii*) **Intrachromosomal, Shift Translocation or Interstitial Transposition.** In this, an interior or interstitial segment of a chromosome that is induced by two breaks is incorporated interstitially into another non-homologous broken chromosome, the latter being induced by a single break. It is also called *intercalation* or *insertion*. If the sequence of genes in the translocated segment is the same as that in the original segment with respect to centromere, it is referred to as *enucentric translocation.* But if the sequence of gene loci in translocated chromosome segment is reversed it is called *dyscentric translocation.*.

2. **Reciprocal Translocation.** When a break occurs at a point where two non-homologous chromosomes touch each other, the broken end of one chromosome may become united with the broken end of second chromosome and that of second chromosome becomes attached to that of first, this is *reciprocal translocation.* Suppose, there are two chromosomes A B C D E F and G H I J K L. These two after reciprocal translocation may produce chromosomes A B C J K L and G H I D E F as shown in Fig. 22.9. The reciprocal translocation is the most common type of translocation.

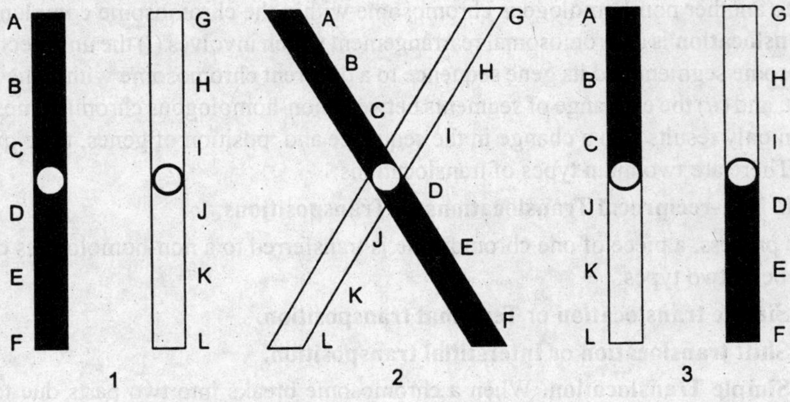

Fig. 22.9 Reciprocal translocation.

If there occurs two breaks in each of the two non-homologous chromosomes, the reciprocal translocation of intercalary segments may be obtained, but it is very rare.

The reciprocal translocation is like crossing over except that it involves exchange between the segments of two non-homologous chromosomes. It is sometimes called "illegitimate crossing over". The origin of translocations is interpreted either according to the breakage-reunion or the exchange model. The unit of translocation may be the chromosome (***chromosome translocation***), the single chromatid (***chromatid translocation***) or a segment of chromatid. The sites of translocations are called *translocation* points. The reciprocal translocation may be either **asymmetrical (*aneucentric*)** or **symmetrical (*eucentric*)**. The asymmetrical translocation gives rise to one dicentric and one acentric chromosomes (Fig. 22.10) and it may lead to the formation of chromosome bridge if the two centromeres of a dicentric translocation product are distributed to opposite poles of spindle during anaphase. In symmetrical translocation, however, the products are *monocentric*.

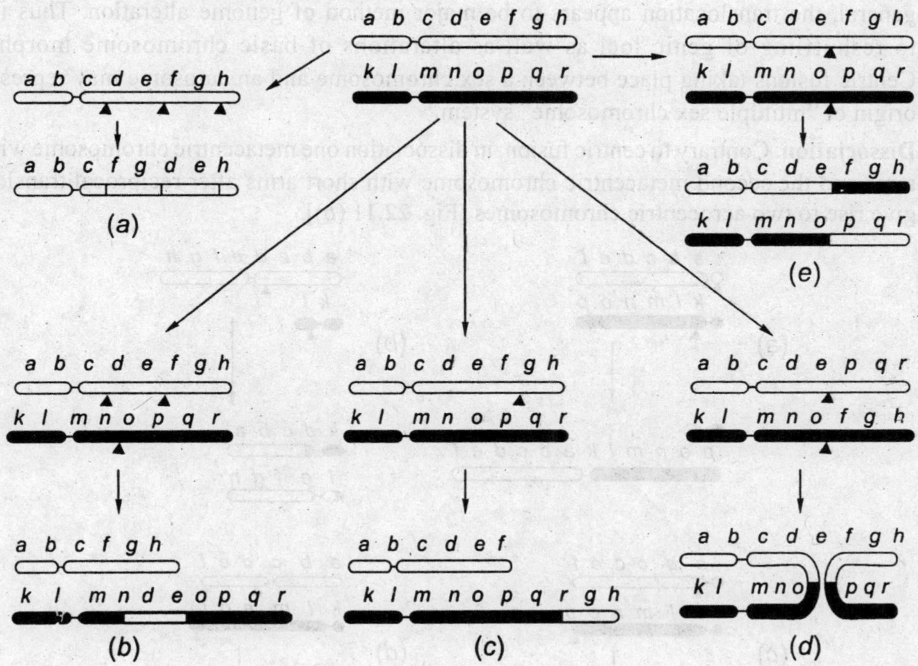

Fig. 22.10 Diagram showing several types of intrachromosomal and interchromosomal translocations. (*a*) intrachromosomal transposition ("shift") of a chromosome segment; (*b*) interchromosomal transposition of a chromosome segment; the sequence d–e is transferred to a heterologous chromosome; (*c*) "terminal translocation" of the segment g–h to a heterologous chromosome; (*d*) reciprocal (asymmetric) translocation (interchange) of segments f–h and p–r; (*e*) reciprocal (symmetric) translocation of segments f–h and p–r; in contrast to an asymmetric interchange the translocation products are monocentric.

Sometime, whole or nearly whole arms of the chromosomes may be transposed or interchanged. This is termed *whole-arm translocation* (Muller, 1940). There are three special cases of whole arm translocation:

(*a*) Centric fusion,

(*b*) Dissociation, and

(*c*) Tandem fusions.

(*a*) **Centric fusion.** Translocation occurring between two sub-telocentric chromosomes, each with sub-terminal centromere and single long arm, may exchange parts in such a way that major part of long arm of one is translocated to the short arm of the other producing one V-shaped long metacentric chromosome (*i.e.*, with median centromere) and one small chromosome with two minute arms. Long V-shaped chromosome has all the vital genetic material and the small chromosome is little more than a centromere *plus* some chromatic material. This type of translocation is called *"centric fusion"* by White (1954) [Fig. 22.11 (*a*)]. Such a translocation may be quite viable and in a heterozygote, one long V-shaped chromosome may synapse normally with two normal unfused one armed homologous chromosomes. The small chromosome is more or less functionless and appears like a supernumerary one which may be lost from any individual but maintained in the population. These supernumerary chromosomes are spare kinetochores. The reverse translocation in them may lead to increase in basic chromosome number. It seems a very likely mechanism for the alteration in the chromosome number, although there is little direct evidence for it. In general, the translocation appears to be major method of genome alteration. Thus it leads to reshuffling of genic loci as well as alterations of basic chromosome morphology. Centric fusions taking place between a sex chromosome and an autosome may represent the origin of "multiple sex chromosome" system.

(*b*) **Dissociation.** Contrary to centric fusion, in dissociation one metacentric chromosome with long arms and the second metacentric chromosome with short arms after reciprocal translocation give rise to two acrocentric chromosomes [Fig. 22.11 (*b*)].

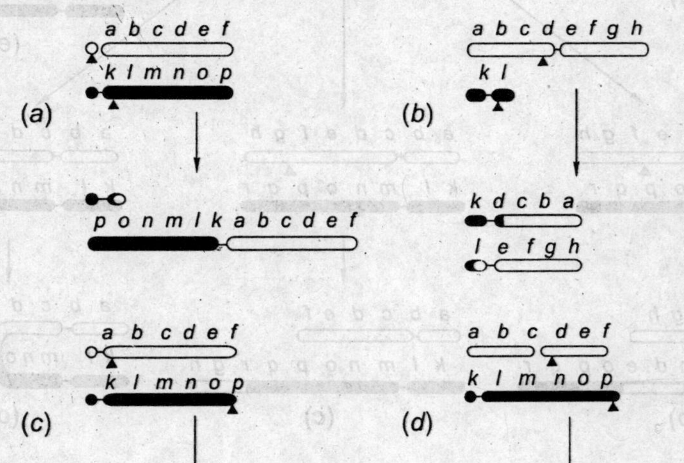

Fig. 22.11 Special types of translocation. (*a*) centric fusion; (*b*) dissociation; (*c*) tandem fusion between two acrocentric chromosomes; (*d*) tandem fusion between one acrocentric and one metacentric chromosome.

(*c*) **Tandem fusions.** This results from a single break in the vicinity of centromere of one chromosome and another break near the end of a second chromosome. If the two chromosomes involved in this process are acrocentrics, then one large acrocentric and one small metacentric chromosomes will result (Figs. 22.11 (*c*), 22.12). But if the interchange takes place between one metacentric and the other acrocentric chromosome, then two acrocentrics, one short and the other large may be formed [Fig. 22.11 (*d*)].

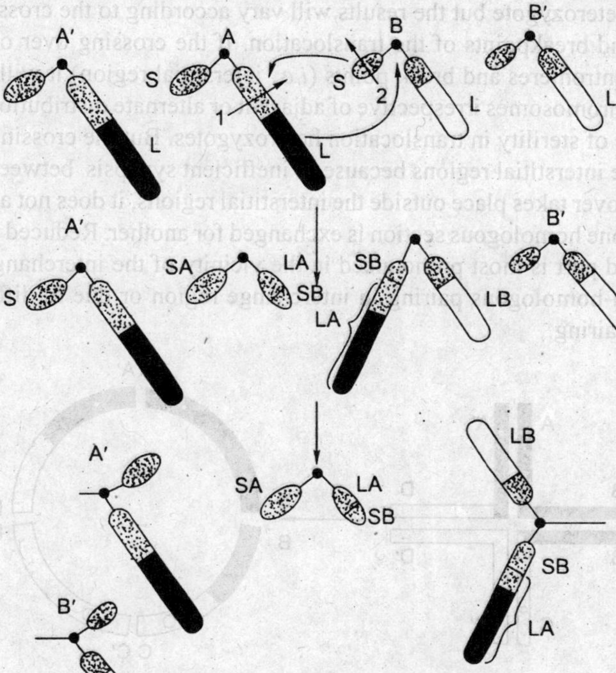

Fig. 22.12 Reciprocal translocation involving two chromosomes each with sub-terminal centromere and a single long arm results one 'V' shaped long armed metacentric chromosome and one small armed chromosome.

Translocations are usually non-lethal in their effects. Like deficiency and duplication, the translocations may also be homozygous or heterozygous provided the aberrations in question are not associated with lethality. The homozygous translocations are characterised by the presence of same gene sequences in the transposed segments of homologous chromosomes. If an organism carrying two pairs of chromosomes with reciprocal translocation is crossed with a normal organism, the offspring will be heterozygous for translocation. Translocation heterozygotes thus possess translocated and normal chromosomes. The following are the important effects or evidences of translocations:

1. *Cytological effects*
2. *Genetical effects*

1. Cytological Effects of Translocations. The chromosomes of homozygous translocations generally behave as do the normal ones from which they arise, except that new linkage groups are formed. If they persist they can give rise to new chromosomal races in the population.

In the individuals heterozygous for a symmetrical reciprocal translocation (structural hybrids) two chromosomes with translocation and two normal chromosomes share a partial homology, but no two are identical. Since synapsis is a matter of homologous regions (gene to gene pairing) and these regions are distributed over four chromosomes, in simple reciprocal translocation heterozygote, the association of all four chromosomes will be formed. Such an association will result in a cross configuration (+) at pachytene i.e., there will develop a group of four associated chromosomes (2 normal + 2 translocated) (Fig. 22.14).

Heterozygosity for translocation reduces the crossing over frequency. The crossing over can take place in any of the four pairing segments of cross-like configuration in reciprocal translocation heterozygote but the results will vary according to the cross over site relative to centromeres and breakpoints of the translocation. If the crossing over occurs in the regions between the centromeres and break points (*i.e.,* interstitial region) it will result in duplicated and deficient chromosomes irrespective of adjacent or alternate distribution patterns. This may be major cause of sterility in translocation heterozygotes. But the crossing over appears to be restricted in the interstitial regions because of inefficient synapsis between the chromosomes. If the crossing over takes place outside the interstitial regions, it does not affect the segregation patterns since one homologous section is exchanged for another. Reduced crossing over within the translocated part is most pronounced in the vicinity of the interchange points. It is so on account of non-homologous pairing in interchange region or due to difficulty in its meiotic chromosome pairing.

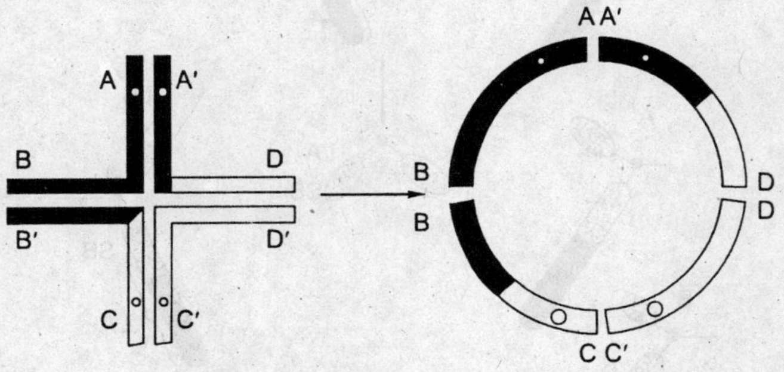

Fig. 22.13 Diagrammatic representation of chromosome behaviour during meiosis in a translocation heterozygote.
A. Pairing of two normal chromosomes with two involving translocation results in the formation of a cross at pachytene.
B. Ring formation by four chromosomes at metaphase I.

The subsequent behaviour of this cross configuration depends upon the frequency and location of the chiasmata and the centromere orientation. If chiasma formation takes place in all four pairing segments, a ring of four chromosomes results. The occurrence of quadrivalent rings has been observed during metaphase in *Datura,* peas, wheat, *Tradescantia* and some other plants as well as animals. If chiasma formation fails in one of the four pairing segments, a chain of four chromosomes results. If chiasma formation takes place in two adjacent or alternate pairing segments, a chain of three chromosomes and one univalent, or two bivalents will result. Rings and chains contain structurally normal and structurally changed chromosomes in alternating sequence. Now the distribution of the four chromosomes in ring or chain configuration at anaphase I of meiosis is determined by the orientation of centromeres. There are two common patterns of distribution, *adjacent* and alternate (Fig. 22.14).

(*a*) **Adjacent distribution.** In this the chromosomes located alternately in the pairing configuration are segregated in such a way that one structurally normal and one translocated chromosome go to one pole and their counterparts, to the opposite pole. In this case both meiotic products are duplicated.

In adjacent distribution there are two events : (*i*) *adjacent 1 distribution* in which the centromeres of neighbouring non-homologous chromosomes segregate to same pole, and (*ii*) *adjacent-2 distribution* in which homologous centromeres migrate to the same pole, but it is very rare in occurrence.

(*b*) **Alternate distribution.** In this the two normal chromosomes go to one pole and the two translocated chromosomes go to the opposite pole during anaphase I and so the gametes-formed are of two kinds; some with normal chromosomes and some with translocated chromosomes.

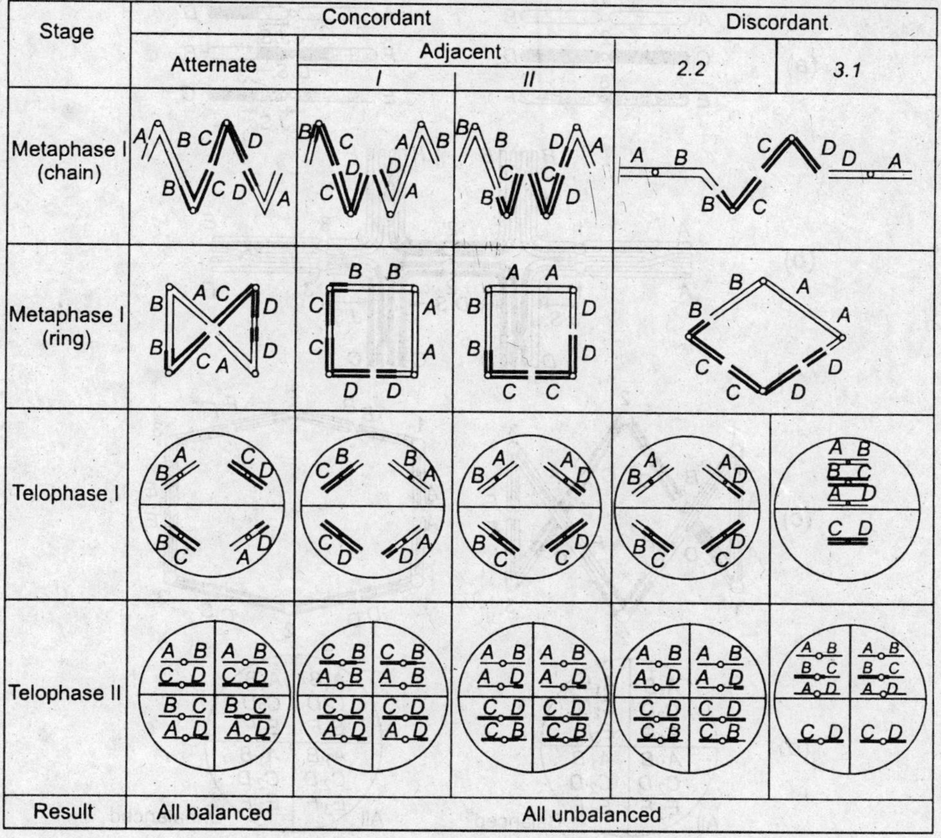

Fig. 22.14 The genetic consequences of various modes of orientation (in metaphase of the first meiotic division) of a chain-of-four or ring-of-four configuration in a karyotype heterozygous for a single reciprocal translocation

Multiple Translocation System

If the arm of one of the two translocated chromosomes is involved in a second interchange with a third non-homologous chromosome, a ring of 6 chromosomes (Fig. 22.15) may be formed at metaphase I. In the same way, a third interchange would give rise to a ring of 8 chromosomes. This process can go on until the entire complement of chromosomes is involved. This produces a translocation complex or complex heterozygote. In *Rhoeo spathacea* all 12 chromosomes are involved in reciprocal translocations and as a result of this a ring of 12 chromosomes is formed during meiosis. *Oenothera* exhibits a similar tendency which varies with the species. In *O. hookeri* normally 7 bivalents are formed but in other species formation of multivalent ring and bivalents may be observed and in some other species, such as *O. lamarkiana* all fourteen chromosomes are linked to form a ring at metaphase I. Translocation complexes of the types described in *Rhoeo* and *Oenothera* are not known in animals. Similar situation has been reported in several other plants exposed to X-rays, gamma-rays and mutagenic chemicals.

2. Genetic Effects of Translocations. The main genetic effects of translocations are as follows:

(*i*) It brings about a qualitative change in the chromosomes structure or linkage group.

(*ii*) It brings about change in the sequences of genes in chromosomes which may eventually produce several abnormalities in body characters. This is *position effect.*

(*iii*) *Semi-sterility*. The translocation heterozygotes are generally semi-sterile because they produce gametes containing duplicated and deficient chromosomes as a result of typical pairing behaviours, crossing over and segregation patterns of chromosomes.

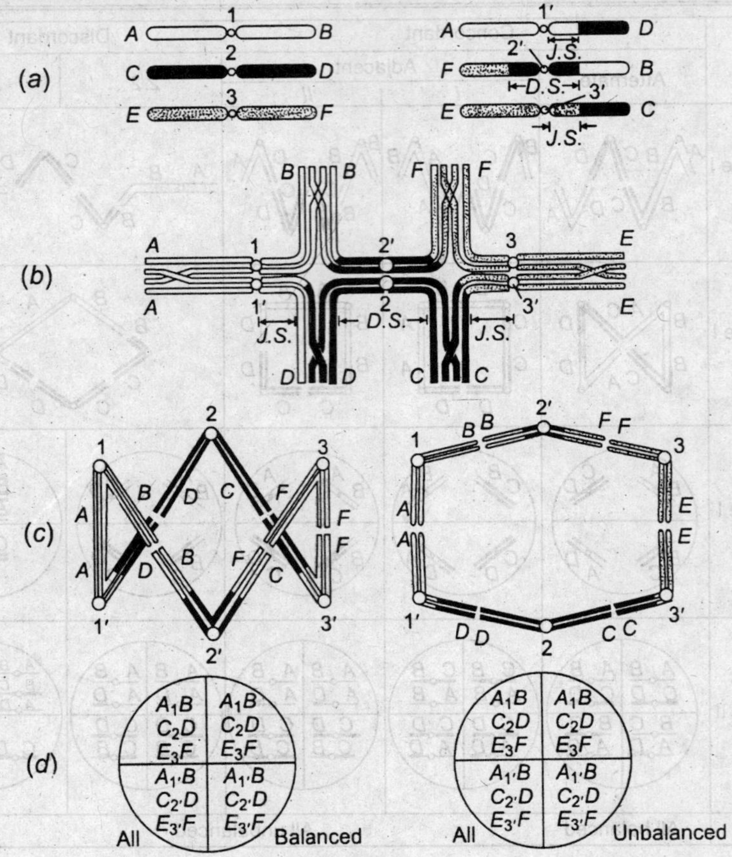

Fig. 22.15 Diagram showing the meiotic consequences of an interchange association of six chromosomes in a karyotype heterozygous for two reciprocal translocations between heterologous chromosomes. (*a*) Left the normal and right the structurally changed chromsomes (I. S. = interstitial segment; D. S. = differential segment). (*b*) The pachytene pairing configuration with the patterns of crossing over and chiasma formation; (*c*) double zig-zag and ring-form orientation of the hexavalent at first meiotic metaphase; (*d*) the genetic constitution of the haploid meiotic products (telophase of second meiotic division).

The translocation is of great importance for the individuals and the species. The most harmful effect of a reciprocal translocation is the semi-sterility it causes. It also causes severe modifications of the normal developmental pattern. In the evening primrose (*Oenothera*) a number of variations are associated with translocations. This was the plant whose variability led De Vries to propose his popular mutation theory.

Inversion

It is an intrachromosomal aberration characterised by inversion or reversal of a chromosome segment and the gene sequence contained therein relative to the standard chromosome or linkage group in question. It occurs in intercalary segment of the chromosome. There is no experimental evidence for the occurrence of inversions in terminal segments of chromosomes. According to the breakage reunion model, the intercalary inversions are formed when two breaks occur in a chromosome, the middle segment between the break points (referred to as *inversion points*) is inverted or rotated through 180° and then reunion of the three segments at the sites of breakage takes place as shown in Fig. 22.16.

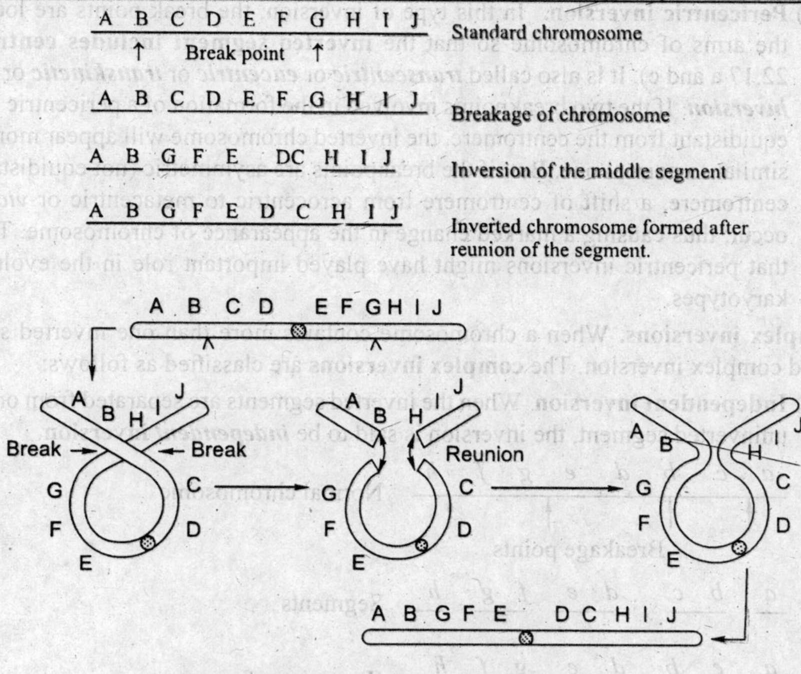

Fig. 22.16 Inversion in a chromosome.

Sometimes inverted segment or a part of it may again undergo inversion. This is called **included inversion**. Thus, in the above example, if the segment, GFEDC in the inverted chromosome undergoes further inversion, the result will be that the chromosome will regain original gene sequence ABCDEFGHIJ.

Types of Inversions

The inversions are classified according to the number of inverted segments within the chromosome and the location of inversion points with respect to each other.

1. Single inversions. When a chromosome contains a single inverted segment, it is called single inversion. Single inversions are classified according to whether or not the inverted segment of the chromosome carries the centromere.

(*i*) **Paracentric inversion**. This is the most common type of inversion which is confined to a single arm of a chromosome. In this, the inverted segment of the chromosome does not **carry centromere**. It is also called ***acentric*** or ***dyscentric*** or ***parakinetic*** or ***asymmetrical inversion*** (Fig. 22.16).

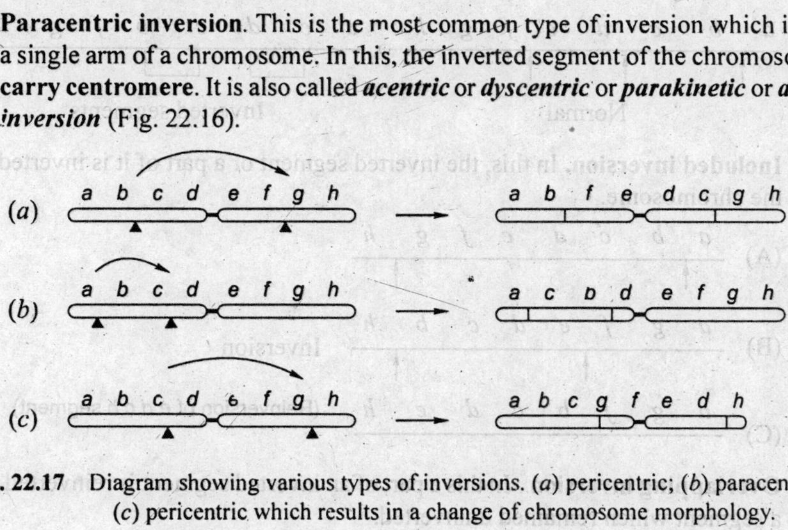

Fig. 22.17 Diagram showing various types of inversions. (*a*) pericentric; (*b*) paracentric; (*c*) pericentric which results in a change of chromosome morphology.

(*ii*) **Pericentric inversion.** In this type of inversion, the break points are located in both the arms of chromosome so that the **inverted segment includes centromere** (Fig. 22.17 a and c). It is also called ***transcentric*** or ***eucentric*** or ***transkinetic*** or ***symmetrical inversion***. If the two breakpoints involved in the formation of a pericentric inversion are equidistant from the centromere, the inverted chromosome will appear morphologically similar to normal one. But, if the breakpoints are asymmetric (not equidistant) from the centromere, a shift of centromere from acrocentric to metacentric or *vice-versa* may occur, thus causing a marked change in the appearance of chromosome. This indicates that pericentric inversions might have played important role in the evolution of new karyotypes.

2. Complex inversions. When a chromosome contains more than one inverted segment, it is called complex inversion. The **complex inversions** are classified as follows:

(*i*) **Independent inversion**. When the inverted segments are separated from one another by uninverted segment, the inversion is said to be ***independent*** **inversion**.

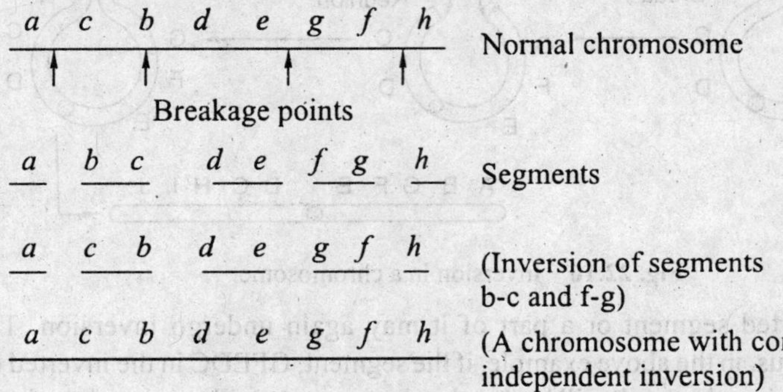

(*ii*) **Direct tandem inversion**. In this, two inverted segments lie adjacent to each other in the chromosome.

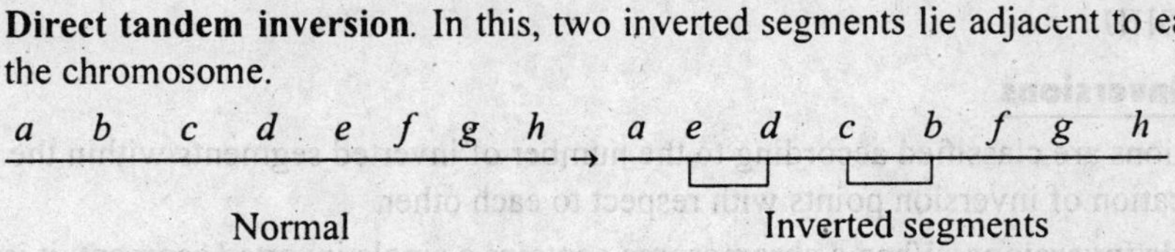

(*iii*) **Reversed tandem inversion**. In this, the two inverted segments are adjacent but mutually interchanged within a chromosome.

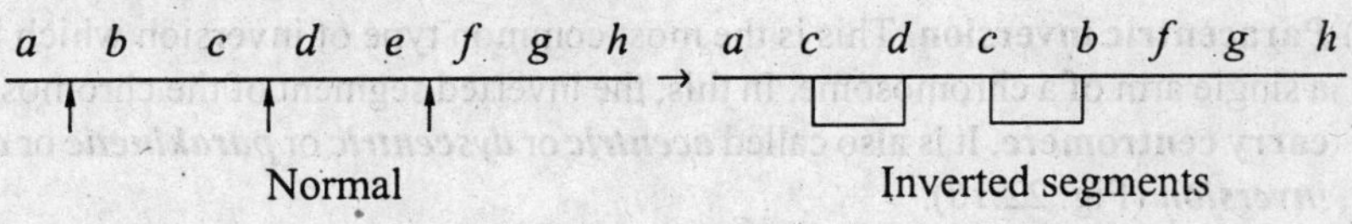

(*iv*) **Included inversion.** In this, the inverted segment or a part of it is inverted again within the chromosome.

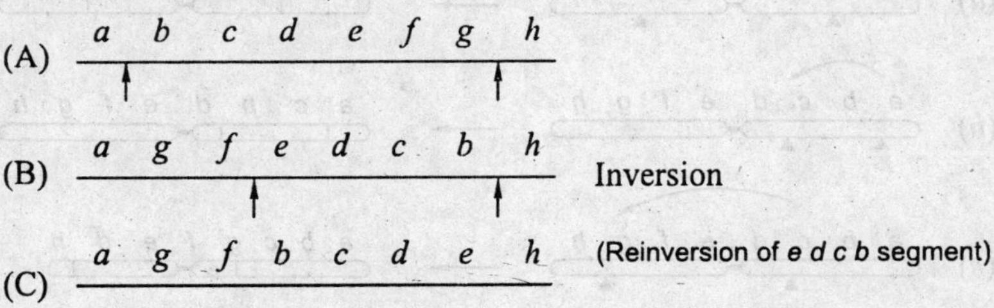

(*v*) **Overlapping inversion**.. In this a part of an inverted segment is reinverted together with a segment which remained uninverted.

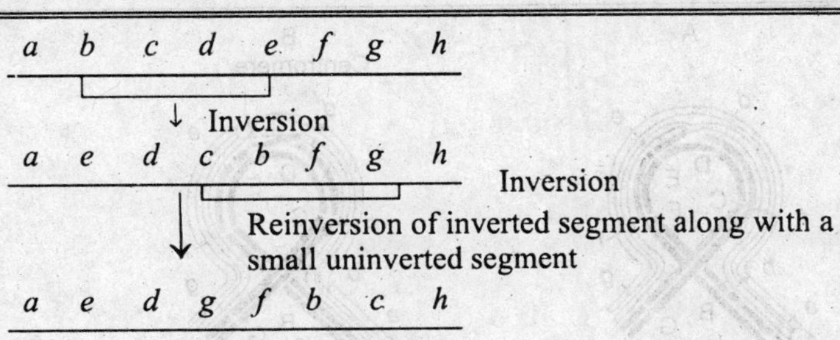

Inversions may be either homozygous or heterozygous. Homozygous inversions have homologous chromosomes with identical inversions. They show normal meiotic pairing and distribution. Heterozygous inversions have one homologue with inversion and the other homologue without inversion (*i.e.,* normal). Inversion heterozygotes show important cytological and genetic effects. Because there is no net loss or gain of genetic material, inversion heterozygotes are perfectly viable. The pairing behaviour of inversion chromosome with standard or non-inverted homologue depends on the length of inversion and the longitudinal relationship of the inverted and uninverted chromosome segments. If the inversion is long, chromosome pairing involves formation of characteristic loop in the normal homologous chromosome and the inversion chromosome pairs with the normal chromosome in such a way that homologous loci pair with each other. The location of the inverted segment, thus, can be recognised cytologically by the presence of an ***inversion loop*** in the paired homologues during meiosis. If the inverted segment is so small that loop formation is not possible either the inverted segment is left unpaired or it may pair with non-homologous segment of normal chromosome. The size of the loop is a function of the size of the inversion—the larger is the inversion, the larger will be the loop.

The crossing over and chiasma formation within and outside inversion loop give rise to secondary structural changes (duplication and deletion) depending upon the type of inversion (paracentric or pericentric), the number of chiasmata and localization of chiasmata. Such structural changes result in meiotic products with unbalanced sets of chromosomes.

Single crossing over within a pericentric heterozygous inversion produces two normal meiotic products and two abnormal products—containing chromosome that are either duplicated or deficient for certain gene loci (Fig. 22.18 B). In plants gametes containing duplication or deficiency are generally not viable. So, pericentric inversion heterozygotes are semi-sterile, although more than 50% viable. In animals, the gametes with duplication and deficiency in chromosomes are usually normal in function, but zygote usually does not survive.

Single crossing over within a paracentric inversion has more complex consequences and it produces one chromosome with two centromere (dicentric chromosome) and one with no centromere (acentric chromosome segment). During anaphase 1 of meiosis dicentric chromosome is pulled as a bridge between two spindle poles and acentric chromosome, because it has no spindle fibre attachment, floats randomly as laggard and is eventually lost (Fig. 22.18 A). The dicentric bridge may break at any point giving rise to duplications and deficiencies in the meiotic products. Remarkably, in *Drosophila* and plant egg cells, the dicentric bridge may remain intact long after anaphase I. Thus, the two daughter nuclei either will be linked by dicentric bridge or will contain the fragments of the bridge if it breaks. The fragment associated with the bridge has the effect of a deficiency and the size of deficiency determines the reduction in fertility. That is, a meiotic product or gamete lacking in one or more genes because of the loss of a segment of chromosome is likely to be non-viable and hence sterile. In general, females heterozygous for paracentric inversion manifest no serious sterility because in most cases the chromosomes are oriented in specific direction during gametogenesis which facilitates exclusion of dicentric and acentric chromosomes from functional gametes.

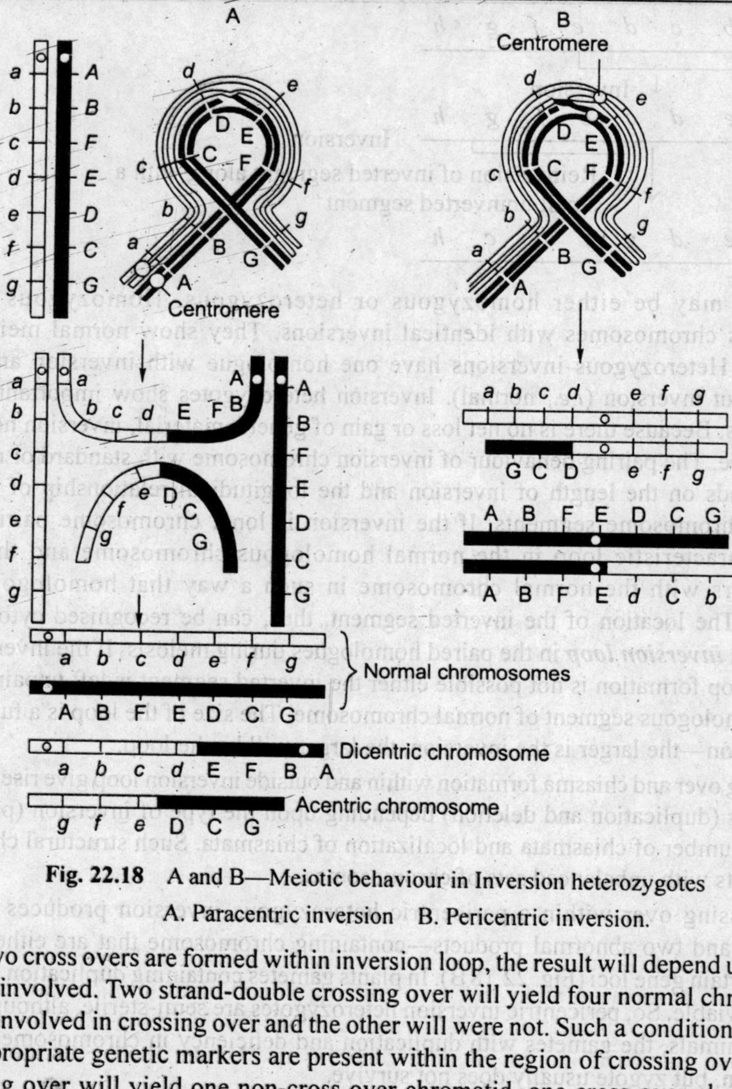

Fig. 22.18 A and B—Meiotic behaviour in Inversion heterozygotes
A. Paracentric inversion B. Pericentric inversion.

When two cross overs are formed within inversion loop, the result will depend upon the number of chromatids involved. Two strand-double crossing over will yield four normal chromatids, two of which will be involved in crossing over and the other will were not. Such a condition can be detected only when appropriate genetic markers are present within the region of crossing over. Three-strand double crossing over will yield one non-cross over chromatid, one cross-over chromatid and two acentric fragments. Four-strand double cross-over would yield two dicentric chromatids and two acentric fragments and deficiency and duplication would presumably lead to inviability of the meiotic products or gametes or would cause death of zygote or embryo if such gametes were involved in fertilization. It is, therefore, evident that paracentric inversions have a drastic effect on the recovery of chromatids involved in crossing over.

Another major effect of inversions is to suppress the recombination of genic loci by crossing over to maintain in the population a specific segment of a chromosome. Crossing over does occur within the inversion, but the crossing over products do not usually contribute to the next generation either because the gametes or zygotes are inviable or because the cross over chromosomes are eliminated from non-functional megaspore in plants or polar bodies in the animals. The presence of recessive lethal gene within inverted segment can be of added advantage in preserving structural heterozygosity because heterozygotes for lethal recessive genes will be viable and homozygous, non-viable. In ClB stock of *Drosophila,* C factor which is a cross-over suppressor is found to be inverted segment of chromosome, l component, a recessive lethal, prevents homozygosity for ClB chromosome and B gene accounts for bar eye. In ClB chromosome C factor is flanked on either side by two marker genes l and B. Muller (1928) was the first to take advantage of cross-over suppressing property of inversion heterozygote to detect sex-linked recessive lethal mutations in *Drosophila* induced by X-rays.

The genetic evidence of inversion thus will be (*i*) suppression of crossing over, and (*ii*) possibly the appearance of mutation owing to position effect.

As mentioned earlier, inversion and translocation involve no loss or gain in the genes as such. Simply they bring about change in the position of some genes and no gene mutation (*i.e.*, no change in the nature of gene) is involved in these cases. The changed positions of genes in the chromosomes may have important consequences since continuous genes sometimes are concerned in the completion of related steps of some repetition biochemical reactions. All such alterations of gene functions due to change in the sequence of genes are referred to as "**position effects**".

Inversion homozygotes can be detected cytologically and genetically in the following ways:

(*i*) By detecting changed linkage relation with the help of genetic linkage studies.

(ii) By detecting the changes in chromosome morphology during mitotic metaphase.

(*iii*) By observing the changes in chromosome bands.

The inversion heterozygotes are detected by the following characteristics:

(*i*) formation of inversion loop during the prophase I of meiosis.

(*ii*) formation of dicentric chromosomal bridge, acentric fragments during anaphase I.

(*iii*) Development of abnormal meiotic products which may be detected by means of tetrad analysis.

(*iv*) Decreased fertility resulting due to production of genetically unbalanced meiotic products or gametes *via* crossing over.

(*v*) Reduced crossing-over frequency

GENE MUTATION OR POINT MUTATION

Under careful genetical observation there appear a few individuals with such characters as have not been previously seen in the population. These characters may breed true through subsequent generations. Some of the mutated characters may be associated with the change in the structure of chromosome or cytoplasm and the others are presumed to be caused by permanent change in the genic material itself. The mutations caused by change of genic materials are called gene mutations. The gene mutation is considered as an inherent property of life. In different groups of species, the rates of gene mutations may vary. It may also vary from gene to gene within the species. According to Beadle (1955), gene mutation may range from one mutation in thousand genes per life cycle to one or less than one in a billion.

Effect of Mutation on trait

The traits are governed by genes. Genes control the traits by controlling the production of specific proteins (enzymes). The general scheme of production of proteins in most organisms is as follows :

1. Gene is a segment of DNA with specific sequence of nucleotides. Different genes show different sequences of nucleotides.
2. DNA (Gene) undergoes transcription to produce a complementary m RNA strand which consists of a sequence of nucleotide triplets or codons.
3. t RNA molecules which form complex with specific aminoacids are provided each with a group of three free nucleotides called anticodon.
4. t RNAs complexed with specific aminoacids are aligned to m RNA forming codon-Anticodon pairs according to base pairing rule.
5. Thus the sequence of codons in m RNA determines the sequence of amino acids in a polypeptide chain (or enzyme) and hence the structure and function of enzyme is directly governed by the

sequence of nucleotides in m RNA which, inturn, depends on the sequence of nucleotides in the segment of DNA from which it is transcribed. The polypeptides or enzymes ultimately effect the trait. Any change in the nucleotide sequence of DNA will result in change of corresponding nucleotide sequence in m RNA. This may affect the alignment of t RNA molecules on m RNA and also the sequence of aminoacids in enzyme molecule. Ultimately the changed structure and properties of enzymes will affect the traits they control. This is ***gene mutation*** or ***point mutation.***

The induction of mutations is referred to as ***mutagenesis***, the organism with changed genes and phenotype is called ***mutant*** and the agents physical or chemical, which induce mutations are termed as ***mutagens***.

Gene mutations or mutations at molecular level can be divided into the following three classes on the basis of molecular changes occurring in gene.

1. Base substitution. 2. Reversion or Reverse mutations, and 3. Frame shift Mutations

Base substitution

When a single base is replaced by a wrong base during replication or a wrong base is incorporated during repair of a DNA segment (gene) it is called ***base substitution.***

Original DNA Strand GAT CAT GAT CAT GAT CAT
↓
Substitution Strand GAT CAT GAT CAT GGT CAT (A is replaced by G)

The consequence of base substitution depends on its location within a gene and not all mutations result in altered phenotype of the organism. ***At protein level*** base substitution may result in the following types of mutations :

(*i*) **Neutral Mutation.** If as a result of base substitution, the mutated codon specifies for the same aminoacid in polypeptide chain as did the original codon, it is called ***neutral mutation***, as for example,

AGG → **C**GG (both triplets code for Aginine)

(*ii*) **Missense mutations.** If the substitution results in a triplet which codes for a different and non functional amino acid, or in other words, it results in a simple change of one amino acid in the polypeptide chain. Such a change is called ***missense mutation***.

(*iii*) **Non-sense mutations.** If a single base substitution modifies a codon to a termination codon, the polypeptide terminates at the point of mutation as for example C A G ⟶ [U] A G, such a mutation is called ***nonsense mutation***. Mutations of this type result in incomplete polypeptide chains which may not be biologically active. Out of 64 codons. 61 code for amino acids and there are three termination codons. The termination codons are UUA, UAG, and UGA.

At DNA level base substitutions are of the following two types :

(*i*) *Transition*

(*ii*) *Transversion*

Transition: Replacement of a purine by another purine base (*i.e.*, A by G or G by A) or pyrimidine by another pyrimidine (*i.e.*, T by C or C by T) is called **transition.** Transitions are the most common types of mutation.

Transversion: If a purine base is substituted by a pyrimidine or *vice versa*, the substitution is called Transversion. Most base pair changes are reversible.

Transversion: A ⟷ C

T ⟷ G

The base pair changes involved in transitions and transversions can be represented as follows :

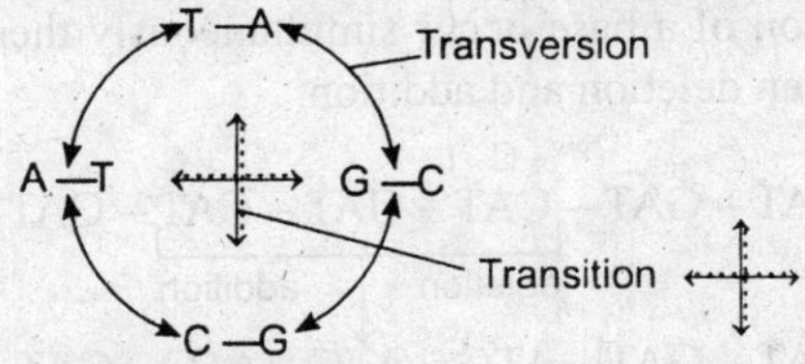

Base pair replacements involved in transitions and transversions

The following eight possible transversions are possible from four purine-pyrimidine pairs. Thus one base pair can undergo one type of transition and two kinds of transversions.

Purine– Pyrimidine $\xleftrightarrow{\text{Transversion}}$ Pyrimidine – Purine

AT → CA   AT → TA   GC → TA   GC → CG

TA → GC   TA → AT   CG → AT   CG → GC

In general, transition mutations code for chemically similar aminoacids while transversions can lead to nonsense mutations and may show greater chances of missense mutations.

Inversion. when an intercalary segment of DNA is removed and inverted or rotated through 180° angles and then reinserted at the site of breakage it results in an *inversion.*

CTA GAT ACT GAT CAT GAT TAC

↓ removed and reinserted after rotating through 180°

CTA GAT TAC TAG TCA GAT TAC

Message is out of frame

Frameshift mutation

Frame shift is mutation in which there is insertion or deletion of one or more nucleotide in a gene but the number of bases lost or gained is not a multiple of three. Both the base substitution and addition mutants alleles are expected to undergo reverse mutation to produce their respective normal or wild type alleles but deletion mutations are not expected to undergo reverse mutations.

The name "frame shift" is derived from the fact that there is shift in reading frame forward or backward by one or two nucleotides. The insertion or deletion of nucleotides results in a new sequence of codons which may code for new amino acids in polypeptide chain. Consequently the protein molecule so formed is entirely different from the wild type and is non functional. If the reading frame shifts by a group of 3 nucleotides the resulting protein is expected to be normal except that it may lack one aminoacid or may contain an extra aminoacid.

Deletion: In a deletion mutation one or more bases are deleted or lost from a gene or nucleotide chain. The removal of even one nucleotide may throw the genetic message out of frame beyond the point of deletion. This will happen on deletion of any number of nucleotides not divisible by three.

Wild type or original reading frame— GAT CAT GAT CAT GAT CAT

–C

After deletion— GAT CAT GAT ATG ATC AT

Message out of frame

Insertion (Addition): Insertion of one or more nucleotides in a gene is referred to as insertion or addition. The genetic message will be disturbed it the number of bases added is not divisible by three.

Wild gene : CAT – GAT – CAT – GAT – CAT – GAT – CAT

↓ + G

CAT – GAT – GCA – TGA – TCA – TGA – TCA – T

Genetic message out of frame

If the deletion and addition of a base occur simultaneously then the message will be out of frame only in the triplets between deletion and addition.

GAT – CAT – GAT – CAT – GAT – CAT – GAT – CAT (– C deletion; + C addition)

GAT – CAT – GAT – ATG – ATC – ATC – GAT – CAT

Message out of frame

TYPES OF MUTATIONS AND THEIR MOLECULAR BASIS

Gene mutations can arise spontaneously or they can be induced. Spontaneous mutations occur in individuals under natural conditions and they occur at quite low but constant rates. Induced mutations are produced when an organism is treated with mutagonic agent or mutagen. Such mutations are recovered at much higher frequencies than spontaneous mutations.

SPONTANEOUS MUTATIONS

Spontaneous mutation was once attributed to the effects of natural backgrounds as for example, cosmic radiations, heat, radioactive substances, naturally occurring base analogues like caffeine, age effects, etc. It has been proved experimentally that sex linked mutations may accumulate as a result of ageing in *Drosophila melanogaster*. Mutagenic agents are also produced in old and decaying seeds. In fact, no mutagens are involved in the production of spontaneous mutation. Both base pair substitution and chromosome aberrations can occur simultaneously. The frequency of spontaneous mutations is usually low ranging from 10^{-7} to 10^{-12} per organism. The rate of detectable spontaneous mutations in an average gene is 1 in 10^{-6}. It should be noted that most methods of estimating the rate of mutations tend to underestimate their frequencies for several reasons. Lethal mutations and micromutations may remain undetected.

Real spontaneous mutations arise from ***tautomerism.*** The ability of a molecule to exist in more than one chemical form is called tautomerism. All the DNA bases (adenine, guanine, cytosine and thymine) have unusual tautomeric forms which are however rare in occurrence (Fig. 22.19).

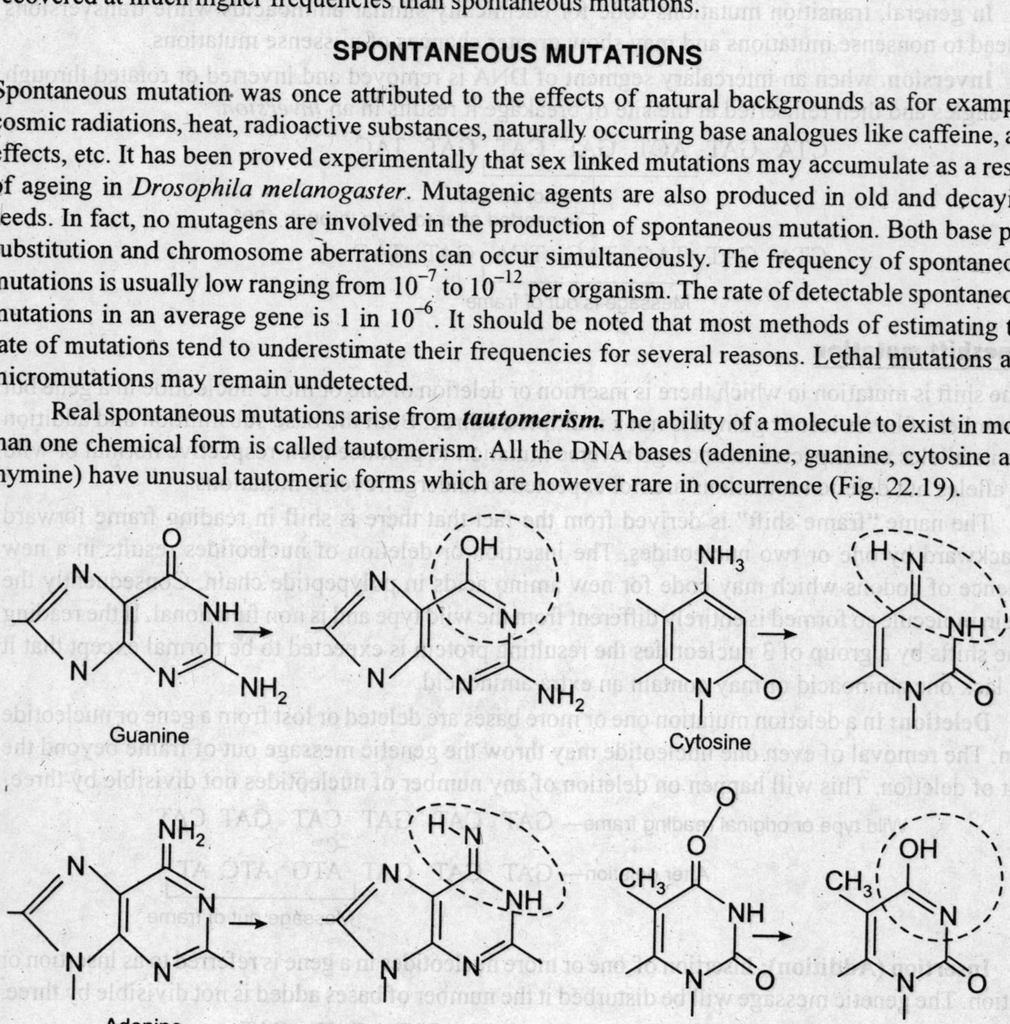

Fig. 22.19. DNA bases and their tautomers

Normally the pairing of bases in DNA is A ↔ T and G ↔ C but tautomeric bases show unusual pairings as for example T ↔ G and A ↔ C.

1. Tautomeric Thymine pairs to normal Guanine or *vice versa*
2. Tautomeric Adenine pairs with normal cytosine or *vice versa*.

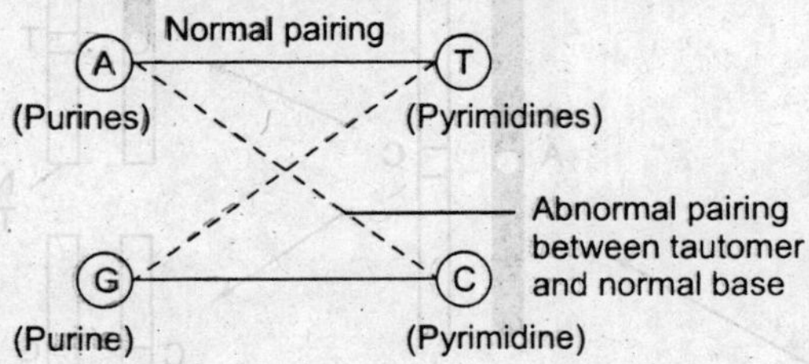

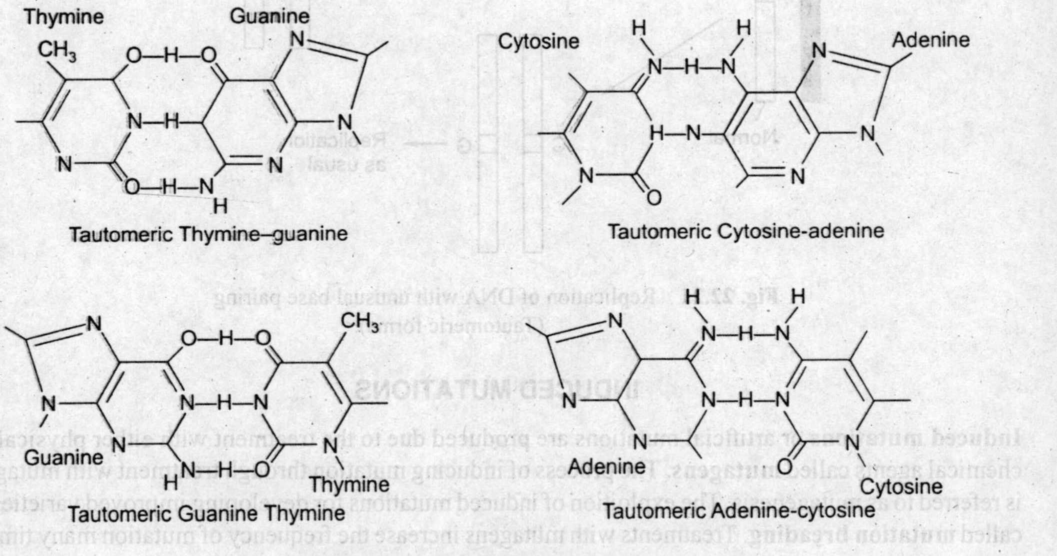

Fig. 22.20. Abnormal pairing of tautomeric bases

The spontaneous mutation can arise due to abnormal base pairing (say A–C or T–G) during DNA replication which may lead to formation of altered codons. It is now thought that the majority of spontaneous mutations occur due to abnormal cellular enzymes.

Watson and Crick (1955) have hypothesized that the occurrence of bases in tautomeric forms possibly provides a mechanism for mutation during DNA replication. If in an old strand of DNA, adenine is in tautomeric condition, it will pair with normal cytosine in the new complementary strand during DNA replication. This may result in a new DNA molecule containing exceptional base pair. The *tautomeric shift is not stable* and in the next replication tautomeric adenine is supposed to return to its normal state and then it will pair with thymine. The cytosine added to complementary strand would then pair with guanine (Fig. 22.21). Thus, two kinds of DNA molecules are formed; some with normal base sequence and some with changed base sequence. The two kinds of DNA molecules formed by transition will eventually be separated into different cells which may result in a recognisable mutation (Fig. 22.21). Mutation which occurs during DNA replication is called **'copy error mutation'**.

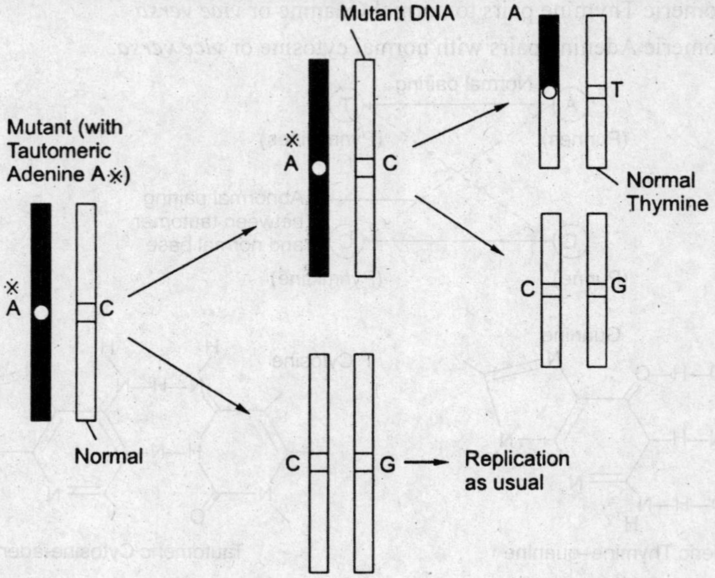

Fig. 22.21 Replication of DNA with unusual base pairing (Tautomeric forms).

INDUCED MUTATIONS

Induced mutations or artificial mutations are produced due to the treatment with either physical or chemical agents called **mutagens.** The process of inducing mutation through treatment with mutagen is referred to as mutagenesis. The exploition of induced mutations for developing improved varieties is called **mutation breading**. Treatments with mutagens increase the frequency of mutation many times.

Factors for Induced mutations (Artificial mutations). The induced mutations are caused artificially by mutagenic agents or mutagens. Artificial mutagenic agents inducing mutations are as follows:

(*i*) Temperature,
(*ii*) Age effect,
(*iii*) Chemical mutagens, and
(*iv*) Physical mutagens.

Temperature. The rate of induction of sex linked lethal genes in *Drosophila melanogaster* increases two or three times if the optimum temperature of the environment is raised by 10°C. However, for several specific loci (particularly some unstable genes tested in both *Drosophila* and maize), this relationship does not hold good. Stability of gene depends not only on its thermodynamic state but also on the genetic environment of the cell.

Age Effect. Mutation rate is influenced by ageing of the organism. It has been proved experimentally that sex linked mutations may accumulate as a result of ageing in *Drosophila.* Mutagenic agents are also produced in old and decaying seeds.

Chemical Mutagens

At present a variety of chemical substances are known which cause mutations in plants and animals. In 1940 for the first time, Auerbach discovered mutagenic property of mustard gas on *Drosophila.* Later on some related substances were also found to produce mutations in every kind of test organisms. In recent years much effort has been directed towards finding chemicals that are even more selective and act on certain genes. Detection of such substances would be of tremendous importance for the theory and practice of genetics. For the research workers they would provide a tool for studying the chemical nature of individuals and for the applied geneticists they would provide means of producing mutations of desired types. Mutagenic chemicals react with the genes directly (*i.e.,* through reaction with cytoplasm). It is also possible and indeed may be frequent cause of mutation. The "gene makes a mistake" when it replicates. Mutagens that are chemically related to the pyrimidines and purines may provoke such mistakes.

Mustard gas, Urethane and **Formaldehyde** produce the same types of mutation as ionising radiations in *Drosophila*. Recently **Diethyl sulphonate (DES)** and **Ethyl methane sulphonate (EMS)** are being most extensively used for inducing mutations in microbes and higher plants.

Quite recently a very potent mutagen **heliotrin** has been found in the common raywort.

In bacteria, at least a proportion of naturally occurring mutation seems to be due to mutagenic purines that are produced in the cells (Automutagenesis). **Caffeine** and some related substances are also used for inducing mutations.

The chemical mutagens can be classified according to their mode of action into the following types:

1. Base analogues which are incorporated into replicating DNA instead of normal bases.
2. Mutagens modifying purines and pyrimidines and agents labilizing bases even in non replicating DNA. These include
   (*i*) Nitrous oxide (HNO_2)
   (*ii*) Hydroxylamine
   (*iii*) Alkylating agents
3. Agents causing distortions in DNA during DNA replication.

Base analogues

A chemical substance resembling a base is called base analogue. Base analogue may be incorporated into new DNA strand instead of normal base during DNA replication. 5-bromouracil (5-BU), 5-chlorouracil (5-CU), 5-iodouracil (5-IU), 2-Aminopurine (2-AP), 2, 6-diaminopurine are the common base analogues and of these 5 BU and 2-AP are commonly used base analogues for mutagenesis.

5- Bromouracil (5-BU). It is a pyrimidine analogue, structurally similar to thymine and pairs with adenine in DNA. Rarely it switches from more stable keto form to the less stable enol state (tautomerisation) and in this form it pairs with guanine.

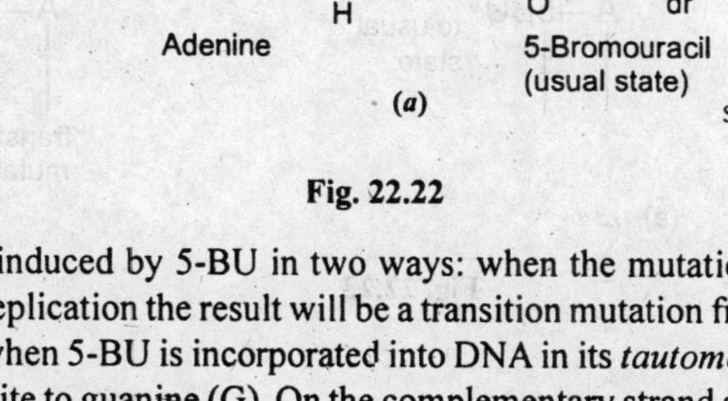

(*a*)

Fig. 22.22

Mutations can be induced by 5-BU in two ways: when the mutation involves *usual form* of 5-BU into DNA during replication the result will be a transition mutation from AT to GC, as shown in Fig. In the second way, when 5-BU is incorporated into DNA in its *tautomeric form* (enol state). This indicates insertion opposite to guanine (G). On the complementary strand and subsequent replication will result in a transition mutation from G–C to AT. [Fig. 22.22(b)]

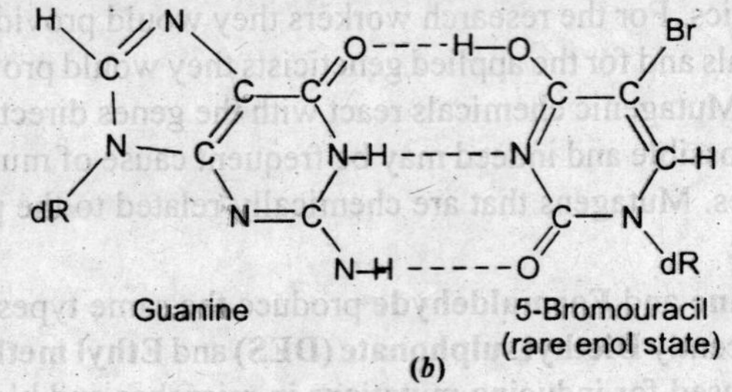

(*b*)

Fig. 22.22

2- Aminopurine. (2, 6-Diaminopurine or 2, 6-DAP)

2-aminopurine is another base analogue like 5-BU. It also exists in two states. In usual state it behaves like adenine and pairs with thymine. In its rare imino state 2-AP behaves like guanine and forms hydrogen bonds with cytosine. Thus 2-AP can induce transition mutations from A–T to GC and from G–C to A–T. Therefore, 2-AP induced mutations can be reverted by 2-AP treatment. 2,6-DAP and 2-AP are less effective as mutagens than 5-BU [Fig. 22.23(a, b)].

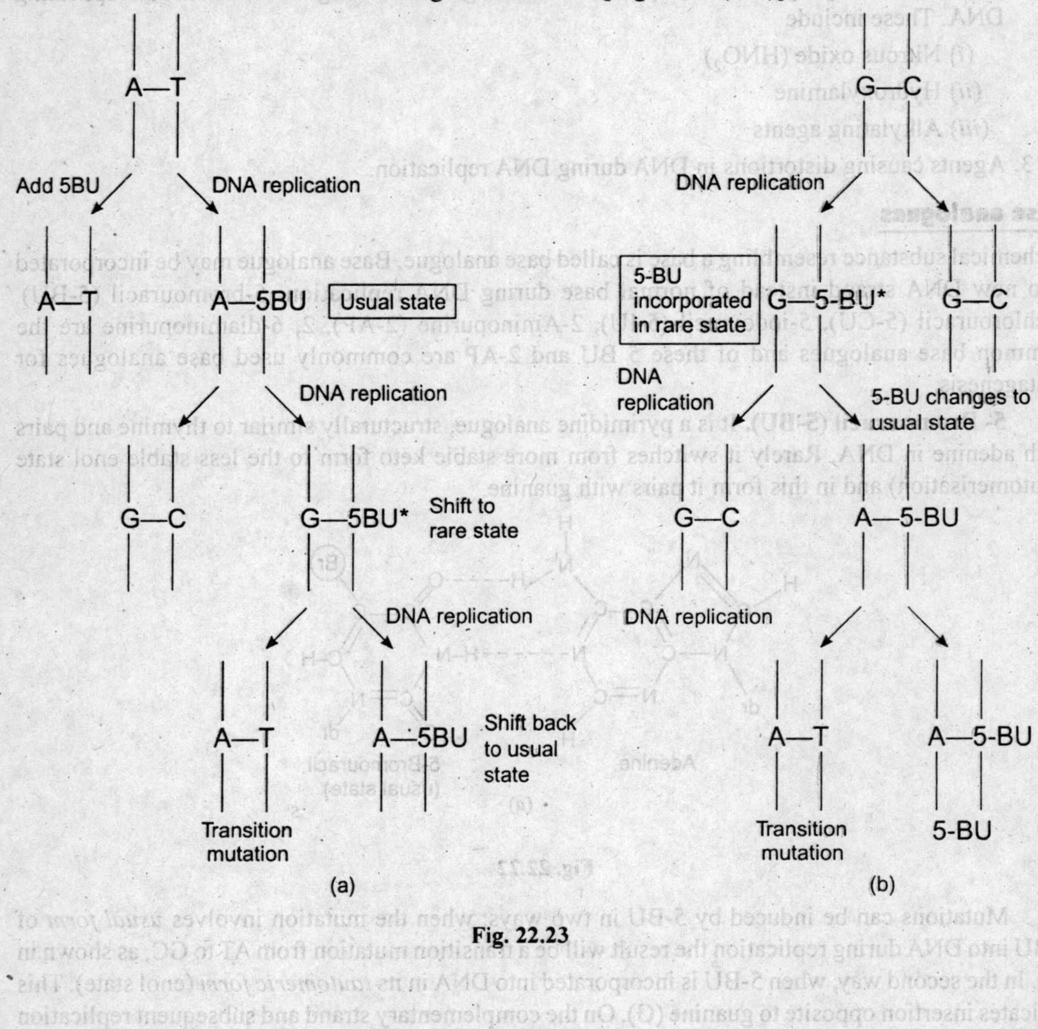

Fig. 22.23

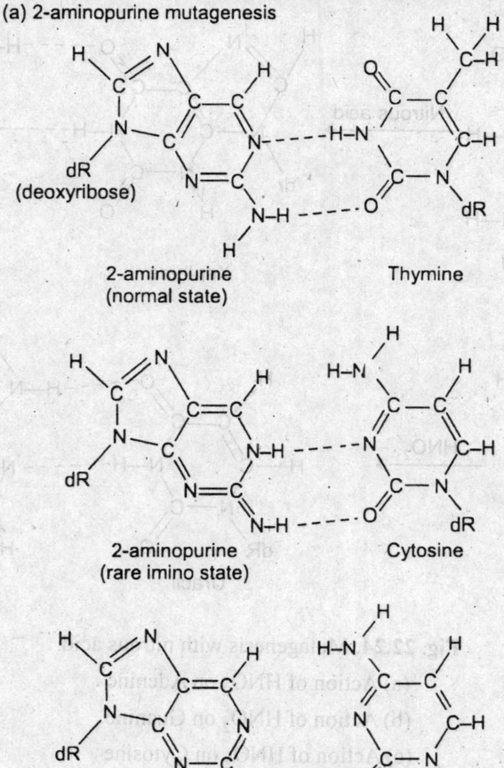

Fig. 22.23c. 2, 6-Diaminopurine and pairing in normal state and Pairing in rare (imino) state

2. Mutagens modifyig Purines and Pyrimidines

(*i*) Nitrous oxide (HNO_2) or NA. Nitrous acid is a dominating agent which removes amino group (NH_2) from the bases. When purines and pyrimidine containing —NH_2 group are treated with nitrous oxide the amino group is replaced by hydroxyl (–OH) group. The order of deamination is adenine → cytosine → guanine. Nitrous acid is potent mutagen which acts on both replicating and non replicating DNA. It removes–NH_2 from 6th C position of cytosine and adenine, 2–NH_2 of guanine is replaced by keto-CO group yielding uracil, hypoxanthine and xanthine respectively. Uracil base pairs with adenine leading to GC–AT transition, Hypoxanthine pairs with cytosine producing AT– GC transition. But Xanthine pairs with cytosine (the same as guanine) hence deamination of guanine does not yield mutation. Nitrous acid, thus produces AT–GC as well as GC–AT transitions. These transitions are produced when DNA treated with HNO_2 undergoes replication.

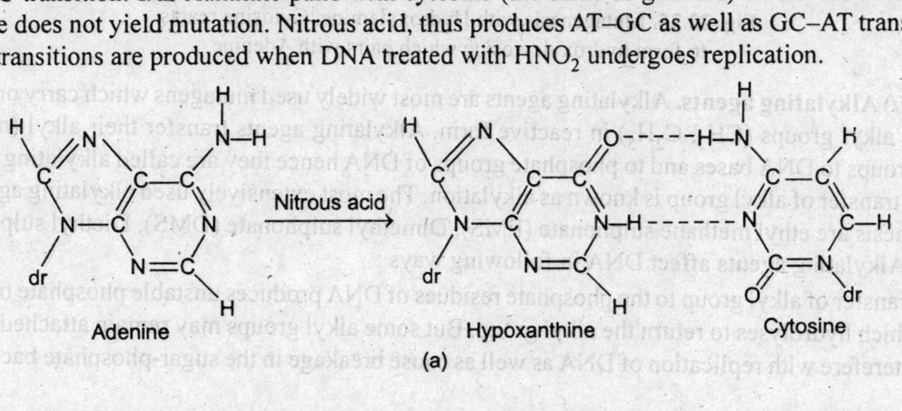

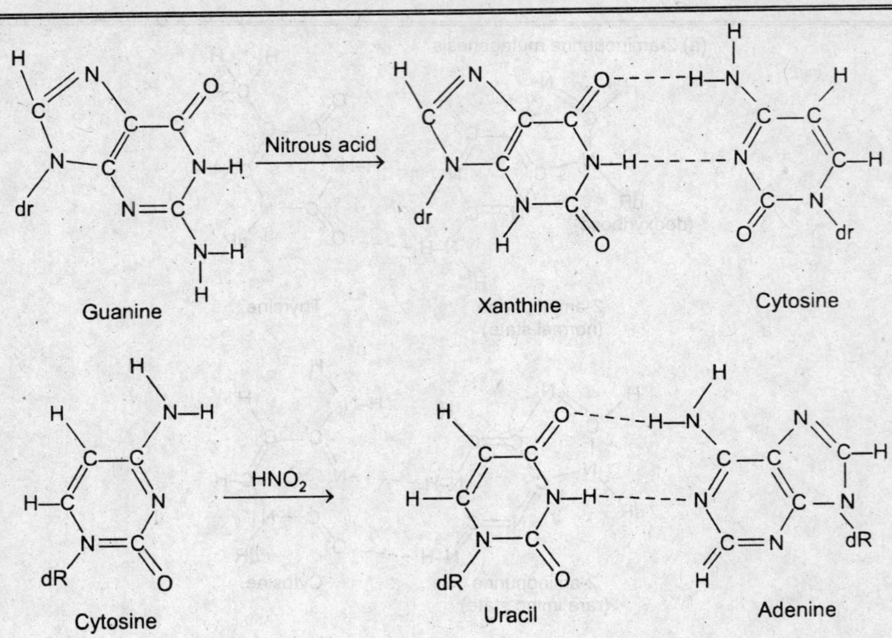

Fig. 22.24. Mutagenesis with nitrous acid.
(a) Action of HNO_2 on Adenine
(b) Action of HNO_2 on Guanine
(c) Action of HNO_2 on Cytosine

(*ii*) **Hydroxyl amine (NH_2OH).** Hydroxyl amine is specific in its action. It reacts mainly with cytosine and causes hydroxylation of cytosine at NH_2 group to form hydroxyl cytosine which pairs with adanine instead of guanine (Fig. 22.25). Hydroxylamine thus induces one way transition mutation from GC to AT.

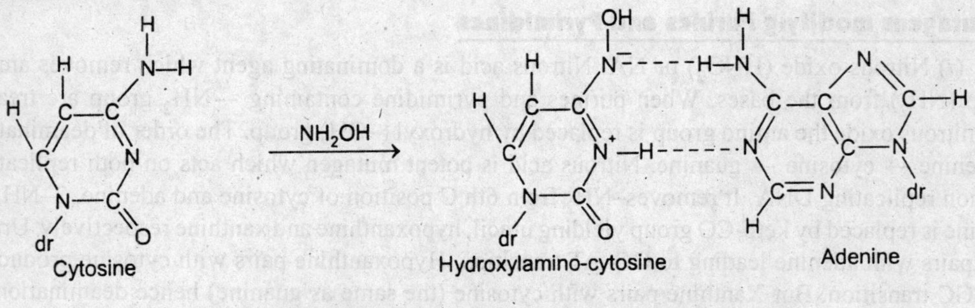

Fig. 22.25. Mutagenesis with Hydroxylamine. Cytosine reacts to form hydroxylcytosine which pairs with Adenine

(*iii*) **Alkylating agents.** Alkylating agents are most widely used mutagens which carry one, two or more alkyl groups (CH_3, C_2H_5) in reactive form. Alkylating agents transfer their alkyl (methyl, ethyl) groups to DNA bases and to phosphate groups of DNA hence they are called alkylating agents and this transfer of alkyl group is known as alkylation. The most extensively used alkylating agents in mutagenesis are ethyl methane sulphonate (EMS), Dimethyl sulphonate (DMS), Diethyl sulphonate (DES). Alkylating agents affect DNA in following ways :

1. Transfer of alkyl group to the phosphate residues of DNA produces unstable phosphate triesters which hydrolyses to return the alkyl group. But some alkyl groups may remain attached which interefere with replication of DNA as well as cause breakage in the sugar-phosphate backbone.

2. Alkyl groups are readily transferred to the 6-oxygen and 7-nitrogen of DNA bases. Out of these alkylation products 7-ethyl guanine is the most common derivative formed which is important for two reasons: (a) other alkylation products are normally repaired through excision-repair process like that for thymine dimers, and (b) 7-ethyl guanine is believed to pair with thymine (in place of cytosine) which ultimately leads to GC → AT transition.
3. During repair of alkylated bases errors may occasionally occur producing transition, transversions or even frame shift mutations.
4. Some difunctional or polyfunctional alkylating agents (with two or more active functional alkyl groups respectively) cross-link DNA strands and or DNA molecules; such cross-link ages prevent DNA replication and lead to chromosome breakage and structural mutations. Such cross-linking of DNA may also produce transition, transversion and frame shift mutations due to errors during repair of affected sites.
5. 7-ethyl guanine often dissociates from deoxyribose and is often removed from DNA strand (*depurination*). The gap in DNA strand so produced may lead to a break in the strand, may interfere with DNA replication or may lead to insertion of wrong base during repair or replication causing mutation.

Agents causing distortions in DNA

Acridine dyes

Certain florescent acridine dyes such as acridine orange, and proflavines cause mutations by insertions or deletions of bases. Crick's work on acridine mutants has provided strong evidence for the genetic code. Like purine bases acridines are planer (flat molecules) and at low concentrations, can be inserted or intercalated between the bases of DNA helix (Fig. 22.26). When it occurs it 'stretches' the distance

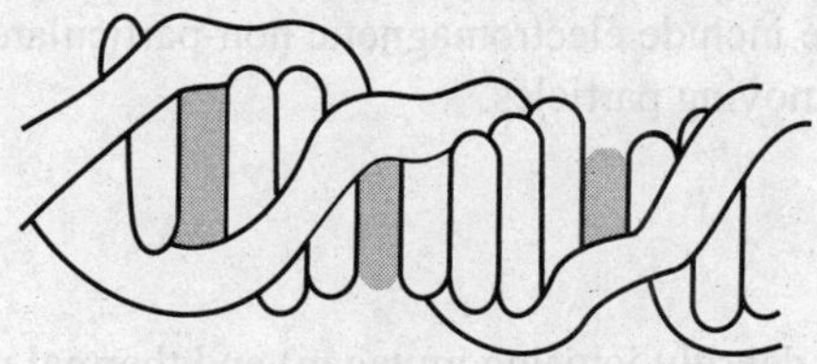

Fig. 22.26. DNA helix with intercalated acridine molecule (shaded)

between adjacent basepairs to 0.68 nm which is precisely double the normal distance. This distorts the DNA strands and result in deletion or insertion of bases during recombination. (Fig. 22.27 a and b).

(*i*) **Intercalation resulting in insertion of bases.** Intercalation of acridine molecule between two bases in template strand results in elongation of DNA molecule. During replication a base X′ is inserted at random in the new chain opposite acridine molecule. During the replication of the strand with X′, a complementary base X will pair with the *inserted base* (*x*′), and in this way a new DNA molecule with additional pair of bases is formed (Fig. 22.27a).

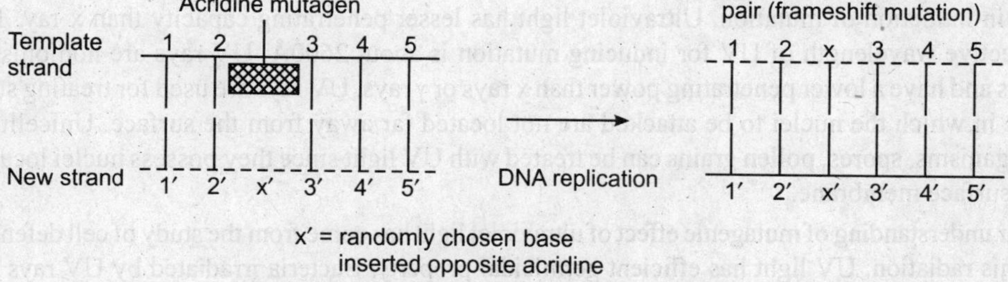

Fig. 22.27. Mutagenesis with acridine molecule

(a) Intercalation resulting in insertion or addition of base

(*ii*) **Intercalation resulting in deletion of base.** Insertion of acridine molecule in DNA strand may block the base in the template strand and does not allow any base to pair with it. So the chain produced is deficient in one base and in the next replication a DNA with a deficient base pair will result (Fig. 22.27 b).

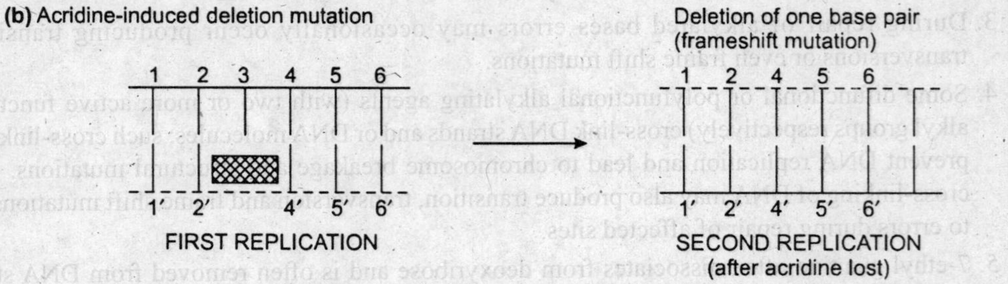

Fig. 22.27b. Mutagenesis with acridine molecule
(b) Intercalation with acridine causing loss of base in the second DNA replication.

Physical Mutagens

These include various kinds of radiations which induce and enhance mutations. H.J.Muller in 1927 demonstrated for the first time that mutations can be artificially induced by treating the flies with x-rays. L.J. Stadler (1928) also demonstrated an increase in the rate of mutations due to x-ray treatment in barley and maize. The mutagenic radiations fall in two categories:

(*a*) *Non-ionising radiations,* such as ultraviolet (UV) light

(*b*) *Ionising radiations.* These include electromagnetic non-particulate radiations and particulate radiations as well as fast moving particles.

Particulate Radiations.

α-rays (densely ionising)

β-rays (sparsely ionising)

Fast neutrons (a common densely ionising mutagen) and thermal neutrons (densely ionising) are used in mutation breeding.

Non-particulate radiations (Electromagnetic radiations): These are as follows:

x-rays (a sparsely ionising radiation commonly used in mutation breeding)

γ-rays or gamma rays (a sparsely ionising radiation used in mutation breeding).

Non-ionising Radiation

Ultraviolet rays. Ultraviolet (UV) rays have a wavelength of 100–3900Å (10–390 nm). When ultraviolet light falls on genetic material it is absorbed and due to increase in the energy level, electrons are *excited.* This is less drastic than the complete ejection of electrons, but it does result in increased chemical activity on the part of affected molecules. This may be the direct effect of UV-rays in induction of mutation. Ultraviolet light has lesser penetrating capacity than x ray. The most effective wavelength of UV for inducing mutation is about 2600Å. UV rays are nonionising radiations and have a lower penetrating power than x rays or γ rays. UV rays are used for treating such materials in which the nuclei to be attacked are not located far away from the surface. Unicellular micro-organisms, spores, pollen grains can be treated with UV light since they possess nuclei located near the surface membrane.

Our understanding of mutagenic effect of ultraviolet light has come from the study of cell defences against this radiation. UV light has efficient germicidal property. Bacteria irradiated by UV rays are killed in direct proportion to the radiation dose so long as the cells are kept in dark. A large proportion of irradiated bacteria survive when they are exposed to visible light. This indicates that there exists a light induced system for repairing damage caused by UV light. The recovery of bacteria exposed to UV light under sunlight also indicates that this radiation induced damage is not itself lethal. There are indications that DNA is the primary target for mutagenic effects of UV radiations. The exposure of DNA and RNA to UV rays causes N-containing bases to become highly reactive instable free radicals. This causes conversion of one base to another (Purine to purine or Pyrimidine to Pyrimidine) changes in mRNA will lead to synthesis of few defective proteins only because of short life of RNA. The substitution in DNA may have lasting effect and all the proteins coded by DNA may be defective. The mutation occurring in germ cells could pass on mutated DNA to the next generation.

The pirmary effect of UV rays appears to be due to the production of pyrimidine dimers in which adjacent thymine or cytosine on the same DNA strand become bonded together by covalent linkages between C—CH_3 and C—CH_3 and CH—CH forming cyclobutane ring.

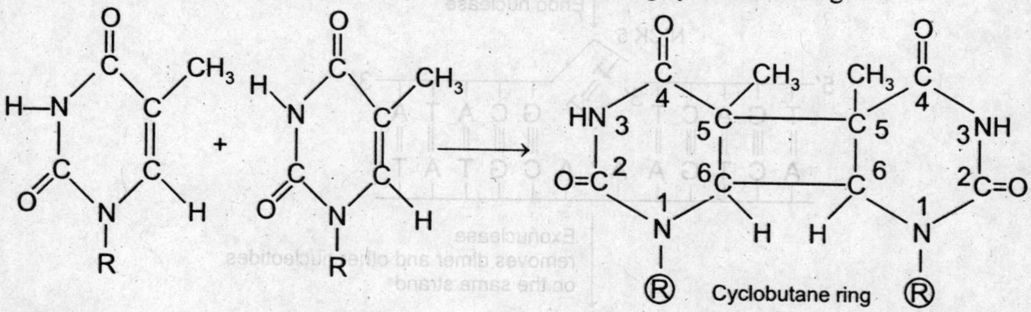

Cross-linking of pyrimidines on different DNA strands

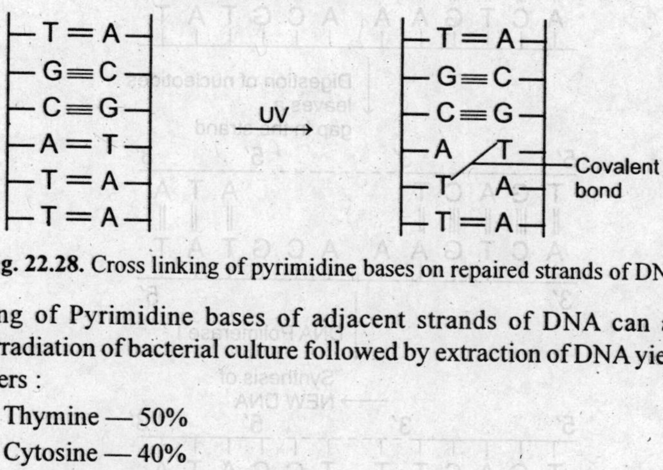

Fig. 22.28. Cross linking of pyrimidine bases on repaired strands of DNA

Cross linking of Pyrimidine bases of adjacent strands of DNA can also form dimers (Fig. 22.28). UV Irradiation of bacterial culture followed by extraction of DNA yielded the following three types of dimers :

Thymine — Thymine — 50%

Thymine — Cytosine — 40%

Cytosine — Cytosine — 10%

Pyrimidine dimer may also develop between adjacent thymine and cytosine. Pyrimidine dimers can not fit into the DNA double helix. This causes distortion of the molecule and failure of DNA replication, leading to lethal effect.

Distortions in DNA molecule due to dimer formation can be corrected by two mechanisms; (I) Excision repair and (II) photoreactivation.

The ***Excision repair mechanism*** was discovered separately by P. Boyce and P. Howard Flanders and R. Setlow and W. Carrier. In this mechanism the first step involves recognition of Dimer in DNA and this is followed by single strand nick (cut) close to damaged DNA by action of endonuclease. A part of DNA strand alongwith pyrimidine dimer is removed by exonuclease enzyme. A second enzyme DNA polymerase inserts correct bases in the gap and a third enzyme polynucleotide ligase joins the inserted bases. The DNA is thus restored to its original condition. Since no light is required for activation in this process, it is called ***dark repair*** (Fig. 22.29).

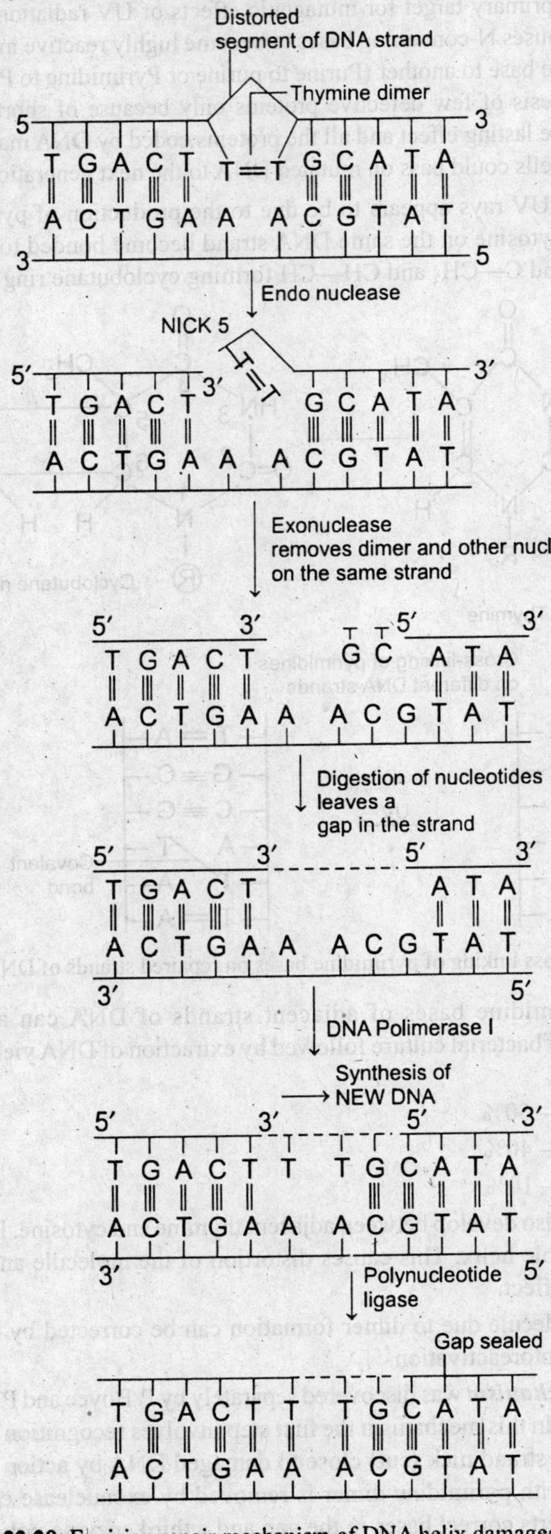

Fig. 22.29. Excision repair mechanism of DNA helix damaged by UV-Rays

In photoreactivation, the pyrimidine dimers are reversed directly to original form. The reaction is dependent on exposure to visible light after irradiation, hence photoreactivation. The reversal process is catalysed by an enzyme called photoreactivating enzyme or photolase which brings about monomerization of dimers when activated by photon of light in the wavelength of 320–370 nm (Fig. 22.30).

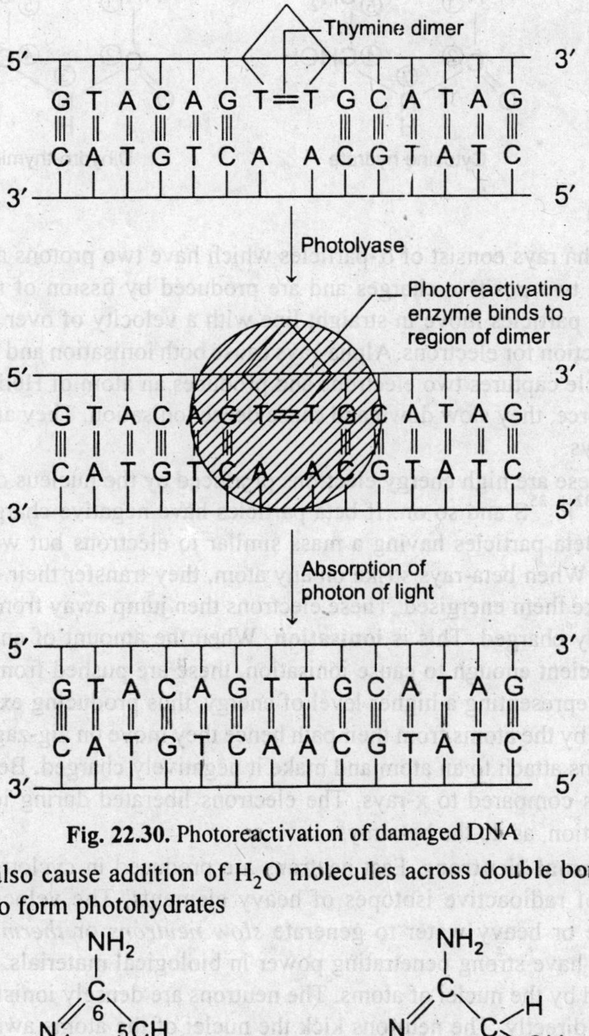

Fig. 22.30. Photoreactivation of damaged DNA

UV radiations also cause addition of H_2O molecules across double bonds to pyrimidines in both DNA and RNA to form photohydrates

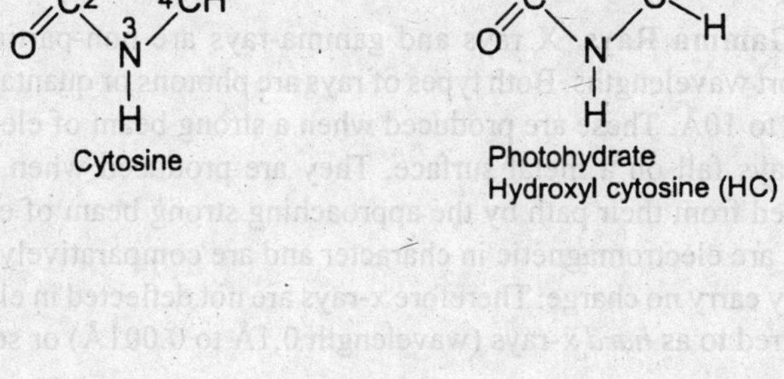

Cytosine

Photohydrate
Hydroxyl cytosine (HC)

(b) Hydration

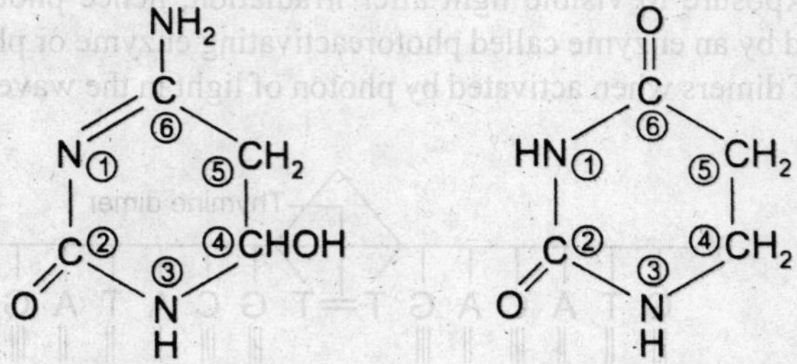

Ionising Radiation

Alpha rays. The alpha rays consist of α-particles which have two protons and two neutrons each. These particles have two positive charges and are produced by fission of radioactive isotopes of heavy elements. The particles move in straight line with a velocity of over 16,000 km per second and have strong attraction for electrons. Alpha-rays cause both ionisation and excitation. After losing energy, each α-particle captures two electrons and produces an atom of Helium. As the α-particles move away from source, they slow down and cause dense ionisation. They are less penetrating than neutrons and beta-rays.

Beta Rays. These are high energy electrons produced by the nucleus of decaying radioactive isotope such as ^{3}H, ^{32}P, ^{35}S and so on. If beta particles have negative charge they are called *beta negative particles*. Beta particles having a mass similar to electrons but with positive charge are called *positrons*(e^{+}). When beta-rays strike on any atom, they transfer their energy to the electrons of that atom and make them energised. These electrons then jump away from their position leaving the nucleus positively charged. This is **ionisation**. When the amount of energy transferred to the electrons is not sufficient enough to cause ionisation, these are pushed from their original orbit to the next outer orbit representing a higher level of energy, thus producing **excitation**. Electrons are liable to be deflected by the atoms from their path hence they move on zig-zag path. After the energy is exhausted, electrons attach to an atom and make it negatively charged. Beta-rays have very little penetrating power as compared to x-rays. The electrons liberated during ionisation also produce ionisation and excitation, as do the beta-rays.

Fast and Thermal Neutrons. Fast neutrons are produced in cyclotrons or atomic reactors as result of decay of radioactive isotopes of heavy elements. The velocity of fast neutrons is reduced by graphite or heavy water to generate *slow neutrons* or *thermal neutrons.* They are neutral particles and have strong penetrating power in biological materials. They move in straight line and are deflected by the nuclei of atoms. The neutrons are densely ionising radiations. They do not cause ionisation directly. The neutrons kick the nuclei of the atoms away which then produce ionisation. As the thermal neutrons are captured by atomic nuclei, the nuclei become unstable and *produce gamma-rays.*

X Rays and Gamma Rays. X rays and gamma-rays are non-particulate electromagnetic radiations of very short wavelengths. Both types of rays are photons or quanta. The x-rays have wave lengths from 0.001Å to 10Å. These are produced when a strong beam of electrons, as for example, a beam of cathode-rays fall on a metal surface. They are produced when the electrons in target substance are deflected from their path by the approaching strong beam of electrons in the form of radiation. These rays are electromagnetic in character and are comparatively more penetrating than cathode rays; but, they carry no charge. Therefore x-rays are not deflected in electric ormagnetic field. x-rays are often referred to as *hard* x-rays (wavelength 0.1Å to 0.001Å) or soft x-rays (wavelength. 10Å to 1Å).

Gamma-rays are produced by decay of some radioactive isotopes e.g. ^{60}Co, ^{14}C, etc. They are the secondary effect of atomic disintegration. After emission of an α and a β particle, the nucleus left behind is in the state of high energy (excited state). The excess energy possessed by it, is released in the form of a quantum or photon of γ-rays.

Gamma-rays are similar to x-rays and light rays but have extremely small wavelengths of 10^{-10} to 10^{-13} m. Gamma-rays have shortest wave lengths, except cosmic rays, like x-rays. They also cannot be deflected even in the strongest electric or magnetic field. They have greater penetrating power than x-rays. They cause ionisation of atom. Both x-rays and gamma rays are harmfull to living tissue.

The non-particulate radiations x-rays and gamma-rays are sparsely ionising than the particulate radiations on the other hand. They are much more penetrating than the particulate radiations. They move deep into the interior of an organism. Once the radiation is passed through an organism there is no longer any radiation energy in its tissues; the energy is dissipated almost instantaneously. Low energy photons transfer all their energy to individual electrons which are then raised to higher energy levels and ejected from the atoms of which they are the part. The initial ejection of electrons (e^-) produces *primary ionisations.* The high energy e^-, in turn, bring about the ejection of other electrons from adjacent atoms. This is *secondary ionisation.* The secondary ionisations are of greater significance than the primary ionisation. This ionising property of low energy photons of x-rays and gamma rays is referred to as *photoelectric effect.* A high energy photon transfers a part of its energy causing ejection of electrons from the atoms (*i.e.,* ionisation). Since it loses energy in repeated ionisations, the wavelength of photon becomes increasingly longer as it travels through the biological material. This is called *Compton scattering effect.* When a high energy photon passes close to the nucleus of an atom it may be completely absorbed. This is accompanied by ejection of a high energy e^- and a high energy positron (e^+) which may produce ionisation and excitation. This is *pair production effect.*

Mechanism of Action of Ionising Radiations

The biological effects of ionising radiations are qualitatively the same. These effects may be both direct and indirect. In direct effect, the radiation energy is directly transferred to an atom or molecule whereas in indirect effect, an atom or molecule receives energy indirectly by transfer from another atom or molecule. Indirect effect is mediated by free radical formation. The free radicals transfer their energy to the other molecule. This indirect effect is important particularly in presence of water, since ionised water molecules produce free radicals. A free radical is extremely reactive. It is an atom or group of atoms containing unpaired electrons. The primary effect of radiations is due to ionisation which results in the formations of free radicals.

(*i*) **Formation of Ion pairs.** When ionising radiations pass through the matter, they eject electrons from the outer orbit of the atoms. With the loss of electrons the atoms become positively charged ions. The ejected electrons move at high speed which, in turn, knock other electrons ejected from other atoms and liberate their energy. The free electrons, at lower energy level, jump to other atoms and make them negatively charged ions as shown below. Oxygen is one of the common electron acceptors. The positively and negatively charged ions, so formed occur in pairs and in this way a cluster or ionic pair is formed along the path of radiation. Ions must undergo chemical reactions to neutralize their charges in order to reach stable configuration. Perhaps during this chemical reaction the ionising radiations cause mutagenesis.

$$A \xrightarrow{\text{Photon}} A^+ + e^- \quad \text{(electron ejected from atom A)}$$

$$B \xrightarrow{+e^-} B^- \quad (e^- \text{captured by atom B})$$

$$A + B \xrightarrow{\text{Photon}} \underset{\text{(ion pair)}}{A^+ + B^-} \quad \text{(overall effect)}$$

(*ii*) **Formation of free Radicals.** Free radicals are neutral molecules with an impaired electron in the outer orbit. They have a very short life of 10^{-10} seconds. The major genetic effect of radiations is that their energy (photon) is involved in the production of free radicals from water. These free radicals then react with DNA to alter its structure.

$$H_2O \xrightarrow{\text{Photon of X-ray or gamma-ray}} H_2O^+ + e^- \text{ (from O atom)}$$

$$H_2O + e^- \longrightarrow H_2O^- \text{ (}e^-\text{ captured by one H atom)}$$

In this way an ion pair H_2O^+ and H_2O^- is formed. In presence of water each ion may form a hydrogen ion and a free radical as shown in the following reactions.

$$H_2O^+ \xrightarrow{\text{Water}} H^+ + OH^\circ \text{ (a free radical)}$$

$$H_2O^- \xrightarrow{\text{Water}} H^\circ + OH^- \text{ (}H^\circ\text{ is a free radical)}$$

H^+ and OH^- ions will combine to form water. H° and OH° are free radicals which are highly reactive. They react and produce H_2O_2 and organic peroxides. The free radicals are removed as follows:

$$H^\circ + OH^\circ \longrightarrow H_2O$$

$$H^\circ + H^\circ \longrightarrow H_2$$

$$OH^\circ + OH^\circ \longrightarrow H_2O_2$$

The presence of O_2 increase the yield of H_2O_2.

$$H^\circ + O_2 \longrightarrow HO^\circ_2$$

$$HO_2^\circ + HO_2^\circ \longrightarrow H_2O_2 + O_2$$

H_2O_2 dissociates to form H_2O and OH°

$$H_2O_2 + H^\circ \longrightarrow H_2O + OH^\circ$$

The formation of peroxides may not be significant in the cells containing peroxidases and catalases. H_2O_2 and organic peroxides are highly reactive and are supposed to be responsible for many biological changes including structural alterations in genetic material. The changes in genetic material results because of the effects of radiation on DNA.

Detection of Mutations

Many mutations can be detected cytologically in the chromosomes, but some particularly gene or point mutations are not so easily detectable. In the past, it remained a matter of dispute whether the visible mutations are in fact distinct from others or they are quantitatively different. It has been suggested, for example that a broken chromosome when rejoins will bear a scar that might alter the function of genes in that place (i.e., it brings about a mutation). Mutations cause many morphological and physiological abnormalities of varying degrees in the organisms. Thus the study of mutation will depend largely upon the methods used to detect the mutant individuals. The method of identification of changes in the morphological characters will be different. A recessive mutation in the gamete of diploid species will yield phenotypically normal offsprings unless the other gamete happens to carry same recessive gene as the result of mutation in the population. Apart from such cases newly arisen recessive gene may be carried in the heterozygotes for many generations till the mating of two such heterozygotes produces a homozygous recessive offspring.

Muller's ClB Method

One important method for studying the mutation rates was devised by H. J.Muller (1927). Muller used a stock of *Drosophila* in which there was an X chromosome carrying the dominant gene "bar" (B) a recessive lethal (1), and an inversion suppressing crossing over (C). This chromosome is termed as ClB chromosome (Fig. 22. 21). All the male flies carrying this chromosome die. So a cross between ClB heterozygote female and a normal male gives equal numbers of normal and bar eyed females and half that number of normal males (*i.e.,* 2 : 1 sex ratio). If the original male transmitted a sex linked recessive lethal, this will be carried by all F_1 female and all F_2 males arising from a bar eyed F_1 female will die. Thus Muller was able to detect the occurrence of mutations in the X chromosomes of the reproductive cells of male *Drosophila* by the failure of these flies to produce any grandson when their daughters carried a known lethal factor and a cross over suppressor in one of the chromosomes.

Many dominant or partly dominant genes do not always give rise to detectable mutants. Many of them produce minor effects that are too small to be observed or they might be attributed to environmental effects. Mutations influencing quantitative characters, for example, grain yield in maize, egg production in poultry, general vigour or intelligence in man are sometimes thought as environmental effects. Presence of lethals in mammals can be detected from the death of embryo in pregnant females.

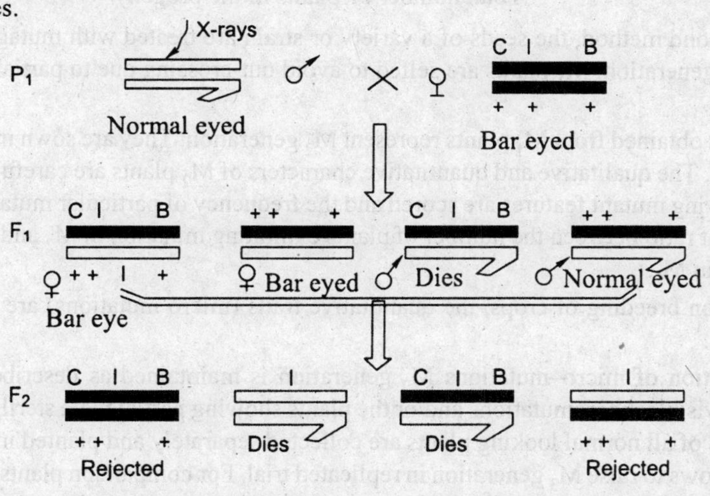

Fig. 22.31 ClB method of detection of mutation in *Drosophila melanogaster*.

A number of other techniques have been evolved to detect mutations in micro-organisms. The most widely used objective method for detecting mutations in fungi (*e.g.,* the bread mould, *Neurospora*) is based on the fact that certain types of mutant spores do not germinate on culture media containing only the bare growth substances—inorganic salts, sugar and vitamin Biotin (*i.e.,* on culture media lacking amino acids etc.). The nutritional requirements of mutants differ from those of the normals. When the spores of fungi are collected and sown on such minimal culture media, only the normal (non-mutants) spores will grow and the mutants will not. The suspension of germinating spores is then filtered from a minimal medium through fine guaze, the germinated spores are trapped because their mycelial web catches in the guaze and the non-germinated mutant spores pass through the guaze. These mutant spores are then plated in a nutritionally enriched medium and the cultures are tested for their nutritional requirements.

A similar method has been evolved to detect mutation in bacteria. This method utilizes the fact that the penicillin kills only growing cells. Hence non-mutant bacteria will be killed if cultured on minimal medium containing penicillin and the antibiotic resistant mutant strains remain alive but do not grow. Such mutants can be recovered by washing and transferring the cells into enriched culture medium.

Detection of mutations in Plants

The techniques for detection of mutations in plants are poorly developed. The following methods are generally followed for detection of mutations in plants :

1. In some plant species, e.g., maize strains homozygous for some recessive genes, as well as those for several dominant alleles are available. The seeds or plants of homozygous strains of dominant genes are treated with mutagens and the M_1 plants (plants treated with mutagens) are crossed with strain having recessive alleles of the same genes (*tester strain*). The treated plants are generally used as female parent because of partial male sterility in some M_1 plants. In the progeny of this cross, plants exhibiting recessive forms of such traits will get one allele from the tester plant and the other recessive allele is supposed to have arisen due to mutation in the mutagen treated parent (with dominant alleles of the genes). Such plants exhibiting recessive forms of traits governed by the concerned genes are scored and the frequency of mutation for a gene is calculated as follows :

   Mutation frequency (%) for a gene

$$= \frac{\text{No. of plants having recessive form of trait governed by a gene}}{\text{Total number of plants in the progeny}} \times 100$$

2. In the second method, the seeds of a variety or strain are treated with mutagen and grown to raise M_1 generation. M_1 plants are selfed to avoid out-crossing due to partial male sterility in M_1 plants.

   The seeds obtained from M_1 plants represent M_2 generation. They are sown in the field to raise M_2 plants. The qualitative and quantitative characters of M_2 plants are carefully recorded. The plants having mutant features are scored and the frequency of particular mutation is calculated as per cent ratio between the number of plants exhibiting mutation in M_2 and the total number of plants in M_2.

3. In mutation breeding of crops, the quantitative traits (micro mutations) are sometimes more important.

   For detection of micro–mutations M_2 generation is maintained as described above. Plants showing visible macromutations and/or the plants showing partial male sterility are discarded. The seeds of all normal looking plants are collected separately and planted in individual plant progeny rows to raise M_3 generation in replicated trial. For comparison plants of parent variety are also grown in separate plots. Data on quantitative traits in M_3 plants are recorded and mean and variance of M_3 progeny are compared with those of parental variety to ascertain if M_3 progeny deviated significantly from the parent variety in respect of given trait.

QUESTIONS

1. What do you understand by variations ? How do the heritable and nonheritable variations differ from each other ?
2. What do you mean by spontaneous mutations and induced mutations ? Discuss the various causes of spontaneous mutations.
3. Define induced mutation and mention the various factors which induce mutations.
4. Discuss briefly the effects of ultra violet radiations on genetic material.
5. Give an account of the effect of various ionising radiations on genetic system and mention briefly the mode of action of ionising radiations.
6. What do you mean by gene or point mutations ? Discuss various biochemical process which induce gene mutations.
7. Describe the different type of chromosomal abnormalities which occur during meiosis.

8. How will you distinguish cytologically the following:
   (*a*) Paracentric inversion and Pericentric inversion, and
   (*b*) Translocation homozygote and translocation heterozygote ?
9. Describe and illustrate how deletion, inversion and reciprocal translocation arise in nature ?
10. Describe the detailed procedure utilized for the detection of sex-linked lethal mutation utilizing CIB method.
11. Describe in brief how mutation takes place at molecular level.
12. Describe different mutagens that induce mutations.
13. Write brief notes on:
   (*i*) Characteristics of mutations
   (*ii*) Excision repair process
   (*iii*) Detection of mutations.

23 CHAPTER

Variations in Chromosome Number

Every cell in the body of living organism is provided with a packet of genetic material (the nucleus) which contains chromosomes. The number of chromosomes in the nuclei of a given species is fixed.

Some species of plants and animals have identical chromosome numbers. In such cases one should not think that these organisms will show similar characters, because it is not the number of chromosomes which differentiates various species from one another, but rather the nature of hereditary material in chromosomes determines the characters of species. In most of the plants and animals the chromosome number ranges between 10 and 50. Numbers above and below this range are comparatively rare. The highest number of chromosomes reported so far is about 1100 found in *Radiolaria,* an unicellular marine protozoan. In plants, chromosome number ranges from two pairs (in a member of compositae, *Haplopappus gracilis*) to several hundreds found in ferns. In *Ophioglossum* 768 chromosomes have been counted in each nucleus. In higher plants and animals, body cells normally have two identical sets of chromosomes (2*n*), *i.e.,* there are two identical members for each type of chromosomes in the nucleus. In 14 chromosomes of pea, for example, there are 7 chromosomes of different shapes, sizes and structures and each kind of chromosome has its own homologue. So, fundamentally there are 7 pairs of chromosomes in pea. The minimum number of chromosomes, though all are different which function as a harmonious and integrated unit, is called **basic chromosome number**. This basic number is denoted by the letter '*x*'. The individuals in which the somatic cells have chromosomes in pairs are called diploid (2*n*). When the reduction division takes place in a diploid cell, the daughter nuclei so formed will have chromosome number that is just half the diploid chromosome number, *i.e.,* each daughter cell now gets one member of each chromosome pair. The chromosome number present in the product of meiosis of a diploid cell is referred to as haploid or gametic number or genome. When two gametes, each carrying haploid number of chromosomes (*n*) fuse they produce a diploid cell. The diploid cell produces the diploid body. In this way, in the life cycle of an individual, the haploid and diploid conditions come regularly one after another and the same chromosomes number is maintained from generation to generation. Several cases, however, are known in which chromosome number becomes changed in the somatic cell. This condition is known as ***ploidy***. The change in the chromosome number involves addition or elimination of individual chromosomes or of complete chromosomal set (genome). On this basis two main classes of ploids can be recognised:

1. Aneuploids or Heteroploids.
2. Euploids.

1. **Aneuploidy or Heteroploidy**. It is a condition in which nuclei contain chromosomes whose number is not true multiple of basic chromosome number or genome. The organisms with such abnormal conditions are called aneuploids.
2. **Euploidy** (GK. *Eu* = true; and *ploid* = unit). This is a condition in which the change in the chromosome number involves additions or elimination of complete chromosome set or genome. The basic chromosome number of euploids are represented by the haploid (*n*). The euploids above the haploid or monoploid level are called polyploids. They may be diploid (2*n*), triploid (3*n*), tetraploid (4*n*), pentaploid (5*n*), hexaploid (6*n*) and so on. In all these cases of euploids the chromosome numbers are exact multiples of original haploid chromosome number (*n*).

ANEUPLOIDY

The individuals carrying chromosome numbers other than somatic or diploid (2*n*), which are not true multiples of basic chromosome number or genome are called *heteroploids* or *aneuploids*. Thus *aneuploidy* results due to an addition of loss of less than a complete set of chromosomes to or from a diploid set. Thus there are two classes of aneuploids:

(*i*) *Hyperploids* which result due to addition of one or more chromosomes to a diploid set, and

(*ii*) *Hypoploids* which develop due to loss of one or more chromosomes from a diploid set.

Causes or Origin of Aneuploids

1. Both the classes of aneuploids arise from an abnormal distribution of chromosomes during anaphase of meiosis I. Because of irregular chromosome distribution at the poles, one daughter cell receives one or more extra chromosomes and the counterpart lacks in one or more chromosomes. This type of abnormal situation arises when one or more pairs of chromosomes (bivalents) fail to separate or disjoin and move to one spindle pole as such, this is called **non-disjunction**. The gametes having extra chromosome, when fuse with normal haploid gametes result in hyperploids and those containing one or more chromosomes less than a haploid number if unite with normal haploid gametes produce hypoploids. Triploid plants are important source of aneuploidy as they produce aneuploid gametes in high frequency. Many univalents are regularly present at Metaphase I due to asynapsis and desynapsis. Consequently most of the gametes are aneuploids.
2. Non-orientation of one or more bivalents at metaphase I.
3. Loss of individual chromosome in meiosis or mitosis leads to the formation of nuclei with hypoploid chromosome number.
4. Irregularities in the segregation of chromosomes during meiosis in polyploids.
5. Multipolar mitosis with irregular chromosome distribution to the daughter cells.

Heteroploids or aneuploids are arranged in several groups. They are listed in Table 23.1 and Fig. 23.1:

Table 23.1 Aneuploids

Aneuploids or Heteroploids $2n \pm$ a few chromosomes					
	Hyperploids (2n + few)		*Hypoploids (2n – few)*		
1.	Trisomic	$(2n + 1)$	Monosomic	$(2n - 1)$
2.	Tetrasomic	$(2n + 2)$	Nullisomics	$(2n - 2)$
3.	Double trisomic	$(2n + 1 + 1)$	Double monosomic	$(2n - 1 - 1)$
4.	Double tetrasomic	$(2n + 2 + 2)$	Double nullisomic	$(2n - 2 - 2)$

Aneuploidy does not include the cases in which numerical variation involves B chromosomes.

1. Monosomic (2*n* – 1)

An organism lacking one chromosome of a diploid complement is called monosomic (2*n* – 1). Suppose in an organism diploid set is formed of 4 pairs of chromosomes. In this, by chance one chromosome

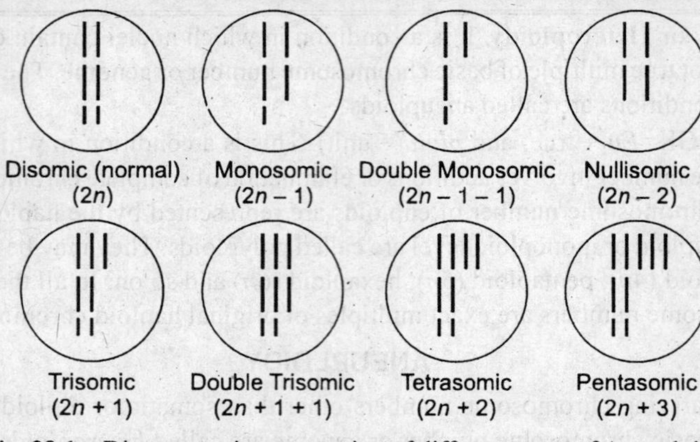

Fig. 23.1 Diagrammatic presentation of different types of Aneuploids.

is missing. So in this case one may find 3 normal pairs of chromosomes and one singlet chromosome. This should, therefore, be called as monosomic. Since monosomics, as well as nullisomic and trisomy affect only one chromosome of the haploid complement, there may be expected as many different types of monosomics as are the chromosomes in the haploid set.

In some plants, monosomics are not viable. In *Nicotiana tabaccum* with basic chromosome number 24, Clausen and his co-workers have found all 24 possible monosomics as expected and have noted that they were much different from one another. In wheat (*Triticum vulgare*) haploid number '*n*' is 21. So one may expect that 21 monosomics of different types should be available. Sears (1948) obtained all the 21 possible monosomics in bread wheat (*Triticum vulgare*), but they did not differ much from the normal diploids.

The monosomics are very rare. However, viable monosomics have been studied in *Drosophila,* tobacco, *Datura,* maize, tomatoes and many other organisms.

Monosomics show abnormal type of meiosis. They have chromosome constitution $2n - 1$, so they produce two kinds of gametes; some with n and others with $n - 1$ chromosomes. The odd chromosome, which has no homologue to pair tends to pass at random to one pole during meiosis I or may divide, as in mitosis, into two chromatids which move to opposite poles. Frequently, however, it acts as a laggard at anaphase and is not included in any of the daughter nuclei and subsequently lost. For this reason the gametes having '*n*' chromosomes occur less frequently than the gametes of $(n - 1)$ type (*e.g.,* wheat produces about 75% $n - 1$ gametes). In plants, the nuclei with a missing chromosome seldom survive. The $n - 1$ gametophytic generation fails, presumably because the complete absence of certain genes that occur on the missing chromosome means the failure of certain biochemical reactions which are accomplished in presence of the enzymes formed by these genes. Thus, monosomic analysis makes it clear what genes are found on the particular chromosome or linkage group.

2. Trisomic (2*n* + 1)

An organism containing two complete genomes *plus* one extra chromosome is termed trisomic $(2n + 1)$. In the normal process of meiosis I, sometimes chromosomal pairs separate in such a way that one member of each pair goes to one daughter nucleus and other member goes to other daughter nucleus. But very rarely one pair of chromosomes fails to disjoin and finally it moves as such to one spindle pole. So half of daughter cells receive an extra chromosome and half of them lose one. In this way gametes with $(n - 1)$ and $(n + 1)$ chromosomes will develop. If by chance the gamete with $(n + 1)$ chromosomes fuses with normal gamete carrying '*n*' chromosomes, the resulting zygote will have $n + 1 + n$ or $2n + 1$ chromosomes and it will produce trisomic organism. Gametes with $(n - 1)$ chromosomes on fusion with gamete carrying '*n*' chromosomes will give rise to monosomic $(2n - 1)$ zygote. Many of the mutations in *Oenothera* noted by De Vries are trisomics. Trisomics are widespread in nature and have been extensively studied in *Datura,* maize, tomatoes, wheat, tobacco and *Drosophila.*

Blakeslee and Belling (1924) announced a mutant type in *Datura* having 25 chromosomes rather than normal 24 chromosomes ($2n = 24$). In that, at first meiotic metaphase one of the 12 pairs was found to have an extra member, *i.e.,* there were 11 normal pairs and one trisome.

The extra chromosome in the trisomic is either completely homologous or partially homologous to a certain chromosomes of the standard complement and, accordingly, the trisomics may be classified into *Primary, Secondary* and *Tertiary* trisomics.

Primary Trisomics. When the extra chromosome is completely homologous to one pair of homologues in the diploid chromosome compliment and, therefore, one chromosome is represented three times in an otherwise diploid complement, it is primary trisomy. Such homology permits formation of a trivalent or one bivalent and rod shaped univalent during prophase of meiosis I. The trivalents may be of different configurations as shown in Fig. 23.3.

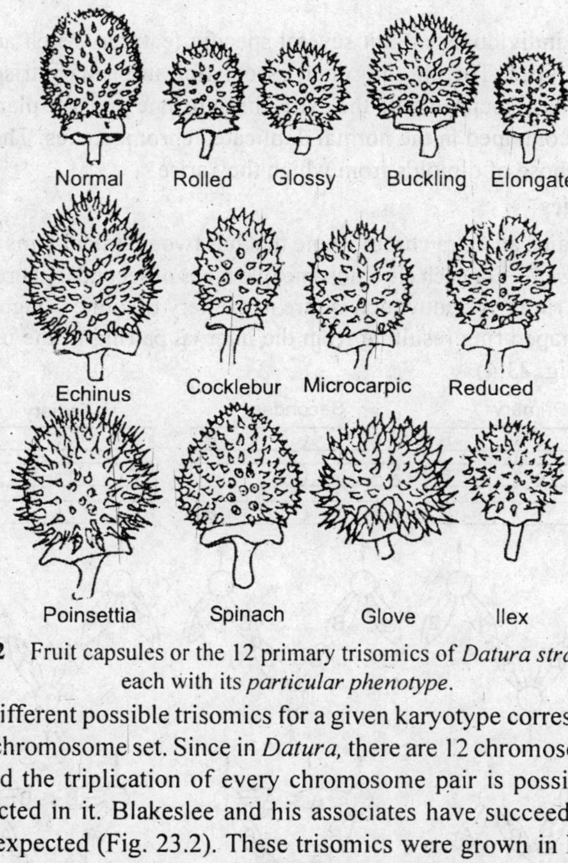

Fig. 23.2 Fruit capsules or the 12 primary trisomics of *Datura stramonium*, each with its *particular phenotype*.

The number of different possible trisomics for a given karyotype corresponds to the number of chromosomes in each chromosome set. Since in *Datura,* there are 12 chromosome pairs with different genic composition, and the triplication of every chromosome pair is possible, 12 distinguishable trisomics can be expected in it. Blakeslee and his associates have succeeded in producing all 12 different trisomics as expected (Fig. 23.2). These trisomics were grown in Blakeslee's garden and all were found to have distinct phenotypic features which could easily be attributed to extra dose of one chromosome to each of the 12 individual sets of chromosomes.

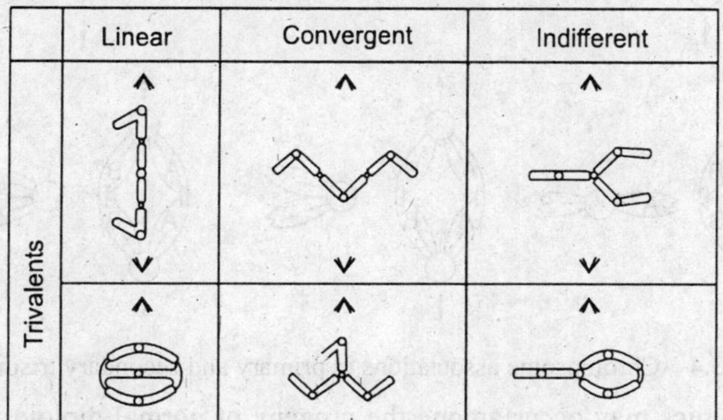

Fig. 23.3 Different types of trivalents formed in trisomic cell.

Trisomics show irregular distribution of chromosomes during anaphase I of meiosis. From a trisomic cell the two kinds of gametes; some with 'n' chromosomes and some with $(n + 1)$ chromosomes, would be expected after meiosis. Recessive genes present on the extra chromosomes in trisomic complexes will express themselves less frequently than in normal diploid plants because comparatively more wild type alleles are present to check their expression. It has been observed by geneticists in *Datura* that extra chromosome enters the offspring through female gamete because the egg tolerates extra dose of one chromosome without any loss of viability. The pollen with extra dose of chromosome are not viable. Only normal pollen with haploid chromosome number succeeds in fertilizing the egg.

Primary trisomic individuals exhibit several specific features which are quite different from those seen in the wild types. The chromosomal association in primary trisomic is shown in Fig. 23.4. The additional traits associated with the extra chromosome give the plants extra dose of all the genes that are already contained in the normal duplicated chromosomes. The viability and fertility are usually lower than those of diploids from which they arise.

(*ii*) Secondary Trisomics

When the trisomic contains an extra chromosome that has two identical arms (isochromosome), it is said to be *secondary trisomic.* For each chromosome two types of secondary chromosomes are possible. At first meiosis either a ring trivalent which is rare in primary trisomic (diagnostic configuration) or a bivalent and one U-shaped ring resulting from the internal pairing of the univalent with identical arms may be formed (Fig. 23.4).

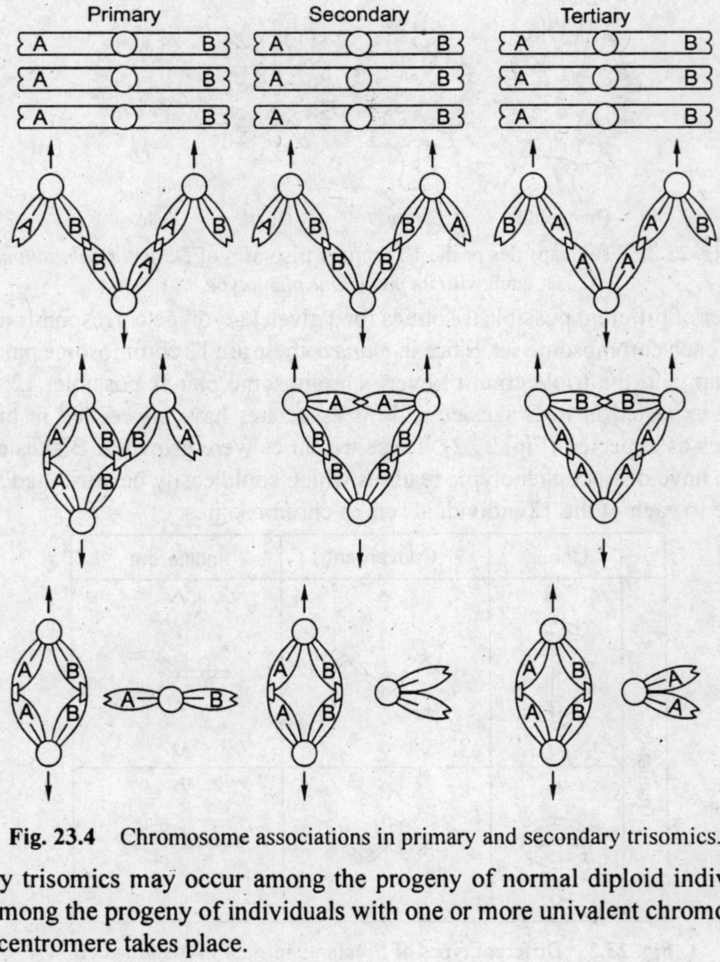

Fig. 23.4 Chromosome associations in primary and secondary trisomics.

Secondary trisomics may occur among the progeny of normal diploid individuals but their best source is among the progeny of individuals with one or more univalent chromosomes in which misdivision of centromere takes place.

Tertiary Trisomics. In these, the extra chromosome is the result of reciprocal translocation between two standard chromosomes. The ends of extra chromosome are homologous with the ends of two different chromosomes. The tertiary trisomics are identified by the presence of a chain of three chromosomes. The tertiary trisomic individual has two chromosome segments represented three times each in an otherwise diploid complement.

The tertiary trisomics occur regularly and possibly exclusively in the progenies of translocation heterozygotes by 3 : 1 segregation of four chromosomes which are partially homologous in such heterozygotes and may form a quadrivalent during prophase I of meiosis.

Segregation of genes included in a primary trisomic association differs from that of genes found in any of the other chromosomes of the complement. For a particular locus (A/a), a trisomic may have the genotype AAA (triplex), AAa (duplex), Aaa (simplex) or aaa (nulliplex). If one of the three chromosomes is not eliminated, selfing or intercrossing of trisome would theoretically result in 25% tetrasomics, 50% trisomics and 25% disomics (diploid) progeny. But this expectation is rarely realized because of irregular meiosis and non-functioning of $n + 1$ gametes.

A regular chromosome distribution at meiosis I (two to one pole and one to the other) in genotypes AAa and Aaa would lead to expected gametic ratio 1 AA : 2Aa : 2A : 1a (dominant : recessive = 5 : 1) and 2Aa : 1 aa : 1A : 2a (dominant : recessive = 1 : 1) respectively. If the male gametes possessing the additional chromosome ($n + 1$) are non-viable as in many plants and chromatid segregation is absent, the following genotype proportions are expected in the progenies of selfed or intercrossed trisomics of genotypes AAa and Aaa (Table 23.2):

Table 23.2. Progenies of selfed and intercrossed trisomics of AAa and Aaa genotypes

Genotype	*AAa*		*Genotype Aaa*		
gametes	♂ 2A	1a	gametes	♂ 1A	2a
♀			♀		
1AA	2AAA	1AAa	2Aa	2AAa	4Aaa
2Aa	4AAa	2Aaa	1aa	1Aaa	2aaa
2A	4AA	2Aa	1A	1AA	2Aa
1a	2Aa	1aa	2a	2Aa	4aa

The observed genotypic ratio in the progenies may deviate considerably from the above theoretical expectation. If instead of chromosome segregation, chromatid segregation takes place, and the crossing over takes place between centromere and locus, then the expected gametic ratio for the genotype AAa will be 6AA: 8Aa:1aa:10A:5a instead of 1AA : 2Aa : 2A :1a.

3. Double Trisomics ($2n + 1 + 1$)

Sometimes an organism receives two complete genomes *plus* two extra chromosomes. This is called double trisomic. The extra chromosomes are homologues of two different chromosomal pairs in normal diploid set. Thus two different chromosome pairs in the normal diploid set will have one additional homologue each with them and thus they form two trisomes. The phenotypic expression of genes on extra chromosomes in trisomics will be deeper than in normal diploids.

4. Tetrasomic ($2n + 2$)

Aneuploid nuclei which have normal diploid genome and one extra pair of chromosomes are called tetrasomics ($2n+2$). In this one pair of chromosomes in the diploid set is reduplicated due to additional chromosome pair. Phenotypic expression of the genes in tetrasomics may be deeper than the effects found for the genes in corresponding trisomics.

In meiosis, the four homologues of tetrasomic set often tend to form a quadrivalent (Fig. 23.5). If disjunction by twos is regular, a fair genetic system will operate. Generally quadrivalents are not formed and disjunction (separation) is not always regular. Nevertheless, tetrasomics are more regular than trisomics in meiosis.

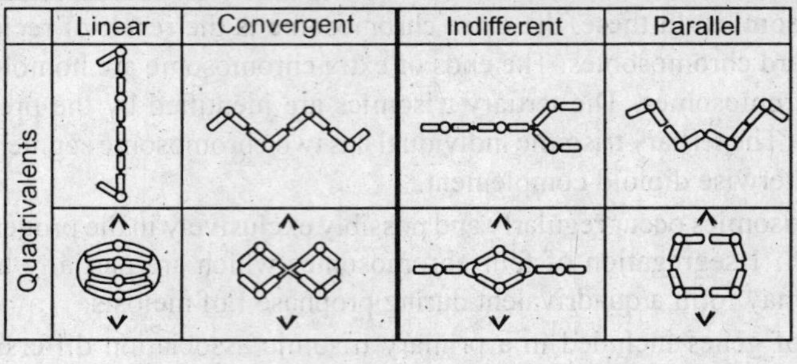

Fig. 23.5 Quadrivalents in tetrasomics.

5. Nullisomic ($2n - 2$)

In these both the homologues of a particular chromosome set are somehow lost so that a chromosomal pair is completely missing from the normal 2n set. The phenotypic expression of such a plant, when compared with that of normal plant, will indicate the number of genes contained in the missing chromosomes and the respective characters they govern. Nullisomics are generally non-viable like the monosomics. In maize and wheat, several nullisomics have been reported. Nullisomics are generally obtained by selfing monosomics. When the gametes containing $n - 1$ chromosomes fuse, they form nullisomics.

EUPLOIDY

The total amount of genetic material in prokaryotic cell or the haploid set of chromosomes of an eukaryotic organism is called genome. The *euploidy* (Gr. *eu.,* even or true; *ploid*, unit or set) refers to a state in which a cell or an individual contains one or more complete chromosome set or sets. The individuals having an integer number of complete chromosome sets or genomes are called *euploids*. Euploids are classified into haploids, diploid and polyploids as detailed in Table 23.3:

Table 23.3. Euploids and their geomomic composition.

Euploids	*Number of genome*	*Description of chromosomes*
1. Haploid or Monoploid	*n*	Contains the basic chromosomes characteristic of the species. Genetically hemizygous.
2. Diploid	$2n$	Two homologous sets.
3. Polyploids:	More than two genomes per cell.	
(*i*) Triploid	$3n$ ($2n$ + n)	Three homologous sets
(*ii*) Tetraploid	$4n$ ($2n + 2n$)	Four homologous sets
(*iii*) Pentaploid	$5n$	Five sets of chromosome
(*iv*) Hexaploid	$6n$	Six haploid sets

Haploids or Monoploids

Monoploids or haploid organisms contain single genome in their cells, *i.e.*, they contain one member of each kind of chromosome. The haploid chromosome number (n) is usually found in gametes of diploid plants and animals and this is unusual for a somatic cell of higher plant. Monoploidy has been reported in several cases, both plants and animals. Viruses and bacteria whose genome consists of single linkage structure are haploids. Majority of lower plants, particularly thallophytes and bryophytes exist in monoploid condition. In higher plants, haploids develop as a result of parthenogenesis of haploid eggs. In *Datura stramonium, Oryza sativa, Oenothera, Triticum, Zea mays* and many other plants, haploidy has been noted. In some animals also, haploids are common. The males of bees, wasps and other hymenoptera are normally haploids.

Characters of Haploids

1. These plants are usually weaker and smaller than the diploids, but in pepper the haploids are as healthy as normal diploid plants.
2. Leaves are generally small.
3. Plants are of low viability.
4. **Cytology of Haploids.** High degree of sterility has been observed in these plants. It is so only because meiosis is irregular. As the chromosomes in haploid set do not have their homologues, the pairing of chromosomes is not possible and the chromosomes are found as univalents at metaphase I of meiosis. So neither the eggs nor the pollens are formed. However, if meiotic process succeeds at all, the univalent chromosomes are found scattered all over the cell. They may constitute restitution nucleus including all chromosomes and may thus give rise to gametes having a complete haploid set of chromosomes. In haploid *Datura stramonium* dyads are produced by microsporocytes which directly enter second meiosis and such haploid spores are functional. Sometimes the distribution of chromosomes to the two poles is irregular and the gametes so formed are highly non-viable (*i.e.,* they die soon). In haploid animals, reduction division does not occur during spermatogenesis. Normal gamete is possible if by chance all the chromosomes pass to one pole during the irregular meiosis.

According to Kimber and Riley (1963) the haploids may be classified as follows on the basis of ploidy status of their parents:

- Euploids
  - Euhaploids (arise from Euploids)
    - Monohaploids (arise from diploids)
    - Polyhaploids (arise from polyploids)
      - → Autopolyhaploid
      - → Allopolyhaploid
  - Aneuhaploids (arise from aneuploids)
    - Disomic haploid ($n + 1$)
    - Nullisomic haploids ($n - 1$)
    - Substitution ($n - 1 + 1$) haploids
    - Addition haploids ($n + 1$)

Monoploids may arise in several ways. In plants, they arise by parthenogenetic development of embryo from unfertilized eggs.

Monoploids and haploids occur spontaneously in maize in low frequencies. They may be induced from pollen grains through callus or embryoid production and by chromosome elimination through interspecific hybridization as in *Hordeum jubatum* × *H. vulgare.* Production of haploids can be induced by using pollen incapable of effecting fertilization such as pollen from a related but incompatible species or the pollen inactivated by exposures to high dosages of x-rays or the pollen made inviable by prolonged storage prior to pollination.

Application of Monoploids in Breeding

Monoploids are generally sterile and are of no use in agriculture directly. They do not undergo segregation and carry a single set of genes. So they are genetically pure. Thus, they can be used experimentally to a good advantage in crop improvement.

(*i*) They are used for developing homozygous diploid lines. In maize, extensive inbreeding is necessary to select lines for hybrid seed production. Monoploids provide a short cut way for getting homozygous diploids which are then tested for their value in the production of hybrid seeds.

The normal diploids can be obtained by treating monoploid tissue with colchicine solution.

(*ii*) They may be useful in isolation of mutants because the mutant allele, even if recessive, expresses itself in M_1 generation due to a single dose of gene in somatic tissues. Haploid mutants can be used for developing homozygous diploid mutant line in a single generation.

(*iii*) Pure autotetraploid varieties of certain economic crops like potato can be produced by doubling the chromosome number of haploids to tetraploid level.

(*iv*) It is easier to get desirable haploid derived diploid lines than to get a desirable zygote derived diploid lines.

Diploids

Most common type of euploid is the diploid which has two complete genomes. As mentioned earlier this chromosome arrangement is nearly universal among animals and common in plants. These organisms are normal in their behaviour, so they are not discussed here in detail.

Polyploids

Polyploidy is commonly met within plant world. In animals, it is indeed rare. About one-half of total species of flowering plants are polyploids (Stabbins, 1950). Among certain families the proportion of polyploids to diploids is higher than 50 per cent. About two-third species of all grasses are polyploids. The cases of polyploidy may be found in many other common plants, *e.g., Chrysanthemum, Solanum, Brassica,* Wheat, *Nicotiana.* In the genus *Chrysanthemum,* the basic number of chromosomes is 9. In this genus species are known with 36, 54, 63, 72 and 90 chromosomes. In wheat (*Triticum sp.*) the basic chromosome number is 7. In this, several varieties are known with 14, 28 and 42 chromosomes which are diploid, tetraploid and hexaploid respectively. In *Nicotiana* with basic chromosome number 12, the varieties with 24, 48 and 72 chromosomes are known. These are all polyploids.

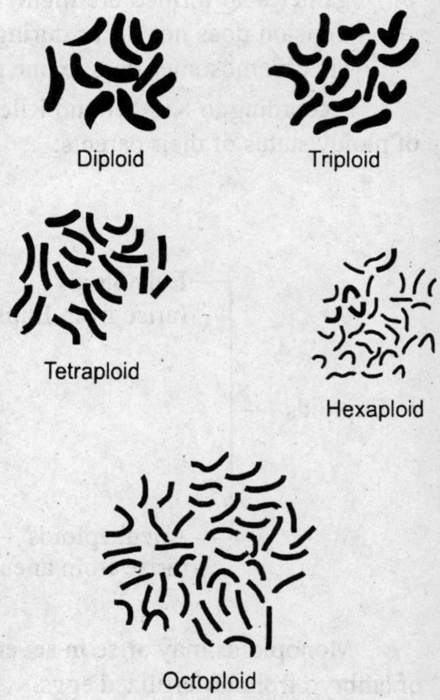

Fig. 23.6 Chromosomes in polyploids of Rose.

Similarly the genus *Solanam*, in which basic chromosome number is 12, consists of several polyploid species and varieties with 24, 48, and 72 chromosomes. In Rose, the basic chromosome number is 5. In this, diploid, triploid, tetraploid, hexaploid and octaploids have 10, 15, 20, 30 and 40 chromosomes respectively (Fig. 23.6).

In animals, the polyploids are rare on account of more delicate sex balance than in plants. In animals, addition of chromosomes generally gives rise to intersexes which are sterile (*i.e.,* do not reproduce). Sterility is virtually always associated with departure from diploid number in animals. Only those animals which can reproduce partheno-genetically (*e.g., Artemia salina*) develop polyploid races. It is so because parthenogenesis enables these animals to escape from the hazards of anomalous gametes.

Two Types of Polyploids

Two factors could account for the origin of polyploids from diploids:

1. Doubling of the chromosome number in reproductive tissue capable of giving rise to gametes.
2. Failure of reduction in the chromosome number during the formation of gametes.

On the basis of their origin, the polyploids are of the following two types.

1. **Autopolyploids**. The individuals containing more than two genomes or sets of homologous chromosomes per cell are called *autopolyploids.* More scientifically, in these organisms, multiple genomes are identical or nearly so. Autopolyploidy arises due to the failure of anaphase resulting in the duplication of genomes. For example, if a diploid species has two identical sets of chromosomes or genomes (AA), the *autotriploid* will have three similar genomes (AAA) and the *autotetraploid* will have four such sets of homologous chromosomes (AAAA) per cell (Fig. 23.7).

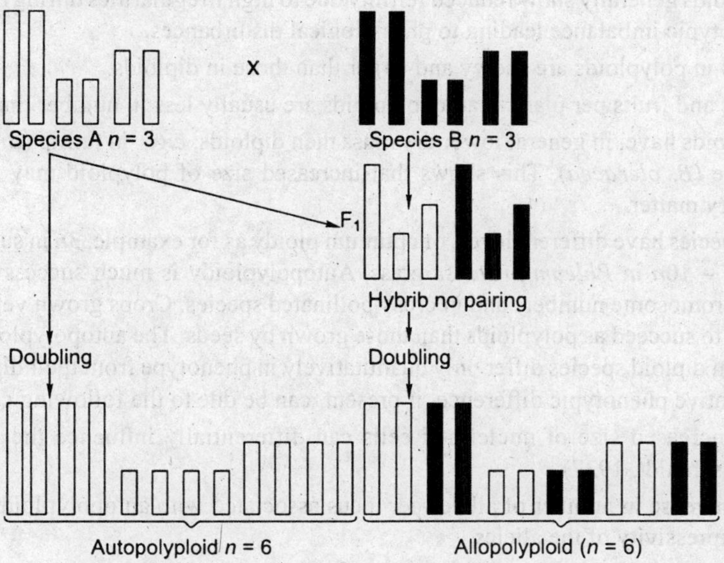

Fig. 23.7 Origin of Autopolyploid and Allopolyploid.

2. **Allopolyploids**. The individuals in which the multiple genomes are not alike are called *allopolyploid.* In allopolyploids, different genomes are derived from two or more distinct species by hybridization and contain the structurally and genetically different sets of chromosomes of the particular parents, each set being present more than once.

Suppose, there are two different species, one with genomes AA and the other with BB. A cross between these two species will produce a hybrid AB. If the genomes A and B of the hybrid are multiplied spontaneously or artificially, so that the individual contains two or more sets of each genome, it is called allopolyploid (Fig. 23.7). If the duplication of different genomes or sets of chromosomes takes place only once, it is referred to as *amphidiploidy* or *allotetraploidy* and the resulting polyploid is called allotetraploid or amphidiploid (*e.g.,* AA, BB). If the multiplication is taking place more than once (*i.e.,* higher multiplication) it is called *allopolyploidy*. It results in *allopolyploids* like *allohexaploids*, *allooctaploids*, etc.

Autopolyploidy

Autopolyploids are not very common in occurrence and are of little evolutionary consequence. Polyploidy appears very seldom in animals but in plants many species are known to be represented by polyploids.

Morphological Features of Autopolyploids

The morphological features of polyploids may vary in different species. Some general features are listed here as follows:

(*i*) In many cases autopolyploidy lead to increase in general vigour and vegetative growth and so they are often larger than their diploid counterparts. In some cases, however, autopolyploids are smaller and weaker than diploids.

(*ii*) Polyploids have large cell sizes than diploids. The guard cells are larger and the frequency of stomata per unit area is lower in the polyploids than in diploids.

(*iii*) Pollen grains in polyploids are larger than in corresponding diploids.

(*iv*) The cells of polyploids contain more chloroplasts than those of diploids.

(*v*) Polyploids are slower in growth and have thicker and larger leaves as compared to diploids.

(*vi*) The flowering occurs later in polyploids than in diploids.

(*vii*) Autopolyploids generally show reduced fertility due to high irregularities during meiosis which causes genotypic imbalance leading to physiological disturbances.

(*viii*) The flowers in polyploids are showy and larger than those in diploids.

(*ix*) The flowers and fruits per plant in autopolyploids are usually less in number than in diploids.

(*x*) Autopolyploids have, in general lower dry mass than diploids, *e.g.,* in turnip (*Brassica rapa*) and cabbage (*B. oleracea*). This shows that increased size of polyploid may not represent increased dry matter.

(*xi*) Different species have different levels of optimum ploidy as for example, 3*n* in sugarbeet (*Beta vulgaris*), 8 – 10*n* in *Phleum pratense* grass. Autopolyploidy is much successful in species with low chromosome numbers and in cross-pollinated species. Crops grown vegetatively are more likely to succeed as polyploids than those grown by seeds. The autopolyploids which are derived from diploid species differ only quantitatively in phenotype from their diploid parents. The quantitative phenotypic difference, if present, can be due to the following causes:

(*a*) The increased size of nuclei and cells can differentially influence the gene activity (Goldschmidt, 1937).

(*b*) The increase in number of alleles per locus associated with autopolyploidy can modify the expressivity of the alleles.

(*c*) If the diploid form is heterozygous for the concerned allelic pairs, an increase in the number of chromosome sets may lead to modifications of the dominance relationships of the alleles.

Cytological Behaviour of Autopolyploids

The cytological behaviour of chromosomes vary with the level of ploidy. In ***auto-triploids***, a variable number of trivalents, bivalents and univalents may be observed in the meiotic cells during prophase I. At anaphase I, separation of chromosomes of trivalents and distribution of univalents are irregular. All possible gamete types with chromosome numbers between haploid to diploid may result which may produce a range of aneuploid progeny, *e.g.,* trisomics, monosomics, tetrasomics, nullisomics, double trisomics and so on. Autotriploids are highly sterile.

In **autotetraploids**, there are four sets of homologous chromosomes. During the prophase of meiosis I, chromosomes of each set are present as quadrivalents or otherwise. 2 bivalents or one trivalent and one univalent or four univalents. The pairing of chromosomes derived from one genome is called autosyndesis. Autotetraploids have somewhat higher fertility because the chances of normal pairing are increased, but the fertility is in all cases lower than that in diploids. In most part, the autopolyploids with even number of haploid sets are more fertile than those with an odd number of sets. In the progeny of autotetraploids, the majority of plants are tetraploids, but some plants showing variable numbers of chromosomes may also be produced. There is a considerable evidence that the fertility of autotetraploids can be improved through selection. This has been achieved in some cases such as maize, rye, rice, bajra etc.

Segregation of Genes in Autopolyploids

The meiotic segregation of genes in autopolyploids is more complex than in diploids. The number of alleles of each gene is represented according to the ploidy level of the individual and gametes containing more than one allele of each gene (homo or heterozygotic) may be produced. According to the number of dominant and recessive genes at a particular locus, the genotype of an autotetraploid may be *quadriplex* and *monoallelic* (AAAA or A^4), triplex and biallelic (AAAa or A^3a), duplex and biallelic (AAaa or A^2a^2), simplex and biallelic (Aaaa or Aa^3) and nulliplex and monoallelic (aaaa or a^4). Autopolyploids show the so called polysomic inheritance. The segregation of genes in autopolyploids is affected by factors which play no essential role in diploid. Among such factors are the number and position of chiasmata in the multivalents, the distance between particular locus and centromere, the behaviour of homologues in multivalent associations during anaphase I and the presence of univalents.

In **autotriploids**, the actual segregation ratios correspond, in principle, with those of trisomics but, as a rule, these ratios are strongly affected by irregularities of chromosome distribution and by chromosome elimination. Therefore, it is difficult to predict segregation pattern of genes in autotriploids.

In **autotetraploids**, if it is assumed that the four homologous chromosomes are distributed to poles in 2 : 2 during anaphase I, theoretical segregation ratios for various autotetraploid genotypes of a locus may be calculated. Theoretically calculated segregation ratios which do not take into consideration, crossing over between the locus concerned and the centromere are referred to as *chromosome segregation.* The segregation ratios which take into account regular crossing over between the concerned locus and centromere and thereby modified segregation of heterozygous loci are referred to as "*chromatid segregation*" (double reduction). When the locus of gene is very close to centromere, crossing over would not occur between the centromere and gene locus. In such a case the two sister chromatids of each chromosome are attached to the same centromere and would move to same pole during anaphase I of meiosis. The relative frequencies of the gamete types and zygote types of different autotetraploid genotypes after selfing and reciprocal crossing, the identical genotypes for pure "*chromosome segregation*" are presented in the following Table 23.4:

Table 23.4 Frequencies of the gamete types and zygote types of autotetraploid genotypes for pure *Chromosome segregation.*

Parent	*Gametes*				*Zygotes*					
genotype	*AA*	*Aa*	*aa*	*Divisor*	A^4	A^3a	A^2a^2	Aa^3	a^4	*divisor*
Quadruplex (AAAA)	1	—	—	1	1	—	—	—	—	1
Triplex (AAAa)	1	1	—	2	1	2	1	—	—	4
Duplex (AAaa)	1	4	1	6	1	8	18	8	1	36
Simplex (Aaaa)	—	1	1	2	—	—	1	2	1	4
Nulliplex (aaaa)	—	—	1	1	—	—	—	—	1	1

If it is assumed that the two sister chromatid segments with identical alleles of one chromosome are distributed after crossing over, the corresponding distribution of chromatids into the same meiotic product will modify expectations of gamete and zygote classes (Table 23.5).

Table 23.5 Segregation ratios of autotetraploid genotypes (pure chromatid segregation).

Parent	*Gametes*				*Zygotes*					
genotype	*AA*	*Aa*	*aa*	*Divisor*	A^4	A^3a	A^2a^2	Aa^3	a^4	*Divisor*
AAAA	1	—	—	1	1	—	—	—	—	1
AAAa	15	12	1	28	225	360	174	24	1	784
AAaa	3	8	3	14	9	48	82	48	9	196
Aaaa	1	12	15	28	1	24	174	360	225	784
aaaa	—	—	1	1	—	—	—	—	—	1

The extent of chromatid segregation actually observed in autotetraploid depends upon the locus of gene and crossing over frequency between the concerned locus and centromere, frequency to multivalent formation, types of orientation and distribution of chromosomes in multivalent associations. The segregation ratios of different genotypes of autotetraploids generally lie between those theoretically expected for pure chromosome segregation and pure chromatid segregation.

Origin of Autopolyploids

The autopolyploidy develops directly or indirectly through chromosome doubling. It may occur spontaneously or may be induced by spindle poisons. It is the result of the formation of a restitution nucleus in somatic cell (*somatic polyploidization*) or in reproductive tissues (*generative* or *gametic polyploidization*).

Some Specific Examples of Autopolyploids

Triploids. Such plants have 3 genomes per nucleus and they are commonly designated as $3n$ plants. Autotriploids occur in a number of plants, *e.g., Oenothera, Datura,* rose, rice and many others. They are usually formed as a result of fertilization between a diploid ($2n$) and a haploid gamete (n), ($2n$ gamete + n gamete = $3n$ zygote) (Fig. 23.8). Diploid gametes are produced by normal tetraploids in meiosis or in sectors of otherwise diploid organisms where automatic doubling of somatic chromosome number (*i.e.,* of $2n$) has taken place. Somatic doubling may be spontaneous or may be induced.

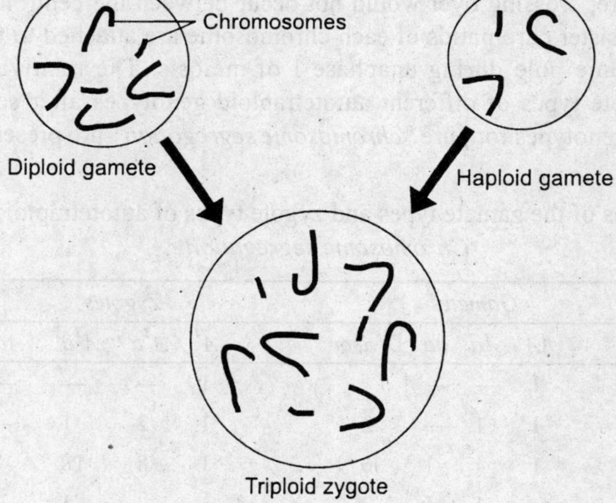

Fig. 23.8 Origin of a triploid (after SRB, Owen and Edgar 1965).

Characters

(*i*) Autotriploids are more vigorous than normal diploids, they are more leafy and show tendency towards perenniality.

(*ii*) Sometimes floral abnormalities may be observed.

(*iii*) Pollengrains, stomatal guard cells and wood cells of xylem in triploids are larger than those of diploids.

(*iv*) Plants are highly sterile and seeds are formed rarely. In nature, seed propagated triploid plants are ordinarily rare because of irregularity at meiosis. Generally the autotriploids occurring in nature are propagated by vegetative means. This is advantageous in horticulture, specially for ornamental plants, *e.g., Chrysanthemum,* roses, Dahlias, etc., and also in the production of seedless fruits. Some of the commercially important fruit plants, as for example, apples, pears, bananas, grapes, oranges, guavas, pineapples are triploid.

Meiosis. In autotriploid it is irregular. The centromeres of three homologous chromosomes have no way to orient themselves so as to give equivalents at the two poles. In this condition, the only possibility is that the components of trivalents separate in such a way that two chromosomes go to one side and third goes to the other. If it is so then the meiotic products may contain either 'n' or '$2n$' chromosomes or any number in between. This is not true irrespective of whether three homologous chromosomes align either as trivalents or as one bivalent and one univalent. Therefore, the gametes arising from triploids, have unbalanced genomes. In fact, probability of formation of haploid and diploid gametes are very rare.

Autotetraploids ($4n$). Autotetraploids have four similar genomes per nucleus. They may originate in one of the following ways:

(*i*) by fusion of two diploid gametes, and

(*ii*) they may result from the duplication of somatic chromosomes ($2n$) followed by failure of mitotic division. The resulting nucleus contains four copies of each chromosomes, instead of the usual two. If this tetraploid perpetuates itself through normal mitosis, the increased chromosome number may become established in a group of cells or tissues within the body of organism. Plants capable of vegetative propagation may be manipulated to produce pure tetraploids.

Characters

(*i*) Autotetraploids usually show marked phenotypic variations, great adaptability and sometimes marked disease resistance. They are commercially more valuable than corresponding diploids,

(*ii*) some are larger and more deeply coloured than diploids and sometimes they bear large seeds and fruits,

(*iii*) they have larger cells with correspondingly larger nuclei than do the diploids,

(*iv*) the stomatal guard cells and epidermal cells are larger and arranged in different patterns in tetraploids than those in diploids,

(*v*) flowers are larger and more showy in tetraploids than those in diploids,

(*vi*) high content of vitamin C (ascorbic acid) have been reported in tetraploids, and

(*vii*) the tetraploids show tendency towards perenniality and may show low fertility.

Meiosis. It may be normal or abnormal. When four homologous align in tetravalents, normal diploid gametes are formed. Sometimes, two bivalent or one trivalent and one univalent may also occur. In these cases, products of meiosis are irregularly formed. Incomplete genomes in gametes may result in total sterility.

Allopolyploidy

Allopolyploids have genomes from two or more genetically distinct species, each genome being present more than once in a nucleus. The occurrence of allopolyploids has been reported in many plant species including some crops, as for example in *Raphanobrassica*, *Triticale*, *Rosa*, *Primula kewensis*, *Spartina townsendii*, *Papaver*, *Iris versicolor*, *Grepis*, etc.

Origin and Production of Allopolyploids

The allopolyploids existing in nature were produced spontaneously by chromosome doubling in the natural hybrids of two distinct species belonging to the same genus or different genera (Fig. 23.9). The chromosome doubling might have occurred in the somatic cells (*Somatic allopolyploidization*) of the hybrid due to irregular mitosis resulting in the formation of allopolyploid sector either in the apical meristem or in the auxillary buds. Such sectors of allopolyploid tissue in the buds would have produced allopolyploid branches. Sexual progeny from such allopolyploid branches possibly resulted in stable natural allopolyploids.

The other alternative way by which allopolyploids could arise might be the formation and fusion of unreduced functional gametes (*generative allopolyploidization*).

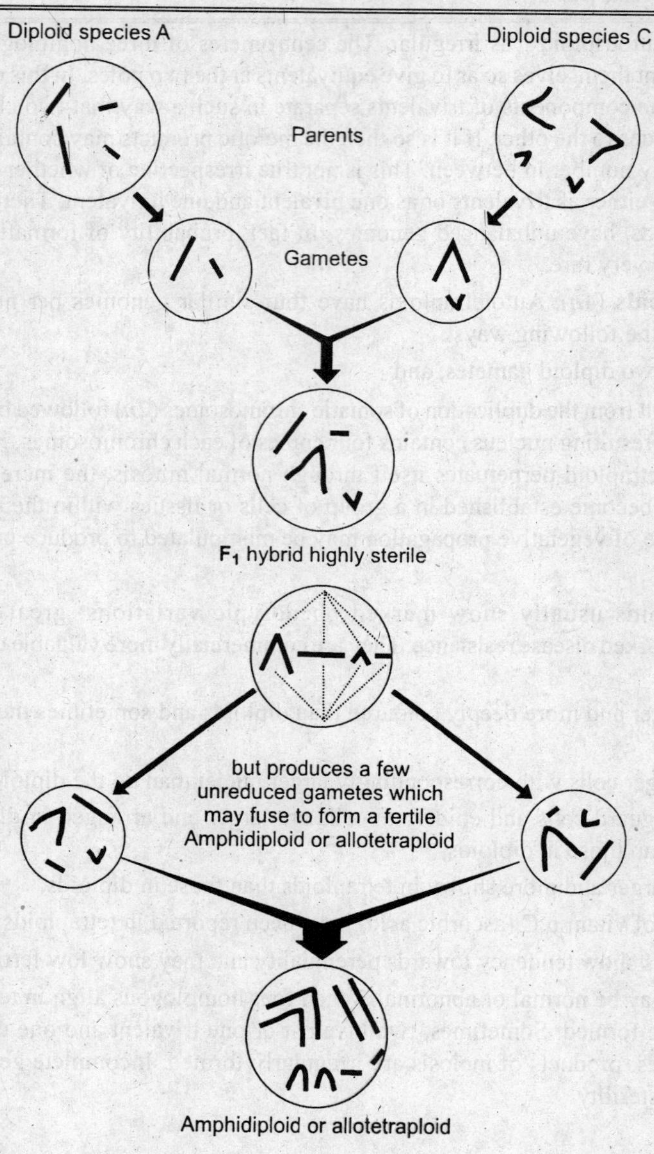

Fig. 23.9 A sequence of events leading to amphi- or allotetraploidy.

In the artificial production of allopolyploids from distant hybrids both the processes, somatic and generative polyploidization have been followed and many important allopolyploids have been developed.

A very important case of allopolyploidy is *Raphanobrassica* which was experimentally evolved by the Russian geneticist G.D. Karpechenko who made intergeneric cross between radish (*Raphanus sativus*) and cabbage (*Brassica oleracea*) (Fig. 23.10). Although these plants were distantly related, they were enough alike to be crossed successfully with each other. Both had 9 pairs of chromosomes. The diploid hybrid *Raphanobrassica* had 18 chromosomes; 9 chromosomes from radish and 9 chromosomes from *Brassica* parent. But it could not perpetuate itself largely because of failure of pairing between unlike chromosomes during meiosis. When chromosomes of sterile *Raphanobrassica* were doubled by artificial means, a fertile polyploid *Raphanobrassica* was produced with 36 chromosomes; 18 of radish and 18 of cabbage. Experimental productionof allopolyploids is achieved by doubling the chromosome number of distant hybrids with the help of colchicine or some other agents. Thus, the homologous chromosomes could pair themselves and the regular meiosis resulted normal gametes, each with 18 chromosomes. This pairing of chromosome is called *autosynapsis,* as actually chromosomes of two genomes are pairing with their respective replicas. This experiment had theoretical significance since it provided a method by which fertile interspecific hybrids could be produced. This also suggests the possibility of incorporating the desirable genotypes of two different species into a new polyploid species.

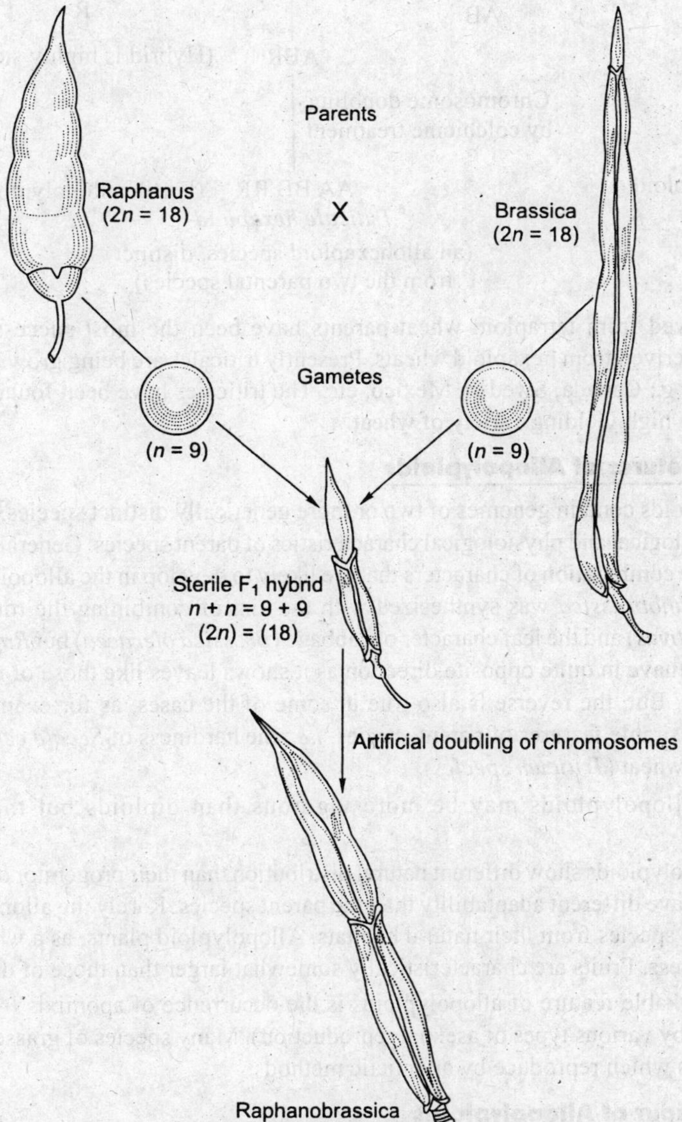

Fig. 23.10 The origin of *Raphanobrassica.*

Since many of the crop species happened to be allopolyploid man thought that he could create new crop spp. at will but the experience with a large number of allopolyploids were disappointing. In fact, it is only Triticale as synthetic allopolyploid that has succeeded as a crop species. *Triticale* is artificially produced allopolyploid produced by crossing tetraploid or hexaploid wheat with rye (*Secale cereale*).

Parents	Tetraploid wheat (*Triticum turgidum*)	X	Diploid rye (*Secale cereale*)
Genomes	AA BB		RR
Gametes	AB	↓	R
F_1		ABR	(Hybrid is highly sterile)
	Chromosome doubling by colchicine treatment	↓	
Amphidiploid		AA BB RR *Triticale hexaploid* (an allohexaploid species, distinct from the two parental species)	(Fertile allopolyploid.)

Triticale derived from tetraploid wheat parents have been the most successful species as compared to those derived from hexaploid wheats. Presently triticales are being grown commercially in some countries, *e.g.,* Canada, Sweden, Mexico, etc. The triticales have been found to be as good in production as any high yielding variety of wheat.

Morphological Features of Allopolyploids

Since the allopolyploids contain genomes of two or more genetically distinct species, they generally combine the morphological and physiological characteristics of parent species. Generally, it is difficult to predict the precise combination of characters that are likely to develop in the allopolyploid species. For example, *Raphanobrassica* was synthesized with the aim of combining the root character of radish (*Raphanus sativus*) and the leaf character of cabbage (*Brassica oleracea*) but *Raphanobrassica* has been found to behave in quite opposite direction as it shows leaves like those of radish and root like that of cabbage. But, the reverse is also true in some of the cases, as for example, *Triticale* has combined the desirable features of parent species, *i.e.,* the hardiness of *Secale cereale* (rye) and yielding potency of wheat (*Triticum species*).

In general, allopolyploids may be more vigorous than diploids but there are many exceptions to it.

Since the allopolyploids show different natural distribution than their progenitor diploid species, it appears that they have different adaptability than the parent species. Rarely the allopolyploids may displace the parental species from their natural habitats. Allopolyploid plants, as a whole, possess a good deal of robustness. Fruits are characteristically somewhat larger than those of diploid parents.

Another remarkable feature of allopolyploids is the occurrence of apomixis (replacement of sexual reproduction by various types of asexual reproduction). Many species of grasses are sexually sterile allopolyploids which reproduce by apomictic methods.

Cytological Behaviour of Allopolyploids

In allopolyploids or amphiploids each chromosome of the original allodiploid or amphihaploid (*e.g.,* AB) is represented twice. The original amphihaploid are sterile hybrids, chromosomes of which behave, more or less, like those of haploids. By formation of allopolyploid, fertile derivatives are obtained.

Meiotic behaviour of allopolyploids is governed by the homology relationships of chromosome sets they possess. There are two types of allopolyploids:

(*i*) Some allopolyploids form only bivalents and no multivalents or univalents are observed during meiosis. In such a case every chromosome finds one identical partner with which it pairs. Such an allopolyploid breeds true for its genotype and there is no segregation.

(*ii*) The other types of allopolyploid are multivalent forming. This may result from:

(*a*) the presence of homologous chromosome sets in more than double dose (*e.g.*, AAAA BB), and

(*b*) the introduction of partially homologous (= homoeologous) chromosome sets.

The segregation of allopolyploids is consequence of behaviour of chromosomes in multivalent associations in both cases. Generally, extreme genetical imbalance occurs among segregants of allopolyploids with numerous complex multivalents in meiosis resulting in reduced fertility.

On the basis of meiotic behaviour of chromosomes of different sets, Stebbins (1945) classified allopolyploids as follows:

1. **Genome allopolyploids**. In these the, chromosomes pairing is exclusively in the form of bivalents during meiosis. Pairing is restricted to exactly homologous chromosomes (homogenetic pairing) which may be consequence of the striking structural dissimilarities between the parental chromosome sets or may result from the activity of specific genes which prevent the pairing of partially homologous chromosomes. The amphihaploid or allodiploid parent of such allopolyploid are usually completely sterile. The fertility is fully restored with the production of the *genome allopolyploid* derivative. Once established, this derivative is usually completely isolated by sterility barriers from its nearest relatives and behaves like a new species.

   In genome allopolyploid, only identical chromosomes of the similar genomic sets pair, behaving as they do in diploid organisms. This type of chromosome pairing is called *autosyndatic pairing* or *autosyndasis.* For this reason allotetraploid with complete autosyndatic pairing is called amphidiploid (=double diploid)

2. **Segmental allopolyploids**. In these, the chromosomes form bivalent and multivalent associations. The parental chromosome sets combined in the diploid hybrid in this case are partially homologous and correspond in a considerable number of segments permitting "heterogenetic pairing". The pairing of partially homologous chromosomes from different genomes is referred to as *allosyndesis.* The term describes pairing of any kind between A chromosome and B chromosome which are homologous in AA BB tetraploid. In allosyndatic pair the chromosomes have one or more small segments that are homologous. Thus, the sterility of diploid hybrid parent of such an allopolyploid is not very pronounced. In contrast to genome allopolyploid, the segmental alloploids show more or less extensive segregation; the segregants being highly disbalanced genetically and sterile. Hence such polyploids are genetically unstable. Through segregation, they may produce stable segregants, which in regard to their chromosome, constitution may be autoploid, genome allopolyploid or stable segmental allopolyploid.

3. **Autoallopolyploids**. In these allopolyploids, chromosomes may form bivalents and multivalents like those in segmental polyploids, because homologous chromosomes are present in more than double dose. Polyploids of this type are possible from hexaploid level upward and combine characteristics of auto and allopolyploidy.

The fertility of allopolyploids can be improved through hybridization and selection. Selection improves chromosome pairing behaviour and through recombinations new gene combinations develop that improve the physiological and cytological behaviour of allopolyploids.

ARTIFICIAL INDUCTION OF POLYPLOIDY

In recent years, a number of methods have been worked out to induce polyploidy in plants. Some of them are described below:

1. **By Radiation.** Polyploidy can be induced in plants in low frequency by exposing their seeds or certain parts, such as vegetative buds and flower buds, to radiations of shorter wavelengths;

ultraviolet rays, x-rays, gamma-rays. Irradiation increases the rate of cell division and also causes the multiplication of chromosome number (somatic doubling of chromosomes).

2. **By Injury.** When the meristematic zones of a plant are injured the cells at the points of injury grow rapidly and form a callus. Callus growth is enhanced by a chemical substance named coumerine which also brings about somatic doubling of chromosomes. Vegetative buds generally developing from callus tissue are polyploid in nature. From injured parts of tomato plants it is possible to produce tetraploids ($4n$) plants.
3. **Regeneration *in vitro*.** Polyploidy is a common feature in the cells of cultured tissue *in vitro*. Some of the plants regenerated from the callus in suspension culture may be found to be polyploids. Polyploids have been developed from callus cultures of *Nicotiana, Datura,* rice and several other species.
4. **By chemical treatment.** A number of chemicals are now known which induce polyploidy in plants. Important among them are colchicine, 8-hydoxyquinolin, acetophenon, nitrous oxide, granosan, chloroform, chloral hydrate, some narcotics and alkaloids, veratin sulphate, acenaphthane, sulphanilamide, ethyl murcury chloride, hexachlorocyclohexane etc. Colchicine ($C_{22}O_6N$) is the best chemical for this purpose.

Colchicine

It is a chemical compound which was first discovered by Pernice in 1889 (Eigsti and Dustin, 1955). Colchicine was first demonstrated by Levan (1938) to be a specific and efficient chemical in creating polyploid restitution nuclei (Eigsti and Dustin, 1955). Colchicine is obtained from the extract of seeds (0.2 – 0.8%) and corms (0.1 – 0.5%) of the *Colchicum autumnale* (member of family Liliaceae). In India it is obtained from *Colchicum luteum* and *Gloriosa superba* (Parthasarthy, 1941 ; Eigsti and Dustin, 1955).

Method of Application of Colchicine

Colchicine treatment is done in one of the following ways:

1. **Seed treatment.** The dry or soaked seeds are soaked in aqueous solution of colchicine of different strength in shallow container to facilitate aeration. Generally, aqueous solutions of 0.25 to 0.5 per cent concentrations are used and for trees 1% solution is preferred. The duration of treatment may vary from 3–24 hours in case of seeds and seedlings white in case of trees it may continue for 2–3 weeks. Since the solution is unstable in water, colchici solution is used fresh each time. Colchicine treatment is given for a definite period which is different for different seeds. After the seeds are soaked in colchicine solution for desired period, they are washed thoroughly in water and then they are sown. Treatment of dry seeds gives better result than soaked seeds in some cases.
2. **Seedling treatment.** Seedlings may be treated in young stage. During treatment, the shoot tips are dipped in 0.2% colchicine solution and root tips are covered with cotton soaked in water and the treatment may be given from 3 to 24 hours and in some cases the treatment should be repeated on 2nd and 3rd day also.
3. **Treatment of growing buds of shoot.** In some cases growing points are treated with 0.1 to 0.5% solution of colchicine which is applied by brush or with a dropper. Sometimes cotton wool soaked in the aqueous solution of colchicine is applied over the growing point of plant. The treatment is repeated once or twice daily for a few days. Alternatively, 0.2 to 0.5% colchicine solution is mixed with lanoline paste and is smeared on the shoot apex. This treatment may be repeated 2-3 times daily for a week.
4. In woody plants, 1.0% colchicine solution may be applied over the growing bud. For proper penetration of colchicine, the solution is applied 2-3 times a day for a week.

At any given time only a small proportion of cells would be in mitotic phase, hence repeated treatment should be given at brief intervals so as to double the chromosome number in good number of cells. There may be a competition between polyploid cells and unaffected diploid cells and some times diploid cells may compete over the polyploid cells and diploid shoots may arise from the treated growing apex.

Effect of Colchicine

When colchicine is applied some irregularities occur in the mitotic division. It has no effect on doubling of chromosomes which proceeds in normal fashion. This chemical is a spindle poison and it simply checks the anaphasic movement of chromosomes to the two poles and consequently mitosis fails to complete. The doubled chromosomes become bounded by a nuclear membrane, thus a restitution nucleus with double chromosome number is formed.

Application of Allopolyploidy in Crop Breeding

The allopolyploidy has been exploited by plant breeders for the following three purposes:

1. utilization as bridging species in transfer of desirable characters from one species to another,
2. for the production of new crop species. Autopolyploidy has contributed to a limited extent in evolution of plant species. Some crops e.g. potato, sweet potato, cabbage, turnip, banana, peanut, coffee are autopolyploids. Autotetraploids are found to be more useful crops than other forms of autopolyploidy. Forage grasses are mostly autopolyploids.
3. for widening the genetic base of existing allopolyploid crop species.

Utilization of Allopolyploids as Bridging Species

An amphidiploid is utilized as bridge in transfer of desirable characters of one species to the other related species (particularly from wild species to cultivated one), when the hybrid between the cultivated species to which the desirable character is to be transferred and the wild species (donor) from which the desirable character is to be transferred is sterile. In such cases, the chromosome number of F_1 sterile hybrid between the recipient species and donor species is doubled to produce amphidiploid which is generally fertile and can be crossed with the recipient species. The progeny of the cross between recipient species and amphidiploid would have $2n$ chromosomes of recipient species and one genome from donor species. This situation renders the progeny to be sufficiently fertile and so the progeny can be back crossed with the recipient species. Such a programme may give rise to alien addition and alien substitution lines which are used for transferring genes or groups of genes or small segments of chromosome to the recipient species.

This can be clear from the following example of synthetic allohexaploid *Nicotiana digulata* as bridge for the transfer of resistance against tobacco mosaic virus from *N. sylvestris to N. tabacum*. The scheme is as follows:

N. sylvestris ($2n = 24$) (Resistant against TMV) X *N. tabacum* ($2n = 4x = 48$)

↓

F_1 Hybrid (Sterile)

↓ Artificial doubling of chromosomes

N. tabacum (Tetraploid) X Synthetic Amphidiploid (*N. digulata*) Back cross

↓

Hybrid

↓ Back crossed to *N. tabacum*

Pentaploid Progeny with $2n$ chromosomes of *N. tabacum* and one genome of *N. sylvestris* and so sufficiently fertile.

The repeated back cross may result in alien addition and alien substitution lines from which materials may be selected with desirable character.

It should be noted that in the use of a species as bridge, the features of amphidiploids are not important and it should be sufficiently fertile to be back crossed with the recipient species.

Use of Allopolyploidy in Creation of New Species

The fact that many species of plants seem to be all natural allopolyploids can lead one to think that allopolyploidy would enable a man to create new species at will and the new species would be superior to many of the existing crop species. But before one thinks on this line one should bear the following facts in mind:

1. allopolyploidy may not necessarily result into a successful species as the allopolyploids may be undesirable and noxious in many cases,
2. the stability of allopolyploids is attained after a long period and so they could hardly be expected to become successful crops within a short period, and
3. the development of several superior characters in synthetic amphidiploid may take a long time and it is not certain that they would develop in the synthetic material.

These facts present a discouraging picture of the possibility for utilization of allopolyploidy in creating new synthetic allopolyploid species.

Triticale and *Raphanobrassica* are the examples of the two most successful synthetic species which have been evolved through allopolyploidy.

Widening of the Genetic Base of Natural Allopolyploids

The genetic bases of some natural allopolyploid species may be narrow and the increase in genetic variabilities by producing new synthetic allopolyploids may be quite useful. *Brassica napus,* for example, has a narrow genetic base and genetic variability can hardly be seen in nature. In such a case, the only recourse available is to synthesize new allopolyploid *B. napus* to widen the genetic base. This can be done by crossing the diploid parents *B. campestris* with *B. oleracea* and subsequently doubling the chromosomes.

Polyploidy and Evolution of Species

The polyploidy has played an important role in evolution of new varieties and species in nature. Angiosperms and Pteridophytes have very high numbers of polyploid species in nature. More than 35 per cent angiosperms are polyploids. According to Manton (1950), among all the plant groups, ferns show highest degree of polyploidy. It is generally noted that with the increase in chromosome number the adaptability and variabilities of species increase progressively. Many mutations in diploids are deleterious or fatal, but in polyploids, where a gene is represented more than twice the chances of appearance of deleterious effects due to mutation are lesser.

Autopolyploidy has limited contribution in the evolution of plant species. Some of our present day crops, many forage grasses and several ornamental plants are polyploids. Some of the autopolyploid crops are listed in following Table 23.6.

Table 23.6. Autopolyploid crop species

Crop Common name	Botanical name	*Somatic chromosome number (2n) of the cultivated form*	*Somatic chromosome (2n) of related wild species*
Potato	*Solunum tuberosum*	48 (4*n*)	2*n* = 24
Sweet potato	*Ipomoea batatas*	90 (6*x*)	
Bananas	*Musa paradisiaca*	33 (3*n*)	22
Coffee	*Caffea arabica*	44 (4*n*)	22, 66, 88
Groundnut	*Arachis hypogea*	40 (4*n*)	
Alfalfa	*Medicago sativa*	32 (4*n*)	14, 16, 32
Tomato	*Lycopersicum esculentum*	48 (4*n*)	24
Black night shade	*Solanum nigrum*	48 (4*n*)	24

Allopolyploidy has played vital role in the evolution of species. A good number of crop species seem to have evolved through allopolyploidy. This is evident from the widespread occurrence of allopolyploids in various genera of plants and their high adaptability in nature. It is estimated that about one-third of the flowering plants are polyploids and a great majority of them are allopolyploids. The importance of allopolyploidy in evolution of species was first pointed out by Winge in 1917. Subsequent studies have supported this contention.

Cytological studies in many species have made it possible to trace the evolutionary history of many allopolyploid species and their diploid parental species have been identified with some degree of certainty. The parental diploid species of allopolyploid is based primarily on the pairing between the chromosomes of the diploid species and allopolyploid species. When the chromosomes of a diploid species pair with some of the chromosomes of the allopolyploid species, the homology between the chromosomes of two species is established which suggests that the diploid species may be one of the parent species of allopolyploid species. But in some cases, owing to considerable chromosome differentiation in diploid and allopolyploid species in course of time, pairing between the chromosomes of the two species might be greatly reduced.

The identification of parental diploid species can also be made by involving the suspected parent species in hybridization and synthesizing an allopolyploid from the interspecific hybrid and then comparing the synthetic allopolyploid with naturally occurring allopolyploid. The synthetic allopolyploid does often resemble the natural allopolyploid species. This has been demonstrated in *Brassica* in which a synthetic allopolyploid developed from the interspecific hybrid of cross *B. oleracea* (cabbage) × *B. campestris* (rape) was found to resemble very closely with *B. napus* (turnip). However, the resemblance between natural and synthetic allopolyploids may not be close in all cases.

The other methods for determining the resemblance between the diploid parents and allopolyploid species are biochemical techniques, such as by electrophoretic analysis of their protein and enzyme patterns and from chromosome banding patterns. It is now possible to reconstruct the probable method of origin of certain species that have arisen through natural hybridization and polyploidization. There are many cultivated allopolyploid species, such as wheat, tobacco, cotton and *Brassica*, whose possible evolutionary history can be inferred with high certainty.

Table 23.7. Some species of wheat, their somatic chromosome number and genomic composition

Species	*Common name*	*Diploid chromosome number and genome*	*Nature*
Triticum monococcum	Einkorn wheat	$2n = 14$ (AA)	Cultivated (C)
T. boeticum	Wild Einkorn wheat	$2n = 14$ (AA)	Wild (W)
T. dicoccum	Emmer wheat	$2n = 4x = 28$ (AABB)	C
T. dicoccoides	Wild emmer	$2n = 4x = 28$ (AABB)	C
T. turgidum	Solid stem wheat	$2n = 4x = 28$ (AABB)	C
T. carthlicum	Persian wheat	$2n = 4x = 28$ (AABB)	C
T. polonicum	Polish wheat	$2n = 4x = 28$ (AABB)	C
T. durum	Durum wheat	$2n = 4x = 28$ (AABB)	C
T. timopheevi		$2n = 4x = 28$ (AAGG)	W
T. americanum		$2n = 4x = 28$ (AAGG)	W
T. spelta	Spelt wheat	$2n = 6x = 42$ (AABBDD)	C
T. aestivum	Common bread wheat	$2n = 6x = 42$ (AABBDD)	C
T. macba		$2n = 6x = 42$ (AABBDD)	W
T. vavilov		$2n = 6x = 42$ (AABBDD)	W
T. compactum	Club wheat	$2n = 6x = 42$ (AABBDD)	C
T. sphaerococcum		$2n = 6x = 42$ (AABBDD)	W

Evolution of Bread Wheat. The origin of bread wheat (*Triticum aestivum*) has been most extensively investigated. The common bread wheat (*T. aestivum*) is a hexaploid species ($2n = 6x = 42$) having three different sets of genomes (AABBDD) which seem to have been contributed by three different diploid species (Sears, Kihara, Percival etc.). The species of wheat *(Triticum)* fall into the following three categories: diploids, tetraploids and hexaploids (Table 23.7).

It is generally believed that the genome A present in the diploid wheats is more or less the same as those present in tetraploid and hexaploid wheats and the genome B of tetraploid Emmer wheats is also more or less similar to genome B found in hexaploid wheats. The pairing behaviour of chromosomes during meiosis in diploid, tetraploid and hexaploid wheats supports this. The chromosome pairing in triploid hybrids between diploid and tetraploid wheats results in 7 II (bivalents) and 7 I (univalents) associations. The pentaploid hybrids ($5n$) resulted from a cross between tetraploid and hexaploid wheats show 14 IIs and 7 Is. Though these facts are generally accepted by the majority of cytogeneticists, yet there is no agreement on the point regarding the diploid parent species which contributed genomes A, B and D to hexaploid wheats. It has been proposed that the source of A genome of common bread wheat is *T. monococcum,* that of B genome is *Aegilops speltoides* and that of D genome is *Aegilops squarrosa.* The possible evolutionary history of hexaploid wheat is as follows:

```
                  T. monococcum          X          Aegilops speltoides
                  2n = 14 (AA)           |             2n = 14 (BB)
                                         ↓
                  F1    2n = 14     A B (Hybrid)  sterile
                                         ↓ spontaneous chromosome doubling

 Aegilops squarrosa      X          AA BB (2n = 28)       Tetraploid
 (DD)    2n = 14         |          (Amphidiploid)        Emmer wheat
                         ↓
 F1               ABD (n = 21)
                         ↓ spontaneous chromosome doubling

                  AABBDD           Hexaploid wheats
                  (Amphidiploid)
                  (2n = 6 × = 42)
```

The genome D present in bread wheat is supposed to have been derived from a grass *Aegilops squarrosa* ($2n = 14$) which grows in the region extending from Armenia to Afghanistan. The different varieties of wheat with higher chromosome numbers have apparently arisen from this type and other related grasses through hybridization followed by chromosome doubling. J. Percival is of the opinion that bread wheat (*T. aestivum*) originated through hybridisation between Emmer wheat ($2n = 28$ chromosomes) and goat grass ($2n = 14$). Experimental results have been obtained by E.S. Mc. Fadden, Sears (1948) and Kihara that definitely established the origin of one type of bread wheat *T. spelta.* These investigators doubled the chromosome numbers of Emmer wheat and goat grass by colchicine treatment (described earlier in this chapter) in order to get diploid gametes from these two plant species. By crossing these two colchicine treated plants they found a fertile hybrid with 42 chromosomes (28 from Emmer wheat and 14 from goat grass). The hybrid resembled bread wheat and was given name *T. spelta.* These researches showed how a moderately useful wheat and a useless grass were combined in nature to produce most valuable variety of wheat. The origin of wheat is still a matter of great speculation and is very confusing even today.

Evolution of Allopolyploid Species of Brassica

The species of Brassica, show a good degree of variation in their somatic chromosome numbers (Table 23.8).

Table 23.8. Somatic chromosome numbers and genomic composition of some species of Brassica

Species of Brassica	*Common name*	*Somatic chromosome number (2n) and genomic set*	*Nature*
Brassica nigra	Black Sarson	16 (BB)	Cultivated (C)
B. oleracea	Cabbage	18 (CC)	C
B. campestris	Rape, turnip rape	20 (AA)	C
B. carinata	Abyssinian cabbage	34 (BBCC)	Wild
B. napus	Rape	38 (AACC)	C
B. juncea	Rai (Indian mustard)	36 (AABB)	C

The possible origin of different polyploid species of *Brassica* has been illustrated in the famous species triangle proposed by N.U. (1935) (Fig. 23.11). According to this NU's triangle, three species

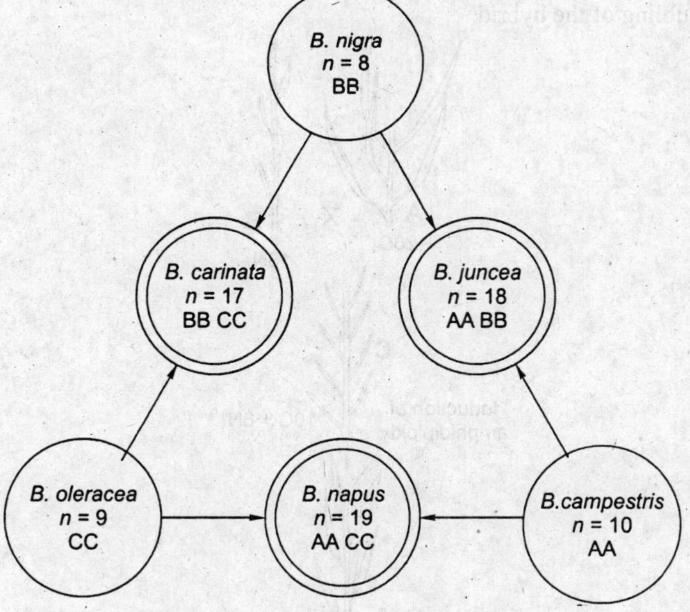

Fig. 23.11 Species triangle of *Brassica*.

B. nigra, B. campestris and *B. oleracea* have contributed genomes for different amphidiploid species and represent the three tips of the triangle. The amphidiploid species are presented midway between their parental species. *B. juncea* is an amphidiploid from a cross *B. nigra* (n = 8) × *B. campestris* (n = 10) (Fig. 23.12), *B. napus* (n = 19) is an amphidiploid from the cross *B.oleracea* (n = 9) and *B. campestris* (n = 10) and *B. carinata* (n = 17) is amphidiploid from a cross *B. nigra* (n = 8) × *B.oleracea* (n = 9).

It has been experimentally possible to get synthetic fertile amphidiploids according to the above scheme which resembled natural amphidiploids.

Evolution of Cotton Species

The genus *Gossypium* is represented by some wild and some cultivated species. The chromosome numbers of some species have been presented in (Table 23.9):

Table 23.9. Somatic chromosome numbers and the genomes of *Gossypium* species

Name of the Species	*Common name*	*2n or somatic chromosome number and genomes*	*(Cultivated/ wild)*
Gossypium arboreum	Asiatic or Desi cotton	26 (A_2A_2)	C
G. herbaceum	Asiatic cotton	26 (A_1A_1)	C
G. thurberi	Wild American cotton	26 (D_1D_1)	W
G. barbadense	Sea island (Egyptian) cotton	52 ($A_2A_2D_2D_2$)	C
G. hirsutum	American upland cotton	52 ($A_1A_1D_1D_1$)	C
G. tomentosum	Wild Hawaii cotton	52	

G. hirsutum, G. barbadense, and *G.tomentosum* ($n = 26$) appear to have arisen through natural hybridization between the diploid species ($n = 13$) from old world or Asiatic cotton and New world or American cotton and subsequent chromosome doubling of interspecific hybrid. A possible origin of *G. hirsutum* is from the cross *G. arborium* $n = 13$ (Asiatic cotton). × *G. thurberi* ($n = 13$) followed by chromosome doubling of the hybrid.

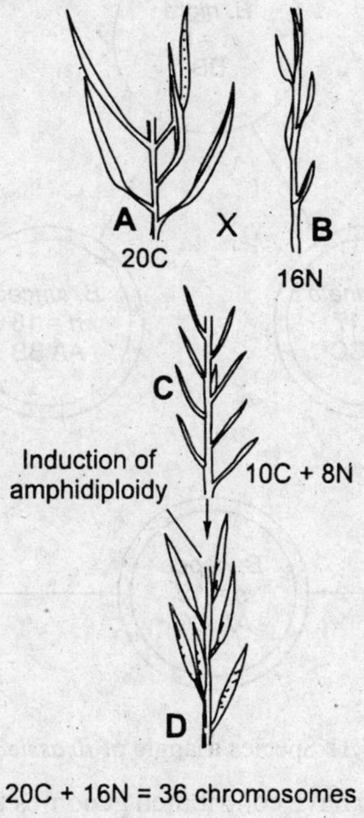

Fig. 23.12 Synthesis of Amphidiploid (D) *Brassica juncea* by a cross between *B. campestris* (A) and *B. nigra* (B).

Evolution of Tobacco

Tobacco (*Nicotiana tabacum*) ($2n = 48$) is probably an amphidiploid from a natural cross *N. sylvestris* ($2n = 24$) × *N. tomentosa* ($2n = 24$). This is evident from the chromosome behaviour during meiosis in the triploid hybrids of the crosses *N. tabacum* × *N. sylvestris* and *N. tabacum* × *N. tomentosa* which formed 12 IIs and 12 univalents during prophase I. This indicates similarity between chromosomes of *N. tabacum, N. sylvestris* and *N. tomentosa*. Morphologically, the amphidiploids of the hybrids from cross between *N. sylvestris* and *N. tomentosa* are similar to *N. tabacum* in many respects. The species *N. tabacum* has undergone considerable differentiation due to gene mutations and modification (possibly due to loss of some duplicated segments) in genomes of the two parents.

QUESTIONS

1. How would you distinguish the following:
   (*a*) Double monosomic and nullisomic, and
   (*b*) Trisomic and nullisomic?
2. What are different kinds of aneuploids?
3. What are different kinds of polyploids? How will you distinguish between allopolyploid and autopolyploid?
4. Describe the cytology of haploids and triploids.
5. Write short notes on:
   (*i*) Genome,
   (*ii*) Evolution of wheat,
   (*iii*) *Raphanobrassica*,
   (*iv*) Amphidiploidy,
   (*v*) Colchicine, and
   (*vi*) Triticale.
6. Discuss the role of Polyploidy in Evolution and Plant breeding.
7. The diploid number of an organism is 12, how many chromosomes would be expected in monosomic, trisomic, tetrasomic, nullisomic, haploid, triploid and autotetraploid?

24 CHAPTER

Genetic Engineering or Recombinant DNA Technology

DNA is a genetic material which contains all hereditary information needed to create an organism. DNA actually does not make organism, it only makes proteins. DNA is transcribed into mRNA and mRNA is translated into protein and the protein then forms organism. The specific regions on the DNA molecule that direct the synthesis of proteins are called genes. By changing the nucleotide sequence in DNA, the protein formation is changed. This leads into either a different protein or inactive protein. Recombinant DNA is the general name for taking a piece of DNA and combining it with another strand of DNA (hence the name recombinant DNA or rDNA). Recombinant DNA is sometimes also referred to as **chimera DNA**. By combining two or more strands of DNA from two different organisms, scientists are able to create a new strand of DNA.

Recombinant DNA Technology (rDNA Tech) or **genetic engineering** is concerned with the manipulation of genetic materials towards desired end in a directed way. It is also known as **gene cloning**. Genetic engineering aims at isolating DNA segments of one organism of interest and recombining that with DNA of second unrelated organisms. In this process the DNA molecules are isolated and cut into pieces by one or more specialised enzymes and the fragments are joined together in a desired combination and restored to a cell for replication and reproduction. **Recombinant DNA**, thus is a composite DNA molecule that results from the physical combination of DNA segments derived from different sources. In other words, the Genetic engineering is biochemical manipulation of genes by which foreign genes or the genes of donor organisms are inserted into the DNA or unrelated recipient organism through the plasmids of bacteria or through bacteriophages.

Historical Account

In nature, the transfer of genetic material occurs through conjugation, transformation and transduction. Recombination is brought about by the following two processes:

(*i*) natural recombination process, and

(*ii*) artificial recombination process.

The **natural recombination** occurs between closely related species and it does not involve unrelated ones. It occurs during meiosis when similar nucleotide sequences in DNAs of homologous chromosomes come closer, break and exchange segments and then rejoin. Thus a new arrangement of genes results from this recombination.

The **artificial recombination** process involves unrelated organisms. The genetic material of one organism is inserted into a specialized DNA molecule of another organism giving rise to recombinant or rDNA.

The history of modern genetic engineering goes back to mid 1970s when it became possible to cut DNA into pieces and transfer a particular piece of DNA containing specific information from donor organism to the DNA of the recipient organism. An yearwise landmarks of the genetic engineering are given here as follows:

1970 **Smith and Nathans** discovered a new class of enzymes, the **restriction enzymes** that act as chemical scissors and cut a DNA molecule into smaller fragments.

1972 **Berg and others** combined DNAs from two viruses to produce what is called **recombinant DNA (rDNA).**

1973 **Cohen and Boyer** inserted recombinant DNAs into host bacteria that reproduced or cloned foreign DNAs. With this, the age of genetic engineering begins.

1977 **Genetech**, one of the first genetic engineering companies starts the biosynthesis of important drugs by rDNA Technology.

1977 **Sanger and Gilbert** independently discovered techniques for rapid sequencing of nucleotides in DNA molecules.

1982 Human insulin produced by rDNA is marketted under the trade name **'Humulin'**.

1983 Tracy Moreno, a ten year old girl born with growth hormone deficiency, grows 5 inches in one year of treatment with engineered hormones. Human growth hormone genes are inserted into mouse embryo producing a giant mouse. For the first time a human gene functions in another animal.

1997 Scientists remove the DNA–containing nucleus from a female's egg and replace it with a nucleus from a different animal of the same species. The scientists then place that egg into the uterus of a third animal. The result, first demonstrated by the birth of a cloned sheep named **Dolly**, is the birth of an animal that is nearly genetically identical to the animal from which the nucleus was obtained. Such an animal is genetically unrelated to the surrogate mother.

2002 Brigitte Boisselier, a biochemist and director of Clonaid, hinted that human cloning experiments were already underway. She was very, very pleased to announce that the first baby clone was born on December 26, 2002 at 11.55 a.m. Another British baby conceived at the Reproductive Genetic Institute was born on February 9, 2002.

2003 The third cloned baby was born on 3rd January, 2003 to a Dutch woman, the head of Raclian sect in Netherlands. Clonaid claims to have produced 13 cloned babies world wide. Claims of cloned babies have yet to be substantiated primarily because of the risk of incarceration. In February, 2003 the Human Cloning Prohibition Act was passed which made it a crime for any one, public or private, to conduct somatic cell nuclear transplantation on fertilized or unfertilized human egg cell for reproductive or therapeutic purposes. The penalty for engaging in such research is US $ 1 million fine and a jail term for 10 years.

Cloning is still in its infancy, but it may pave the way for improved domestic animals and medical products.

Procedure of Genetic Engineering

The gene cloning or recombinant DNA formation requires the following:

1. DNA fragments to be cloned or **target gene** sequence of desired type.
2. Restriction endonucleases for cutting DNA molecule into fragments.
3. Cloning vector.
4. DNA ligase enzyme for splicing DNA segments.
5. Prokaryotic or Eukaryotic cells to serve as host.

The major steps involved in the genetic engineering are as follows (Fig. 24.1):

1. **To break open the living cells.** Several methods are available to break open the living cells. One of the popular methods involves mechanical shearing, the cells in a blender and then treating them with a detergents.

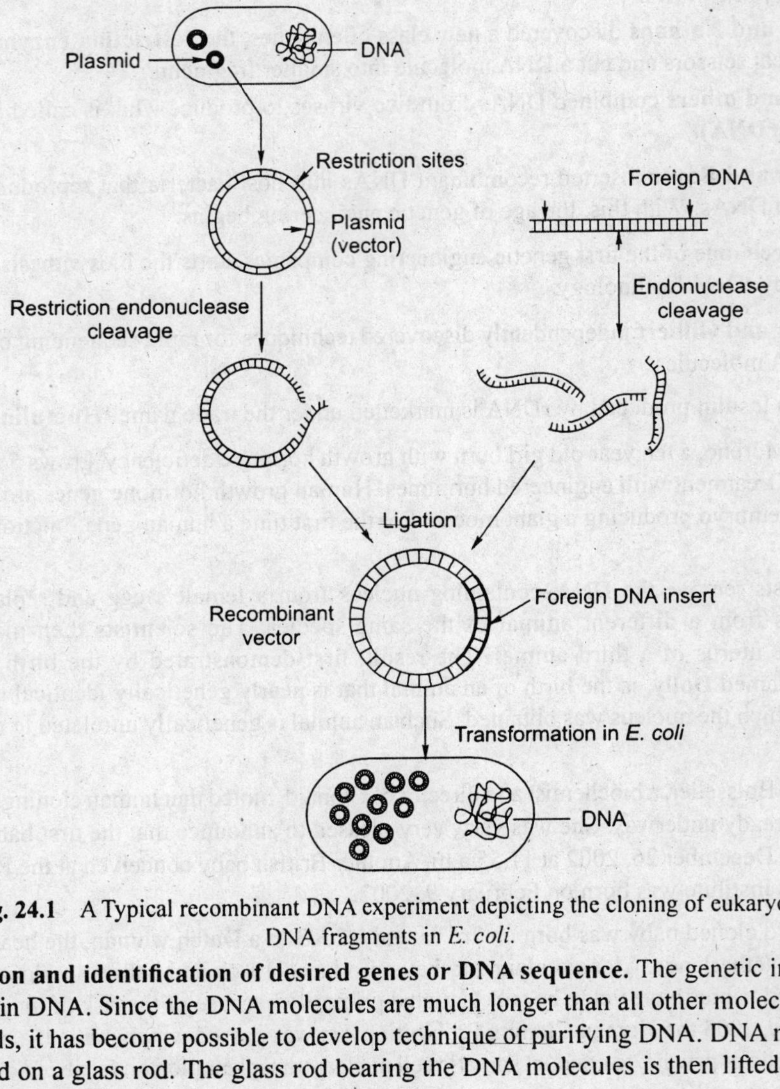

Fig. 24.1 A Typical recombinant DNA experiment depicting the cloning of eukaryotic DNA fragments in *E. coli*.

2. **Isolation and identification of desired genes or DNA sequence.** The genetic information is stored in DNA. Since the DNA molecules are much longer than all other molecules found in the cells, it has become possible to develop technique of purifying DNA. DNA molecules are spooled on a glass rod. The glass rod bearing the DNA molecules is then lifted out from the mixture of broken cells.
3. **Cutting of the DNA molecules into segments containing specific genes from the rest of DNA.** DNA is cut into gene size segments with the help of molecular scissors, called restriction endonucleases. The restriction endonucleases recognise specific base sequences in DNA molecule and make two cuts, one in each strand. Such an action results in generating 3′ hydroxyl and 5′ phosphate terminals.

   Several enzymes are used as biological tools in genetic engineering, some of them are as follows:

   Exonucleases. These enzymes act upon genome and digest the base pairs on 5′ or 3′ ends of a single stranded DNA.

Restriction endonucleases. Restriction enzymes used in recombinant DNA technology are capable of making cuts in the DNA molecules by two distinct arrangements. One type of cleavage results in **blunt end DNA fragments** and the other type of cleavage results in **cohesive or sticky**

ended molecule. The blunt end DNA fragments are formed when the restriction enzymes act on the line of symmetry *i.e.,* they cleave both DNA strands precisely at opposite points (Fig. 24.2A). The cohesive ended molecules are produced when the enzyme acts around the line of symmetry *i.e.,* two strands are not cut on opposite points and the cuts are staggered as illustrated below (Fig. 24.2 B).

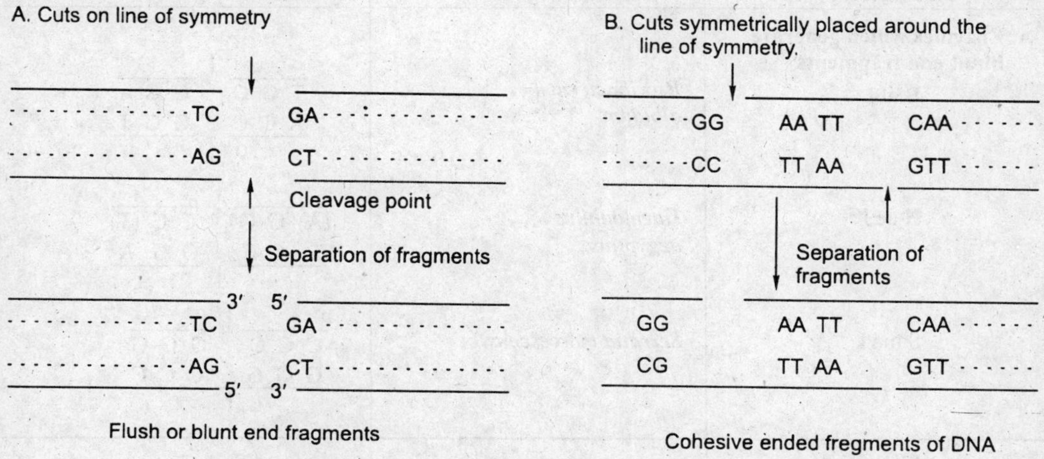

Fig. 24.2 A-Formation of non-sticky DNA fragments by restriction endonuclease.
B-Formation of sticky ended fragments by restriction enzyme.

The sequences recognised by restriction enzymes are often **palindromes** (*i.e.,* inverted repetitive sequences which are symmetrical. Restriction endonucleases have the following important features:

1. These enzymes make breaks in palindromic sequences, each consisting of 4–6 complementary bases as illustrated below (Fig. 24.3).

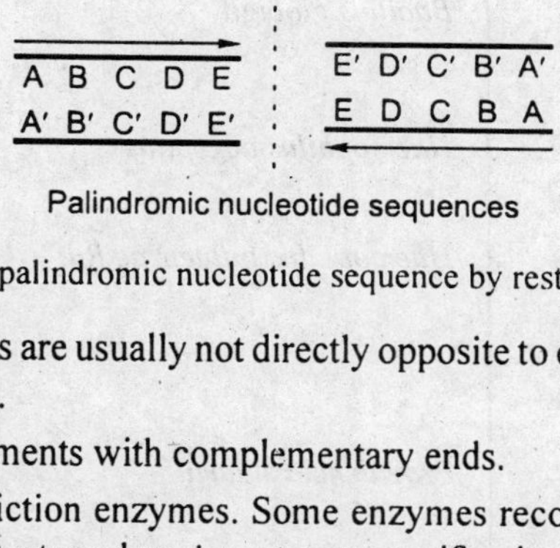

Fig. 24.3 Formation of palindromic nucleotide sequence by restriction enzyme.

2. The breaks in two DNA strands are usually not directly opposite to each other but are staggered cuts forming cohesive strands.
3. The enzymes yield DNA fragments with complementary ends.

There are two groups of restriction enzymes. Some enzymes recognise a specific nucleotide pair sequence and then cut the DNA at random *i.e.,* at non-specific site away from the recognition site. The other group of enzymes cleave the DNA at specific site. All restriction enzymes are sequence specific and thus the number of cuts they make in a DNA molecule depends upon the number of times, the specific sequences are repeated in DNA molecules. Different endonucleases found in different organisms recognise different nucleotide sequences and they cut DNA at different cleavage sites. The restriction enzymes are named according to the name of organisms from which they are derived. The first letter of the genus together with first two letters of species name of the organism and then first letter of strain in capital form the enzyme name. Roman number at the end of enzyme name indicates the order of their discovery. Some restriction endonucleases, their source, recognition sequences and sites of cleavage indicated by arrows are presented in Table 24.1.

Table 24.1. Some restriction endonucleases, their source, target sequences and cleavage sites

Name of restriction endonuclease	*Source (Micro-organism)*	*Target sequences and cleavage sites*
A. Enzymes which generate blunt end fragments		
B al I	*Brevibacterium albidum*	T G G ↓ C C A A C C ↑ G G T
H ae I	*Haemophilus aegyptius*	(A) G G ↓ C C (T) T C C ↑ G G A
S ma I	*Serratia marcescens*	C C C ↓ G G G G G G ↑ C C C
Enzymes which generate cohesive restriction fragments		
Eco RI	*Escherichia coli*	G ↓ A A T T C C T T A A ↑ G
B am HI	*Bacillus amylolique faciens H.*	G ↓ G A T C C C C T A G ↑ G
B gl III	*Bacillus globigii*	A ↓ G A T C T T C T A G ↑ A
H ae II	*Haemophilus aegyptius*	Pu ↓ C G C G Pu Py G C G C ↑ Py
H indl III	*Haemophilus influenzae Rd.*	A ↓ A G C T T T T C G A ↑ A
P st I	*Providencia stuartii*	C T G C A ↓ G G ↑ A C G T C
S al I	*Streptococcus albus G.*	G ↓ T C G A C C A G C T ↑ G
T aq I	*Thermus aquaticus*	T ↓ C G A A G C ↑ T
		Arrow indicates cleavage site in DNA strands

Since each restriction enzyme recognizes unique base sequence of 4–6 nucleotides irrespective of source of DNA, the pieces obtained from DNA molecule of one organism have the same cohesive ends as the pieces produced by the same enzyme acting on the DNA molecule from another organism. Since the restriction enzymes recognize unique base sequence, the number of cuts made in DNA molecule from an organism by a particular enzyme is limited. A typical bacterial DNA molecule which contains around three million base pairs is cut into several hundred to several thousand fragments, and nuclear DNA of mammals may be cut into more than a million fragments. A set of fragments generated by a single restriction enzyme can be detected easily by gel – electrophoresis of enzyme treated DNA and particular DNA segments can be isolated by cutting out the portion of the gel containing the fragments and then removing DNA fragment from the gel. Of the special interest are the smaller DNA molecules such as viral DNA, DNA of Plasmids which may have only 1 – 10 or even none cleavage sites for particular enzyme. Plasmids are especially valuable in recombinant DNA technology because they have single cleavage site for a particular restriction enzyme.

The restriction endonucleases are found in nearly every microorganism, but more in bacteria. No similar enzymes have been identified in a few eukaryotes examined so far. About 975 restriction endonucleases have been isolated from different species of bacteria which recognise more than 100 different nucleotide sequences. They are now available commercially in pure form and most of these enzymes recognize specific nucleotide sequences. Some restriction enzymes, however, are not specific in their action and they do not cut DNA strands at specific sites. In bacteria these endonucleases act as chemical weapon against invading viruses. If any foreign DNA molecule enters the cell, that is recognized, cut into pieces and made ineffective. But there is a possibility that the bacterial DNA itself may be cut into pieces by restriction endonucleases which would amount to self-destruction. To prevent this, there is a protective mechanism available in bacterial cytoplasm and in each restriction site one or more nucleotides of corresponding sequences in the genome of bacterium are protected by methylation at an adenine or cytosine residue.

An important event in the study of restriction enzyme with electron microscope was that the fragments produced by many restriction enzymes were spontaneously circularised (Fig. 24.4). The circles could be relinearised by heating but after circularisation if they are treated with DNA ligase, the circularisation becomes stable. The Eco R_1 enzyme discovered in the laboratory of Dussoix and Boyer was able to cleave a circular DNA molecule to form a linear duplex.

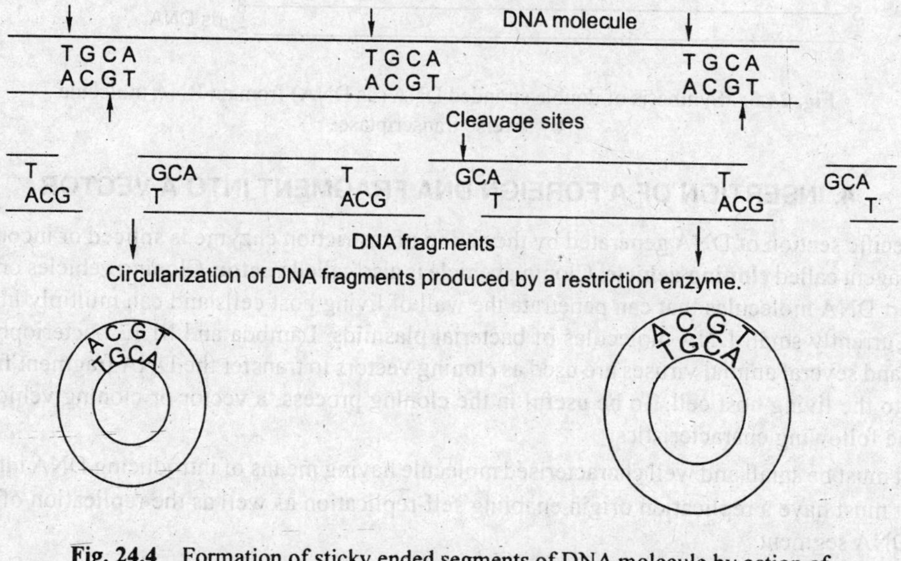

Fig. 24.4 Formation of sticky ended segments of DNA molecule by action of restriction endonuclease and circularization of fragments.

The desired DNA or gene sequence of one organism to be added to the DNA of another organism is called **foreign DNA**. Foreign DNA may be a fragment of DNA molecule which is enzymatically isolated or a copy DNA or cDNA or ready made genes procured from gene banks. Foreign DNA can be procured from a variety of sources depending on the aims and scope of cloning experiment. Copy DNA or cDNA is synthesized by using messenger RNA as a template with the help of an enzyme called **reserve transcriptase**. Since identification and characterisation of DNA sequences on its genome are rather more difficult than using a messenger RNA if it is in pure form complementary DNA or cDNA is synthesized by using mRNA which carries transcript for coding protein. Normally, the flow of genetic information proceeds from DNA → mRNA → Protein.

In 1960 Temin and Vallimore independently demonstrated reverse flow of genetic information *i.e.,* complementary DNA or cDNA can be synthesized from single mRNA in presence of enzyme *reverse transcriptase*. Using this method a particular gene combination is isolated (Fig. 24.5). c DNA is single stranded DNA (ss DNA). ssDNA is separated from mRNA – DNA complex by alkali hydrolysis. ssDNA is then converted into double stranded DNA (ds DNA) using DNA polymerase of bacterial origin. The two strands of DNA so obtained are still linked by a covalent bond. The bond is broken by S_1 nuclease in order to make free two DNA strands. Poly A or Poly T tails are now added to DNA strands with the help of enzyme *terminal transferase* in presence of ATP or TTP respectively (Fig. 24.5). Rapid nucleotide sequencing technique are now available to use a synthesized gene for cloning.

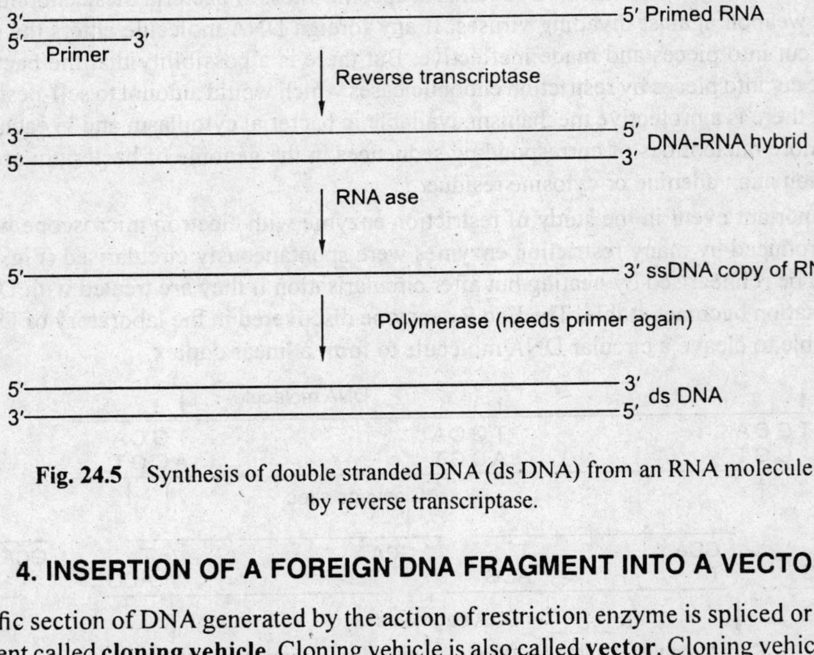

Fig. 24.5 Synthesis of double stranded DNA (ds DNA) from an RNA molecule by reverse transcriptase.

4. INSERTION OF A FOREIGN DNA FRAGMENT INTO A VECTOR

The specific section of DNA generated by the action of restriction enzyme is spliced or incorporated into an agent called **cloning vehicle**. Cloning vehicle is also called **vector**. Cloning vehicles or vectors are short DNA molecules that can penetrate the wall of living host cells and can multiply inside the cells. Currently small DNA molecules of bacterial plasmids, Lambda and M 13 bacteriophages of *E. coli* and several animal viruses are used as cloning vectors to transfer the DNA fragment from test tube into the living host cell. To be useful in the cloning process, a vector or cloning vehicle must have the following characteristics:

1. It must be small and well characterised molecule having means of introducing DNA into a cell.
2. It must have a replication origin enabling self-replication as well as the replication of foreign DNA segment.
3. It should be capable of selection of hybrid molecules by a straight forward assay preferably by the growth of a host cell on a solid culture medium.

The DNA fragment generated by action of restriction endonucleases has to be joined with the vector, either a plasmid or a phage, before that is cloned into a bacterium. The plasmid is separated from main chromosome by ultracentrifugation. The plasmid ring is opened by restriction enzyme and the gene is combined with the vector. The splicing process produces a **chimeric DNA** or **hybrid DNA molecule** containing foreign DNA and the DNA of cloning vehicle. Such a DNA is also called **recombinant DNA molecule**. Once the foreign DNA is integrated into cloning vehicle, the chimeric DNA so formed is *introduced* or *transferred* into a host cell. Usually the host cells used for cloning are single celled organisms such as bacteria or yeasts. The host cell that has received recombinant DNA or chimeric DNA is then said to be **transformed** and the introduction of chimeric DNA into the host cell is called **transformation**. Gene cloning is essentially the insertion of a specific piece of foreign DNA in a cell in such a way that inserted DNA is replicated and handed on to daughter cells during cell division. The host cell must accept the vector with foreign genes, get it incorporated into its genome and start transcribing that gene.

Plasmids as Vector

Plasmids are small, double stranded, extrachromosomal circular DNA possessing limited number of restriction sites. They are present in the bacterial cytoplasm and yeasts. The plasmid that has been used extensively in cloning is pBR 322 (Fig. 24.6) which has 4363 base pairs. They replicate independently of the bacterial chromosome and carry genes for fertility, antibiotic resistance, ability to ferment sugars and also genes for production of bacteriocins, haemolysins etc., a property that helps in the selection of transformed cells. They differ in their size and genes contained in their DNA. Plasmids can be categorized on the basis of number of copies per cell into (*i*) **Relaxed plasmids**, and (*ii*) **Stringent plasmids**. They can also be classified into **conjugative** and **non-conjugative** depending on whether or not they carry a set of transfer gene or **tra** genes that promote bacterial conjugation. Generally **conjugative plasmids** are stringent plasmids of high molecular weight which are present in limited number (1 – 3 copies per cell) and the **non-conjugative plasmids** are relaxed type that are small with low mol. wt. and are present in multiple copies (20 – 25 copies per cell). Plasmids that are popular in cloning contain about 6,500 base pairs and have a single site where they can be cleaved by a restriction endonuclease. Besides, there are some other plasmids which are larger enough and are difficult to handle. They are generally not used in gene cloning. Some plasmids are capable to integrate themselves into the bacterial chromosome and these are called **episomes**, *e.g.,* F factors of *E. coli.* Sometimes after cleavage by a restriction enzymes and during insertion of foreign DNA fragment into the plasmid, a few important genes are destroyed *e.g.,* **Bam HI** restriction enzyme destroys **ter-r gene** (resistant against tetracycline). In general, the plasmid vehicles have a different density than host DNA and thus, they can be purified easily by sedimentation. Thus to clone a

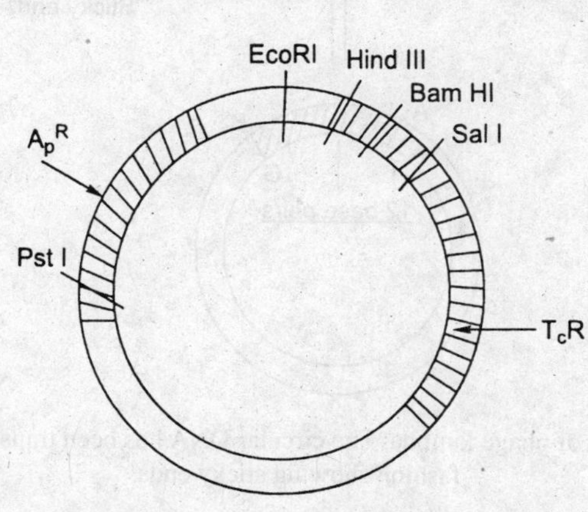

Fig. 24.6 A pBR 322 Plasmid.

piece of foreign DNA, the DNA is integrated or spliced into the plasmid DNA and the chimeric or recombinant DNA is used to transform a host bacterium such as *E. coli*. The plasmid with foreign DNA then replicates within the bacterium when it grows and divides and thus the piece of foreign DNA becomes cloned.

Bacteriophages as Vectors

Bacteriophages or their derivatives are also used as vector or cloning vehicles. Bacteriophage Lambda (λ) is one of the best understood elements in bacterial genetics.

A **bacteriophage** is a virus that infects bacteria. These are commonly called **phages**. Bacteriophages are more complicated than plasmids. Unlike plasmids, the phage vectors are required for cloning of large DNA fragment and, therefore, gene bank or genomic libraries can be constructed. Phage contains a head and a long tail attached to the head which are made up of proteins. Inside the proteinaceous sheath, there is a linear double stranded DNA molecule. Phage DNA contains genes coding for sheath proteins and it has an origin of replication. Phages, like plasmids, lack in the machinery needed to make proteins and hence they reproduce inside living bacterial cells. When a phage comes in contact with a bacterial cell, the phage tail sticks to the cell wall of bacterium *E. coli* and it injects its linear DNA into the cytoplasm of host cell. DNA of Lambda phage is a linear duplex molecule of about 48.5 Kb pairs possessing some 60 genes. Functionally related genes are clustered together on the linear map (Fig. 24.7). Gene cluster on the left of linear map code for head and tail proteins, those in the middle are involved in Lysogenic phage *i.e.,* in recombination and the genes on the right of central region are concerned with regulation and prophage immunity to infection, DNA synthesis and lysis of host cell.

Fig. 24.7 Genetic map of the phage lambda.

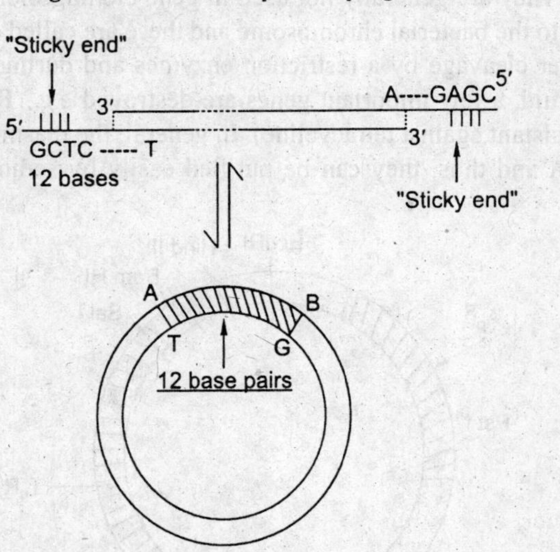

Fig. 24.7a DNA of phage lambda. The circular DNA has been transformed in a linear fashion showing sticky ends.

When present inside the host cell, the phage DNA may be either integrated into bacterial chromosome or it may remain in free state in the cytoplasm. The **replication cycle of phage** is accomplished in two ways: (*i*) Lytic process, and (*ii*) Lysogenic process. In the **lytic process**, the linear double stranded DNA molecule injected inside the bacterial cell becomes circular through the single strand of 12 nucleotides called cos site [Fig. 24.7 (*a*)] which replicates by a rolling circle mechanism and produces several genomes. In Lambda phage which replicates in lytic phase, much of the central region is not essential for phage growth and that may be deleted or substituted without impairing the infectious growth cycle. At the same time phage DNA also directs the synthesis of proteins required to produce empty heads where DNA molecules are packed. Eventually, tail is also formed and attached to the head and progeny phage particles are released out of bacterial cell (Fig. 24.8). At early stage of infection some phages produce proteins that destroy the bacterial DNA, chopping that into individual nucleotides. Lambda phage replicates in lytic phage.

In the **lysogenic** process the phage DNA becomes integrated into the bacterial chromosome and replicated along with the host chromosome with no detrimental effect to the host (Fig. 24.8).

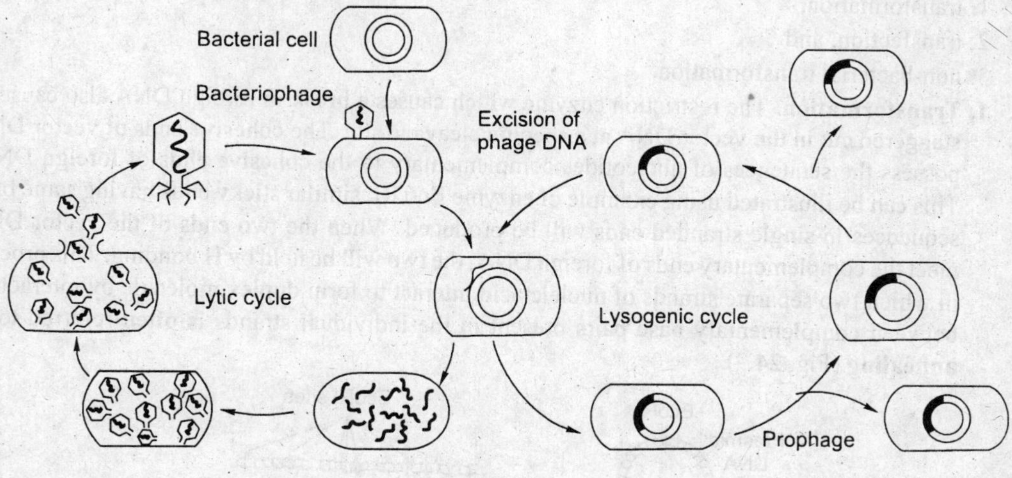

Fig. 24.8 The life style of a temperate phage. Infection of *E. coli* with phage can result in either phage multiplication and lysis or lysogenization of the bacterium.

The following are the advantages of phage cloning systems over plasmids:

(*i*) Foreign DNA can be packed *in vitro* into phages and transduced into bacterial host cell with better efficiency.

(*ii*) Long foreign DNA molecules upto 25 kPB in length can be inserted into phage DNA.

(*iii*) Screening and storage of recombinant DNA is more easy in phages than in plasmids.

M-13 Phages

M 13 Vector is a filamentous single stranded DNA phage. M-13 phage can infect male bacterial cell through pili. Once inside the host cell, phage DNA is replicated, transcribed and translated into viral gene products. The plus (+) strand of virus is used to synthesize complementary (–) strand giving rise to a double-stranded copy of M-13 which is known as the **parental replication form (RF)**. The replication form serves as template from which several (100 – 200) copies of RF molecules are synthesized. Single stranded DNA phages **harm the** bacterial cell, instead the phage particles are released continuously from the cell without any harm to it. No packaging of viral DNA is needed with this system. M 13 particles continue to infect the surrounding male bacteria and a *plaque* of slow growing cells is formed. Thus with M 13 system recombinant phage can be distinguished from the non-recombinant on the basis of plaque colour.

Cosmid Vectors

Cosmids (cos + plasmid) are artificially synthesised vectors containing plasmid DNA with a fragment of λ DNA including the cos site packaged in a phage particle. Cosmids consist essentially of a circular double stranded DNA with antibiotic resistant genes, several restriction sites and cos sites of λ phage. The cosmid HC 79 for instance, consists of 1.7 kb DNA fragment consisting of cos region of λ phase inserted into any one essential region of pBR 322. Foreign DNA can be cloned into any one of the restriction sites on pBR 322 and product after ligation is packaged *in vitro*.

The cosmid infects the host which is selected for growth on antibiotic containing medium and the colonies formed are analysed for the presence of plasmid DNA. Recombinant cosmids, unlike λ vector, are maintained like plasmids in the host cell. Recombinant cosmids are more stable and are easy to store. Larger DNA fragments can be cloned in the cosmids. Thus they combine the best features of cloning vector with maximum capabilities.

Methods of Recombinant DNA Formation

The recombinant DNA is formed by the following three methods:

1. transformation,
2. transfection, and
3. non-bacterial transformation.

1. **Transformation.** The restriction enzyme which causes a break in foreign DNA also causes a staggered cut in the vector DNA at a specific cleavage site. The cohesive ends of vector DNA possess the sequences of nucleotides complementary to the cohesive ends of foreign DNA. This can be illustrated using example of enzyme Eco RI, similar sticky ends having same base sequences in single stranded ends will be produced. When the two ends of the vector DNA meet the complementary ends of foreign DNA, the two will be held by H bonding. This process in which two separate strands of nucleic acid interact to form duplex molecule by interaction between complementary base pairs present in the individual strands is often referred to as **annealing** (Fig. 24.9).

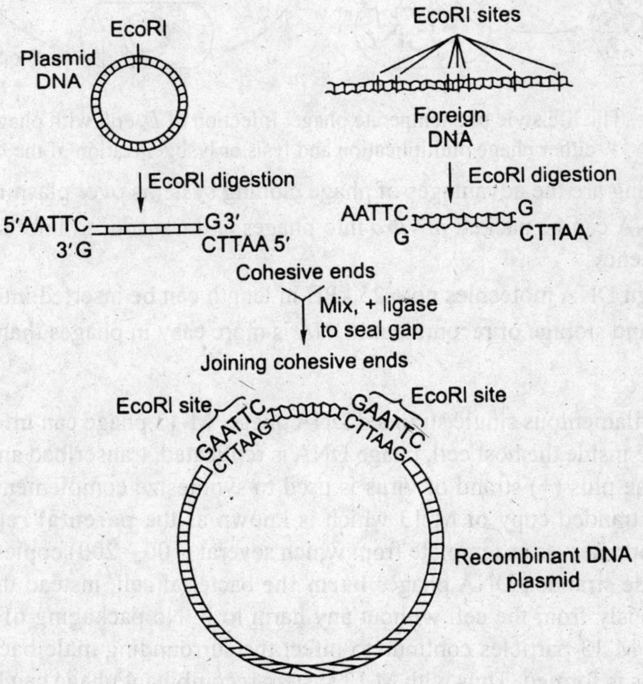

Fig. 24.9

Foreign DNA fragment with non-sticky ends are first made stricky by addition of a homopolymer tail to the 3′ hydroxyl end. Homopolymer tail is essentially a sequence of the few similar nucleotides, *e.g.*, poly A (AAAAA.....) or poly T (TTTTT.....). Now the complementary nucleotide sequences of DNA fragment and those of vector are covalently linked together with the help of enzyme terminal deoxynucleotidyl transferase (Figs. 24.10 and 24.11).

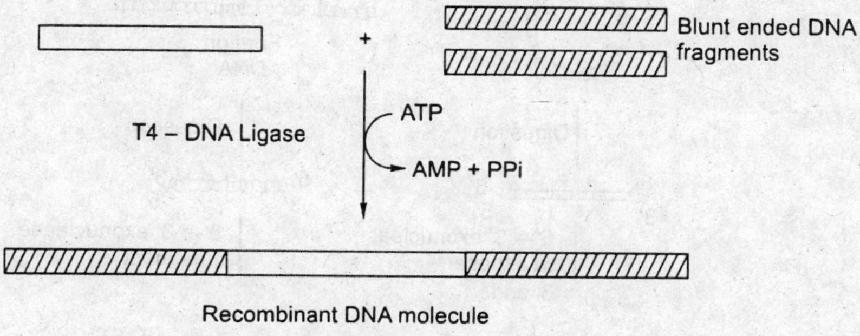

Fig. 24.10 Integration of blunt ended DNA fragments with vector DNA.

Enzyme **DNA ligase** seals and stabilizes the cut ends of two DNA molecules by catalysing phosphodiester bonds between 3′O H and 5′ phosphate of two adjacent nucleotides. The main advantage of this technique is that the foreign DNA segment can be easily retrieved from the cloned copies of chimeric DNA by recleavaging using the same enzyme. At the same time, there are two disadvantages of this technique (*i*) the two ends of the cleaved vector DNA or of foreign DNA may join end to end and they may be recircularised before insertion, (*ii*) the recognition site particularly in the segment to be cloned may not lie at a convenient position.

In this process, the insert may contain a selectable marker which allows for identification of recombinant molecules. An antibiotic marker is often used. So a host cell without a vector dies when exposed to a certain antibiotics and the host with the vector will live because it is resistant.

Transformation of *Escherichia coli* with relatively large pieces of chimeric DNA can be facilitated by treating the cells with $CaCl_2$ at low temperature to make them permeable. With this procedure the transforming DNA enters the cell intact and the host bacterium remains viable. The entry of rDNA into *E.coli* is also facilitated by suddenly raising the temperature to 42°C for 2 – 5 minutes. At this temperature, the cell membrane of *E. coli* becomes permeable to macromolecules. Strains of *E. coli* have now been evolved which are permeable to macromolecules. Then during the growth of transformed *E. coli* cell, the recombinant DNA of plasmid replicates. The bacterial cell contains several plasmids in it and some of them are recombinant type and the others are just the plasmid vehicles. In such cases it may not be easy to determine whether a clone of the transformed *E. coli* cell carries a recombinant DNA molecule or just the plasmid vehicle.

The selection and isolation of transformed host depends on their resistance to antibiotics, colour change or some other characteristics. Different vectors have different properties to make them suitable to different applications. Some properties can include symmetrical cloning sites, size, and high copy number. For selection of transformed cells on the basis of their resistance to antibiotics, it is important to incubate the cells in a medium without antibiotic for about an hour to allow the plasmid antibiotic resistant genes to be expressed. The cells can then be plated on a solid medium containning antibiotic for the selection of colonies with recombinant DNA. Plasmids possessing antibiotic resistance marker will form a colony and those without resistance marker will be killed. This procedure for detecting insertion is called **insertional inactivation**. In strain pBR-322, there are two different antibiotic markers—tet-*r* gene responsible for resistance to tetracycline and amp-*r* gene for resistance to ampicillin. One of the two genes becomes inactivated during restriction enzyme cleavaging and insertion of foreign DNA. Now if the cells are plated on a medium containing ampicillin, all surviving colonies must be amp-r resistant.

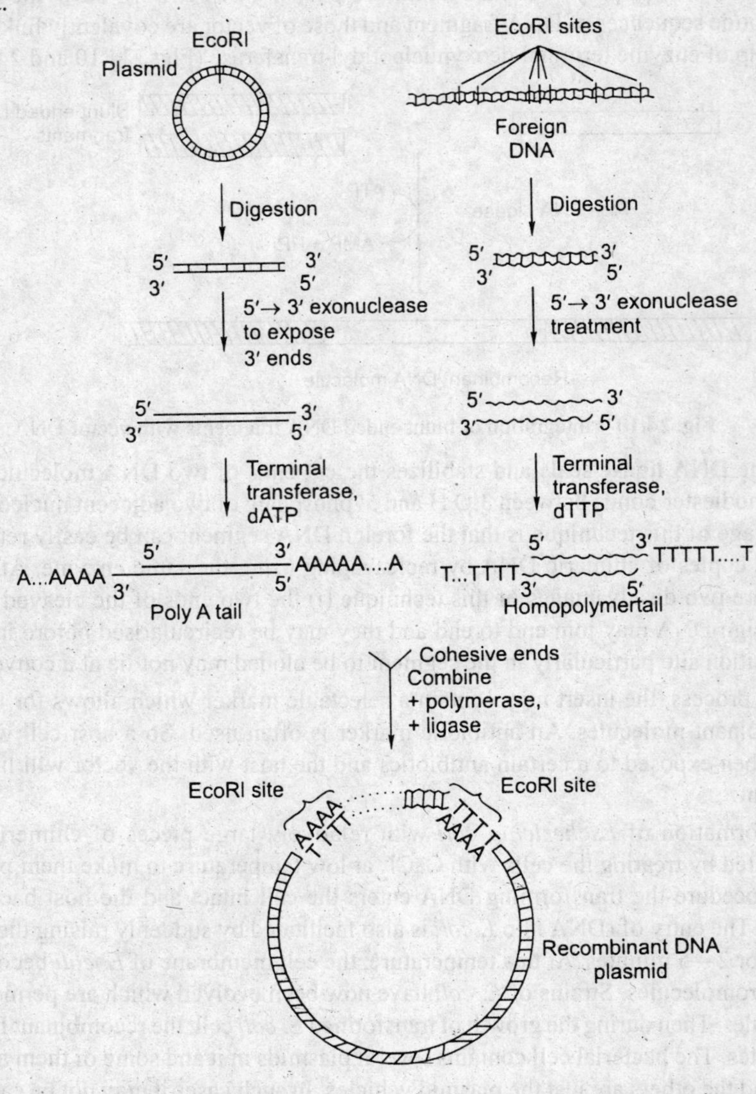

Fig. 24.11 Non-sticky ends of DNA fragments are digested by exonuclease to expose 3′ ends of fragments followed by addition of homopolymer tails to the ends and subsequent ligation.

2. Phage Introduction or Transfection. Phage introduction is the process of transfection which is equivalent to transformation excepting that a phage is used instead of bacterial plasmid. *In vitro* packaging, a vector uses lambda or M 13 phages to produce phage plaques which contain recombinant DNA. The recombinant DNA can be identified using various selection methods. For the first time bacteriophage was used to transfer the foreign DNA into *E. coli* cells. If the vector is bacteriophage, its replication in bacterial host would result in phage particles, each carrying an identical copy of target gene. When phage vectors are used, a population of cells is infected with the viruses and virus replication proceeds spontaneously. Eventually, the phage DNA containing the target gene is inserted into the bacterial chromosome where it will replicate as though it was a part of normal chromosome. Thus these vectors along with target genes are introduced into a bacterial host. This is usually achieved with the enzyme DNA ligase. It is important for ligation that vector DNA and the target DNA have been cut by the same restriction endonuclease.

3. **Non-Bacterial Transformation.** This process is similar to transformation. The only difference between the two is that non-bacterial transformation does not use bacteria such as *E. coli* as the host. In microinjection the DNA fragment is injected directly into the nucleus of the cell being transformed.

   In biolistics, the host cells are bombarded with high velocity microprojectile gun such as particles of gold or tungsten that have been coated with DNA (Fig. 24.12).

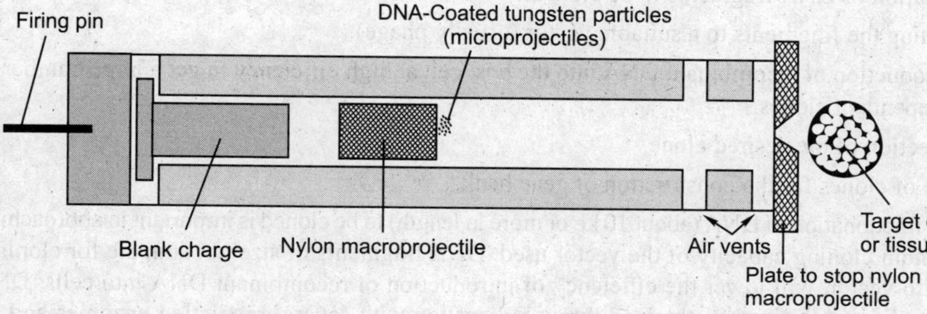

Fig. 24.12 Diagrammatic representation of the structure of a microprojectile gun.

5. MULTIPLICATION OF HOST CELL AND CLONING OF GENES

The host cell containing foreign DNA in addition to its own DNA is allowed to multiply forming a clone having millions of identical cells. Each member of the clone contains in addition to its normal DNA segment of foreign DNA joined to a DNA of cloning vehicle. By this process a piece of genetic information can be transferred or introduced into DNA genome of recipient species where it could never occur naturally, hence a new organism can be created.

In general, simply cloning a piece of DNA is not enough. The information in DNA must be converted into a useful product. To make a product useful the information in DNA is usually transferred from the gene to the site where a new protein molecule is synthesized through gene expression. A number of problems may be associated with the expression of cloned genes in the recipient cell.

Recombinant DNA works when the host cell expresses protein from recombinant gene. A significant amount of recombinant protein will not be produced by the host unless expression factors are available in the cell. Protein expression depends upon the gene surrounded by a collection of signals which provide instructions for the transcription and translation of the gene by the cell. These signals include the initiators, promoters, the ribosome binding site and the terminator. All bacterial genes contain such signals but the signals associated with eukaryotic genes are not recognised in bacteria. One solution to this problem is to insert the eukaryotic DNA into the middle of bacterial gene, thus using the existing regulatory signals. But the product of translation will then be a hybrid protein with additional aminoacid sequences (derived from bacterial gene) at either end. Expression vectors, in which the foreign DNA is inserted, contain these signals. The signals are species specific. In the case of *E. coli* these signals must be *E. coli* signal as *E. coli* is unlikely to understand the signals of human promoters and terminators.

Problems are encountered if the gene contains introns or it contains signals which act as terminators in bacterial host. This results in premature termination and the recombinant protein may not be synthesized properly and folded correctly, or it may even be degraded. Production of recombinant proteins in eukaryotic systems generally takes place in yeast and filamentous fungi. The use of animal cells is difficult due to the fact that many need a solid support surface, unlike bacteria and have complex growth needs. However, some recombinant proteins are too complex to be produced in bacteria and so eukaryotic cells must be used.

Gene Bank or Genome Bank

Gene bank is also called **genomic library**. It refers to the complete collection of cloned DNA fragments which comprises the entire genome of an organism (Dahl *et. al*, 1981). Gene bank is developed by a short gene experiments where genome of a cell is cloned in the form of random and unidentified clones. The cloned DNA are produced by the following method:

1. Isolation of DNA fragments to be cloned.
2. Joining the fragments to a suitable vector (usually phage).
3. Introduction of recombinant DNA into the host cell at high efficiency to get a large number of independent clones.
4. Selection of the desired clone.
5. Use of clones for the construction of gene bank.

For fractionation of DNA (about 10 kb or more in length) to be cloned is important in approaching the maximum cloning capacity of the vector used. DNA fragments of size not suitable for cloning, if ligated to vector, will lower the efficiency of introduction of recombinant DNA into cells. DNA fragments of suitable size are obtained through digestion with several restriction enzymes and by gel electrophoresis. These fragments are cloned without any attempt to select particular sequences. Bacteria containing recombinant DNA are plated on culture media to give rise to individual colonies. If the size of the fragments is small, more colonies are needed to be grown.

cDNA, Clone Bank or cDNA Library

A complementary copy of DNA developed on mRNA molecule is known as copy DNA or cDNA. The cDNA molecule inserted into a suitable vector, transforms a bacterial cell in such a way that it does not disrupt the normal functions of the host cell. The transformed bacterial cell containing a plasmid with cDNA is known as **cDNA clone**. It is difficult to get cDNA from double stranded DNA molecule. Therefore, cDNA clones have been prepared from mRNA sequences of eukaryotic cells. The procedure for obtaining cDNA to develop a library is illustrated in Fig. 24.13. An eukaryotic cell may contain thirty thousand different mRNA sequences. The cDNA clone bank is defined as a population of bacterial **transformants**, each containing a plasmid with single cDNA insert and with sufficiently large number of individual transformants so that every mRNA molecule is represented at least once in the bacterial population (Williams, 1981). A complete bank is one that contains about 5,000 to 10,000 clones of the sequences. A desired clone is screened, identified, isolated and used as and when required. The benefit of cDNA clone bank are as follows:

1. Preparation of cDNA clone is easy in RNA viruses for example, influenza virus, reovirus etc.
2. Screening of cDNA is easy as it contains less number of cDNA prepared from mRNA of the eukaryotic cells that may contain between 10,000 to 30,000 mRNA molecules hence there is a possibility of getting 10,000 – 30,000 DNA fragments.
3. As the genes in prokaryotes do not contain **introns** (the intervening non-transcribed regions of DNA sequence which interrupt the transcript of most of the genes), the DNA clones possess an uninterrupted copy of mRNA molecules. It may be attributed to the facts that the prokaryotes produce splicing enzymes for ligation process just to remove the sequences of interons. In eukaryotes, however, the interons split the transcribed DNA into many fragments which are known as *exons*.
4. Following hybridization process, the nucleotide sequence of the gene interrupted by introns may be determined from a genomic clone by comparing it with mRNA transcripts. Therefore, a cloned cDNA is a most suitable probe for hybridization.

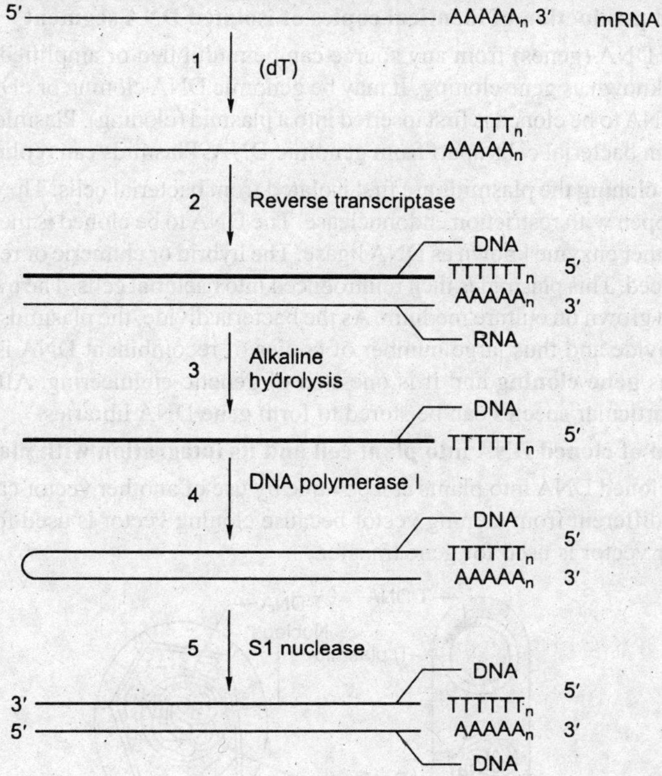

Fig. 24.13 Formation of complementary DNA or cDNA on mRNA of known base sequence.

GENE TRANSFER OR GENETIC ENGINEERING IN PLANTS

Different techniques are evolved for the transfer of gene into microorganisms, plants and animals. One of the techniques which is commonly used in plants is described here.

Gene transfer or genetic engineering in plants involves the following steps:

1. Isolation and identification of desired DNA/genes.
2. Cloning and production of identical copies of isolated DNA segment.
3. Introduction of cloned DNA into plant cells and its integration with plant DNA.
4. Expression of introduced genes in the plants.

1. Isolation and Identification of desired DNA/gene

Two types of DNA can be isolated:

(*i*) Genomic DNA

(*ii*) Complementary DNA (cDNA)

(*i*) **Genomic DNA.** Total DNA isolated from the cells is called genomic DNA. It is isolated by breaking the walls of the cells by physical or biochemical methods. Finally the DNA is separated by ultracentrifugation.

(*ii*) **Complementary DNA (cDNA).** In this case first the mRNA is isolated and then DNA is synthesized on mRNA template by the process of reverse transcription. The base sequence of this DNA is complementary to mRNA base sequence. Hence, it is known as complementary or cDNA.

2. Cloning and production of identical copies of isolated DNA segment

Fragment of DNA (genes) from any source can be multiplied or amplified more than million fold. This is known as gene cloning. It may be genomic DNA cloning or cDNA cloning. In this technique, DNA to be cloned is first inserted into a plasmid (cloning). Plasmids are small circular DNA found in bacterial cells apart from genomic DNA. Plasmids can replicate independently.

During gene cloning the plasmids are first isolated from bacterial cells. They are purified. They are then cut open with restriction endonuclease. The DNA to be cloned is then joined to plasmid DNA by another enzyme known as DNA ligase. The hybrid or chimeric or recombinant plasmid is thus produced. This plasmid is then reintroduced into bacterial cells. The transformed bacterial cells are then grown on culture medium. As the bacteria divide, the plasmids containing foreign DNA also divide and thus large number of copies of recombinant DNA is produced. This is referred to as **gene cloning** and it is one step of genetic engineering. After cloning various genes of a particular species can be stored to form gene/DNA libraries.

3. Introduction of cloned DNA into plant cell and its integration with plant DNA

Transfer of cloned DNA into plants can be done by use of another vector called **gene transfer vector**. It is different from cloning vector because cloning vector is used for gene cloning but gene transfer vector is used for gene transfer.

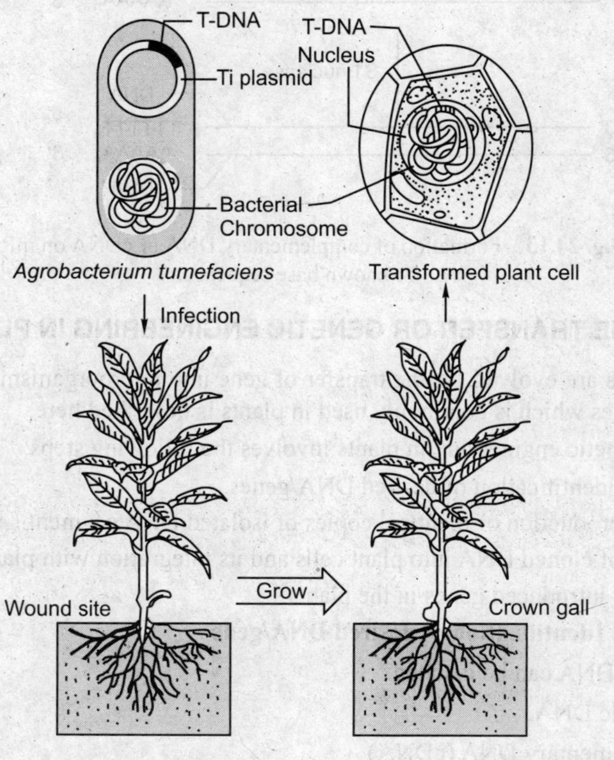

Fig. 24.14 Crown gall formation on dicotyledonous plants after infection with the bacterium *Agrobacterium tumefaciens*.

In dicotyledonous plants, the soil borne pathogen, bacterium, *Agrobacterium tumefaciens* is used as gene transfer vector. The bacterium causes crown gall disease in many species of dicotyledonous plants. The bacterium enters the plants through wound and after infection a callus or tumour is formed near the wounded area at the juncture of root and stem. This bacterium contains, apart from its genomic DNA, a large plasmid called *Ti plasmid* (Tumour inducing plasmid). The plasmid contains tumour inducing genes. When the bacterium infects plant cells, a part of the Ti Plasmid with recombinant DNA gets incorporated into DNA of the plant cell. Thus, the foreign DNA becomes part of plant DNA and is called transfer or T DNA (Fig. 24.14).

The desirable donor gene/DNA after cloning is first incorporated into Ti Plasmid of *Agrobacterium*. This DNA is called transfer DNA, or T DNA or recombinant DNA. The bacterium is then introduced into the plant at cut end.

Bacteria enter the plant tissue and a part of its plasmid having T DNA gets integrated into plant DNA. The cells of callus thus contain foreign DNA from donor species. These cells are known as **transformed or transgenic cells**. Through tissue culture technique plants can be generated from the callus containing transgenic cells (Fig. 24.15). These plants will now contain additional DNA from donor species. These plants are called **transformed plants** or **genetically modified (GM) plants** or **transgenic plants**.

Fig. 24.15 Undifferentiated callus growing on synthetic medium.

4. **Expression of Introduced Genes in Plants**

The recombinant DNA works when the recipient cell expresses protein from recombinant genes. The introduced genes should be transcribed, processed and translated in the plant, *i.e.*, the desirable character of donor species should be expressed in the recipient plant species.

Recent Trends in Genetic Engineering

Research work in plant genetic engineering is advancing at good pace throughout the world including India. Techniques have now been developed which allow direct DNA uptake by plants without using *Agrobacterium*. Foreign DNA can be directly introduced into the plant cells:

1. By use of polyethylene glycol (PEG) into the naked protoplast.
2. DNA can also be introduced directly into nucleus of recipient cell by **microinjection** under microscope with the help of micromanipulator.
3. Technique of electroporation has been used for DNA transfer. In this method, plant protoplasts are electrically permeated in the presence of Vector DNA by discharging about 350 V through a protoplast suspension.
4. Very recently DNA bombardment technique has been used for DNA transfer. In this technique DNA is bombarded into cells by microbullets coated with foreign gene.
5. Very recently Laser beam has been used to make holes into the recipient cells and DNA can be introduced through the holes.

APPLICATION OF GENETIC ENGINEERING

Genetic engineering has many practical applications. It has opened up new possibilities in the field of agriculture, food processing, medicine, energy and industry. Recombinant DNA technology has revolutionized the production of numerous chemicals on a commercial scale such as enzymes and drugs as well as numerous human growth hormones, interferons, vaccines etc.

1. Application in Agriculture

An important application of recombinant DNA technology is to alter the genotype of crop plants to make them more productive, nutritious, rich in proteins, disease resistant, and less fertilizer consuming. Recombinant DNA technology and tissue culture techniques can produce high yielding cereals, pulses and vegetable crops. Some plants have been genetically programmed to yield high protein grains that could show resistance to heat, moisture and diseases. Some plants may even develop their own fertilizers and some have been genetically transformed to make their own insecticides. Through genetic engineering some varieties have been produced that could directly fix atmospheric nitrogen and thus there is no dependence on fertilizers. Scientists have developed transgenic potato, tobacco, cotton, corn, strawberry, rape seeds that are resistant to insect pests and certain weedicides.

Bacterium, *Bacillus thurenginesis* produces a protein which is toxic to insects. Using the techniques of genetic engineering, the gene coding for this toxic protein called **Bt** gene has been isolated from bacterium and engineered into tomato and tobacco plants. Such transgenic plants showed resistance to tobacco horn worms and tomato fruit worms. These genotypes are awaiting release in USA.

There are certain genetically evolved weed killers which are not specific to weeds alone but kill useful crops also. Glyphosate is a commonly used weed killer which simply inhibits a particular essential enzyme in weeds and other crop plants. A target gene of glyphosate is present in bacterium *Salmonella typhimurium*. A mutant of *S. typhimurium* is resistant to glyphosate. The mutant gene was first cloned to *E. coli* and then recloned to *Agrobacterium tumifaciens* through its Ti Plasmid. Infection of plants with Ti plasmid containing glyphosate resistant gene has yielded crops such as cotton, tobacco, maize, all of which are resistant to glyphosate. This makes possible to spray the crop fields with glyphosate which will kill the weeds only and the genetically modified crops with resistant genes remain unaffected. Recently Calogene, a biotech company, has isolated a bacterial gene that detoxifies the side effects of herbicides. Transgenic tobacco plants resistant to 7 MV mosaic virus and tomato plants resistant to Golden mosaic virus have been developed by transferring virus coat protein genes into susceptible plants. These are yet to be released.

The gene transfer technology can also play significant role in producing new and improved variety of timber trees.

Several species of microorganisms have been produced that can degrade toxic chemicals and could be used for killing harmful pathogens and insect pests.

For using genetic engineering techniques for transfer of foreign genes into host plant cells, a number of genes have already been cloned and complete libraries of DNA and mt DNA of pea are now known. Some of the cloned genes include: (*i*) Genes for phaseolin of french beam, (*ii*) few phaseolin leg haemoglobin for soybean, (*iii*) genes for small sub-unit RUBP carboxylase of pea, and (*iv*) genes for storage protein in some cereals.

Efforts are being made to improve several agricultural crops using various techniques of genetic engineering which include:

(*i*) transfer of nitrogen fixing genes (nif genes) from leguminous plants into cereals.

(*ii*) transfer of resistance against pathogens and pests from wild plants to crop plants.

(*iii*) improvement in quality and quantity of seed proteins.

(*iv*) transfer of genes for animal proteins to crop plants.

(*v*) Elimination of unwanted genes for susceptibility to different diseases from cytoplasmic male sterile lines in crop like maize, where cytoplasmic male sterility and susceptibility are located in mitochondrial plasmid

(*vi*) improvement of photosynthetic efficiency by reassembling nuclear and chloroplast genes and by the possible conversion of C_3 plants into C_4 plants.

(*vii*) Development of cell lines which may produce nutritious food in bioreactors.

2. Application in Medicine

Genetic engineering has been gaining importance over the last few years and it will become more important in the current century as genetic diseases become more prevalent and agricultural area is reduced. Genetic engineering plays significant role in the production of medicines. Microorganisms and plant based substances are now being manipulated to produce large amount of usefull drugs, vaccines, enzymes and hormones at low costs. Genetic engineering is concerned with the study of inheritance pattern of diseases in man and collection of human genes that could provide a complete map for inheritance of healthy individuals.

Gene therapy by which healthy genes can be inserted directly into a person with malfunctioning genes is perhaps the most revolutionary and most promising aspect of genetic engineering. The use of gene therapy has been approved in more than 400 clinical trials for diseases such as cystic fibrosis, emphysema, muscular dystrophy, adenosine deaminase deficiency.

Gene therapy may some day be exploited to cure hereditary human diseases like haemophilia and cystic fibrosis which are caused by missing or defective genes. In one type of gene therapy new functional genes are inserted by genetically engineered viruses into the cells of people who are unable to produce certain hormones or proteins for normal body functions. Introduction of new genes into an organism through recombinant DNA technology essentially alters protein makeup and finally the body characteristics.

Vaccines

Recombinant DNA Technology is also used in production of vaccines against diseases. A vaccine contains a form of an infectious organism that does not cause severe disease but does cause immune system of body to form protective antibodies against infective organism. Vaccines are prepared by isolating antigen or protein present on the surface of viral particles. When a person is vaccinated against viral disease, antigens produce antibodies that acts against the viral proteins and inactivate them. With recombinant DNA technology, scientists have been able to transfer the genes for some viral sheath proteins to vaccinia virus which was used against small pox. Vaccines produced by gene cloning are contamination free and safe because they contain only coat proteins against which antibodies are made. A few vaccines are being produced by gene cloning, *e.g.,* vaccines against viral hepatitis, influenza, herpes simplex virus, virus induced foot and mouth disease in animals.

Hormones

Until recently the hormone insulin was extracted only in limited quantities from pancreas of cows and pigs. The process was not only costly but the hormone sometimes caused allergic reactions in some patients of diabetes. The commercial production of insulin was started in 1982 through biogenetic or recombinant DNA technology and the medical use of hormone insulin was approved by Food and drug administration (FDA) ofUSA in 1982.

The human insulin gene has been cloned in large quantities in bacterium *E. coli* which could be used for synthesis of insulin. Genetically engineered insulin is commercially available as **humilin**.

Lymphokines

Lymphokines are proteins which regulate immune system in human body. α-**Interferon** is one of the examples. Inteferon is used to fight viral diseases such as hepatitis, herpes, common colds as well as cancer. Such drugs can be manufactured in bacterial cell in large quantities.

Lymphokines can also be helpful for AIDS patients. Genetically engineered interleukin-II, a substance that stimulates multiplication of lymphocytes is also available and is being currently tested on AIDS patients.

Somatostatin

A fourteen aminoacid polypeptide hormone synthesized by **hypothalamus** was obtained only in a small quantity from a human cadavers. Somatostatin used as a drug for certain growth related abnormalities appears to be species specific and the polypeptide obtained from other mammals has no effect on human, hence its extraction from hypothalamus of cadavers. Genetic engineering technique has helped in chemical synthesis of gene which is joined to the pBR 322 plasmid DNA and cloned into a bacterium. The transformed bacterium is converted into somatostatin synthesising factory. ADA (adenosine deaminase) deficiency is a disease like combined immune deficiency which killed the bubble boy David in 1984. The children with ADA deficiency die before they are two years old. Bone marrow cells of the child after removal from the body were invaded by a harmless virus into which ADA gene has been inserted.

Erythropoetin, a genetically engineered hormone is used to stimulate the production of red blood cells in people suffering from severe anaemia.

Production of Blood Clotting factors

Normally heart attack is caused when coronary arteries are blocked by cholesterol or blood clot. Plasminogen is a substance found in blood clots. Genetically engineered tissue plasminogen activator (tPA) enzyme dissolves blood clots in people who have suffered heart attacks. The plasminogen activator protein is produced by Genetech company which is so potent and specific that it may even arrest a heart attack underway.

Cancer

Cancer is a dreaded disease. Antibodies cloned from a single source and targetted for a specific antigen (monoclonal antibodies) have proved very useful in cancer treatment. Monoclonal antibodies have been tagged with radioactive elements or cytotoxins like Ricin from castor seed to make them more deadly. Such antibodies seek cancer cells and specifically kill them with their radioactivity or toxin.

3. Energy Production

Recombinant DNA technology has tremendous scope in energy production. Through this technology it is now possible to bioengineer energy crops or biofuels that grow rapidly to yield huge biomass that can be used as fuel or can be processed into oils, alcohols, diesel, or other energy products. The wastes from these can be converted into methane. Genetic engineers are trying to transfer gene for cellulase to proper organisms which can be used to convert wastes like sawdust and cornstalks first to sugar and then to alcohol.

4. Application in Industries

Genetically designed bacteria are put into use for generating industrial chemicals. A variety of organic chemicals can be synthesised at large scale with the help of genetically engineered microorganisms. Glucose can be synthesised from sucrose with the help of enzymes obtained from genetically modified organisms. Now-a-days with the help of genetic engineering strains of bacteria and cyanobacteria have been developed which can synthesize ammonia at large scale that can be used in manufacture of fertilizers at much cheaper costs. Microbes are being developed which will help in conversion of cellulose to sugar and from sugar to ethanol.

Recombinant DNA technology can also be used to monitor the degradation of garbage, petroleum products, naphthalene and other industrial wastes. For example bacterium *Pseudomonas fluorescens* genetically altered by transfer of light producing enzyme called *luciferase* found in bacterium *Vibrio fischeri*, produces light proportionate to the amount of its breaking down activity of naphthalene which provides way to monitor the efficiency of the process.

Maize and soybeans are extensively damaged by black cutworm. *Pseudomonas fluorescens* is found in association with maize and soybeans. *Bacillus thuringiensis* contain a gene pathogenic to the pest. The pest has, over the years, not only become dangerous to the crops but has developed resistance to a number of pesticides. When the gene from *B. thuringiensis* (**Bt**) was cloned into *Pseudomonas fluorescence* and inoculated into the soil, it was found that genetically engineered *Pseudomonas fluorescens* could cause the death of cutworms.

CONTROVERSIES AND HAZARDS OF GENETIC ENGINEERING

Although recombinant DNA technology has played tremendous roles in many aspects, yet it may bring ecological catastrophs and disturb balance of nature. It is difficult to assess the nature's destruction in ones' lifetime. It is beyond man's control as it has vast transformational potential. Sometimes genetic manipulations may be responsible for extensive genetic erosion and destruction of plant germplasm. Sometimes recombinant DNA technology may release in dangerous materials in the environment. The toxic chemicals produced due to this technology can be used as agents for mass destruction. Accidently unnatural microorganisms may be produced which can destroy the natural environment and thus they can be used in biological warfare. Possibility is also there that a microbe with a specific degree of resistance might be transformed into a pathogenic form.

As far as the reaction of public towards the use of recombinant DNA genetic engineering is concerned, it is mixed. The production of medicines through the use of genetically engineered organisms, in general, has been welcomed. But the critics of recombinant DNA fear that disease causing organisms used in some recombinant DNA experiments might generate extremely infectious strains that could cause epidemics worldwide. To prevent such an occurrence, The National Institute of Health (NIH) in USA has passed regulations restricting use of pathogenic organisms in recombinant DNA experiments.

According to Animal rights, the production of transgenic animals is harmful to other animals and the interbreeding between genetically engineered and non-engineered fishes may change the characteristics of wild fishes in unpredictable and undesirable way. Some experts fear that the engineered fishes may compete with wild fishes for food and they may even replace wild form in some areas.

The critics of recombinant DNA question the use of genetically engineered bovine somatotropin (BST), a hormone which is injected to increase the yield of milk in cows. In Canada BST cannot be sold legally since 1999 because BST causes health problems for cows. Canadian scientists found that BST could cause mastitis or infection of udder in cows, lameness and infertility. Nevertheless, a large percentage of dairy cows are treated with BST in USA.

Genetically modified crops or transgenic plants also present controversial issues. Allergens can be transferred from one food crop to another through genetic engineering. Brazil nuts contain an allergenic protein. In an attempt to increase nutritional value of soybeans, a genetic engineering company transferred a gene from Brazil-nut into soybean plant which directed increased synthesis of an essential aminoacid methionine. However, when scientists found that genetically engineered soybeans caused allergy in people sensitive to Brazils nuts, the project was cancelled.

The critics of genetic engineering warn that genetically modified food crops could result in some unforeseen problems. They quote a report on 1999 study that found that genetically engineered corn produced pollen that killed caterpillar larvae of monarch butterfly in laboratory. As a precaution the Environmental Protection Agency in United States passed a regulation in January 2000 to reduce potential risks posed by Transgesic corn crop.

Many developing and European countries have expressed concern about the health and environmental hazards associated with genetically engineered food imported from United States and other countries. 130 nations have prepared the protocol of Biosafety in 2000. Once ratified, the treaty will require exporting countries to notify importing countries that the products contain genetically modified organisms including seeds, food crops, cattle and fruit trees.

It is, therefore, necessary that any genetically engineered product must undergo proper evaluation before being allowed into the market. India has a fairly rigorous system of evaluation that is one of the most stringent in the world. The products are evaluated for environmental and food safety. The nodal agencies for approval of genetically modified products are the Department of Biotechnology (DBT) and the Ministry of Environment through various committees. Some of the main committees for the purpose are Review Committee on Genetic Manipulation (RCGM), Monitoring cum Evaluation Committee (MEC) and Genetic Engineering Approval Committee. Several recombinant DNA products in health care are already approved and available to consumers and several products are presently under trials.

QUESTIONS

1. What do you mean by Recombinant DNA? Give an illustrated account of gene cloning.
2. What are restriction endonucleases and DNA ligases? Discuss characteristics of endonucleases.
3. Explain the formation of blunt ended and cohesive ended DNA fragment.
4. Write short notes on:
   (*i*) Plasmid vector,
   (*ii*) λ phase, and
   (*iii*) Cosmids.
5. Discuss the lytic and lysogenic cycles of bacteriophage and explain the mechanism of transfection.
6. Discuss the transformation or plasmid cloning.
7. Discuss the insertion of blunt ended and sticky ended fragments of foreign DNA into Plasmid.
8. How the recombinant DNA is formed?
9. Give an illustrated account of development of transgenic plants.
10. Discuss briefly the practical utility of DNA biotechnology in Agriculture.
11. How the recombinant DNA technology has been exploited in the field of human health?

25

CHAPTER

Micropropagation of Plants (Tissue Culture)

Plant cell culture is a rapidly developing technology which holds promise of restructuring agricultural and forestry practices. The tissue culture technique greatly reduces the labour and space requirements for producing new varieties and can markedly enhance propagation rates. **Theory of totipotency** provides that it should be possible to produce an organism from any one nucleated cell of the body, since all the information needed to specify an organism is contained in its DNA. Consequent upon the pioneering research, plant tissue culture has found wide application in agriculture, horticulture, forestry and plant breeding.

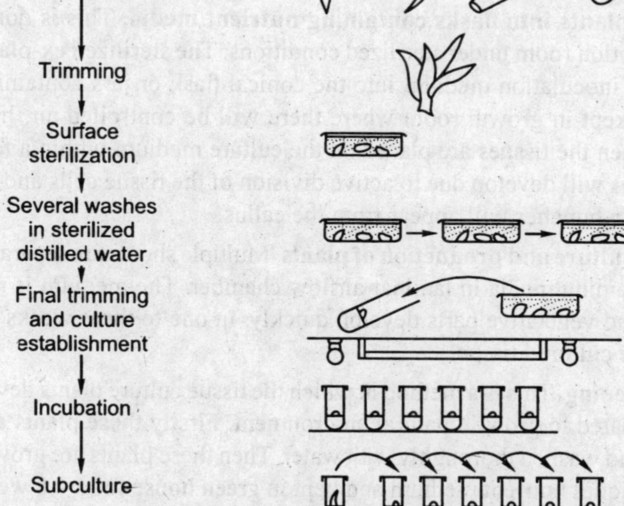

Fig. 25.1 Process of plant tissue culture.

Nature's method of self-preservation in plant is mainly by seeds. This method is also the most economical one and, therefore, universally used for plant propagation. However, in several plants seeds are either not formed or they become inviable at an early stage or seeds are produced in small numbers. Besides, several plants are cross pollinated and produce a promising progeny that shows wide variation. Under such circumstances horticulturists use vegetative methods of propagation from underground stems, tubers, coms, rhizomes bulbs, etc. by cutting, layering, budding and grafting.

Plant tissue culture is being used to supplement the existing methods of sexual and asexual reproduction. The application of this technique is based on the observation that a single shoot tip when cultured on an appropriate nutrient medium, develops large number of shoots, identical to parents. This method of clonal propagation is similar to the rooting of cuttings except the cuttings used for tissue culture are much smaller often less than a centimeter in length (Fig. 25.1). Unlike the conventional methods where one cutting produces only one plant, by tissue culture thousands of plants can be produced in a short period from a single vegetative bud as in *Chrysanthemum*, gerbera, turmeric, teak, eucalyptus, etc. The success of clonal multiplication in plants depends on the following:

1. **Selection of mother plant.** Selection of mother plant with all desirable characters like high yield, good quality and disease resistance plays crucial role in tissue culture.
2. **Collection of Ex-Plants.** A cut portion of plant organ or tissue used for initiation of an *in vitro* culture is called an **ex-plants**. These can be shoot tips, root tips, leaves, tuber etc. Shoot tips are generally preferred.
3. **Sterilization of Ex-Plants.** The ex-plants should be washed thoroughly with sterilizing agents, such as sodium hypochloride, mercuric chloride solution and then washed with sterile distilled water before transferring to the nutrient media.
4. **Preparation of culture media.** The appropriate composition of the nutrient medium largely determines the success of the cultures. In general, the media contains inorganic salts, vitamins, growth regulators (auxins, cytokinins, gibberellins), variety of carbohydrates, organic supplements such as yeast, malt extracts, coconut water and agar boiled in sufficient quantity of water. Then it is sterilized in an autoclave and stored in media room. Now ready-made culture media are available with the scientific suppliers and they can be directly used for the purpose.
5. **Transfer of ex-plants into flasks containing nutrient media.** This is done by experienced person in inoculation room under sterilized conditions. The sterilized ex-plants are transferred with the help of inoculation medium into the conical flask or jars containing culture media. Such flasks are kept in growth room where there will be controlled air, light, humidity and temperature. When the tissues are placed on the culture medium within a few days a mass of cells called **callus** will develop due to active division of the tissue cells and within 3 – 6 week multiple shoots in bunches will appear from the callus.
6. **Method of sub-culture and production of plants.** Multiple shoots are separated and transferred to the flask containing media in laminar airflow chamber. The medium is modified in such a way that roots and vegetative parts develop quickly. In one to three weeks young plants may develop from the cultured tissue.
7. **Method of hardening.** This is a method in which the tissue culture plants developed in artificial media are habituated to grow in natural environment. Firstly these plants are taken out from nutrient media and washed thoroughly with water. Then these plants are grown in netted plastic pots filled with liquid nutrient medium and kept in green house for 6 – 8 weeks. This is called **Primary hardening**. Afterwards the plants are transferred to polybags filled with potting mixture and grown under shaded house for 6 – 8 weeks. This is called **Secondary hardening.** After secondary hardening the plants are suitable for growing in farmer's fields.

SUMMARY CHART OF TISSUE CULTURE OF BANANA

Selection of Elite Plants

↓

Selection of Daughter Sword Sucker

↓

Surface Sterilisation of Suckers AND Dissection
of Shoot Tips under Aseptic Conditions
↓
Inoculation of Shoot Tip on Nutrient Media and
Incubation in Controlled Microclimatic Conditions
↓
Establishment of Cultures and Origin of New Shoots
↓
Multiplication of Shoots
↓
Elongation or Shooting / Rooting
↓
Primary Hardening of Rooted Plants in Green House
↓
Secondary Hardening in Shade House
↓
Planting in the Field

Application of Tissue Culture Technique:

1. Large number of UNIFORM plants can be produced round the year, independent of season.
2. Techniques are also available to produce hybrid plants with combination of several desired characters (genetic manipulation) by tissue culture.
3. The plants so produced are uniform in growth similar to mother plant.
4. Where existing methods of propagation such as budding/grafting etc. are slow, complicated and costly.
5. Micro propagation is carried out under aseptic laboratory conditions and, therefore, plants so produced are always free from seed borne diseases, however, subject to selection of healthy mother plant.
6. To preserve/store large number of germplasm of elite and endangered plants in small place without losing its originality.
7. Limited space is required to produce and maintain plants *i.e.,* about 20 sft space is needed to produce 3,000 banana plants as against one hectare area by traditional method.
8. Plants of desired sex can be produced through tissue culture.

Disadvantages of Tissue Culture:

1. High investment.
2. Requires advanced level of skills and technology.
3. Needs commercial protocols (controlled environment and nutrient media).
4. Plants produced are comparatively costly.
5. Plants produced by tissue culture methods are observed to have poor photosynthetic activity because of abnormal leaf anatomy, malfunctioning of stomata, lack of epicuticular waxes and, therefore, mortality in the field is high. These plants, therefore, must be exposed through the hardening process under green houses before planting in field conditions, to acclamatise them to open field conditions.
6. If the production process is not observed scientifically, the plants so produced may lack in quality *e.g.,* with increased cycles of sub-culturing, many times genetically aberrant plant may emerge.

Future Prospects of Tissue Culture in India:

Tissue culture technique can be profitably employed for the production of elite clones and disease free propagules of desired plant type and a new era of *green revolution* in agriculture, horticulture and forestry can be brought in. In India present consumption level of tissue culture plants is negligible but the anticipated demand for tissue culture plants will be high with the commercialization of agriculture. The first commercial tissue culture laboratory in India was set up by A.V. Thomas group of companies in Cochin. Since then, several industrial houses have set up tissue culture units. These are Indo-American Hybrid Seeds, Bangalore, Harrison Malayalam in Chennai, Nath Seeds, Aurangabad, Godrej Plant Biotech, Hyderabad, Hindustant Agri-Genetics, Hyderabad, Proagro Seeds, Hyderabad, A.G. Biotech, Hyderabad, Agrindia Biotechnologies, Gaziabad. Jain Biotech Ltd., Jalgaon, Dharampur Sugar Mills Ltd., New Delhi, International Pulp and Fibres Inc., Gujarat, Kalptaru Irrigation System Ltd. Pune etc. These units are engaged in production of tissue culture plants of fruit crops, vegetable crops, ornamental plants, forest trees and plantation crops.

QUESTIONS

1. Discuss the importance of plant tissue culture. What are the demerits of tissue culture?
2. Give an illustrated account of the procedure for the culture of plant tissue in laboratory.

26 CHAPTER

Modern Concept on the Origin of Life

In an International conference on *the origin of life on earth* in Moscow, the Russian Biochemist A.I. Oparin (1968) stated that life is a special form of matter in motion arising in an orderly manner at a certain stage in the evolution of matter.

According to Korn and Korn (1971), a living system is a physicochemical system capable of elaborate polymeric synthesis directed in specific patterns by information contained in a polymeric form (nucleic acid) which is capable of changing to other controlled patterns at more efficient levels of environmental exploitation.

Origin of life is one of the most fascinating problems in biology. Most of the biologists of today believe that life might have originated on the earth sometime in the remote past from non-living substances. But how did it so happen? Nobody could reply this question properly and none has succeeded in creating living organism from non-living matters. The explanations which some early biologists and philosophers have proposed in this regard are baseless and rather highly fantastic. It is only in the twentieth century the biologists started thinking over the subject deeply and have analysed the problem scientifically and some of them have provided plausible conclusions. Modern biologists do not believe in discontinuous origin of life. They are of the view that life was eventual outcome of long process of chemical evolution which took place during the early stages of the earth's history. The **modern concept** or the **naturalistic theory** suggests that there was no trace of life on this planet about two billion years ago. The earth originated sometime about 4.5 billion years ago either as a part shot out from the sun or in the form of interstellar gas cloud. Urey (1952) postulated that the earth cloud originally contained chiefly hydrogen, methane, inert gases, a large number of iron compounds, ammonia, silicates etc. Most authorities believe that the earth originally was a very hot and homogeneous gaseous ball. That in due course of time cooled down due to cosmic cooling effect and as a result of the cooling three zones *viz.*, Core (radius 2500 km. from the centre), mantle (about 2900 km. above the core surface) and solid crust (about 9 to 40 km. above the mantle) became differentiated (Fig. 26.1). The crust is the superficial solid zone of earth sphere, the mantle is the middle zone which is still in molten condition and earth core is the central part that is most possibly in vaporized state. The conditions on the primitive earth were not suitable for life as free oxygen was not available in the earth's atmosphere. All the oxygen was in combined state, either in the form of water or as oxides of other elements. It is perhaps about 2.5 billion years ago that conditions consistent with life developed.

In the cooling process several physical changes took place in the earth and as a result of these changes, oceans and mountains were formed. Side by side, innumerable continuous chemical reactions were going on. Water which was formed first was not in the liquid state but remained in vaporized state for very long time. The water vapour accumulated and became condensed to form dense cloud around the earth. The water vapour on cooling precipitated in the form of rain drops. Rain drops on reaching the hot surface of earth became instantaneously vaporized and the vapour again accumulated to form clouds. Continuous and prolonged raining made the surface of earth more cold. Then the rain water accumulated in the low areas and formed oceans on the earth. Countless volcanoes and fissures continuously gushed methane, steam, ammonia and perhaps carbon dioxide which accumulated around the earth and formed earth's first atmosphere. That air contained four main elements of life, carbon, hydrogen, oxygen and nitrogen which were in the form of gases deadly to present day life. Moreover, the atmosphere was flooded with short wave cosmic radiations or ultraviolet radiations and was stabbed by incessant lightning. This is the brief history of origin of earth and its atmosphere.

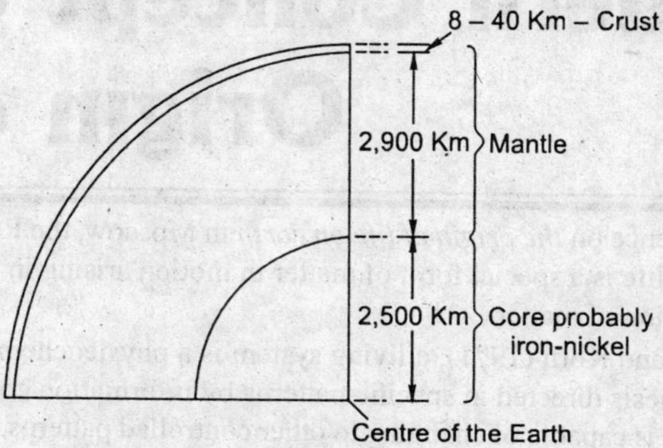

Fig. 26.1 Formation of core, mantle and crust in the earth (sectional view).

Now the question arises as to how life originated on this earth in this elemental turmoil.

In the year 1924, A.I. Oparin, a Russian academician and biochemist, putforth the concept that living matters might have originated from the non-living or inorganic matters. The same view was also expressed by J.B.S. Haldane and R. Bentner. Oparin published a fascinating book *The Origin of Life* in 1936, in which he stated that in the first place carbon must have been present in the atmosphere of primitive earth in the form of hydrocarbons and not in the form of CO_2. Hydrocarbons might have united with other compounds to produce a variety of organic compounds, such as, alcohols, aldehydes, carboxylic acids and so on, and some of carboxylic acids could react with ammonia to form amino acids, the basic molecules of life.

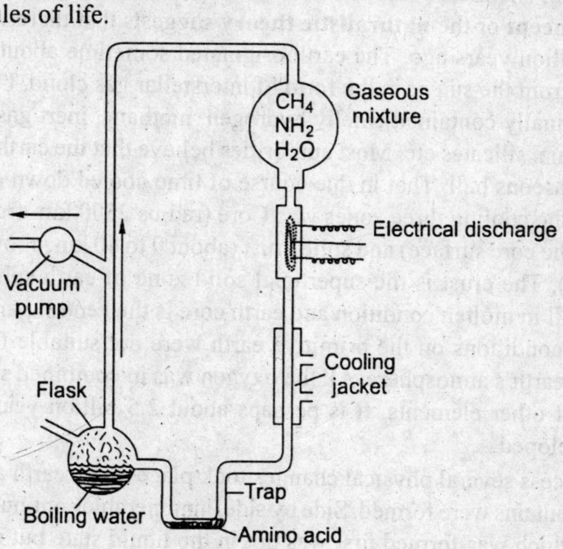

Fig. 26.2 Miller's Experiment.

Dr. Haldane, an English biochemist, wrote that the first formed substances must have been stored in the primitive oceans before life originated in them. He called such compound laden oceans "a hot dilute soup". He also suggested that on exposure to ultraviolet radiations the inorganic compounds should have been converted into organic compounds, such as, amino acids, polypeptides (proteins) and so on.

At one time, scientists believed that organic compounds could be produced only by living system, but since 1950 the scientists are striving to demonstrate experimentally that basic molecules of life could have emerged under primitive sets of conditions on earth.

Dr. Melvin Calvin of California University irradiated a mixture of CO_2 and water vapour using cyclotron and obtained some organic compounds particularly formic acid, oxalic acid and succinic acid with one, two and three carbon atoms respectively. These are metabolic products of living organisms. Dr. Harold Urey, atomic scientist, in those days at the University of Chicago reasoned that if hydrogen, ammonia and methane were probable constituents of primitive atmosphere of earth, what would happen if these gases were placed in a flask and intermittent electric discharge was passed into it! In 1953, Stanley Miller, a student of Dr. Urey, performed one such experiment (Fig. 26.2) in which he exposed a mixture of water vapour, methane, ammonia and hydrogen gases to continuous electric sparking for a week and at the end of the experiment he found sugars, aldehydes, amino acids, such as glycine, alanine and a number of more complex organic compounds. These experiments provided clue that such complex organic compounds were formed in the atmosphere as a whole in the remote past during continuous rains and lightning. Cosmic radiations might have made the course of reactions more easy by supplying tremendous energy required in the processes. These substances entered into solution. Life came into being as weak organic liquor (Fig. 26.3). Now at this juncture after knowing the probable ways in which amino acids and other complex organic molecules were formed, one would question as to how proteins were formed from these amino acids and how the proteins would have given rise to first living cells?

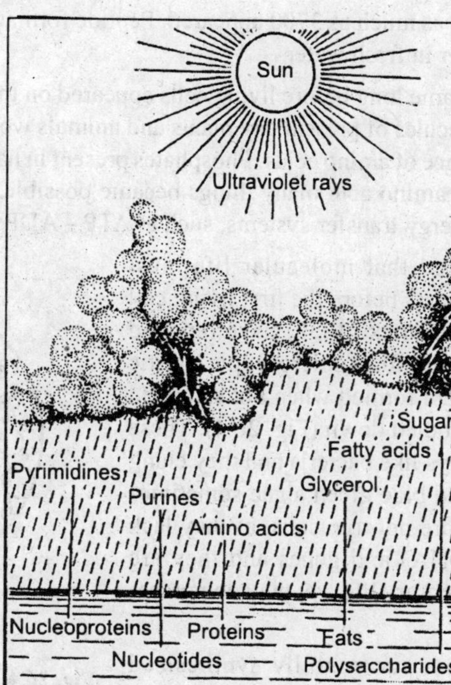

Fig. 26.3 Formation of organic chemicals in primitive ocean.

These two questions are puzzling the scientists all over the world and satisfactory explanations for them are yet to come out. Dr. Sidney Fox of the Institute of Molecular Evolution, Miami Florida, and Dr. George Wald of Harvard University have recently tried to supply some reasonable explanations for the above questions.

Blume (1957) and Dr. George Wald have pointed out that under pre-biological conditions, pools containing dissolved amino acids might have dried up because of high temperature of primitive earth and that primitive amino acids could have formed long polypeptides or primitive proteins under the influence of solar energy. The following reaction illustrates the formation of polypeptides from amino acids.

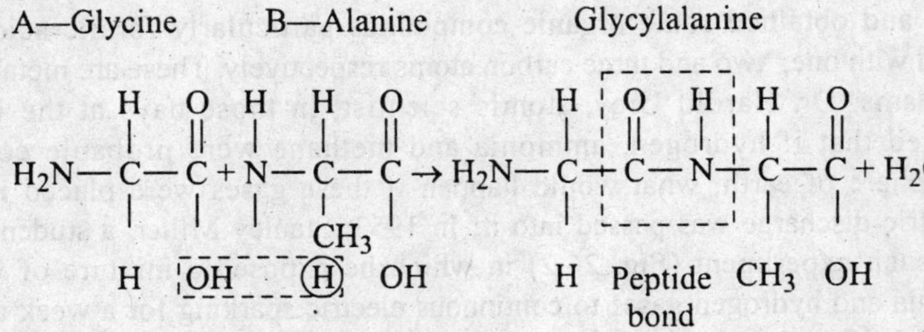

Dr. Fox and Harada (1958) have demonstrated some such reactions in the laboratory by exposing a solution of amino acids at 170°C for evaporation. What they actually found in the experiment was that many amino acid molecules became linked up forming long thread-like sub-microscopic structures, the proteinoids. At least twenty different amino acids are known to take part in the synthesis of life proteins. These aminoacids unite in different ways to form various types of body proteins.

In 1963, one simple experiment was conducted to demonstrate formation of polypeptide chains. When oxygen was bubbled through artificial sea water containing dissolved amino acids, polypeptide chains with molecular weights as much as 3700 appeared. Peptide formation by this means possibly occurred in sea water but never in fresh water.

The primitive proteins came long before living cells appeared on the earth. It is presumed that precisely ordered protein molecules of present day plants and animals would have taken many years in acquiring the definite sequence of amino acids. Phosphates present in natural bodies of water, fresh or salty, when combined with amino acid many things became possible. Nucleotides were formed leading to developments of energy transfer systems, such as ATP – ADP and to nucleotide chains.

Dr. Calvin has estimated that molecular life must have evolved about 2 billion years before the first living cell appeared. Fox, Harada and Kendrick (1959) have reported that a large number of small uniform microscopic elastic spherules with firm outline were formed when hot saturated solution of proteinoides was cooled down (Fig. 26.4). These spherules behaved in many ways like bacteria and they clung together in chains as one celled blue green algae sometimes do. Although these spherules are not true cells, as they lack Deoxyribose nucleic acid (DNA) and genes and they are simpler than any other contemporary life, yet they do possess many cellular properties.

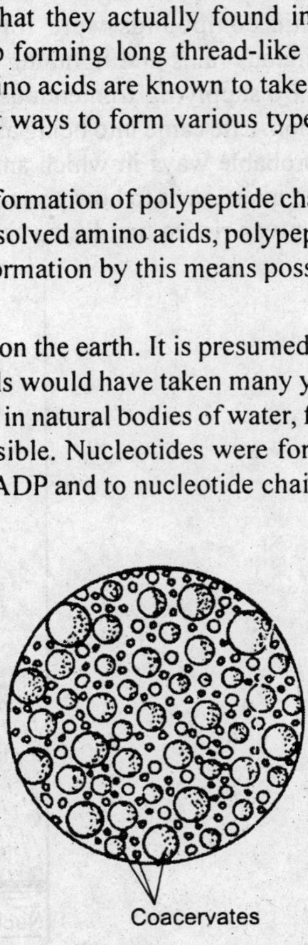

Fig. 26.4 Small uniform microscopic spherules like present-day bacteria or blue green algae

S.W. Fox (1956–57) has successfully synthesized uredo-succinic acid, one of the precursors of nucleic acid, by his thermal experiments. Life of proteins in water is very short. So one can expect that spherules of proteinoids could have persisted for a few weeks only. Among the most effective substances which check the decomposition of proteins are nucleic acids. Perhaps, some nucleic acid precursors such as uredo-succinic acid might have prevented the decomposition of proteins in the primitive pools.

Dr. Fox (1957, 1959) has putforth an interesting hypothesis on the basis of results of his thermal experiments. In this, he suggested that in the pockets of compressed gases rising through the fissures of earth crust, simple compounds combined to form biological staples such as amino acids, vitamins, pyrimidines and carbohydrates and these staples likewise recombined to form biological macromolecules, such as, proteins, polysaccharides and nucleic acids involving loss of water and addition of energy. He further suggested that if the gases escaped through fissures in the hot oceans, the proteinoids would have formed the first cell like microsystems containing nucleic acid precursors and proteins. In fact, life began with the coupling of nucleic acid polymerisation with amino acid polymerisation. In this way primitive 'protein-nucleic acid system' bounded by a membrane or the cell wall would have been formed automatically and rapidly from the gases escaping out through the fissures in the earth crust.

Porphyrins which supplied molecular framework for such vital compounds as cytochromes, haemoglobin and chlorophyll must have been formed almost immediately after acetic acid and simple amino acid glycine first accumulated in the primeval ocean, a big step toward a living system. Porphyrins are extremely stable at high temperatures and at the same time they form very stable complexes with all kinds of metals. The porphyrins might have been metabolically well active in the earliest and only partly living structures. They are coloured compounds, they are able to trap the energy of visible light and make it available for chemical conversions and photocatalytic reactions.

It seems reasonably certain that during an enormous period of time, one to two billion years, the conditions outside the first self reproducing units must have approached those found now only within the living protoplasm. During this long period of genesis of life no line could be drawn between the living and the non-living.

The solar energy reaching the earth surface underwent highly significant qualitative changes. The atmosphere of primitive earth was possibly without free oxygen. The short wave ultraviolet rays (UV-rays) amounting to 5% of the total radiant energy, would have penetrated the atmosphere throughout and would have decomposed methane, ammonia, water and molecular hydrogen into reactive radicals. In this way traces of oxygen started accumulating in the atmosphere. The oxygen of primitive atmosphere absorbed the energy of radiations of short wavelengths, like UV-rays, X-rays and gamma rays and consequently a part of oxygen, particularly at high atmospheric level became excited to form *Ozone layer*. Then the UV-rays of solar radiations were progressively absorbed by ozone layer. Thus, the very source of energy responsible for the initiation of ozone became gradually cut off from the lower atmospheric levels by virtue of its own deeds. That marked the beginning of next phase of chemical and biological evolution.

In the course of chemical evolution probably some proteins combined with nucleic acids and formed complex molecules which acquired the property of synthesising the molecules like themselves. Such hypothetical autocatalytic biological macromolecules would have had some of the properties of the viruses or free genes.

The genes arose in such an environment where all the materials required for their multiplication and chemical activities were present. When the protobionts or primitive cells were formed spontaneously, the DNA-RNA-Protein system became established and with this a variety of control mechanisms were also generated. Replication of DNA and division of primitive cells became integrated.

The primitive cells at some definite stage of growth automatically divided to form daughter cells. The newly formed cells resembled in all respects to their parents (reproduction property).

The first living cells which originated in the oceans were heterotrophic in nature. The organic molecules present in the primitive oceans were source of energy and food for the growth and reproduction of those cells (Fig. 26.5).

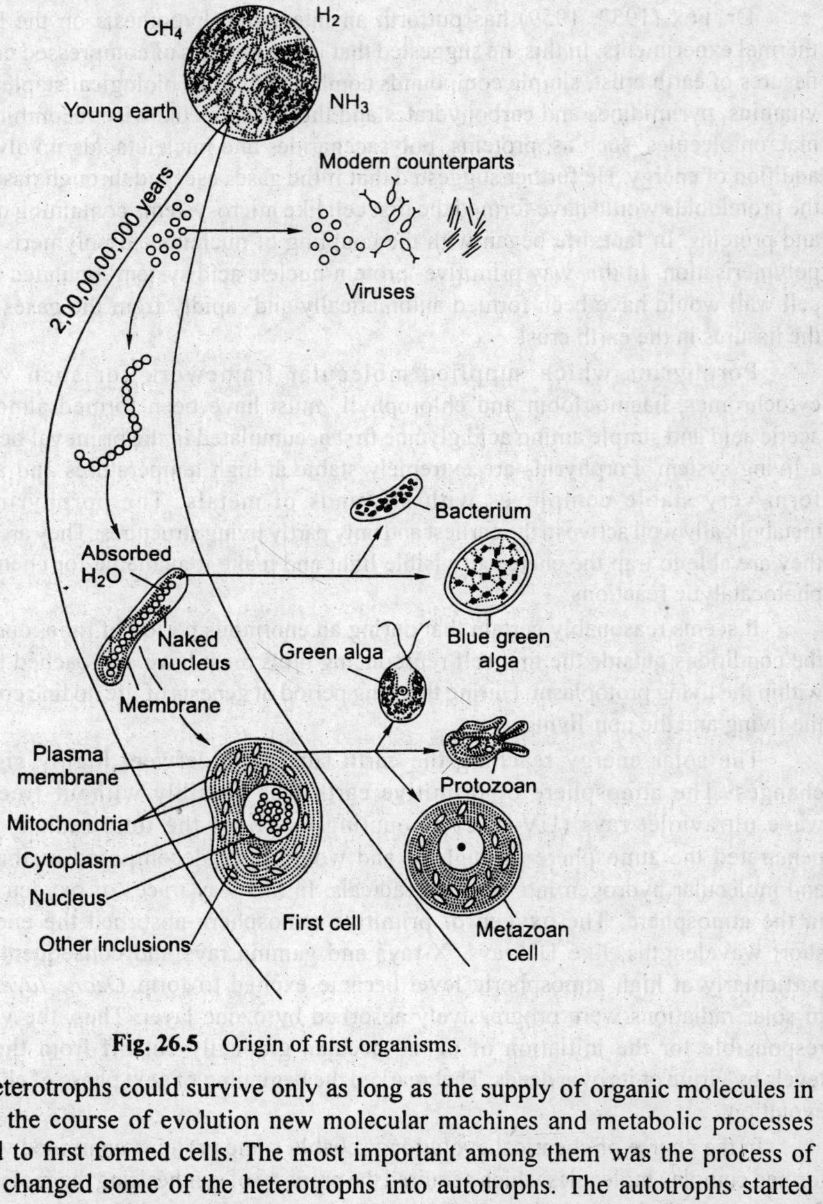

Fig. 26.5 Origin of first organisms.

The primitive heterotrophs could survive only as long as the supply of organic molecules in the oceans existed. In the course of evolution new molecular machines and metabolic processes must have been added to first formed cells. The most important among them was the process of *photosynthesis* which changed some of the heterotrophs into autotrophs. The autotrophs started synthesising carbohydrates from free carbon dioxide and water liberating large quantity of oxygen. In this way, the photosynthetic machinery provided a new source of food for life and changed the atmosphere into an oxidising one.

The history of cellular forms of life could be traced back to the pre-cambrian period of geologic time, some 3000 million years ago. During the most part of that long period, the only inhabitants of the earth were simple microscopic prokaryotic organisms, many of which could be compared with modern bacteria. The conditions under which those organisms lived were entirely different from those of today. Genetic variations made some individuals better adapted which survived and reproduced in a given environment. The emergence of new forms of life through this process of natural selection was influenced by the changes in the physical environment. The development of oxygen generating photosynthetic process and accumulation of oxygen released as a by-product of photosynthesis in the atmosphere affected a new cycle of biological adaptation. In due course of time eukaryotes, a new type of cellular forms emerged in which genetic material was aggregated in a distinct nucleus having a well defined nuclear membrane. Such nucleated cells were highly organised and were capable of sexual reproduction, the process whereby the genetic variations of the parents can be passed on to the offsprings. Gradually the large complex multicellular forms of life came into being.

The first evidence of the occurrence of fossil microscopic plants resembling the modern blue-green algae and bacteria was reported by A. Tylor (1954). In the recent years such micro fossils have been collected from a number of fossiliferous localities in the world and they throw sufficient light on the early history of origin of life. Much information has been obtained from the fossil remains of early microorganisms, and their size, shape, and degree of morphological complexity are among the most easily recognised features.

Comparison of the metabolism and biochemistry of prokaryotes and eukaryotes provide strong evidence that the latter group arose only after the accumulation of oxygen in the atmosphere. The microfossils present a variety of morphological complexities such as:

(*i*) Branched filaments made up of cells resembling modern fungi or green algae.
(*ii*) Fossils of unicellular algae containing intracellular membranes and small dense bodies.
(*iii*) A group of 4 spore-like cells in a tetrahedral fashion resembling some algae.
(*iv*) Spiny cells or algal cysts resembling some modern eukaryotic organisms.
(*v*) Highly branched filaments of large diameter similar to those of eukaryotic algae.
(*vi*) Spheroidal microfossils having two layered walls.
(*vii*) Tetrahedral group of 4 small cells resembling spores of some green algae.

Cyanobacteria are of particular interest as they might have been responsible for the development of an oxygen-rich atmosphere. Like higher plants, cyanobacteria carried out photosynthesis in which energy of sunlight was used to make carbohydrates from water and CO_2 and molecular oxygen was released as a by-product.

Although the biochemistry of the cyanobacteria differs from that of green eukaryotic plants, it suggests that the group originated during the time of fluctuating O_2 concentration.

Early prokaryotes were capable of fixing nitrogen. The enzyme nitrogenase necessary for fixation of nitrogen was produced in the thick walled cells resembling heterocysts of some blue-green algae. It is likely that the capability for N_2-fixation in primitive prokaryotes developed early in the pre-cambrian and in an environment where fixed nitrogen was in short supply. Some nitrogen–fixing organisms became adapted to an anaerobic habitat. By the time eukaryotes appeared, O_2 and fixed nitrogen were enough in the atmosphere, so the eukaryotes never produced enzymes required for N_2- fixation.

In all the bacteria photosynthesis is anaerobic process and O_2 is not evolved as a by-product. Anaerobic photosynthesis evolved in primitive bacteria early in the Pre-cambrian when there was no O_2 in the atmosphere. In due course of time, such bacteria gave rise to the first organisms capable of aerobic photosynthesis, the precursors of modern cyanobacteria. The anaerobic bacteria became adapted to habitats where there was less light and lower concentration of oxygen. Many photosynthetic bacteria occupy such a habitat even today. There are still such organisms living today as are closely related to the earliest forms of life. Thus, it appears that oxygen must have been freely present in the early atmosphere when first eukaryotic cells appeared probably 1400 to 1500 million years ago. Evidence comes from the study of minerals which contain oxygen. Modern biochemistry, geology and minerology have further supported the history of pre-cambrian life. The most primitive forms of life where probably spheroidal prokaryotes comparable to modern bacteria of *Clostridium* group. In the beginning those prokaryotes derived their energy from the minerals which were organic in nature, but of non-biological origin.

The appearance of aerobic photosynthesis resulting in the increase in the oxygen content introduced a change in the global environment during the mid pre-cambrian. This led to the disappearance of many anaerobic organisms. With the development of citric acid cycle and release of energy from foodstuff, the dominance of aerobic organisms in the biological community was confirmed.

The diversity of eukaryotic forms suggests that some forms of sexual reproduction might have evolved. Within the later 400 million years, the rapid diversification of eukaryotes had led to the emergence of multicellular forms of life and some of them were similar to modern plants and animals.

Mutations must also have played good role in the evolution of primeval life from time to time. The periodical mutations or changes in the cells provided it with new enzymatic compounds which controlled the inheritance of characters. Thus, the new individuals possessing these new physiological characters would have had better enzymatic reactions and better buffering or new sources of food.

In summary, the origin of life may be seen as taking place in six stages:

(*i*) Anaerobic phase rich in hydrogen and with energy entering the system as ultraviolet rays, ionising radiations, potential chemical energy and local heat which was responsible for the initiation and accumulation of organic substances of the sort now universally utilized by living cells in building up their own vital constituents.

(*ii*) Anaerobic phase poor in hydrogen when slow accumulation of molecular oxygen blocked the penetration of short-wave ultraviolet rays. Under the influence of short-wave ultraviolet rays some of the oxygen particularly at high atmospheric level assumed a state of excitation and became ozone which progressively absorbed short-wave ultra violet rays. Thus the very source of energy which initiated and promoted the most important reactions became cut off from the lower atmosphere by virtue of its own activity. Then the visible light reaching the earth atmosphere and organic chemicals were the main source of energy and the organic substances became more and more complex and diverse.

(*iii*) Simple anaerobic organisms distinguishable from their surroundings such as purple bacteria came into existence in a way difficult to imagine.

(*iv*) Depletion of accumulated organic substances from the environment by anaerobic micro-organisms of the early period with anaerobic photosynthesis utilizing visible light as the energy source. Thus, intense competition was set in among the organisms, as a result of which one particular kind of microorganisms outgrew and outmultiplied all other kinds, setting the stage for further evolution.

(*v*) In the next stage, chance combination of porphyrins and manganese-protein complexes probably led to the formation of chlorophyll and then the photosynthesis of the kind typical of green plant cells became dominant with accompanying liberation and accumulation of free oxygen into the atmosphere and water caused gradual disappearance of anaerobic organisms except those that escaped into oxygen-free region in the soil and elsewhere.

(*vi*) In the sixth stage, oxygen itself became an evolutionary force which caused extinction of most anaerobic microbes and at the same time enabled achlorophyllous descendants of green cells as animal organisms to make use of the vast energy potential of the biological oxidation of organic substances.

The spontaneous origin of life in present time seems impossible, yet all those events, which occurred in the past, are quite possible at present.

Why spontaneous origin of life seems impossible in the present time? There are two main reasons for this which are as follows:

1. The accumulation of organic molecules seems impossible in present set of conditions because bacteria, moulds and living things of other kinds will attack on these substances and will make them disappear soon.
2. The presence of free oxygen in the atmosphere will cause quick oxidation of organic molecules.

The above two factors present in the modern biological environment were not available in the prebiotic times.

If the modern theory of origin of life is correct then one can hope that conditions for the origin of life must exist on some other planets also and there the probability of existence of life can be expected. Wherever life is possible it should appear and ramify into a wide range of varieties, some of them may be quite similar to those which are found on this earth but others may be quite dissimilar to the forms of life existing on this planet.

QUESTIONS

1. Discuss the experimental evidences which suggest that organic matter could be synthesized in an abiotic environment. What kinds of simplest life in your opinion might have originated first in the past?
2. Discuss the conditions which should have been present in the past to make origin of simplest life possible. If those conditions are made available, can the life be developed in the modern times? If not, why?
3. Discuss the Oparin view on the origin of life.
4. Write short notes on:
   (*a*) Miller's experiments.
   (*b*) S.W. Fox's contribution to the problem of origin of life.

27

CHAPTER

Evolution

The term evolution means unrolling or unfolding (Latin word '*evolve*' = unfolding). It is defined as the process of gradual and orderly change from one condition to another. Such gradual and orderly changes are taking place in all the living, as well as non-living things. The evolution, therefore, may be of two types:

1. inorganic evolution, and,
2. organic evolution.

1. **Inorganic evolution.** It explains the changes that are taking place in non-living objects. The planets, stars, the earth's topography, chemical compounds and chemical elements which form the inorganic world have undergone several changes in the past and shall remain in changing state in the future also. This is inorganic evolution. It is important in the study of geology, astronomy, chemistry and physics.
2. **Organic evolution.** It is concerned with living world. The modern living world (organic world) including both plants and animals has evolved gradually and in orderly manner from primeval ancestors through a process which we speak as *organic evolution.* **Organic evolution refers to the continuous process by which living beings have come to their present forms and functions.**

According to doctrine of organic evolution, the present living objects on the earth have descended from the early simpler organisms by the process of gradual modifications and changes. These modifications and changes are brought about by interactions among organisms and their environments. Through gradual modifications, new individuals with better adaptive features appear. To this we call **natural selection**.

The process of evolution is continuous and it will never come to an end. There are many good proofs in nature to suggest that evolution is taking place and it is taking place on different patterns.

Evolution may speed in either *progressive or retrogressive* way. ***Progressive evolution*** leads the simple forms towards the complex structural and physiological organizations. There are, however, many examples among plants and animals in which evolutionary transformations lead them from high complexity towards decreased complexity. Thus, structurally more complex organic forms produce simplified individuals. Such a type of evolution is called ***retrogressive evolution***. The possible derivation of fungi from algal ancestors as a result of loss of chlorophyll, the development of structurally simplified types of flowers from more elaborate and more complex flowers, change from autotrophic nature to parasitic mode of nutrition in parasitic *Cuscuta* due to loss of chlorophyll are some of the important examples of retrogressive evolution in the plant kingdom. In the evolution of different groups of plants and animals, many changes occur frequently which follow a similar trend and culminate into a more or less similar morphological organization, although the plants which are subjected to evolutionary changes may be genetically only very distantly related. Such an evolution is known as *parallel evolution.* Parallel evolution may run either in progressive direction or in retrogressive direction. Independent evolution of vessels among angiosperms, in some gymnosperms and some of the pteridophytes is one example of parallel evolution. Parallel evolution proceeding in different plant groups in almost similar way and in nearly similar conditions sometimes produces forms which resemble morphologically one another so closely that it becomes difficult to distinguish them. This striking condition is sometimes called convergence and this type of parallel evolution is called ***convergent evolution***. The development of hydromorphic and xeromorphic characters in plants of very distantly related groups of angiosperms may be quoted here as examples of convergent evolution.

At this juncture one may believe that evolution is taking place in a continuous and orderly fashion and it produces a series of modifications in the individuals but one confusion may develop in mind and that is—why stages or steps of evolution have never been seen? The only satisfactory answer to this is that the processes of organic evolution are very slow and they require thousands and millions of years in completion. Unfortunately, the age of man is so short that one cannot see the happenings of many million years in his life time. It is only by seeing the similarities and dissimilarities in the structural organizations and behaviour of organisms in plant and animal kingdoms, the possible trends of evolution are interpreted.

In brief, the modern concept of organic evolution can be summarized in the following points:

(*i*) The theory suggests that the living world has undergone vast modifications in the remote past, it is still changing, and presumably it will go on changing in the future also.

(*ii*) The evolution in the past has resulted different types of organisms among which many have become extinct and many of them are still existing.

(*iii*) Evolution which yields new groups of plants and animals advances very slowly and it requires millions of years to produce new taxa.

(*iv*) Evolution proceeds either on progressive or on retrogressive lines. Both these processes of evolution are proceeding simultaneously in nature.

THEORIES ON EVOLUTION OF ORGANIC FORMS

Several theories have been proposed from time to time to account for the existence and evolution of various kinds of animals and plants on the earth. Among these approaches, some are mechanistic and the others are materialistic. The mechanistic views are based on the belief that life originated from inorganic matter governed by the natural laws, whereas the materialistic views hold that life originated from organic compounds. The mechanistic views are of historical importance now and the materialistic theories seem to be more plausible and scientific. The following theories have been proposed on the origin and evolution of life:

1. Theory of eternity of present conditions,
2. Theory of special creation or creationism,
3. Cosmozoic theory,
4. Theory of catastrophism and spontaneous creation, and
5. Modern theory of organic evolution.

1. Theory of Eternity of Present Conditions

This theory suggests that various living forms have existed in this universe without undergoing any transformation or change and will presumably continue to exist unchanged in future also. This old theory does not stand on scientific grounds, hence it has no validity at present.

2. Theory of Special Creation or Creationism

Until the middle of 19th century life was believed to have been created by supernatural power either once or at successive intervals and each species was presumed to have been created separately. The theory of special creation is based on the descriptions of the Bible. It states that the world was made by God in six natural days. On the first day, the basic materials were made out of nothing. On the second day and third day the ancestors of all the living plants suddenly came into existence. On the fifth day the ancestors of all the living animals were created and finally on the sixth day God created man. This theory was strongly advocated by Father Suerez and John Milton. Linnaeus also supported this theory. This theory was based on religious faith, hence unscientific and could not be subjected to experimental study.

3. Cosmozoic Theory

Cosmozoic theory was proposed by Richter (1865). According to this theory, the living material (protoplasm) might have been carried accidentally to the earth surface from other planets. But this idea was discarded on account of the following two reasons:

(*i*) It does not explain as to how the ultimate source of protoplasm evolved.

(*ii*) The protoplasmic mass must not have survived in the extreme cold and intensive lethal radiations of interplanetary space during its migration.

Recently in 1961, some structures have been discovered from meteorites (rock particles of stars). They seem to contain fossils of microorganisms, possibly of some algae. This simply suggests the possibility of life on other planets also but still there is no evidence that life could be transported safely from one planet to other through meteorites.

4. Theory of Catastrophism and Spontaneous Creation

Empedocles (5th century B.C.) propounded the theory of spontaneous creation or abiogenesis which states that life has originated repeatedly from non-living materials spontaneously. Anaximander and Anaximenes believed that life appeared from small seeds or *'spermaia'* which came to earth along with rain drops. Prominent supporters of spontaneous origin of life were Aristotle, William Harvey, Von Helmont and John Needham. The following are the refutations of the theory of spontaneous creation:

(*i*) Francesco Redi (1668) experimentally disapprove this view and showed that the meat if placed under a screen so as to check flies from laying eggs on it never developed maggots spontaneously.

(*ii*) Lazzaro Spallanzani (1765) demonstrated that even microbes failed to develop spontaneously in a broth, a liquid culture medium, boiled and sealed in a jar. The supporters of this theory argued that heating had spoiled the air in the jars in which spontaneous generation failed to occur. Spallanzani had no answer to this and his experiments thus remained inconclusive.

The **theory of catastrophism** was propounded by Baron Georges Cuvier (1769 – 1832) in view of the discovery of fossils of certain animals whose existence antedated some of the living types. He believed that the old forms of life disappeared due to some revolution or catastrophes and then some new forms came into existence from time to time.

D. Orbringy also postulated that new creations followed catastrophes. The creations of new types in past correspond to the principal geologic periods.

5. Theory of Organic Evolution

The doctrine of evolution is strictly opposed to the older theory of special and separate creation of each plant and animal. The morphological similarities found in plants and animals of different periods in the earth's history, the existence of numerous intermediate forms bridging the gaps between species of one period and those of other, the increasing complexity of structure in both plants and animals through geological periods and the actual demonstrable present day origin of new plants and animals from older types, all support the idea that living organisms on this earth are changeable and they are lineal descendants of pre-existing forms. New species developed due to modification of pre-existing forms.

In the present century notable workers like H.F. Osborn (1857 – 1935), A.S. Romer, G. Simpson etc. have made valuable contributions in vertebrate palaeontology supporting organic evolution.

Among population geneticists notable workers like Haldane, Fisher and Wright have provided mathematical evidence that the natural selection can lead to genetic changes in population. Finally, the various views on evolution were merged into one broad, coherent theory—the modern synthetic theory of evolution.

Historical Account. The development of thinkings regarding evolution started several hundred years ago. Early Greek thinkers like Thales (624 – 548 B.C.), Anaximander (588 – 548 B.C.), Empedocles (495 – 435 B.C) etc., had vague notions about evolutionary processes, because they had no idea about biology. Aristotle believed that the individuals were created by God.

With renaissance, the ideas about evolution quickened and from fourteenth century onward many people have formulated theory of organic evolution in their own way.

Long before renaissance, man had discovered dead remains of plants and animals from the earth. Some of the fossils resembled parts of modern plants and animals and others were quite unlike any known forms. Leonardo da Vinci (1452 – 1519) pointed out for the first time that some of the fossils were undoubtedly remains of animals that had lived in the remote past.

Buffon (1707 – 1788) was the first biologist who discarded the idea of special creation and unchangeableness of animals and plants. He concluded that the animals were plastic and had changed from time to time under the influence of changing environment.

Lamarck (1744 – 1829) proposed for the first time the theory of evolution. He also gave us the first conception of phylogenetic tree or the tree of life.

Charles Darwin (1809 – 1882) was beyond doubt the foremost figure in the history of organic evolution. Since Darwin's time, the theory of organic evolution has been more and more generally accepted. There is no doubt that it is the only logical way whereby the creation can be interpreted and understood. The methods in 'operation' are, however, not definitely known but we may rest assured that the process has been in accordance with great natural laws, some of which still remain unknown.

EVIDENCES SUPPORTING ORGANIC EVOLUTION

The sources of evidence supporting the doctrine of organic evolution have come from different branches of biology. These are outlined as under:

1. Evidences from palaeontology,
2. Evidences from comparative morphology,
3. Evidences from taxonomy,
4. Evidences from comparative physiology and biochemistry,
5. Evidences from Embryology–Doctrine of recapitulation or Biogenetic laws,
6. Evidences from biogeography (distribution of organisms in space), and
7. Evidences from genetics and cytology—Experimental breeding and hybridization.

1. Evidences from Palaeontology

Palaeontology, the science of fossils, provides the strongest and direct evidence in support of the theory of organic evolution. Fossils are dead remains of plants and animals preserved in the rocks of the remote past.

Process of Fossilization. Fossils are found in the sedimentary or stratified rocks. They were formed from those organisms or fragments of organisms which became entombed (covered) under the sediments before they had a chance to disintegrate completely. Complete preservation is very rare and incomplete preservation is directly proportional to resistance of tissues of various disintegrating fragments. It is very seldom that a plant or animal may be found preserved in one piece with its several parts attached and all the tissues intact. Generally different parts become detached before fossilization. Leaves, seeds and fruits become detached from the twigs on which they grew and stems break loose from the roots, and further a proportion of softer tissues may be decayed before the fossilization comes into play. The study of fossils is mainly, therefore, of the unconnected parts, **some of which are**

usually not completely preserved. Often the parts of many different kinds of plants are mixed together indiscriminately. Pieces of plants and animals are brought by wind and rainfluries into rivers, lakes or shallow sea where they settle at the bottom and then they are covered by sedimentary deposits. These organic parts are usually preserved by sedimentary beds of the earth light-heartedly. It is for us to sort them as best as we can. Hard parts of animals such as bones, shells and teeth are preserved unaltered. Very rarely the whole organism may be preserved as such.

Types of fossils. Fossils are of the following three types:

(*i*) Petrified fossils or petrifactions,

(*ii*) Moulds of animals and plants, and

(*iii*) Compression and impression.

(*i*) **Petrified fossils.** Sometimes, fossilized materials are converted into stones because original protoplasmic substance of each cell is replaced by mineral particles. The important mineral compounds which replace the cytoplasmic components of fossilized materials are iron oxide, iron pyrites, sulphur, silica, malachite, etc. Petrified fossils show nearly exact external forms, internal structure and sometimes substances of original plants or animals often in great detail.

(*ii*) **Natural moulds of plants and animals.** These are formed in mud and sand which become hardened into slates or stones. In these fossils, central portion of the material is not preserved at all. The cavity (mould) formed as a result of disintegration of central tissue is subsequently filled up with hard minerals and mud, thus forming an incrustation or cast of the original structure. Soft bodied animals, tree trunks, stumps, fruits and seeds are preserved in this way. Many fossil plants and animals are known only from their casts and impressions they have left. Casts are valuable fossils, because they give good idea about external shapes of the organisms.

(*iii*) **Compression and impression.** These are found in sedimentary deposits. The external form of materials is modified by vertical pressure of overlying sediments in which the material is entombed. In compression, original shape of materials is changed and the material becomes flattened. If the plant part is cylindrical it would be changed to elliptical shape. If rock containing compressed fossils is broken into two halves in such a way that one half contains fossils and the other half bears only impressions or imprints of the fossils, the half with fossils will be called positive rock piece and the other half having imprints of the fossils will be called negative piece or impression piece.

The study of fossils shows that the animals and plants have appeared on this earth in a series of progressive complexity of organizations. If the evolution has taken place, the students of geology and palaeontology may be inclined to think that they should be able to get a complete series of animals and plants representing all their changes which took place in them in the past. If such a complete geological record existed it would undoubtedly be conclusive evidence in favour of evolution, but it is not the case and there is imperfection in the geological record. It is in fact rarely ever complete. There are, however, a few cases in which record are so complete that the entire evolutionary history of the group can easily be traced.

Geological Time-Table

The geological time-table is a calendar of the earth's past history which shows that:

1. None of the organisms of the past were exactly similar to those found today.
2. All fossils did not appear at the same time but in different periods.
3. The older forms are relatively simpler and comparable, at least, with unspecialized member of the living phyla.
4. There has been a gradual progress from the simpler to complex forms in time.
5. A new type appearing in one period became predominant in the latter period and that finally disappeared. Though there have been extinctions of many large groups, once a major phylum was established that has persisted with more or less number of species living today.
6. The mammals among the animals and angiosperms in plants are the latest products of evolution.

Table 27.1 Geologic Time Scale

Eras and year of their duration	*Major divisions of era*	*Periods and their beginning*	*Epochs or Notes*	*Plants*	*Animals*	*Outstanding features*
Cenozoic from 60,00,000 years ago-up to date	Quaternary	20,00,000 years ago	Recent	Dominance of herbs	Civilized Man	
			Pleistocene	Extinction of many trees and increase of herbs.	Extinction of mammals.	Ice Age glacial cold
	Tertiary	Late Tertiary	Pliocene	Evolution of herb, reduction of forest.	Man appeared	Development of plains and grasslands, Mid climates becoming cooler and drier.
			Miocene	Forest reduction.	Culmination of mammals.	
		Early Tertiary 60,00,000 years ago	Oligocene	World wide distribution of tropical forests.	Disappearance of primitive mammals, and evolution of higher mammals and birds.	
			Eocene	Evolution of modern Flowering plants	Modern birds and marine mammals appear.	
Mesozoic (12,50,00,000 years)	Late mesozoic	Upper cretaceous		Angiosperms dominate. Dwindling Gymnosperms	Rise of primitive mammals.	Last great spread of Inland seas
		Middle cretaceous		Rapid development of Angiosperms	Extinction of great reptiles.	Mountain Building at close of period.
		Lower cretaceous (12,50,00,000 years ago)		Rise of angiosperms Cycads & conifers dominants.		Mild do cool climates.
	Early Mesozoic	Jurassic 15,70,00,000 years ago		Rise of angiosperms. Caytoniales, conifers and cycads dominant. Cordiatales disappeared	Primitive birds and flying reptiles (Pterodactyles) Dinosaurs & higher insects dominant.	Continents low Inland seas restricted, Mild climates.
		Triassic 18,50,00,000 years ago		Dominance of higher gymnosperms. Disappearance of seed ferns.	First mammals. Rise of giants reptiles (Dinosaurs).	Elevated continents, Extensive deserts.

Contd on next page

Eras and year of their duration	*Major divisions of era*	*Periods and their beginning*	*Epochs or Notes*	*Plants*	*Animals*	*Outstanding features*
Paleozoic (36,80,00,000 years)	Later Paleozoic	Permian 22,30,000,00 years ago		First cycads & conifers. Reduction of pteridophytes.	Rise invertebrate Reptiles.	Uplift of Continents, Elimination of Inland seas. Cold & dry.
		Pennslyvanian (upper carboniferous) (27,10,00,000 years ago)		Dominance of fossil lycopods, Equisetales, ferns, Pteridosperms & cordiatales.		Formation of coal in Swamps, shallow Inland seas. Warm humid climates.
		Mississippian. (Lower carboniferous) (30,90,00,000 years ago)		Dominant lycopods, Horse-tails, ferns & seed, ferns. Early coal deposit.	Rise of Reptiles & insects. dominance of fishes.	Mountain Building Extensive Inland Warm climates.
	Middle Paleozoic	Devonian. (35,40,00,000 years ago)		Early land plants-Psilophytales Equisetales, ferns, seed ferns, lycopodiales. First Forests.	Rise of amphibians dominance of fishes.	Mountain Building Extensive Inland Seas, heavy rainfall.
		Silurian 38,10,00,000 years ago		First known land plant. (Algae dominant).	Lung fishes and air breathing animals and corals.	Continents low with greatest extention of Inland seas.
	Early Paleozoic	Ordovian 44,80,00,000 years ago		Marine algae	Many groups of invertebrates	Warm mild climates.
		Cambrian 555 million years ago		Marine algae	Invetebrates. Molluscs, Echinoderms, Brachypods	Spreading of Inland seas. Climates mild.
Proterozoic 90,00,00,000 years		15,00,000,000 years ago		Bacteria, Algae and Fungi.	Worms, crustaceans and brachiopods.	
Archeozoic (55,00,00,000 + years)		20,00,000,000 years ago		Organisms very simple.	No fossils.	

The geological approach to the subject of palaecontology is mainly from the standpoint of the correlation of rock formations to evolutionary events. Fossils are markers of the geological times. Rocks in widely separated localities containing similar floras and faunas are generally thought to be approximately of the same age. The largest divisions of geological time are called eras. The eras are recognised on the basis of presence of major groups of plants and animals in their rocks. Eras are divided into periods and periods are divided into epochs. A brief summary of geological time scale and succession of various representative groups of plants and animals at different times have been given in the Table 27.1.

Evolution of Horse. Perhaps the most thoroughly known phylogeny of any vertebrates is that of horse (Fig. 27.1). The history of the horse covers a period of 60,000,000 years. The first horse appeared in the Eocene period about 6 million years ago as a small horse like animal which was about 12″ high and had four toes and a remain of 5th toe. That extinct animal is named ***Eohippus***. The second stage is represented by ***Orohippus*** which was a little more advanced than *Eohippus,* having lost the remains of 5th digit of the foot.

The next stage is represented by the three toed horse called ***Mesohippus*** which made its appearence in the Oligocene period. The height of animal was double that of an *Eohippus.* The teeth had also become more complex.

Mesohippus was later replaced by Merychippus in the Miocene period. ***Merychippus*** represented a transitional stage between modern horse and *Mesohippus*.

Pliohippus, the first single toed horse, appeared in Pliocene. It was about 40 inches in height. The modern horse ***Equus*** first appeared in the upper Pleistocene and now represents the final stage of evolution of horse. This is illustrated by a series of steps in Fig. 27.1.

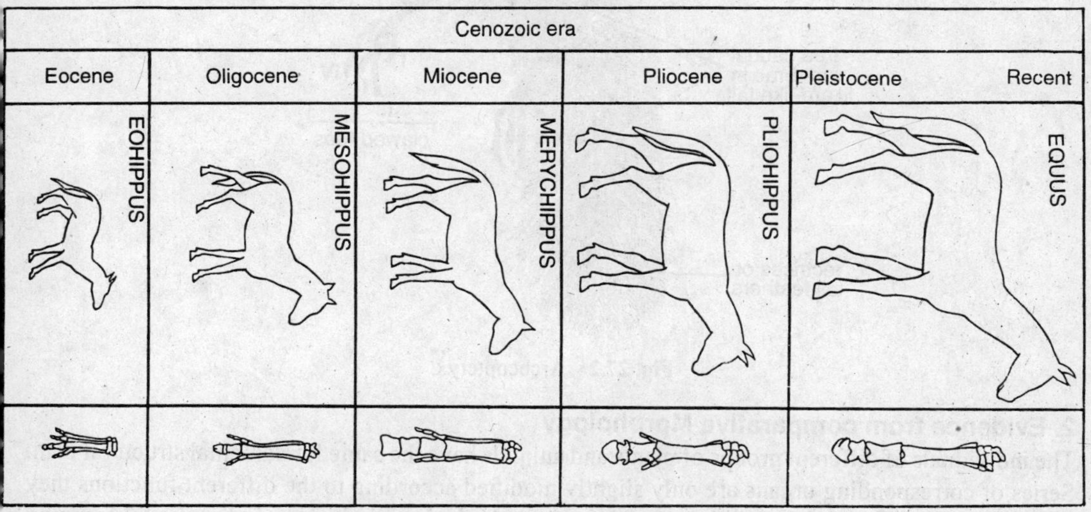

Fig. 27.1 Evolution of horse in different epochs.

The history of evolution of horse in wider sense is merely an example of the evolution of a genus. But, many other fossils are known which clear our notion about the evolution of different classes of vertebrates and fill in the gaps between two different groups. Fossil genus *Archeopteryx,* for example, links birds with reptiles (Fig. 27.2).

Similar such examples are known in palaeobotany also. For example, Psilophytales group in pteridophyta forms a connecting link between non-vascular and vascular plants and similarly seed ferns (pteridosperms) definitely form a connecting link between ferns and cycadales.

Fossil record, however, is imperfect. This imperfection may be due to the following reasons:

(*i*) Incomplete preservation of the organic parts. In the process of fossilization hard parts are preserved with little change but the softer parts are incompletely preserved.

(*ii*) Fragmentary nature of the fossilized organic parts.

(*iii*) Even if the organisms are preserved well, they may be destroyed by physical agencies, such as, high pressure, strong heat, volcanoes and so on.

(*iv*) Some organisms which might have been in imperfect stage of formation might not have been fossilised.

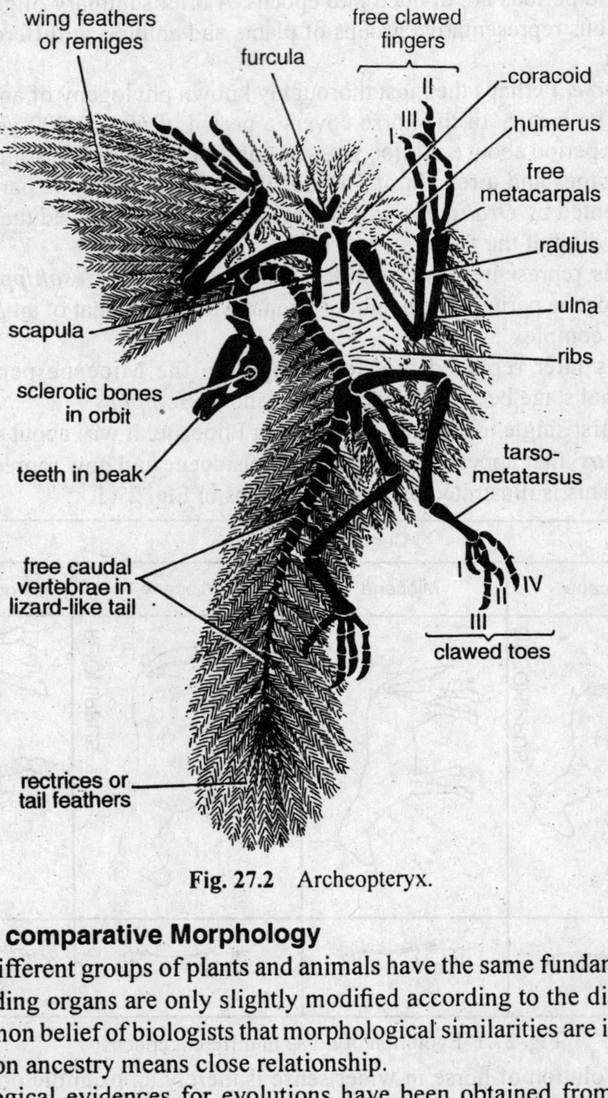

Fig. 27.2 Archeopteryx.

2. Evidence from comparative Morphology

The individuals of different groups of plants and animals have the same fundamental structural plan. Series of corresponding organs are only slightly modified according to the different functions they perform. It is a common belief of biologists that morphological similarities are indicators for common ancestry and common ancestry means close relationship.

The morphological evidences for evolutions have been obtained from the studies of floral morphology, seed structure, wood anatomy of plants and from the comparative studies of organ systems in animals. Some of the morphological evidences in support of organic evolution are enlisted below.

(*i*) Among animals, studies on organ systems make it evident that structures in question show a common fundamental structural plan. Only minor changes in the structure have brought alterations in the morphology of organs. Vertebrae in all higher vertebrates, for example, show the same fundamental structural plan. The vertebrae originate from similar embryonic rudiment. In less evolved vertebrates, vertebrae are simple but in highly evolved members, these are modified to a high degree but the basic structural plan remains unaltered.

(*ii*) The appendages of crustacea are built fundamentally on the same bimarous pattern and so they are considered to be serially homologous.

(*iii*) The forelimbs of all vertebrates are built up on the same pattern as they all have the same bones namely humerus, radius, ulna carpels, (Fig. 27.3). These bones differ in different animals because they are modified and specialized according to different modes of life, such as, swimming walking, jumping, flying, climbing, digging, etc. Such structures are said to be homologous organs.

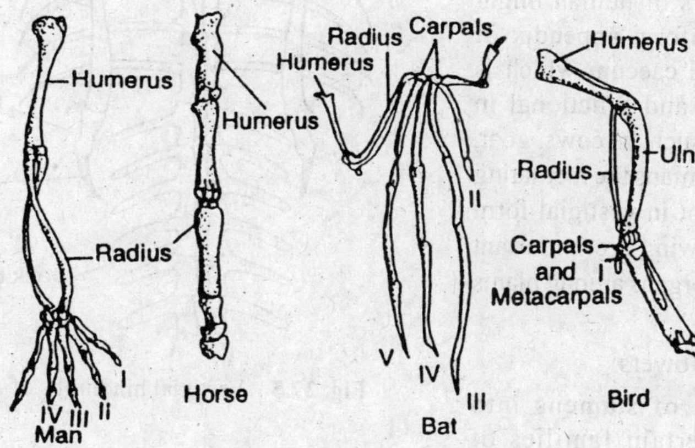

Fig. 27.3 Homology in structure of forelimbs in vertebrates
1. Human forelimb, 2. Horse forefoot,
3. Forelimb of bat, 4. Forelimb of bird.

(*iv*) **Similarities in other organ system of vertebrates.** Practically every organ system presents some basic structural similarities among the vertebrates. The alimentary canal, nervous system, excretory system and reproductive system of the vertebrates are built on similar generalized plans. They show progressive changes from fish to mammals. Heart is two chambered in fishes, three in amphibians and completely four chambered in birds and mammals.

Homology and Analogy. Organs that are similar in their structural plan and in their embryonic developments are called homologous organs. Wings of birds, bat and *Pterodactyle* are homologous because structurally and embryologically they are similar (Fig. 27.4). Structures which are similar in functions but different in morphology and origin are called analogous organs. The phylloclade of *Ruscus* and cladodes of *Asparagus* are similar in functions and origin, so they are homologous, but both of them are analogous to leaves, because they are modified stems. In plants, the flowers are supposed to be homologous to a branch and its various parts, such as, sepals, petals, stamens and carpels are homologised with leaves of a twig. The homologous organs are some of the strongest morphological proofs in support of the theory of organic evolution.

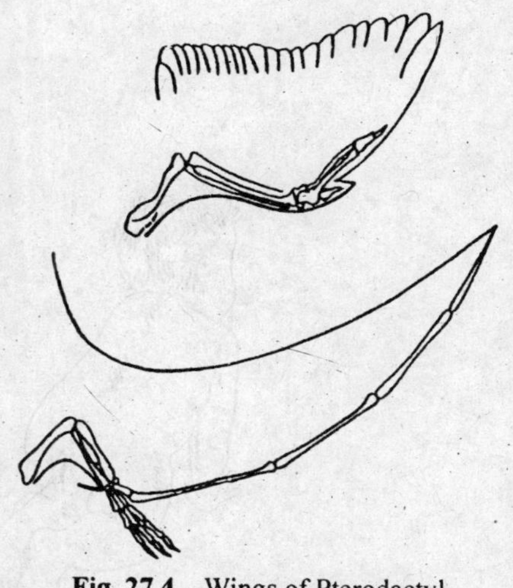

Fig. 27.4 Wings of Pterodactyl, bird and bat.

Vestigial organs. Vestigial organs are the structures in body that are found in reduced condition and serve no function. They, however, resemble and correspond to the same organs which are well developed and useful in other animals. The Kiwi, a flightless bird, has wings that are reduced to structures called *halters*. In some of the snakes (pythons), vestiges of the pelvic girdles are found (Fig. 27.5). In man, there are approximately ninety vestigial organs. Presence of tail like structure in some rare children is a case of vestigial organ. This suggests that ancestors of human might have had tails. Vermiform appendix of human is degenerated caecum which is very well developed and functional in herbivorous animals, such as, cows, goat, rabbits. In birds and humans the nictitating membranes are present in vestigial form (Fig. 27.7). The following are important examples of vestigial organs among plants (Fig. 27.6):

1. Staminodes in flowers.
2. Specialization of stamens into nectaries in certain families of angiosperms.
3. Rudimentary leaflets in apricot (Rosaceae).
4. Reduction of leaflets into glands in apricot.

The existence of vestigial organs can be explained only if it is admitted that being useless they are disappearing. These evidences suggest that the animals and plants are changing and becoming modified.

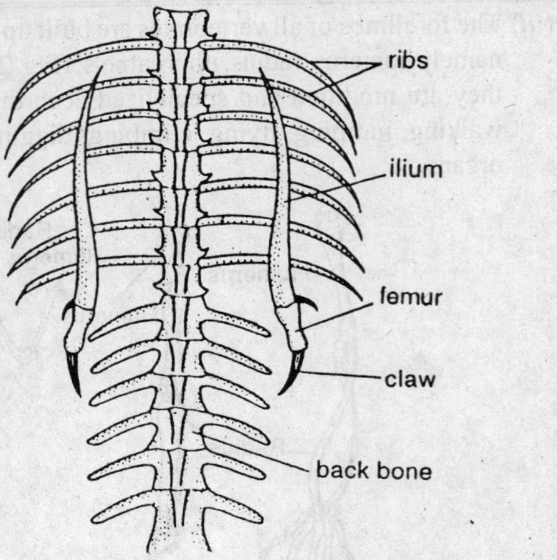

Fig. 27.5 Vestigial hindlimbs of *Python* (Snake).

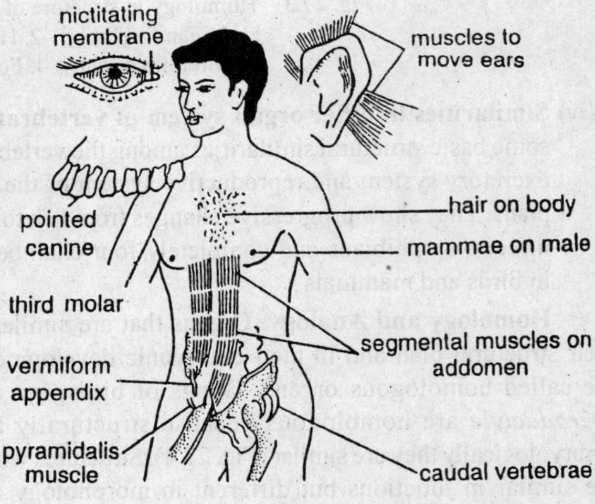

Fig. 27.6 Vestigial organs of man.

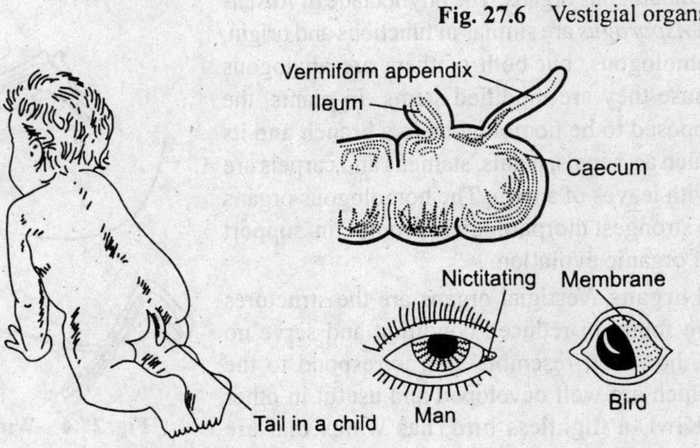

Fig. 27.7 Some vestigial organs in man and bird.

3. Evidence from Taxonomy

Taxonomy, a science concerned with the description, naming, classification and phylogeny of plants and animals, also furnishes many evidences in support of the theory of organic evolution. Evidence from taxonomy is based upon existence of (*i*) border line or intermediate taxonomic groups, and (*ii*) series of plants and animals from simple to complex. Although different individuals belonging to particular phyla or groups have common ancestry, they possess common fundamental characters. All birds, for example, share many features in common. As one goes down to the lower classificatory units, one finds that the species of a genus differ only in some minor characters. Classification in graded series strongly supports affinities of plants and animals by lineal and common descent. Some forms of life form connecting link between two closely related groups, as for example, fossil group *Cycadofilicales* (seed ferns) stood intermediate between ferns and cycads. This group shared some characters of ferns on one hand and some characters of cycads on the other. This suggests that ferns were probable ancestors of cycads and related forms. Such intermediate forms indicate the trends through which a group of organisms modified into another group. The existence of such intermediate and transitional forms is a very strong evidence of evolution. Intergrading species are formed as a result of active evolution.

The individuals of some major groups of plants and animals are simple and their organs are relatively unspecialized, but the other members of the same groups, though they retain some primitive features, are complex. The more complex plants and animals may thus be traced back to simple beginnings.

4. Evidence from Comparative Physiology and Biochemistry

Basic physiological and biochemical differences and similarities in the individuals of particular plant or animal group also furnish important information regarding its relationships and ancestry. The following facts of physiology and biochemistry support the theory of organic evolution.

(*i*) **The protoplasm.** The basic substance of living body contains nearly the same elements compounded from roughly the same proportions of proteins, carbohydrates, fats, water and other substances. Much the same is true with respect to the chemistry of chromosomes. It appears that throughout the entire living world the chromosomes consist of basic proteins and nucleic acids. The nucleic acids are rather uniform.

(*ii*) **Enzymes and Hormones.** In a particular group of plants or animals, similar enzymes and hormones occur. This is specially true of some of the digestive enzymes in animals. In almost all the green plants belonging to diverse groups photosynthetic enzymes are the same. This indicates seemingly a common ancestry of all green plants.

(*iii*) **Metabolites.** The presence of characteristic chemical substances, such as aromatic volatile oils in the members of Rutaceae and Labiatae, resins in conifers and so on, suggests that the members are closely related.

Excretory product analysis. Analysis of the urinary wastes of animals provides some evidence of evolutionary relationship. Nitrogenous excretory waste depends upon the activity of particular kinds of enzymes which are determined by genes and have been selected during the course of evolution. The waste products of the metabolism of purines, adenine and guanine are excreted by man and other animals in the form of uric acid, allantoin, urea and ammonia. Vertebrates' evolution shows successive loss of enzymes for degradation of uric acid. In the early stages of development in chick embryo first ammonia is excreted, then urea and finally uric acid. The enzyme uricase is present in the early stage of chick embryo which catalyses the first step in the degradation of uric acid but later on it disappears.

(*iv*) **Hematological characters.** Crystals of hematine obtained from Haemoglobin of all the species of genus share many characteristics in common while crystals obtained from the members of different classes have characteristics which are peculiar to them. Thus, it is possible on the basis of examination of hematine crystals alone to find out relationship among the various classes of vertebrates.

(*v*) **Comparative serology.** Serum precipitation tests based on "antigen-antibody reactions" in the animals show many basic similarities in physiological and chemical properties of sera of different groups of animals and establish physiological relationships among them, for example, serum precipitation tests of mammals show a much greater chemical relationship between man and ape than between man and dog. Sera of mammals are more sharply distinguished from those of other vertebrates.

(*vi*) **Blood groups.** The blood of human has been classified into four well known groups 'A', 'B', 'AB' and 'O' on the basis of antigen-antibody reaction. Red blood cells may carry antigen 'A' or 'B' or both 'AB' or 'O' (none) and person's blood group is named accordingly. Such series of blood groups have been found in many other animals also. It has been experimentally proved that bloods of horse and ass are similar in their properties. This suggests a relationship between them.

(*vii*) **Biochemistry and Recapitulation phosphogens.** Muscle contraction in animals is brought about by energy rich adenosine triphosphate (ATP). ATP when breaks down yields energy for the contraction and then a second energy rich compound called a phosphogen breaks and releases energy for the resynthesis of ATP.

ATP $\rightarrow$ ADP + Phosphate + Energy

Phosphocreatine + ADP $\rightarrow$ creatine + ATP

(Phosphogen).

In the muscles of vertebrates, the phosphogen is always a specific compound, creatine phosphate, whereas in most invertebrates it is argenine phosphate. Argenine and creatine are very closely related chemicals and in vertebrates the former is actually used in the synthesis of the latter. Hemichordates, the most primitive chordates, have both the phosphogens, a condition found elsewhere only in certain echinoderms. In the embryos of shark, argenine is abundant but becomes more restricted in the adult. Thus, it is possible that creatine metabolism in vertebrate is a biochemical recapitulation which is comparable to embryological recapitulation. Some interesting cases of recapitulations are seen in fishes and amphibians.

5. Embryological Evidences (Biogenetic Laws)

Comparative studies of the embryology show great similarity in different groups of plants and animals. Pea and bean, although quite dissimilar in adult stage, show similar embryology. Monocots and dicots, which are quite different from each other, show similar pattern of embryo development in the beginning and it is believed that one of the two cotyledons of dicot embryo disappeared for ever, giving rise to monocot embryo. In this way, monocots might have evolved from dicots.

Embryos of all multicellular animals develop from single cell (zygote) and all pass through a common series of stages, such as blastula, gastrula, and so on. The nearer the relationship of the adult structures, the greater is the similarity in the courses of embryo development. The embryos of fishes, reptiles, birds and mammals are all alike in the early stage of development (Fig. 27.8). In the embryogenic stages they all possess two chambered heart, aortic arches, gills and gill slits which are exactly like those of the fish embryo. Adults of those animals do not retain these characteristics. In the course of evolution, the embryonic gills were closed, the aortic arches became modified into the carotid and the other arteries and the heart became three chambered in amphibians and reptiles, then finally four chambered in birds and mammals. All these evidences indicate that amphibia, reptiles, birds and mammals have descended from some common fish-like aquatic ancestors.

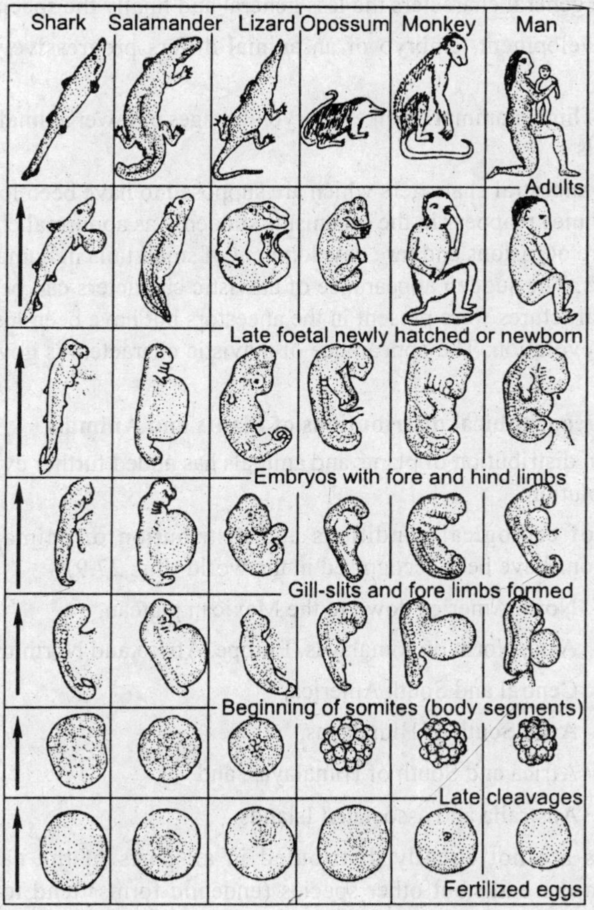

Fig. 27.8 Comparative embryology from fish to man.

It is now evident that during embryonic development all the plants and animals pass through the similar stages, their ancestors went through in the long course of evolution. Thus in brief, it can be said that the embryonic development or ontogeny is recapitulation of the ancestral history or phylogeny or in other words, *"ontogeny repeats phylogeny"*. This is known as '**Biogenetic law**' or "**Doctrine of recapitulation**". Ontogeny is the development of the individual and the phylogeny is the development of the species or race. The law was propounded by E. Haeckel who was of the opinion that each of the embryogenic stages corresponds to the ancestral adult and hence provides direct evidence to the lines of descent. The fertilized egg of higher animals can be compared with the single celled flagellate ancestors and the blastula can be compared with a colonial protozoan or some blastula-like animal which has been postulated to be ancestor of all metazoa. Recapitulations do occur but not in the ways Haeckel has interpreted, for the resemblances are chiefly in the embryos. Protonema stage in mosses can be cited as an example of recapitulation in the plant. This stage apparently indicates the algal ancestry of mosses. Another example of recapitulation is found in the living cycads in which the sperms are ciliated. This condition is common in ferns and bryophytes.

Von Boer's principles. It seems that in formulating his Biogenetic laws Haeckel started from Von Boer's principles of embryonic differentiation which are perhaps more sound, more correct and better guides to the embryological evidence for organic evolution. Von Boer's principles are as follows:

(*i*) General characters appear before specialised characters in the development.

(*ii*) From the more general characters the less general and finally the special characters appear.

(*iii*) During the development, embryo of an animal differs progressively from those of other animals.

(*iv*) Young stages of higher animal are like embryonic stages of lower animals but are not like adults of those animals.

Atavism. The ancestral characters which are supposed to have been lost during the course of evolution may sometimes reappear in the organisms suddenly as abnormal characters. Large canine teeth in human on rare occasions and rare development of small tails in human babies are examples of atavistic characters. The sudden appearance of atavistic characters can be explained only if it is supposed that these structures were present in the ancestors but have been lost in the modern forms during the course of evolution. The appearance of atavistic characters is governed by principles of genetics.

6. Evidences from Geographical distributions of Plants and Animals

Study of geographical distribution of plants and animals has added further evidence in favour of the theory of organic evolution.

On the basis of ecological conditions and distribution of animals, the following six zoogeographical regions have been recognised in the world (Fig. 27.9):

(*i*) Nearctic — North America down to the Mexican plateau,

(*ii*) Palearctic — Asia, North of Himalayas, Europe, Africa and North of Sahara desert,

(*iii*) Neotropical — Central and South America,

(*iv*) Oriental — Asia, South of Himalayas,

(*v*) Ethiopian — Africa and South of Himalayas, and

(*vi*) Australian — Australia and associated Islands.

The organisms are not equally distributed in all parts of the earth. Many types are widely distributed on the earth but other species (endemic forms) tend to be restricted to their specific ranges. Places with similar climatic conditions in different regions of the world do not always possess the same type of flora and fauna, for example, elephants occur in Africa and India but not in Brazil. Climatic conditions of desert region in south-western United States and western Africa are similar but animals and plants in the two places are quite different. South America and African regions both are mostly tropical and crossed by equator and they have similar habitats yet they have dissimilarities in fauna more greatly than similarities. Succulent Euphorbias are common in African deserts but rare in the deserts of United States. Again, representatives of a particular group of plants or animals may occur in widely separated places, but not in the regions between them. This uneven distribution of plants and animals on the earth cannot be explained except through evolution. Geographical barriers, such as oceans, rivers, mountains, etc. isolating groups of animals and plants give rise to divergent forms.

Generally it is seen that the floras and fauna in the regions that are isolated for only a short period of geologic time contain similar but not identical plants and animals. The longer the period of isolation of one region from another, the greater is the degree of differences among the plants and animals of the two regions. These facts lead to the conclusion that the existing plant and animal species are descendants of the extinct forms which were of more generalized type. When they were dispersed away from places of their origin they became modified to meet the new environment. Populations representing intermediate evolutionary changes possibly disappeared. This is the only possible explanation of the uneven distribution of plants and animals in different parts of the world.

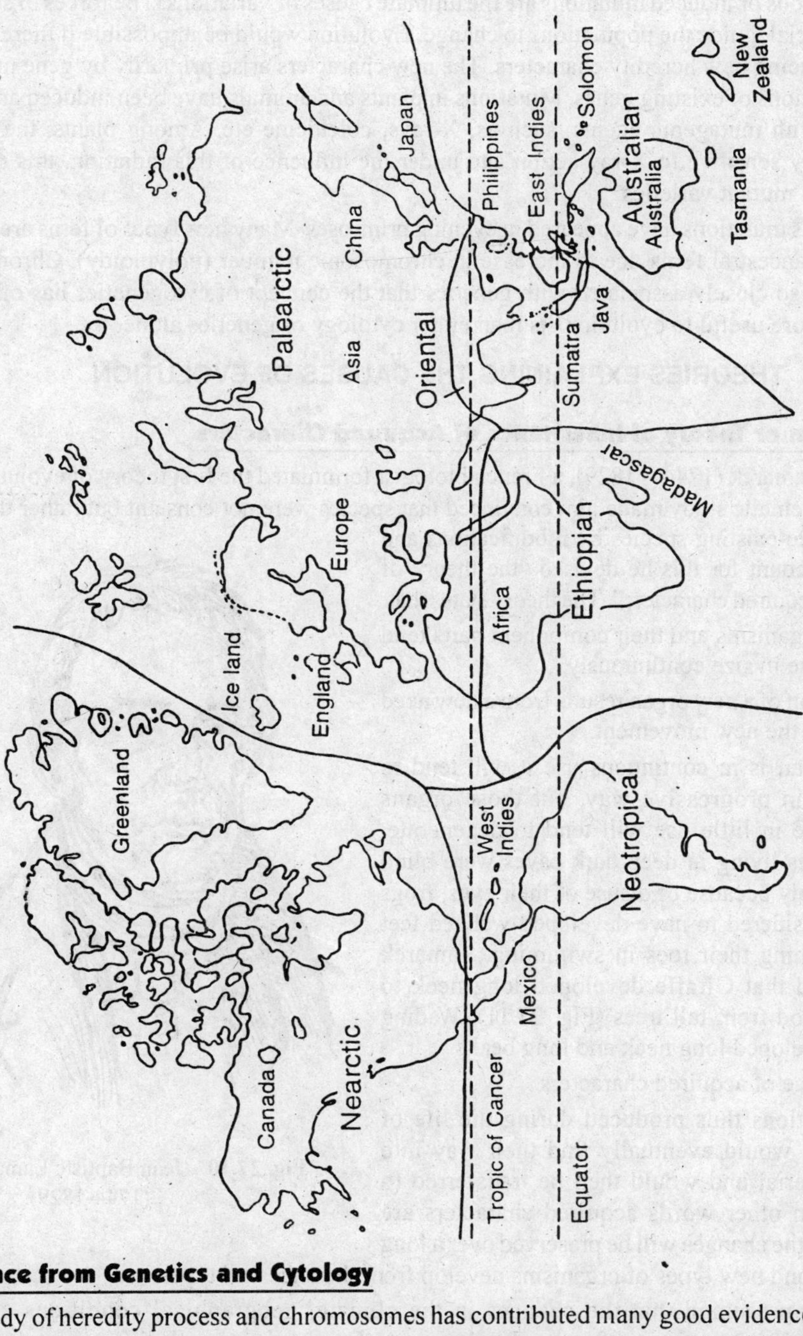

Fig. 27.9 Different zoogeographical regions of the world.

Evidence from Genetics and Cytology

The study of heredity process and chromosomes has contributed many good evidences which support the concept of unceasing change in the living organisms. Cytogenetical evidences in support of organic evolution are final. In fact, the evolutionary relationship is clearly reflected in the karyotype or chromosomes morphology of different species. By selective breeding man has been able to develop different new types of domestic animals and new varieties of plants which are so much different from the wild (common) forms that they can be classified almost as separate species. In the study of hereditary mechanism of plants and animals biologists have observed the frequent, sudden and unpredictable developments of new types from the parents by mutations.

Spontaneous or induced mutations are the ultimate causes of variations. The forces of selection, natural or artificial, guide the populations to change. Evolution would be impossible if there were no means of producing new heredity characters. The new characters arise primarily by gene mutations and recombinations of existing genes. Mutations in plants and animals have been induced artificially by treatment with mutagenic agents, such as, X-rays, colchicine etc. Among plants, the tobacco is specially very sensitive to X-ray action and under the influence of this radiation, this plant has produced many mutant varieties.

Numerous mutations have appeared in evening primroses. Many new types of ferns are evolved from previous ancestral ferns due to increase in chromosome number (polyploidy). Chromosomal differences are so closely associated with genetics that the concept of cytogenetics has often been proved to be more useful to evolutionists than either cytology or genetics alone.

THEORIES EXPLAINING THE CAUSES OF EVOLUTION

1. Lamarckism or Theory of Inheritance of Acquired Characters

Jean Baptiste Lamarck (1744 – 1829), a French biologist formulated the first theory of evolution (Fig. 27.10). His systematic study made him convinced that species were not constant but rather they were derived from pre-existing species by modifications and changes. To account for this he devised "the theory of inheritance of acquired characters". The theory states that:

(*i*) Living organisms and their component parts tend to increase in size continuously.

(*ii*) Production of a new organ results from a new need and from the new movement.

(*iii*) If an organ is in continuous use it will tend to develop in progressive way, but those organs which are in little use will tend to degenerate, *e.g.*, fishes living in deep dark caves were blind presumably because of disuse of their eyes, frogs were considered to have developed webbed feet by stretching their toes in swimming. Lamarck presumed that Giraffe developed long neck to secure food from tall trees (Fig. 27.11). Wading birds developed long neck and long beaks.

(*iv*) Inheritance of acquired characters.

Fig. 27.10 Jean Baptiste Lamarck (1744-1829).

Modifications thus produced during the life of the individuals would eventually find their way into hereditary material and would then be transferred to offsprings or in other words acquired characters are inherited. Thus the changes will be preserved over a long period of time and new types of organisms develop from the ancestral types.

Lamarck presumed that the changes in the climate, geographical conditions and food requirements were main causes for the development of new organs and disappearance of older ones. The existence of vestigial organs could not be explained unless it is admitted that because of continuous disuse they are gradually degenerating.

Objections. Lamarck's theory found little support and was rejected on the following grounds:

(*i*) His first principal suggestion that the organs have tendency to increase in size, although illustrated in many groups of animals is far from universal truth. Many organisms do not show any tendency towards gigantism and in many cases size reduction has been a prominent feature of evolution.

Fig. 27.11 Evolution of modern long necked Giraffe from primitive short necked Giraffe.

(*ii*) The second principle that new organs arise to fulfil new needs of organisms is incorrect, since the desire of an animal alone cannot lead to the formation of new organs. The idea that organs in continuous use will tend to develop in a progressive way and those in little use will tend to degenerate, appears meaningless in practice and has been disapproved by many experiments.

(*iii*) The idea that the acquired characters are inherited has not been demonstrated experimentally, *e.g.*, the powerful arm muscles of blacksmith are not inherited by his child. The germinal material during the developmental processes in plants and animals cannot be easily modified by environmental modifications.

August Weismann (1892), the author of the *"Theory of Continuity of Germplasm"* suggested that reproduction in animals was accomplished not by the ordinary body cells (somatoplasm) but by germplasm which was transmitted essentially unchanged from generation to generation.

Neo-Lamarckism

Although Lamarck's doctrine of inheritance of acquired characters was strongly refuted and rejected, yet a few biologists accepted the theory in modified way. The prominent among them were Cope (1840 – 1897), Giard (1846 – 1908), Packard Spencer and Mc Bridge. According to Neo-Lamarckism, the adaptation is the universal feature of living beings. It arises as a result of interaction between the structure, function and environment. The changed or fluctuating environmental conditions can alter the habit and structure of the organisms. So, the organisms acquire new adaptation in response to new environmental conditions and consequently variations among plants and animals may result. These variations become established gradually in the heredity of the race. This is the modified version of Lamarckism or Neo-Lamarckism because it does not follow the general perfecting tendency in evolution and stresses mainly on the direct action of environment on organic structure. According to some Neo-Lamarckians, the fur development on the skin of some animals as adaptation against cold weather is the consequence of changed environment from warmer to colder state. On the other hand, if the environment comes to the normal state again, the fur would also disappear. Neo-Lamarckians have discarded natural selection as the sole mechanism of evolution. They are of the opinion that the interaction between the structure, function and environment is the only cause of evolution. Today no evolutionist supports the Neo-Lamarckism.

2. Evolution by Natural Selection (Darwinism)

Charles Robert Darwin (1809 – 1882) was born in England (Fig. 27.12). His father Robert was an eminent physician. After his early education, Darwin studied medicine at the University of Edinburg, but he was of academic taste. Then his father advised him to become clergyman which he liked.

He also studied Theology in Cambridge University. Later on he developed interest in natural history and in collection of natural objects. John Steven Henslow, a botanist at Cambridge developed in him interest in plants. He was a keen observer of plants, animals and geological specimens on the advice of Dr. Henslow. After graduation, he accepted job of naturalist aboard the HMS Beagle exploration ship for five years. The ship Beagle left Plymouth on December 27,1831 and visited several Islands of Atlantic ocean, coasts of South America and some Islands including Galapagos in South Pacific. During that voyage period he observed and collected an enormous amount of materials and he was impressed with the striking adaptations found in the animals and plants he collected and with the progressive changes in the different species of given genus. In Galapagos Island he observed several species of individual genera which resembled other species of the mainland. This observation led Darwin to reason that the species may become modified. When he returned home in 1836, he spent two years in writing a book, *The Voyage of the Beagle.*

Fig. 27.12 Charles Darwin.

He was married in 1838 and later Darwin couple were blessed with two daughters and five sons. Darwin acted as secretary of Geological Society from 1838 to 1841. While he was working on the concept of evolution, Darwin was surprised to receive from Alfred Russel Wallace, another English Naturalist, then in Malaya, an essay for his opinion. The essay contained all the salient points of the theory that Darwin was giving shape. Darwin first hesitated in giving any comment on Wallace idea but on the insistence of the geologist Charles Lyell and the Botanist Joseph Hooker, Darwin prepared an abstract of his conclusion for joint publication with Wallace. Both these observers of nature read their paper in 1858, before Linnaean Society, London. Darwin published his first book, *The Origin of Species,* in 1859.

Darwin in his life time wrote more than a dozen books besides numerous articles and letters. The publication of the theory of evolution caused great furor as the basic ideas contained in that were against the ideas of the Bible. But progressive missionary supported him. Darwin and Wallace's theory of evolution is based essentially on three observations and two conclusions which are as follows:

Observation 1

Overproduction free from environmental pressure. Every species tends to multiply in geometric progression. In other words, a population doubling its number in one year should increase four times its size in the second year and eight times in the third year. Simple plant of *Amarathus* may form 20,00,000 seeds. Mushroom in one breeze may scatter two billion spores and so on. Thus the number of offsprings produced by plants and animals is always greater than that which could actually survive in nature.

Observation 2

The size of the population remains constant over long period of time.

Conclusion I. *Competitions or struggles for existence.* Not all the sperms and eggs will form zygotes, not all zygotes will develop into adults and not all adults will survive and reproduce. If there is overproduction of species, there must start a struggle for existence among members of same species (intraspecific competition) or among members of different species (interspecific competition) or against the environment changes (environmental struggle).

Observation 3

Variation. Not all members of a species are alike (*i.e.*, there exist considerable individual variations in them).

Conclusion II. *Survival of the fittest.* In the struggle for existence, therefore, individuals having several favourable variations will have better chance of survival and less fit species will be destroyed by the physical or biotic factors. This is known as "natural selection".

Thus, according to Darwin and Wallace, environment is the principal cause of natural selection. Under the prolonged and continued selective influence of the environment, the organisms accumulate many new favourable variations. The variations probably caused evolution of new species from ancestral species.

In order to explain the inheritance of characters, Darwin proposed *The Hypothesis of Pangenesis* in 1866 which assumed that all the organs and perhaps all the cells in the body of an individual produced minute particles or pangenes which carried information about them. The pangenes travelled from the organs through blood stream to the gonuds where they were assembled to form sex cells or gametes—the eggs and sperms. The gametes, therefore, contained pangenes from each and every organ and they transmitted them to the next generation. In this way, environmental changes would produce modified organs which, in turn, would produce modified pangenes that would transmit the change to the next generation.

Criticisms

Galton (1875) on the basis of several experimental proofs refuted the hypothesis of pangenesis.

Many objections were raised against the theory of natural selection. Some of them are as follows:

(*i*) Natural selection is regarded as a cause of the evolution of new species but it is not the sole cause because it is now known that varieties arise due to mutations. Moreover, the mutations depend purely on chance. Further, the variations of small magnitudes were not taken into consideration in the explanation of natural selection.

(*ii*) Darwin's theory did not explain as to how the variations arose in the individuals.

(*iii*) It is not always the case that useful variations are selected. A useful organ can develop to such an extent (over-adaptation) that some time it may be lethal for the individuals. An example of over-adaptation is found in *Yucca* flowers in which stigmas are so peculiarly shaped that special insect can only affect natural pollination in them. In absence of that moth or insect *Yucca* flowers fail to form seeds. If natural selection had operated favourably on useful variations then non-useful variations should not have been selected. This fact is opposed to Darwin's theory. It is also probable that over-adaptation might have been an important factor in the extinction of many kinds of plants and animals during the course of organic evolution.

(*iv*) It has not been explained that the useful and non-useful variations are inherited.

(*v*) New variation would be lost by 'dilution' as the individuals bred with others. Although the phenotypic expression of a gene may be altered when it exists in association with certain other genes, yet the gene itself is not altered and is transmitted to successive generations as such.

(*vi*) Darwinism indirectly accepts the Lamarckian principle of inheritance of acquired characters which cannot be explained in the light of recent concept of genetics.

Neo-Darwinism

This is modification of Darwinism in light of cytogenetics. The main contributors of this theory are T.H. Huxley, S. Wright, Th. Dobjhansky, Karl Gegenbaur, Ernst Haeckel, Weismann and G.L. Stebbins. Darwin's theory was criticised for several defects. It did not explain the origin of variations and could not distinguish somatic (non-heritable) from germinal (heritable) variations. Darwin failed to explain why even harmful variations were selected or preserved leading ultimately to extinction of the species, *e.g.*, horns of Irish elks. The supporters of this theory believed that natural selection has accounted every thing that has undergone evolution.

Weismann and his supporters discarded Darwin's theory except its principal element of natural selection. This made clearcut distinction between somatic and germinal variations in their theory of germplasm but they could not appreciate the role of mutations in evolution. Darwin was of the opinion that the mutation resulted mainly by single force, *i.e., natural selection.* According to Neo-Darwinism adaptations result from multiple forces and the natural selection is one of them. The theory also holds that the characters are not inherited as such but there are *character determinant* which control the development of the particular characters. The ultimate character results due to the interaction of the determinants as well as the interaction between organisms and the environment during development.

This theory gives better mechanism of evolution and states that

(*i*) the sources of variations are mutations and gene recombinations through sexual reproduction.

(*ii*) chromosomal organization and variation affect genetic linkage and produce variations in gene pool of cross fertilizing populations.

It is remarkable that variations only provide the *raw material* and not the direction of evolution which is the function of natural selection. The limits to which the natural selection can guide the population are set by reproductive isolation only. Neo-Darwinism has received great support from recent studies of population genetics.

De Vries Mutation Theory

Hugo De Vries (1884 – 1935), a Dutch botanist, propounded the mutation theory in 1901 to explain the origin of new species in the living organisms (Fig. 27.13). According to this theory, new species arise not by gradual accumulation of small variations in the organisms over a long stretch of time but by appearance of some permanent sudden and unpredictable changes (mutations). Fully established new species may continue to live side by side with the parental forms. In the course of his studies on the evening primrose (*Oenothera lamarckiana*) De Vries marked the sudden appearance of new characters. Those newly developed characters were inheritable. He called such sudden hereditary changes as **mutations** and believed that some of his mutants produced in a single step were actually new species. Mutations are some time called discontinuous variations because they are not connected by a series of intermediate forms. In this way, Darwin's theory differs basically from the mutation theory of evolution because the theory of natural selection emphasized the importance of continuous variation in forming an unbroken gradually merging series.

Fig. 27.13 Hugo De Vries.

Oenothera lamarckiana has 14 chromosomes but De Vries found that some of the offsprings had 15, 16, 20, 22, 24, 27, 28, 29 and 30 chromosomes. The new individuals differed in flower size, shape, arrangement of buds and size of seeds, etc. To those new abnormal types, De Vries called **elementary species**. He believed that those differences were due to mutations and they furnished sufficient materials for evolution.

De Vries concluded his theory in the following ways:

(*i*) New elementary species appear suddenly and attain full constancy at once.

(*ii*) The same new species are produced in large number of individuals. This increases the chance of selection by nature.

(*iii*) Mutability is something fundamentally different from fluctuating variability.

(*iv*) Mutations take place in nearly all directions and may involve any character.

Evolution is based upon the genetic variations and changes in gene frequency. The ultimate source of all genetic variations is mutation whether produced due to change in the molecular organisation of genes or in the linear sequence of genes on the chromosomes or in the number of chromosomes. Recombination is another source of genetic variation. Some interesting facts highlighting the role of mutation in evolution are outlined below:

1. Mutations may be spontaneous or induced. A mutation arising in a natural population is spontaneous. Haemophilia in man, white eyes and vestigeal wings in *Drosophila*, resistance of bacteria to penicillin and resistance of flies to DDT etc. are spontaneous mutations.
2. Mutations may be induced by X-rays, Gamma-rays, UV-rays and chemicals like mustard gas, formaline and colchicine.
3. Mutations arising spontaneously are not directed by environment. The environmental influence can affect the mutation rate but cannot cause a particular mutation to take place. The environment plays important role in the selection of mutants.
4. Mutations are relatively, though not always, persistant. A mutant gene may later mutate again into a new allele or it may change back to original condition.
5. Generally, mutations are deleterious and rarely they are advantageous but only useful mutations are selected. The organisms with useful mutations are better adapted species in course of evolution.
6. Beneficial mutations spread through population rapidly. If they are dominant, they are selected at once resulting in rapid evolutionary change. Recessive mutant genes spread very slowly.
7. The chromosomal mutations are important in genetics of the individuals. Such mutations result in an increase in the amount of genetic variabilities which are important from evolution point of view. Addition of new genes by insertion into the individual chromosome may sometimes be important in promoting evolutionary change.

Criticisms

(*i*) The mutations observed by De Vries in *Oenothera* were actually due to chromosomal rearrangement and segregation, rather than the gene mutations. What De Vries called as new species were actually hybrids of the old species.

(*ii*) The mutation theory does not give any proper direction to the evolution unless natural selection operates on them.

(*iii*) It is well known that most of mutations are not useful but usually lethal and, therefore, they would eliminate the species rather than propagate it.

(*iv*) If mutations are only accidents then mutant animals may not be necessarily adapted to their environments.

(*v*) Genetic studies have shown that small variations produced mutations and gene recombinations provide the *raw material* on which natural selection operates. The natural selection acts to short out individuals with favourable variations from population for survival in the struggle for existence and only the *favoured individuals* of the population lead to the origin of new species in course of evolution.

Modern Synthetic Theory of Evolution

The various views on evolution have now been merged into one broad coherent theory, **the modern synthetic theory**.

The modern theory of evolution is a synthesis of Mendel's and Darwin's observations. In this theory, Mendelian population is described as *object of evolution*. The heritable variations which appear among the individuals of such a population either by sexual recombinations or by mutations are described as the *raw materials* of evolutionary process, and natural selections acting on the heritable variations of population have been described as *mechanism of evolution*. The principles involved in the evolution of new species according to modern theory, may be grouped under two main heads:

1. Isolation.
2. Hybridization.

1. Isolation. In order to bring new species into existence, population should necessarily be segregated into smaller units. This segregation of population into small units is referred to as isolation. Isolation may be brought about by geographical barriers, such as, high mountains or large bodies of waters over which plants cannot pass (*geographical isolation*) or by factors restricting or eliminating interbreeding or natural population or subsequent reproduction of the hybrids between them (*reproductive isolation*). By isolation, plants reach in different types of environments. The isolated species develop in divergent lines and become quite distinct. Mutations also arise in the isolated species. Such changes result new races and eventually new species in the population: Evolution is essentially a change in the genetic compositions of populations. To be ranked as separate species two populations must differ in many genes. It is now obvious that if one population has to become different from another, then the two must be distantly separated over a longer period of time so as to prevent unrestricted interbreeding.

2. Hybridization. Natural selection in the modern view of evolution is a creative force. The species hybridization is believed to be important in the production of new forms of plants and animals, although the role which it plays is generally believed to be less important than isolation. Hybridization is possible between two different sexually reproducing species. Sexual reproduction involves not only the fusion of two kinds of gametes but also the entire mechanism of independent assortment of homologous chromosomes and segregation and recombination of genes which comprise the essentials of Mendelian laws of heredity in its modern meaning. The combination of genes from different individual increases the number of variations upon which the natural selection operates. Sexual reproduction does not induce struggle for existence or survival of the fittest in the population. Organisms usually struggle rather very rarely. Indeed animals prefer to avoid struggle and similarly the plants have never been seen engaged in the struggle. Moreover, natural selection does not eliminate the unfit individuals from population. The fit individuals may be the mightiests in the population, but they might happen to be sterile so the natural selection cannot operate upon them.

Evolution at Molecular Level

The recombination is of great importance in the evolution. Single mutational change may be lost or passed on without great impact on population but if its effect is modified and enhanced by recombination an unending contribution to variation is begun. Variation is the raw material for evolutionary change. Recombination is its principal source, says Savage, J.M., of the University of Southern California. Mutation alone has relatively little effect on variation without the pervasive impact of recombination.

Gene mutations, although often expressed in visible characteristics or physiological activity of the organism, are ultimately caused by rearrangement, addition or deletion of specific nitrogenous bases in DNA. Changes in the order of bases result in the modifications in the polypeptide chain produced under the control of gene and these modifications will result in production of modified enzyme or the protein or suppression of some enzyme feature. The modified enzyme may ultimately express itself by affecting a change in the physiology of development of the organism. Since a change in one of the bases may theoretically occur at any point along the length of one strand of DNA and incredible number of mutations are possible. Some mutations involve either addition or deletion of one or many bases along the DNA strand. The possibility for mutation along the extraordinarily long DNA molecules that form the axis of chromosome provides more than adequate basis for variation and, therefore, raw materials for future evolutionary changes.

QUESTIONS

1. List the different theories available to explain the organic evolution. Discuss in brief their merits and demerits.
2. What do you mean by organic evolution? Discuss the palaeontological, embryological and anatomical evidences which support the theory of organic evolution.
3. Give an outline of different postulates of Lamarckism and discuss their merits and demerits.
4. Discuss important postulates of Darwinism and demonstrate that some of these postulates still hold good.
5. How does mutation theory explain the organic evolution?
6. What modifications in the mutation theory were considered necessary in view of the present knowledge?
7. What is Neo-Darwinism?
8. Explain the evolution at molecular level.
9. What do you mean by the modern synthetic theory of evolution? Discuss the salient features of synthetic theory of evolution.

28 CHAPTER BIOSTATISTICS

Biostatistics or Biometry

Statistics (German-**Statistik**, Latin-**Statista**) is a science which deals with the collection, compilation and systematic analysis of observed data for drawing valid conclusions and making numerical predictions for future policy planning. It is concerned with the systematic study of the relationship of numerical data. Statistics is also concerned with the study of random variables and it helps in reaching to a logical decision from the state of uncertainty.

The data regarding the qualitative and quantitative characters of the organisms are essential components of experimental biology. The statistical analysis of observed data and interpretation of results of analysis are much helpful in drawing some definite conclusions about biological problems. The branch of statistics which deals with the mathematical methods in analysis and interpretation of biological problems is called **Biostatistics**, **Biometry** or **Quantitative Biology**.

The science of biological statistics was begun by Sir Francis Galton in 1911. Biostatistics was in developing phase when Mendel published the details of his breeding experiments. The other pioneer worker in the field of biometry besides Sir Francis Galton (1822 – 1911), was L.A.J. Quetelet (1796 – 1874). The real advancement of biostatistics took place in the first half of the twentieth century. Karl Pearson (1857–1936) and his co-workers established a school of Biometry in the University of London. Tippet (1927), Professor Sir Ronald, A. Fisher (1947, 1954), Fisher and Yates (1938), S. Wright, J.B.S. Hadane, K. Mather have made commendable contributions to the field of biometry.

Biostatistics has wide application in various fields of biology and the statistical methodologies have now become essential part of researches in Genetics and plant breeding, ecology, agriculture, forestry, biogeography, taxonomy, medicine and other disciplines of life science. Therefore, the knowledge of biostatistics is necessary for the biologists, agricultural scientists and geneticists for the following reasons:

1. It helps in understanding the nature of variability,
2. It helps in deriving general laws from small samples, and
3. It helps in making predictions for future planning.

Biometric study involves two types of statistics; (*i*) **Sample statistics,** and (*ii*) **Test statistics**. **Sample statistics** explains the distribution of data which originate from samples. The data are collected, systematically analysed and interpreted for drawing reasonable and valid conclusions. Sampling statistics is utilized to calculate population parameters such as means, standard deviation etc. It also indicates whether the results obtained from the analysis or computation of data are statistically significant or insignificant.

Test statistics is used to test hypotheses about sample data. In statistical tests one is required to choose more data for analysis which are obtained by conducting experimental design. The various biometrical principles such as probability, binomial expansions, variance, various tests of significance as *t*-test, F-test, Z-test are considered for drawing general conclusions for population from small samples.

SAMPLING METHODS

The total number of individuals belonging to a particular species in a particular area from which all possible data pertaining to an experiment are drawn is called population. In other words, the population is the total set of actual or possible variables. A population may be finite or infinite and the variables continuous or discrete. In biological experiments it is difficult to study each and every individual of entire population. Therefore, it is advised to select a part of population with the objective of study. That is called **sample**. The selected sample must have relationshipwith population. Data originate from samples. The process of collecting samples is called **sampling**. The number of individuals considered in a sample is basically referred to as the ***sample size***.

The sampling is done with the following objectives:

(*i*) Research objectives,

(*ii*) Experimental designs, and

(*iii*) Practical considerations.

If the sampling process is correct, the conclusions drawn from the data will be truly representative. The accuracy of the result depends largely on the sampling process and the size of the samples in relation to population. It is, therefore, important that the sampling methods must be correct and the sample size must be large enough to give reliable data about the population under study.

Sampling Designs

The sampling process can be grouped under the following two categories:

1. Random sampling designs.
2. Non-random sampling designs.

1. Random Sampling Designs. In random sampling all the individuals of population have equal and independent chance for being included in the sample. Random samples are characterised by the ways in which they are selected. Random is not used in the sense of "haphazard" or "hit or miss". To ensure randomness in sample selection one adopts either the ***lottery system*** or consults the ***tables of random numbers***.

Lottery Method. In this method, all items of the population are numbered and named on separate paper slips of identical shapes and sizes. The slips are folded and rolled and then mixed up in a container or beaker. This is followed by blindfold selection of rolled paper slips in desired number. For example, suppose we want to take a sample of 10 plants out of a population of 100 plants, the procedure is to write the number of plants from one to hundred on separate slips of paper, fold and roll those 100 slips, mix them thoroughly in a beaker or container and then select any 10 slips at random.

Table of Random Numbers. Several standard tables of random numbers are available which have been tested extensively for randomness. Some of the standard Tables are mentioned here:

(*i*) Tippett's Table (1927),

(*ii*) Fisher and Yates Table (1938), and

(*iii*) Rao, Mitra and Matthai's Table (1966).

The advantages of random sampling are as under:

(*i*) It provides essentially unbiased estimates and has measurable precisions.

(*ii*) It does not depend upon the detail information.

(*iii*) With the help of probability sampling it is possible to evaluate the relative efficiency of various sampling designs. However, there are certain limitations to probability sampling. These are as follows:

(*a*) It requires high level of skill and experience for use.

(*b*) It is a time taking process.

(*c*) It is a costly sampling process.

2. **Non-random Sampling Process.** It is a process of sample collection without randomization. The sampling is made at regular intervals and the selection of samples is done on the basis of consequences or investigators' judgement. The most important difference between random and non-random sampling is that the pattern of selecting samples can be ascertained in random sampling while in case of non-random sampling, there is no way of knowing pattern of variability of the process. There are three types of non-random samplings:

   (*i*) Judgement Sampling,

   (*ii*) Convenience Sampling, and

   (*iii*) Quota Sampling.

(*i*) **Judgement Sampling.** In this type of non-random sampling, the choice of sample items depends on the discretion of investigator. For example, suppose for estimating grain yield if a sample of 10 plants is to be selected from a population of 100 plants, the investigator would select 10 plants which in his opinion are the representatives of population under study. This sampling technique is often used in solving business and economic problems. Though this method of sampling is easy and simple yet it has the following limitations:

(*a*) It is not a scientific method,

(*b*) It involves risk of wrong conclusions, and

(*c*) It cannot be recommended for general use because of possible subjectiveness.

(*ii*) **Convenience Sampling.** In this type of non-random sampling, a fraction of population under study is selected neither by probability nor by judgment but by convenience. Convenience sampling is done by interview. The result of convenience sampling can hardly be representative of population and hence they are generally biased and unsatisfactory.

(*iii*) **Quota Sampling.** This is a type of judgement sampling. In this technique, quota are set up according to specified characteristics such as so many in each age-group, so many in each of several income groups and so on. This sampling technique is used in public opinion studies. It provides satisfactory results if the interviewers are properly trained and they follow the instructions carefully.

Selection of Samples

Samples are of two types:

1. Qualitative samples.
2. Quantitative samples.

1. Qualitative Samples. These samples are selected on the basis of quality characters of individuals. The quality characters are considered when two samples or objects are to be compared, *e.g.*, colours of fruits, size of fruits (small or big), heights of plants (tall or dwarf).

2. Quantitative Samples. These samples represent the quantities and the data are expressed in numerical figures, such as the measured heights of plants, total number of fruits per plant, total number of seeds per plant, 100 or 1,000 seed weight, number of days for fruit maturity and so on. Numerical measurements provide definite information about a particular population.

Selection of Appropriate Sampling Method

Sampling method to be adopted in a particular experiment has important bearing in biometry and the size of sample is important matter to be considered while adopting a sampling technique. Some statisticians are of the opinion that the size of sample should be 5% of the population size while others hold the view that sample size should be at least 10% of the size of population. The following two points should be considered while deciding the appropriate size of the sample:

1. The size of sample should increase as the variation in the individual items increases.
2. The larger the size of sample, the greater will be the accuracy.

The sampling technique should have the following merits:

(*i*) It should be less time taking.
(*ii*) It should involve less cost.
(*iii*) It should provide more detailed information.

The sampling process has, no doubt, several advantages yet there are limitations with it which are mentioned below:

1. Sample survey must be done carefully otherwise the results obtained may not be accurate.
2. It generally requires the services of experts.

Sampling and Non-sampling Errors

If the inferences are drawn about the population on the basis of a few observations or if proper care is not taken in selecting samples, some errors in result arise and the results obtained may not be accurate. These errors are called sampling errors. The errors arising only due to ascertainment are called non-sampling errors. In other words, these errors arise due to factors other than the inductive process of making inferences about population from samples. Sampling errors are of following two types:

(*i*) Biased errors, and
(*ii*) Unbiased errors.

The biased errors arise from the bias committed during selection of samples or ignoring certain members of population during sampling process. The unbiased errors arise due to chance of differences between such members of population as are included in the sample and those not included in the sample. The reasons for biased errors may be:

(*i*) faulty selection process,
(*ii*) faulty method of analysis, and
(*iii*) defective method of data collection.

Non-sampling errors may occur at any stage of planning and arise due to the following reasons:

(*i*) Defective method of interview or data collection,
(*ii*) Errors due to non-response from the unit during interviews,
(*iii*) Defective tabulation of data,
(*iv*) Errors in data processing such as coding, bunching, verification, etc., and
(*v*) Errors committed during interpretation of data and presentation of results.

Sometimes, the non-sampling errors may be big and deserve proper attention of the investigator. Chances for sampling errors decrease with the increase in sample size whereas non-sampling errors tend to increase with the increasing sample size. The non-sampling errors should be controlled and reduced to a level at which they do not spoil the final results.

QUESTIONS

1. What is meant by biostatistics? Discuss the scope of biostatistics.
2. What are the objectives of sampling? Discuss the different types of sampling techniques.
3. Write brief notes on sampling and non-sampling errors.

29 CHAPTER

Data Presentation

Data collected for the purpose of statistical analysis are first compiled or systematically arranged. It is essential because haphazard raw data as such do not express any fact. There are several methods of presenting data in systematic manner, some of which are as follows:

1. Tabular form.
2. Graphic or diagrammatic form.
3. Text presentation.
4. Pictorial presentation for comparing large populations.
5. Cartographic presentation through maps etc.

Here only tabular and graphic methods of data presentation will be discussed.

Tabular Presentation of Data

In this method, the collected data are presented in Table form, *i.e.,* data are arranged in systematic form in rows or columns. The data table with its title should bear the details or important features. There are two basic types of tables:

(*i*) reference/general/source table.

(*ii*) summary/text/analytical table.

The following format shows functional parts of a statistical table (Table 29.1).

Table 29.1. Format for Data Table: Title Complete and Compact

S. No.	*Column head*	*Column head*	*Column head*	*Column head*	*Column head*	*Column head*
1	2	3	4	5	6	7
						Body field

Source note: (*Source from where data are obtained*).

Footnote, if any: (*Statement clarifying some specific entries*).

Guidelines for Preparing Data Tables

For proper presentation of data in table form data should be arranged in systematic manner in alphabetical, geographical order or in order of magnitude. Some of the guidelines for good presentation are outlined below:

1. If comparisons are to be made, data should be recorded in adjacent columns or rows.
2. The size and shape of Table must be based on preliminary layouts so that data can be arranged properly.
3. The Table number and title of the table should be at the top and preferably at the centre.
4. A zero quantity or non-avoidable data should be shown by letter n.a. (not available) or by a dash (–).
5. Use of abbreviation or expressions etc. should be avoided in Tables and headings.
6. Ditto mark (") should not be used.

Graphic Presentation of Data

The presentation of data in graphic form is complementary to the tabular presentation. A graph or chart is a visual form of presenting data. The graphic presentation of data may be in the following forms:

(*i*) Line graph or curve form,

(*ii*) Geometric forms, such as, bars, rectangles, circles (Pie diagrams), Squares and cubes, and

(*iii*) Pictographs (*i.e.,* charts consisting of pictures and cartographs (or maps) to indicate data geographically.

Of the above graphic forms of data presentation, the charts and graphs are very effective sources for drawing attention to the broad and salient features. The commonly used charts, and graphs for presenting and comparing data are described below:

(*i*) Line chart,

(*ii*) Bar charts, Histograms and frequency polygones, and

(*iii*) Pie chart or circular charts.

(*i*) **Line Chart.** A line chart is presented through simple lines. It is commonly used when the distribution is not continuous or the sample points are distantly separated. In this the values of different classes are represented by equally spaced vertical or horizontal lines on a common scale. An example of line chart prepared from the following data table of cotton production in some States of India during the period 1992-93 is given in Fig. 29.2.

States	*Production in Lakh tonnes*
Andhra Pradesh	3.5
Karnataka	15.5
Maharashtra	17
Madhya Pradesh	6.9
Gujarat	21.0
Rajasthan	7.0

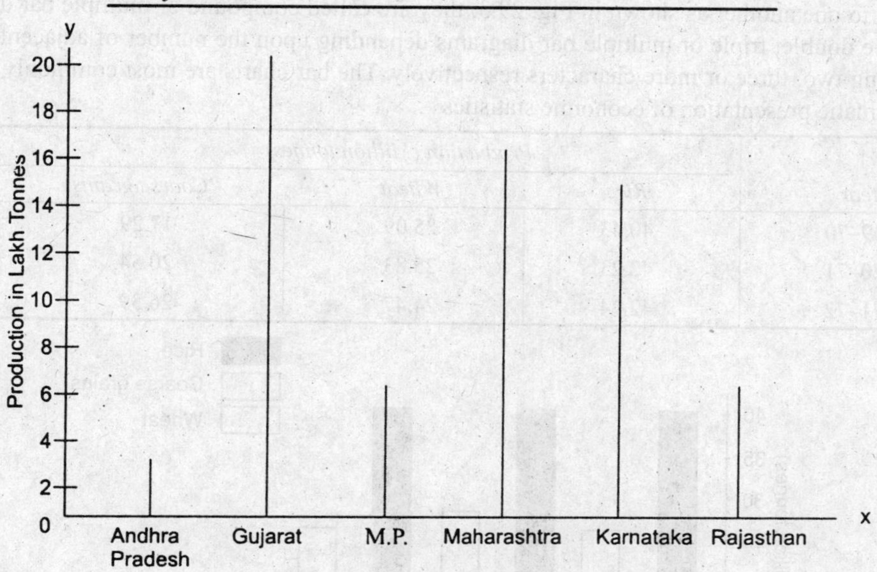

Fig. 29.1 Line Chart.

(*ii*) **Bar Chart and Histograms.** A bar chart is graph that consists of a number of rectangles of variable heights but of equal width.

These rectangles are called bars. In this class intervals are represented on horizontal axis and the frequency starting from zero are indicated on vertical axis on a common scale. Now the magnitudes of frequencies are marked above the respective class intervals and rectangles are made over them. An example of bar chart prepared from the following data table is given in Fig. 29.2.

Months	VO_2 consumed ml/kg/h
Jan- 98	77
Feb-98	84
Mar-98	94
Apr-98	95
May-98	110
Jun-98	115
Jul-98	105
Aug-98	100
Sep-98	100
Oct-98	88
Nov-98	84
Dec-98	74

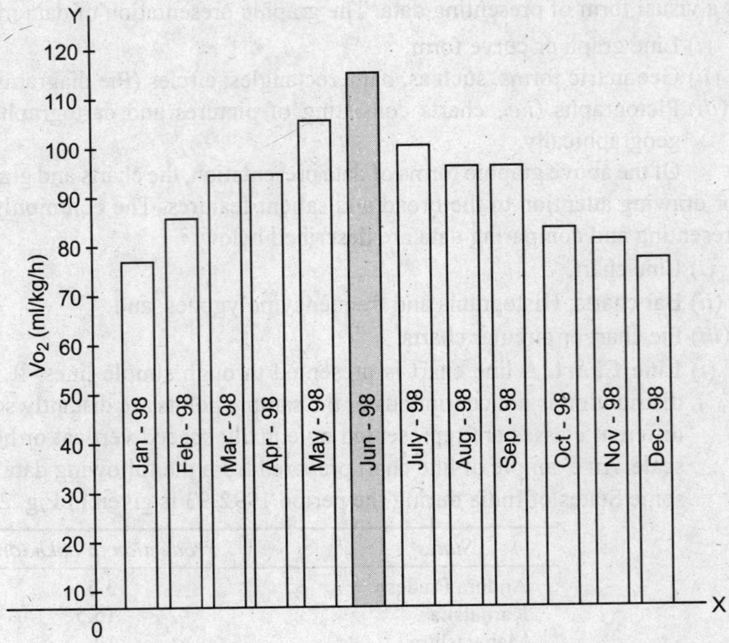

Fig. 29.2 Simple Bar Chart or Histogram.

When for the purpose of comparison, bar charts depicting two or more characters are made adjacent to one another as shown in Fig. 29.3, they are called compound or multiple bar diagrams. It may be double, triple or multiple bar diagrams depending upon the number of adjacent bars for comparing two, three or more characters respectively. The bar charts are most commonly used for diagrammatic presentation of economic statistics.

Year	*Production (Million tonnes)*		
	Rice	*Wheat*	*Coarse grains*
1969–70	40.43	25.09	17.29
1970–71	42.23	25.83	20.54
1971–72	42.74	24.47	26.39

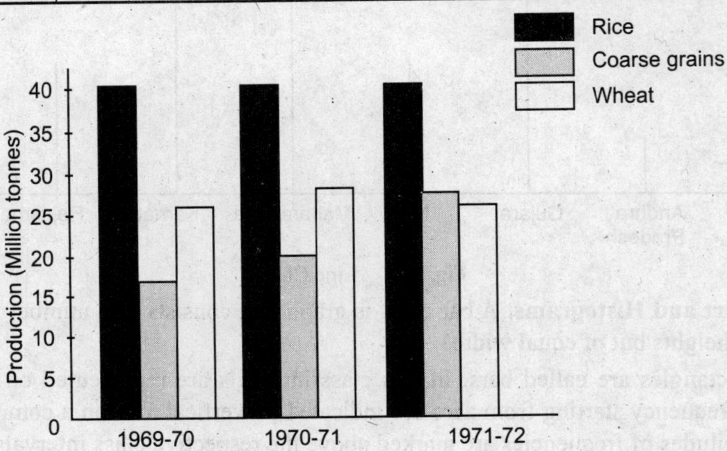

Fig. 29.3 Multiple Bar Chart.

Histograms

These are used for continuous variables. While presenting the data in the form of histogram, the true class intervals are represented on the horizontal axis which need not begin at zero. On the vertical axis the frequency per unit scale is marked. Now above each class interval its frequency is marked on vertical axis which begins at zero and then rectangles are drawn over the classes whose heights are proportional to the magnitudes of frequencies of various classes. The frequency per class interval provides a quick visual appreciation of the number in particular classes. Figure 29.4, illustrates the histogram prepared from the following data table:

Class interval e.g., seed yielding capacity	*Frequency e.g., No. of plants*
0 – 10	30
10 – 20	40
20 – 30	70
30 – 40	75
40 – 50	90
50 – 60	90
60 – 70	20
70 – 80	50
80 – 90	20
90 – 100	15

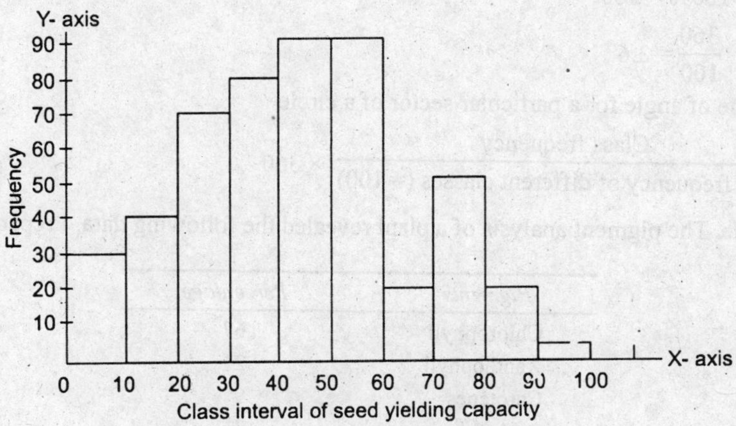

Fig. 29.4 Histogram showing seed yielding capacity and their frequency of plants.

Frequency Polygon. When the frequency mid points of rectangles are joined by straight lines a polygon is obtained. That is called **frequency polygon**. A frequency polygon can also be obtained by putting a dot at the same heights as in rectangles over the mid points of various class intervals and then the dots are joined by straight lines. Vertical lines can be made from horizontal axis to the dots above the mid points first and the last class intervals (Fig. 29.5). In this, there is no need of making rectangles over different class intervals.

Class intervals	0 – 10	10 – 20	20 – 30	30 – 40	40 – 50	50 – 60	60 – 70	70 – 80	80 – 90
Frequency	3	4	7	8	9	9	2	6	2

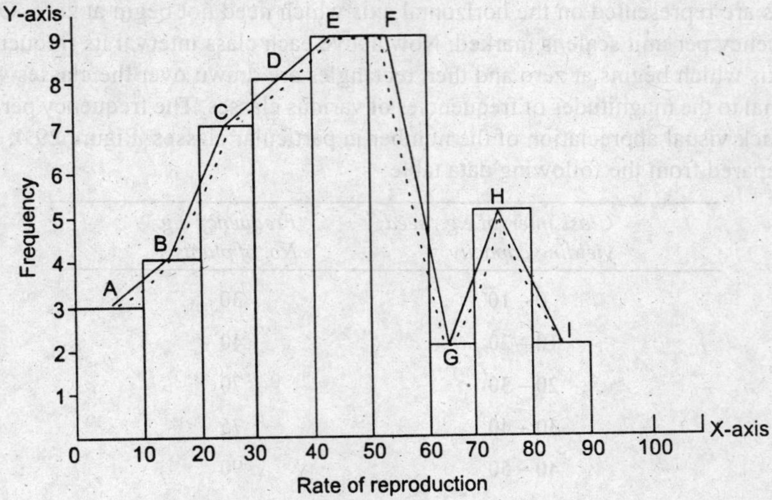

Fig. 29.5 Frequency polygon and frequency curve (Dotted line) showing rate of reproduction and their frequency.

(*iii*) **Pie Chart or Circular Chart.** When the discrete data of any particular character are given in percentages they can be diagrammatically presented in the form of a circle. A circle has 360° angle on its radius. So the area of circle to be covered by a particular data is calculated in the following way:

Circle = 100% = 360°

$$\text{So } 1\% = \frac{360}{100} = 3.6°$$

The value of angle for a particular sector of a circle

$$= \frac{\text{Class frequency}}{\text{Total frequency of different classes } (= 100)} \times 360$$

Example. The pigment analysis of a plant revealed the following data. Prepare a pie chart for the pigments.

Pigments	*Percentage*
Chlorophyll	67
Xanthophyll	21
Carotenes	9
Anthocyanines	3

The above data can be presented in the form of pie chart as follows (Fig. 29.6).

Calculation of Angles:

Chlorophylls	67 × 3.6°	= 241.2°
Xanthophylls	21 × 3.6°	= 75.6°
Carotenes	9 × 3.6°	= 32.4°
Anthocyanines	3 × 3.6°	= 10.8°

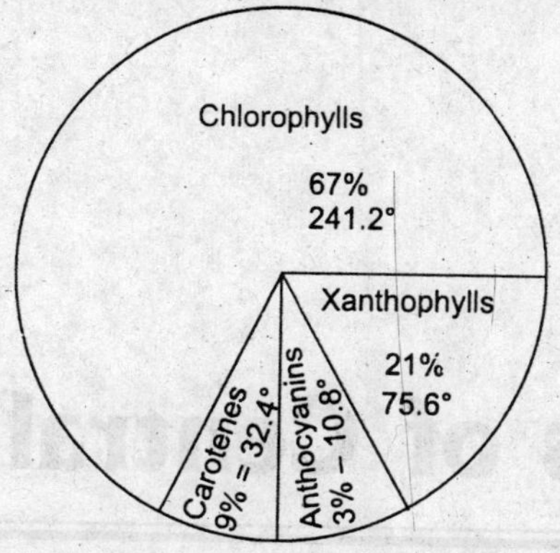

Fig. 29.6 Pie chart or circular chart.

QUESTIONS

1. Describe various methods for presenting statistical data.
2. Describe different types of graphical presentation of data.
3. How would you prepare a pie chart from the given percentages of different characters?
4. How would you prepare frequency polygones from the given data?

30 CHAPTER

Measures of Central Tendency

It is commonly seen that a variety of characters such as plant height, days to flower, grain yields of a crop in different plots, eye colour of insects are not identical and they show some degree of variation. The need for statistical analyses arises from these variations. The characters which show variation of some sort are called **variates** or **random variables**. The variables are of two types:

1. Continuous variables.
2. Discrete or Discontinuous variables.

1. **Continuous variables.** The characters which can theoretically take any value in a given range, finite or infinite. The grain yield, 1,000 grain weight, plant height, etc., are continuous variables. The yield of a rice variety, for example, may vary from 106 kg to 108 kg in different plots. This means that the per plot yield for that variety may be 106 kg or 106.5 kg or 106.8 kg or 107.3 kg, 107.7 kg or 108.0 kg and so on. So the seed yield is a continuous variable.
2. **Discrete or Discontinuous variables.** These variables do not show values on continuous scales, for example, the number of insects with particular eye colour, the number of diseased or healthy plants in a plot and so on. The scale of measurement is in discrete unit, often whole number, *e.g.,* the number of insects with red eye in a population can never be suggested to be 12.5, it would be either 12 or 13, the whole number. The number of individuals representing any variable or class is referred to as **frequency**.

Determination

The records of observations are called *determinations*. There are two types of determinations for different variables; (*i*) Measurements, and (*ii*) counts. The continuous variables are measured, *e.g.,* yield of milk is measured in terms of weight or volume and grain yield per plant is measured in terms of weight. The discontinuous or discrete variables are determined by counting, as for example, the number of seeds per ear is determined by counting the number of seeds in invididual ears, the egg production of an insect is determined by counting the number of eggs. The countings yield integers (whole numbers) and discrete variables cannot take up fractional values between the integers.

For developing a general theory, the variables are denoted by letter symbols, say X denotes a variable to be studied. X at one time may be used to denote grain yield per plot and on other occasion X may denote the number of diseased wheat plants per plot. Further, X_1, X_2, X_3, X_4 may represent the grain yields of the first, second, third and fourth plot respectively and in general, let X_i (read as x subi) denote the grain yield of *ith* plot where *i*th may stand for any number say 1, 2, 3, ..., 20.

A sign of operation used continuously in biometrics is the sign of summation which is denoted by Σ, a Greek letter sigma. Thus the sum of $X_1 + X_2 + X_3 + X_4 + X_5$ is written as $\sum_{1}^{5} X_i$ which is read as the sum of all the X_s from 1 to 5 or summation X_i, i running from 1 to 5. The first value is written below Σ sign and the last above Σ. The lower and upper values are called the limits of summation.

The numerical data or quantitative values of characters or variables centre around a particular value. This tendency of numerical data of any character is referred to as **central tendency**. Thus, the central tendency refers to some central value of data. Generally, this value is located somewhere in the middle of a series of quantitative values and represents the important characteristic of frequency distribution. This is why this is called *statistical mean* or the *representative value* or *measure of central tendency* of numerical data.

The measures of central tendency are of the following two types:

(*i*) Mathematical measures or Arithmetic mean.

(*ii*) Positional measures including,

(*a*) Median, and

(*b*) Mode.

Arithmetic Mean

The mean is a measurement of unit most frequently used to describe a frequency distribution of same type. The arithmetic mean or average is calculated by dividing the sum (total) of all the individual values of data series by the total number of items.

$$\text{Arithmetic mean or } \overline{X} = \frac{X_1 + X_2 + X_3 + X_4 + X_5 \ldots\ldots\ldots + X_n}{n}$$

$$\text{or } \overline{X} = \frac{\Sigma_X}{n}$$

where Σ_X is sum of all *Xs* from 1 to *n* and *n* = number of items.

Example. The heights of 10 plants are found to be 15, 21, 17, 19, 21, 16, 14, 17, 15 and 13 cm. Calculate the arithmetic mean of the plant height.

Solution:

$$\text{Arithmetic mean } \overline{X} = \frac{\Sigma_X}{n} \text{ i.e., } \frac{\text{Sum of the values of all items}}{\text{Number of items}}$$

$$\Sigma_X = X_1 + X_2 + X_3 + X_4 + X_5 + X_6 + X_7 + X_8 + X_9 + X_{10}$$

$$= 15 + 21 + 17 + 19 + 21 + 16 + 14 + 17 + 15 + 13 = 168 \text{ cm}$$

$$n = 10$$

$$\overline{X} = \frac{168}{10} = 16.8.$$

So the arithmetic mean of the plant height is 16.8 cm or the average plant height is 16.8 cm.

In case of a large number of observations, *i.e.*, more than 10, the arithmetic mean can be calculated by preparing a frequency distribution table. Frequency is the number of individuals representing a class.

I. *When the data are discrete and unclassified but tabulated, the arithmetic mean can be calculated in the following way:*

Item	X_1	X_2	X_3	X_4	---------	Xn
Frequency	f_1	f_2	f_3	f_4	---------	fn

$$\text{Arithmetic mean } (\overline{x}) = \frac{f_1x_1 + f_2x_2 + f_3x_3 + f_4x_4 + \ldots\ldots\ldots f_nx_n}{f_n}$$

$$\text{or } (\overline{x}) = \frac{\Sigma fx}{\Sigma f} = \frac{\Sigma fx}{n}$$

Example. The seed yield (in gms) per plant for 30 maize plants was found to be 110, 115, 112, 108, 110, 120, 115, 112, 108, 115, 121, 115, 110, 112, 108, 121, 110, 110, 115, 110, 115, 112, 110, 115, 112, 112, 115, 110, 112 and 110. Calculate the average yield per plant from the given data.

Solution: From the given data it is clear that there are some figures which appear more than once, for example seed yield 110 g/plant has been recorded in nine plants. So the frequency of plants with 110 g. seed yield in this case is 9. The frequencies of other values can also be calculated in the same way and the data can be tabulated as follows:

Class *Seed yield per plant* ×	*Frequency (f)* *No. of plants representing*	$f \times X$
108	3	324
110	9	990
112	7	784
115	8	920
120	1	120
121	2	242
Total	30	$\Sigma fx = 3380$

$$\text{Arithmetic meam } (\bar{x}) = \frac{\Sigma fx}{n} = \frac{3380}{30} = 112.66 \text{gm/plant.}$$

So the average yield per plant is 112.66 gms.

II. *When the data of continuous series are classified*, then only the frequency of individuals falling in the particular class appears. In such cases for calculating the arithmetic mean, the mid value of each class is taken into consideration and the subsequent steps of computation are like those of discontinuous series.

Example 1: Calculate the arithmetic mean from the following data:

Class	10 – 20	20 – 30	30 – 40	40 – 50	50 – 60
Frequency	6	10	12	18	4

Solution:

Class	*Mid value of class* (×)	*Frequency (f)*	*fx*
10 – 20	15	6	90
20 – 30	25	10	250
30 – 40	35	12	420
40 – 50	45	18	810
50 – 60	55	4	220
Total		$\Sigma f = 50$	$\Sigma f.x = 1790$

$$\text{Arithmetic mean } (\bar{x}) = \frac{\Sigma fx}{\Sigma f} = \frac{1790}{50} = 35.8.$$

Example 2: Calculate the arithmetic mean of flowers per plant from the following data:

Class (flowers/Plant)	0 – 10	10 – 20	20 – 30	30 – 40	40 – 50
Frequency (No. of plants)	3	10	15	12	8

Solution:

Class	*Mid value of class* (×)	*Frequency (f)*	*fx*
0 – 10	5	3	15
10 – 20	15	10	150
20 – 30	25	15	375
30 – 40	35	12	420
40 – 50	45	8	360
Total		$\Sigma f = 48$	$\Sigma fx = 1320$

$$\text{Arithmetic mean } (\bar{x}) = \frac{\Sigma fx}{\Sigma f} = \frac{1320}{48}$$

$$= 27.5$$

So the average number of flowers per plant is 27.5 or 28.

Short Cut Method of Determining Arithmetic Mean

Arithmetic mean can be calculated by short cut method in the following way:

(*i*) The value of any item in a data series is taken as assumed mean (*A*). That value may or may not be necessarily present in the series. Generally the item whose frequency is highest is taken as assumed mean.

(*ii*) From value of each item of the series is now subtracted the value of assumed mean, which gives deviation ($dx = X - A$). The deviation values are recorded along with their positive or negative sign.

(*iii*) The deviation value of each item is multiplied with its respective frequency (*i.e.*, *dxf*) and the sum of all the products ($\Sigma\ dxf$) is determined.

(*iv*) The sum of all the products ($\Sigma\ dxf$) is divided by the number of items and the value so obtained is added to assumed mean which gives arithmetic mean.

$$\text{Arithmetic mean } (\bar{x}) = \text{A} + \frac{\Sigma fdx}{\Sigma f}$$

$$= \text{A} + \frac{\Sigma fdx}{n}$$

In case the data are classified, the mid-value of each class is considered to calculate the deviation. But if the data are not classified and non-tabulated, the following formula is used to calculate the arithmetic mean $\bar{x} = \text{A} + \dfrac{\Sigma dx}{n}$.

Example. The heights of tomato plants are presented in the following table. Calculate the average plant height by short cut method.

Class Plant height (cm)	25	35	45	55	65	75	85	95
Frequency (No. of plants)	3	4	7	20	35	9	2	1

Solution: Suppose assumed plant height mean (*A*) = 65.

Height class (x)	*Frequency [No. of plants (f)]*	*Deviation* $dx = X - A$	*fdx*	
25	3	25 – 65 = – 40	– 120	
35	4	35 – 65 = – 30	– 120	
45	7	45 – 65 = 20	– 140	= – 580
55	20	55 – 65 = – 10	– 200	
65	35	65 – 65 = 0	00	
75	9	75 – 65 = + 10	+ 90	
85	2	85 – 65 = + 20	+ 40	= + 160
95	1	95 – 65 = + 30	+ 30	
	$\Sigma f = 81$		$\Sigma fdx = - 420$	

$$\text{Arithmetic mean } (\bar{x}) = A + \frac{\Sigma fdx}{\Sigma f}$$

$$= 65 + \frac{-420}{81}$$

$$= 65 + (-5.18) = 65 - 5.18$$

$$= 59.82 \text{ cm.}$$

So the average plant height of tomato is 59.82 cm.

Example. Calculate the arithmetic mean of plant height from the following data:

Plant height class (in cm)	20 – 25	25 – 30	30 – 35	35 – 40	40 – 45	45 – 50	50 – 55
Frequency (No. of plants)	5	10	20	40	30	20	5

Solution:

Class (Plant height)	*Frequency [No. of plants (f)]*	*Mid-value of class (x)*	*Deviation dx = X – A*	*fdx*	
20 – 25	5	22.5	22.5 – 37.5 = – 15	– 75	
25 – 30	10	27.5	27.5 – 37.5 = – 10	– 100	= – 275
30 – 35	20	32.5	32.5 – 37.5 = – 5	– 100	
35 – 40	40	37.5 = A	37.5 – 37.5 = – 00	= 00	
40 – 45	30	42.5	42.5 – 37.5 = + 5	= + 150	
45 – 50	20	47.5	47.5 – 37.5 = + 10	= + 200	= + 425
50 – 55	5	52.5	52.5 – 37.5 = + 15	= + 75	
	$\Sigma f = 130$			$\Sigma fdx = + 150$	

$$\text{Arithmetic mean } (\bar{x}) = A + \frac{\Sigma fdx}{\Sigma f} = 37.5 + \frac{150}{130}$$

$$= 37.50 + 1.15 = 38.65.$$

So the arithmetic mean of plant height is 38.65 cm.

Merits of Arithmetic Mean

1. It is a mathematical mean.
2. It is easy to calculate.
3. It is always definite.
4. It establishes a simple relation among the values of data series.

Demerits of Arithmetic Mean

1. It cannot be determined by simple visual observation of data.
2. It cannot be presented by chart or graph.
3. It is not suitable for qualitative studies.
4. Sometimes it does not show representation of series.

Median

When the values of all items of a series are arranged in increasing (ascending) or decreasing (descending) order it is usually called an array and the middle item of an array is called **median**. The median divides the series into two groups; one group in which the values of items are less than the middle value and the other group in which the values of the items are greater than the middle item. Median is denoted by **Me** or **Mdn**.

The methods of calculating the median are comparatively simple. The value of median is not affected by change in extreme values. If the number of data in a series is odd, the median is the middle value. But if the number of data in a series is even, the median is the average of the two middle values.

Methods of Determining Median

1. *For unclassified and untabulated data:* In order to calculate the median, the data are first arranged in increasing or decreasing order and then the following formula is used:

(*i*) **For odd number of data** in the series

$$\text{Me} = \frac{n+1}{2}, \text{ where } n = \text{number of items or data.}$$

Example. The heights (in cm) for 9 plants are given below. Find out the Median Height — 67, 65, 70, 68, 62, 63, 64, 63, 66.

Solution: The height measurements can be arranged in ascending order as follows:

62, 63, 64, 65, 66, 67, 68, 69, 70.

The number of data in the above array $n = 9$.

$$\text{Median} = \frac{n+1}{2} = \frac{9+1}{2} = 5\,(\textit{i.e.}, \text{ value of 5th data in the array})$$
$$= 66$$

(*ii*) **For even number of data in the series.**

The median is calculated as follows:

$$\text{Me} = \frac{\frac{n}{2}\text{th item} + \frac{n}{2}+1}{2}$$

Suppose, for example, there are 10 data in a series.

$$\text{So Me} = \frac{\left(\frac{10}{2}\right)+\left(\frac{10}{2}+1\right)}{2} = \frac{5+(5+1)}{2} = \frac{5+6}{2}$$

$$= \frac{\text{Value of fifth data + Value of sixth data of the array}}{2}.$$

Example. The number of flowers recorded on 10 plants are: 15, 10, 8, 12, 13, 7, 11, 14, 9, 16. Find out the median value of flowers per plant.

Solution. The given numbers of flowers on 10 plants can be arranged in ascending order as under:

7, 8, 9, 10, 11, 12, 13, 14, 15, 16.

Total number of data in the above array $n = 10$ (even number).

$$\text{Me} = \frac{\frac{n}{2} + \frac{n}{2} + 1}{2}$$

$$= \frac{\frac{10}{2} + \frac{10}{2} + 1}{2} = \frac{\text{5th value + 5 + 1 value}}{2}$$

$$= \frac{11+12}{2} = \frac{23}{2} = 11.5 \text{ flowers.}$$

Example. Calculate the median of the following series of data obtained by measuring the heights of 16 plants: 9, 10, 10, 8, 9, 7, 8, 11, 7, 12, 14, 12, 11, 14, 15, 13.

Solution: The given data of plant heights are arranged in ascending order as follows:

7, 7, 8, 8, 9, 9, 10, 10, 11, 11, 12, 12, 13, 13, 14, 14, 15

$n = 16$ (even number)

$$\text{Me} = \frac{\frac{n}{2} + \frac{n}{2} + 1}{2} = \frac{\frac{16}{2} + \frac{16}{2} + 1}{2} = \frac{\text{Value of 8th item} + \text{Value of 9th item}}{2}$$

$$= \frac{10 + 11}{2} = 10.5$$

So the median height = 10.5 cm

2. For grouped data.

(*i*) *Discontinuous or Discrete series of data.* To calculate the median for discrete grouped data, first of all the cumulative frequency of whole series is obtained. The value of data against $\frac{n+1}{2}th$ cumulative frequency will be the median **for odd number** of data and the mean of values against $\frac{n}{2} + \frac{n}{2} + 1th$ cumulative frequencies will be the median for series containing **even number of data.**

Example. Calculate the median of the following data obtained by counting the number of flowers on 19 plants.

Class (No. of flowers/Plant)	1	2	3	4	5
Frequency (No. of plants)	3	4	6	3	3

Solution:

Class (No. of flowers/Plant)	*Frequency*	*Cumulative frequency*
1	3	3
2	4	7
3	6	13
4	3	16
5	3	19

$n = 19$ (odd number)

$$\text{Median (Me)} = \text{Value of data against } \frac{n+1}{2} \text{ cumulative frequency}$$

$$= \text{Value against } \frac{19+1}{2} \text{ cumulative frequency}$$

$$= \frac{19+1}{2} = \frac{20}{20} = 10.$$

Since cumulative frequency 10 is included in 13 which represents the class value 3, therefore, for cumulative frequency 10, the class value will be 3 which is the median.

Example. Calculate the median for the following data recorded for heights (in cm) of 80 plants.

Class (Plant height)	119	120	121	122	123	124	125
No. of plants	4	9	14	18	15	13	7

Solution:

Class [Plant height (in cm)]	*Frequency (No. of plants)*	*Cumulative frequency (Cf)*
119	4	4
120	9	13
121	14	27
122	18	45
123	15	60
124	13	73
125	7	80

$n = 80$ (even number)

$$\text{Median (Me)} = \frac{\text{Class values against } \frac{n}{2}\text{th} + \frac{n}{2}+1 \text{ th cumulative frequency}}{2}$$

$$= \frac{\frac{80}{2}\text{th} + \frac{80}{2} + 1 \text{ cumulative frequencies}}{2}$$

$$= \frac{\text{class values against } c.f.\ 40 \text{ and } 41}{2}$$

The class values for cumulative frequencies 40 and 41 are included in the class value of cumulative frequency 45 which is 122. Therefore, Median (Me) = $\frac{122 + 122}{2} = 122$.

(*ii*) **For classified grouped data.** The median is determined in the following way:

(*a*) first, the cumulative frequency of all the classes are obtained from the given frequencies.

(*b*) median class value is determined which is $\frac{N}{2}$ *th* class.

(*c*) then that class is ascertained whose cumulative frequency preceeds that of median class (c.f.).

(*d*) the median is calculated by the following formula.

$$\text{Median (Me)} = L + \frac{\frac{N+1}{2} - F}{f} \times i$$

$$= \text{L} + \frac{\frac{N}{2} - CF}{f} \times i$$

where L = lower limit of median class

N = sum of all the frequencies ($N = \Sigma f$)

cf = cumulative frequency of class preceding that of median class

f = frequency of median class

i = class interval of the median class (*i.e.*, upper limit – lower limit or $L_2 - L_1$).

Example. The number of seeds produced by 55 plants of a plot are given in the following table. Calculate the median seed number of a plant.

Class interval	40 – 50	50 – 60	60 – 70	70 – 80	80 – 90	90 – 100	100 – 110	110–120
No. of plants	5	9	9	15	8	4	3	2

Solution:

Class	*Frequency (f)*	*Cumulative frequency (c.f.)*
40 – 50	5	5
50 – 60	9	14
60 – 70	9	23 = *c.f. of the class preceding that of median class*
70 – 80	15 (*f*)	38
80 – 90	8	46
90 – 100	4	50
100 – 110	3	53
110 – 120	2	55
	$\Sigma f = 55$	

To find out median class we use formula $\frac{n}{2}$.

So, Median class $= \frac{55}{2} = 27.5$

Now, the median class value falls within the cumulative frequency of class interval 70 – 80, therefore, the median class is 70 – 80.

The width or class interval $(i) = L_2 - L_1$

$= 80 - 70 = 10$

Now the median is calculated by the following formula

$$\text{Me} = \text{L} + \frac{\frac{n}{2} - cf}{f} \times i$$

$$= 70 + \frac{27.5 - 23}{15} \times 10$$

$$= 70 + \frac{4.5}{15} \times 10 = 70 + \frac{45}{15}$$

$$= 70 + 3 = 73.$$

While calculating the median for classified grouped data the following facts must be kept in mind:

(*i*) Class intervals must be equal for all classes. If not equal, they should be rearranged allowing equal interval as shown below:

Class	10 – 15	15 – 17.5	17.5 – 20	20 – 30	30 – 35	35 – 40	40 – 50
Frequency	10	17	19	27	28	30	40

Because the class intervals are not equal in different classes, the above data are rearranged as follows:

Class	10 – 20	20 – 30	30 – 40	40 – 50
Frequency	46	27	58	40

(*ii*) The classes should be presented by exclusive method (for example 10 – 20, 20 – 30, 30 – 40 ----and so on). If the classes are presented in inclusive manner then they should be changed to exclusive one by subtracting 0.5 from the lower limit and adding 0.5 to the upper limit as examplified below:

Inclusive presentation of classes

50 – 54, 55 – 59, 60 – 64, 65 – 69, 70 – 74

↓ ↓ ↓ ↓ ↓

49.5 – 54.5, 54.5 – 59.5, 59.5 – 64.5, 64.5 – 69.5, 69.5 – 74.5 exclusive presentation.

Similarly, if the classes are 0 – 9, 10 – 19, then they will be changed to 0 – 9.5, 9.5 – 19.5.

Merits of Median

1. It is calculated easily and located exactly.
2. It is not affected by abnormally large or small values.
3. Its size cannot be changed much by adding a few more items.
4. Median can be used in quantitative measurements.

Demerits of Median

1. The median of two or more series cannot be calculated by using the median of the component series.
2. It may not be represented in central data.
3. It cannot be used where weightage is given to some items.

Mode

Mode is another unit of measurement of central tendency which is used to obtain a quick estimate of central tendency. **Mode** is defined as the particular value which appears again and again for maximum number of times in a set of observations. In other words, mode is that variate or item in the series which shows highest frequency. It is denoted by Mo or Z. Sometimes, in a series of data all the variates have equal frequency. In such a series there is no mode. In some series more than one data have equal frequency. In such series there may be more than one mode. Such a series is said to be multimodal series. That series is analysed properly to get the genuine mode.

Method of Determining Mode

For different types of data, the methods of computation of Mode are different.

1. **For non-grouped data.** If the values of all the items are given separately then the mode is the value of that item which shows largest frequency.

Example. Find out the mode size of shoe from the following numbers of shoes 2, 7, 4, 1, 7, 7, 6, 7, 7, 6, 5, 3, 2, 2, 7, 7, 5 Mode = 7.

2. **For discrete series of data.** Mode of discrete series of data is determined easily because the figure with highest frequency is mode of the series.

Example. Determine the mode from the following data regarding the number of flowers per plant.

Number of flowers/plant	1	2	3	4	5	6	7	8	9	10
No. of plants	4	7	10	19	35	25	20	5	3	1

Solution: The above Table of data distribution indicates that 5 flowers per plant is represented 35 times out of 129 plants which is the largest frequency. Therefore, the value of Mode is 5 flowers per plant.

For grouped data, the mode can be calculated by identifying the class with highest frequency by grouping method which is detailed as under:

1. Grouping Method. When in a given series two or more classes show same frequency, mode can not be determined by inspection method. Sometimes the frequencies of neighbouring classes affect the mode determination in such a way that a class other than the true mode class with highest frequency is treated as mode. In order to avoid such mistakes, grouping method is used to determine mode which is the best method. Grouping method is outlined below:

Grouping Table. For this, table is made in which first column is meant for observed items or classes and the next 6 columns are meant for frequency. In the first frequency column, the frequency of each item or class is mentioned. In the second frequency column, the given frequencies are grouped in pairs starting from first and the combined values of paired frequencies are mentioned.

In the third frequency column, the groups of two frequencies each are made starting from the frequency of the second item or class and the combined values are noted against different groups.

In the fourth frequency column, groups of three frequencies each are made starting from the first frequency and the combined values of three frequencies in different groups are recorded.

In the fifth frequency column, the frequencies of first column are made into groups of three each starting from the second frequency and combined values of 3 frequencies of different groups are recorded.

In the sixth frequency column, the frequencies of first column are bracketed into groups of three each starting from the third frequency and the combined value of three frequencies is noted against each group.

2. Analysis Table. This is made to record the items or classes represented by highest frequency of different columns.

3. Determination of Mode. The value of item or class which is represented by highest in the analysis table is mode.

The grouping method for determining mode is illustrated by taking the example of the following data:

Example. Find out the mode of the following series.

No. of fruits per plant	30	31	32	33	34	35	36	37
No. of plants	3	9	5	4	8	7	6	2

(*i*) *Grouping Table*

Number of fruits per plant	*Frequency Column*					
	1	*2(2-2)*	*3(2-2)*	*4(3-3)*	*5(3-3)*	*6(3-3)*
30	3	12		17		
31	9		14		18	
32	5	9				17
33	4		12	19		
34	8	15			21	
35	7		13			15
36	6	8				
37	2					

(*ii*) *Analysis Table.*

	Items or class with highest frequency							
Frequency column	*30*	*31*	*32*	*33*	*34*	*35*	*36*	*37*
1		31						
2					34	35		
3		31	32					
4				33	34	35		
5					34	35	36	
6			32	33	34			
Total		2	2	2	4	3	2	

From the above analysis table it is clear that 34 fruits/plant appeared four times which has the highest frequency. Therefore, the mode (Mo) = 34.

(*iii*) *For Classified Data.* First of all the class with highest frequency, *i.e.,* Mode class is identified either by inspection method or grouping method and then the mode is calculated with the help of the following formula:

$$\text{Mode (Mo)} = L_1 + \frac{fm - f_1}{2fm - f_1 - f_2} \times (L_2 - L_1)$$

Since $L_2 - L_1 = i$

$$\text{Mo} = L_1 + \frac{fm - f_1}{2fm - f_1 - f_2} \times i$$

where, L_1 = Lower limit of mode class

fm = Frequency of mode class

f_1 = Frequency of class preceding the mode class

f_2 = frequency of the class succeeding mode class

i = Width of the mode class (*i.e.,* upper limit – lower limit or $L_2 - L_1$ of mode class)

Let us take the following classified data table for calculation of mode here.

Class	40 – 50	50 – 60	60 – 70	70 – 80	80 – 90	90 – 100	100 – 110	110–120
Frequency	5	9	9	15	8	4	3	2

From the above data table it is apparent that the class 70 – 80 has highest frequency, *i.e.,* 15. Therefore, 70 – 80 is the mode class. The mode can now be calculated with the help of the following formula.

$$\text{Mo} = L_1 + \frac{fm - f_1}{2fm - f_1 - f_2} \times i$$

Here, $L_1 = 70, fm = 15, f_1 = 9, f_2 = 8$ and $i = 10$

$$\therefore \text{Mo} = 70 + \frac{15-9}{2\times15-9-8} \times 10$$

$$= 70 + \frac{6}{30-9-8} \times 10$$

$$= 70 + \frac{6}{30-17} \times 10 = 70 + \frac{6}{13} \times 10$$

$$= 70 + \frac{60}{13} = 70 + 4.61$$

$$\text{Mo} = 74.61.$$

Example. Calculate the mode of data presented in the following Table for 220 plants of a local variety of wheat.

No. of seeds per plant	50 – 60	60 – 70	70 – 80	80 – 90	90 – 100	100 – 110	110–120
No. of plants scored	7	24	18	24	15	22	10

Solution:

(*i*) Grouping Table.

Class	*Frequency Column* 1	2	3	4	5	6
50 – 60	7	31		49		
60 – 70	24		42		66	
70 – 80	18	42				57
80 – 90	24		39	61		
90 – 100	15	37			47	
100 – 110	22		32			
110 – 120	10					

(*ii*) Analysis Table.

	Classes with highest frequencies				
1	60 – 70		80 – 90		
2		70 – 80	80 – 90		
3	60 – 70	70 – 80			
4			80 – 90	90 – 100	100 – 110
5	60 – 70	70 – 80	80 – 90		
6		70 –80	80 – 90	90 – 100	
Total	3	4	5	2	1

The analysis table reveals that class 80–90 has the highest frequency. Therefore, it is mode class.

Now, the mode can be calculated with the help of the following formula:

$$Mo = L_1 + \frac{fm - f_1}{2\,fm - f_1 - f_2} \times i$$

Where, $L_1 = 80,\ fm = 24,\ f_1 = 18,\ f_2 = 15,\ i = 10$

$$\therefore \text{Mo} = 80 + \frac{24 - 18}{2 \times 24 - 18 - 15} \times 10$$

$$= 80 + \frac{6}{48 - 33} \times 10 = 80 + \frac{60}{15}$$

$$= 80 + 4 = 84$$

Mo = 84.

Merits of Mode

1. It is based on all values and can be subjected to further algebric treatments.
2. It represents group as well as extreme items and is not affected unless they are in modal class.

Demerits

1. It requires intricate calculation.
2. It cannot be used for further calculations and interpretations.

QUESTIONS

1. Define Mean, Median, and Mode.
2. Discuss various methods of determining median of unclassified and tabulated data.
3. Discuss the methods for determining Mode from the classified data.
4. Define and explain Arithmetic mean, and mention its merits and demerits.
5. Calculate mean and median from the following data:
   (*i*) 10, 8, 20, 22, 18, 39
   (*ii*) 17, 16, 19, 21, 21, 19, 23, 23, and
   (*iii*) No. of seeds per fruit 8, 7, 3, 15, 13, 17, 16, 18, 20.
6. When Mean $(\overline{x}) = 58$ and $\Sigma \overline{x} = 30$, compute N (No. of observations).
7. Calculate mean, median and mode in the following ungrouped and grouped series.
   No. of seeds per plant in *convolvulus* species:
   52, 55, 35, 39, 45, 49, 48, 52, 33, 48, 47, 50, 51, 50, 53, 53, 55, 50, 53, 54, 48, 49, 33, 55, 50, 30, 50, 51, 53, 50, 53, 52, 49, 51, 50, 50, 50, 51, 44, 58, 59, 57, 59, 60, 51, 58.
8. Calculate, median and mode from the following data:

Class interval	16 – 20	21 – 25	26 – 30	31 – 35	36 – 40	41 – 45	46 – 50
Frequency	4	4	9	7	13	3	2

9. Calculate harmonic mean of a series 1, 5, 10, 15, 20, 25.
10. Find median from the following data recorded on number of seeds per fruit:
    11, 5, 0, 42, 19, 27, 36, 52, 21, 30, 50.

31 CHAPTER

Measures of Dispersion or Spread

Various methods of data presentation such as tables, graphs, histograms on one hand give an idea about the kinds of variation available in frequency distribution and the measures of central tendency and on the other hand provide an idea about the nature of clustering of data. Mean is an important measure of central tendency which represents the values of whole series, but that does not show the relationship between the closeness and clustering of data around the central value. Two different series of data may show same mean value but this does not mean that the two series are identical. If the mean values of the two data series remain same, it may be possible that data in one series are widely dispersed from the mean value and those in the other series are closely clustered around the mean value. The lack of uniformity in the values of data of a series is referred to as **dispersion**. In other words, *dispersion is the measure of extent to which the individual items or data vary*. The limit to which numerical data tend to spread about mean value is referred to as variation or dispersion of data. Thus, the knowledge of the measures of dispersion is helpful in the following ways:

1. It helps in knowing the limit to which the mean represents the frequency distribution of a series. If the data are close to the mean value, the dispersion will be small and the mean represents the data of a series satisfactorily.
2. Unless the measure of dispersion is known, reliable conclusion about frequency distribution of a series on the basis of its mean cannot be drawn.

In this way, mean and measures of dispersion are complementary to each other.

MEASURES OF DISPERSION

The various measures of dispersion or measures of spread are as follows:

1. Range.
2. Deviation:
   (*i*) mean deviation, and
   (*ii*) standard deviation.
3. Variance.
4. Co-efficient of variation.

RANGE

This is the simple measure of dispersion. The range of distribution is the difference between the largest and the smallest data or items of a series. The range of data provides certain amount of variability.

Suppose that the heights (in cm) of ten individual wheat plants in a plot are: 70, 50, 60, 90, 80, 65, 40, 75, 55 and 85.

The range of plant height (R) = Highest value – Lowest value

$$\text{or } R = H - L$$
$$= 90 - 40 = 50$$

i.e., plant height ranges from 40 cm to 90 cm.

$$\textbf{Co-efficient of Range} = \frac{\text{Highest value} - \text{Lowest value}}{\text{Highest value} + \text{Lowest value}}$$

Example. Find out the range and co-efficient of range for the following series of data of plant heights (in cm): 17, 15, 12, 16, 10, 18, 20, 14, 11.

Solution. Range (*R*) = Highest value – Lowest value.

$R = H - L$

Here, $H = 20, L = 10$

So, $R = 20 - 10 = 10$ cm.

$$\textbf{Co-efficient of Range} = \frac{H-L}{H+L} = \frac{20-10}{20+10} = \frac{10}{30} = \frac{1}{3}$$

Co-efficient of $R = \frac{1}{3}$.

Example. Calculate the range and co-efficient of range of plant height from the following data:

Plant height (in cm)	10 – 15	15 – 20	20 – 25	25 – 30	30 – 35	35 – 40	40 – 45	45 – 50
Number of plants	5	10	20	40	35	30	25	15

Solution. Range (R) = $H - L$

Here $H = 50$ and $L = 10$

$\therefore$ Range = 50 – 10 = 40 cm

$$\textbf{Co-efficient of } \boldsymbol{R} = \frac{H-L}{H+L} = \frac{50-10}{50+10} = \frac{40}{60} = \frac{2}{3} = 0.66$$

Merits of Range

1. It represents the two extremes of observations.
2. Very easy to calculate.
3. It indicates two extremes between which the data or items of the series are dispersed or spread.

Demerits

1. Range cannot be considered as reliable measure of variability because it is based on only two extremes of observation, *i.e.,* highest and lowest values.
2. It is extremely influenced by the variation in sampling.
3. Range is much sensitive to size of sample or population.
4. It is not based on frequency of data.

DEVIATION

Mean Deviation

Deviation refers to the difference between the calculated means (arithmeatic mean, median and mode) and the value of individual data. **Mean deviation** is the mean of the deviations of all data of a series from the real mean value. Theoretically, the mean deviation is a true measure of dispersion. In the calculation of mean deviation all the values (+ and –) of deviation are taken as positive.

When the mean deviation is divided by the mean of the series, the value so obtained is called co-efficient of mean deviation.

Computation of Mean Deviation

1. When the data are unclassified and untabulated–

(*i*) Mean deviation $(\delta\overline{x})$ from arithmetic mean $= \frac{\Sigma (d\,\overline{x})}{n}$,

(*ii*) Co-efficient of Mean deviation or Co-efficient of $\delta\overline{x} = \frac{\delta\overline{x}}{x}$,

(*iii*) Mean deviation from Median or $\delta M = \frac{\Sigma (dm)}{n}$,

(*iv*) Co-efficient of Mean deviation from Median (Co-efficient of δM) $= \frac{\delta M}{M}$,

(*v*) Mean deviation from Mode (*Mo* or *Z*) or $\delta Z = \frac{\Sigma (dZ)}{n}$, and

(*vi*) Co-efficient of deviation from Mode $(\delta Z) = \frac{\Sigma (dZ)}{Z}$.

Where, $d\overline{x}$ = deviation from arithmetic mean

dM = deviation from Median

dZ = deviation from Mode

n = number of items or data in a series

$\overline{x}$ = Arithmetic mean

M = Median

Mo or Z = Mode.

Example. Calculate the mean deviation and co-efficient of mean deviation from the arithmetic mean from the following data on seed weight:

4.8, 5.5, 6.5, 7.1, 7.3, 7.5, 8.0 and 8.5.

Solution:

S. No.	*Seed Weight (x)*	*Arithmetic Mean*$(\overline{x})$	*Deviation from the Arithmetic Mean (all in +)* $d = x - \overline{x}$
1	4.8	$\overline{x} = \frac{55.2}{8}$	4.8 – 6.9 = 2.1
2	5.5	= 6.9	5.5 – 6.9 = 1.4
3	6.5		6.5 – 6.9 = 0.4
4	7.1		7.1 – 6.9 = 0.2
5	7.3		7.3 – 6.9 = 0.4
6	7.5		7.5 – 6.9 = 0.6
7	8.0		8.0 – 6.9 = 1.1
8	8.5		8.5 – 6.9 = 1.6
$n = 8$	55.2		$\Sigma d\overline{x} = 7.8$

Mean deviation from Arithmetic Mean $d\overline{x} = \frac{\Sigma d\overline{x}}{n} = \frac{7.8}{8} = 0.975.$

Co-efficient of mean deviation from Arithmetic Mean $= \frac{d\overline{x}}{\overline{x}} = \frac{0.975}{6.9} = 0.141.$

2. **When the data are ungrouped and tabulated.** The mean deviation of ungrouped data is calculated with the help of the following formulae:

(*i*) $\delta\overline{x} = \dfrac{\Sigma f d\overline{x}}{n}$,

(*ii*) $\delta M = \dfrac{\Sigma f dM}{n}$, and

(*iii*) $\delta Z = \dfrac{\Sigma f dZ}{n}$.

where, f = frequency.

3. **When Data are Classified and Tabulated**. The mean deviation of Tabulated and Classified data is also calculated with the help of same formulae as are used for ungrouped data.

Example. 100-seed weight of maize recorded for 100 plants are presented in the following Table. Calculate the mean deviation and co-efficient of mean deviation from the data table:

100-Seed weight (in gm)	10 – 20	20 – 30	30 – 40	40 – 50	50 – 60
Number of plants	10	25	42	15	8

Solution:

100-Seed weight class	*Mid value of the class (x)*	*Frequency (f) of the class*	*fx*	$\overline{x} = \dfrac{\Sigma fx}{n}$	*Deviation* $d\overline{x} = x - \overline{x}$	
10 – 20	15	10	150	$\dfrac{3360}{100}$	15 – 33.6 = 18.6	186
20 – 30	25	25	625	= 33.60	25 – 33.6 = 8.6	215.0
30 – 40	35	42	1470		35 – 33.6 = 2.6	109.2
40 – 50	45	15	675		45 – 33.6 = 12.6	189.0
50 – 60	55	8	440		55 – 33.6 = 22.6	180.8
		$n = 100$	$\Sigma fx = 3360$		$\Sigma f\overline{x} = 65.0$	$\Sigma f d\overline{x} = 880$

$$\delta\overline{x} = \frac{\Sigma f d\overline{x}}{n} = \frac{880}{100} = 8.8$$

$$\text{Co-efficient of } \delta\overline{x} = \frac{\delta\overline{x}}{\overline{x}} = \frac{8.8}{33.6} = 0.261.$$

Standard Deviation

The best and most important measure of dispersion is standard deviation which was first worked out by Karl Pearson (1833). It is the positive square root of mean of deviations of individual values of a data series from the arithmetic mean of the series. In other words, the square of standard deviation is equal to mean of the square of deviations of individual observations from the arithmetic mean. It is also called mean square deviation from the mean and is denoted by a Greek symbol σ.

Let us suppose that n is the total number of observations (Say, X_1, X_2, X_3, X_4 -------, X_n) and the sum of the values of all observations is S.

Thus the mean $(\overline{x}) = \dfrac{S}{n}$.

The deviations of individual observations can now be expressed as follows:

$X_1 - \overline{x}, X_2 - \overline{x}, X_3 - \overline{x}, X_4 - \overline{x} \,.........\, X_n - \overline{x}$.

The square of standard deviation SD^2 or σ^2

$$= \frac{(X_1-\bar{x})^2 + (X_2-\bar{x})^2 + (X_3-\bar{x})^2 + (X_4-\bar{x})^2 \ldots\ldots\ldots (X_n-\bar{x})^2}{n}$$

$$\therefore \text{S.D. or } \sigma = \sqrt{\frac{(X_1-\bar{x})^2 + (X_2-\bar{x})^3 + (X_3-\bar{x})^2 + (X_4-\bar{x})^2 \ldots\ldots\ldots (X_n-\bar{x})^2}{n}}$$

If the values of observed data are equal, the mean deviation will be zero which represents its lowest value. The value of standard deviation will increase with the increase in deviations of individual data from their arithmetic mean. To make the standard deviation comparable, co-efficient of standard deviation is calculated which is the ratio between standard deviation of observation series and its mean, *i.e.*, standand deviation (S.D. or σ) = $\sqrt{\frac{\Sigma d^2}{n}}$.

$$\text{Co-efficient of S.D.} = \frac{\text{Standard Deviation}}{\text{Arithmetic Mean}} = \frac{\sigma}{\bar{x}}.$$

Methods of Calculating Standard Deviation

Generally, the following three methods are used for calculating standard deviation:

1. Direct Method.
2. Short Cut Method.
3. Step Deviation Method.

1. Direct Method. In this method, first of all arithmetic mean ($\bar{x}$) of the series is calculated. The deviations of individual values from the mean are calculated ($d = X - \bar{x}$) which may be either positive or negative number. The value of d^2 is always a positive figure. Then the standard deviation (S.D. or σ) is calculated as follows:

(*i*) *when the data are unclassified*, the following formula is used to calculate the standard deviation:

$$\text{S.D. or } \sigma = \sqrt{\frac{\Sigma d^2}{n}}$$

where n = number of data.

Example. Calculate S.D. from the given data collected on 10 plants: 55, 58, 60, 62, 68, 65, 66, 72, 75, 69.

Solution. Arithmetic mean $= \frac{650}{10} = 65.0$

S. No.	*Data*	*Mean* $\bar{x}$	*Deviation* $d = X - \bar{x}$	d^2	Σd^2
1	55	$\frac{650}{10}$	55 – 65 = – 10	100	358
2	58	= 65.0	58 – 65 = – 7	49	
3	60		60 – 65 = – 5	25	
4	62		62 – 65 = – 3	9	
5	68		68 – 65 = + 3	9	
6	65		65 – 65 = 0	0	
7	66		66 – 65 = + 1	1	
8	72		72 – 65 = + 7	49	
9	75		75 – 65 = + 10	100	
10	69		69 – 65 = + 4	16	

$$\text{S.D. or } \sigma = \sqrt{\frac{358}{10}} = 6.0,$$

(*ii*) when the data are ungrouped and tabulated, the S.D. is calculated with the help of the following formula:

$$\text{S.D. or } \sigma = \sqrt{\frac{\Sigma fd^2}{n}}, \text{ where } f = \text{frequency of data.}$$

(*iii*) if the data are classified and tabulated then also the same formula is used in the calculation of S.D. as is used for ungrouped data. But, the difference here is that, in case of classified data, the mid-values of different classes are used to calculate the mean $(\bar{x})$ and then the deviation (d) of each class is calculated by subtracting the mean from mid-value of each class, *i.e.*, $d = (X - \bar{x})$. The deviation squares (d^2) are then multiplied by the frequency (f) of the respective classes to get (fd^2). The standard deviation is calculated with the help of the following formula S.D. or $\sigma = \sqrt{\frac{\Sigma fd^2}{n}}$.

Short Cut Method for Calculating S.D.

The method involves the following steps:

1. Any value of an unclassified series or mid-value of any class in case of classified series is treated as assumed mean (A).
2. The deviation (dx) of individual item or class is calculated by subtracting the assumed mean from the individual data (in case of unclassified series) or mid-value (X) of each class (in case of classified series) $dX = X - A$.
3. The deviation values are squared to get dx^2 and then for ungrouped data and for classified data separate formulae are used to calculate standard deviation:

   (*i*) Formula for calculation of S.D. of ungrouped data

$$\sigma = \frac{\sqrt{\Sigma dx^2}}{n} - \left(\frac{\Sigma dx}{n}\right)^2$$

$$\text{or } \sigma = \frac{1}{n}\sqrt{n\Sigma dx^2 - (\Sigma dx)^2}.$$

   (*ii*) Formula for calculation of S.D. of tabulated classified data:

$$\sigma = \frac{\sqrt{\Sigma fdx^2}}{n} - \left(\frac{\Sigma fdx}{n}\right)^2$$

$$\text{or } \sigma = \frac{1}{n}\sqrt{n\Sigma fdx^2 - (\Sigma fdx)^2}.$$

Example. Find out the standard deviation of the following observations regarding the number of flowers per plant

3, 2, 1, 5, 7, 9, 11, 13, 14, 15.

Solution. Arithmetic mean $= \frac{80}{10} = 8$

$$\therefore (\bar{x}) = 8$$

S. No.	No. of flowers per plant	Deviation (d) $=(x-\bar{x})$	Square of deviation (d^2)
1	3	$3-8=-5$	25
2	2	$2-8=-6$	36
3	1	$1-8=-7$	49
4	5	$5-8=-3$	9
5	7	$7-8=-1$	1
6	9	$9-8=+1$	1
7	11	$11-8=+3$	9
8	13	$13-8=+5$	25
9	14	$14-8=+6$	36
10	15	$15-8=+7$	49
$n=10$	$S=8$		$\Sigma dx^2=240$

$$\text{S.D.}^2 \text{ or } \sigma^2 = \frac{240}{10} = 24$$

$$\text{S.D.} = \sqrt{24} = 4.899.$$

Example. Calculate the S.D. of the following data on the heights (in cm) of 10 plants: 73, 75, 80, 42, 57, 65, 52, 42, 47, 67.

Solution. Suppose that the assumed mean of the given series is 57.

S. No.	Height in cm (x)	Deviation dx = (X – A)	dx^2
1	73	$73-57=16$	256
2	75	$75-57=18$	324
3	80	$80-57=23$	529
4	42	$42-57=-15$	225
5	57 (A)	$57-57=0$	00
6	65	$65-57=+8$	64
7	52	$52-57=-5$	25
8	42	$42-57=-15$	225
9	47	$47-57=-10$	100
10	67	$67-57=10$	100
$n=10$		$\Sigma dx=30$	$\Sigma d^2=1848$

$$\text{S.D.}(\sigma) = \frac{1}{n}\sqrt{n\,\Sigma dx^2-(\Sigma dx)^2}$$

$$= \frac{1}{10}\sqrt{10\times1848-(30)^2}$$

$$= \frac{1}{10}\sqrt{18480-900} = \frac{1}{10}\sqrt{17580}$$

$$= \frac{1}{10}(132.58) \therefore \text{S.D.} = 13.258.$$

Example. Calculate the standard deviation of data presented in the following table:

Height of plants of mutant population (in cm)	50–60	60–70	70–80	80–90	90–100	100–110	110–120	120–130	130–140	140–150	150–160
Frequency	5	7	18	26	35	25	40	20	16	6	2

Solution. Suppose 105 is the assumed mean of plant height.

Class (Plant height in cm)	*Mid-value of clas (x)*	*Frequency (f)*	*Deviation dx = x − A*	*f.dx*	*dx²*	*f.dx²*
50 – 60	55	5	55 – 105 = – 50	– 250	2,500	12,500
60 – 70	65	7	65 – 105 = – 40	– 280	1,600	11,200
70 – 80	75	18	75 – 105 = – 30	– 540	900	16,200
80 – 90	85	26	85 – 105 = – 20	– 520	400	10,400
90 – 100	95	35	95 – 105 = – 10	– 350	100	3,500
100 – 110	105 = A	25	105 – 105 = 00	00	00	00
110 – 120	115	40	115 – 105 = 10	400	100	4,000
120 – 130	125	20	125 – 105 = 20	400	400	8,000
130 – 140	135	16	135 – 105 = 30	480	900	14,400
140 – 150	145	6	145 – 105 = 40	240	1,600	9,600
150 – 160	155	2	155 – 105 = 50	100	2,500	5,000
		n = 200		Σ f.dx =−320		Σ f.dx² =94,800

$$\text{S.D. } (\sigma) = \sqrt{\left(\frac{\Sigma fdx^2}{n}\right) - \left(\frac{(\Sigma fdx)}{n}\right)^2}$$

$$= \sqrt{\frac{94800}{200} - \left(-\frac{320}{200}\right)^2}$$

$$= \sqrt{474 - 2.56} = \sqrt{471.44}$$

$$\sigma = 21.71.$$

Merits of Standard Deviation

1. Standard deviation is mathematically computed and reliable. Therefore, it is used in quality research work.
2. It is based on every individual data of a series.
3. It is least affected by sudden deviations of any type.
4. It is a definite measure of dispersion.

Demerits of Standard Deviation

1. Method of its calculation is complicated and so inconvenient.
2. In this measure, higher values are given more importance than the lower values which affect the S.D. value.

VARIANCE

It is very useful measure of dispersion which is commonly used for population data. The arithmetic mean of the squares of deviations obtained from the mean is referred to as **variance**. It is based on standard deviation and the square of the standard deviation is termed as **variance**

$$V = \sigma^2 \text{ or } (\text{S.D.})^2 = \frac{\Sigma (X - \bar{x})^2}{n}$$

$$\text{or } V = \frac{\Sigma d^2}{n} \text{ or } \frac{\Sigma f.d^2}{n}.$$

Thus, if the spread of a series is known, its standard deviation can be determined:

$$\sigma = \sqrt{variance}.$$

Co-efficient of Variation

Co-efficient of variation is a relative term which measures the relative magnitudes of variations present in observations and is related to the magnitude of their arithmetic mean. Since the standard deviation is independent measure of dispersion obtained from a single series of data, the comparison of variability of two different series of data is not possible directly by standard deviations. To make it possible Karl Pearson used co-efficient of variation. **Co-efficient of variation** is defined as the ratio of standard deviation to arithmetic mean and is expressed in percentage. Co-efficient of variation or $CV(\%) = \frac{\sigma}{\bar{x}} \times 100$, where σ = standard deviation of a series $\bar{x}$ = arithmetic mean of series.

Thus, when standard deviation of a series is compared to arithmetic mean of the series in terms of percentage, we get co-efficient of variation. It expresses the relative variability of each data series. The co-efficient of variation can be used to compare the homogeneity, consistency and stability of data of two or more series. The data series with high value of co-efficient of variation is said to be unstable and the series with low values of co-efficient of variation are stable and they show good degree of homogeneity of data.

Example. Compare the relative variability of two samples with the following data:

Sample A–Arithmetic mean $(\bar{x}) = 30$ and S.D. = 1.5

Sample B–Arithmetic mean $(\bar{x}) = 20$ and S.D. = 2.5

Solution. (*i*) For sample A, co-efficient of variation (CV) $= \frac{\text{S.D. or } \sigma}{\text{mean}} \times 100$

$$= \frac{1.5}{30} \times 100 = 5.0\%.$$

(*ii*) For sample B, $CV = \frac{\sigma}{\text{mean}} \times 100$

$$= \frac{2.5}{20} \times 100 = 12.50\%.$$

Example. Calculate standard deviation, variance and co-efficient of variation from the following classified data regarding number of pods per plant recorded on 50 plants in a plot:

Number of pods/plant	1 – 3	3 – 5	5 – 7	7 – 9	9 – 11	11 – 13	13 – 15
No. of Plants	2	6	7	16	11	5	3

Solution.

Class (No. of pods per plant)	*Mid-value of class (x)*	*No. of Plants (f)*	*dx = (x – A)*	*dx²*	*f.dx²*	*f.dx*
1 – 3	2	2	2 – 8 = – 6	36	72	– 12
3 – 5	4	6	4 – 8 = – 4	16	96	– 24 = – 50
5 – 7	6	7	6 – 8 = – 2	4	28	– 14
7 – 9	8 = A	16	8 – 8 = 0	00	00	00
9 – 11	10	11	10 – 8 = 2	4	44	22
11 – 13	12	5	12 – 8 = 4	16	80	20 = 60
13 – 15	14	3	14 – 8 = 6	36	108	18
Total		n = 50			Σ f.dx² = 428	Σ f.dx = 10

Arithmetic mean $(\bar{x}) = A + \frac{\Sigma f.dx}{n}$

$$\bar{x} = 8 + \frac{10}{50} = 8 + 0.2 = 8.2$$

$$\text{S.D. or } \sigma = \sqrt{\frac{\Sigma fdx^2}{n} - \left(\frac{\Sigma fdx}{n}\right)^2}$$

$$= \sqrt{\frac{428}{50} - \left(\frac{10}{50}\right)^2} = \sqrt{8.56 - 0.04}$$

$$= \sqrt{8.52} = 2.92$$

$$\text{Variance } = \sigma^2$$

$$= \left(\sqrt{8.52}\right)^2$$

$$\text{Co-efficient of variation } = \frac{\sigma}{\bar{x}} \times 100$$

$$\text{or C.V.} \quad = \frac{2.92}{8.2} \times 100$$

$$\text{or C.V.} \quad = \frac{292}{8.2} = 35.61\%.$$

Standard Error of the Mean

If instead of considering one sample, several samples are taken from the same population, the standard deviations of different samples may vary. This variation is measured by standard error or standard error of mean. Standard error is a measure of dispersion of different hypothetical means around the grand mean. It is a statistical constant which measures the dispersion of sample means around population mean. Standard error is calculated by the following formula:

$$\text{Standard Error (S.E.)} = \frac{\text{Standard deviation or S.D. or } \sigma}{\sqrt{n}}$$

Where S.D. or σ = Standard deviation of individual sample

n = Total number of observations.

Standard error thus gives an idea as to how a mean of any sample differs from the true mean of population. Therefore, while making a statement on some measurements one should state the average ± standard error, say for example, the height of plants of particular species measures 105 ± 0.13 cm. Here, 105 cm is the average height and 0.13 is the standard error.

Sometimes instead of standard error, probable error (P.E.) is used which is 2/3 or 0.6745 time the standard deviation

$\therefore$ P.E. = 2/3 or 0.6745σ.

Example. In a sample plot of 1 square metre, the height of 100 plants was measured and the standard deviation was calculated to be 5.0. What will be the standard error of the sample?

Solution.

$$\text{Since, S.E.} = \frac{\sigma}{\sqrt{n}}$$

$$\text{Therefore, S.E.} = \frac{5.0}{\sqrt{100}}$$

$$= \frac{5.0}{10} = 0.50.$$

QUESTIONS

1. Define Dispersion. Point out various measures of dispersion. Discuss range and its merits and demerits.
2. Explain mean deviation. How would you calculate mean deviation and co-efficient of mean deviation from arithmetic mean of the given data?
3. What is standard deviation? Discuss the methods for calculating standard deviation and point out the merits and demerits of standard deviation.
4. What do you mean by variance and co-efficient of variation?
5. What do you mean by standard error?
6. Find out the range of the following two series:

   Series I 70, 15, 89, 62, 56, 72, 86.

   Series II 18.2, 19.5, 16, 23, 21, 13, 19.
7. Calculate the arithmetic mean and standard deviation of the following frequency distribution:

Variable	15	10	5	20	25	30	40	45	50	60
Frequency	6	4	2	6	10	10	10	6	4	2

8. A score of 50 wheat plants measured with a maximum height of 50 cm., the mean was 24 and $S = 10$. If a score of 100 plants measured 100 cm each, find out the standard deviation of the second score.
9. Calculate the variance of data: 3, 4, 1, 7, 6, 7, 6, 5, 6.
10. Calculate mean deviation of data: 1, 2, 3, 4, 8, 10, 12, 18, 20, 22.
11. Calculate mean deviation of the following data:

    Water content in fruit: 63, 62, 62, 67, 64, 68, 71, 72, 71, 68, 69, 68, 66, 62.

32 CHAPTER

Probability

Probability means possibility or likelihood for the occurrence of a specific event, as for example, the likelihood of winning the cricket match. If several events are taking place, the possibility for the occurrence of any specific event can be predicted. Thus, if a coin is tossed 100 times, it is likely that it will fall "*head*" 50 times and "*tail*" 50 times because chances for turning up of head and tail are equal, *i.e.,* 1:1. When an experiment results in two or more different results, the experiment is called **trial** and the results obtained from the experiment are called events.

The likelihood of any specific event may be influenced by several factors. For example, the result of an examinee depends on his performance, the type of question paper (easy or hard), evaluation of answer books by the examiners etc. Contrary to such events, there are some events whose occurrence is not affected by any factor other than chance, as for example, when a coin is tossed, either head or tail turns up and similarly, the sex of a new born baby is either male or female.

When several events occur at a time, then the probability of occurrence of a specific event will be equal to number of times that event happens divided by the total number of opportunities for it to happen or total number of trials.

$$P = \frac{1}{n}$$

Where P = Probability of any event, and

n = total number of trials.

Value of p does not exceed 1 and all the values of P lie between 0 and 1. In the same way, the probability of failure of any specific event may be calculated as follows:

$$P = \frac{\text{No. of cases not favourable}}{\text{Total number of trials}}$$

If the total number of trials is represented by n and the number of times any specific event (A) happens is m, then

$$P(A) = \frac{m}{n}$$

If m represents the number of favourable cases out of total number of events (n), the number of failures or non-favourable cases ($\overline{A}$) will be $n - m$, and then the probability of the failure of event $P(\overline{A})$ will be

$$P(\overline{A}) = \frac{n-m}{n}$$

or $$P(\overline{A}) = \frac{n}{n} - \frac{m}{n}$$

or $P(\overline{A}) = 1 - P(\overline{A})$ $\left[\text{Since } \frac{n}{n} = 1 \text{ and } \frac{m}{n} = P(A)\right]$

and thus $P(\overline{A}) + P(\overline{A}) = 1$.

If a cubical die having six sides or faces mentioned 1, 2, 3, 4, 5 and 6 is rolled, any one of the 6 different faces may turn up. Here, the group of six possible results is said to be exhaustive and the rolling of die is called a trial. Here the probability of rolling a "*four*" on a die in a single trial or $P(4)$ will be 1/6, because, the die has six faces. As each face is likely to turn up, the average result should be one *four* for every six trials.

Example. Find out the probability of birth of male child to an expecting woman.

Solution. There are two conditions in this case, *i.e.,* the new born may be either a male child or a female child. So $n = 2$, and each condition stands one chance out of two. Therefore, the probability for a boy would be

$$P(B) = \frac{1}{n} = \frac{1}{2}$$

and similarly, probability for a girl $P(G) = \frac{1}{n} = \frac{1}{2}$.

Types of Events

The events are of several types, some of which are defined here as under:

1. **Simple Events.** When a single event occurs at a time, this is called simple event, as for example, taking out a single card from the stack of playing cards.
2. **Compound Events.** When two or more events occur simultaneously. They are called compound events, as for example, turning up of heads when two coins are tossed simultaneously.
3. **Mutually Exclusive Events.** When two events are so related that the occurrence of one prevents the occurrence of the other or *vice-versa* in a trial, they are said to be *mutually exclusive* events. For example, tossing of a coin is a trial which provides an equal opportunity for the head or tail to turn up and in the trial only one, either head or tail can occur but not both. Similarly, a new born baby can be a boy or a girl. Thus, the occurrence of boy excludes that of girl and *vice-versa.* Therefore, these events are mutually exclusive events. The probability of mutually exclusive events is determined by the following simple formula:

   $$P = \frac{1}{n}$$

   When two mutually exclusive events are such that the possibility of occurrence of one event is neither less nor more than the possibility for the occurrence of other event, they are said to be ***equally likely events***. For example, when a coin is tossed, the possibilities for turning up of head or tail are equal and thus, the two are equally likely events.
4. **Independent Events.** When two or more events are so related that the occurrence of one does not affect the possibility of the occurrence of remaining event(s), they are said to be independent events. For example, when two coins A and B are tossed together or separately, the head or tail of coin A does not affect the possibility of the occurrence of head or tail of coin B. Thus, turning up of head or tail of coin A is independent of that of coin B.

   If coin A and B are tossed together, one may obtain any of the following combinations of events:

   (*i*) Head of both coins, (*ii*) Head of coin A and tail of B, (*iii*) Tail of coin A and head of B, and (*iv*) Tail of both A and B.

Table 32.1 Probability of different combinations of head and tail of coin *A* and coin *B* tossed either simultaneously or separately.

S. No.	*Coin A*	*Coin B*	*Probability*
1	*H*	*H*	$\frac{1}{2}\times\frac{1}{2}=\frac{1}{4}$
2	*H*	*T*	$\frac{1}{2}\times\frac{1}{2}=\frac{1}{4}$
3	*T*	*H*	$\frac{1}{2}\times\frac{1}{2}=\frac{1}{4}$
4	*T*	*T*	$\frac{1}{2}\times\frac{1}{2}=\frac{1}{4}$

The combination of head of coin *A* and tail of coin *B* and tail of coin *A* and head of coin *B* may be grouped together since both of them have head and tail and the only difference between the two is the particular coin in which head or tail appears first. The condition of getting one head and one tail on tossing the two coins is satisfied by both these conditions. Therefore, the possibility of getting one head and one tail will be the sum of the probabilities of these two combinations $\left(\frac{1}{4}+\frac{1}{4}=\frac{1}{2}\right)$. This is so because such combinations of independent events behave as mutually exclusive events.

Similarly, when three coins are tossed simultaneously, the following eight combinations of head and tail are possible: HHH, HHT, HTH, THH, HTT, THT, TTH and TTT. Each of these combinations will have the probability of $\frac{1}{8}=\left(\frac{1}{2}\times\frac{1}{2}\times\frac{1}{2}\right)$. These eight combinations ignoring the sequence of occurrence of head (H) and tail (T) can be grouped in the following four classes:

(*i*) One combination of three heads (HHH)

(*ii*) Three different combinations of 2 head and one tail– (HHT, HTH,THH).

(*iii*) Three different combinations of one head and two tails– (HTT, THT, TTH)

(*iv*) One combination of three tail–TTT.

Thus, the possibilities of getting 3 heads, 2 head and one tail, one head and two tail and three tail on tossing three coins simultaneously will be $\frac{1}{8},\frac{3}{8},\frac{3}{8}$, and $\frac{1}{8}$ respectively. The number of possible combinations of head and tail on tossing 4, 5, 6, and 7 coins together will be 16, 32, 64 and 128 respectively. This indicates that the number of different combinations increases with the increase of the number of independent events. Increase in the number of independent events by one will double the number of possible combinations.

5. **Dependent Events.** When the occurrence of one event affects the occurrence of other and *vice-versa* such related events are said to be dependent events.

6. **Sample Space or Probability Space.** All the possible results of an experiment or trial form a set which is referred to as sample space or probability space. For example, a woman gives birth to a child that may be a male or female. Therefore, the sample space = M,F. If two women give birth to one baby each, the probability space (S) may be MM, MF, FM, FF.

7. Conditional Events. When two events are so related that one occurs either before or after the other, they are said to be conditional events. Suppose, for example, there are two events E_1 and E_2 which may be defined as follows:

(*i*) E_1/E_2 which means that E_1 occurs after E_2.

(*ii*) E_2/E_1 which means that E_2 occurs after E_1.

Mathematical Probability and Statistical Probability

When the probability is calculated on the basis of reasoning even before any trial or experiment that is called mathematical or *apriori* probability. In practice, the actual probability calculated by trials may not coincide with *apriori* probability, for example, when a coin is tossed one gets the head (H), though the mathematical probability of occurrence of head is only 1/2 in the trial $P(H) = 1$ or $P(H) = 0$. But if the coin is tossed 10 times, the occurrence of head (H) can be 0 or 1 or 2 or 3 -- or 10, the extreme cases being very rare. Suppose H occurred 4 times out of 10 trials. Now consider the occurrence of H as the favourable event, the occurrence of H 4 times out of 10 trials gives the relative frequency 4/10 for H (The *apriori* probability is 1/2).

But if the number of tosses (trials) is increased, it is likely that the ratio of the occurrence of favourable event becomes more close to 1/2. In general, if favourable event, say head (H), occurs n time out of N trials, the relative frequency of favourable event will be n/N. The limit of this relative frequency as N becomes infinitely large is known as **statistical probability.**

$$P = \frac{Lt}{N-\infty} = \frac{n}{N}$$

Rules or Theorems of Probability

The concept of probability is of paramount importance in statistics as it provides the basis for all the tests of significance. It is usually estimated on the basis of the following two basic rules of chances:

1. addition rules or addition theorem.
2. multiplicative rule or multiplication theorem.

1. Addition Rule of Probability. This rule is applied when two events are **mutually exclusive,** *i.e.*, both events do not occur simultaneously. Suppose that E_1 and E_2 are mutually exclusive events and the occurrence of one event precludes the occurrence of the other event. The probability of occurrence of either E_1 and E_2 is the sum of the probabilities of the individual events.

$P(E_1)$ or $P(E_2) = P(E_1) + P(E_2)$

This addition rule of probability can be extended to any number of mutually exclusive events as follows:

$P(E_1$ or E_2 or E_3 ....or $E_n) = P(E_1) + P(E_2) + P(E_3) + \ldots\ldots. P(E_n)$

If a coin is tossed, the probability of the occurrence of one face or that of other face is equal to the probability of one face *plus* the probability of the other face, *i.e.*,

$$P(H) \text{ or } P(T) = P(H) + P(T) = \frac{1}{2} + \frac{1}{2} = 1.$$

Problem. In a bunch of 8 flowers, four are red and the others are white. Find out the probability that (*i*) one flower drawn from the bunch at random is red, (*ii*) two flowers drawn from the bunch at random are of the same colour.

Solution.

(*i*) As all the 8 flowers of the bunch have equal chance for being drawn and there are 4 red flowers out of 8, the probability of selection of a red flower in a random chance is $\frac{4}{8} = \frac{1}{2}$.

(*ii*) The two flowers selected at random can be both red or both white. So the required probability = Probability of selecting two red flowers + probability of selecting two white flowers because both are mutually exclusive and equally likely events. The number of ways in which two flowers of the same colour (red or white) can be selected from 4 flowers of one colour is six because if red flowers are numbered R_1, R_2, R_3, R_4, and white flowers are numbered W_1, W_2, W_3, W_4, the possible combinations are $R_1 R_2, R_1 R_3, R_1 R_4, R_2 R_3, R_2 R_4$ and $R_3 R_4$ of red flowers and $W_1 W_2, W_1 W_3, W_1 W_4, W_2 W_3, W_2 W_4, W_3 W_4$ of white flowers.

In general, the number of combinations of two from *n* different things ($^nC_2$) is calculated by the formula:

$$^nC_2 = \frac{n}{1} \times \frac{n-1}{2}.$$

The number of combinations of 3 from *n* different things is determined by the formula:

$$^nC_3 = \frac{n}{1} \times \frac{n-1}{2} \times \frac{n-2}{3}.$$

So the probability of getting at random two flowers of red colour from 4 is $^4C_2 = \frac{4}{1} \times \frac{3}{2} = 6$ and the number of combinations of 2 from 8 different things is $^8C_2 = \frac{8}{1} \times \frac{7}{2} = 28.$

Therefore, the probability of drawing two flowers of red colour is $\frac{6}{28}$ and the probability of getting two flowers of white colour at random is $\frac{6}{28}$ and then the required probability of drawing two flowers is $\frac{6}{28} + \frac{6}{28} = \frac{12}{28} = \frac{3}{7}.$

(*iii*) **Multiplicative Rule of Probability.** This theorem states that *the probability of two independent events occurring simultaneously is the product of individual probabilities of those events.* If $P(E_1)$ and $P(E_2)$ are the probabilities of independent events E_1 and E_2, then the probability of the two events happening together is calculated as follows:

$P(E_1) + P(E_2) = P(E_1) \times P(E_2).$

This theorem can be extended to any number of independent events as follows:

$P(E_1, E_2, E_3 \text{-------} E_n) = P(E_1) \times P(E_2) \times P(E_3) \times \text{-----} \times P(E_n).$

Problem. In a flock of 16 birds, the ratio of black feathered birds to white birds was 5:3. If three birds are drawn at a time at random, find out the probability that:

(*i*) They are all black,
(*ii*) They are all white, and
(*iii*) They all are either black or white.

Solution. Since the ratio of black birds to white birds is 5:3, out of 8 birds, the ratio of black to white out of 16 birds will be 10 black and 6 white.

Now the possible number of combinations of three birds from 16 birds will be

$$^{16}C_3 = \frac{16}{1} \times \frac{15}{2} \times \frac{14}{3} = 560.$$

(*i*) The number of ways 3 black birds can be selected out of 10 black birds will be:

$$^{10}C_3 = \frac{n}{1} \times \frac{n-1}{2} \times \frac{n-2}{3} = \frac{10}{1} \times \frac{9}{2} \times \frac{8}{3} = 120.$$

So, the probability of getting three black birds in a draw is:

$$\frac{^{10}C_3}{^{16}C_3} = \frac{120}{560} = \frac{3}{14}.$$

(*ii*) The number of ways in which three white birds are selected in a draw out of 6 white birds is:

$$^{6}C_{3}=\frac{n}{1}\times\frac{n-1}{2}\times\frac{n-2}{3}=\frac{6}{1}\times\frac{5}{2}\times\frac{4}{3}=20$$

So, the probability of getting 3 white birds in a draw at random from a flock of 16 birds is:

$$\frac{^{6}C_{3}}{^{16}C_{3}}=\frac{20}{560}=\frac{1}{28}.$$

(*iii*) In a draw of three birds at a time, the probability of getting all three birds of either black colour or of white colour will be the same as of individual probability of black birds and white birds.

$$P=\frac{^{10}C_{3}}{^{16}C_{3}}+\frac{^{6}C_{3}}{^{16}C_{3}}=\frac{3}{14}+\frac{1}{28}=\frac{7}{28}=\frac{1}{4}.$$

PROBABILITY DISTRIBUTION

The characters which show variation are referred to as **random variables** or **variates**. They may be discrete or continuous. On the basis of nature and probability of occurrence of random variables, the following three types of probability distribution have been formulated:

1. Normal probability distribution.
2. Binomial probability distribution.
3. Poisson probability distribution.

Normal Probability Distribution

The normal form of distribution was worked out by De Moivre (1733) which was further developed by Gauss (1809) and Leplace (1812). For this reason normal probability distribution is named after these workers. Gauss noted that the errors of observations on any character follow normal distribution. Therefore, the normal probability distribution is also referred to as **Gaussian law of error**. The term *normal* does not stand for normal trend of any character nor it stands for the norm of distribution of any character. In fact, normal distribution is the distribution of characters showing continuous

Wt. of fishes	15	17	19	21	23	25	27	29	31	33	35
Frequencies	9	11	13	15	17	19	17	15	13	11	9

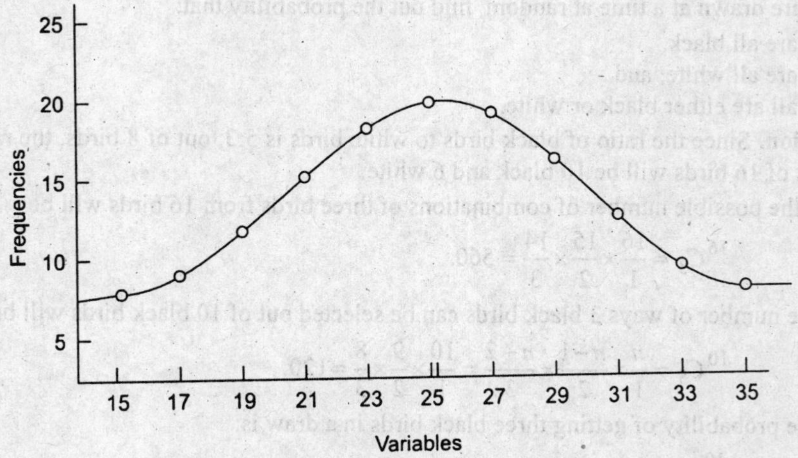

Fig. 32.1 Normal Distribution Curve.

variation. In binomial distribution $(p + q)^n$, if the value of power n is infinite and all the frequency points are linked by a line on graph paper, a simple curved line will be obtained. That is called normal frequency distribution curve (Fig. 32.1). The shape of normal distribution is unimodel, symmetrical and the ends of the curve tail off to the base. Graphically total area under the curve is unity. The algebric equation of normal distribution curve is $P_{(x)} = \frac{1}{\sigma\sqrt{2\pi}} e - \frac{(x-\mu)^2}{2\sigma^2}$ where μ = population mean and σ = standard deviation when frequency of a character is N and distribution is normal, then the frequency $F_{(xi)} = \frac{N}{\sigma\sqrt{2\pi}} e - \frac{(xi-\mu)^2}{2\pi\sigma^2}$.

Binomial Distribution of Probability

Binomial probability distribution has been worked out by James Bernoulli. The law of probability of coincidence of independent events states that the chance or probability of the simultaneously occurring two or more independent events is the product of the individual probabilities of the events. If two independent events E_1 and E_2 occur simultaneously and their probabilities are p and q respectively, then the probability of their joint occurrence will be $p \times q$. For example, suppose if 10 coins are tossed together one may get 5 heads and 5 tails, 4 heads and 6 tails, 3 heads and 7 tails most frequently, other results less after and either 10 heads and no tails or *vice-versa* rarely. When the number of respective heads and tails of different tosses of coins are plotted on graph paper, a symmetrical graph (curve) would be obtained. This is called symmetrical or normal probability curve. Here if P stands for probability of head, q stands for probability of the tail and n stands for the number of coins, the probability of head (p) and the probability of tail (q) are both equal to $\frac{1}{2}$. In other words, for one coin, the distribution of probability will be or $(p + q)^2$.

If two coins are tossed together, the following probability can be expected:

Coin I	*Coin II*	*Result*	*Probability*
H	H	HH	$p \times p = p^2$
H	T	HT	$p \times q = pq$
T	H	TH	$q \times p = pq$
T	T	TT	$q \times q = q^2$

So the probability distribution = $p^2 + 2pq + q^2 = (p+q)^2$.

This is an expansion of general binomial $(p + q)^n$. For different values of n, the following expansion of $(p + q)^n$ can be obtained:

$$(p+q)^1 = p+q$$

$$(p+q)^2 = p^2 + 2pq + q^2$$

$$(p+q)^3 = p^3 + 3p^2q + 3pq^2 + q^3$$

$$(p+q)^4 = p^4 + 4p^3q + 6p^2q^2 + 4pq^3 + q^4$$

$$(p+q)^5 = p^5 + 5p^4q + 10p^3q^2 + 10p^2q^3 + 5pq^4 + q^5$$

$$(p+q)^n = p^n + p^{n-1}q + p^{n-2}q^2 + \text{--------} q^n$$

The complete expansion of the binomial $(p + q)^n$ involves two steps. The first step involves the determination of various terms of expansion and the second step involves the determination of co-efficient of these terms. The first term obtained on simplification of binomial $(p + q)^n$ is $p^n q^0$ or p^n (as $q^0 = 1$). In the subsequent terms, the index of p sequentially decreases by 1 and that of q sequentially increases by 1 till it reaches $p^0 q^n$ or q^n. Simplification of $(p + q)^n$ in this way yields the following expansion:

$$(p+q)^n = p^n + p^{n-1}q + p^{n-2}q^2 \text{---------} + q^n.$$

For example $(p+q)^5 = p^5 + p^4q + p^3q^2 + p^2q^3 + pq^4 + q^5$.

In the above expansion each term has an appropriate co-efficient. The co-efficient of the term can be calculated by the following general formula:

$$\text{Co-efficient} = \frac{n!}{S!t!}$$

where n = index of binomial, S = index of p in the term and t = index of q in the term and ! = factorial, *e.g.*, $n! = 1\times2\times3\times\text{---------}\times n$, $S! = 1\times2\times3\text{------}\times S$, $t! = 1\times2\times3\times\text{-------}\times t$ (in other words, $3! = 1\times2\times3$, $4! = 1\times2\times3\times4$, $5! = 1\times2\times3\times4\times5$ and so on).

The set of terms written along with their appropriate co-efficients constitutes the complete binomial expansion.

Calculation of Probabilities using Binomial expansion

The binomial expansion can be used only in the case of two mutually exclusive events. The binomial expansion appropriate for a problem has the same index as are the number of independent events in the problem. One of the two mutually exclusive events is represented by p and the other is represented by q. Indices of p and q in a term are the same as the numbers of mutually exclusive events they represent. The probability of a given combination of mutually exclusive events is obtained by simplifying the term of binomial expansion which is appropriate for that combination. The steps for calculation of probability of occurrence of a combination of two exclusive events are as follows:

(*a*) Selection of binomial appropriate for the problem. The binomial appropriate has the same index as are the independent events in the problem.

(*b*) Expansion of binomial along with the co-efficients of the various terms of binomial.

(*c*) **Selection of appropriate term.** In the term appropriate to a combination of independent events, the index of p is equal to the number of one exclusive event in combination while that of q is equal to the number of other mutually exclusive event.

(*d*) **Simplification of selected term.** For simplification p and q are replaced by the probabilities of the events they represent.

QUESTIONS

1. What is meant by events? Describe different types of events.
2. What is the difference between mathematical and statistical probability. Discuss the multiplicative rules of probability?
3. Describe the different types of probability distribution.
4. What do you mean by Binomial distribution of probability?

33

CHAPTER

Tests of Significance

In biometrics, the size of sample affects the reliability and acceptability of the result. The larger the number of observations in an experiment, the greater shall be the reliability of result. The samples having less than 50 observations are said to be *small size sample* and those having 50 or more than 50 observations in the experiments are treated as **big size samples**. An important use of biological statistics in the field of genetics, agriculture and other biological disciplines is for testing whether a particular ratio of observations may be taken to fit an expected ratio. Let us consider a monohybrid experiment in which the ratio of dominant to recessive is expected to be 3 : 1 but actual ratio obtained from the observations of experiment is 2.84 : 1. Such deviations of observed ratios from the expected ratios are found to occur in all such experiments. It is, therefore, necessary to test whether the deviation is within the reasonable limit. Statisticians have developed several tests of hypotheses for this purpose, which are called tests of significance. The tests of significance used especially **for small size samples** are as follows:

1. t-test or student's t test
2. F-test or variance ratio test and
3. Fisher's *Z* test or *Z*-test.

For large size sample an important test of significance is used which is called X^2 (chi square) test.

Student's *t*-test or *t*-test

It is one of the simplest tests used for drawing conclusions or interpretations for small samples. This test was worked out by W.S. Gosset (pen name "Student"). *t*-test is used to test the significance of means of two samples drawn from a population, as well as the significance of difference between the mean of small sample and hypothetical mean of population (expressed in terms of standard error).

(*I*) Application of t-test for assessing the significance of difference between the sample mean and population mean

The computation of *t*-value involves the following steps:

(*i*) **Null Hypothesis.** First of all, it is presumed that there is no difference between the mean of small sample and the population mean (μ) or hypothetical mean. Thus,

Null hypothesis (Ho): sample mean $(\overline{X})$ = population mean (μ) or Ho = $\overline{X} = \mu$

(*ii*) **Test statistics.** *t*-value is calculated by the following formula:

$$t = \frac{(\overline{X} - \mu)}{S/\sqrt{n}}$$

where, $\overline{X}$ = Sample mean, μ = population mean
n = number of units in the sample
s = standard deviation of sample

For small samples

$$s = \sqrt{\frac{1}{n-1}\sum(X-\overline{X})^2} \quad \text{or} \quad \sqrt{\frac{1}{n-1}\sum d^2}$$

(*iii*) **Degree of freedom.** It is one less than the number of units in the sample.
Degree of freedom (*d. f.* or *v*) = No. of units in the sample – 1
= $n-1$

(*iv*) **Level of significance.** Any level of significance can be considered to test the hypothesis but generally 1% (=0.01) or 5% (= 0.05) levels of probability is considered for testing the hypothesis.

(*v*) **Table value of *t*.** After calculating the value of *t* (= *t* – cal) with the help of above formula, the value of *t* is noted from Fishers and Yates *t*-table at given degree of freedom and 5% level of significance. Then after the calculated value of *t* is compared with the table value of *t*.

Table 33.1 STUDENT'S *t*-DISTRIBUTION

Degrees of freedom	*Value of P* 0.10	0.05	0.02	0.01	0.002	0.001
1.	6.314	12.71	31.82	63.66	318.3	636.6
2.	2.920	4.303	6.965	9.925	22.33	31.60
3.	2.353	3.182	4.541	5.841	10.21	12.92
4.	2.132	2.776	3.747	4.604	7.173	8.610
5.	2.015	2.571	3.365	4.032	5.893	6.869
6.	1.943	2.447	3.143	3.707	5.208	5.959
7.	1.895	2.365	2.998	3.499	4.785	5.408
8.	1.860	2.306	2.896	3.355	4.501	5.041
9.	1.833	2.262	2.821	3.250	4.297	4.781
10.	1.812	2.228	2.764	3.169	4.144	4.587
11.	1.796	2.201	2.718	3.106	4.025	4.437
12.	1.782	2.179	2.681	3.055	3.930	4.318
13.	1.771	2.160	2.650	3.012	3.852	4.221
14.	1.761	2.145	2.624	2.977	3.787	4.140
15.	1.753	2.131	2.602	2.947	3.733	4.073
16.	1.746	2.120	2.583	2.921	3.686	4.015
17.	1.740	2.110	2.567	2.898	3.646	3.965
18.	1.734	2.101	2.552	2.878	3.610	3.922
19.	1.729	2.093	2.539	2.861	3.579	3.883
20.	1.725	2.086	2.528	2.845	3.552	3.850
21.	1.721	2.080	2.518	2.831	3.527	3.819
22.	1.717	2.074	2.508	2.819	3.505	3.792
23.	1.714	2.069	2.500	2.807	3.485	3.767
24.	1.711	2.064	2.492	2.797	3.467	3.745
25.	1.708	2.060	2.485	2.787	3.450	3.725
26.	1.706	2.056	2.479	2.779	3.435	3.707
27.	1.703	2.052	2.473	2.771	3.421	3.690
28.	1.701	2.048	2.467	2.763	3.408	3.674
29.	1.699	2.045	2.462	2.756	3.396	3.659
30.	1.697	2.042	2.457	2.750	3.385	3.646

(*vi*) **Test criterion.** If the calculated value of t (t-cal) irrespective of the positive (+) or negative sign (–) is less than the tabulated value of t at respective degree of freedom and at 5% or 0.05 level of probability, then the Null hypothesis is correct, *i.e.*, difference between the sample mean and population mean is insignificant and so the hypothesis is acceptable.

But, if the calculated value of t (t – cal) is greater than the tabulated value of t (t-tab) at given degree of freedom and at 5% level of probability, then the observed difference is considered statistically non-significant and so the null hypothesis is incorrect and rejected. In such a case the observed data are not according to the expected data or in other words, the sample under test does not represent the population with μ as its mean.

(*II*) ***t*-test for assessing the significance of the difference between the means of two samples drawn from the same population**

t-test is also applied to test the significance of the difference between the arithmetic means of two samples drawn from the same population. The procedure of the test is as follows:

(*i*) **Null hypothesis.** In this, first of all it is presumed that there is no difference in the standard deviations of the two samples under test, *i.e.*

$\mu_1 = \mu_2$

where μ_1 and μ_2 are the standard deviations of the sample I and sample II respectively.

(*ii*) **Test statistics.** Next, the value of t is calculated by the following formula:

$$t = \frac{(\overline{X}_1 - \overline{X}_2) - (\mu_1 - \mu_2)}{S\sqrt{\frac{1}{n_1} + \frac{1}{n_2}}}$$

where, $\overline{X}_1$ = arithmetic mean of sample I

$\overline{X}_2$ = arithmetic mean of sample II

n_1 = No. of observations in sample I

n_2 = No. of observations in sample II

S_1 = Standard deviation of sample I

S_2 = Standard deviation of sample II

Since according to hypothesis $\mu_1 = \mu_2$

$$\text{therefore,} \quad t = \frac{X_1 - X_2}{S\sqrt{\frac{1}{n_1} + \frac{1}{n_2}}} = \frac{\text{difference between the sample means}}{\text{standard error of difference between the means}}$$

$$S = \frac{\sqrt{n_1 S_1^2 - n_2 S_2^2}}{n_1 + n_2 - 2}$$

$$\text{or} \quad \frac{\sqrt{\sum(X_1 - \overline{X}_1) + \sum(X_2 - \overline{X}_2)}}{n_1 + n_2 - 2}$$

(*iii*) **Degree of freedom.** (*d.f.*) = $n_1 + n_2 - 2$.

(*iv*) **Level of significance.** The level of significance is generally considered at 5% (0.05) level of probability.

(*v*) **Tabulated value of *t*.** Value of t is recorded from the Fisher and Yates table at the given degree of freedom and at 5% level of significance.

(*vi*) **Test criterion.** At last, the calculated value of t (| t | - cal) is compared with the table value of t at the given degree of freedom and 5% level of significance.

(*a*) If the calculated value of t-exceeds the table value, the observed difference is considered statistically significant and hence null hypothesis is rejected.

(*b*) If the calculated value of t (| t | - cal) is less than the table value of t, the differences are not significant. Therefore, Ho hypothesis is accepted and the samples represent the population well.

F-test or Variance Ratio Test

F-test was first worked out by G.W. Snedecore. This is based on F distribution and is used to test the significance of difference between the standard deviations of two samples. Snedecore calculated the variance ratio of the two samples and named this ratio after R. F. Fisher. Therefore, the test is referred to as F-test or variance ratio test. The procedure for F-test is as follows:

(*i*) **Null hypothesis.** In this, it is presumed that the ratio of variance of two samples is not significant.

(*ii*) **Test statistics.** Value of F is determined by the following formula:

when $S_1^2 > S_2^2$ –

$$F = \frac{S_1^2}{S_2^2}$$

and when $S_1^2 < S_2^2$ –

$$F = \frac{S_2^2}{S_1^2}$$

Where S_1 is the standard deviation of sample I and S_2 is the standard deviation of sample II.

Value of S_1^2 and S_2^2 are calculated by the following formula:

$$S_1^2 = \frac{\sum (X_1 - \overline{X}_1)^2}{n_1 - 1} \quad \text{and} \quad S_2^2 = \frac{\sum (X_2 - \overline{X}_2)^2}{n_2 - 1}$$

Where $\overline{X}_1$ = arithmetic mean of sample I and $\overline{X}_2$ is the arithmetic mean of sample II.

n_1 = number of observations or data of sample I and n_2 = the number of observations or data for sample II.

Degree of freedom. (*d.f.*) for sample I (υ_1) = $n_1 - 1$ and d.f. for sample II (υ_2) = $n_2 - 1$

The level of significance: Generally the level of significance is considered either at 1% (0.01) or 5% (0.05).

The tabulated value of *F*. The value of F for a sample with greater spread at given *d.f.* is located in the Table from left to right and for the other sample with low spread at respective degree of freedom is located in the same table from top downward. Where the two meet each other that value of given level of significance is F-value.

Now if the calculated value of F is less than the tabulated value of F then null hypothesis is true and accepted, *i.e.*, the difference between the standard deviations of two samples is not significant and so the two samples under test might have been drawn from the same population.

Table 33.2(*a*). 1 Per cent Points of Variance-Ratio (F) Distribution

$F_2\backslash F_1$	1	2	3	4	5	6	7	8	9	10	12	15	20	30	00
1.	4052	4999	5403	5625	5764	5859	5929	5982	6022	6056	6106	6157	6209	6261	6366
2.	98.50	99.00	99.17	99.25	99.30	99.33	99.36	99.37	99.39	99.40	99.42	99.43	99.45	99.47	99.50
3.	34.12	30.82	29.46	28.71	28.24	27.91	27.67	27.49	27.35	27.23	27.05	26.87	26.69	26.50	26.13
4.	21.20	18.00	16.69	15.98	15.52	15.21	14.98	14.80	14.66	14.55	14.37	14.20	14.02	13.84	13.46
5.	16.26	13.27	12.06	11.39	10.97	10.67	10.46	10.29	10.16	10.05	9.89	9.72	9.55	9.38	9.02
6.	13.75	10.92	9.78	9.15	8.75	8.47	8.26	8.10	7.98	7.87	7.72	7.56	7.40	7.23	6.88
7.	12.25	9.55	8.45	7.85	7.46	7.19	6.99	6.84	6.72	6.62	6.47	6.31	6.16	5.99	5.65
8.	11.26	8.65	7.59	7.01	6.63	6.37	6.18	6.03	5.91	5.81	5.56	5.52	5.36	5.20	5.86
9.	10.56	8.02	6.99	6.42	6.06	5.80	5.61	5.47	5.35	5.26	5.11	4.96	4.81	4.65	4.31
10.	10.04	7.56	6.55	5.99	5.64	5.39	5.20	5.06	4.94	4.85	4.71	4.56	4.41	4.25	3.91
11.	9.65	7.21	6.22	5.67	5.32	5.07	4.89	4.74	4.63	4.54	4.40	4.25	4.10	3.94	3.60
12.	9.33	6.93	5.95	5.41	5.06	4.82	4.64	4.50	4.39	4.30	4.16	4.01	3.86	3.70	3.36
13.	9.07	6.70	5.74	5.21	4.86	4.62	4.44	4.30	4.19	4.10	3.96	3.82	3.66	3.51	3.17
14.	8.86	6.51	5.56	5.04	4.69	4.46	4.28	4.14	4.03	3.94	3.80	3.66	3.51	3.35	3.00
15.	8.68	6.36	5.42	4.89	4.56	4.32	4.14	4.00	3.89	3.80	3.67	3.52	3.37	3.21	2.87
16.	8.53	6.23	5.29	4.77	4.44	4.20	4.03	3.89	3.78	3.69	3.55	3.41	3.26	3.10	2.75
17.	8.40	6.11	5.18	4.67	4.34	4.10	3.93	3.79	3.68	3.59	3.46	3.31	3.16	3.00	2.65
18.	8.29	6.01	5.09	4.58	4.25	4.01	3.84	3.71	3.60	3.51	3.37	3.23	3.08	2.92	2.57
19.	8.18	5.93	5.01	4.50	4.17	3.94	3.77	3.63	3.52	3.43	3.30	3.15	3.00	2.84	2.49
20.	8.10	5.85	4.94	4.43	4.10	3.87	3.70	3.56	3.46	3.37	3.23	3.09	2.94	2.78	2.42
21.	8.02	5.78	4.87	4.37	4.04	3.81	3.64	3.51	3.40	3.31	3.17	3.03	2.88	2.72	2.36
22.	7.95	5.72	4.82	4.31	3.99	3.76	3.59	3.45	3.35	3.26	3.12	2.98	2.83	2.67	2.31
23.	7.88	5.66	4.76	4.26	4.94	3.71	3.54	3.41	3.30	3.21	3.07	2.93	2.78	2.62	2.26
24.	7.82	5.61	4.72	4.22	3.90	3.67	3.50	3.36	3.26	3.17	3.03	2.89	2.74	2.58	2.21
25.	7.77	5.57	4.68	4.18	3.85	3.63	3.46	3.32	3.22	3.13	2.99	2.85	2.70	2.54	2.17
26.	7.72	5.53	4.64	4.14	3.82	3.59	3.42	3.29	3.18	3.09	2.96	2.81	2.66	2.50	2.13
27.	7.68	5.49	4.60	4.11	3.78	3.56	3.39	3.26	3.15	3.06	2.93	2.78	2.63	2.47	2.10
28.	7.64	5.45	4.57	4.07	3.75	3.53	3.36	3.23	3.12	3.03	2.90	2.75	2.60	2.44	2.06
29.	7.60	5.42	4.54	4.04	3.73	3.50	3.33	3.20	3.09	3.00	2.87	2.73	2.57	2.41	2.03
30.	7.56	5.39	4.51	4.02	3.70	3.47	3.30	3.17	3.07	2.98	2.84	2.70	2.55	2.39	2.01
40.	7.31	5.18	4.31	3.83	3.51	3.29	3.12	2.99	2.89	2.80	2.66	2.52	2.37	2.20	1.80
60.	7.08	4.98	4.13	3.65	3.34	3.12	2.95	2.82	2.72	2.63	2.50	2.35	2.20	2.03	1.60
120.	6.85	4.79	3.95	3.48	3.17	2.96	2.79	2.66	2.56	2.47	2.34	2.19	2.03	1.86	1.38
00	6.63	4.61	3.78	3.32	3.02	2.80	2.64	2.51	2.41	2.32	2.18	2.04	1.88	1.70	1.00

If the calculated value of F is, however, greater than the table value of F (F - cal > F - tab.) the null hypothesis is rejected and the difference between the standard deviations is significant which means that the two samples under test cannot be supposed to be parts of the same population.

Example. Prove that there is no significant difference between the standard deviations of the following two samples and they are drawn from the same population:

	Sample I	*Sample II*
No. of observed data	12	10
Standard deviation	5.34	3.65

Solution. Standard deviation of sample I is greater than that of sample II. Here $n_1 = 12$, $S_1 = 5 = 34$ and $n_2 = 10$, $S_2 = 3.65$

Table 33.2(*b*). 5 Per cent Points of Variance-Ratio (F) Distribution

$F_2\backslash F_1$	1	2	3	4	5	6	7	8	9	10	12	15	20	30	00
1.	161.4	199.5	215.7	224.6	230.2	234.0	236.8	238.9	240.5	241.9	243.9	245.9	248.0	250.1	2543
2.	18.51	19.00	19.61	19.25	19.30	19.33	19.35	19.37	19.40	19.40	19.41	19.43	19.45	19.46	19.50
3.	10.13	9.55	9.28	9.12	9.01	8.94	8.89	8.85	8.81	8.79	8.74	8.70	8.66	8.62	8.53
4.	7.71	6.94	6.59	6.29	6.26	6.16	6.09	6.04	6.00	5.96	6.91	5.86	5.80	5.75	5.63
5.	6.61	5.79	5.41	5.19	5.05	4.95	4.88	4.82	4.77	4.74	4.68	4.62	4.56	4.50	4.36
6.	5.99	5.14	4.76	4.53	4.39	4.28	4.21	4.15	4.10	4.06	4.00	3.94	3.87	3.81	3.67
7.	5.59	4.74	4.35	4.12	3.97	3.87	3.79	3.73	3.68	3.64	3.57	3.51	3.44	3.38	3.23
8.	5.32	4.46	4.07	3.84	3.69	3.58	3.50	3.44	3.39	3.35	3.28	3.22	3.15	3.08	2.93
9.	5.12	4.26	3.86	3.63	3.48	3.37	3.29	3.23	3.18	3.14	3.07	3.01	2.94	2.86	2.71
10.	4.96	4.10	3.71	3.48	3.33	3.22	3.14	3.07	3.02	2.98	2.91	2.85	2.77	2.70	2.54
11.	4.84	3.98	3.59	3.36	3.20	3.09	3.01	2.95	2.90	2.85	2.79	2.72	2.65	2.57	2.40
12.	4.75	3.89	3.49	3.26	3.11	3.00	2.91	2.85	2.80	2.75	2.69	2.62	2.54	2.47	2.30
13.	4.67	3.81	3.41	3.18	3.03	2.92	2.83	2.77	2.71	2.67	2.60	2.53	2.46	2.38	2.21
14.	4.60	3.74	3.34	3.11	2.96	2.85	2.76	2.70	2.65	2.60	2.53	2.46	2.39	2.31	2.13
15.	4.54	3.68	3.29	3.06	2.90	2.79	2.71	2.64	2.59	2.54	2.48	2.40	2.33	2.25	2.07
16.	4.49	3.63	3.24	3.01	2.85	2.74	2.66	2.59	2.54	2.49	2.42	2.35	2.28	2.19	2.01
17.	4.45	3.59	3.20	2.96	2.81	2.70	2.61	2.55	2.49	2.45	2.38	2.31	2.23	2.15	1.96
18.	4.41	3.55	3.16	2.93	2.77	2.66	2.58	2.51	2.46	2.41	2.34	2.27	2.19	2.11	1.92
19.	4.38	3.52	3.13	2.90	2.74	2.63	2.54	2.48	2.42	2.38	2.31	2.23	2.16	2.07	1.88
20.	4.35	3.49	3.10	2.87	2.71	2.60	2.51	2.45	2.39	2.35	2.28	2.20	2.12	2.04	1.84
21.	4.32	3.47	3.07	2.84	2.68	2.57	2.49	2.42	2.37	2.32	2.25	2.18	2.10	2.01	1.81
22.	4.30	3.44	3.05	2.82	2.66	2.55	2.46	2.40	2.34	2.30	2.23	2.15	2.07	1.98	1.78
23.	4.28	3.42	3.03	2.80	2.64	2.53	2.44	2.37	2.32	2.27	2.20	2.13	2.05	1.96	1.76
24.	4.26	3.40	3.01	2.78	2.62	2.51	2.42	2.36	2.30	2.25	2.18	2.11	2.03	1.94	1.73
25.	4.24	3.39	2.99	2.76	2.60	2.49	2.40	2.34	2.28	2.24	2.16	2.09	2.01	1.92	1.71
26.	4.23	3.37	2.98	2.74	2.59	2.47	2.39	2.32	2.27	2.22	2.15	2.07	1.99	1.90	1.69
27.	4.21	3.35	2.96	2.73	2.57	2.46	2.37	2.31	2.25	2.20	2.13	2.06	1.97	1.88	1.67
28.	4.20	3.34	2.95	2.71	2.56	2.45	2.36	2.29	2.24	2.19	2.12	2.04	1.96	1.87	1.65
29.	4.18	3.33	2.93	2.70	2.55	2.43	2.35	2.28	2.22	2.18	2.10	2.03	1.94	1.85	1.64
30.	4.17	3.32	2.92	2.69	2.53	2.42	2.33	2.27	2.21	2.16	2.09	2.01	1.93	1.84	1.62
40.	4.08	3.23	2.84	2.61	2.45	2.34	2.25	2.18	2.12	2.08	2.00	1.92	1.84	1.74	1.51
60.	4.00	3.15	2.76	2.53	2.37	2.25	2.17	2.10	2.04	1.99	1.92	1.84	1.75	1.65	1.39
120.	3.92	3.07	2.68	2.45	2.29	2.17	2.09	2.02	1.96	1.91	1.83	1.75	1.66	1.55	1.25
00	3.84	3.00	2.60	2.37	2.21	2.10	2.01	1.94	1.88	1.83	1.75	1.67	1.57	1.46	1.00

Null Hypothesis. Here it is presumed that there is no difference between the standard deviations of sample I and sample II

$$F = \frac{S_1^2}{S_2^2} = \frac{(5.34)^2}{(3.65)^2} = \frac{28.5156}{13.3225} = 2.14$$

d.f. for sample I = $n_1 - 1 = 12 - 1 = 11$

d.f. for sample II = $n_2 - 1 = 10 - 1 = 9$

Suppose that the considered level of significance is 5% = 0.05.

Therefore, the tabulated values of F at 11 and 9 degrees of freedom and 5% level of significance is between 3.14 and 3.07 $= \frac{6.21}{2} = 3.10$.

So the calculated value of F (= 2.14) is less than the tabulated value of F(= 3.10) at 11 and 9 degrees of freedom and 5% level of significance. So the Null hypothesis is accepted, *i.e.*, there is no significant difference between the standard deviations of the two samples under test and they might have been drawn from the same population.

Example. The data regarding the effects of two nutrients A and B on the increase in height of plants after 10 days of application have been presented in the following table. For nutrient A a sample of 12 plants and for nutrient B a sample of 10 plants were considered. With the help of F-test prove that there is no significant difference in the variances of the two samples at 5% level of significance.

Nutrient	*Increase in height (in cm) after 10 days of application*											
Nutrient A	22	13	7	15	12	14	18	8	23	21	17	10
Nutrient B	16	10	6	17	13	12	08	14	15	9		

Solution. Table for calculation of variances.

	Sample for Nutrient A			*Sample for Nutrient B*		
S. No	*Height increase* X_1	$d_1(X_1 - \overline{X}_1)$	d_1^2	*Height increase* X_2	$d_2(X_2 - \overline{X}_2)$	d_2^2
1.	22	22 – 15 = 7	49	16	16 – 12 = 4	16
2.	13	13 – 15 = – 2	4	10	10 – 12 = – 2	4
3.	7	7 – 15 = – 8	64	6	6 – 12 = – 6	36
4.	15	15 – 15 = 0	0	17	17 – 12 = 5	25
5.	12	12 – 15 = – 3	9	13	13 – 12 = – 1	1
6.	14	14 – 15 = – 1	1	12	12 – 12 = 0	0
7.	18	18 – 15 = 3	9	8	8 – 12 = – 4	16
8.	8	8 – 15 = – 7	49	14	14 – 12 = 2	4
9.	23	23 – 15 = 8	64	15	15 – 12 = 3	9
10.	21	21 – 15 = 6	36	9	9 – 12 = – 3	9
11.	17	17 – 15 = 2	4			
12.	10	10 – 15 = – 5	25			
	$\Sigma X_1 = 180$		314	$\Sigma X2 = 120$		120
	$\overline{X}_1 = \dfrac{180}{12} = 15$			$\overline{X}_2 = \dfrac{120}{10} = 12$		

Null hypothesis. It is presumed that there is no difference in the variances of the samples for nutrient A and B.

$$\text{Variance of sample for nutrient } A \ (S_1)^2 = \frac{\sum d_1^2}{n_1 - 1}$$

$$= \frac{314}{12-1} = \frac{314}{11} = 28.55$$

$$\text{Variance of sample for nutrient } B \ (S_2)^2 = \frac{\sum d_2^2}{n_2 - 1}$$

$$= \frac{120}{10-1} = \frac{120}{9} = 13.33$$

When $S_1^2 > S_2^2$,

$$F = \frac{S_1^2}{S_2^2} = \frac{28.55}{13.33} = 2.14$$

Degree of freedom for $A = n_1 - 1 = 12 - 1 = 11$

Degree of freedom for $B = n_2 - 1 = 10 - 1 = 9$

Level of significance 5% or 0.05

Now, the tabulated value of F at 5% level of significance and 11 and 9 degree of freedom $= \frac{3.14 + 3.07}{2} = 3.10.$

When the calculated value of F (= 2.14) is compared with the tabulated value of F (= 3.10) at 10 and 9 degrees of freedom and 5% level of significance, it is apparent that F-cal is less than F-tab. Therefore, there is no significant difference in the variances of the samples for nutrient A and B at 5% level of significance. Hence, null hypothesis is correct and accepted and the two samples appear to be drawn from the same population.

Fisher's Z-test or Z-test

Z-test is based on the normal probability distribution and is used for testing the significance of several measures. The relevant test statistics is worked out and compared with its probable value (to be read from the table showing the area under normal curve) at a given level of significance in order to judge the significance of measures concerned. Z-test is generally used to compare the mean of large sample to some hypothetical mean for population.

$$Z = \frac{\overline{X} - \mu}{\sigma / \sqrt{n}}$$

where, $\overline{X}$ = Sample mean

μ = Hypothetical mean of population

σ = Standard deviation

n = Total number of observations

X^2-Test (Chi-Square Test)

X^2 square test (named after Greek letter *x* pronounced as *ki*) is a statistical method of testing significance which was worked out by Karl Pearson. Any biological study is based on a limited number of individuals which constitute a sample. A sample is a small part of population at random. If a sample is drawn from population at random, each individual of the population is given equal opportunity to be included in sample and so the properties of samples will reflect the properties of population of which they are the part. Thus, the various statistics like mean, ratios, variance calculated from a random sample are estimates of those of parent population. The mean ratios variance of sample may be close but seldom equal to those of parent population.

In a cross between tall and dwarf pea plants Mendel found 787 tall and 277 dwarf plants out of 1064 plants in F_2 generation that yielded a ratio of 2.84 tall: 1 dwarf which was close to but numerically slight different from the expected ratio 3 Tall : 1 Dwarf from the imaginary population. Such deviations are bound to occur in all biological situations like this. Thus, there should be some objective criteria to decide if a set of observed data is in accordance with specified or expected ratio. In other words, it is applied to test goodness of fit of the frequency of observed data with the expected or specified ratio. X^2-test is generally applied to enumeration data and rarely to measurement data and it involves the following steps:

1. **Formation of a Null hypothesis.** In this, it is presumed that the observed data are not different from the specified or expected ratio, *i.e.*, the deviation of observed data from the expected ratio is not real but is due to chance only.
2. **Test statistics or computation of value of X^2.** After the formation of null hypothesis, the deviation of observed data from the expected frequencies of different classes of data will be considered due to chance only. Calculation of the value of X^2 from observed data is made in the following way:

(*i*) The observed frequencies of different classes are arranged sequencially *e.g.*, Tall 287 and Dwarf 277 in F_2 generation of Tall × Dwarf.

(*ii*) *The expected frequencies* of different classes are calculated on the basis of expected ratio (as stated in the null hypothesis) and the total number of observed values. For example, the expected ratio of Tall to Dwarf is 3 : 1 and the total number of observations is 1064 plants. This ratio specifies that out of every 4 (3 + 1), 3 (*i.e.*, 3/4) will be tall and one (*i.e.*, 1/4) will be dwarf plants. Therefore, out of total 1064 plants, three-fourth (or 1064 × 3/4) = 798 should be tall and one-fourth (*i.e.*, 1064 × 1/4) = 266 should be dwarf plants.

Likewise in a dihybrid experiment involving Tall and red flowered × Dwarf, white flowered, the expected ratio for F_2 plants is Tall Red: Tall, White: Dwarf, Red : Dwarf White :: 9 : 3 : 3 : 1. If the total number of plants in F_2 generation is 556, the distribution of frequencies for different classes will be calculated in the following way:

Class	*Ratio*	*Expected Frequency*
Tall, Red	9 or 9/16	9/16 of 556 = 556 × 9/16 = 312.75
Tall White	3 or 3/16	3/16 of 556 = 556 × 3/16 = 104.25
Dwarf Red	3 or 3/16	3/16 of 556 = 556 × 3/16 = 104.25
Dwarf White	1 or 1/16	1/16 of 556 = 556 × 1/16 = 34.75
Total = 9 + 3 + 3 + 1 =16		

(*iii*) In the next step deviation of observed frequency from the expected frequency of each class is determined by subtracting the expected frequency from the observed frequency of the respective class (deviation = observed frequency – Expected frequency or $d = o - e$). The value of deviation may have either positive (+) or negative (–) sign. Thus, in the above example of monohybrid cross, the deviation value for tall plants is 787 – 798 = – 11 and that for Dwarf will be 277 –266 = +11.

(*iv*) Now the deviation value (*d*) of each class is squared to get all d^2 values in positive.

(*v*) The importance of a deviation depends on its magnitude as well as on the value of expected frequencies from which the deviation has occurred. Suppose that the magnitudes of deviations in two classes of data are equal, say 11 in monohybrid cross Tall × Dwarf but the expected frequencies of the two classes are different say 798 for one class (Tall) and 266 for the other class (Dwarf). Thus there is disparity in the importance of deviations due to unequal expected frequencies of different classes of data. In order to remove this disparity, the deviation squares are divided by expected frequencies of the respective classes $\frac{(o-e)^2}{e}$. All the numbers (quotients) so obtained are based on the expected frequency of one and are, therefore, equal in importance.

(*vi*) **X^2 value.** The sum (total) of all quotients of different classes yields the calculated value of *chi square. Thus Chi Square may be defined as the total of the quotients obtained by dividing the deviation squares of different classes with their respective expected frequencies.* This is expressed as follows:

$$X^2 = \sum \frac{(o-e)^2}{e} \text{ or } \sum \frac{d^2}{e}$$

where, Σ = summation

o = observed frequency of the class

e = expected frequency of the class

d = deviation or difference between the observed value and expected frequency value

(*vii*) **Degree of freedom.** It is one less than the total number of classes and is expressed as $d.f. = n - 1$ where n = No. of classes.

Table 33.3 The X^2 Distribution

Degrees of freedom	*Accept* 0.99	← 0.95	*Value of P* 0.05	→ 0.01	*Reject* 0.001
1.	0.000157	0.00393	3.841	6.635	10.83
2.	0.0201	0.103	5.991	9.210	13.82
3.	0.115	0.352	7.815	11.34	16.27
4.	0.297	0.711	9.488	13.28	18.47
5.	0.554	1.145	11.07	15.09	20.51
6.	0.872	1.635	12.59	16.81	22.46
7.	1.239	2.167	14.07	18.48	24.32
8.	1.646	2.733	15.51	20.09	26.13
9.	2.088	3.325	16.92	21.67	27.88
10.	2.558	3.940	18.31	23.21	29.59
11.	3.053	4.575	19.68	24.72	31.26
12.	3.571	5.226	21.03	26.22	32.91
13.	4.107	5.892	22.36	27.69	34.53
14.	4.660	6.571	23.68	29.14	36.12
15.	5.229	7.261	25.00	30.58	37.70
16.	5.812	7.962	26.30	32.00	39.25
17.	6.408	8.672	27.59	33.41	40.79
18.	7.015	9.390	28.87	34.81	42.31
19.	7.633	10.12	30.14	36.19	43.82
20.	8.260	10.85	31.41	37.57	45.31
21.	8.897	11.59	32.67	38.93	46.80
22.	9.542	12.34	33.92	40.29	48.27
23.	10.20	13.09	35.17	41.64	49.73
24.	10.86	13.85	36.42	42.98	51.18
25.	11.52	14.61	37.65	44.31	52.62
26.	12.20	15.38	38.89	45.64	54.05
27.	12.88	16.15	40.11	46.96	55.48
28.	13.56	16.93	41.34	48.28	56.89
29.	14.26	17.71	42.56	49.59	58.30
30.	14.95	18.49	43.77	50.89	59.70

(*viii*) **Table value of X^2.** It is obtained from X^2-table (Table 3) at given degree of freedom and 1% (0.01) or 5% (0.05) level of probability. In Chi Square table maximum values of Chi Square at different probability levels and at different degrees of freedom obtained purely due to chance are listed which serve as points of reference while deciding whether the calculated value of X^2 is due to real or chance deviation. The value of X^2 depends on the variables, degree of freedom and probability.

(*ix*) **Level of significance.** In X^2 table probability is listed in top most row and is designated by *P*. The maximum values of X^2 at different probability levels are recorded against different degrees of freedom.

The value of Chi Square as given in the X^2 table against 1 degree of freedom at 5% (0.05) probability level is 3.84. This means that out of a large number of similar experiments with one degree of freedom, 5% or 0.05 experiments will show a X^2 value 3.84 or less than that only due to chance. Consequently in 0.95 or 95% of experiments X^2 value 3.84 or of lower magnitudes will be due to real deviation of observed data from the expected data.

(*x*) **Test of Hypothesis.** After determining the X^2 value from observed data, the calculated value of X^2 is compared with table value of X^2. There are two ways in which a conclusion may be drawn regarding the validity of null hypothesis:

(*a*) If the calculated value of X^2 is less than the table value at 5% probability against given degree of freedom, the deviations of observed frequencies from the expected frequencies are not significant and are accepted to be purely due to chance. Therefore, null hypothesis is accepted and it is inferred that the observed data are in accordance with the expected ratio.

(*b*) If the calculated value of X^2 is equal or greater than the table value of X^2 at 5% level of probability and given degree of freedom, then the deviations of observed data from expected value are statistically significant. In such a case null hypothesis is rejected and it is concluded that observed data are not in accordance with the expected ratio.

The following example will clarify the procedure of X^2-test.

Example. In a cross between tall and dwarf plants Mendel found 787 tall and 277 dwarf plants in F_2 generation. Using X^2-test find out the goodness of fit.

Solution. *Null hypothesis:* It is presumed that observed data are in ratio of 3 Tall: 1 Dwarf in F_2 generation.

	Description	*Classes of Plants in F_2*		*Total*
		Tall	*Dwarf*	
1.	Observed frequencies (*o*)	787	277	1064
2.	Expected frequency according to 3 : 1 ratio	1064 × 3/4 = 798	1064 × 1/4 = 266	1064
3.	Deviation ($o - e$)	787 – 798 = –11	277 – 266 = 11	00
4.	Square of deviations	$(-11)^2 = 121$	$(11)^2 = 121$	
5.	Squares of deviation divided by expected value or $\frac{(o-e)^2}{e}$	$\frac{121}{798} = 0.152$	$\frac{121}{266} = 0.455$	$\Sigma\frac{(o-e)^2}{e} = 0.607$
6.	Calculated X^2-Value	$\Sigma\frac{(o-e)^2}{e} = 0.607$		

Degree of freedom = number of classes – 1

= n – 1 = 2 – 1 = 01

Considered level of significance: 5% or 0.05.

Table value of X^2 at 1 degree of freedom and 5% level of probability obtained from X^2 table is 3.84.

Test of significance. On comparing the calculated value of X^2 with table value of X^2, it is clear that the calculated value (0.607) is less than the table value (3.84). So the deviation of observed data from the expected ratio is purely due to chance. Hence the null hypothesis is accepted and it is concluded that the observed data are in accordance with the expected ratio, *i.e.*, they show goodness of fit with the expected ratio.

Example. In a dihybrid cross between pea plants with round seeds and yellow cotyledons and those with wrinkled seeds and green cotyledons Mendel found 35 round, yellow; 108 round, green, 101 wrinkled, yellow and 32 wrinkled, green seeds in F_2 generation. Find out the goodness of fit of observed data with expected F_2 ratio 9 : 3 : 3 : 1 by X^2-test.

Solution.

Null hypothesis. It is presumed that the observed data are in accordance with the expected F_2 ratio 9 : 3 : 3 : 1.

	Classes of F_2 seeds				
Description	*Round, Yellow*	*Round, Green*	*Wrinkled, Yellow*	*Wrinkled, Green*	*Total*
Observed frequency (*o*)	315	108	101	32	556
Expected frequency (*e*) according to 9 : 3 : 3 : 1 ratio	$556\times\frac{9}{16}$ = 312.75	$556\times\frac{3}{16}$ = 104.25	$556\times\frac{3}{16}$ = 104. 25	$556\times\frac{1}{16}$ = 34.75	556
Deviation (*d*) = (*o* – *e*)	315 – 312.75 = 2.25	108 – 104.25 = 3.75	101 – 104.25 = – 3.25	32 –34.75 = – 2.75	0
Square of deviation	$(2.25)^2$ = 5.06	$(3.75)^2$ = 14.06	$(-3.25)^2$ = 10.56	(– 2.75)2 = 7.56	
$\frac{d^2}{e}=\frac{(o-e)^2}{e}$	$\frac{5.06}{312.75}$ = 0.016	$\frac{14.6}{104.25}$ = 0.135	$\frac{10.56}{104.25}$ = 0.101	$\frac{7.56}{34.75}$ = 0.217	0.469
$X^2=\Sigma\frac{(o-e)^2}{e}$	= 0.469				

Degree of freedom. *d.f.* = number of classes – 1 = *n* – 1

d.f. = 4 – 1 = 3

Considered level of probability. 5% or 0.05.

Table value of X^2. The value of X^2 recorded in X^2 table (Table 3) at 3 *d.f.* and 5% probability level is 7.815.

Test of significance. The calculated value of X^2 from the given data is less than the table value of X^2 at 3 *d.f.* and 0.05*P* indicating that the deviations are due to chance only and, therefore, null hypothesis is accepted and it is concluded that the data are in agreement with the expected ratio 9 : 3 : 3 : 1, *i.e.*, observed F_2 data show goodness of fit with the expected F_2 ratio 9 : 3 : 3 : 1.

Yates Correction

In the cases where only two classes exist and the frequency of one of the two classes is less than 50, Yates correction is applied to increase the precision of X^2-test. Yates correction is made by deducting or subtracting 0.5 from the deviations of the two classes. In the subtraction process the + or – sign of the deviation value is ignored. Thus the corrected deviations are 0.5 less than the actual deviation values. The subsequent steps in calculation of X^2 from the corrected deviations and drawing conclusions are the same as discussed earlier. The following example will illustrate the procedure of X^2 test involving Yates correction.

Example. In a cross between round seeds and wrinkled seeds, Mendel found 165 round seeds and 35 wrinkled seeds in F_2 generation. State:

(*i*) What is the expected ratio in this case?
(*ii*) Null hypothesis.
(*iii*) Degree of freedom.
(*iv*) Correction required in this case.
(*v*) Carry out X^2-test to ascertain if the observations agree with the expected ratio.

Solution.

1. The data presented in this problem indicate that the cross is of monohybrid nature and the expected in F_2 generation is 3 : 1.
2. **Null hypothesis.** It is presumed that the observed F_2 data are in agreement with expected F_2 ratio 3 : 1.
3. **Degree of freedom (d.f.)** = number of classes – 1
   = $n - 1$ or $2 - 1 = 01$
4. Since the frequency of one of the two classes, *i.e.*, Wrinkled seeds is 35, Yates correction could be applied in this case.
5. Calculation of X^2 value from the observed data after Yates correction.

Description	*F^2 classes*		*Total*
	Round Seeds	*Wrinkled Seeds*	
1. Observed frequency (o)	165	35	200
2. Expected frequency (e) according to 3 : 1 ratio	$200 \times 3/4 = 150$	$200 \times 1/4 = 50$	200
3. Deviation (d) = ($o - e$)	$165 - 150 = +15$	$35 - 50 = -15$	0
4. Corrected deviation (0.5 less from original deviation)	$15 - 0.5 = 14.5$	$15 - 0.5 = -14.5$ (– sign of original deviation ignored)	
5. Square of corrected deviation	$(14.5)^2 = 210.25$	$(-14.5)^2 = 210.25$	
6. Square of corrected deviation divided by expected frequency	$\frac{210.25}{150} = 1.416$	$\frac{210.25}{50} = 4.205$	5.621
7. $X^2 = \Sigma \frac{(o-e)^2}{e} = 5.621$			

Considered level of probability 5% or 0.05 P. The value of X^2 recorded in X^2 table at 1 *d.f.* and 0.05 *P* level is 3.84.

Since the calculated value of X^2 from the observed F_2 data is 5.621 which is more than the table value of X^2 recorded at 1 *d.f.* and 5% level of probability. This indicates that the deviations of observed data from expected frequencies are real and significant. Therefore, the null hypothesis is rejected and it is concluded that observed data are not in agreement with the expected F_2 ratio 3 : 1.

QUESTIONS

1. What is student's t-test? How is t-test applied in assessing the significance of the difference between the means of two samples drawn from same population?
2. Discuss F-test and Z-test.
3. Discuss the various statistical tests for determining the test of significance.
4. Define deviation and test of significance.
5. Calculate t-test for the variates 3, 2, 8, 3, 7, 5, 4, 6, 4.
6. In a cross between Tall (TT) and Dwarf (tt), 1574 tall and 554 dwarf plants were obtained in F_2. Suggest if 3 : 1 ratio is suitable or not.
7. In a cross between black male and grey female *Drosphila*, the offsprings obtained were 25 black, 35 grey. Calculate X^2 and give your inference on the ratio of black to grey offsprings 1 : 1.
8. There are two factors X and Y apparently showing dominance. Suppose in F_2, the progeny were $XX = 494$, $XY = 162$, $xY = 106$, $xy = 54$ out of 816 individuals. Test the hypothesis that F_2 segregation ratio is 9 : 3 : 3 : 1.
9. What is X^2 test? Mention the hypothesis on which it is based.
10. Calculate X^2 value for two samples 15 : 25 and 240 : 260 on the basis of 1 : 1 hyphosis (at $d.f. = 1$. X^2 value of 0.05 P is 3.84).

34 CHAPTER

Correlation

Sometimes two or more events are interrelated, *i.e.,* any change in an event may affect the other events. If such changes are expressed in the form of numerical data and they appear to be interdependent they are said to be correlated. For example, the weight of human body increases with the increase in height and age. Here the age and body weight are two separate characters but they are interdependent, so they are correlated. Thus, the purpose of correlation is to study the relationship between the data of one series and those of the other series. If two or more variables are so related that an increase in one variable may cause change in the other variables they are said to be correlated. According to Bowley, when two quantities are so related that the fluctuations in one are in sympathy with the fluctuations in the other so that an increase or decrease in one is found in connection with an increase or decrease or inversely of the other and the greater the magnitude of the change in one, the greater the magnitude of the change in other, the two quantities are said to be correlated.

Types of Correlation

Depending upon the direction and proportion of changes in the variables and the number of data series, the correlation may be of the following types:

1. Positive and negative correlations.
2. Linear and non-linear correlations.
3. Simple, multiple and partial correlations.

1. Positive and Negative Correlations. When the changes take place in same direction in two variables or data series, the correlation between them is said to be positive and direct. For example, if increase in one variable causes increase in the other variable or a decrease in one variable causes decrease in the other variable, the two variables show **positive correlation**. But when the changes in two variables occur in opposite directions, the correlation is said to be **inverse** or **indirect** or **negative**, *i.e.,* if increase in one variable may cause decrease in the other or vice-versa, the two variables show negative or inverse correlation.

When the increase or decrease in the value of one variable does not affect the other variable, the correlation is said to be **zero**. Sometime two variables show positive correlation with each other but later they show negative correlation. Then the correlation between two such variables is said to be **curvilinear**.

2. Linear and Non-linear Correlations. When the proportion of changes in the two variables or data series is fixed, *i.e.,* the dispersions in two series are equal, the correlation is said to be **linear**. If two such variables are plotted on $O - X$ and $O - Y$ axes of graph paper, the changes in two will result in straight line graph. Contrary to it, if the change rates are not fixed, the correlation graph will be a curved line and this type of correlation is called **non-linear** or curvilinear.

3. **Simple, Multiple and Partial Correlation.** The correlation existing between two variables or data series is said to be **simple correlation**. Of the two series, one which causes change in the other is called **independent** or **subject series** and the other which is affected is called **dependent series**.

When the correlation involves three or more variables or data series that is called **multiple correlation**. Out of several correlated series, one is *dependent series* and the remaining two or more series are *independent* or *subject series* which jointly affect the dependent series.

When the correlation exists between two or more variables at a time, the correlation is considered between any two of them keeping the interest of other series constant, such a correlation is called **partial correlation**. For example, there is a multiple correlation among seed production, number of fruits per plant and weight of seeds where the fruit number per plant and the weight of seeds, both jointly affect the seed production. If the correlation is studied only between seed, production and the number of fruits per plant keeping the seed weight constant, this correlation is said to be partial correlation.

Degree of Correlation

The degree of correlation is expressed by the value of coefficient of correlation which ranges between + 1 and – 1. The value of coefficient of correlation directly indicates the degree of correlation as detailed in the following table:

Value of Correlation Coefficient	*Degree of Correlation*
+ 1 value	Perfect positive correlation
– 1 value	Perfect negative correlation
Value between 1 and 0.75	High degree of correlation
Value between 0.75 and 0.5	Moderate degree of correlation
Value between 0.5 and 0	Low degree of correlation
0 (zero)	No correlation

Method of Determining Correlation

The following methods are generally used to determine simple correlation:

1. Graphic method.
2. Scatter diagram or Dotogram method.
3. Karl Pearson's method.
4. Spearman's ranking method.
5. Coefficient of correlation by concurrent deviation.

1. **Graphic Method.** When the values of dependent series are plotted on $O-X$ axis and independent series are plotted on $O-Y$ axis of graph paper, a linear or non-linear graph will be obtained which will simply indicate the direction of correlation and not the numerical magnitude. If the graph lines of two independent series move in upward direction from left to right, the correlation is **positive**, but if the graph line of one series moves upward from left to right and that of the other independent series moves downward from left to right, they show **negative** correlations. If the values of two data series do not show either positive or negative trend then it should be inferred that there is no correlation.

2. **Scatter Diagram or Dotogram Method.** This method is more or less similar to graphic method. In this method, the values of independent data series are plotted on $O-X$ axis and those of dependent series on $O-Y$ axis and then the pairs of values are plotted on the graph paper. In this way, graphs of dots are obtained. These dots are scattered in different forms. Therefore, the graphs are called scatter diagrams or dotograms. The patterns of scatter diagrams indicate the direction and magnitude of correlation. The scatter diagram may indicate the following conditions:

   (*i*) If the dots of the two series are advancing in a definite direction like a current, this condition indicates that the data series are definitely correlated.

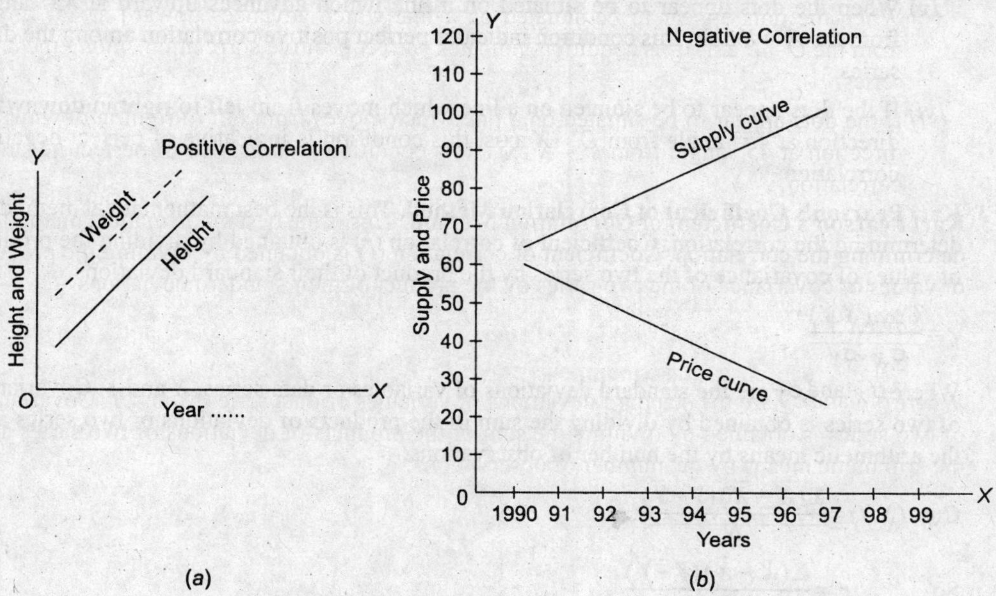

Fig. 34.1 (*a*) Positive correlation (*b*) negative correlation.

(*ii*) When the arrays of dots advance from left to right in upward direction, the correlation is definitely positive [Fig. 34.2 (*c*)].

(*iii*) When the scatter diagram advances from left to right in downward direction, the correlation is negative [Fig. 34.2 (*a*)].

(*iv*) When the dots are not in definite arrays and are scattered haphazardly, this condition indicates that there is no correlation between the data series [Fig. 34.2 (*b*)].

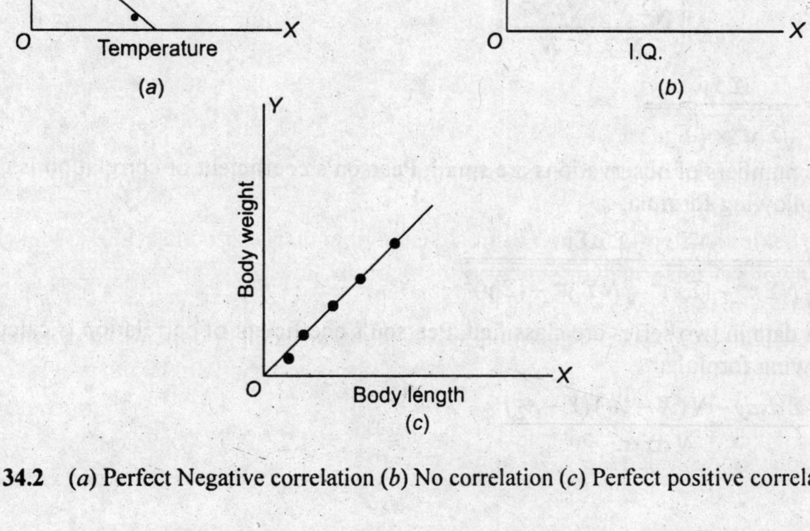

Fig. 34.2 (*a*) Perfect Negative correlation (*b*) No correlation (*c*) Perfect positive correlation

(*v*) When the dots appear to be situated on a line which advances upward at 45° angle from the $O-X$ axis, this condition indicates perfect positive correlation among the data series.

(*vi*) If the dots appear to be situated on a line which moves from left to right in downward direction at 45° angle from $O-X$ axis, this condition is indicative of perfect negative correlation.

3. Karl Pearson's Coefficient of Correlation Method. This is the best mathematical method of determining the correlation. Coefficient of correlation (r) is obtained by dividing the product of values of covariance of the two series by the product of their standard deviations.

$$r = \frac{Cov.(X.Y)}{\sigma_X . \sigma_Y}$$

Where σ_X and σ_Y are the standard deviations of variables or data series, X and Y. Covariance of two series is obtained by dividing the sum of the products of deviations of two series and the arithmetic means by the number of observations.

$$\text{Cov. }(X.Y) = \frac{\Sigma(X-\bar{X})(Y-\bar{Y})}{N}$$

So, $$r = \frac{\Sigma(X-\bar{X})(Y-\bar{Y})}{N.\sigma_X . \sigma_Y}$$

Where $\bar{X}$ = The arithmetic mean of series X

$\bar{Y}$ = Arithmetic mean of series Y

σ_X = Standard deviation of series X

σ_Y = Standard deviation of series Y

N = Number of observations

If $X-\bar{X}=x$ and $Y-\bar{Y}=y$

$$r = \frac{\Sigma xy}{n \sigma x \sigma y}$$

Since $\sigma x = \frac{\sqrt{\Sigma x^2}}{N}$ and $\sigma y = \frac{\sqrt{\Sigma y^2}}{N}$

Therefore $$r = \frac{\Sigma xy}{N\left(\frac{\sqrt{\Sigma x^2}}{N} . \frac{\sqrt{\Sigma y^2}}{N}\right)}$$

or $$r = \frac{\Sigma xy}{\sqrt{\Sigma x^2} \times \sqrt{\Sigma y^2}}$$

If the numbers of observations are small, Pearson's coefficient of correlation is calculated by the following formula:

$$r = \frac{N\Sigma xy - \Sigma x . \Sigma y}{\sqrt{N\Sigma x^2 - (\Sigma x)^2}\ \sqrt{N\Sigma y^2 - (\Sigma y)^2}}$$

If the data in two series are classified, Pearson's coefficient of correlation is calculated by the following formula:

$$r = \frac{\Sigma dxdy - N(X - A_1)(Y - A_2)}{N \sigma x \sigma y}$$

Where, A_1 = assumed mean of series X

A_2 = assumed mean of series Y

dx = deviation from assumed mean of series – X or $(X - A)$

dy = deviation from assumed mean of series Y or $(Y - A_2)$

σ_x = standard deviation of series X

σ_y = standard deviation of series Y

The above formula can also be expressed as follows:

$$r = \frac{N\Sigma dxdy - \Sigma dx \Sigma dy}{\sqrt{N\Sigma dx^2 - (\Sigma dx)^2}\sqrt{N\Sigma dy^2 - (\Sigma dy)^2}}.$$

Example. Calculate Pearson's coefficient of correlation from the following data:

Series *X*—18, 20, 22, 27, 27, 26, 27, 29, 28, 26.

Series *Y*—23, 27, 28, 28, 29, 30, 31, 33, 35, 36.

Solution.

S. No.	*Series X*			*Series Y*			
	Observation (*X*)	*Deviation* $dx = X - \bar{X}$	dx^2	*Observation* (*Y*)	*Deviation* $dy = Y - \bar{Y}$	dy^2	*dx dy*
1	18	18 – 25 = – 7	49	23	23 – 30 = – 7	49	49
2	20	20 – 25 = – 5	25	27	27 – 30 = – 3	9	15
3	22	22 – 25 = – 3	9	28	28 – 30 = – 2	4	6
4	27	27 – 25 = 2	4	28	28 – 30 = – 2	4	– 4
5	27	27 – 25 = 2	4	29	29 – 30 = – 1	1	– 2
6	26	26 – 25 = 1	1	31	30 – 30 = 0	0	0
7	27	27 – 25 = 2	4	30	31 – 30 = 1	1	2
8	29	29 – 25 = 4	16	33	33 – 30 = 3	9	12
9	28	28 – 25 = 3	9	35	35 – 30 = 5	25	15
10	26	26 – 25 = 1	1	36	36 – 30 = 6	36	6
10	$\Sigma X = 250$		$\Sigma dx^2 = 122$	$\Sigma Y = 300$		Σdy^2	$\Sigma dxdy$
Mean	$\bar{X} = \frac{250}{10} = 25.$			Mean $\bar{Y} = \frac{300}{10} = 30$		= 138	=105 – 6 =99

$\bar{X} = 25$

$\bar{Y} = 30$

$$\sigma_X = \frac{\sqrt{\Sigma dx^2}}{n} = \frac{\sqrt{122}}{10} = 3.49$$

$$\sigma_Y = \frac{\sqrt{\Sigma dy^2}}{n} = \frac{\sqrt{138}}{10} = 3.71.$$

Pearson's coefficient of correlation:

$$r = \frac{\Sigma dx\, dy}{n\sigma_X \sigma_Y}$$

$$= \frac{99}{10\times3.49\times3.71} = \frac{99}{129.479}$$

$r = 0.76.$

Example. Calculate Pearson's coefficient of correlation from the following data series:

Series X	63	66	67	67	71	69	68	59
Series Y	66	64	65	66	67	68	70	71

Solution. Suppose that the assumed mean for series X $(A_1) = 68$ and the assumed mean for series Y $(A_2) = 67$.

S. No.	*Series X* Observation (X)	$dx = X - A_1$	dx^2	*Series Y* Observation (Y)	$dy = Y - A_2$	dy^2	$dx\,dy$
1	63	$63 - 68 = -5$	25	66	$66 - 67 = -1$	1	5
2	66	$66 - 68 = -2$	4	64	$64 - 67 = -3$	9	6
3	67	$67 - 68 = -1$	1	65	$65 - 67 = -2$	4	2
4	67	$67 - 68 = -1$	1	66	$66 - 67 = -1$	1	1
5	71	$71 - 68 = 3$	9	$\boxed{67 = A_2}$	$67 - 67 = 0$	0	0
6	69	$69 - 68 = 1$	1	68	$68 - 67 = 1$	1	1
7	$\boxed{68 = A_1}$	$68 - 68 = 0$	0	70	$70 - 67 = 3$	9	0
8	59	$59 - 68 = -9$	81	71	$71 - 67 = 4$	16	-36
$n = 8$		$\Sigma dx = -14$	$\Sigma dx^2 = 122$		$\Sigma dy = 1$	$\Sigma dy^2 = 41$	$\Sigma dx\,dy = 15-36 = -21$

Coefficient of correlation

$$r = \frac{n\Sigma dx\,dy - \Sigma dx\,dy}{\sqrt{n\Sigma dx^2 - (\Sigma dx)^2}\,\sqrt{n\Sigma dy^2 - (\Sigma dy)^2}}$$

$$r = \frac{8\times-21-(1\times-14)}{\sqrt{8\times122-(-14)^2}\,\sqrt{8\times41-(1)^2}}$$

$$= \frac{-168+14}{\sqrt{976-196}\,\sqrt{328-1}} = \frac{-154}{\sqrt{780}\times\sqrt{327}}$$

$$= \frac{-154}{\sqrt{255060}} = \frac{-154}{505.03}$$

$r = -0.305.$

4. **Spearman's Ranking Method.** Professor Charls Spearman worked out a method for determining the correlation in which the values of all data of a series are assigned ranks in decreasing or increasing (ascending) order. In this ranking process, the highest value is given rank 1 and the next higher value is given rank 2 and so on. In some series the values of two or more data are similar. In that case, the mean of the ranks will be equally shared by those data, as for example, in one series there are two observations; one at S. No. 3 and the other at S. No. 10 of 67 each. In ranking process 67 at S. No. 3 and 67 at S. No. 10 instead of being ranked 6 and 7 respectively are ranked at 6.5 (mean of rank 6 and rank 7). In the same way if there are three or more data in a series as have got same value, all those data will share the rank which will be the mean of their ranks. The number or frequency of the data with similar value is indicated by **m**.

In the next step, the difference between the ranks (D) of respective data of the two series are obtained ($D = R_1 - R_2$) which may be positive or negative figures. Then after, the values of D^2 and sum of D^2 ($= \Sigma D^2$) are determined.

For two such series as are taking in data with similar values, the following formula is used to determine the coefficient of correlation by ranking method (Symbolised by Rho = ρ):

$$\rho = 1 - \frac{6 \Sigma D^2}{N(N^2 - 1)}$$

For determining correlation coefficient by ranking method of two such series as have got 2 or more data of similar values, the following Spearman's formula is used:

$$\rho = \frac{1 - 6[\Sigma D^2 + \frac{1}{12}(m_1^3 - m_1) + \frac{1}{12}(m_2^3 - m_2) + \frac{1}{12}(m_3^3 + m_3) ...]}{N(N^2 - 1)}$$

Where D = difference between the ranks of respective data ($R_1 - R_2$)

N = total number of data in a series

m = frequencies of different data of same value in the series.

Example. Calculate coefficient of correlation for the following series by Spearman's ranking method:

Series X	64	65	67	63	68	62	68	70	66	67	69	71
Series Y	65	68	68	66	69	66	71	68	65	67	68	70

Solution.

S. No.	*Series X*	*Series Y*	*Ranks of series X* (R_1)	*Ranks of series Y* (R_2)	$D = R_1 - R_2$	D^2
1	64	65	10	11.5 m_5	– 1.5	2.25
2	65	68	9	5.5 m_3	3.5	12.25
3	67	68	6.5 m_2	5.5 m_3	1.0	1.00
	63	66	11	9.5 m_4	1.5	2.25
5	68	69	4.5 m_1	3	1.5	2.25
6	62	66	12	9.5 m_4	2.5	6.25
7	68	71	4.5 m_1	1	3.5	12.25
8	70	68	2	5.5 m_3	– 3.5	12.25
9	66	65	8	11.5 m_5	– 3.5	12.25
10	67	67	6.5 m_2	8	– 1.5	2.25
11	69	68	3	5.5 m_3	– 2.5	6.25
12	71	70	1	2	– 1.0	1.00
$n = 12$						$\Sigma D^2 = 72.5$

$$\rho = \frac{1 - 6[\Sigma D^2 + \frac{1}{12}(m_1^3 - m_1) + \frac{1}{12}(m_2^3 - m_2) + \frac{1}{12}(m_3^3 - m_3) + \frac{1}{12}(m_4^3 - m_4) + \frac{1}{12}(m_5^3 - m_5)]}{N(N^2 - 1)}$$

Here the frequencies of m_1 data is 2, that of m_2 is 2, that of m_3 is 4, $m_4 = 2$ and $m_5 = 2$

$$\therefore P = 1 - \frac{6\,[72.50 + \frac{1}{12}(2^3 - 2) + \frac{1}{12}(2^3 - 2) + \frac{1}{12}(4^3 - 4) + \frac{1}{12}(2^3 - 2) + \frac{1}{12}(2^3 - 2)]}{12(144 - 1)}$$

$$=1-\frac{72.50+\frac{1}{12}(6)+\frac{1}{12}(6)+\frac{1}{12}(60)+\frac{1}{12}(6)+\frac{1}{12}(6)}{2(144-1)}$$

$$=1-\frac{72.50+0.5+0.5+5.0+0.5+0.5}{2(143)}$$

$$=1-\frac{79.50}{286}$$

P = 1 – 0.278 = 0.722 (coefficient of correlation), *i.e.*, the correlation between the two data series is moderate.

Example. Calculate the coefficient of correlation by Spearman's ranking method and indicate the degree of correlation in the following two data series:

Series X	19	28	26	23	27	13	26	31	21	22
Series Y	21	28	23	25	31	17	22	29	27	20

Solution.

S. No.	*Series X*	*Series Y*	*Rank of data in series X* (R_1)	*Rank of data in series Y* (R_2)	$D = R_1 - R_2$	D^2
1	19	21	9	8	+ 1	1
2	28	28	2	3	– 1	1
3	26	23	4.5	6	– 1.5	2.25
4	23	25	6	5	+ 1	1
5	27	31	3	1	+ 2	4
6	13	17	10	10	0	0
7	26	22	4.5	7	– 2.5	6.25
8	31	29	1	2	– 1	1
9	21	27	8	4	4	16
10	22	20	7	9	– 2	4
$n = 10$						$\Sigma D^2 = 36.50$

$$\text{Coefficient of correlation } (P)=1-\frac{6\Sigma D^2}{N(N^2-1)}$$

$$P=1-\frac{6\times 36.50}{10(100-1)}$$

$$=1-\frac{6\times 36.50}{990}$$

$$=1-\frac{7.3}{33}\text{ or }1-0.22$$

$$=+0.78.$$

The value of coefficient of correlation + 0.78 indicates high degree of positive correlation.

Correlation coefficient by Concurrent Deviation

This method is used to indicate whether the correlation is in positive or negative direction especially in the data series characterised by short-term fluctuations of data. Correlation coefficient by concurrent deviation is calculated as follows:

1. First of all, the direction of deviation [positive (+) or negative (–)] of each observation in respect of preceding data are marked for different series in separate columns. If the value of a data is greater than that of the preceding data of the series, the direction of deviation is marked + and if it is less, then the direction of deviation will be marked –.
2. Next, the deviation signs of respective data of the two series are multiplied (+ × + = +, + × – = – and – × – = +) and the products are recorded in a separate column.
3. The total number of positive signs in the column for product of deviation signs is recorded which is called concurrent deviation (= *C*).
4. The coefficient of correlation (*RC*) by concurrent deviation is determined by the following formula:

$$RC = \pm\frac{\sqrt{\pm(2C-N)}}{N}$$

Where, C = total number of + signs in the column for products of two deviations

N = number of observations in a series.

The following example will illustrate the process:

Example. Calculate the co-efficient of correlation of the following two data series by concurrent deviation methed:

Series *X*	50	51	53	51	50	54	55	56	61	62	63
Series *Y*	73	71	75	76	79	80	83	84	86	85	91

Solution.

S. No.	*Series X*	*Deviation from preceding data*	*Series Y*	*Deviation from preceding data*	*Product of deviation signs of respective data in two series*
1	50		73		
2	51	+	71	–	–
3	53	+	75	+	+
4	51	–	76	+	–
5	50	–	79	+	–
6	54	+	80	+	+
7	55	+	83	+	+
8	56	+	84	+	+
9	61	+	86	+	+
10	62	+	85	–	–
11	63	+	91	+	+
	$N = 10$		$N = 10$		$C = 6$

$$RC = \pm\frac{\sqrt{\pm(2C-N)}}{N}$$

$$= \pm\frac{\sqrt{\pm 2\times6-10}}{10}$$

$$= \pm\sqrt{\pm\frac{12-10}{10}} = \sqrt{\frac{2}{10}}$$

$RC = +0.477.$

Therefore, the series *X* and series *Y* show positive correlation.

QUESTIONS

1. Define and explain correlation and correlation coefficient with suitable examples.
2. Fruit weight and water content percentage of twelve fruits of same species are given in the following table. Find out the correlation coefficient from the given data:

Fruit weight in g	27.2	31.5	32.6	29.2	28.3	30.3	27.9	31.8	32.4	28.6	31.8	27.1
Water content percentage in fruit	62.6	64.2	67.0	68.6	71.0	72.6	71.0	69.7	67.0	64.5	62.0	60.7

3. The height of plant and weight of plant in 7 experimental plots are presented in the following table. Find out the correlation coefficient of the two variables:

Plant height in cm	11.7	13.9	15.5	17.8	18.5	19.2	21
Weight of plant in g.	7.10	12.42	15.35	23.20	28.45	32.25	39.84

4. Find out the coefficient of correlation (r) from the following mean of deviation of two variables X and Y:

   X → – 4, – 3, – 2, – 1, 0, 1, 2, 3, 4

   Y → – 3, – 3, – 4, 0, 4, 4, 1, 2, – 2, – 1

35 CHAPTER

Regression

Regression means to return or come back. It is used in the cases where the values of data of two or more series tend to return toward the mean of the series. Sir Francis Galton observed that tall fathers produced tall sons and the dwarf fathers produced dwarf sons. The average height of tall sons was found to be less than the average height of fathers. Similarly, the average height of dwarf sons exceeded the average height of their dwarf fathers. This tendency of height in human beings to move towards the average height was called *regression* by Galton.

In biological studies it is important to know how the variation in one variable affects the variation in other variable, as for example, the yields of wheat vary with the change in the amounts of fertilizers applied as well as the change of plant density, and similarly, the weights of animals vary with the change of protein contents in feed. In such problems, it may be possible to distinguish the variable of primary interest and one or more variables of secondary interest which affect the main one. The main variable is called **dependent variable** which is denoted by Y and the secondary variables are called **independent variables** which are denoted by X.

Thus, the change in the yield of wheat is dependent variable (Y) and the change in the amounts of fertilizers applied is independent variable (X) which affects the yield potency.

Regression Analysis

The purpose of regression analysis is to predict the value of one character or variable (say Y) from the known value of the other variable (say X). The value of variable to be predicted is *dependent variable* and the known variable is called the *independent variable*. The regression analysis is done by (*i*) Regression lines and (*ii*) Regression coefficient. Regression lines explain the mean relationship between X and Y variables.

Kinds of Regression Analysis

In general, there are two types of Regression analysis:

1. **Simple Regression.** It involves two variables, one of which is independent and the other dependent. It may be classified into *Linear* and *non-linear* or curvilinear regressions.
2. **Multiple Regression.** It involves more than two variables. In this, the number of independent variables may be two or more than two.

A linear regression problem is one in which the same change in dependent variable Y can be expected for a change in the independent variable X irrespective of the value of X. In this case, a mathematical function representing a straight line graph can be used to describe the relationship between the two variables. In other words, the two variables show linear relationship when the change in independent variable X causes a change in dependent variable Y. For illustrating linear regression, let us consider the following data recorded on 7 animals:

Age of animals in months (X)	0	2	3	4	6	8	12
Weights of animals in kg (Y)	4	18	14	16	20	28	36

$$\Sigma Y = 126 \qquad \Sigma X = 35$$

$$Y = 18 \qquad X = 5$$

The relationship between dependent variable (weight) and independent variable (age) may be illustrated in the form of *scatter diagram* by plotting weights (Y) against ages (X) Fig. 35.1.

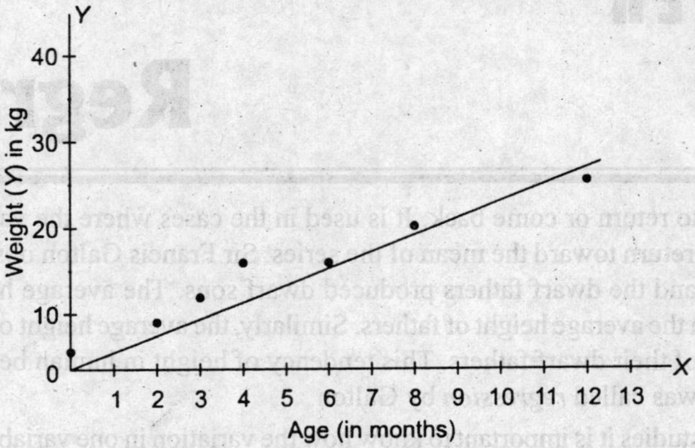

Fig. 35.1

It is evident from the scatter diagram that there is a linear relationship between weight and age, *i.e.*, weight is affected by age or in statistical terms, there is a linear regression of weight on age. The line passing through the majority of the dots or the line of best fit is referred to as **regression line**. If the regression line is straight the regression is said to be **linear regression** and if the regression line is curved then it is called **curvilinear** or **non-linear regression**.

When two variables X and Y are plotted on a scatter diagram, two **lines of best fit** can be drawn which pass through the plotted points. These lines are called regression lines and two equations based on these 2 lines are called regression equations.

Regression Lines

The composition of regression lines is based on the least square assumptions. Regression analysis is generally based on the summation of squares of deviations of observed values of Y from the lines of the best fit. A line of regression gives the best average value of one variable from any given value of the other variable.

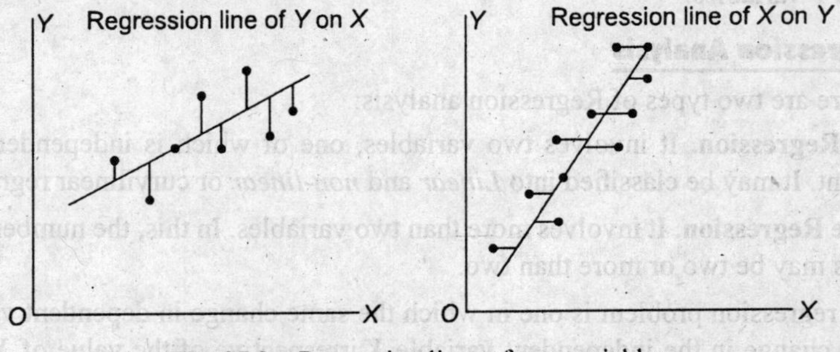

Fig. 35.2 Regression lines of two variables.

In regression analysis, there are usually two regression lines to show the average relationship between X and Y variables. It means that if there are two variables X and Y, then one line represents regression of Y upon X and the other shows the regression of X upon Y (Fig. 35.2). On these lines if the value of one variable is known, the corresponding value of variables on the other axis can be obtained. When the regression lines are nearer to each other then there is a high degree of correlation between X and Y. Another explanation for two regression lines is that regression lines are the lines of best fit which are made on the basis of assumptions of least squares of deviations of observed values. Accordingly, the lines of best fit are those which represent the minimum values of deviation squares of observed values. Deviations can be measured in two ways: (*i*) horizontally, and (*ii*) vertically. To get the minimum of two deviations, two lines are essential. That line on which the sum of the squares of horizontal deviations is minimum is called regression line of X on Y and the line on which the sum of the squares of vertical deviations becomes minimum is called regression line of Y on X.

If $X = O$, both the variables are independent and they will cross each other at right angle. When the regression lines intesect each other at the point of means of X and Y, and if a perpendicular line is drawn from that point to the X axis, it will touch the axis on the mean value of X. Similarly, if another perpendicular line is drawn from that point to the Y axis, it will touch the Y axis at the mean value of Y (Fig. 35.3).

The lesser the correlation, the greater will be the divergence of angles. The steepness of the regression lines indicates the extent of correlation. Closer the correlation, the greater is the steepness of regression lines X and Y or Y and X.

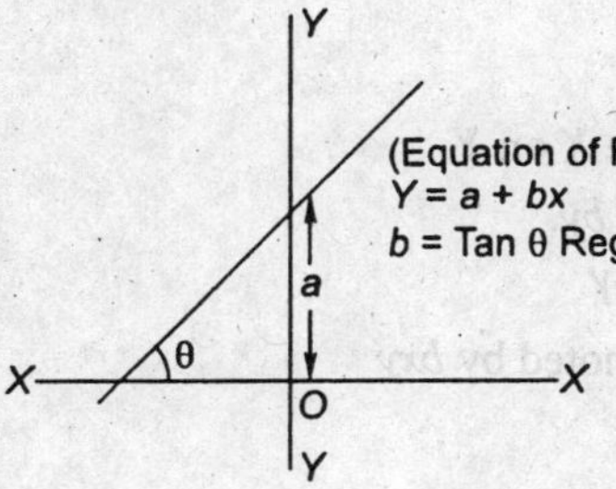

Fig. 35.3

Regression Equations

The mathematical expression of relations between two variables, one dependent and the other independent is called **regression equation**. Regression equations are the algebric expressions of regression.

The regression equation of Y on X indicates the changes in the values of X for changes given in Y. Likewise regression equation of X on Y indicates the changes in the values of Y for the changes given in X.

Since there are two regression lines, there are two regression equations too. These regression equations are discussed here as under:

1. **Regression Equation of *Y* on X.** Regression equation of Y on $X = yx = a + b$ where X and Y are variables and '*a*' and '*b*' are unknown constants. Constant a is distance between the point of origin and the point where the regression line touches Y axis. Constant b is the steepness or slope of line and is also called **coefficient of regression**. It indicates that for every unit change in X, there will be two unit change in Y. Constants '*a*' and '*b*' can be obtained by the following formulae:

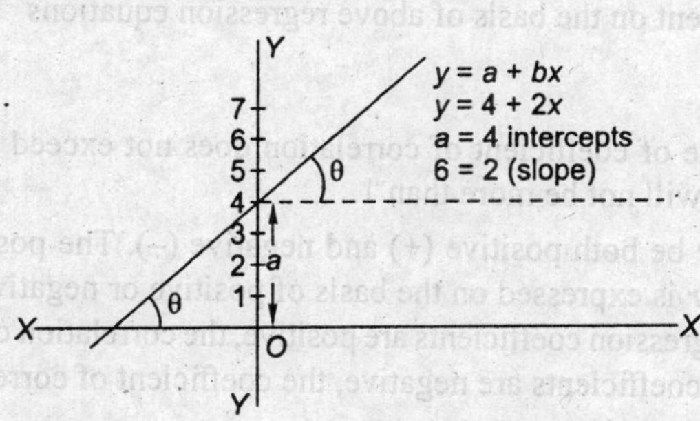

Fig. 35.4 Position of a and b from the equation $y = a + bx$.

$a=\bar{Y}-b\bar{x}$ where $\bar{X}$ and $\bar{Y}$ are the means of X and Y respectively.

$$b=\frac{\Sigma x.y}{\Sigma x^2}=\frac{\Sigma X.Y-\frac{\Sigma X.\Sigma Y}{N}}{\Sigma X^2-\frac{(\Sigma X)^2}{N}}$$

where $x=(X-\bar{X})$ and $y=(Y-\bar{Y})$, *i.e.*, the deviations of X and Y respectively.

If σ_x is the regression of X series and σ_y is the regression of Y series and the coefficient of correlation is r, then

$bxy=r\times\frac{\sigma \text{ of } Y \text{ series}}{\sigma \text{ of } X \text{ series}}$ or $r.\frac{\sigma x}{\sigma y}$ and regression line of Y on X

$$Y-\bar{Y}=byx(X-\bar{X})$$

or $Y-\bar{Y}=r\frac{\sigma y}{\sigma x}(X-\bar{X})$

2. Regression equation of Regression line *X* on *Y*.

Regression equation of X on $Y = x = a + by$.

Where b = slope of regression line X on Y

or coefficient of regression of X on Y denoted by bxy

$$b \text{ or } bxy=r\times\frac{\Sigma xy}{\Sigma y^2} \text{ or } r\times\frac{\sigma x}{\sigma y}$$

and $a = \bar{X}$ = Mean of series X

$\bar{Y}$ = Mean of series Y

$x=(X-\bar{X})$ and $y=(Y-\bar{Y})$

If σ_x = standard deviation of series X,

σ_y = standard deviation of series Y, and

r = coefficient of correlation between X and Y.

3. Regression Equation of *X* on *Y*.

It is represented as follows:

$$X-\bar{X}=bxy(Y-\bar{Y})$$

$$X-\bar{X}=r\frac{\sigma x}{\sigma y}(Y-\bar{Y})$$

Correlation coefficient on the basis of above regression equations

$$r=\sqrt{bxy-byx}$$

Since the magnitude of coefficient of correlation does not exceed 1, the value of regression co-efficients *bxy* and *byx* will not be more than 1.

The value of r may be both positive (+) and negative (–). The positive or negative value of coefficient of correlation r is expressed on the basis of positive or negative value of two regression coefficients. If both the regression coefficients are positive, the correlation coefficients will be positive and if the two regression coefficients are negative, the coefficient of correlation will be negative.

If the arithmetic means of deviations of two series is not considered and assumed mean is considered in place of that, the values of two regression coefficients will be expressed as follows:

$$bxy = \frac{\eta \Sigma\, dxdy - \Sigma\, dx . \Sigma\, dy}{\eta \Sigma\, dy^2 - (\Sigma\, dy)^2}$$

and then $bxy = \dfrac{\eta \Sigma\, dxdy - \Sigma\, dx . \Sigma\, dy}{\eta \Sigma\, dx^2 - (\Sigma\, dx)^2}$

where dx = deviation from the consumed mean (A) or $dxy = (X - A)$

and dy = deviation from assumed mean (B) or $dy = (Y - B)$.

Regression Equation through Least Square Method

Regression equation can be derived through least square method. Linear regression equations are the equations of those lines of best fit which are made on the basis of least square method. In this method 'a' and 'b' are determined by regression equations.

For regression equation of regression line X on Y ($X = a + by$)

$$\Sigma x = na + b\Sigma y$$

$$\Sigma xy = a\Sigma y + b\Sigma y^2$$

and for the regression equation of regression line

Y on $X = (Y = a + bx)$

$$\Sigma y = (na + b\Sigma x)$$

$$\Sigma yx = a\Sigma x + b\Sigma x^2$$

It is noteworthy here that in both the regression equations value of a and b are different.

Regression Coefficient

Prediction of the value of one character from the knowledge of the other character is done by finding out another constant called regression coefficient. The significant of regression coefficient is to know whether there is a linear relationship between X and Y at population level tested by t analysis. Regression co-efficient of Y for one unit of X and regression coefficient of X for one unit of Y are found out by one of the following three formulae:

Formula I. When the correlation coefficient (r) is known, regression coefficient can be derived in the following ways:

(*i*) Regression coefficient of X on Y

$$bxy = r \times \frac{\sigma \text{ of } X \text{ series}}{\sigma \text{ of } Y \text{ series}} \text{ or } r \times \frac{\sigma X}{\sigma Y}$$

(*ii*) Regression coefficient of Y on X

$$byx = r \times \frac{\sigma \text{ of } Y \text{ series}}{\sigma \text{ of } X \text{ series}} \text{ or } r \times \frac{\sigma Y}{\sigma X}$$

(*iii*) Multiplying both equations–

$$bxy \times byx = r \times \frac{\sigma X}{\sigma Y} \times r \times \frac{\sigma Y}{\sigma X}$$

$$bxy \times byx = r^2$$

$$\text{or } r = \sqrt{bxy . byx}.$$

It means that square root of the product of regression coefficients of X on Y and Y on $X = r$.

Formula II. If means are already calculated the regression coefficients are derived in the following way:

$$bxy = \frac{\Sigma xy}{\Sigma y^2} \text{ or } \frac{\Sigma(X-\bar{X})(Y-\bar{Y})}{\Sigma(Y-\bar{Y})^2}$$

$$byx = \frac{\Sigma xy}{\Sigma x^2} \text{ or } \frac{\Sigma(X-\bar{X})(Y-\bar{Y})}{\Sigma(X-\bar{X})^2}.$$

Formula III. If means of the two series are not calculated, the following simple and direct method may be adopted to calculate regression coefficient:

$$bxy = \frac{\Sigma XY - \left(\frac{\Sigma X . \Sigma Y}{N}\right)}{\Sigma Y^2 - \frac{(\Sigma Y)^2}{N}}$$

$$\text{and } byx = \frac{\Sigma XY - \left(\frac{\Sigma X . \Sigma Y}{N}\right)}{\Sigma X^2 - \frac{(\Sigma X)^2}{N}}.$$

Here the X and Y are the original measurements of two parameters and not the deviations.

Solved Problem

Problem I. In a study on the effect of *water temperature* on the *content of chlorophyll* of a plant, the following data were obtained. Find out the type of relationship and its strength between these two parameters by applying analysis and correlation techniques.

Temperature (in °C) X	24.2	24.6	25.4	26.2	27.0	28.1	28.8	29.7	30.6	31.3
Chlorophylla content in mg/l Y	52	64	72	81	92	101	106	116	122	130

Solution.

Temp. (°C) X	*Chlorophyll – a content, in mg/litre Y*	*XY*	*X²*	*Y²*
24.2	52	1258.4	584.64	2704
24.6	64	1574.4	605.16	4096
25.4	72	1828.8	645.16	5184
26.2	81	2122.2	686.44	6501
27.0	92	2484.0	729.00	8464
28.1	101	2838.1	789.61	10201
28.8	106	3052.8	829.44	11236
29.7	116	3445.2	882.09	13456
30.6	122	3733.2	939.36	14886
31.3	130	4069.0	979.69	16900
$\Sigma X = 275.9$	$\Sigma Y = 936$	$\Sigma XY = 26406.1$	$\Sigma X^2 = 7668.59$	$\Sigma Y^2 = 93686$

$n=10, \ \bar{X}=27.59, \ \bar{Y}=93.6$

$$b=\frac{\Sigma XY-\bar{X}.\Sigma\bar{Y}}{\Sigma X^2-\bar{X}\Sigma X}=\frac{26406.1-27.59\times 93.6}{7668.59-27.59\times 275.9}=\frac{581.86}{56.509}=10.296$$

$$a=\bar{Y}-b\bar{X}=93.6-(10.296\times 27.59)=-190.58$$

$$r=\frac{\Sigma XY-\bar{X}\Sigma Y}{\sqrt{(\Sigma X^2-\bar{X}\Sigma X)(\Sigma Y^2-\bar{Y}\Sigma Y)}}$$

$$r=\frac{26406.1-27.59\times 936}{\sqrt{(7668.59-27.59\times 275.9)(93686-93.6\times 936)}}=0.99$$

$$r^2=0.98,\ t=\frac{\sqrt{n-2}}{\sqrt{1-r^2}}=\frac{\sqrt{10-2}}{\sqrt{1-0.98}}=20.21 \qquad \textit{Ans.}$$

t at 5% level from the t table = 2.306.

High positive value of r = 0.99 shows strong correlation between temp X and chlorophyll a content Y in a plant.

If a graph is prepared putting the temperature data on X axis and chlorophyll a content on Y axis, the value of chlorophyll a on regression line can be predicted at any temperature within naturally occuring limits.

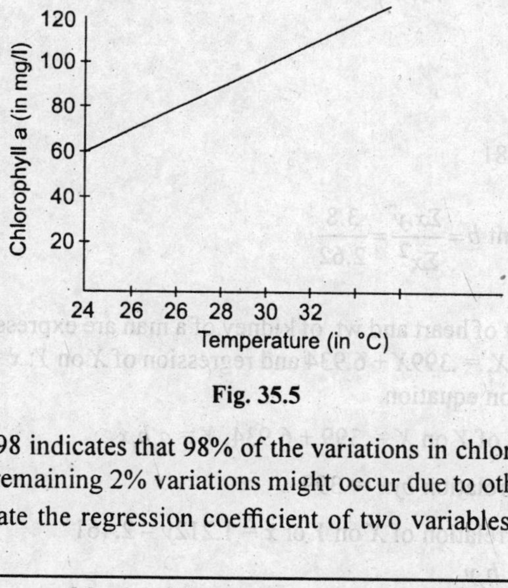

Fig. 35.5

The value of r^2 = 0.98 indicates that 98% of the variations in chlorophyll a are influenced by temperature changes and remaining 2% variations might occur due to other factors.

Problem II. Calculate the regression coefficient of two variables X and Y presented in the following table:

X	4.5	4.0	5.2	4.6	5.2	5.2	4.3	4.0	4.5	5.5
Y	5	5	6	6	6	7	4	4	5	6

Solution. $\Sigma X=47$

$\Sigma Y=54$

$\Sigma X^2=223.52$

$\Sigma Y^2=300$

$\Sigma X.Y=257.6$

(*i*) $\Sigma x^2=\Sigma X^2-\dfrac{(\Sigma X)^2}{N}=223.52-\dfrac{(47)^2}{10}=223.52-220.9$

(*ii*) $\Sigma y^2 = \Sigma Y^2 - \frac{(\Sigma Y)^2}{N} = 2.62$

$= 300 - \frac{(54)^2}{10}$

$= 300 - 291.6$

$= 8.4$

(*iii*) $\Sigma xy = \Sigma XY - \left(\frac{\Sigma X.\Sigma Y}{N}\right)$

$= 257.6 - \frac{47 \times 54}{10}$

$= 257.6 - 253.8 = 3.8$

(*iv*) Correlation coefficient $(r) = \frac{\Sigma x \times y}{\sqrt{\Sigma x^2 \times \Sigma y^2}}$

$$= \frac{\Sigma XY - \frac{\Sigma X.\Sigma Y}{N}}{\sqrt{\left(\Sigma X^2 - \frac{\Sigma X^2}{N}\right)\left(\Sigma Y^2 - \frac{\Sigma Y^2}{N}\right)}}$$

$$= \frac{3.8}{\sqrt{2.62 \times 8.4}}$$

$$= \frac{3.8}{\sqrt{22.01}} = \frac{3.8}{4.69} = 0.81$$

Regression coefficient $b = \frac{\Sigma x.y}{\Sigma x^2} = \frac{3.8}{2.62}$

$= 1.45$

Problem III. Weight of heart and wt. of kidney of a man are expressed as X and Y respectively and the regression of Y on X, $= .399X + 6.934$ and regression of X on Y: $x = 1.212Y - 2.461$. Find out the coefficient of correlation equation.

Regression equation of Y on $X = .399 + 6.934$, $Y = a_1 b_1 x$

$\therefore$ Coefficient of correlation $byx = .399$

$\therefore$ Coefficient of correlation of X on Y or $x = 1.212y - 2.461$

$X = a_2 + b_2 y$

$b_2 = bxy$

$\therefore$ Coefficient of correlation $bxy = 1.212$

$\therefore$ Correlation coefficient between X and $Y = r = \sqrt{bxy.byx}$

$r = \sqrt{1.212 \times .399}$

$= \sqrt{.483588}$

$r = 0.695.$

QUESTIONS

1. Define regression. Discuss regression lines and regression equations.
2. What do you mean by regression equation?
3. Discuss regression coefficient. How is it derived?
4. Calculate the regression coefficient from the following data on the number of branches and the number of fruit clusters:

No. of branches	6	7	8	9	10	11	12	13	14	15
No. of clusters	13	14	15	17	16	20	19	18	21	22

5. Considering the following data of two varieties draw a scattered diagram from the data provided:

X	1	2	3	4	5	6
Y	6	4	3	5	4	2

36 CHAPTER Plant Breeding

Man is dependent directly or indirectly on plants for his basic necessities– food, clothings, and shelter. Man has been cultivating and conserving plants according to his needs since the time immemorial and even now various agricultural crops are grown for food grains, fibres, oil seeds, fodder and feed for domestic animals and for other plant products of commercial importance. The farmers try to increase the production of crops in various ways such as by selecting good varieties of crops for cultivation, application of manures and fertilizers, proper irrigation, disease and pest control practices, removal of weeds and so on. But the good environmental conditions can enhance the yield of any crop upto a certain limit. The production, however, can be increased further only by developing such new plant varieties as may utilize the environmental conditions more efficiently and to a greater extent than the existing varieties. This is possible only by changing the heredity of existing varieties.

The present-day plants and animals are products of natural and artificial breeding. Plant breeding is concerned with the production of improved new crop varieties that are superior to existing varieties in many characters. Plant breeding also helps in introducing many superior qualities originally found in different plants of particular crop into a single plant.

J. M. Poehlman (1959) defined plant breeding as "*the art and science of changing and improving the heredity of plants*".

According to D. C. Smith (1966), plant breeding is the art and science of improving the genetic patterns of plants in relation to their economic uses.

Objectives of Plant Breeding

The main aims and objectives of plant breeding are to improve the qualities of plants so that they are much desirable agronomically and economically. The specific objectives vary greatly depending upon the crops under consideration. The main objectives of plant breeding have been listed here as follows:

1. To evolve new varieties of crops which have better yielding potency than the local crop varieties. To increase the yield of desired plant products such as, grains, fodder, fibres, oils and so on.
2. To increase the qualities of crops with regard to size, colour, shape, taste, nutrients contents, milling, keeping and cooking etc. of grains, vegetables, flowers and fruits and many special qualities as high sugar contents in sugar crops, strong, long and fine fibres in fibre crops, high protein contents in pulses, appealing flavour, high sugar and starch content and large size of fruits in fruit crops. To evolve such varieties of crops as are suitable to producers and consumers both and can meet the requirements of the consumers to a good extent.

3. To produce varieties that are resistant to fungal, bacterial diseases, insects and pests.
4. To change the duration of crop maturation according to need. Synchronous maturity is highly desirable in certain crops like mung bean (*Vigna radiata*), pigeon pea (*arhar*) etc.
5. To develop varieties as may utilize fertilizers and irrigation efficiently and to a greater extent.
6. To obtain suitable varieties for particular agroclimatic region.
7. To develop varieties with a wide range of adaptability.
8. To change the growth habit of crops as needed. The development of plant varieties that are tolerant to moisture stress and salts would be useful in increasing the production in rainfed areas and saline soils respectively. Winter hardiness would be desirable in certain situations.
9. To produce new plant varieties for new seasons and new areas.
10. To eliminate toxic substances from the consumable parts to make them safe for human consumption. For example, khesari (*Lathyrus sativa*) seeds have a neuro-toxin, βN-oxalyl amine alanine (BOAA) which causes paralysis. Similarly, erucic acid found in *Brassica* seeds is harmful to human health. Elimination of such toxic substances would increase the nutritional value of these crops.

Historical Account

The history of plant breeding in India can be traced back to the Vedic period (1500 to 800 B.C.). Among the oldest crops grown in this country were wheat, sugarcane, rice, cotton, etc. The excavations of Mohanjodaro and Harappa (in undivided India) which existed over 5000 years ago stand good witness to the same. European tasted sugarcane hundreds of years after it was grown in India. Crops were maintained at high standards. Cotton grown in India in ancient period was of fine qualities and muslin of Dacca (undivided India) was famous all over the world. The Dacca muslin was so fine that it could be compared with the spider's web and also with the finest quality cloth of modern time and the story went that a full length piece of 40 yards of the muslin could be wrapped over a thumb or packed into a match box.

Plant breeding began when man first chose certain plants for cultivation. The process of bringing a wild species under human management is referred to as *domestication*.

The early plant breeders improved their crops by selecting only superior type of healthy plants in mass and collecting their seeds for cultivation purpose. During selection time they considered only external characters of plants and seeds. Actually, they had no idea about the scientific principles behind the methods they followed. In the early days of human civilization there prevailed a general feeling in the people that better crop yield could be achieved by pleasing God.

Migration of human population from one place to another was also responsible for the movement of cultivated plant species. The introduction of plant species or varieties into a new area forms an integral part of plant breeding today.

Besides selection and plant introduction, cross breeding was also in practice in ancient days. Babylonians and Assyrians pollinated date palms artificially as early as 700 B.C.

Camerarious developed for the first time the plant hybrid by a cross between hop plant and hemp. Thomas Fair Child (1717) obtained artificial hybrid by cross pollinating red flowered and pink flowered species of sweet williams. Linnaeus (1707-1778) believed that new plant species developed by natural hybridization. Joseph Koelreuter made over 500 different crosses during 1760s and 1770s in Germany.

Carl Gaertner published an account of his 10,000 breeding experiments in 1849. Similar hybridization experiments were performed by Thomas Andrews Knight in 1780, John Goss and Alexander Seton in 1820, French botanist Charles Naudin, and Mendel.

Le Conteur and Shireff in 1840 used individual plant selection and progeny test to develop suitable cereal varieties. Vilmorine (1856) developed progeny test and employed the method successfully in improvement of sugarbeets (*Beta vulgaris*). Nilsson and his associates at Svalof, Sweden developed a detail method of *individual plant selection* around 1900 and in 1903 Johannsen proposed *pureline theory* that provided the genetic basis for individual plant selection.

Plant breeding on scientific basis started developing only after the rediscovery of Mendel's laws of heredity in 1900. Recent discoveries in the field of cytogenetics have provided very useful tool to the plant breeders and have enabled them to achieve wonderful results. G.H. Shull has made a noteworthy contribution on breeding in maize (*Zea mays*). He found that inbreeding produced a considerable loss of vigour (inbreeding depression). But when the weak inbred lines were crossed the resulting hybrids were found to be more vigorous than the original variety. These observations led to the production of hybrid varieties in several crops like maize, jowar (*Sorghum bicolor*), bajra (*Pennisetum americanum*). The science of plant breeding is very much advanced now and has been associated with other disciplines of modern sciences such as, genetics, cytology, taxonomy, plant physiology and biochemistry, plant pathology, bacteriology, entomology, biometrics, agronomy etc.

QUESTIONS

1. Define plant breeding and discuss the aims and objectives of plant breeding.
2. What do you know about the history of Plant breeding?

37 CHAPTER

Pollination Control

In flowering plants, the transfer of pollen grains from the anthers to the stigmas of the pistil is referred to as pollination. Pollination is of two types:

1. **Self-pollination or Autogamy.** When the pollen grains of a flower fall on the stigma of the same flower or on the stigmas of the other flowers of the same plant, it is referred to as **self-pollination** or **autogamy**.
2. **Cross pollination or Allogamy.** When the pollen grains from the flowers of one plant are transferred to the stigmas of the flowers of another plant, it is called **Cross pollination** or **Allogamy**.

Self-Pollination

The majority of annual crop plants are self-pollinated. Self-pollinated species are believed to have evolved from cross pollinated ancestors. These species normally have hermaphrodite flowers. In autogamous plants self-pollination is not cent per cent and cross-pollination may take place up to 5%. The various mechanisms which promote self-pollination are as follows:

1. **Cleistogamy.** In this case the hermaphrodite flowers do not open at all and under such condition only self-pollination can be possible. Since the flowers do not open, foreign pollen do not reach the stigma. Some varieties of *Triticum* species, oats (*Avena* species), barley (*Hordeum vulgare*) are cleistogamous.
2. **Chasmogamy.** When the flower buds open only after self-pollination has occurred, the condition is called **Chasmogamy**. In some cereals such as wheat, barley, rice, oats, flowers do open but some degree of cross pollination may also occur.
3. In some plants pollination takes place after the flower bud opens but the stigma is surrounded by anthers which ensures self-pollination. Examples are *Solamum melongena, Lycopersicum esculentum* etc.
4. In some plants the flower buds do open, but stamens and stigmas are covered by two or more petals which ensures self-pollination, as for examples, in Legumes (*Vigna radiata* or mung bean, *V. mungo* or urd, Soybean or *Glycine max*), Pea or *Pisum sativum*, gram or *Cicer arietinum*.
5. In some plant species, the receptive stigmas elongate through staminal columns. This ensures self-pollination.

Self-pollination leads to a rapid increase in homozygosity. Therefore, self-pollinated species are highly homozygous. These species do not show inbreeding depression but may exhibit considerable heterosis. Therefore, the aim of self-pollination in plant breeding is generally to develop homozygous varieties.

Cross Pollination

Many crop plants are naturally cross pollinated but self-pollination up to 5 – 10% may also be possible. In some self-pollinated crops cross pollination may exceed 5% and may reach 50%. Such species are referred to as *often cross pollinated* species, as for example, arhar (*Cajanus cajan*), safflower (*Carthemus tinctorius*), cotton etc. In such species, breeding methods suitable for both self- and cross- pollinated species may be applied.

Cross pollination promotes heterozygosity in a population. Cross pollinated species show considerable amount of heterosis. Economically useful hybrids or synthetic varieties of some crops are developed through cross pollination.

In cross-pollinated species, the transfer of pollen from one plant to the stigmas of the other plants is brought about by means of wind (Anemophily), by water (Hydrophily), and by insects (Entomophily). The various mechanisms promoting cross pollination in plants are listed below:

1. **Unisexuality or Dicliny.** Decliny is a condition in which the flowers are either female (pistillate) or male (staminate). There are two conditions for dicliny:

   (*a*) Monoecy and (*b*) Dioecy.

   In **Monoecy**, the male and female flowers occur on the same plant either in the same inflorescence for example, in castor (*Recinus communis*), bananas (*Musa paradisiaca*), coconut, mango, etc. or in separate inflorescences *e.g.*, in maize.

   In **Dioecy**, the male and female plants are separate, *e.g.* papaya (*Carica papaya*), dates (*Phoenix dactilifera*), spinach (*Spinacea oleracea*), *Asparagus officinalis, Canabis indica*. In some plants hermaphrodite flowers also develop in addition to male and female flowers.

2. **Dichogamy.** In some plants, the flowers are hermaphrodite but stamens and pistils mature at different times. This condition ensures cross pollination.

   (*a*) **Protandry.** In this condition, the stamens of flowers mature before pistils, *e.g.*, in maize, sugarbeet, etc.

   (*b*) **Protogyny.** In this, pistils of the flowers mature before the stamens, *e.g.*, bajra.

3. In alfalfa (*Medicago sativa*), the stigma is covered by waxy film. Honey bees destroy the waxy film and make the stigma receptive and ultimately cross pollination is effected.

4. In some species two or more devices either singly or in combination may promote cross pollination, *e.g.*, maize in which monoecy and protandry both are exhibited.

5. **Self-incompatibility.** The failure of the pollen grains of flower to affect fertilization in the same flower or other flowers in the same plant is referred to as self- incompatibility. Several species of *Brassica*, *Nicotiana*, radish and grasses show self-incompatibility. There are two types of self-incompatibility:

   (*a*) Sporophytic incompatibility.

   (*b*) Gametophytic incompatibility.

   Selfing in both the cases results in failure of seed setting. Self-incompatibility is effective in promoting cross-pollination.

6. **Male sterility.** Formation of sterile or non-functional pollen grains in the anthers of hermaphrodite flowers is referred to as male sterility. Male sterility is less common in natural populations. Male sterility may be due to (*a*) genetic, (*b*) cytoplasmic, and (*c*) cytoplasmic-gentic causes. Male sterility is of much importance in production of hybrid seeds.

Method of Determining Mode of Reproduction of a Species

The mode of reproduction of any species is determined in the following steps:

1. In the first step critical examination of the flowers of the species is made to detect the mechanisms like dioecy, monoecy, protogyny, protendry, cleistogamy, etc., which clearly indicate the mode of pollination.
2. The second step involves recording of seed setting of a plant under isolation. This is done by growing individual plants at long distance (space isolation) or by growing the plants in cages or

by bagging the flowers. The failure to set seeds in isolation proves that the species concerned is cross pollinated and the setting of seeds in the plants under isolation indicates that the species is self-pollinated.

Finally if the selfing or inbreeding causes loss of vigour or inbreeding depression, the species may be cross-pollinated. Self-pollinated species do not show inbreeding depression.

Determination of Percentage of Cross Pollination

In order to determine the degree of cross pollination in a species, two strains, one homozygous strain with dominant characters and the other homozygous recessive strain are grown in mixed stand in such a way that plants of recessive strain are surrounded by those of dominant strain to provide plenty of pollen grains. The seeds of recessive strains are collected. The percentage of seeds carrying dominant allele represents the percentage of cross pollination. Since the frequency of cross pollination varies with the variety, weather and location, such study should be conducted for two or more years at serveral locations and on several varieties of species.

In order to determine whether a species is apomictic, a recessive strain (as pistillate parent) of that species is crossed with a dominant strain (male parent). If the progeny of the cross includes sufficiently high frequency of recessive individuals, the species is most likely apomictic but for this every care should be taken to check self-pollination.

Importance of Mode of Reproduction in Breeding

The knowledge of mode of reproduction and pollination has the following importance in plant breeding programme:

1. Modes of reproduction and pollination determine the **genetic constitution** of species.
2. They help in **pollination control**.
3. They determine the **stability of varieties**.

1. **Genetic constitution of the species.** The cross pollinated species are heterozygous and show loss of vigour or inbreeding depression. The self-pollinated species, on the other hand, show homozygosity and do not show inbreeding depression. The cross pollinators consist of one genotype (hybrid varieties) or many genotypes (open pollinated) and synthetic varieties but self-pollinators have plants of single genotype. The asexually reproducing species are similar to cross pollinated species and they represent an added advantage because any desirable plant at any stage of breeding programme may be directly utilized as a variety.
2. **Pollination control.** The flower structure and mode of pollination determine the ease or difficulty in making crosses. Self-pollinated crops show natural selfing while in the cross pollinated ones, the flowers have to be pollinated by hand and protected from foreign pollen.
3. **Stability of varieties.** The self-pollinated varieties are homozygous and fairly stable in their genetic constitution and natural cross pollination is negligible in them. They can be used in cultivation for several years provided precaution is taken to remove offtypes from the population from time to time. Similar is the case with asexually reproducing species. In cross pollinated species enough precautions are to be taken to check contamination by foreign pollen. Seeds of hybrid varieties are replaced every year due to inbreeding depression. In synthetic varieties, the seeds should be replaced after a few years.

SELF-INCOMPATIBILITY

Self-incompatibility refers to the failure of pollen grains of a flower to fertilize the same flower or other flowers of the same plant, *i.e.*, the flowers do not set seeds on selfing. Self-incompatibility may be due to the following reasons:

1. Either pollen grains fail to germinate on the stigma or if they germinate, their pollen tubes fail to penetrate the stigma, as for example, in radish, cabbage, rye etc.
2. Pollen grains germinate, pollen tubes enter the style but they grow too slowly to affect fertilization before the flowers drop, *e.g.*, in *Petunia violacea, Nicotiana sanderae*.

3. Sometimes fertilization is affected but the zygote or embryo abort at very early stage, *e.g.*, in some strains of cocao and sugarbeet.

Self-incompatibility appears to be a biochemical reaction but precise nature of this reaction is not clearly understood. The genetic control of self-incompatibility is simple.

Lewis (1954) has classified self-incompatibility as follows:

Self-incompatibility
- Heteromorphic system
- Homomorphic system
  - Sporophytic control
  - Gametophytic control

Heteromorphic System

Some species of plants bear flowers of different morphology, as for example, *Primula*, buck wheat, and sweet potato. There are two types of flowers; **pin and thrum**. **Pin type** flowers have short stamens and long styles and **Thrum** flowers have long stamens and short styles (Fig. 37.1). Pin and Thrum flowers are self-incompatible in themselves but compatible mating is possible between thrum and pin flowers reciprocally.

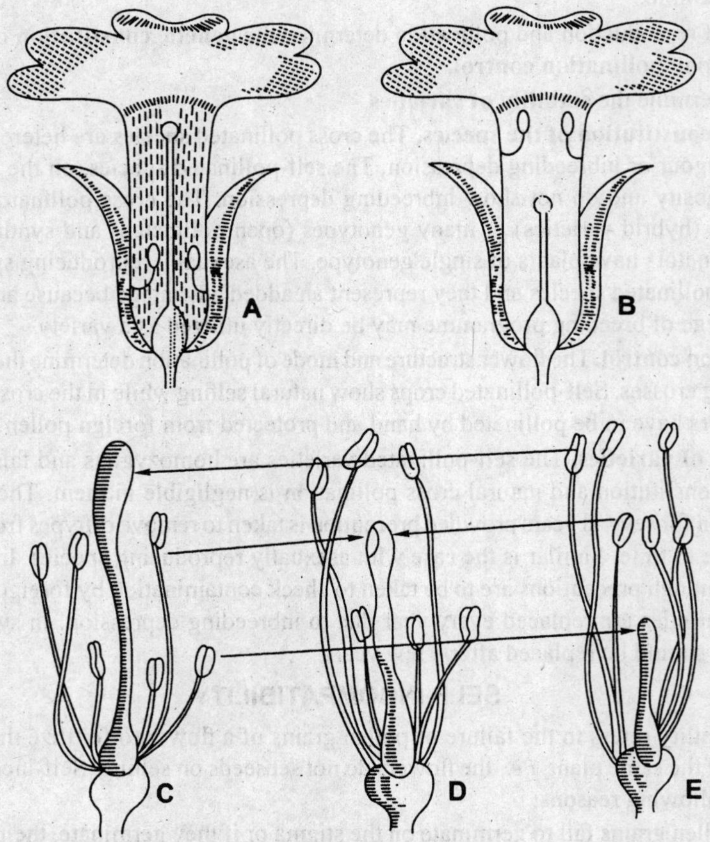

Fig. 37.1 Heterostyly in *Primula*. A. Pin type flower, B. Thrum type flower, C, D, E – Pollination in heteromorphic system.

This character is governed by a single gene *s*. The genotype *ss* (homozygous recessive) is responsible for pin character and genotype *Ss* (heterozygous dominant) is responsible for thrum character. The allele *S* is dominant over *s*. The compatibility reaction of pollen is determined by the genotype of parent plant. The system is, therefore, **heteromorphic sporophytic system**. All the pollen grains of pin flowers have *s* genotype and are self-incompatible in reaction whereas in thrum flowers, half of the pollen grains have *S* genotype and remaining half have *s* genotype. The allele *S* is dominant over *s* in determining incompatibility as well as length of style. Pollen of an incompatible group would not normally grow on the stigma of same incompatibility group and so unable to effect self-fertilization. Such a mating is said to be incompatible. The mating between pin (*ss*) and thrum (*Ss*) plants would produce pin and thrum progeny in equal numbers.

Phenotype	Pin flowers	Thrum flowers
Genotype	*ss*	*Ss*
Gametes	*s*	*S* and *s*

Mating Reaction

Mating phenotypes	*Genotypes*	*Progeny*	
		Genotype	*Phenotype*
Pin × Thrum	*ss* × *Ss*	1*Ss* : 1*ss*	1 Thrum : 1 Pin
Thrum × Pin	*Ss* *ss*	1*Ss* : 1*ss*	1 Thrum : 1 Pin
Thrum × Thrum	*Ss* *Ss*	Incompatible mating	
Pin × Pin	*ss* *ss*	Incompatible mating	

Homomorphic System

In this system, self-incompatibility is not associated with the morphological differences between flowers but the incompatibility reaction of pollen may be governed by the genotype of plant on which it is produced (*sporophytic control*) or by its own genotype (*gametophytic control*).

1. Sporophytic Control. In this system incompatibility reaction of pollen is governed by the genotype of sporophytic plant on which it is produced and not by the genotype of the pollen and that of style is governed by its own genotype. It was first reported by Hughes and Babcock in *Crepis foetida* and by Garstel (1950) in *Parthenium argentatum*. The sporophytic incompatibility occurs in radish (*Raphanus sativus*), diploid *Brassica* etc.

Genotypes of sporophytes:	S_1S_2	S_1S_3	S_1S_4	S_2S_3	S_2S_4	S_3S_4
	↓	↓	↓	↓	↓	↓
Genotypes of gametes	S_1,S_2	S_1,S_3	S_1,S_4	S_2,S_3	S_2,S_4	S_3,S_4
Incompatibility reaction of pollen grains:	All S_1	All S_1	All S_1	All S_2	All S_2	All S_3
Incompatibility reaction of styles:	S_1	S_1	S_1	S_2	S_2	S_3

Letters within parenthesis, say (S_1), (S_2) denote incompatibility reactions of styles and pollen

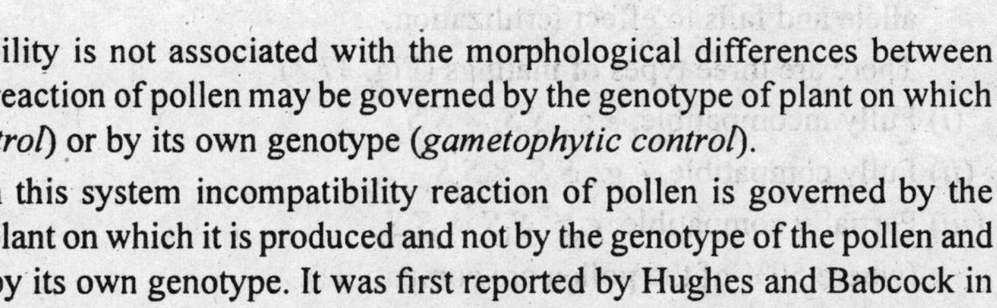

S_1S_2 Selfed → Complete incompatibility; $S_1S_2 \times S_1S_4$ → Complete incompatibility; $S_1S_2 \times S_2S_3$ → Complete incompatibility

Fig. 37.2 Sporophytic system of incompatibility.

In sporophytic system, self-incompatibility is determined by a single gene S with multiple alleles (S_1, S_2, S_3, S_4 and so on) which show complete dominance in a sequence $S_1 > S_2 > S_3 > S_4$ ----etc. In sporophytic system S alleles may exhibit dominance, individual action or competition and so there exist many incompatibility relationships (Fig. 37.2). Lewis has suggested the following characteristics of this system:

(*i*) There may be frequent reciprocal differences.

(*ii*) Incompatibility may occur in female parent.

(*iii*) Homozygotes are normal part of this system.

(*iv*) Incompatibility group may contain two genotypes.

(*v*) A family can consist of three incompatible groups.

Incompatibility reaction of pollen grains is governed by the genotype of the parent plant on which they are produced and the reaction of style is controlled by its own genotype. The incompatibility alleles or S alleles show complete dominance in the sequence $S_1 > S_2 > S_3 > S_4$ ---- etc.

2. Gametophytic System of Incompatibility. East and Mangelsdorf described this system in *Nicotiana sanderae* in 1925. In this system, incompatibility reaction is governed by genotype of pollen itself and that of style is determined by its own genotype. It is determined by a single gene S with multiple allelic series. In some cases two genes with multiallelic series are also known and generally two alleles do not show dominance, *i.e.*, they show codominance. Incompatibility reaction of pollen is determined by its own genotype and not by the genotype of its parent plant (sporophyte). Pollen tube grows very slowly in the style containing same S allele and fails to effect fertilization.

There are three types of matings (Fig. 37.3):

(*i*) Fully incompatible, *e.g.*, $S_1S_2 \times S_1S_2$

(*ii*) Fully compatible, *e.g.*, $S_1S_2 \times S_3S_4$

(*iii*) Partially compatible, *e.g.*, $S_1S_2 \times S_2S_3$ (where 50% of the pollen are compatible).

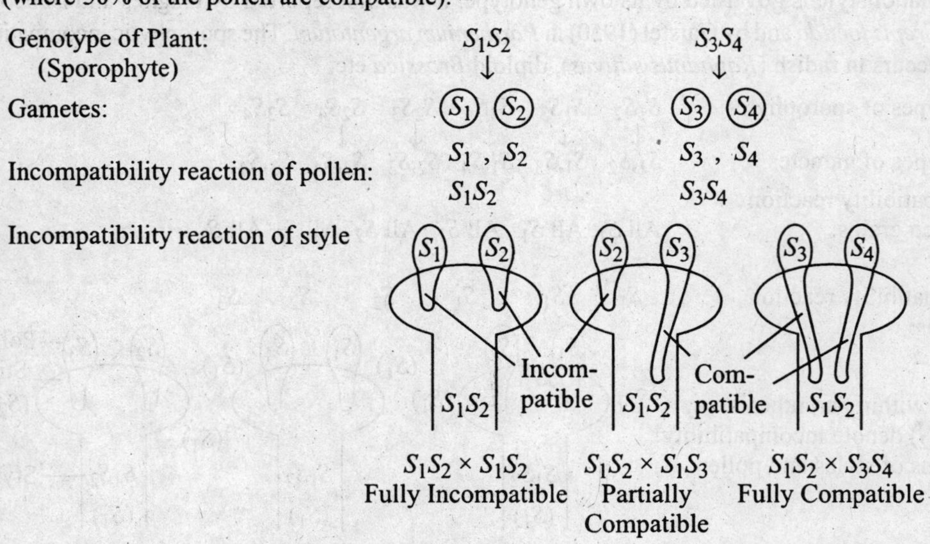

Fig. 37.3 Gametophytic system of incompatibility.

There are some cases which contain an allele for self-fertility S_f and the pollen with S_f allele does not show incompatibility reaction. Thus a plant of genotype S_fS_1 on selfing produces progeny

S_fS_f and S_1S_f. Allele S_f may be induced by irradiating the pollen used for selfing. Another allele $S_{f'}$ is also known which retards growth of S_f pollen tube and causes self-incompatibility. Gametophytic system of incompatibility is reported in Pineapple, rye grass, *Trifolium*, coffee, etc.

MALE STERILITY

Plants which are unable to produce functional pollen but possess normal fertile female structure are referred to as *male sterile*. Thus, *male sterility is characterised by non-functional pollen grains.* Male sterility occurs possibly due to mutations and is classified into the following three groups on the basis of inheritance pattern:

1. Genetic male sterility.
2. Cytoplasmic male sterility.
3. Cytoplasmic-genetic male sterility.

Genetic Male Sterility

In some crops such as barley, corn, sorghum, sugarbeet, genetic male sterility is governed by specific *male sterility gene 'ms'* which may arise spontaneously or may be induced. Generally the male sterility gene (ms) is a recessive chromosomal gene and the dominant allele Ms produces male fertility. This shows Mendelian pattern of inheritance. Male sterile line may be maintained by crossing it to heterozygous male fertile plant (Back cross). Such a cross results in male sterile and male fertile plants in 1 : 1 ratio as shown here:

Parents: ms ms (Pure male sterile) X $M_s M_s$ (Pure male fertile)

↓

F_1: Ms ms (Heterozygous male fertile)

↓ Selfing

F_2: 1 $M_s M_s$ (Male fertile) — 2 Ms ms (Male fertile) — 1 ms ms (Male sterile)

Maintenance of Male Sterile Line

♀ ms ms (Male sterile line) X Ms ms (Back cross) (Male fertile F_1 heterozygote)

♂ Ms ms Male fertile (seeds not collected) — 1 : 1 — ♀ ms ms Male sterile (seeds harvested)

♂ Ms ms (Male fertile hetrozygous plant) X ms ms ♀ (sibmating) (male sterile)

♂ Ms ms Male fertile (seeds are not harvested) — 1 : 1 — ♀ ms ms Male sterile (seeds are harvested and the male sterile line is maintained by sibmating indefinitely)

Utility of Male Sterile Lines in Plant Breeding

In certain crops hybrids are found much superior to their parents in respect of yield. The hybrid production involves emasculation (removal of anthers) of flower buds of female parent and their subsequent pollination. The emasculation is a tidious process which involves high cost. So, in breeding process if the population of male sterile line is available, it can be utilized as female parent. In practice such male sterile lines are interplanted with a homozygous male fertile (Ms Ms) pollinator. Male sterile population may occasionally have a few male fertile plants. Such plants must be identified and removed (roguing) before pollen shedding. Pollen dispersal by hand should also be done for good seed set in the female line. This considerably reduces the labour and cost of hybrid seed production. In some countries it is being successfully utilized for hybrid seed production of certain crops on commercial scale. In USA it is being successfully practised in castor breeding.

Cytoplasmic Male Sterility

Cytoplasmic male sterility is determined by cytoplasmic factors. Since the bulk of cytoplasm of zygote is contributed by the egg cell and the male gamete contributes negligible or no cytoplasm, sterility factors present in the cytoplasm of egg cell will be transmitted to the progeny which would always be male sterile. The occurrence of cytoplasmic male sterility has been reported in many plant species including some crop plants. Important crop plants in which male sterility is controlled, at least in part, by cytoplasmic factors are sugarbeet (Owen, 1945), maize (Rhoades, 1933), sugarcane (Raghavan, 1951), sorghum or jowar (Stephens, 1935), Bajra (*Pennisetum typhoides*) (Menon, 1957), Onions (Jones and Clarke, 1943).

Cytoplasmic male sterility may be transferred easily to a given strain by using that strain as pollinator in successive generations of back-cross programme. Since the cytoplasmic male sterile lines have nuclear genotype similar to that of male fertile pollinator, such a male fertile line is known as the *maintainer line* as they are used for maintaining male sterile line.

A case of extrachromosomal inheritance of male sterility was discovered in maize by Rhoades (1933). He noticed a male sterile plant in maize which produced only male sterile progeny when fertilized with pollen from normal male fertile plants. The F_1 male sterile plants were then back crossed repeatedly with normal fertile parent until all the chromosomes from male sterile plants had been replaced by those from male fertile plant (Fig. 37.4). Even then male sterility persisted which indicated that male sterility was not controlled by nuclear genotype.

In rare cases, male sterile plants produced a few fertile pollen grains. When reciprocal crosses were made between male sterile parent (with fertile pollen) and normal male fertile (female), the progeny was found to be male fertile. Such cases confirmed maternal inheritance of male sterility.

Dhawan and Paliwal (1964) have also demonstrated differences between the hybrids of reciprocal crosses involving two strains of maize Sikkim primitive-2 and another strain from Colorado. When Sikkim primitive-2 was used as female parent in the cross, the progeny showed little vigour and poor yield, but when Colorado strain of maize was used as seed parent, the hybrids were more vigorous and they showed high yield potency. These differences in the hybrids of reciprocal crosses indicate that yield and vigorosity are governed by female cytoplasm and this can be accepted as a good case of extrachromosomal inheritance.

Utility of Cytoplasmic Male Sterility in Plant Breeding

Cytoplasmic male sterility has some value in production of hybrid seeds in ornamental plants or in plants in which vegetative parts are of economic value. Since the hybrid progeny of cytoplasmic male sterile plants would be male sterile, it is practically of no use in the production of hybrid seeds of such plants in which seeds are of no economic value.

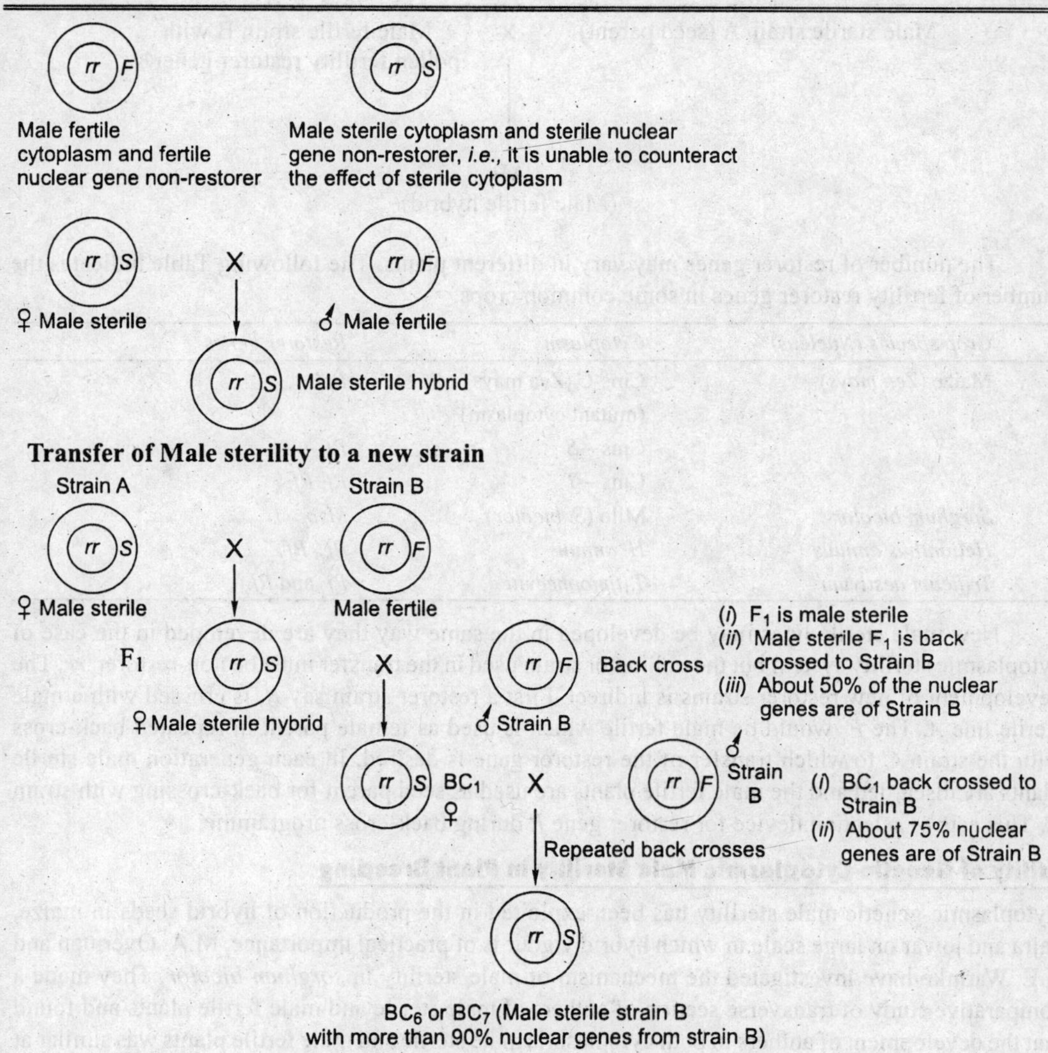

Fig. 37.4 Cytoplasmic male sterility.

Cytoplasmic—Genetic Male Sterility

Male sterility is not always a cytoplasmic effect. When the cytoplasmic male sterility is expressed in a male sterile line which is known to possess nuclear gene *R* for restoring the fertility of pollen, it is referred to as Genetic-cytoplasmic male sterility. The nuclear gene which restores male fertility in male sterile line is referred to as **restorer gene**. The expression of cytoplasmsic male sterility depends on the nature of nuclear genotype with respect to restorer genes. The plants would be male sterile in presence of male sterile cytoplasm only if the nuclear genotype is *rr* (pure recessive). The restorer genes in dominant condition (*RR* or *Rr*) suppress the effect of cytoplasmic male sterility genes. Thus expression of male sterility depends on an interaction between nuclear and extranuclear genes. A single dominant nuclear gene can restore pollen fertility even if extrachromosomal male sterile gene is present. Strains or varieties with pollen fertility restoring genes if crossed to another male sterile plants (seed parent), fertility may be restored in them.

Stephen and Holland (1954) have demonstrated this fact in *Sorghum*.

Male sterile strain A (seed parent) X Male fertile strain B with pollen fertility restorer gener®

AB
(Male fertile hybrid)

The number of restorer genes may vary in different plants. The following Table indicates the number of fertility restorer genes in some common crops.

Crop species (Nucleus)	*Cytoplasm*	*Restorer genes*
Maize (*Zea mays*)	Cms-*C* (Zea mays) (mutant cytoplasm)	Rf_3
	Cms—*S*	Rf_4
	Cms—*T*	Rf, Rf_2
Sorghum bicolor	Milo (*S. bicolor*)	*Mso*
Helianthus annuus	*H. annuus*	Rf_1, Rf_2
Triticum aestivum	*T. timopheeviti*	Rf_1 and Rf_2

New male sterile lines may be developed in the same way they are developed in the case of cytoplasmic sterility system but the pollinator strain used in the transfer must be non-restorer, *rr*. The development of new restorer strains is indirect. First a restorer strain say *R*, is crossed with a male sterile line *A*. The F_1 would be male fertile which is used as female parent in repeated back-cross with the strain *C* to which transfer of the restorer gene is desired. In each generation male sterile plants are discarded and the male fertile plants are used as seed parent for back-crossing with strain *C*. This acts as selection device for restorer gene *R* during back-cross programme.

Utility of Genetic-Cytoplasmic Male Sterility in Plant Breeding

Cytoplasmic-genetic male sterility has been exploited in the production of hybrid seeds in maize, bajra and jowar on large scale in which hybrid vigour is of practical importance. M.A. Overman and H.E. Warmke have investigated the mechanism of male sterility in *sorghum bicolor*. They made a comparative study of transverse section of anthers of male sterile and male fertile plants and found that the development of anthers in both cytoplasmic male sterile and male fertile plants was similar at the time of meiosis. After that, the tapetal cells, which provide nutrients to the developing microspores became thicker, more vacuolated and underwent cytoplasmic fusion in male sterile anthers whereas in normal fertile plants, the tapetal cells were disintegrated and their contents were released which served as nutrient medium for developing microspores. Thus, abnormal behaviour of Tapetum was the main cause of male sterility in *sorghum*.

QUESTIONS

1. Discuss briefly the pollination mechanisms in crop plants. How would you determine the mode of reproduction in a species of plant?
2. What do you mean by self-incompatibility? Discuss in brief homomorphic and heteromorphic systems of incompatibility.
3. Define male sterility. Discuss different types of male sterility in brief.
4. Write brief note on:
   (*i*) practical utility of male sterile plants, and
   (*ii*) maintenance of genetic male sterile line.

38

CHAPTER

Methods of Crop Improvement

METHODS OF CROP IMPROVEMENT

The principal aim of plant breeding is to evolve improved plant varieties that have many good qualities in them and are superior to existing varieties in several respects. The crop improvement has been achieved through several methods such as (*i*) *plant introduction and acclimatization*, (*ii*) *simple and honoured methods of selection*, (*iii*) *hybridization*, (*iv*) *mutation breeding*, and (v) *breeding for disease resistance*.

INTRODUCTION AND ACCLIMATIZATION OF CROPS

Plant introduction is the process of introducing plants into a new locality with different climate from their natural growing places. In other words, introducing a genotype or a group of genotypes of plants into a new place where they have not been grown earlier is **plant introduction**. The plant introduction may be **intercontinental, intercountries, inter-states** or **inter-districts**. The introduction is followed by acclimatization. The process of acclimatization refers to adaptation or adjustment of the population of introduced plants for a number of generations in a new locality with changed agroclimatic conditions.

Purpose of Introduction

Introduction of plants is done for several purposes as listed below:

1. **To introduce new crop plants in an area.**
2. **For direct use in agriculture, forestry and industries.** Many exotic varieties of food crops, grasses, trees and medicinal plants have been introduced in India. Several varieties of wheat have been introduced in India from Mexico, some varieties of rice have been introduced from Japan and Philippines. Germplasm of many varieties of crops, medicinal plants and forest trees are introduced on exchange or gift basis from foreign countries to India.
3. **For Aesthetic Purpose.** Several ornamental herbs, shrubs, forest trees and lawn gasses are imported from foreign countries as well as from other states and grown in the gardens, parks and forests for their aesthetic values and decoration purposes.
4. **For Genetic Improvement of Economic Crops.** Several promising varieties of crop plants and their wild relatives are introduced from different countries and other states and they are utilized in hybridization programme for transferring useful genes from exotic varieties into the local or indigenous varieties.
5. **To Protect from Diseases and Pests.** Sometimes, certain economically important crops are introduced in new areas to protect them from diseases and pests. As for instance, rubber plant

(*Hevea*) was introduced in Malaya from South America to save that from a leaf disease and coffee was introduced in South America from Africa to prevent loss due to leaf rust.

6. **For Scientific Studies.** Sometimes collection of plants are made from different parts of the world for the purpose of studies on biosystematics, evolution and origin of species.

History of Plant Introduction

The majority of crop plants have moved from their primary centres of origin to new areas primarily due to the activities of man. Plant wealth of almost all the nations to a larger extent are the result of plant introduction. Most of the crops were introduced in pre-historic time. For several centuries the agencies of plant introduction were settlers, traders, travellers, invaders, pilgrims, explorers, etc. and the introductions were made either knowingly or unknowingly. Mung, mustard, pear, apple and walnut (*Juglans regia*) were introduced from the Central Asia Centre of Origin, Sesame, jowar, pigeon pea, Asian cotton (*Gossypium herbaceum*) and finger millet (*Eleusine coracana*) were introduced in India from Africa in pre-historic period. Muslim invaders brought cherries and grapes from Afganistan to India. Portuguese introduced chillies, potatoes, sweet potato, maize, groundnut, cashewnut (*Anacardium occidentale*) and tobacco in sixteenth century. Tea (*Camelia sinensis*), litchi and loquat from China, cabbage, cauliflower and other vegetables from Mediterranean, mahogany timber tree from West-Indies were introduced in India in the last quarter of 18th century by East India Company. A number of botanic gardens, various agricultural and horticultural research stations, forest departments have introduced crop plants, medicinal plants, horticultural plants, forest trees and several other plants of economic values from foreign countries to India. Quinine and *Hevea* rubber trees were introduced in India from South America by Kew Botanic Gardens, England.

Centres of Origin of Cultivated Plants

It is evident from the study of geographical distribution of plants that the species of plants are not uniformly distributed throughout the world. The part of earth between 20°S and 37°N including South East Asia, China, Indochina, India, Malaya, Archipelago, South West Asia, Tropical Africa, Avicenia, Central America, South America, Mexico abound in different kinds of vegetation and the species show maximum diversity of forms in these regions. The regions or areas having species with maximum diversity of forms were termed the **Primary centres of origin of cultivated crops** by N.I. Vavilov in 1936. According to the concept of centres of origin of cultivated plant species put forth by Vavilov, the then Director of Institute of Plant Industries, Leningrad, the cultivated crops of the world originated in these primary centres of origin. In other words, the places where the crop species show maximum diversity of forms, genetic variability and heritability and the primitive ancestral species are in abundance are referred to as primary centres of origin. In the central region of the primary centres of origin occur dominant genes and in peripheral part, there occur recessive genes. These places display continuous variability in a particular crop species. In these primary centres of origin, plants occur in wild state. Later, when the crop species move to different places primarily due to activities of man hybridization and mutations occur in wild forms and the useful variations or changed wild forms are selected and protected by man. The wild forms contain dominant genes and their cultivated forms possess recessive genes.

Besides the primary centres of origin, there are some regions where some crop species show considerable diversity, although, they did not originate there and only recessive genes are prevalent there. These regions are referred to as **secondary centres of origin** of those species. The secondary centres of origin of crop plants are not inhabited by closely related wild species. For example, Oat (*Avena sativa*) shows maximum variability in Spain and several species occur over there, yet all related species do not occur there. Therefore, Spain cannot be primary centre of origin but it is the secondary centre of origin of Oats.

J.R. Harlen in 1948 planned an expedition to Turkey for the exploration of plants and in the course of journey he found tremendous plant diversity in some small patches of land to which he named **microcentres**. From such microcentres some precious plants may be obtained and utilized in experimental studies relating to biosystematics and origin of cultivated species.

Nikolai Ivanovich Vavilov (1887–1943), a famous Russian Botanist recognised the following centres of origin of cultivated species and grouped them into **Old World Centres of Origin** and **New World Centres of Origin**.

Old World Centres of Origin

The old world centres of origin include the following centres:

1. **The China Centre of Origin.** This centre is considered to be the oldest and the largest independent centre of origin of cultivated crops in the world. It includes mountaineous parts of Central and Western China and neighbouring tracts. 136 independent opp have been tested from this centre and improved species are Jowar (*Sorghum species*), soybeans (*Glycine max*), opium poppy (*Papaver somniferum*), tea (*Camelia sinensis*), radish (*Raphanus sativus*), brinjal (*Solanum melongena*), *Cucumis* species, buckwheat (*Fagopyrum esculentum*) etc., are considered to have originated in the China Centre of Origin. This centre is also considered to be the place of origin of apple, plums (*Prunus divaricata*), peach (*Prunus persica*), orange (*Citrus nobilis*), apricot (*Prunus armenia*), etc. This is the secondary centre of origin of sesame (*Sesamum indica*), turnip (*Brassica rapa*), Rajma (*Phaseolus vulgaris*), cowpea (*Vigna anguiculata*), beans and maize (*Zea mays*).

2. **The Hindustan Centre of Origin.** The Hindustan Centre of Origin has been divided in the following two centres of origin by Vavilov (1935).

2-A. Indo-Burma Centre of Origin. Excepting Punjab and North Western parts, whole of India and Burma are included in this centre. It is considered to be the centre of origin of rice (*Oryza sativa*), Sugarcane (*Saccharum officinarum*), pigeonpea (*Cajanus cajan*), chickpea (*Cicer arietinum*), Mung (*Vigna radiata*), Cowpea (*Vigna anguiculata*), Jowar (*Sorghum vulgare*), bean (*Dolichos lablab*), lettuce (*Lactuca indica*), *Cucumis sativus*, oat (*Avena sativa*), Yams (certain species of *Dioscorea*) etc. All tropical fruits such as guava (*Psidium guayava*), Jamun (*Eugenia jumbolana*), Mango (*Mangifera indica*) are considered to have originated in this centre. Deshi cotton (*Gossypium arboreum*), Jute (*Corchorus sp.*), *Susbania aegyptiaca*, indigo (*Indigofera sp.*), black pepper (*Piper nigrum*), arecanut, some citrus species, coconut (*Cocos nucifera*), cardemom etc., also originated in this centre of origin.

2-B. The Siam-Malaya-Java Centre of Origin. It includes Malaya, Java, Sumatra, Philippines, Indochina etc. Several important plants like turmeric (*Curcuma domestica*), ginger, black pepper, hemp (*Canabis sativa*), cardemom, banana (*Musa paradisiaca*), some species of citrus, and several fruits are thought to have originated in this centre of origin.

3. **The Central Asia Centre of Origin.** It is also known as Afganistan Centre of Origin. This centre includes Northwest India (Punjab, Kashmir), Afganistan, Russia, Northwest Frontier Province, Tadjikistan, Uzbekistan and adjoining areas. Important crops like Russian wheat (*Triticum aestivum*), club wheat (*Triticum compactum*), sunn hemp (*Crotolaria juncea*), carrot, Pea (*Pisum sativum*), mung, radish (*Raphanus sativus*), onion (*Allium cepa*), garlic (*Allium sativum*), spinach (*Spinacea oleracea*), broad bean (*Vicia faba*), brassicas, linseed, and some fruits like apple (*Pyrus malus*), almond (*Pyrus amygdalus*), grapes (*Vitis vinifera*), pistacia nut (*Pistacia vera*), apricot (*Prunus americana*), pear (*Pyrus communis*) are supposed to have originated in this centre. It is the secondary centre of origin of rye (*Secale cereale*).

4. **Near East or Persian Centre or Asia Minor Centre of Origin.** This centre includes Asia minor, whole of Transcaucasien Iran, highlands of Turkmanistan, etc. Nine species of *Triticum* including *T. monococcum*, *T. durum*, *T. turgidum*, *rye*, alfalfa (*Medicago sativa*), persian clover (*Trifolium resupinatum*), carrot (*Dancus carota*), cabbage (*Brassica oleracea*), *Avena* species, *Lactuca sativa* and fruits like pomegranate (*Punica granata*), pear, almond, chestnut (*Castenea sp.*), water melon and *Cucumis melo* are supposed to have originated in this centre. This is the secondary centre of origin of coriander (*Coriandrum sativum*), black mustard (*Brassica nigra*), rape (*B. campestris*), leaf mustard (*B. japonica*), turnip (*B. rapa*).

5. **Mediterranean Centre of Origin.** This centre is considered to be the place of origin of some cultivated species of wheat such as durum wheat (*Triticum durum*), emmer wheat (*T. dicoccum*), oat (Species of *Avena*), barley (*Hordeum vulgare*), lentil (*Lens culinaris*), several species of lathyrus, pea (*Pisum sativum*), chick pea, broad bean (*Vigna faba*), clover (*Trifolium species*), vetch (*Vicia sativa*), mustard (several species of *Brassica*), onion, garlic, beet (*Beta vulgaris*), *Asparagus officinalis*, lavender, Pepermint (*Mentha* species).
6. **Abyssinian Centre.** Some important crops such as barley, emmer wheat, lentil (*Lens culinaris*), linseed (*Linum usitatissimum*), bajra (*Pennisetum americanum*), jowar (*Sorghum bicolor*), chickpea, sem (*Dolichos lablab*), pea, safflower (*Carthemus tinctorius*), sesame, castor (*Ricinus communis*), okra (*Abelmoschus esculentus*), coffee (*Coffea arabica*) are considered to be originated in this centre of origin. It is the secondary centre of broad bean.

New World Centres of Origin

Expeditions in new world were made in 1932 and the following centres of origin were recognised:

7. **The South Mexican and Central American Centre of Origin.** It is also known as the Mexican Centre of origin. It includes South Mexico and Central America. Maize (*Zea mays*), rajma (*Phaseolus vulgaris*), lima bean (*Phaseolus lunatus*), sweet potato (*Ipomoea batatas*), tobacco (*Nicotiana tabacum*), *Gossypium hirsutum*, *G. purpureascens*, chillies (*Capsicum annuum*), pumpkins (*Cucurbita melanosperma*), melons (*Cucumis melo*), papaya (*Carica papaya*), arrow-root (*Maranta arundinacea*), Cocoa (*Thebroma cacao*) and some other crops originated from this centre of origin.
8. **The South American Centre of Origin.** It includes high mountaineous regions of Peru, Bolivia, Columbia, part of Chile and Brazil, whole of Paraguay and Ecuador. Many cultivated species of Potatoes, maize, lima bean, peanut (*Arachis hypogea*), tomatoes, egyptian cotton (*Gossypium barbadense*) and quinine (*Cinchona calysaya*), cassava (*Manihot Atilissima*), rubber (*Hevea sp.*) originated in this centre.

   Recently one more centre namely USA Centre of Origin has been introduced in which two crops sunflower (*Helianthus annuus*) and Jerusalem artichoke (*Helianthus tuberosus*) are believed to have originated.

Procedure for Plant Introduction and Acclimatization

Plant materials are introduced following a definite procedure. This involves the following steps:

(*i*) Collection of plants from other sources,
(*ii*) Despatch of material,
(*iii*) Plant Quarantine,
(*iv*) Cataloguing,
(*v*) Evaluation of introduced material,
(*vi*) Multiplication and distribution to growers.

1. Methods of Germplasm Collection

The sum total of hereditary material or genes present in a species is known as **germplasm** of that species. Collection of a large number of genotypes of a species and its relatives is called **germplasm collection** or **gene bank**. When the collections are large and include genotypes from all over the world, it is called **world collection**. Germplasm collections furnish the richest source of variability. The crop inprovement ultimately depends upon the variability of germplasm to be utilized in breeding programme.

The desired material is demanded directly from the concerned agency or authority of the foreign country through plant introduction organisation of the country. The needed material can be had easily from Food and Agricultural Organization (FAO) which maintains world catalogues of genetic stocks of wheat, rice and some other crops. The material can also be obtained through Embassies of friendly countries or through many international agencies existing in the country either on exchange basis or as gift or against payment.

For the introduction of plant materials the following four methods are adapted:

(*i*) **By personal visit.** In this method the breeders or their assistants visit the places of availability themselves to collect the plants of their choice.

(*ii*) **By expeditions.** In this method a team of trained persons plan visit to different places to collect desired plants. Such types of expeditions are arranged by private bodies, Government agencies of the country, or joint committee of two or more governments. The expeditions are of the following two types:

(*a*) *Ad-hoc Expeditions*. The expeditions organised and financed by the government of one country are called *ad-hoc* expeditions. Such expeditions have been organised by Government of India to Burma, Sumatra, Java and Indochina under "*Spontaneum Expedition Scheme for Sugarcane*" improvement to collect several varieties of 'Kaans' (*Saccharum spontaneum*). The introduced varieties were maintained at Sugarcane Research Institute, Coimbatore and they were involved in cross hybridization with sugarcane (*Saccharum officinarum*) to incorporate disease resistant genes of *S. spontaneum* into sugarcane. This played very significant role in the improvement of sugarcane.

(*b*) *Co-operative Expeditions*. The expeditions organised and funded jointly by the goverments of two or more countries are called co-operative expeditions. Expeditions arranged by Food and Agriculture Organisations (FAO) belong to this category.

(*iii*) **By Correspondence.** In this method correspondence is made with the Government or Plant Research Stations of the concerned friendly country and request made for the supply of plant materials (seeds or cuttings, etc.). Though this method is quick and cheap yet there is little possibility for the supply of pure and desired plant material. Cotton, berseem, linseed and sugarcane seeds cannot be obtained through letter post.

(*iv*) **Plant Collection on Exchange Basis.** In this method of plant collection, the government or the plant breeding stations or individual plant breeders first decide about the plants to be exchanged and then the exchange is made on reciprocal basis. This method of plant collection is quick, involves less expenditure of money and makes available genuine plant material.

The export and import of certain plant species are restricted or prohibited. Almost all countries restrict exchange of germplasm of one or more plant species. India restricts the export of propagating materials of blackpepper, jute and tea. Introduction of the following is prohibited in India: plants and seeds of cocoa and other *Theobrome* species from Africa, West Indies and Srilanka, sugarcane from Australia, Fiji, Newguinea, West Indies, and Philippines, sunflower from Argentina and Peru, unginned cotton and Mexican jumping beans, plants and seeds of rubber plant from West Indies and America.

2. Despatch of Materials

The exporting agency, before sending the materials, cleans the material, makes it free from contaminations, treats it with fungicides, pesticides, etc. and then packs it up. Cuttings are enclosed in damp cotton wool and then packed inside polyethene bags. Labels bearing reference number are placed inside and outside the package for the purpose of exporting seeds and pollen grains. Polyethene tubes are also used. The sender mails the materials generally by sea mail. Import of plants by air is restricted and cannot be done in India without prior permission from Plant Protection Adviser to the Government of India.

3. Plant Quarantine

Past experience has made it clear that the control of undesired organisms could not be possible in spite of big expenditures on the process. In fact, some of the plant introductions proved to be more costly and harmful than being beneficial. Therefore, strict and effective measures are must to check the entry of unwanted agents.

Quarantine literally means to keep introduced materials in isolation for forty days to prevent spread of diseases, etc. But it is not always necessary or desirable to keep every introduction in isolation for forty days. Sometimes it is desirable that the materials received at entry place should be screened and treated with pesticides within a few hours and sent to the addressee. Some introduced plant materials, however, may be kept in isolation for months or year. All the introduced plant material is thoroughly inspected for contamination with pathogens, insect pests and weeds. The materials suspected to be contaminated are fumigated or given treatments to make them free from contamination. If it is necessary, the introduced materials are grown in isolation and kept under observation for diseases, insect pests and weeds. The entire process involving inspection, fumigation or treatment with pesticides and growing the introduced materials in isolation is known as **quarantine** and the rules prescribing them are known as **quarantine rules**. Now several countries are applying stringent measures to check the entry of exotic diseases, insect pests and weeds. The scientists of plant quarantine department destroy the pathogens, insects and weeds by treating the imported materials with fungicides, insecticides and weedicides at the place of entry and then they issue *Phytosanitary certificates* (import permission). They also keep the records of quarantine measures applied. The phytosanitary certificate is attached to the package and then the package is sent for delivery. Introductions suspected to be contaminated or those not conforming to the quarantine rules are destroyed by NBPGR or returned to the senders. Certain plant materials which are prohibited for import in India cannot be demanded from foreign countries. The quarantine measures are exercised by the NBPGR at prescrbed ports of entry, *e.g.,* Delhi, Chennai, Mumbai, Kolkata etc. The Division of Mycology and Plant Pathology and Division of Entomology, IARI, also assist NBPGR in quarantine activities.

4. Cataloguing

The material received is examined, identified, numbered, catalogued, and then grown in isolated areas under local conditions and tested for acclimatization.

In cataloguing, besides giving entry number, name of the species and variety, place of origin, adaptation and characteristics are also recorded. The introduced plant materials are classified into three groups:

(*i*) *Exotic collections* are given prefix **'EC'**.

(*ii*) *Indigenous collections* are designated as **'IC'**.

(*iii*) Indigenous wild collections are marked **'IW'**.

5. Evaluation

The performances of introduced materials are evaluated at different sub-stations of NBPGR. The evaluation and maintenance of those crops are done by the concerned Central Research Institutes, as for example, rice, sugarcane, potato, tabacco, tea, etc. In many cases the introduced materials are grown in isolation to observe diseases, pests and weeds that might have been introduced along with them. Susceptibility to diseases and pests is evaluated under conditions that favour their spread. The performance of introduced material in new environment is referred to as **acclimatization**. The extent of acclimatization is determined by—

(*i*) mode of pollination,

(*ii*) range of genetic variability in original population, and

(*iii*) duration of life cycle of the crop.

Cross-pollination is more helpful in acclimatization than self-pollination. The greater the genetic variability the more is the extent of acclimatization. The duration of life cycle of crop is important because annuals will produce several generations by the time a perennial completes its life cycle. Each generation produces new gene combinations, facilitating thereby acclimatization. Mutations also help in acclimatization.

6. Multiplication and Distribution

The individuals with promising features are selected from the introduced population, multiplied and released as varieties after necessary trials. Introductions having desirable characters are maintained for future use which are utilized in hybridization programmes. They are supplied to other workers on request.

Types of Plant Introduction

Plant introductions are of two types:

1. **Primary Plant Introduction.** The promising varieties of cultivated plants have been introduced in India from the foreign counties from time to time and some of the exotic. varieties have been proved to be superior to indigenous or local varieties. Such varieties have been adapted for cultivation as such. When the introductions are made for changing wild species into cultivated ones or when the introduced genotypes are released as such as new varieties for cultivation in an area with new or changed agroclimatic conditions they are called **primary introductions**. Some examples of primary introductions are listed here as under:

   (*i*) *New Crop Species*. Several important crops grown in India are introductions. Potatoes, maize, groundnut, chillies, coffee, rubber, guava, grapes, pine apple, papaya, etc. and some ornamentals like Goldmohr, *Phlox*, *Salvia*, Aster, etc. are introductions. Soybean and sugarbeet were introduced in 1960. Oil palms and jojoba were recently introduced in India which are presently in trial stage.

   (*ii*) *Directly Released Varieties*.

   Varieties of wheat. **Ridley** variety of wheat developed in Australia is grown in Uttar Pradesh, Punjab, Himachal Pradesh. Mexican Dwarf wheat varieties **Sonora 64** and **Lerma roso** introduced in India were released for cultivation directly.

   Oat *(Avena sativa)*. Several Australian varieties of oat such as Kent, Flaming, Gold, Green graph, Mountain etc. and Algerian variety 'Grey Algerian' introduced in India have been released directly as varieties for cultivation.

   Rice (Oryza sativa). Paddy varieties China-10 introduced from China and IR-0, IR-8, and IR-28 introduced from International Rice Research Institute (IRRI), Philippines and the variety Taichung Native-1 developed in Taiwan and introduced in India were directly released as varieties for cultivation which are very successful in several rice growing belts of the country.

   Pea (Pisum sativum). 'Bonneville' and 'Early Badger' varieties of Pea introduced from America are doing well in India.

   Tomato (Lycopersicum esculentum). 'Sious' variety of tomato introduced from America is directly released for cultivation in India.

   Several introductions of vegetable crops such as cowpea, cauliflower, onion, lettuce, water melon have been relased directly as varieties for cultivation in India.

2. **Secondary Introductions.** If the introduction of wild species or cultivated varieties is made for developing genetically improved variety by selection or cross hybridization, it is referred to as **secondary introduction**. Secondary introductions are more important than the primary ones.

1. **Varieties Selected from Introductions.** Many cultivated varieties of crops have been developed through selection directly from the exotic varieties introduced in India. Kalyan sona and Sonalika varieties of wheat were selected from introductions from CIMMYT, Mexico. The varieties 'Jamnagar Giant', 'Improved Ghana' of Bajra, 'Pusa Lal', 'Pusa sunahari' varieties of sweet potato, 'Pusa Basmati' variety of cowpea, and 'Japanese white' and '40-days' varieties of radish have been selected from introducted materials.

2. **Varieties Developed through Hybridization.** Several introductions have been involved in hybridization with indigenous crops for their improvement. All dwarf wheat varieties are derived from crosses with Mexican dwarf wheats. The majority of dwarf rice varieties possess dwarfing genes Dee-geo-woo-gen either from Taichung Native or IR-8. 'Pusa Ruby' tomato variety was developed from a cross between Meeruty and Sioux, an American variety. Hybrid maize, jowar (*Sorghum bicolor*) and bajra (*Pennisetum americanum*) varieties have one parent (the male sterile or female parent) which is introduction. All the recently released varieties of sugarcane have been derived from introduced noble canes (*Saccharum officinarum*).

Plant Introduction and Quarantine

The cultivation of introduced plant materials has often created serious problems of diseases, insect pests and weeds. Man himself has violated the prohibitary measures and has been responsible for the spread of several diseases, insects and weeds in new areas where they were not known to exist earlier. Several diseases, pests and weeds have found their ways from one country to another along with introduced plant materials and have caused tremendous damage and other problems in India. Some noxious weeds like *Parthenium, Phylaris minor* which were not known in India a few years back, have entered our country along with imported food grains. Now they are spread all over the country and creating problem in agriculture.

Plant Introduction Agencies in India

A centralized plant introduction agency was initiated at Indian Agricultural Research Institute (IARI), New Delhi in 1946. In 1956 that unit was converted into *Plant Introduction and Exploration Organization.* Subsequently in 1961 it was made an independent *Division of Plant Introduction*. In 1976, this division was reorganised as *National Bureau of Plant Introduction* under Indian Council of Agricultural Research (ICAR). In 1977, its name was changed to *National Bureau of Plant Genetic Resources* (NBPGR). It has its head office at New Delhi. The nature of activities of the Bureau remained the same but the scope and the activities have increased considerably. It is responsible for introduction and maintenance of the germplasm of agricultural and horticultural plants. NBPGR has the following four stations for testing the introduced plant materials:

1. **Simla (Himachal Pradesh).** It represents the north hilly temperate zone, approximately 2300 m above sea level.
2. **Amaravati (Maharashtra).** It represents the central zone or mixed climatic zone.
3. **Jodhpur (Rajasthan).** It represents arid zone.
4. **Kanya Kumari (Tamil Nadu).** It represents the tropical zone.

NBPGR is the central agency for the introduction and export of germplasm and it is assisted in its activities by the various Central Research Institutes of ICAR. The activities of NBPGR are listed here as under:

1. Introduction of germplasm from other countries as required.
2. Arranging explorations within and outside the country to collect valuable germplasm.
3. Screening and quarantine of introduced plant materials.
4. Testing, multiplication and maintenance of germplasms obtained from various sources at its sub-stations or Central Research Institutes of ICAR.
5. Supply of germplasm to scientists and institutions on request. If germplasms are not available in the country, they are procured from outside and then supplied.
6. Maintain a record of plant name, variety name, propagating materials, special characteristics, source, date and other relevant information about the materials received.
7. Supply of germplasm to the counterparts in other countries or their agencies.

8. To publish its exchange and collection lists and catalogues. News letter containing such lists is published by Food and Agriculture Organisation (FAO).
9. To set up natural gene sanctuaries of plants where genetic resources are endangered.
10. Improvements of certain medicinal and aromatic plants.

Demerits of Plant Introduction

Plant introduction, no doubt, helps in crop improvement and it is an easy method of crop improvement, yet slight carelessness in handling introduced materials may cause several problems which are listed below:

1. **Diseases.** Some diseases of foreign origin may enter along with the introduced materials and cause damage not only to introduced crop but also to other local crops.
2. **Weeds.** Some introduced plants in new environment behave as dangerous weeds and create nuisance, as for example, *Lantana camera* was introduced in India as ornamental plant but soon that spread as problem weed along the rail tracks and in orchards. *Argemone mexicana, Eicchornea crassipes* (water hyacinth), *Phylaris minor, Parthenium* etc., are some of the instances of weeds which were introduced in India from other countries along with imported plant materials.
3. **Insect pests.** Several insect pests found their way in India alongwith introduced plants, as for example, wooly aphis of apple, fluted scales of citrus, rubber moth from Italy were introduced in India alongwith introduced plant materials.
4. Sometimes the introduced materials do not prove to be beneficial and thus the money and labour spent on their maintenance go waste.

SELECTION

Selection of good quality plant from their population has been in use from the remote past. The selection is of the following two kinds:

1. Natural selection,
2. Artificial selection.

Natural Selection

This is a continuous and autonomous process which is influenced by natural factors. According to Darwin's theory of natural selection, over-production of plants creates struggle for existence in them and the environment induces in them many new variations that make certain plants most suitable and others less adapted to live in the prevailing environmental conditions. Finally, the natural selection comes in operation due to which suitable and most fit varieties survive well and less fit and weak plants do not persist and eventually they are eliminated from the environment (survival of the fittests). Because of natural selection, the plant species are gradually adjusted towards advantageous direction. The cultivated crops have probably been evolved from wild types through natural selection. Certain climatic and regional races are the results of this process.

Artificial Selection

Farmers and plant breeders choose certain individual plants of good quality from mixed populations for the purpose of taking qualitative as well as quantitative yield from them. This is known as artificial selection. In other words, artificial selection is a process which involves the selection of certain individual plants from a mixed plant population having both desired and undesired qualities. The following are three artificial methods of plant selection:

1. Mass selection,
2. Pure-line selection, and
3. Clonal selection.

1. **Mass Selection.** Mass selection is the oldest and simplest method of the crop improvement which is still in use by the farmers. This method is indirect and unsystematic. This method is used in those varieties which are cross-pollinated and self-defective. In this process, fruits of plants that are superior in desired traits (elites or best and healthy plants) are selected from the fields either during or after harvesting and they are thrashed together. The mixture of seeds so obtained is called 'mass' and hence the name of process *mass selection.* Sometimes seeds with extreme characters (best looking and most healthy ones) are selected directly. In the next year, the mixture of seeds is sown *en masse* and new crop is raised and again the same practice of selection is applied. This process is repeated till there reaches a state of complete uniformity in the plants with regard to desired characters. This process takes about 8 years. The general procedure for evolving a variety by mass selection process is outlined as under:

   (*i*) In the **first year**, desired plants, heads or seeds possessing all the possible desired characters are collected in pretty good number from the pre-determined places or fields.

   (*ii*) In the **second year**, the composite seeds of each mass selection are sown in the isolated fields and the progeny grown, and the plants allowed to self or cross-pollinate naturally. The elites are selected and others with undesired characters are removed.

   (*iii*) In the **third, fourth** and **fifth years**, the selected mass is subjected to repeated yield trials and the yield of elites is compared with the yield of old variety.

   (*iv*) In the **sixth, seventh** and **8th years** the selected varieties are sent to different regional research stations and grown there to determine the adaptability and yield in different regions.

   (*v*) At the end of 8th year the variety is multiplied, named, recommended and released for general cultivation.

There are two divergent opinions on the question whether the mass selection should be practised under best cultural conditions or under normal or even unfavourable conditions. According to Hallet (1869), mass selection should be practised under most favourable cultural conditions. But, Rimpau (1867) suggested that it should be practised under ordinary or even unfavourable cultural conditions with minimum supply of water, nutrients, etc.

Limitations

1. It is applicable to crops that show some degree of heterozygosity. The heterozygosity is associated with cross-pollination, this is why mass selection is practised in cross-pollinated crops.
2. The application of mass selection is much limited to self-pollinated plants because they are pure and homozygous.
3. There is no check on cross-pollination in the selected plants. This can cause further heterozygosity.
4. Superior variety cannot be separated easily from the slightly inferior variety.

2. **Pure-line Selection.** Johanssen (1903) was the first to introduce the term "pure-line". Johanssen grew seeds of bean (*Phaseolus vulgaris*) separately and selfed them repeatedly. In this way he made purity in the individuals. Since progeny became more and more homozygous by repeated selfing, the chances of variability became least. Johanssen defined pure-line *as a group of descendants selected among progeny of single genetically pure and self-fertilized individuals.* Hayes, Immer and Smith (1955), Sinnot, Dunn and Dobhzansky (1950), Poehlman (1959), Darlington and Mather (1952) all have defined pure-lines in more or less similar ways.

   In simple terms, the pure-line can be defined as a group of plants obtained from a single homozygous (genetically uniform) and self-fertilized plant. Before Johanssen developed this method of selection in 1903 and started his systematic experiment with bean, Vilmorin in France had also started single plant selection.

Pure-line selection is also called 'individual plant selection', 'inbred selection', 'progeny selection', single line selection (because single plant is selected, inbred and its progenies are maintained separately) and also as pedigree selection (this system is named after Pedigree, a Swedish seed station of Svalof).

General Procedure. The method of production of a variety by the pure-line selection consists in testing the progenies of a single individual plant from the mixed population of naturally self-pollinated crops. It takes nearly 9 to 10 years to produce a variety. In general, the procedure is outlined below:

(*i*) In the **first year**, 50 or more plants or heads of desired quality are selected out from mixed population of crop at the time of harvesting. The seeds from individual plants are picked separately and numbered.

(*ii*) In the **second year**, 25-30 seeds of individual plants are grown in individual rows. Every tenth row is seeded to some standard variety for comparing the selected population with it. The undesired plants as well as diseased plants, are removed from the rows. The selected plants are tested for disease resistance by inoculating the suspension of spores of pathogens on the plant body. The progenies of rows are harvested and seeds of plants in each row composited together and composite produce of each row is maintained separately which is treated as experimental strain.

(*iii*) In the **third year**, each experimental strain of second year is grown in 4 or 5 rows instead of one row as in the second year. The superior strains are maintained and undesired and diseased plants are removed from the field. The seeds of each strain are composited separately after harvesting.

(*iv*) In the **4th, 5th and 6th** years seeds of each strain are sown in small seven rows-plots. Every 5th plot is seeded to some standard strain or local strain for comparing the selected and local strains. The inferior and diseased plants are removed and superior ones are maintained. The yield trials are made during these years and finally two or three superior strains are selected as improved strains.

(*v*) In the **7th year** seeds of finally selected strains are multiplied and sent to various research stations.

(*vi*) In the **8th, 9th** and **10th years,** they are grown at different regional research stations to determine the comparative yield and adaptability under different cultural conditions. The breeder then finally selects one (or at the most two) and give to it variety name and releases it for general cultivation in the recommended area.

Advantages of pure-line selection

(*i*) Pure-lines may be used as a new improved varieties which are uniform in appearance and performance.

(*ii*) Pure-lines may be used as a parent in cross-breeding programmes.

(*iii*) By this method different stains are easily purified from self-pollinated crops.

(*iv*) This process is easier than hybridization.

This process is difficult because–

(*i*) it requires much labour and time, and

(*ii*) it requires careful attention and thorough observation.

3. Clonal Selection. In some plants seeds are either lacking or they are of low viability. Such conditions enforce the plants to use vegetative organs for their propagation. Further, in some fruit trees as orange, mango, apple, etc., there exist wide heterozygosity and high degree of polyploidy. If they propagate by seeds, they may exhibit many variations that may affect the purity of race and may also result in low production. Many crops, such as, bananas, potato, sugarcane, onion, garlic, turnip, grapes, ginger, *Colocasia* propagate through their vegetative part. A group of plants that are derived from the vegetative organs of a single plant is known as **clone**. In simple terms, a clone may be defined *as a progeny of a single plant obtained by vegetative propagation.*

The chief characteristics of a clone are as follows:

(*i*) A clone is obtained by vegetative propagation of single plant and it propagates vegetatively in future generations.

(*ii*) The plants of a clone are genetically alike and since in a clone genetic constancy is ensured, it is immaterial whether the parent is homozygous or heterozygous.

(*iii*) The clones retain their original characters after many years of vegetative propagation, *i.e.*, they are stable.

The clonal selection is concerned with the selection and propagation of best individuals of clones from mixed population of vegetatively propagated crops.

General procedure. Superior clones are selected on the basis of their phenotypic characters from mixed population of a local as well as introduced varieties of crops. Selection of individual is made always among the clones and never within a clone because all the members of a clone are genetically identical.

The units of selection, as well as method involved, are different in different crops. Some of them are listed here:

1. Pieces of underground stem—	*Colocasia,* yam, banana, pineapple, *Aloe, Agave*, potato, *Chrysanthemum,* onion, garlic, etc.
2. Stem Cutting—	Sugarcane, sweet potato, grapes, betel, pepper, rose, jasmine, Croton, *Clerodendron, Thespesia,* etc.
3. Runners—	many grasses
4. Bulbils—	*Dioscorea*
5. Grafts and buds—	*Mango, Citrus, rose, apple,* etc.
6. Roots—	Sugar beet, turnip, Dahlia, etc.

The selected units are grown by vegetative methods and then they are compared with the normal variety. The diseased and poor quality clones are removed and best clones are selected and sent to different regional research stations where they are grown for determining their yield and adaptability under the particular set of conditions for three continuous years. Finally best clones are given variety names and multiplied, recommended and released for general cultivation.

Since in the course of vegetative multiplication variations do not arise in the clones, the characters of clones remain stable. Nevertheless, the clones can be improved by the following methods:

(*i*) By inducing mutations and chromosomal changes in somatic cells which eventually produce mutant buds. The somatic buds are then grown vegetatively. If found suitable, the clones of buds are maintained.

(*ii*) **By Hybridization.** Improvement of clone is also possible in those cases where cross-breeding can be done between the different varieties of clonal populations and viable seeds can be produced. The hybrids obtained from cross-breedings are studied and compared with standard variety. If found suitable, they are named, multiplied, recommended and released for general cultivation.

In sugarcane, which is propagated vegetatively by stem cutting, it was believed that seeds were totally non-viable. Thus, nobody could even think about introducing any improvement in its clones. In 1886, it was noticed that some of the seeds in certain clones were viable but for very short period after maturation and their viability was lost within a few weeks. This discovery provided great scope for improvement of sugarcane and very significant breeding work has been done at the Imperial Sugarcane Station of Coimbatore with spectacular success. Many new and excellent varieties of sugarcane could be produced by crossing cultivated varieties of canes with wild species and with other genera like *Sorghum* (Jowar) and bamboos.

Importance of clonal selection—Selection of clones is important because–

(*i*) it is the only method for improving clonal crops,

(*ii*) it offers opportunity for exploiting desirable bud mutations, and

(*iii*) it maintains the purity of race in heterozygous state.

HYBRIDIZATION

Hybridization can be defined briefly as the method of producing new crop varieties by crossing two genetically different parents. According to the nature and relationship of plants to be crossed, the process is divided into the following types:

1. **Inter-varietal hybridization**. In this, cross is made between plants of two different varieties which belong to the same species. It is sometimes called intra-specific hybridization. Many new varieties of cultivated crops have been evolved by this method.
2. **Intra-varietal hybridization.** In this, plants to be crossed have different genotypes but belong to the same variety.
3. **Inter-specific or Intra-generic hybridization.** In this, cross is made between two different species of same genus. Inter-specific crosses have been successfully tried in wheat, cotton, tobacco, mustard, *Luffa*, etc.
4. **Inter-generic hybridization**. In this, plants belonging to two different genera are crossed. *Raphanobrassica*, sugarcane-*Sorghum*, sugarcane-bamboo, wheat-rye (Triticale), wheat-*Aegilops* hybrids are results of intergeneric crosses.

Objects of Hybridization

1. To evolve a variety having all the desirable characters, such as, high yielding capacity, good quality, resistance to diseases and drought, adaptability to particular place and climate, etc.
2. To produce useful variations by introducing recombinations of characters.
3. To produce and utilize hybrid vigour.

Techniques for Hybridization

The following informations are essential before making crosses:

1. Details of male parent.
2. Details of female parent.
3. Whether plants are unisexual or bisexual.
4. Time of anthesis (opening of flowers) and harvesting of plants.
5. Whether flowers are self or cross-pollinated.

The various steps involved in the operation of hybridization programmes are described below:

1. **Selection of Parents.** Desirable male and female parents are selected from the available materials keeping in mind all the important characters to be combined.
2. The parents are grown in isolated places and are self-pollinated repeatedly in order to bring homozygosity in desired traits. The parents to be crossed should be grown keeping in mind time they take in their maturation and flowering. Both the parents should mature at the same time.

3. Preparation of Female and Male Parents

A. Female parent

(*a*) **Emasculation.** The majority of the crops bear bisexual flowers and, therefore, for making crosses in normally self-pollinated crops, emasculation is essential. Removal of anthers before they dehisce and shed their pollens from bisexual flowers of female parent is known as emasculation. It is done in order to prevent self-fertilization and, therefore, it is usually performed before opening of the flower buds (anthesis). In plants with unisexual flower, the female flowers are bagged before their stigmas become receptive. In hermaphrodite, protandrous and protogynous flowers, the emasculation is done before the anthers shed their pollen.

There are several methods of emasculation:

(*i*) In the case of large flower buds, it is easily performed by opening the flower bud by means of sterilized forceps and fine needle and then removing the anthers. The instruments are sterilized before use by dipping them in rectified spirit. The emasculation should be done without causing injury to other floral parts, such as, sepals, petals, and pistil (Figs. 38.1 and 38.2).

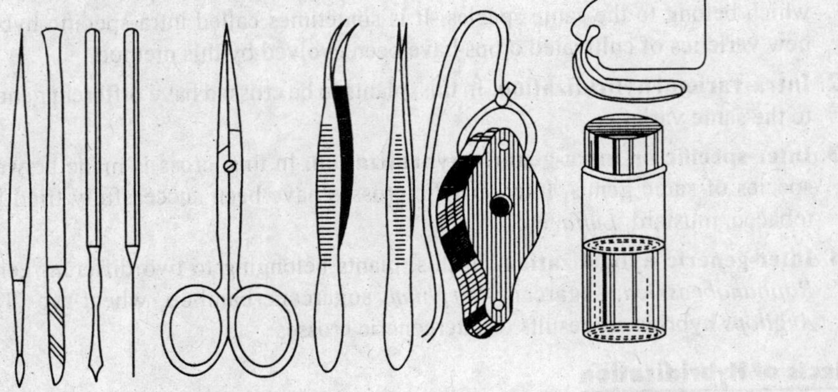

Fig. 38.1 Instruments and the materials required in emasculation.

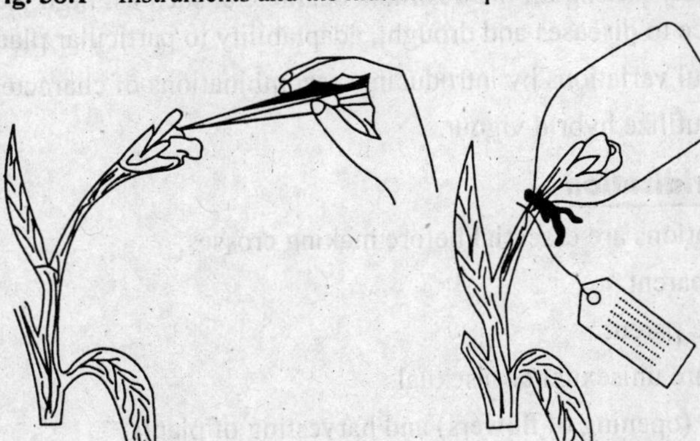

Fig. 38.2 Emasculation of flower bud.

a. An immature flower bud is opened and anthers are being removed by forcep.

b. Emasculated flower bud is enclosed inside a bag and a label is tagged to the bag.

(*ii*) **Emasculation by Hot Water Treatment**. In the cases where the flowers are small hermaphrodites and are crowded in dense inflorescences, as for example, in bajra, jowar, rice, etc., emasculation by forceps and needle is rather very difficult and it requires special skill and technique. In such cases the gynoecium can withstand a temperature at which anthers are killed. The emasculation is performed in one operation by dipping the whole inflorescence in hot water at temperature between 45°C and 50°C for a definite period (1 to 10 minutes). The emasculation in some cases is carried out also by dipping inflorescences into cold water or into alcohol for definite period (Fig. 38.4).

(*iii*) In some self-pollinated crops, emasculation is eliminated by the use of male sterile plants in which anthers are sterile. Male sterility can be induced artificially by spraying 2, 4 D, naphthalene acetic acid (NAA), maleic hydrazide (MA) and tri iodobenzoic acid on the immature flower buds.

The following precautions should be taken during emasculation:

(*i*) Immature flower buds should be selected but that should not be too young. In mature buds there is possibility of self-pollination and in very young buds parts are not well differentiated, they may be injured.

(*ii*) Besides the emasculated bud, all other buds, flowers and fruits should be removed in order to avoid confusion and mistake.

(*iii*) Sufficient number of buds should be emasculated to minimise the chance of failure.

(*iv*) Generally emasculation should be done in the evening when stigma is not receptive, anthers do not dehisce and buds can heal at night the injury done to them during emasculation.

(*v*) As far as possible anthers should be removed by holding their filaments between forceps so that they may not get burst or broken during the operation.

(*vi*) Perfect sterilization of fingers and instrument is necessary before emasculation of each bud so that pollen of the bud previously emasculated may not be transferred to the next bud.

(*vii*) As far as possible, no other part of the flower should be injured, except those which are necessary.

(*viii*) After emasculation buds should be left in their natural form.

(*ix*) After emasculation and before bagging, the stigma must be carefully examined with a hand lens for ascertaining that no pollen grain is present on the stigma.

(*x*) During emasculation, anthers should be collected and counted to make sure that all of them have been removed from the flower bud.

(*b*) **Bagging.** After emasculation flower buds are enclosed in bags of convenient sizes that are made up of thin butter paper, muslin cloth, plastics, parchment, glassine, celophane or polyethene. The bags are tied by threads, wire, pins, etc. This process is known as bagging. Emasculated flower buds are bagged to check cross-pollination of emasculated flowers with undesired foreign pollens.

B. Male Parent

(*i*) Irrespective of the fact that the flowers are unisexual or bisexual, the unopened flower buds of male parent are covered as in the case of female parent in order to avoid the contamination of pollen grains and to make sure that pollen of the bagged flowers are of the same plant and not of any other plant.

(*ii*) **Collection and Storage of Pollen**. Pollen grains for the purpose of crossing are collected from bagged male flowers. Pollens or anthers are collected in petridishes or in paper bags just after dehiscence and they are placed in vials or capsules for future use.

4. Cross Pollination

It is performed by removing bags of emasculated flowers and dusting or brushing the stigmas of pistils with the collected pollens of male parent. In wheat, whole anther is inserted between lemma and palea with the help of forceps. After crossing female flowers are bagged again. In Bajra and Jawar, pollen and penicle parents are grown side by side and crossing is done by enclosing male penicle and emasculated female penicles in one bag. This ensures automatic crossing.

5. Labelling

The emasculated and crossed flowers are bagged, tagged and labelled properly. The labels are tagged to the bags with help of threads. The labelling card should bear the following details.

1. Ref. No..........
2. Date of emasculation..........
3. Date of crossing..........
4. Details of male parents..........

The above entries should be made soon after the pollination.

Breeders keep complete and uptodate records of their works in a separate field record book.

6. Collection of Hybrid Seeds

After maturity of seeds, the crossed heads are collected along with their labels separately in envelops. In the next season, the seeds of each head are sown separately to raise F_1 generation. All the plants of F_1 are genetically similar and may look exactly alike. They may sometimes show hybrid vigour, *i.e.,* increased growth, size, yield or function over the mean of parents.

7. Final Steps

The final step consists in the selection, testing, naming and releasing the variety.

Hybridization Technique in Some Crop Plants

1. Wheat. For emasculation such spikes are selected in which anthesis will occur in coming one or two days. One-third of the upper and lower spikelets are removed. Awns should also be removed with the help of scissors in such a way that small proportions of glumes, lemma and palea are also removed. Each alternate spikelet in the centre should also be removed. The two flowers in each spikelet, which are not fully developed can be conveniently removed. Each of the fertile flowers should be opened carefully by putting fine-pointed forceps between the lemma and palea and the three stamens removed carefully. Care should be taken that the stamens have not attained maturity and they are not dehisced during emasculation. The maturity of anthers is indicated by their colours. Green and white anthers are treated as immature ones and yellow should be treated as mature. The anthers generally burst before extrusion. The emasculated spikes are bagged and labelled.

Pollination is done next morning and may be repeated on second day also depending upon the weather conditions and receptivity of stigmas.

For collecting the pollen from the pollen parent upper portions of the spikelets are cut and the spike is exposed to sunlight for 10 to 15 minutes, so that the anthers burst and liberate their pollen grains. Pollen grains are collected on a paper by tapping the ear with hand. The pollen from the male parent is dusted over the stigmas of female parent. After pollination, the spikes are again bagged and labelled properly.

With a slight modification, the same method is followed for barley and oats also.

2. Rice. Three methods are followed for emasculation in rice:

(*a*) **Ordinary method.** With the help of forceps the lemma and palea are gently opened out and the stamens are removed by scissoring. This is done in 1-2 hours before normal time for anthesis. When sufficient number of flowers have been emasculated, the remaining spikelets are removed to avoid selfing. The ear head is labelled and enclosed in a muslin bag (Fig. 38.3).

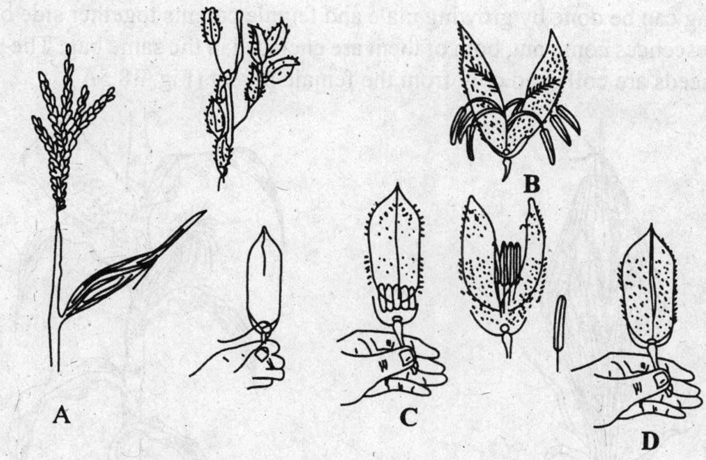

Fig. 38.3 Emasculation in rice.

A. A flower.
B. Palea and lemma spaced away by forcep.
C. Male and female parts of flowers, and one anther removed.
D. Emasculated flower.

(*b*) **Black Paper Method**. In this method penicle is enclosed in a black paper cover 1½ to 2 hours before blooming. This raises the temperature inside the bag. At high temperature the anthers are extruded but they do not dehisce then. This takes about 15-30 minutes. The extruded anthers are scissored off.

(*c*) **Hot Water Treatment.** Jordon (1938) reported that pollen viability was lost when the penicles of rice were dipped into water at 43–44°C (104–111.2°F) for 10 minutes. Stigmas were unaffected. A thermos flask is used and penicles are dipped into water in the morning before normal blooming (Fig. 38.4).

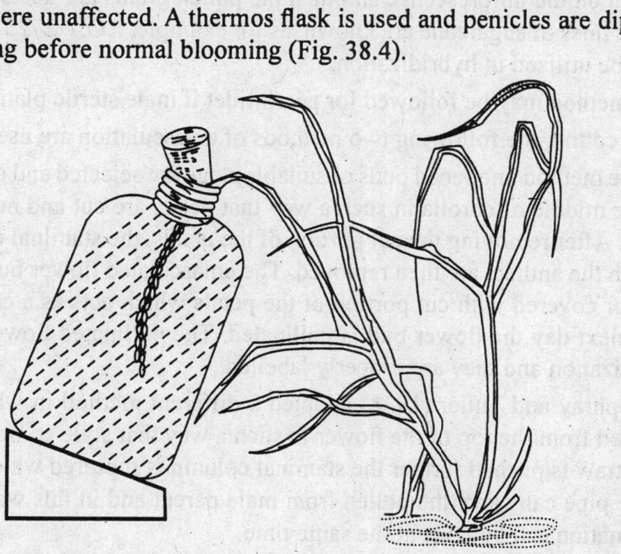

Fig. 38.4 Rice penicle is dipped in hot water to make pollen grains inactive.

The emasculated penicles are bagged and next morning when stigmas become receptive the pollen from the bagged penicles of desired male parents is dusted over them. Then bagging and labelling is done.

3. Sugarcane. In sugarcane, the flowers are minute and so the ordinary method of emasculation can not be practised. For breeding purposes such female parent is used which is male sterile. The crossing can be done by growing male and female parents together side by side and when their inflorescences come out, both of them are enclosed in the same bag. The pollination takes place and seeds are collected only from the female parent (Fig. 38.5A).

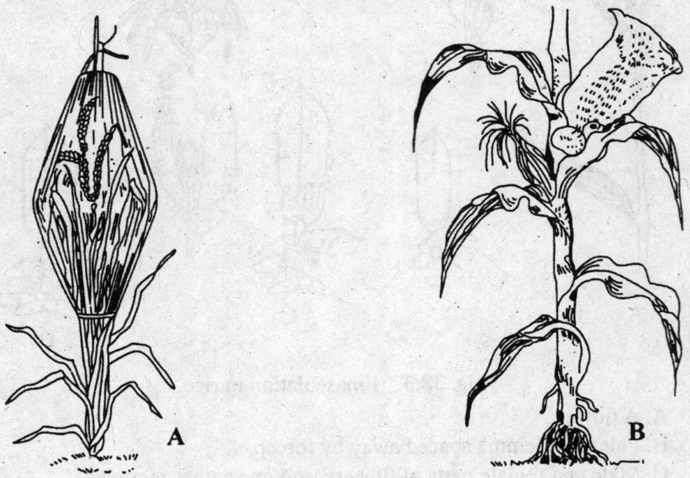

Fig. 38.5 A. Emasculated and Bagged flower of sugarcane.
B. Female inflorescence of maize covered with bag.

The other method for crossing may be that only the female parent which is male sterile is grown and when it comes to maturity, the pollen grains from male parent are dusted over its inflorescence which is then bagged. In certain cases it has been seen that the pollen grains do not germinate probably because of lack of food material. In such cases 10% sugar solution is first sprayed on the inflorescence and then the pollen grains are dusted over stigmas. Several male sterile lines of sugarcane are known, as for example, P.O.J. 2725 and Villai Var. Co. 281, which can be utilized in hybridization.

The same method may be followed for pearlmillet if male sterile plants are available.

4. Cotton. In cotton, the following two methods of emasculation are used:

(*a*) In one method unopened buds of suitable stage are selected and circular incision is made in the middle of corolla in such a way that petals are cut and no injury is caused to the style. After removing the cut portion of the petals, the staminal column is exposed from which the anthers are then removed. The emasculated flower bud is enclosed in a paper bag or covered with cut portion of the petals which acts as a cap over the gynoecium. The next day the flower bud is pollinated. The pollinated flowers are kept covered till fertilization and they are properly labelled.

(*b*) Humphray and Tuller (1938) adopted a different method in which a drinking straw is pushed from the top of the flower in such a way that style comes inside the tube. When the straw is pushed further the staminal column is ruptured which is then removed. The same pipe can carry the pollen from male parent and in this way the emasculation and pollination can be done at the same time.

5. Pea, Gram, Pigeon pea and Mung. These are strictly self-pollinated crops. For emasculation, the flower buds of suitable size in which anthesis has not yet occurred are selected. The wings are bent in front of the standard petal, the keels are opened carefully and then all the stamens are removed by forceps. In doing so, the care is taken that no injury is caused to sepals, petals and stigma as far as possible. The stamens are counted to make sure that all of them have been removed. The emasculated flower buds are bagged. Anther dehiscence commences between 9 a.m. and 12 noon. The emasculated flower buds are cross pollinated next morning at 9 a.m.

6. **Groundnut.** The flower buds which are likely to bloom next morning are emasculated. In the emasculation process, the floral parts are opened out carefully as in pea and 8 fertile and 2 sterile stamens are removed with the help of forceps. The stigma and style are covered by keel. No covering is necessary in this crop as it is always self-pollinated and the floral parts protect the stigma from foreign pollen.

   Dehiscence of anther takes place normally at about 5 a.m. Next day, the extended receptive stigma of emasculated flower bud is pollinated between 7 a.m. and 11 a.m. and labelling is done. 5-7 days after pollination successful crosses develop gynophore.

7. **Dolichos lablab.** In flat beans the anthesis takes place two days prior to opening of the flowers. Buds emasculated and pollinated immediately before dehiscence of anthers are found to be most successful in setting. The standard is split with the help of pointed needle and bent down. The wing petals are similarly held. The keel petals are split up and undehisced stamens are removed. The stigma is examined with a lens to ensure that it is free of pollen. For pollination a flower bud with dehisced anthers is chosen. The standard and wings are cut and the keel tube with mass of pollen inside is tapped on the stigmas of emasculated flowers. This method of pollination is superior as it affords natural conditions. After pollination and labelling, bagging are done.

8. **Castor.** Since the plant is monoecious and the flowers are unisexual, emasculation is easy and this consists in removal of the male flowers which are borne at the base of the inflorescence. In some cases, male flowers are intermingled with female flowers. In such cases all the male flowers should be carefully removed from the inflorescence. After pollination the female flowers are labelled and bagged.

9. **Chilli.** Since the flowers open in the morning, emasculation is carried out the previous evening. The petals are opened out and stamens cut and removed. One easy method of emasculation is to gently pull out the corolla. The stamens are epipetalous and, therefore, they are removed alongwith the corolla. The emasculated flower buds are covered with paper bags.

   The next day, the stigma of emasculated flower bud is pollinated with pollen from desired male parent. Then the bud is labelled and bagged.

10. **Maize.** In maize, the flowers are unisexual and cross pollinated and hence hybridization is very easy. There is no need of emasculation. The entire tassel of male plant is removed (detasselled). The tip of female inflorescence is cut and bagged about 2-3 days prior to the emergence of silky hairs. The varieties to be crossed are grown side by side in isolated land (Fig. 38.5B).

11. **Sorghum and Other Grasses.**

    (*i*) For mass emasculation in sorghum, generally hot water treatment is adopted. A tin containing water at 45°C is mounted on tripod. The panicle is dipped in hot water for about 10 minutes. By hot water treatment, the pollen grains are killed. The temperature and duration of treatment for different plants may be worked out by preliminary experiments.

    (*ii*) The second method is the mechanical removal of anthers. In this method, a plant in which head is just emerging is selected and all the flowers are removed except a few near the upper and lower ends of only upper branches of the panicle. Now the emasculation is done by holding the flower and pressing the glumes laterally and removing the three stamens. After emasculation the panicle is bagged in a glassine bag.

Pollination is done when stigmas protrude out of the glumes. The pollen grains are collected from bagged panicles of male parent by ticking the bag with fingers on the pollen panicle.

After pollination the female panicle is again bagged with brown paper bag.

Methods of Hybridization

This step consists handling of F_1 and subsequent generations by different selection methods. The methods of hybridization are different for self- and cross-pollinated crops.

I. Hybridization in self-pollinated crops. Many crop plants, such as, wheat, barley, oat, rice, cotton, tobacco, potato, peas and beans are self-pollinated. The different selection methods of hybridization in self-pollinated crops are as follows:

(*a*) Pedigree method.

(*b*) Bulk method.

(*c*) Back cross method.

(*d*) Multiple cross method.

(*a*) **Pedigree method.** In this method, desirable hybrids of F_1 population are selected on the basis of desired characters. The seeds from each selected plant are collected and grown separately in rows and F_2 is raised. Again in F_2, best performers are selected and the product from these plants are sown separately in rows to raise F_3 and the same process is repeated up to F_6 until it becomes homozygous and fairly uniform. Undesirable varieties are discarded during the selection process. In this way, a complete ancestral record of each progeny is maintained. Finally the progeny of F_6 are tested in bulk and then named, recommended and transferred to the farmers.

(*b*) **Bulk method**. In this method hybrid plants are grown in bulk. In F_2, plants with many desirable traits are selected and bulked together to form F_3. Seeds obtained from selected F_2 are sown *en masse* and in F_3 and subsequent generations same method of selection is repeated and the process is continued up to F_6 or F_7 till the homozygosity is obtained. Finally plants with superior qualities are named and released for cultivation after conducting yield trials at different stations.

(*c*) **Back-cross method.** In this method, F_1 hybrids are crossed with one of the parents. The object of this cross is to transfer a particular quality of one parent into another lacking it. Disease resistance, drought resistance, all such characters are introduced into susceptible crops with many good qualities. This method is used in the improvement of both self and cross-pollinated plants. Suppose, a good variety A (recurrent or recipient parent) lacks some desirable characters which are present in an inferior variety B (donor variety). Now the desirable character, say, disease resistance of B is transferred to A by crossing A with B. The F_1 hybrids, instead of allowing self-pollination in them, are back crossed with the recurrent parent A. The hybrids of first back cross (BC_1) are again crossed with recurrent parent A and this back crossing is repeated for several generations (upto BC_6 generation) till desirable character is obtained. At the end of BC_6 the selected plants are selfed to make the population homozygous with regard to new character. The process can be shown in the following way:

A ♀ × B (disease resistance) ♂
↓
A × F_1 hybrid
↓
A × BC_1 hybrid
↓
A × BC_2 hybrid
↓
A × BC_3 hybrid
↓
BC_6 (Plants with all good qualities of A + Disease resistance)

The method of back-crossing is regarded as the best because:

(*i*) the required characters of a plant can be brought easily into the desirable variety, and

(*ii*) fertility is established and sterility can be minimised after repeated back-crossing.

(*d*) **Multiple cross method.** In this method, a series of bridge crosses are made and the desirable monogenic characters, which are found scattered in many pure line parents, are combined together in one variety. Suppose there are 8 varieties, each with one good quality. Then the qualities of these 8 varieties are combined into one by the following method:

I A × B, C × D, E × F, G × H → Single crosses

↓ ↓ ↓ ↓

II F_1 Hybrid (AB) × F_1 Hybrid (CD) F_1 Hybrid (EF) × F_1 Hybrid (GH) → Double crosses

↓ ↓

III Hybrid (ABCD) × Hybrid (EFGH) → Multiple cross.

↓

Hybrid (ABCDEFGH)

The hybrids of multiple crosses are selfed and F_2 is raised. Further breeding is carried out either by pedigree or by bulk method of selection.

II. Hybridization in cross pollinated crops. Common crops which are cross pollinated are maize, rye, cucurbits, fruit trees and forage crops. In these crops, the desirable characters, originally found scattered in different purelines (inbreds) are combined in the following ways:

(*a*) Single cross
(*b*) Double cross
(*c*) Three way cross
(*d*) Top cross (or Inbred × variety)
(*e*) Synthetic cross.

(*a*) **Single cross.** It is a cross between two inbreds, *e.g.*, A × B or C × D. The hybrids are directly distributed to farmers for cultivation purpose. If there are many inbreds, the number of possible single crosses can be calculated by the formula:

No. of single crosses $= \frac{x(x-1)}{2}$, where x = No. of inbreds. If there are 4 inbreds, the possible single crosses will be $\left(\frac{4\times 3}{2}\right) = 6$.

(*b*) **Double cross.** It is a cross between the F_1 hybrids of two different single crosses, each involving two different inbreds.

I single cross A × B ↓ F_1 hybrid (AB) × II single cross C × D ↓ F_1 hybrid (CD)

↓

Double cross hybrid (ABCD)

If there are many inbreds, the possible number of double crosses can be calculated by the following formula:

No. of double crosses $= \frac{x(x-1)(x-2)(x-3)}{8}$, where x is the number of inbreds.

(*c*) **Three way cross.** This is a cross between F_1 hybrid of a single cross and an inbred which is used as male parent, *e.g.*, [(A × B) × C].

(*d*) **Top cross.** It is a cross between an inbred and open pollinated variety.

Variety × inbred

(*e*) **Synthetic cross.** In this a number of inbreds are crossed in order to combine different desirable characters into one variety.

INBREEDING DEPRESSION AND HYBRID VIGOUR

Cross breeding or hybridization between different unrelated strains generally results in F_1 hybrids which are healthier, larger in size, more luxuriant in growth and yield and show more vigour and fertility than either parent. This superiority of hybrids over their parents is called **hybrid vigour**.

Inbreeding Depression

The cross-pollinated species and asexually reproducing species are highly heterozygous and these plants lose their increased vigour and other superior qualities if selfing or inbreeding is allowed in them. **Inbreeding** is mating between genetically related individuals, *i.e.*, brother-sister mating or sib-mating. The mating between the individuals of different genetic constitution is known as **out breeding.** The reduction in growth, vigour and fertility due to inbreeding is referred to as **inbreeding depression**. Inbreeding reduces heterozygosity and leads to homozygisity in the progeny. Thus, there is a direct correlation between the homozygosity and degree of inbreeding. Selfing reduces the heterozygosity by a factor of 1/2 in each generation and the degreees of inbreeding increases in the same proportion. Several harmful effects develop due to inbreeding. In human societies marriages between closely related individuals have been prohibited. An extreme example is Hindu Society in which marriages between the individuals related by descent or ancestry are prohibited howsoever, distant they may be. The harmful effect of inbreeding on fertility was noticed in seventeenth century. Koelreuter (1761-66) conducted his hybridization experiments on *Nicotiana* and suggested the following facts:

(*i*) The vigour of hybrid is related to the degree of genetic dissimilarity of parents, and

(*ii*) Hybrid vigour is of particular significance in evolution.

Darwin (1876) published the account of his experiment on self and cross pollination in maize in his book entitled '*Cross and Self fertilization in Vegetable Kingdom*' and concluded that the progeny obtained from self-fertilization is weaker than those derived from out crossing or cross fertilization. Subsequent studies on inbreeding in maize by East (1906) and on some other crops by Shull (1909) and others also made it clear that in cross-pollinated and asexually propagated species inbreeding showed harmful effects.

Effects of Inbreeding

Some of the harmful effects of inbreeding are summarized as under:

1. **Appearance of Lethal Alleles.** Inbreeding results in segregation of harmful recessive alleles and the lines carrying them usually do not survive. Such characteristics include chlorophyll deficiencies, rootless seedlings, abnormal flowers.

2. **Vigour Loss.** Inbreeding results in loss of vigour in progeny, *i.e.*, general reduction in the size of various plant parts.

3. **Reduction in Fertility.** Inbreeding adversely affects the reproductive ability in many plants or lines and the inbred lines reproduce so poorly that they can hardly be maintained and utilized in future breeding programmes.

4. **Rapid Segregation of Population into Distinct Lines.** Inbreeding causes rapid segregation of phenotypes and increases the homozygosity and random fixation of various allels in different lines which differ in their genotype and phenotype.

5. **Increase in Homozygosity.** Repeated inbreeding causes increase in homozygosity of different lines and after 7 or 8 generations of selfing, the lines become almost uniform (more than 99 per cent homozygous).

   The lines which are almost homozygous due to repeated inbreeding and are maintained through close inbreeding are known as **inbred lines**. Inbred lines are maintained in homozygous state by strict inbreeding or selfing.

6. **Reduction in Yield.** Inbreeding leads to reduction in yield. In maize, the best inbred lines yield about half as much as their open pollinated parent varieties. In Alfalfa and carrot the loss in yield due to inbreeding is much greater. In onion and cucurbits, however, the reduction in yield is very small.

Degree of Inbreeding Depression

The degree of inbreeding response differs in different species and may range from very high to very low or even zero. Accordingly, it may be grouped into the following four categories:

1. High inbreeding depression
2. Moderate inbreeding depression
3. Very low inbreeding depression
4. No inbreeding depression.

1. **High Inbreeding Depression.** In some plants, as for example, Alfalfa (*Medicago sativa*), carrot (*Daucus carota*) the loss in vigour and fertility due to inbreeding is so great that a very few lines survive after 3 or 4 generations of inbreeding and they show greatly reduced yield, generally less than 25 per cent of the open pollinated varieties.
2. **Moderate Inbreeding Depression.** Some plant species such as jowar (*Sorghum bicolor*), bajra (*Pennisetum americanum*), maize show moderate inbreeding depression. Selfing of progeny results in many weak, lethal or sub-lethal types in population which are eliminated but a good proportion of population can be maintained under self-pollination.
3. **Low Inbreeding Depression.** Some plants such as onion (*Allium cepa*), cucurbits, rye (*Secale cereale*), sunflower (*Helianthus annuus*), sunn hemp (*Crotolaria sativa*) show very little loss in vigour and fertility due to inbreeding and some of the inbred lines yield as much as their open pollinated parent varieties.
4. **No Inbreeding Depression.** The self-pollinated species do not show inbreeding depression, although they show some degree of hybrid vigour or heterosis.

HYBRID VIGOUR OR HETEROSIS

Hybrid vigour is a hereditary process by which two genetically different individuals when crossed with each other produce F_1 hybrids which are superior to the best parent or mean value of the two parents in respect of size, growth, yield and fertility. The superiority of F_1 hybrids over parents is referred to as **hybrid vigour**. G.H. Shull in 1914 used the term **heterosis** for hybrid vigour and this term was unanimously accepted in 1922. The term heterosis was used for the growth and vigour arising due to heterozygosity and thus, the terms hybrid vigour and heterosis were used for the same process and were synonyms.

There has been considerable confusion due to use of various terms for describing heterosis and also due to the use of heterosis in different contexts.

Brubaker (1964) suggested that the term heterosis should be used for the genetical expression of beneficial effects of hybridization. But, Powers (1944) suggested that **heterosis** should be used for the apparent expression of one or more characters in the hybrid as may be superior or inferior to that of any of the two parents. According to this view heterosis and hybrid vigour are not synonyms because the term hybrid vigour is used for increased size or vigour of the hybrid as compared to that of their parents, whereas the heterosis may be of two types: (*a*) **positive heterosis** or superiority of hybrids, and (*b*) **negative heterosis** or inferiority of the hybrids over their parents. Whale (1924) has tried to explain the difference between hybrid vigour and heterosis and suggested that the superiority of hybrid over better parent should be termed hybrid vigour and the mechanism by which superiority or hybrid vigour is developed should be termed heterosis. Now the use of the term heterosis is preferred to hybrid vigour because plant breeders compare the F_1 hybrids with their parents and in this comparison F_1 hybrids may be found to be either superior or inferior to their parents.

Although heterosis refers to both superiority and inferiority of the hybrids over their parents, in plant breeding generally the superiority of the hybrid over parents is taken into consideration to define heterosis. Usually the following two criteria are used for comparing the hybrids with their parents:

(*i*) Superiority of hybrids over the superior or better parent.

(*ii*) Superiority of hybrids to the mid-parent value or the mean value of the two parents.

Realising the above criteria Witzer and associates (1968) and Fonseka and Paterson (1968) used the term **heterobeltiosis** for the superiority of hybrids over the better parent.

Thus, *the superiority of the hybrids over the mean value of the two parents is termed heterosis* and *the superiority of the hybrids over the better parent is called heterobeltiosis.*

For clarity and simplicity, it is desirable to avoid use of many terms to describe similar but somewhat different situations.

General Aspects of the Expression of Heterosis

1. Hybrid vigour is a normal process in plants and animals but in cross-pollinated plants it is more conspicuous than in self-pollinated ones.
2. It is a natural process and occurs more automatically in heterozygous plants than in homozygous plants.
3. Heterosis does not manifest equally in all hybrids of a species. From the studies on inter-varietal crosses in maize, it became clear that some of the hybrids showed heterosis while others did not. Crosses between distinct types or genetically diverse varieties exhibited greater heterosis than those involving closely related varieties.
4. Heterosis is governed by genetic factors.
5. Heterosis shows maximum manifestation in F_1 hybrids and in F_2 and subsequent generations it decreases gradually because the genes producing heterosis segregate.
6. Heterosis is directly related to specific combining ability of the parents. Heterosis shows maximum manifestation in the hybrids of such parents as have got high specific combining abilities.
7. In some agricultural crops, heterosis is large enough to be used in production of hybrid varieties, *e.g.*, in maize, jowar, bajra, wheat, barley, rice, cotton, sunflower, brinjal, onion, tomato, tobacco, brassica, castor, radish, carrot, cucumber etc.

Effects of hybrid vigour. The different effects of heterosis are as follows:

1. Heterosis results in the increase in size of the plants.
2. It increases the yield of the crop.
3. It increases the number and size of fruits and seeds.
4. It increases the fertility of plants.
5. It increases the vitality of plants.
6. It results in more efficient germination and growth.
7. It induces early flowering and fruit setting.

8. It increases the disease resisting power of plants.
9. It makes the plants more suitable for dry conditions.

The non-beneficial heterosis shows effects that are quite reverse of the positive ones.

Causes of Heterosis

Many hypotheses have been advanced by different workers to explain the causes of heterosis. These have correlated the phenomenon with genetic and physiological causes.

Genetical causes of Heterosis. Two hypotheses based on experimental works have been put forth by various workers to account for the heterosis. These are as follows:

(*i*) **Dominant factors hypothesis.** This was advanced by Bruce (1910). According to this theory, the increased vigour results due to larger number of favourable dominant genes in hybrid than in parents. The cross breeding generally results in the accumulation of large number of dominant genes in the hybrids and finally hybrid vigour. For example, a cross between pure-line AAbbcc and inbred aa BB CC gives a genotype Aa Bb Cc. When all the three dominant genes are favourable in the hybrid, it will show increased vigour than the parents. It is important to note that maximum vigour is seen in F_1 hybrids and this vigour goes on gradually decreasing in F_2, F_3, F_4, F_5 and other subsequent generations because of segregation of factors.

(*ii*) **Overdominance hypothesis.** This hypothesis was proposed independently by G.H. Shull (1903) and East (1908). According to exponents of this hypothesis, the heterozygote is superior to homozygote and hybrid vigour increases with the increase in the amount of heterozygosity. In simple terms, heterozygous $A_1 A_2$ is superior to $A_1 A_1$ and $A_2 A_2$. The genes A_1 and A_2 govern different functions, therefore, the heterozygote is more vigorous than the inbreds.

Physiological Causes of Heterosis. The heterosis can be attributed to the following physiological causes:

(*i*) **Cytoplasmic-nuclear interaction hypothesis.** Some prominent breeders like Leivis, A.F. Shull, Michaelis and many others have opined that heterosis or hybrid vigour might be caused by interaction of cytoplasm and nucleus. It is due to the effect of changed nucleus and relatively unchanged cytoplasm upon each other.

(*ii*) **Greater Initial Capital (Potential) hypothesis.** According to this hypothesis, hybrid vigour is due to an increased initial embryo size or due to the so-called greater initial capital. This idea was proposed by Ashby (1930).

Importance and Practical utility of Hybrid vigour

This phenomenon results in hybrids that have better characters (*i.e.*, large size, high productivity, early maturation, resistance to diseases, pests and drought than the inbreds). Many ornamentals and fruit trees, valuable vegetables, good quality cereals are results of cross breeding of inbreds.

Hybrid vigour has been extensively utilized in the improvement of both plants and animals. The earliest work on heterosis has been done on maize. The self-fertilized progeny in this crop is inferior in size, vigour and yield as compared to open pollinated plants. The F_1 hybrids from different inbred lines are found to be much superior to their parents in yield and other characters (Fig. 38.6). The hybrid maize varieties have led now to such a phenomenal increase in yield that almost all the varieties of maize cultivated in USA are hybrids. In other parts of the world including India, a great majority of high yielding varieties of maize presently under cultivation are developed from four double crosses. The high yielding maize cultivars propagated in India are Ganga 101 [a double cross (CM 103 × CM 104) × (CM 201 × CM 105)], Ganga safed hybrid-2, a high starch product of three way cross (CM 400 × CM 300) × CM 601, Ganga 3, Ganga 5, EH 238, EH 450, Ganga hybrid 89 which is a product of double cross [(CM 116 × CM 107) × (CM 202 × CM 117)], Himalayan hybrid 123, a product of double cross [(CM 202 × CM 205) × (CM 113 × CM 112)], Deccan 101, a double cross hybrid [(CM 202 × CM 206) × (CM 114 × CM 115)], VL Makka 54, a double cross hybrid [(CM 107 × CM 108) × (CM 203 × CM 204)].

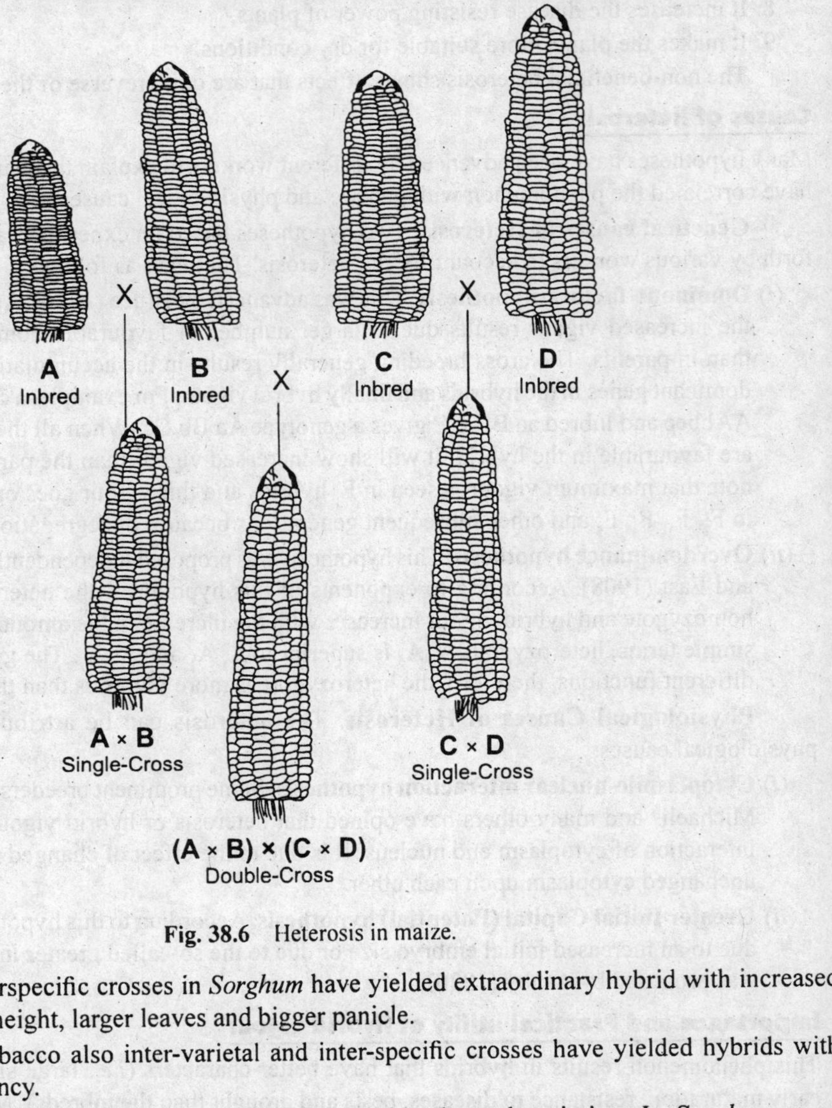

Fig. 38.6 Heterosis in maize.

Similarly, interspecific crosses in *Sorghum* have yielded extraordinary hybrid with increased stem thickness and height, larger leaves and bigger panicle.

In rice and tobacco also inter-varietal and inter-specific crosses have yielded hybrids with increased yield potency.

In some cases, however, the vigour in the hybrids showed variation. In *Sorghum* some inter-specific hybrids showed positive heterosis but in some other cases the hybrids were found to be either intermediate between the two parents or they were inferior to parents. Heterosis is commonly used in the form of hybrids or synthetic varieties in cross-pollinated or often cross-pollinated crops.

In India much work has been done on heterosis breeding in a number of self-pollinated crops of economic value such as sugarbeet, brinjal, lady's finger, cotton, tomato, radish, onions, cabbage, maize, sugarcane, rice, *Sorghum*, cucumber, squashes, coconut, sunflower etc. Hybrid vigour has led to the improvement of the plants in many traits such as height, size, productiveness, earliness, viability, resistance to diseases and pests, drought, etc. Many ornamentals, fruit trees and vegetables, good quality cereals are the results of cross breeding of inbreds. They are heterozygous for most of their desirable genes and they possess hybrid vigour to a marked degree. Since inbreeding in them leads to the appearance of detrimental traits, many of them are propagated vegetatively by cuttings, grafting, budding etc. Hybridization produces better and valuable plants and the vegetative propagation preserves valuable traits and vigour.

Crossing of dissimilar plants does not always lead to positive heterosis. In wheat and barley mostly there is self-pollination and inbreeding in them does not result in loss of vigour. Attempts have been made and are being made to utilize heterosis in breeding of wheat and barley, but so far no success has been achieved.

MUTATION BREEDING

Mutation sometimes produces many important qualities in the plants. Therefore, it can be used in changing the genotypes and phenotypes of plants so as to produce strains of desired type. Mutation breeding is, thus, concerned with the induction of mutations in the plants with the help of mutagenic agents (mutagens) and their utilization for production of new superior varieties.

In plants, mutations can be induced artificially by certain mutagenic agents such as atomic radiations of various sorts such as ultraviolet rays, X-rays, gamma rays (γ-rays), sudden heat shocks and mutagenic chemicals. Different mutagenic agents and methods of their application have been given in the chapter on Mutations.

The history of mutation breeding in India can be traced back to 1935, but large scale work started when Gamma gardens of Bose Research Institute, Kolkata and IARI, New Delhi were established in 1959 and 1960, respectively. At present, mutation breeding work is being carried out at several places in India. The prominent places are Bhaba Atomic Research Centre (BARC), Trombay, Tata Institute of Fundamental Research, Mumbai, Bose Research Institute, Kolkata, Indian Agricultural Research Institute, New Delhi and National Botanical Research Institute, Lucknow. Breeders carrying out researches at various Institutions, Universities and Departments generally get their materials irradiated from these places. A Radioisotope Laboratory setup at IARI, New Delhi, has provided new momentum to mutation breeding programme in India.

Methods of Inducing Mutations

(*i*) **Treatment of plant materials with physical mutagens.** For inducing mutations, the plant materials such as seeds, bulbs, rhizomes and buds of perfectly established varieties are irradiated with x-rays, γ-rays, UV rays under controlled conditions of moisture content, temperature, atmospheric pressure and oxygen level. The materials are exposed to different doses of mutagenic rays, say at 5Kr, 10Kr, 15Kr, 20Kr, 25Kr, 30Kr, 35Kr, 40Kr, 45Kr, 50Kr and so on, and for each treatment many seeds, say 500 to 2000 or more are exposed. Dry seeds are less responsive to irradiation than the seeds pre-soaked in water.

The control and treated seeds are germinated in petridishes and germination percentage is recorded. The germinated seeds are then transferred to experimental plots and emergence of seedlings in field is recorded. The particular radiation dose which reduces the germination to 50% is termed as lethal dose 50% or L.D. 50. For different crops fifty per cent lethal doses are different. The increase in the radiation dose brings exponential increase in chromosomal aberrations and linear increase in gene mutations.

(*ii*) **Treatment of plant materials with chemical mutagens.** Chemical mutagens such as Ethyl methane sulphonate, Ethyleneimine or Methyl methane sulphonate are used in solution form at different concentrations. The mutation frequency depends on dose (concentration x time), temperature, pH and some other factors. For inducing mutations seeds, either dried or pre-soaked in distilled water for a few hours are placed in freshly prepared solutions of chemical mutagens at different concentrations, say at 0.025%, 0.05%, 0.1%, 0.2%, 0.3%, 0.4%, 0.5% and so on, for two to six hours or even longer period. To avoid dissociation of chemical mutagen, the treatment is generally given at low temperature and the acidity of solution is controlled by use of buffer solutions. The treated plant materials are washed in running water for 2-3 hours and then placed in petridishes for recording the germination percentage. The germinated seeds are finally transferred to experimental plots.

Handling of Mutant Materials

The populations raised from the treated materials are designated as M_1 generation (First mutant generation). The populations raised from the X-ray irradiated materials are specifically designated as X_1, X_2, X_3 and so on. The plants in M_1 generation are critically studied for their morphology and reproductive behaviour. Primary effects of mutagens are manifested in colour, shape, venation of leaves, branching and flowers etc. The abnormal plants in each treatment are labelled and their seeds are collected separately. Ten seeds from a single M_1 plant in cereals may be sufficient. If M_1 generation is too large then one or two seeds per M_1 plant will be enough for raising M_2 generations. Every tenth to twentieth row should be sown by control material.

Segregation of mutations may occur in M_2 generation, so the selection of plants with desired characters can be made from M_2 populations. Two to five best possible plants for a particular desired characters should be selected from one row. The selected plants are selfed and the seeds obtained from such plants are grown to raise M_3 generation. Selection of best plants is again made in M_3 generation.

Use of Mutant Genes

(*i*) **Direct use**. When certain true breeding mutant lines with desired genes are established, then suitable breeding methods are employed to obtain desired genotype. In autogamous species, mutants selected in M_3 generation can be used directly as genotype if they breed true for a particular character. *Sharbati Sonora,* a single gene mutant in wheat is a selection from mutant population of *Sonora*-64 variety. In cross-pollinated populations there occurs a greater degree of heterozygosity, so the selected mutants should be allowed to inbreed for six or seven generations. If they breed true, then they can be used as mutant genotypes.

(*ii*) **Incorporation of mutant genes in self-fertilized crops.** The desired genotypic recombination after M_3 generation can be obtained from autogamous species in one of the following ways:

(*a*) The mutant is crossed with parental variety,

(*b*) Two or more mutants from the same parental variety are involved in pedigree programme,

(*c*) Back-crossing mutant as donor parent with a different variety as recipient parent.

(*iii*) **Incorporation of Mutant genes in Allogamous species.** In cross fertilized species, the desired genotype can be achieved through hybridization by the mutants followed by inbreeding and recurrent selection. Though mutation breeding in cross-pollinated crops is difficult, yet mutagenesis increases variability to widen the range for selection. Further, it can be used in developing male sterile lines.

BREEDING FOR DISEASE RESISTANCE

Plants are attacked by a large number of diseases which are caused by fungi, bacteria, viruses, insects, nematodes etc. The diseases can be controlled by two methods:

(*i*) By phytopathological measures, and

(*ii*) By growing disease resistant varieties of plants.

The phytopathological measures are used to prevent or eradicate the pathogens, but the disease resistant varieties resist the attack of the pathogens. The disease resistance in plants may be either due to certain morphological features that enable the plants to resist the attack of disease causing organisms or due to its protoplasmic resistance which may be caused often by certain chemical compounds produced by the plant body in response to certain environmental conditions.

The disease resistant varieties can be evolved by various breeding methods such as introduction, selection, hybridization, budding and grafting, and mutation breeding. These methods have already been described earlier in this chapter.

QUESTIONS

1. Describe the aims and objectives of plant breeding.
2. Discuss in brief the history of plant breeding in India.
3. What are the different methods of crop improvement and when and where are they used in particular crop? Which is the oldest and which is the latest method of crop improvement?
4. Define artificial selection. What are its different types?
5. In what type of crop plants the mass selection, pure-line selection and clonal selection are applied? Discuss by giving examples of crop plants in each case.
6. What do you mean by hybridization? Does hybridization bring the improvement in crops by creating entirely new genes not originally found in the parents?
7. List the different steps involved in hybridization procedure for production of new variety and discuss each step in brief.
8. What are the different hybridization methods for self- and cross pollinated crops?
9. Discuss the method of mutation breeding.
10. Explain the following terms.
    (*i*) Clone selection,
    (*ii*) Pure-line selection,
    (*iii*) Introduction and acclimatization,
    (*iv*) Emasculation, and
    (*v*) Hybrid.
11. What do you mean by centres of origin? Discuss in brief different centres of origin of plants.
12. Discuss in brief the procedures of plant introduction.
13. Discuss the purpose and importance of plant quarantine in plant introduction.
14. What do you mean by inbreeding depression? Discuss the effects of inbreeding in crop plants.
15. Define hybrid vigour. Discuss in brief the various effects and causes of heterosis in plants.
16. Give a brief account on the practical utility of heterosis breeding in crop plants.

39 CHAPTER

Breeding Work on Some Important Crops

BREEDING OF WHEAT
(*Triticum Species*)

Wheat is the most important food crop used by man. The origin of cultivated wheat is still a controversial subject. From the available records, it appears that wheat originated in South West Asia. It is a general belief that Aryans brought wheat to India and Afghanistan and from the centre of origin this crop spread to European countries. In the United States, wheat was first cultivated in the 17th century along the Atlantic coast.

Distribution

Wheat is grown extensively throughout the world. Among the major wheat growing countries, the maximum area under wheat is in Soviet Russsia followed by USA, China and India. In regard to yield per hectare, the United Kingdom ranks first followed by West Germany.

India is one of the major wheat growing countries and it stands fourth in acreage and seventh in yield. According to one estimate in the year 1974-75 the total area under wheat cultivation in this country was 18,107 thousand hectares and the total production being 24,235 thousand metric tonnes. The major wheat producing States in India are U.P., Punjab, Haryana, Bihar, Madhya Pradesh, Rajasthan, Maharashtra and Gujarat. Though the maximum acreage and production of wheat are in U.P., Punjab ranks first in respect of per hectare yield, followed by Haryana.

Four species of wheat; *Triticum aestivum* (Common bread wheat), *T. durum* (Macaroni wheat), *T. dicoccum* (Khapli or emmer wheat) and *T. sphaerococcum* grow in India. *Triticum aestivum* and *T. durum* are under extensive cultivation, *T. dicoccum* is grown in small scattered pockets in Maharashtra, Karnataka. Andhra Pradesh and Tamil Nadu and *T. sphaerococcum* has practically gone out of cultivation. The common bread wheat (*T. aestivum*) is most extensively cultivated in Punjab, Haryana, U.P., Bihar, Madhya Pradesh and Rajasthan and *T. durum* is grown in certain parts of Maharashtra, Madhya Pradesh and parts of U.P., Rajasthan and West Bengal.

The scientific wheat improvement work was initiated by Sir Albert Howard and Mrs. Howard in 1904 at the Agricultural Research Institute, Pusa (Bihar). That work resulted in the development of varieties N.P. 4 and N.P. 52. In the year 1936, the Pusa Institute was moved to New Delhi and work on breeding of wheats is still being continued there in the Indian Agriculture Research Institute. Dr. B.P. Pal (1935) reoriented the programme, so as to develop rust resistant varieties and soon varieties like N.P. 710, N.P. 718, N.P.. 770, N.P. 798 and N.P. 809 became quite popular among the farmers. Dr. Ram Dhan Singh also produced many varieties like C.591, C.518, C.273, C.281 in Punjab. Since then research work on wheat has been carried out in different States of India and as a result of which a number of new varieties were developed which showed superiority in yield and disease resistance over the old varieties.

In the year 1962, scientists at I.A.R.I., New Delhi geared up a study programme to know the different barriers responsible for the stagnation and instability in the production of wheat and they found the following causes responsible:

1. Most of the varieties were tall and hence they were lodged when the fields were irrigated at the advanced stage of crop maturation.
2. High atmospheric temperature and dry soil conditions during March checked the seed growth, as the irrigation was not advisable because of fear of lodging.
3. Rains, strong wind and hail storm during March made the crop lodged which consequently delayed the maturation and checked the seed growth.
4. Most of the varieties were susceptible to rusts and smuts and the disease problem became more severe when the crops were irrigated at advanced stage of crop maturation. This analysis led to the conclusion that dwarf varieties were essential.

Pollination in Wheat

The inflorescence of wheat is called ear, a compound spike in which the main axis bears two opposite rows of lateral spikelets and a simple terminal spikelet. In *Triticum monococcum*, the terminal spikelet is absent. The spikelets are sessile, alternately arranged and each with a pair of glumes. Blooming normally starts several days after the spike emerges. The flowering proceeds in the basipetal succession and it continues throughout the day with 2 to 3 days required by a spike to finish blooming. When the glumes open anthers protrude from them and the pollen is shed outside. Wheat is a self-pollinated crop. Entry of the foreign pollen while the flower is open, may result in a small amount of cross pollination (normally less than 1%). For cross pollination, flowers of female parent are emasculated and then the pollens collected from male parent are dusted over the stigma. Emasculation is done by clipping back the glumes and removing the anthers with fine pointed tweezers. Flowers are emasculated one or two days before pollination. Emasculated and pollinated spikes are bagged. Cytoplasmic male sterility and genetic male sterility are available in common wheats. Male sterile lines are utilized to eliminate emasculation. Inbreeding is ensured by covering the spike with butter paper bag prior to flowering.

Aims of Wheat Breeding

Breeding of wheat in India is carried out with the following objectives:

(*i*) to evolve a high yielding variety,

(*ii*) to produce wheat that makes good *chapatis* and stores better under primitive conditions of grain storage,

(*iii*) to evolve varieties resistant to rusts ; Brown rust (*Puccinia triticina*), black rust (*P. graminis tritici*) and yellow rust (*P. glumarum*), smuts, leaf blight, ear cockle, etc,

(*iv*) to evolve climatically suitable and prolific varieties, and

(*v*) recent trend in wheat breeding has been to evolve dwarf wheat varieties both for increasing yield potential of this crop through effective use of water and fertilizers and to remove some of the factors responsible for instability and stagnation in yield.

Methods of Wheat Breeding

The breeding work on wheat in India has reached to a very advanced stage and now this country is recognised as one of the top wheat breeding countries of the world.

Attempts at improving wheat in this country have been made by—

1. introducing new varieties from other countries,
2. selection,

3. crossing certain varieties in order to combine many good qualities of different other varieties into one variety, and
4. recently mutation breeding work has also been attempted and many promising varieties with high yield potentials have been evolved.

1. **Introduction and Acclimatization.** Several introductions of wheat varieties have been made into India but most of them have not been found much beneficial. Most of the foreign varieties imported by Howards were found to be late maturing under Indian growing conditions. The varieties "Ridley" introduced into India from Australia is grown extensively. Most of the introduced strains have been used extensively in hybridization, programmes as a source of genes for disease resistance and dwarfness. Through hybridization genes for disease resistance and dwarfness have been transferred to adopted types and several new varieties have been developed.

   In 1962, scientist at I.A.R.I., New Delhi, realized the importance of "Norin" genes in dwarfing the variety. These genes were discovered in Norin wheat variety in Japan. Strain Norin-10, a Japanese wheat was brought to the U.S.A by Dr. Solman in 1946. Three dwarfing genes with additive effects have been identified so far in this strain. With the Norin dwarfing genes, two dwarf varieties, Gaines, a dwarf winter wheat and a dwarf spring wheat variety were developed by Dr. Ovella Vogal at Washington (U.S.A) and Dr. N.E. Borlaug and associates at CIMMYT (International Centre for Wheat and Maize Improvement), (Mexico) respectively. Seeds of these were obtained from Rockefeller Research Foundation and Mexican Ministry.

   Dr. N.E. Borlaug, the Nobel Prize Winner in Peace for the year 1978, and wheat breeder, visited India and despatched to I.A.R.I. 100 kgs each of 4 dwarf and semi-dwarf Mexican varieties of spring wheat : Lerma Rojo-64 A, Sonora 63, Sonora 64 and Mayo 64 and small samples of 613 promising selections. The materials were grown at I.A.R.I., New Delhi, Ludhiana, Kanpur, Pant Nagar, Pusa, Bhowali and Wellington (Nilgiris hills) and were studied for their acclimatization capacity and reactions to different rusts. Those four strains were tested in 1964-65 under the All India Coordinated Wheat Trials and were also grown in farmer's fields under the National Demonstration Project with a view to demonstrate the new findings to farmers of the country. Lerma Rojo 64 A and Sonora 64 were approved and recommended by the Central Variety Release Committee of Government of India for cultivation in irrigated areas in 1965.

   Seeds of S-227, S-307, S-308 and S-331 obtained from Mexico in 1963 were also subjected to trials in subsequent years and all were found to be of very high yield potentials. In 1967, these varieties were approved for general cultivation. A great majority of wheat varieties now grown in India are dwarfs. They are resistant to lodging, fertilizer responsive and high yielding. They are mostly resistant to rusts and other major diseases of wheat owing to the presence of resistance genes in their genotypes. These varieties are photo-insensitive and some of them are suitable for late sowing. This has made possible the cultivation of wheat in some non-traditional areas like West Bengal.

2. **Selection.** When the wheat improvement programme was initiated by Howards at Pusa, several varieties with outstanding yield and qualities were developed by selection from local types. Prior to the initiation of wheat improvement work in India, various mixed types of wheat were being grown in different regions under local names. In Punjab, Punjab type 9, Punjab type 11 and Punjab type 3 varieties were selected from the local types. Some of the other varieties selected from the local types were N.P. 4, N.P. 6, N.P. 12, 9-D, 8-A, K-13, K-54, C-46 (U.P.), A.O. 90 (M.P), etc.

3. **Hybridization.** Breeding work on wheat in India was carried extensively by Dr. B.P. Pal who was responsible for the release of several N.P. (New Pusa) varieties of wheat. Most of

the important varieties of wheat developed in India in the last 50 years have been developed by hybridization. Different types of hybridization techniques have been used to develop new varieties, such as, by back cross method composite crossing involving several selected varieties which offered an opportunity to bring together innumerable genetic combinations. Some of the varieties evolved through hybridization are as follows:

Inter varietal hybrids:—N.P.165, N.P.710, (N.P. 52 × N.P. 165), N.P. 718 (N.P. 52 × N.P. 165), N.P. 784, N.P. 785, N.P. 786, N.P. 790, N.P. 797, N.P. 798, N.P. 799, N.P. 809 (Democrat × C. 518) × (Spalding's prolific × N.P. 114) × E. 220, N.P. 825, N.P. 828, C. 281 (C. 591 × N.P. 4), C. 285 (C. 228 × B. 256 G-Kenya), R.S. 31-1 (Pb. C 591 × Jaipur red local).

Inter-specific hybrids—Niphad 4 (Motia × Khapli × N.P.4).

T. dicoccum:—Inter-varietal hybrids—N.P. 201 (N.P. 200 × E. 274).

T. durum:—Inter-varietal hybrids—Hyb. 38 (A.O.13 × E. 220.) N.P. 412 (Ekadania 69 × Gazra).

4. **Mutation Breeding.** Dr. M.S. Swaminathan, formerly Director General of International Rice Research Institute, Manila (Philippines) did pioneering work in mutation breeding on wheat and the variety Sharbati Sonora was released under his direction at I.A.R.I., New Delhi and for that he was awarded the Raman Magasway Award in 1978. Irradiation and chemical mutagens are being used to induce mutation in wheat. N.P. 836 was selected from x-ray irradiated N.P.799 at I.A.R.I., New Delhi.

New Varieties

The following promising varieties of wheat have been recently released by the Central Varietal Release Committee, Government of India which have brought break through in the production mark of wheat in the country. Besides, many others are in advanced stages of testing.

1. **Kalyan Sona.** This is selection of S. 227 (F_n-K58-N-N10B) Gabo 55, which is double dwarf wheat resistant to loose smut and hill bunt diseases. Grains are amber, hard and medium size. It grows well under low fertility, irrigated as well as rainfed conditions. It possesses good *chapati* and bread-making property.
2. **Sonalika (S. 308).** This is amber grained single gene dwarf derived through selection from S.308 (Maxican Cross II-53-388 × An) × (Yt. 54 × N.10B). L.R. III 8427. This variety is early maturing which is resistant to all the three wheat rusts. Grains are bold, semi-hard and amber.
3. **Lerma rojo 64.** It is a single dwarf wheat with wide adaptation and resistance to yellow and black rusts. It is derived from a cross (Yaque 50 × N. 10B) × L 52/Lr.2. Grains are red, soft and medium bold.
4. **Safed Lerma.** It is a semi-hard single gene dwarf derived from the Mexican cross (Y. 50 × N. 10.B) (L.52) LR3 (material back crossed thrice to Lerma Rojo-64 A).
5. **Chhoti Lerma.** It is white seeded two gene dwarf variety selected from the Mexican cross LR 64 (Sib) × HURA. The variety is highly resistant to lodging and also resistant to all the three rusts.
6. **Sonora 64.** It is a double dwarf wheat derived from a cross Yaktana 54 – N 10B × Yakui 54, having hard, red grain of good quality and is suitable for late sowings.
7. **Sharbati Sonora** evolved by Gamma irradiation at IARI, New Delhi, in 1967. It is high yielding, dwarf, amber grained wheat.
8. **Shera (H.D. 1925).** It is medium and white grained two gene dwarf variety evolved from a cross between Sonora-64 and L.R. 64 A.
9. **Raj. 911 (Durum wheat).** It is two gene dwarf medium maturing variety evolved through selection from V-0229 (CIMMYT). Grains are bold white.
10. **Moti.** It is three gene dwarf variety with early maturing property. Grains are white and medium sized.

11. **Janak (H.D. 1982).** It is a sharbati colour bold grained variety evolved from a cross between E 557 and H 845.
12. **Arjun (H.D. 2009).** It is two gene dwarf variety evolved from a cross between Lerma Rojo-64 × Nai 60. It is less susceptible to rusts and early maturing variety. Grains are white and medium sized.
13. **Pratap (H.D. 1981).** It is bold grained single gene dwarf variety derived from a cross between E. 557 and H. 845. It is susceptible to rusts, medium maturity and suited to dry farming regions of North-West India.
14. **H.D. 2402.** It is a bread wheat double dwarf genotype of medium maturity duration which was released in 1985 for irrigated normal sown conditions. It is now identified for irrigated and very late sown conditions (sown upto mid-January) in north western plain zone (NWPZ) and NEPZ.
15. **H.D. 2204.** A double dwarf durum wheat with high degree of resistance to all the three rusts. It was bred at IARI, New Delhi from a cross HD 2092 × H.D.1962-E4870-K65. The grains are amber, hard to semi-hard, medium bold, ear fully awned, fusiform, brown and glabrous. The grains are as good as those of *Kalyan Sona*. Besides the timely sown conditions, it does well under late sown as well as under limited irrigated conditions.
16. **C. 306.** It is desi sharbati colour grained variety evolved through a cross (Regent 1974 × Chz 3) C 591^2 (P 19 × C281). It is recommended for dry farming regions.
17. **U.P. 319.** It is sharbati coloured, medium grained three gene dwarf variety derived from a cross No. 'S' (Sonora-64. K1. Rend) 8156. It is susceptible to rust and is early maturing variety.
18. **N.P. 836.** It is produced at I.A.R.I., New Delhi, by Gamma irradiation of N.P. 799 wheat. It is awned and high yielding variety.
19. **K. 816.** It is a tripple gene dwarf, late sowing and early maturing variety with shining seeds.
20. **K. 852.** It is one gene dwarf variety selected from S. 308. It is susceptible to rust under excessive manuring and irrigation. Grains are medium sized and shining.
21. **K. 65.** It is a 'desi' variety evolved through a cross N.P. 773 × C591.
22. **K. 68.** It is also a 'desi' variety evolved from a cross N.P. 773 × K. 13. Its grains are large and sharbati colour.
23. **Kanpur Kundan or K. 101.** (M.P Hybrid 65 × Hybrid 11). It is short maturing variety with shining and amber coloured seeds, yielding 15% greater than K.68. This variety is resistant to rust and *Alternaria*.
24. **K-7410 Shekhar.** It is double dwarf wheat with good adaptation to early timely sown conditions. It does well under recommended 4-6 irrigations as well as under limited one or two irrigation. This variety was bred at C.S. Azad University of Agriculture and Technology, Kanpur from K 812 × Kalyan Sona. Its grains are amber bold, hard lustrous and of good quality. It is resistant to black rust and moderately to brown rust in the field. Ears are oblong, fully awned and brown in colour. It is recommended for cultivation in Eastern U.P, Bihar, West Bengal, Orissa and Assam.
25. **Mukta (H. 1. 385).** It is a white and bold grained single gene dwarf variety evolved from a cross between hybrid 633 and N.P. 842.
26. **H.I. 617.** It is a tall conventional wheat with fully awned white glabrous glumes which was bred at IARI Regional Research Station, Indore (M.P.) by selection from maintenance (Pedigree not known). It is adapted for cultivation in Madhya Pradesh, Gujarat, South-Eastern Rajasthan (Kota and Udaipur divisions) and Bundelkhand area of U.P. Grains are amber, hard, lustrous and very attractive.

27. Shailaza (H.S. 1138-6-4). It is a sharbati coloured medium grained single gene dwarf variety evolved from a cross between E4870 and S.308. It is recommended for northern hilly regions of India.

28. Girija (H.S. 1097-17). It is also sharbati coloured medium grained but two gene dwarf variety evolved from a cross between Cajeme 60 and E 4717. It is recommended for northern hilly regions.

29. V. 401 and V. 404. Both these varieties are light coloured and medium grained and are recommended for northern hilly regions.

30. VL-421. This double dwarf wheat variety was evolved from a cross Sonora 64 × Y50 E-Gaba at Vivekanand Parvatiya Krishi Anusandhansala, Almora (U.P.). It is adapted for cultivation in Hills of Jammu and Kashmir, Himachal Pradesh, Uttaranchal and West Bengal. It shows high degree of resistance to all the three rusts and loose smut under field conditions but is susceptible to hill bunt and powdery mildew. It is medium early to medium late in maturity, easy to thresh. Grains amber semi-hard and medium sized. Head awned with white glabrous glumes.

31. W.H. 147. It is a double dwarf wheat with excellent yielding potency under natural conditions and either free from rusts or with high degree of resistance to rusts. It was bred at Haryana Agricultural University, Hissar from cross (E 4870 × C 308) × (S 339 × P.V. 18). The grains are amber, hard, medium in size and lustrous. Ears are awned, white, glabrous and elevated. The variety does well under limited irrigations and low inputs and is adapted for M.P., Gujarat, South-East Rajasthan and Bundelkhand area of U.P.

32. I.W.P. 72. It is a double dwarf wheat resembling Kalyan Sona but differing from Kalyan Sona by its high resistance to brown rust with amber coloured, hard and medium sized grains of high quality. It was developed at IARI, New Delhi from a cross E 6056 × K. Sona 2. Ears are fully awned, brown and glabrous, slightly earlier in maturity than Kalyan Sona and C 306. The variety does well in timely sown, rainfed conditions and is recommended for Punjab, Haryana, Rajasthan Western U.P., Delhi, North Western part of M.P., and Jammu region of Jammu & Kashmir.

33. HUW-12. A double dwarf wheat with compact head with a tendency for clubbing was evolved from N.P. 876 × Cno-66 at Benaras Hindu University, Varanasi. It is characterised by thick stem, white and glabrous ear, amber coloured, hard, medium sized and lustrous grains; medium late in maturity and it shows resistance to black rust and moderately susceptible to brown rust. It shows good performance in timely sown, irrigated and good fertility conditions. It is recommended for cultivation in eastern part of U.P., Bihar, Orissa, West Bengal and other Eastern States.

34. H.W. 657. A single dwarf wheat with high degree of resistance to black and brown rusts was bred at IARI, New Delhi from Timgalin × K 65. The variety is characterised by fusiform fully awned ears with glabrous glumes, amber, hard and medium sized grains. It shows good performance in timely sown rainfed and low fertility conditions and is recommended for cultivation in Maharashtra, Andhra Pradesh, Karnataka and the plain area of Tamil Nadu.

35. H.D. 4672 (Malavratna). A high yielding disease resistant, semi-dwarf quality Durum wheat variety evolved jointly by the scientists of IARI, New Delhi, and the Regional Research Station, Indore has been identified by 38th Wheat Workshop in August, 1999 for rainfed and limited water availability conditions of central zone (M.P., U.P., and Rajasthan) evolved after a gap of 30 years. It is much superior to A-9-30-1 and sujata varieties in yield.

BREEDING OF POTATO
Solanum tuberosum, Family Solanaceae

The potato is one of the important food crops in India. The potato is native of South America and has been under cultivation in Andes and Chilean highlands of that continent for over a thousand years. In India it is an introduced crop. In our country, potato cultivation was started in the beginning of 17th century. Probably, this crop was introduced in India in the beginning of seventeenth century by English businessmen from some advanced Western countries. In India the total area under potato cultivation is nearly 5,04,000 hectares (3½ lakh hectares in plains and rest in the hills) with an annual production of about 42,33,000 tonnes. U.P. is the top potato-growing State and about 45% of the total area under potato cultivation is in U.P. alone. The area and the production are on the increase.

In U.P., the potato improvement programme was initiated in 1882. In 1887, many varieties were introduced from foreign countries. Some of the varieties tested for their yield potential at Botanical Garden of Sharanpur in 1887 were American rose, Hedge Kidney, New hundredfold, River Royal, Ashleaf Kidney, New hundred milk white, early good rib and also many other indigenous varieties. Many of the foreign varieties were found poor yielding. In India, large scale scientific work for improving potato varieties was started in 1935 at I.A.R.I., New Delhi and now most of the cytogenetical and breeding work on this crop are carried out at the Central Potato Research Institute, Simla. A large number of potato varieties have been produced by Dr. Pushkar Nath from Simla Institute.

History of Central Potato Research Institute

The Government of India, following the recommendations of the Imperial Commission of Agriculture, initiated a scheme for improvement of the potato crop in the Nilgiri Hills. This scheme was replaced by a new scheme "Potato Breeding in Northern India" in 1935 which was later on extended to whole of the country with a station at Simla under the overall control of Indian (then Imperial) Agricultural Research Institute (IARI). The breeding and genetical work could be effectively undertaken only in the high hills, as only two Indian varieties 'Phulwa' and 'Darjeeling Red Round' flowered in plains and also because production of healthy breeder's seeds free from degenerative effects could also be undertaken in hills only, the headquarter of the Institute was transferred to Simla in 1956 on the recommendations of the expert committee appointed by the Government of India. Simultaneously, on the recommendations of the same committee five potato experimental trial centres at Jullundur, Babugarh, Rajgurunagar, Ootacamund and Shillong were set up during 1956-60 to give a fillip to the regional research problem. The earlier headquarter stationed at Patna too was converted into a regional station. In 1964, a wart testing station at Darjeeling was established to tackle the serious potato wart disease problem, peculiar and confined only to this region. In 1967, a foundation seed station at Jagu (Simla Hills) with a unit located in the plains at Jullundur and in 1971 a third unit at Daurala under this scheme were started for the production of breeder's seed.

In 1970, an *All India Coordinated Potato Improvement Project* was initiated, with seven main centres at Simla (coordinating Unit), Jullundur, Patna, Kanpur, Shillong, Ootacamund and Rajgurunagar (Maharashtra) and seven sub-centres at Ranchi, Bhubaneshwar, Kalyani, Hassan, Shopian (Jammu and Kashmir), Deera and Chhindwara. Six of the seven main centres under the project are also the regional stations of Central Potato Research Institute. Through this project a unified and collaborated programme of potato research and development is undertaken actively involving the Agricultural Universities, State departments of agriculture and other collaborating agencies. The overall administrative control of the Central Potato Research Institute, Simla rested in the Ministry of Food and Agriculture, Government of India till April,1966 when the Indian Council of Agricultural Research took over its control along with other commodity and crop institutes.

—From Scientific Reports of the C.P.R.I.
Simla (H.P.) 1971-73

Cytogenetics

The basic chromosome number of this species is generally considered to be 12. A basic chromosome of 6 has also been suggested but no wild species with chromosome number $n = 6$ is known which can be regarded as progenitor of the present species. Five different levels of polyploid $2n$, $3n$, $4n$, $5n$ and $6n$ are known in tuber bearing *Solanum*. *Solanum tuberosum* was first considered to be an autotetraploid but according to recent view it is more probably a segmental alloploid derived possibly from crosses between cultivated diploid *S. spontaneum* and the wild diploid species *S. sparsipilum*.

The genetics of potato has been reviewed by M.S. Swaminathan and Howard, as well as by Howard. Induced polyploids may be produced by placing colchicin solution in the leaf axil of decapitated shoot or by covering the eyes of potato with lanolin paste containing colchicin. Haploid plants may be useful in genetics and breeding studies. The doubling of chromosome number of the haploids will result in genetically homozygous diploid clones. The haploid plants are of rare occurrence. Triploids can occur when tetraploid *S.tuberosum* is pollinated by pollen from a diploid species.

Botany

The potato flower contains five sepals, five petals and bilocular ovary with single style and bilobed stigma. The petals are united and stamens are epipetalous and bear straight anthers which form a close column or cone around the style. In some varieties flower buds fall at premature stage. Pollen production in the majority of commercial varieties is very poor and in some, pollens are not formed. Fruits are berries often called seed balls (Fig. 39.1). In some clones fruit balls are not formed. Flowers in the majority of the cultivated clones open in early morning but in a few other cases, they continue to open throughout the day. Self-pollination is a rule and cross-pollination by wind or insects occurs infrequently. Male sterility and self-incompatibility have been noted in some varieties.

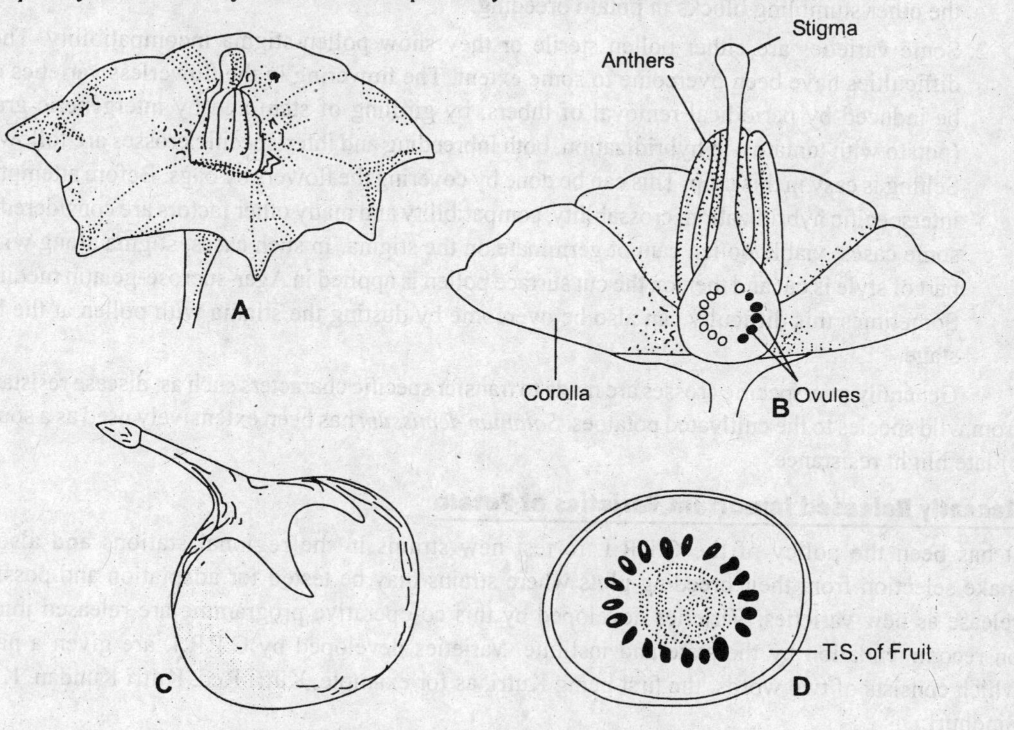

Fig. 39.1 Potato flowers (A & B) and fruits (C & D).

Asexual reproduction in potato is by tubers. The sprouts arise by germination of buds in the eyes of tuber. Vegetative reproduction is also possible by rooting stem cuttings.

Aims of Potato Breeding

Breeding of Potato in India is carried out with the following objectives:

1. High tuber yield.
2. Early maturity.
3. Good storage and cooking quality.
4. Tuber flesh colour, under water weight of tubers, starch content.
5. Attractive skin colour.
6. Resistance to fungal, bacterial, leaf roll virus diseases and frost.

Crossing Techniques

The flower-buds are emasculated in the evening. Flower buds that will open the next morning are selected and rest of the buds and flowers are removed. The petals of the selected flower buds are pushed apart and the five stamens removed with pointed forceps. The emasculated flower-buds along with one or two leaves are bagged. Emasculation is omitted in self-sterile lines. Next morning, fully mature anthers from male parent are collected in a petridish and pollen collected on pen-knife or forceps is dusted over the stigma of the emasculated flower-buds. Flowers are bagged again after pollination.

In the hybridization of this crop, there are two major difficulties:

1. Some varieties do not flower due to photoperiodic effects and in some, flower buds fall before opening. Many cultivars show poor flowering and low fertility. Crossing work with primitive cultivars or wild relatives is hampered by several crossing barriers and by difference in ploidy levels. High degree of heterozygosity and low heritability of many important characters are the other stumbling blocks in potato breeding.
2. Some varieties are either pollen sterile or they show pollen-stigma incompatibility. These difficulties have been overcome to some extent. The flowering in the flowerless varieties can be induced by periodical removal of tubers, by girdling of stem and by intergeneric grafts (potato with tomato). In hybridization, both inbreeding and inter-specific crosses are followed. Selfing is easy in this crop. This can be done by covering the flowers by bags. Before attempting interspecific hybridization, crossability, compatibility and many other factors are considered. In some cases, viable pollen cannot germinate on the stigma. In such cases, stigma along with a part of style is cut and then on the cut surface pollen is applied in Agar-sucrose-gelatin medium. Sometimes this difficulty can also be overcome by dusting the stigma with pollen at the bud stage.

Generally interspecific crosses are made to transfer specific characters such as, disease resistance from wild species to the cultivated potatoes. *Solanum demissum* has been extensively used as a source of late blight resistance.

Recently Released Important Varieties of Potato

It has been the policy of the C.P.R.I. to test new strains in the regional stations and also to make selection from their breeding plots where strains may be tested for adaptation and possible release as new varieties. Varieties developed by this co-operative programme are released jointly on recommendation by the state and institute. Varieties developed by C.P.R.I. are given a name which consists of two words; the first being Kufri, as for example, Kufri Red, Kufri Kundan, Kufri Sindhuri.

At present, nearly 500 varieties are on record, but here only most promising and high yielding popular strains that are released recently are listed:

Variety name	*Pedigree*	*Important Features of the Varieties*
1. Satha	Selection	Yield low, but early maturing, becomes ready in about 2 to two and half months, grown mostly in Bihar. Tubers small, white, smooth.
2. Phulwa	Introduced in India from European countries, a century ago.	Tubers are small to medium, smooth, white, round, flesh yellow coloured, keeping quality good. Late maturing variety, grown in North India.
3. Gola	Selection	White, pear shaped tubers, eyes set deeply, flesh white, early maturing.
4. Up-to-date	Disease free clones selected from Military special, an introduced variety from Rangoon.	Early maturing, tuber large, smooth, oval and flattened, flesh white. It is most popular variety under cultivation in India and is particularly suited for Uttar Pradesh, Himachal Pradesh, Punjab, Madhya Pradesh, Bihar and Maharashtra.
5. Darjeeling Red Round	Selection clone	Wide adaptability, tuber round, medium sized, eyes deep red coloured. Keeping quality poor. It matures slightly earlier than Phulwa.
6. Craig's Defiance (Released in 1953).	Introduced from U.K. in 1953	Suited for cultivation in hills of H.P. and Punjab. Tubers are large, long, smooth, white with fleet eyes and white flesh.
7. Kufri red (Released in 1957)	Selection from Darjeeling red round	Excellent yield in plains of U.P., Bihar and W.B. Tubers are smooth, red, round and flesh white. Keeping quality good.
8. Kufri Alankar	Kennebec × ON 2090	An early tuberising, photo-insensitive, high yielding variety with oval to oblong tubers having white flesh and floury texture.
9. Kufri Chamatkar	Phulwa × Ekishirazu	A medium maturing variety with uniform sized, smooth and shining tubers.
10. Kufri Jeevan	M 109 × 698 D.	A late maturing, high yielding variety with oval white tubers. It is resistant to late blight and *Cercospora* infections.
11. Kufri Jyoti	3069 a (4) × 2814 a (1)	A medium maturing, widely adaptable variety with oval white tubers. It is resistant to late blight.
12. Kufri Khasi-Garo	Taborky × 698 D	A medium maturing variety with oval to round white tubers, good resistance to late blight, and moderate resistance to early blight and viruses. Recommended for growing in hilly regions of Assam, Tripura, Manipur, etc.
13. Kufri Moti	3460 × M 109-3	A medium maturing high yielding variety, immune to late blight, suitable for cultivation in Nilgris Hills.
14. Kufri Naveen	3070 d (4) × 692-D	A late maturing variety with oval white tubers possessing high resistance to late blight and moderate resistance to early blight and *Cercospora*, suitable for growing in Khasi Hills.
15. Kufri Sheetman	Phulwa × Craig's Defiance	A frost resistant medium maturing variety with oval white tubers. Suitable for growing in the plains of North India.
16. Kufri Lovkar	Sarkov × Adina	A high yielding variety suitable for growing in Deccan Plateau of Maharashtra in rainy and Rabi seasons.
17. Kufri Kuber (O.N. 2236) (Released in 1958)	(*S. cartilobum* × *S. tuberosum* × *S. andigena*)	Good for cultivation in the plains and hills of Punjab, Bihar, Maharashtra. It is early maturing and high yielding variety. Tubers are medium sized, oval, flattened and with white flesh.

	Variety name	Pedigree	Important Features of the Varieties
18.	Kufri Kisan (P.S. 454) (Released in 1958)	Up-to-date × O.N. 2091.	It gives high yield under long day conditions. Tubers irregular, white and medium or large sized, pink eyes deeply set, and flesh white. It has very good keeping quality.
19.	Kufri Kundan (Hybrid-8) (Released in 1958 by C.P. R.I., Simla)	K. Siraju × Catadine	Suited to heavy and wet soil of hills, good performance in dry conditions. Tubers are medium sized, smooth surfaced, flat eyed and with yellow flesh, keeping quality excellent.
20.	Kufri safed	Selection from Phulwa	Late ($4\frac{1}{2}$ to $5\frac{1}{2}$ months) variety, yield very high in the plains of U.P., M.P., Delhi, Punjab. Tuber medium sized, smooth and white, deeply set eyes, keeping quality good.
21	President	Introduced from England	It is a medium maturing ($4\frac{1}{2}$ months) variety, tubers white round, grown mostly in Nilgiri Hills. It is resistant to blight disease.
22	President Kufri clone	Clonal selection from President (C.P.R.I., Simla)	Performance better than President.
23.	Kufri Sindhuri (C. 1411)	(Kufri red × Kufri Kundan)	Tubers red, medium sized, eyes fleet. Resistant to frost. Medium keeping quality.
24.	Kufri Chandramukhi (A-2708) (Released in 1968)	(S.-4485 × Kufri Kuber)	Matures in $2\frac{1}{2}$ months in plains. It is frost resistant.

BREEDING OF PADDY OR RICE
(Genus. *Oryza*) 2*n* = 24.

Rice is one of the important and oldest food crops in India. It is a staple food of the natives of Eastern, Southern, and South Western parts of the country. This crop is grown practically in all the states of India. The principal rice producing states are Bihar, West Bengal, Orissa, Andhra Pradesh, Tamil Nadu, Kerala, M.P., U.P., Maharashtra, Assam, and several others. The total area under rice cultivation is between 78 and 80 million acres. Over 85% of the world rice is grown in China, India, Japan, Pakistan, South-Eastern Asia and the adjacent Islands of Pacific. China and India produce about 50% of the total world rice supply.

Origin of Rice

Rice has been cultivated in China and India for several thousand years. The cultivated species of rice (*Oryza sativa*) is supposed to have originated in south and south-east tropical Asia because in these regions there exists a great diversity of the cultivated forms. The only other cultivated species of rice, *Oryza glaberrina* is indigenous to West Tropical Africa. As regards the progenitors of cultivated rice, a number of divergent views have been advanced, but general consensus of opinion favours that it originated from the species complex known as *Oryza perennis.* Polyphylatic origin of *Oryza* species from several ancestral types is generally accepted. In addition to the two cultivated species, there exist several wild species of genus, the exact number of which is 23 (according to some workers) or 28 (according to some others). From India cultivated rice reached to Egypt, Europe, Africa, Australia and America and from China it found its way to Korea and Japan.

Varieties of Rice

On the basis of geographical adaptation and morphological features, the cultivated rices of the world have been grouped into three varietal groups namely, *indica, japanica* and *javanica*. The distinguishing features of these three varietal groups are as follows:

Indicas	*Japanicas*	*Javanicas*
1. Profuse tillering	Moderate tillering	Low tillering
2. Broad, light-green leaves, usually sensitive to photoperiod.	Leaves narrow, deep green, often sensitive to photoperiod.	Leaves broad, stiff, light green, and insensitive to photoperiod.
3. Have grain dormancy.	Do not have grain dormancy.	Have little dormancy.
4. Rarely responsive to nitrogenous fertilizers.	Responsive to N. fertilizers.	Moderately responsive to N. fertilizers.
5. Susceptible to shattering.	Resistant to shattering.	Resistant to shattering.
6. Typical awnless.	Awnless to long awned.	Typically long awned.
7. Thin and short hairs on glumes.	Dense and long hairs on glumes.	Long hairs on the glumes.
8. Include varieties from India, Southern China, Taiwan, Sri Lanka, Java and other regions.	Include varieties from Japan, Korea and Northern China.	Include small number of varieties from Indonesia, which are referred to as "blue variety".

Paddy crop has wide range of physiological adaptability and more than a thousand varieties differing in morphological and agricultural characters are under cultivation. The varieties differ in their season of growth, maturation period and adaptability to different conditions of soil; alkalinity, salinity and acidity of soil, depth of standing water, temperature and altitudes. Varieties of paddy are generally classified as early, medium and late maturing. In some provinces, they are classified as autumn, winter, and spring varieties and of these 90% of the total production is obtained from winter crop. Rice is grown in widely different regions, from deeply submerged areas to high altitudes. The varieties growing in submerged areas with one to 2 m deep water are called shallow water varieties. Varieties growing in the areas where the depth of water is two to five metres are called *deepwater* varieties or *floating varieties.* Stems of deep water varieties elongate with the rising level of water. The varieties adapted to high altitude and those grown on dry land are called *high altitude varieties and upland varieties* respectively.

Botany and Genetics

The inflorescence is panicle which bears single flowered spikelets. The flower consists of a lemma and a palea, the structures which form the hull or husk of paddy (Fig. 39.2). The outer glumes are usually obscured and about one-fourth the length of lemma and palea. Flowers have 6 stamens. The flowers open between 8 A.M. and 11 A.M. The flowers in a panicle bloom over a period of 7 to 10 days. Anthers dehisce at the time the flowers open. The blooming proceeds from apex to base. The flowers are usually self-pollinated. Natural crossing varies from 0 to 3%.

Crossing Technique. Several methods have been developed for emasculating the rice flowers. Before emasculation, immature as well as previously opened flowers are removed from the panicle and only those flowers are left which will bloom the following day. In the conventional method of emasculation, the glumes are separated with a pair of forceps and six stamens are removed. The emasculation and pollination may be facilitated, by cutting of the buds of florets. The emasculation is done in the afternoon and pollination in the following morning. The emasculated flowers are bagged with butter paper bags.

Hot water may be used to open the florets of rice. In the hot water treatment, the rice panicle is immersed in water contained in thermos flask at 42 to 44° centigrade for a period of 5 to 10 minutes. Panicle in the second or third order are chosen as female parents. Flowers already opened as well as the immature florets are removed an hour or so before treatment. The thermos bottle may be supported on a trough like holder at an angle of 35 degrees. This treatment causes the florets to open in normal manner and stamens may be removed without injury to the stigma.

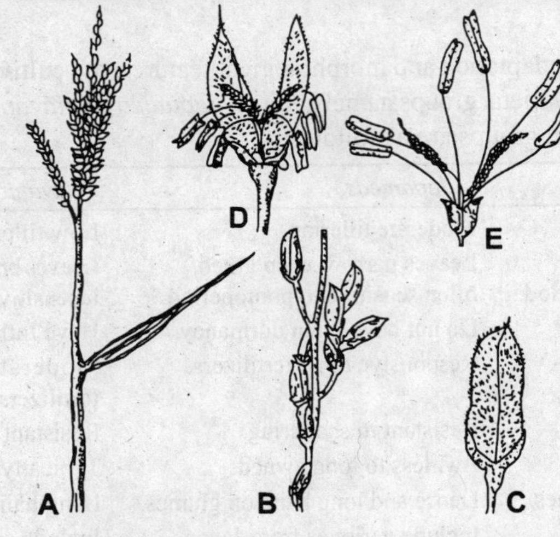

Fig. 39.2 Inflorescence of rice (A, B) and floral parts (C, D, E).

Pollination may be made within a period of 30 minutes before the glumes naturally close. Florets nearing the blooming stage are opened and the turgid mature anthers are taken with forceps and pollen grains are dusted over the stigma. Pollens of rice do not normally remain viable for more than few minutes.

Vegetative Propagation. Plants can be vegetatively propagated by separating the tillers and planting them as small individual plants.

Cytogenetics. The somatic chromosome number ($2n$) of cultivated rice is 24. Some wild species are tetraploids $2n = 4X = 48$. Some cytological and genetical evidences indicate that it is a secondary polyploid and the basic chromosome number is 5. The presence of duplicate genes and secondary association of chromosomes during the meiosis clearly indicate that *Oryza sativa* is aneuploid. The haploid chromosomes in this theory may be denoted as *a, b, c, d, e,* and *a', b', c', d', e'*. According to this hypothesis, a species with haploid set of 5 chromosomes will produce hybrid with chromosomes *a, b, c, d, e, a', b', c', d', e'*. Through meiotic abnormalities two chromosomes (*a', b'*) were duplicated in the hybrid and this was followed by doubling of chromosomes to produce progeny of *O. sativa.*

Interspecific crosses in rice are generally accompanied by high degree of partial or even complete sterility in the hybrid, which is determined by the lack of affinity between the parents. The sterility is also very common in the crosses between *japanica* and *indica* varieties as well as in intervarietal crosses in *Oryza sativa* species. The reasons for this may be segregation and recombination of specific genes, chromosomal abnormalities and irregularities and cytoplasmic influences.

Mutations have been common and these must have caused the evolution of a large number of varieties grown throughout South and South-East Asia. Certain chemical mutagens have been found to be effective in producing polyploid strains but such forms have been found tillerless and are less fertile than the corresponding diploids.

Aims of Rice Breeding

Some aspects of rice breeding are as follows:

1. High yield, good table qualities (non-glutinous) and good response to fertilizers.
2. Resistance to drought and to blast diseases (*Piricularia oryzae*), stem rot (*Helminthosporium sigmoidium*), brown spot (caused by *Helminthosporium oryzae*), bunt of rice (caused by *Neovossia horrida*), false smut (caused by *Ustilaginoidea virens*), viruses and also resistance to insects and pests.
3. Earlyness of crop.
4. Flood resistance.

5. Resistance to lodging and short stature of plants.
6. Resistance to alkalinity and salinity of soil.
7. Non-shedding of grains from the panicle.
8. Dormancy of seeds.
9. Combining the cooking quality and aroma acceptable to consumers.
10. Insensitive to photoperiod.

Breeding and Achievements

In India, investigations have been carried out at a number of research centres. Dr. K. Ramiah, an Indian rice Breeder of International repute, Dr. Richharia and Ghose *et al.* have contributed much in the field of breeding. Major part of the genetical work on this crop has been done at Coimbatore. At present, work on the improvement of rice is being carried out at the Central Rice Research Institute, Cuttack.

Central Rice Research Institute, Cuttack (India)

CRRI was established by the Government of India in 1946 to conduct fundamental research on an All-India basis on all aspects of rice production, breeding, genetics, disease and insect problems. The institute was transferred to ICAR in 1965. The institute maintains 15,000 entries of rice germplasm. An All-India Coordinated Rice Improvement Project (AICRIP) was organised in 1965 for coordinating the breeding and other researches on rice conducted by the State and Central Government organisations in the country. The Rockefeller Foundation in India is co-operating in this coordinated effort to accelerate the rice breeding programme in India. The popular varieties of rice presently under cultivation, like Jaya, Padma, Sona, Vijaya, Bala, Vani, Sakti are some of the notable achievements of CRRI.

International Rice Research Institute, Philippines

It was established on the campus of College of Agriculture, University of Philippines, Los Banos, in 1960 through the joint efforts of Rockefeller and Ford Foundations. The institute has initiated vigorous rice breeding programmes in order to develop high yielding, photoperiod insensitive strains which may be utilized either directly or as parent material in rice breeding programme throughout S.E. Asia. The institute maintains world collection of tropical and subtropical rice germplasm consisting of some 42,000 entries. The dwarf varieties of rice were first developed by IRRI.

Methods of Rice Breeding

The attempts at improving the rice of this country have been made by the following methods:

1. By selection of promising local strains.
2. By introduction and acclimatization of new varieties from foreign countries like Japan, China, U.S.A., Philippines, Java, etc.

   The development of dwarf rice (*O. sativa*) varieties has revolutionised the rice cultivation. These varieties were derived from 'Dee-geo-woo-gen', an early maturing dwarf variety of *Japanica* rice from Taiwan. 'Taichung Native-1 developed in Taiwan and IR-8 developed at IRRI, Philippines were introduced for the first time in India in 1966 and grown extensively for a few years but later several superior Dwarf varieties of rice like Jaya, Ratna etc., developed in the country replaced the introduced varieties. The Dwarf varieties are resistant to lodging, responsive to fertilizer application, high yielding and photo-insensitive. Many of the photo-insensitive varieties were adopted for cultivation in certain non-traditional areas like Punjab. Even in traditional areas where only one crop was possible in an year, it has now become possible to rotate early maturing rice varieties with wheat.
3. By crossing certain varieties in order to combine certain desirable characters found in different parental stocks into one variety.

   Breeding work on rice has been carried out in almost all rice growing States of this country and many improved varieties have been evolved which give 20-50% higher yield than local varieties. At present the number of high yielding varieties is more than 500.

The hybridization programme involving *O. indicus* and *O. japanica*, which is still under way, has resulted in very promising materials with new genetic variability. Hybridization has been done on intervarietal and interspecific levels.

In rice breeding, Taichung Native-1 and IR-8 have contributed in some way or the other in the development of many high yielding varieties in India. The cross Taichung Native 1 × T. 141 has yielded two superior varieties of rice; 'Jaya' and 'Padma', Padma has shorter maturation period and finer grains than Jaya. The other popular varieties of rice developed through hybridization and selection, include Bala, Karuna, Cauveri, Krishna, Ratna, Sabarmati, etc.

4. Mutation breeding and polyploidy have also been applied quite recently for specific purposes. A semi-dwarf rice variety; 'Jagannath' has been developed from Tall cultivated variety T 141 through mutagenesis with Gamma-rays. The variety is resistant to lodging, higher yielding, more responsive to fertilizer application than parent variety T 141. Another rice variety 'CRM 13-324' was also developed through mutagenesis. It has reduced height, shorter maturity period, higher yielding capacity than the parent variety.

Some of the good varieties with high yield potentials are given below:

Selection. Adt. 1, Adt. 3, Adt. 5, 10, Co. 4, Co. 6, Co. 10, Co. 14, Mtu. 1, Mtu. 3, Mtu. 6, Mtu. 9, Mtu. 13, Ptb. 10, S.P.-1, S.P.-2, Type-1 (a selection from variety Ram Jewan of Dehradun), Type 3 (a selection from the variety Chaul of Rampur), Type A 64 (a selection from the variety Hansraj of Unnao) etc.

Intervarietal Hybrids

T.K.M. 6, S. 1086, S. 1088, S. 1089, Adt. 20, Adt. 25, T. 136, T. 137, L. 12 (L.N. 41 × C.P.I), P. 502, R. 575 (R.J. 100, × C.P.I.)

Interspecific Hybrids

Several species of *Oryza* have been involved in crosses with *O. sativa* for improving cultivated rice varieties, especially for transferring disease resistance genes from wild races to cultivated ones. *O. longistaminata, O. nivara, O. perennis, O. latifolia, O. minuta, O. longiglumis, O. ridleyii, O. australiensis, O. alta, O. punctata* and some others have been tested at I.R.R.I., and several Rice Research Institutes all over the world and genes resistant to blast and blight diseases, and insect pests have been identified in them. Further researches on developing resistant varieties utilising these wild species are underway:

Co. 31, I.J.-1, Adt. 27 (GEB 24 × Norin-8).

Varieties Produced by Mutation Breeding

P. 500-28 (from T-1145) was developed at Bose Research Institute, Kolkata. It is high yielding variety.

Some Recently Released High Yielding Varieties

Ambemohar-157, Varangal 487, Andrew sail, Basmati-307, B.R. 46, Co. 4, Co. 26, F.R. 13 A, Hr. 35, Keresail, Mtu. 1, Patnai S. 1092, Adt. 27, Taichung-65, Taivan-3, I.R. 5, I.R. 49, Nagina-22, Sudha, N.S.J. 98, N.S.J. 200, K-22, Kasi, T. 136, T. 9, T. 100, T. 21, T. 23, T. 26, T. 36, T. 3, N.U. 10, Varieties Co. 25, Co. 26, Hybrid Kashmere 60, M. 42 are highly resistant to blast (*Piricularia oryzae*) and Co. 24, T. 141 is resistant to *Helminthosporium.* Var. S.P.-268 is salt resistant and is good for cultivation in Orissa. Dudhalchi, Jaisuria, Chakia 59, are flood resistant varieties which are suitable for cultivation in U.P. under deep water. Work on rice is still in progress and many new varieties are under trial at different Research Stations.

The names of some other varieties, their parentage and grain characters are given in the following Table:

S. No.	Varieties	Parentage	Duration in days	Grain type
1.	Taichung Native	1 Tsaiyuenchang × Dee Geo Woo-gen	110	Short bold
2.	I.R. 8	Peta × Dee Geo Woo-gen	138	Medium bold
3.	Sarjoo 49	T 65 (TN 1 × Taichung 65)	110 – 115	Short bold
4.	Jaya	TN 1 × T. 141	133	Medium bold
5.	Padma	T. 141 × TN 1	110	Short medium
6.	Cauveri	TN 1 × TKM 6	107	Short medium
7.	Kanchi	TN 1 × Co 29	112	Short medium
8.	Krishna	Gele 24 × TN 1	115	Short slender
9.	Bala	N 22 × TN 1	106	Short round
10.	Sabarmati	Basmati 370 × TN 1	109	Medium slender
11.	Jamuna	Basmati 370 × TN 1	110	Medium slender
12.	Pankaj	Peta × Tangkai Rotan	146	Medium bold
13.	Jagannath	Induced mutant of T 141	157	Medium bold
14.	IR 20	IR 262 × TKM 6	140	Medium slender
15.	Saket 3	HR 12 × TN 1	99	Long slender
16.	Ratna	TKM 6 × IR 8	120	Long slender
17.	Vijaya	T 90 × IR 8	145	Medium slender
18.	IR 24	IR 8 × IR 127-2-2	125	Long slender
19.	I.E.T. 1039	T 90 × IR 8	139	Long slender
20.	I.E.T. 1991	Geb 24 × TN 1	130	Long slender
21.	Pusa 2–21	I.R. 8 × TKM 6	110	Short bold
22.	C.R. 44–35	TKM × IR 8	125	Long slender
23.	Improved Sabarmati	TN 1 × Basmati 370	110	Long slender
24.	I.R. 26	IR 24 × TKM 6	125	Medium slender

BREEDING OF COTTON
(*Gossypium species*)

India is one of the important cotton growing countries in the world. It grows the largest acreage of cotton than any country in the world but in total production it stands far behind U.S.A and Russia. In India cotton is grown in all the states but about 90% of the acreage is in peninsular or Southern India. The total area under cotton cultivation in India is about 8 million hectares and it is grown mainly between 8 and 32° latitudes and 70 and 80 degree longitudes.

The genus *Gossypium* is represented by four cultivated species namely *G. hirsutum, G. barbadense L., G. herbaceum L.* and *G. arboreum L.* and thirty diploid short fibred or lintless wild species. The wild species are scattered around the world. *G. hirsutum* (upland cotton) is most important cultivated cotton species which accounts for most of the world's cotton fibre.

In India, all the four species of cotton namely *Gossypium arboreum, G. herbaceum* (desi cottons), *G. hirsutum* and *G. barbadense* (American cottons) are in cultivation.

Of the total cotton crop, nearly 53% is shared by *hirsutum* cotton, 28% by *arboreum* cotton and 19% by *herbaceum* cotton, while barbadense cotton is grown only in a few thousand hectares. Because of outstanding merits of *hirsutum* cotton both qualitatively and quantitatively, more emphasis has been laid on breeding of this cotton than the other cottons.

The cotton growing areas of our country are divided into four regions, *viz.*, Northern, Central, Southern and Eastern regions. The names of important varieties of cotton grown in different regions are given below:

Region	States	Species	Recent varieties under cultivation
Northern	Punjab, Rajasthan, U.P.	*G. hirsutum* *G. arboreum*	216 F, Indore I, Ganga Nagar, 320 F, LL 54, H. 14231 R, Raniben
Central	Gujarat, Maharashtra, Madhya Pradesh	*G. herbaceum* *G. hirsutum* *G. arboreum*	Kalyan, Digvijai, Devi Raj., 134-Co., 2-M, C.J. 73, Vinar, Buri 147, Gaorani 22, G. 46, Daulat.
Southern	A.P., Mysore, Tamil Nadu	*G. hirsutum* *G. arboreum* *G. herbaceum* *G. barbadense*	MCU 2, MCU 3, 216 F; Parbhani-American, K6, K7, Jayandhar, Laxmi, Sel 69, Westerns I, Cocandas 2, N-14, Nandium, Adonicum, Andrews.
Eastern	Assam, Tripura, Manipur	*G. arboreum*	D. 46-2-1

Cotton is indigenous to South and South-East Asia. It has been cultivated in Indus valley for more than 5,000 years. Relics of Mohenjodaro period indicate a high degree of art of spinning and weaving of cotton at that time. It is also indigenous to America.

Cytogenetics

The genes *Gossypium* has several genomic groups; six groups of diploids (AA, BB, CC, DD, EE and FF) and one group of tetraploids (AA DD). Each genome has 13 chromosomes. *G. thurberi* (wild American cotton) is a diploid species with D_1D_1 genomes and $2n = 26$ chromosomes. All tetraploids have one genome of 13 chromosomes similar to that of cultivated old world's diploid cotton and another set of 13 chromosomes similar to that of new world's cotton species.

The somatic chromosome numbers of the different cotton species grown in India are given here as under:

1. *Gossypium arboreum* $(2n = 26)\ A_2A_2$ Diploid species (Desi or Asiatic cotton with coarse and short fibre).
2. *Gossypium herbaceum* $(2n = 26)\ A_1A_1$ Asiatic cotton
3. *G. hirustum* $(2n = 52)\ A_1A_1D_1D_1$ (New world or upland cotton) Tetraploid species with fine long staples.
4. *G. barbadense* $(2n = 52)\ A_2A_2D_2D_2$ (Egyptian cotton)
5. *G. thurberi* (wild American cotton, Diploid chromosomes species with D_1D_1 genomes with $2n = 26$)

The chromosomes in *G. hirsutum* and *G. barbadense* (New world cotton species) are comparatively smaller in size than the chromosomes in *G. herbaceum* and *G. arboreum* (Old world species of cotton). Each of the tetraploid species of new world cotton has 26 large and 26 small chromosomes. This indicates that the tetraploid New World species are probably allopolyploids which originated by hybridization between Old World and New World diploid species (Fig. 39.3). A possible origin of *G. hirsutum* is from a cross between Asiatic cotton *G. arboreum* ($2n$ = 26, AA) and American wild cotton *G. thurberi* ($2n$ = 26, D_1D_1) followed by chromosome doubling in the interspecific hybrid. According to more recent scheme *G. hirsutum* ($2n$ = 52) originated from a cross *G. herbaceum*, Var. *africanum* × *G. raimondii* followed by chromosome doubling of interspecific hybrid.

The incorporation of germplasm from wild species with either AA or DD genome set would appear easy yet many cotton breeders do not recommend interspecific hybridization programme for improvement of cultivated cotton. The incorporation of wild germplasm is really difficult because of (*i*) finely tuned genetic system controlling lint quality-strength and texture, (*ii*) hybrid breakdown among interspecific hybrids, (*iii*) highly negative correlations between fibre strength and high yield and (*iv*) reduced chiasma frequency in many of the interspecific hybrids.

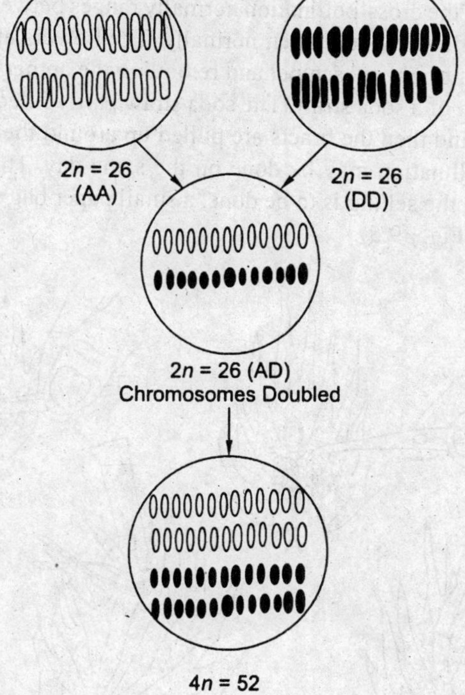

Fig. 39.3 Origin of New World cotton from old-world cotton through Natural hybridization and amphidiploidization.

One method of utilizing wild cotton species is to hybridize diploid species, treat F_1 hybrids with colchicine and then cross the amphidiploids with cultivated tetraploid species. This is followed by back-crossing with tetraploid.

Several male sterile genes-ms_1, ms_2 and so on and sometimes haploid plants are frequently met within cotton. One gene ms_4 is dominant for male sterility. Cytoplasmic male sterility has been obtained in cotton plants with *G. anamalum* cytoplasm. Introducing cytoplasmic male sterility to cultivars from wild species is another potential use of wild cotton germplasm. The nucleus of *G. hiroutum* has been incorporated into the cytoplasm of wild species. Male fertility restorer gene are known in some cases.

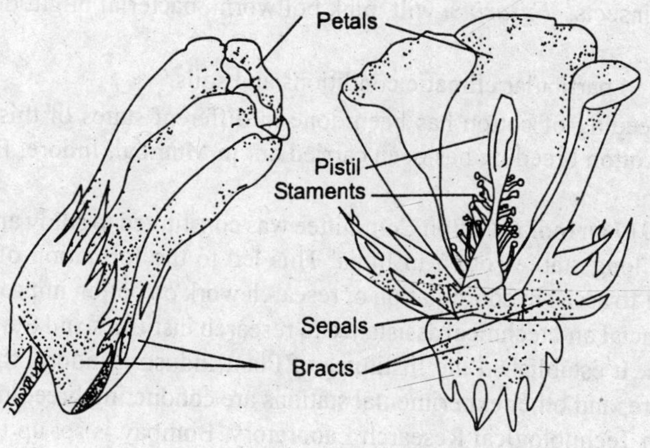

Fig. 39.4 Cotton flower.

Pollination Techniques. Flowers are hermaphrodite with fused stamens forming staminal tube (Fig. 39.4). The amount of the cross-pollination normally ranges between 5 and 25%. The crosses are made one day before the flower would open normally. The flowers are emasculated by cutting the corolla with small scissors or curved scalpel and removing the anthers. Pollen is collected from the male parent in short length of a soda straw. The soda straw partly filled with anthers is slipped down over the exposed stigma and then the bracts are pulled up around the soda straw and wired tightly. The emasculation and pollination may be done on the same day. The emasculated and pollinated flowers are bagged. When the selfing is to be done, a small paper bag should be placed over the bud in the afternoon it opens (Fig. 39.5).

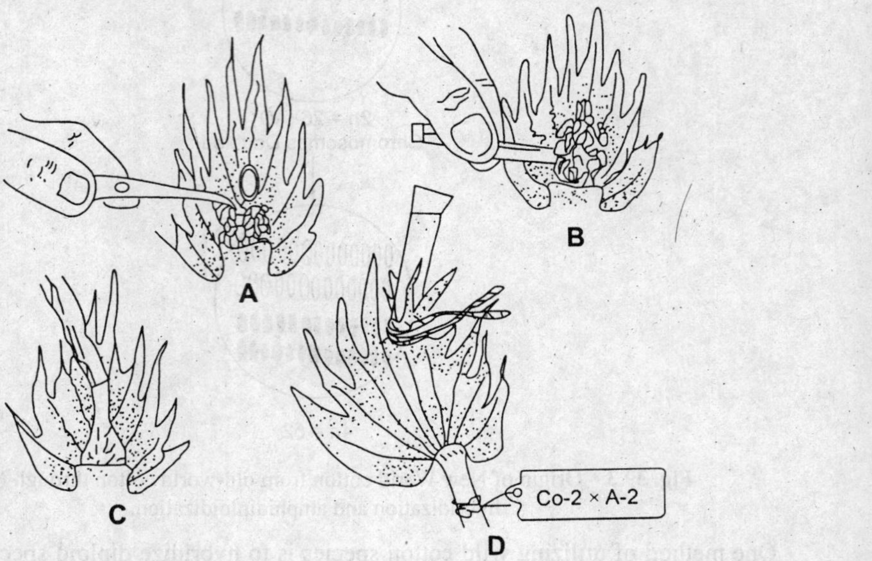

Fig. 39.5 Different steps in crossing of cotton flower.

Objectives of Cotton Breeding

The different objectives of cotton breeding are as follows:

1. Early maturity.
2. Fibre quality : Staples of long and fine type with tensile strength.
3. High ginning percentage.
4. Resistant to insects, *Fusarium* wilt, pink bollworm, bacterial blight or angular leaf spot, root rot, etc.
5. Adaptability to particular climatic conditions and soils.

Although breeding of cotton has been done in different states of this country, the extensive research work on cotton breeding has been carried out in Mumbai, Indore, Punjab, Coimbatore and IARI, New Delhi.

In the year 1917 an Indian Cotton Committee was constituted to study and encourage the growth and production of long staple cotton in India. This led to the formation of Indian Central Cotton Committee in 1921 to assist in coordination of research work on cotton improvement. The committee also provided financial and technical assistance to research institutes and State experimental stations working on cotton. It established the Institutes of Plant Industry, Indore, the PIRRCOM Research Station, Coimbatore, and other experimental stations are conducting breeding and other research on cotton. The Cotton Technological Research Laboratory, Bombay is set up to assist cotton breeders by fibre testing and evaluating the spinning qualities of new varieties of cotton. Since 1965, the research of the Indian Central Cotton Committee is being integrated into that of ICAR, New Delhi.

Methods of Cotton Breeding

The methods employed for improving the cotton varieties are as follows:

1. Introduction and acclimatization.
2. Selection.
3. Hybridization.
4. Mutation breeding
5. Use of polyploidy.

1. Introduction and Acclimatization. The work for improving cotton varieties in India was started in 1790. At that time Indian cotton was exported mainly to England. East India Company with a view to capture the cotton market started extensive trials of introduced varieties mainly of *hirsutum* origin towards the end of the eighteenth century. In the 19th century, many foreign varieties of cotton were introduced in this country from America, Brazil, Egypt, Sea Island and other countries but only a few of them were found to be useful.

Besides *hirsutum* introduced from America, several others have been introduced from Africa. A strain of *hirsutum* introduced from Uganda has been used to improve the staple length of Cambodian cotton in Tamil Nadu State. Recently one strain of *G. hirsutum* has been introduced from Russia. That is being used in breeding experiment in Tamil Nadu. Andrews variety of *G. barbadense* has been introduced into Tamil Nadu from Sea Island.

2. Selection. Selection of high yielding and disease resistant plants is the cheapest and quickest method of breeding. In cotton, both mass selection and pure line selection processes have been used. Dharwar-American variety of cotton was developed through mass selection of an introduction of *Gossypium hirsutum* from U.S.A. The variety 'Dodahatti Local' was developed from Dharwar-American by mass selection.

Two varieties, Gohari A-26 and Broach Desi 8 (B D 8) were selected from a variety of *G. herbaceum* by pure line method. B D 8 was wilt resistant and has high yielding and good spinning value but it had a low ginning out turn (34%). An improved variety 'Buri 107' was isolated by pure line selection from 'Buri' variety of *G. hirsutum.* Cambodia- Coimbatore 2 (CC2) and MCU 1 (Madras Combodia Uganda) were also developed through pure-line selection. A *Gossypium hirsutum* variety, MCU 1 was selected from the variety Coimbatore 4 (Co4) for having acceptable level of resistance to Black-arm under field conditions. Gadag 1 was developed by pure-line selection from Dharwar-American which was superior in yield and other agronomic features to parent variety but was susceptible to red leaf blight disease.

3. Hybridization. Many varieties of cotton have originated either from natural hybridization or by artificial hybridization. Various types of hybridization, such as intervarietal and interspecific hybridization, back crossing, heterosis breeding have been used for improvement of cotton. Several promising varieties of cotton have been developed in India by hybridization. To remove the susceptibility to red leaf blight, Gadag 1 was crossed to Cambodia Coimbatore 2 (CC2) and variety Laxmi was selected which was better than Gadag 1 in ginning out turn, fibre properties, earliness and resistant to red leaf blight.

Two cotton varieties 170-Co2 and 134-Co 2M, were developed by back cross method. *G. hirsutum* variety Dharwar-American 2-6-5 was crossed with *G. arboreum* variety Gaorani. 6. The F_1 was sterile; it set only a few seeds which gave rise to tetraploid plants perhaps as result of back cross to the *G. hirsutum* parent. The progeny from F_1 were back crossed to *G. hirsutum* varieties Dharwar-American (First back cross), Cambodia Coimbatore (CC2) (Second back cross) and Meade (Third back cross). From the back cross progeny '170-Co2' and '134 Co2M' varieties were selected which had staple length of about 2.7 cm and were widely cultivated in Gujarat both under irrigated and rainfed conditions.

The characters other than disease resistance have been transferred by back cross method. Good examples of this type available in cotton are *G. hirsutum* varieties Vijapla, Vijay, Digvijay and Kalyan. BD 8 has been used as recurrent parent in a back cross with 'Goghari A 26' which has a high ginning out turn (42-47%). High ginning out turn was transferred from Goghari A-26, the non-recurrent parent and a variety 'Vijay' was developed. Vijay was further improved by back cross method and the characters like better fibre length and early maturity were transferred to it from 1027 AL-F (non-recurrent parent) and the variety Digvijay was developed from the back cross programme.

Heterosis breeding has also been utilized in developing many commercial cotton varieties. H_4, Varalakshmi, J.K. Hy 1 and CPH 2 are the hybrid cotton varieties. H_4 first hybrid cotton variety obtained from a cross between two *G. hirsutum* strains is widely cultivated in India. The variety Varalakshmi is an interspecific hybrid between *G. hirsutum* × *G. barbadense.* Both the Hybrid cotton varieties, H_4 and Varalakshmi produce extra long staple of high quality. In additions a male sterile line 'Gregg' is being used for hybrid seed production. All India Coordinated Cotton Improvement Project is extensively using this male sterile line to reduce the cost of hybrid seed production.

4. **Mutation Breeding**. Irradiation breeding has also been attempted for improving cotton varieties. A variety Indore-2 was developed from Malwa Upland 4 by x-ray treatment at Plant Industry, Indore, M.P. in the year 1950. x-ray treatment of the variety *Mescilla acala* caused 40-50% increase in hair density which renders the plants resistant to jassids.
5. **Use of Polyploidy**. The polyploidy has been found to be of good use in facilitating interspecific crosses between tetraploid American cotton (n = 26) and diploid Asiatic cotton (n = 13). In certain cases if the hybrids are found to be sterile, amphidiploidy makes them fertile.

Improved Varieties

The species-wise names of different improved strains are listed below:

G. arboreum. This group includes varieties which are predominantly coarse and short stapled, though a few of them are medium stapled and fairly fine. They are grown in almost all cotton growing states in India. Improved varieties include Virnar (N.R. 5 Jarila), Jarila, Pratap, Malvi 9, Maljari (Malvi-9 × Jarila), H. 420 (*Bani* × *Cernuum*) × Com 170, Co 2 or Deviraj (*G. hirsutum* × *G. arboreum*). 134-Co. 2-M or Devitej (*G. hirsutum* × *G. herbaceum*), I.S.C. 67 (*G. hirsutum* × *G. arboreum*), Bhog, Daulat, Cocanadas 2, Gaorani 46. K. 2, K. 5, K-7, AK-231, AK-235, R, R. 18, and C. 520, Shyamali, G-27, 231-R.

G. herbaceum. This group includes varieties that are generally much finer and longer in staple length than **arboreum** group. They are cultivated in Tamil Nadu, Andhra Pradesh, Karnataka, and Maharashtra. The improved varieties include Kalyan (Wagad 8 × Seg. 22-3-1 × Wagad 8), Vijay (B.D. 8 × Goghari A. 26 × BD. 8), Digvijai (Vijai × 1027 ALF × Vijay), Vijapla (Vijay × Vijay × 1027 ALF), Wagad 8, Jayawant, Jayadhar, Westerns-1, Suyodhar S. 69. H-1 Suyog.

G. hirsutum. This group includes varieties with medium to long staple length. Its fibres are much finer than those of other groups. The varieties are grown largely in Punjab, Uttar Pradesh, Rajasthan, Andhra Pradesh, Tamil Nadu and Maharashtra.

Gujrat-67, Hybrid 4, MCU-4, MCU-5, B-100, Deviraja Krishna are important varieties. Other varieties include Co. 4 selected from the original *hirsutum,* Cambodia bulk introduced in 1906. A-12, a *hirsutum* type from Uganda), Laxmi (Gadag 1 × CC 2), MCU 2 a selection from Co. 4 × M C U 1. M C U 5, one of the most promising varieties developed in India was released in 1968. It has a staple length of 1-3/16 and is suitable for growing in the irrigated tracts of Tamil Nadu, Andhra Pradesh and Maharashtra. It has replaced earlier varieties MCU1, MCU2, MCU3 etc.

M. A. 2, H. 190, Indore (produced by mutation breeding from x-rayed material of Malwa Upland-4 by Plant Industry, Indore). L.SS. Buri 0394, 320 F, and H.14.

G. barbadense, Andrews, Sea Island, Sujata, Co. pusa, Egyptian, and P.R.S.-72 are excellent varieties with long staple length. Sujata is high spinning egyptian type with extra long staple length suitable for irrigated cotton tracts in Tamil Nadu.

Disease Resistant Varieties

Suyog, Vijay, Kalyan, Virnar, Jarila, H. 420, Jayawant, Jayadhar, and L.S.S varieties are wilt disease resistant.

Some other important varieties

Name-Variety	*Parentage*	*Yield per ha.*	*Maturity*	*Fibre length*
Pramukh (*G. hirsutum*)	Reselection of M_4 from Sindh. Released in 1965	10 Q.	190 days	23.63 mm
Shyamli (*G. arboreum*)	35/1 × C.J. Released in 1965	12 Q.	175 days	19.3 mm
Lohit (*G. arboreum*)	Selection from local sanguireum. Released in 1968	12 Q.	190 days	17.5 mm

BREEDING OF SUGARCANE (Genus *Saccharum*)

Sugarcane is one of the prominent cash crops and is the main source of sugar in tropical Asia. It is grown on large scale in India, Pakistan, Mauritius., Cuba, Brazil , Australia, Fiji, Argentina, Hawai, West Indies and Java (East Indies). In India, about 2.5 million hectare area comes under sugarcane cultivation. Sugarcane is cultivated mostly in tropical belts across North India from Assam to Punjab with about 60% of the total area under sugarcane cultivation in U.P.

Botany

Sugarcane (Genus *Saccharum*) belongs to the family Gramineae (now-Poaceae).

Saccharum officinarum, (the noble cane) *S. sinense* and *S. barberi* are the cultivated species of sugarcane and *S. robustum* and *S. spontaneum* are the wild species. The present cultivated sugarcane clones are mostly complex hybrids of these species and it would be difficult to classify them into particular species. Besides, there are several other species of this genus which have little or no sugar and so they are of no practical value in breeding work. *Saccharum officinarum*, the noble cane is susceptible to most of the diseases and is not adapted to adverse conditions of soil and climate. *S. barberi* is very much suited to subtropics. It is immune to sereh disease but susceptible to mosaic disease. *S. sinense* is similar to *S. barberi* in reaction to mosaic and also in vigour. *S. spontaneum* is immune to mosaic and sereh diseases but susceptible to smut, downy mildew and leafspots. *S. robustum* is similar to *S. spontaneum* but shows susceptibility to mosaic diseases.

The flowering in sugarcane is very rare except in tropical areas. It is affected by the ecological situations. Warm nights, high humidity and rainfall favour flowering and cool weather and high altitude inhibit flowering. Flowering is undesirable in commercial sugarcanes. However, flowering, capacity to produce fertile pollen and true seeds are important and necessary in breeding of sugarcane in order to get genetic recombinations. Flower opens in early morning between 5 a.m. and 6 a.m. and about 7-14 days are required for complete flowering of the arrow. Flowering starts in basipetal succession. Cross pollination normally occurs in sugarcane. Flowers show a wide range of fertility and production, from male sterility to high pollen productivity and self-sterility to complete self-fertility.

Seeds of sugarcane are of extremely small size, often poorly developed and invisible. The seeds and floral structures along with long silky hairs at the base of spikelets are generally called fuzz or fluff (Fig. 39.6). The fuzz is carried away by wind.

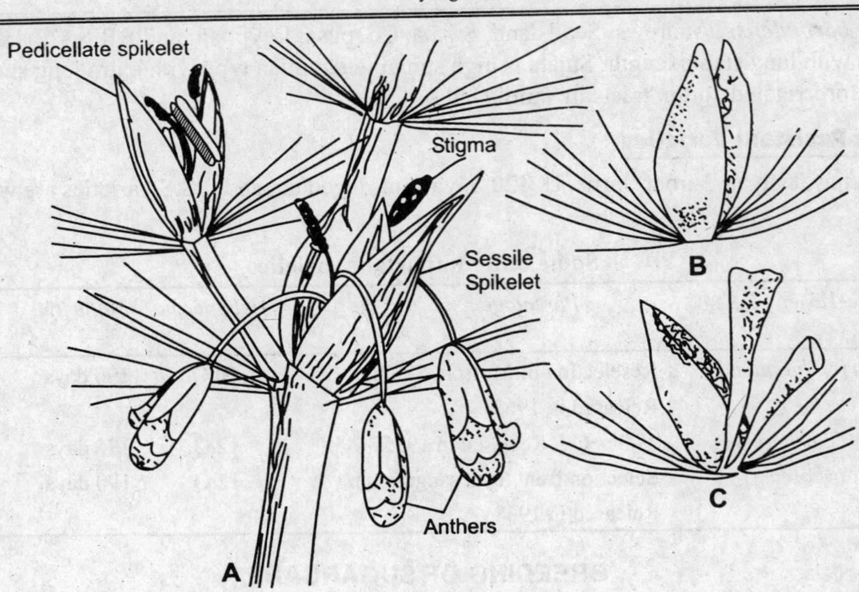

Fig. 39.6 Inflorescence and flowers of sugarcane.

Sugarcane is a vegetatively propagated crop and it is propagated by stem cuttings which are called *Setts* or *Seedcanes.* Each sett has one or more buds. When the setts are placed in moist soil, several roots develop from the nodes and the buds grow into *primary stems*, from the basal nodes of which *secondary stems or tillers* arise.

Origin and Cytogenetics

It is believed that *Saccharum sinense and S. barberi* commonly found in Northern India originated in Northern India itself by natural hybridization with *S. officinarum* and *S. spontaneum. S. officinarum* originated possibly from wild *S. robustum* in New Guinea region. Later on, it migrated from the centre of origin and hybridized with *Erianthus maximus, Sclerostachya fusca* and other related species and got modified.

The present sugarcane clones in cultivation are mostly complex hybrids with high degree of ploidy. The chromosome numbers may vary in different clones studied. The number may be modified by the appearance of aneuploids within the species. The most common basic chromosome numbers are 8 and 10. The chromosome numbers of different species of sugarcane and allied species are given in the following Table:

Species	*Somatic chromosome number (2n)*
Saccharum officinarum	80
S. barberi (Saretha group) Indian canes	90, 92
S. barberi (Sunnabile group)	82, 116
S. barberi (Nargori group)	107, 124
S. barberi (Mungo group)	82
S. sinense (Indian canes)	116, 118
S. spontaneum (Kans or wild canes)	40 to 128
S. robustum (wild canes)	60 to 148
Sclerostachya fusca	30
Narenga porphyrocoma	30
Erianthus species	60

Species	Somatic chromosome number (2n)
Sorghum versicolor (wild *Sorghum*)	10
S. bicolor (Jawar)	20
S. halepense (Johnson grass)	40
S. durra	20

Interspecific crosses can be made among the different species of sugarcane belonging to the genus *Saccharum*. The increase in chromosome number are common when species of *Saccharum* are hybridized. Modern varieties have very high aneuploid chromosome numbers, ranging from $2n = 100$ to 125. Since vegetative propagation is practised for commercial production, fertility is not required by hybrid plants. The increase in chromosome number is important for adding desired characters from wild species because there is little gene exchange between *S. officinarum* and other species of the genus. *S. spontaneum* × *S. officinarum* crosses have yielded plants with as many as 160 chromosomes.

Selfing and Crossing techniques. The selfing may be ensured and cross pollination checked by covering the arrow with a lantern which consists of bamboo framework or cage covered with a closely woven cloth or polyethene bag. The lanterns when placed over the arrows are supported by bamboo pole or scaffoldings. Shading by lantern may be necessary during hot periods to keep the temperature low.

For cross pollination, the arrows of female parent are enclosed within a lantern and pollens from male parent are dusted over the arrows of the female parent. Due to the small size and large number of flowers in an arrow, emasculation is not possible. Sometimes, the arrows of two parents are enclosed within the same lantern. The flowers open between 5 and 6 a.m. and pollens are dusted over the female parent arrows soon afterwards. Since flowering continues normally over the period of 7-11 days, the pollination process must be repeated daily.

Frequent shaking of the enclosed arrows together helps pollination. For crossing purpose detached arrows are generally used in one of the following ways:

(*i*) Sugarcane stalks containing arrow heads are cut and the cut ends of stalks are immersed in weak acid solution containing 150 ppm SO_2, 75 ppm H_3PO_4, 37.5 ppm H_2SO_4 and 37.5 ppm HNO_3. The cut stalk immersed in acid solution is transported to crossing block in upright position. The acid solution is changed fortnightly.

(*ii*) In the second method, the sugarcane stalk is marcotted prior to flowering stage or when arrow formation starts. Marcotting is done by wrapping a polyethene strip containing wet soil mixture around a node of the cane about 2 nodes above the soil. Rooting occurs within 10-12 days. Marcotted stalks are transported to the crossing block and replanted in soil near their counterparts.

Varieties

The term "variety" in cultivated sugarcane refers to a particular clone developed by propagation from seedcanes or setts. The improved sugarcane clones are identified throughout the world by letters identifying the sugarcane breeding stations where the particular clones in question are selected, *e.g.*, Co. 213 where letters "Co" stands for Coimbatore, a city in India where the sugarcane breeding institute is located, and 213 identifies the specific clone bred at Coimbatore Station. Letters identifying clones from some of the principal sugarcane breeding stations throughout the world are as follows:

B = Central Sugarcane Breeding Station, Barbados, British West Indies.

Co = Sugarcane Breeding Institute, Coimbatore, India.

Cp = United States Department of Agriculture, Canalpoint Florida.

M = Mauritius Sugar Industry Research Institute.

H = Hawaiian Sugar Planters Association, Honolulu, Hawaii.

N = South Africa Sugar Association, Natal, S. Africa.

P.O.J. = Jawa Sugar Experiment Station, Pasuruar, Jawa.

Q = Bureau of Sugar Experiment Station, Brisbane, Queensland, Australia.

Central Sugarcane Breeding Research Institute, Coimbatore, Tamil Nadu

It was established in Coimbatore, India, in 1912. In the initial phase the work was concentrated on development of improved cane varieties for northern tropical regions of India where most of the sugarcane is grown. The reason for establishing the institute at Coimbatore was that sugarcane flowers and bears seeds more freely in that area rather than in sugarcane producing areas of North India. Crosses are made, selection of seedlings and preliminary testing of clones are carried out at Coimbatore before selected clones are sent to the state sugarcane station for testing and evaluation. Sub-stations of Sugarcane Breeding Institute have been established at Karnal, Punjab (in the western subtropical regions) and Cannanore, Kerala (on the peninsular west coast).

Indian Institute of Sugarcane Research (IISR), Lucknow

This institute was established in Lucknow in 1952 to carry out research on breeding and improvement of sugarcane.

In India, at present, there are about 40 research stations maintained by the State Governments.

In U.P., the State Council of Sugarcane Research established at Shahjahanpur is the main sugarcane breeding centre. This station has its sub-stations at Muzaffar Nagar, Kanpur, Lucknow and Gorakhpur where clonal testing and selections are done. Until recently, the clones developed at Coimbatore were tested at Shahjahanpur and the selections were sent to sub-stations for trial in different agroclimatic conditons. The selected clones were finally designated as Co S. Now Shahjahanpur Research Station is independently carrying out breeding work on sugarcane.

Breeding Objectives

The main objectives in the sugarcane breeding are:

1. High yield. The height and thickness of cane, and tillering ability of the clone contribute to the high yield.
2. High sugar content in the juice.
3. Delay in flowering,
4. Disease resistance, insect resistance and lodging resistance. Maximum loss of sugarcane is caused by cane borer and other insects. Thus, the main objective before the sugarcane breeders is to develop such varieties as are resistant to insect pests and high yielding. Lodging resistance can be induced by promoting healthy and well developed root system and incorporating the characters from *S. spontaneum*.
5. High adaptability in different conditions as frost, drought and other environmental adversities. Resistance to frost is required in North India. The characters from *S. barberi* and *S. sinense* may be transferred to *S. officinarum*. The resistance to frost can be transferred to *S. officinarum* from *S. barberi, S. sinense* and *S. spontaneum*.
6. Breeding of early maturing varieties. In U.P. the majority of sugar mills are closed by the end of March. But most of the varieties mature by that time and thus the mills stop accepting sugarcane by the end of March. Secondly, the farmers in U.P. have poor land resource and they harvest the sugarcane by the start of December and sow wheat or some other crops in the same fields. Thus, the development of such sugarcane varieties is highly desirable and beneficial as are early maturing.

Method of Sugarcane Breeding

Sugarcane is vegetatively propagated crop. Hence, *clonal selection and hybridization* have been the principal breeding methods. Besides, the *introduction and acclimatization* play very important role in supplying sources of breeding materials. Mutation breeding has also been resorted to quite recently to increase genetic variabilities.

Introduction and Germplasm Collection

The indigenous sugarcane of north India belonging to the species *S. barberi* and *S.sinense* are thin stemmed and poor in sugar content. The tropical noble canes of *S. officinarum* introduced from New Guinea have thicker stems and higher sugar content, are less tolerant to frost and drought and are grown as chewing canes in Northern India only in a small area. They have been grown in India and used extensively in crosses with indigenous canes of India. C.A. Barber, T.S. Venkatraman and others at the Sugarcane Breeding Institute, Coimbatore transferred the desirable characters like thicker stems, higher sugar contents, disease resistance from wild species to *S. officinarum* and other Indian canes. This is commonly known as ***noblisation of Indian canes.***

The Sugarcane Breeding Institute at Coimbatore and Cannanore (Kerala State) and the United States Department of Agriculture at Canalpoint, Florida, are maintaining large world collections of sugarcane clones. Sugarcane Breeding Institute, Coimbatore maintains over 2,000 clones of sugarcane of which about one-half are hybrids developed at Coimbatore or elsewhere in India and the remainder are clones of foreign commercial hybrids of various species. Nearly 700 clones are of *S. officinarum*. All the sugarcane varieties have been derived from introduced noble canes, *Saccharum officinarum*.

Clonal Selection

It is used to isolate desirable clones from genetically mixed populations which may be native or improved populations or inbred populations or hybrid populations. Isolated clone superior for particular characteristics can be used for breeding. The wild *S. spontaneum* is not suitable to be utilised directly for cultivation.

Inbreeding has been used to concentrate genes for desirable characteristics, such as sugar content in sugarcane varieties. It may also be used to increase homozygosity of genes for more simply inherited characters.

Hybridization. The common procedure for developing new varieties of sugarcane involves hybridization between clones followed by selection of desirable clones from the hybrid population. Most of the sugarcane varieties presently grown in India are complex hybrids between Indian and introduced canes, *S. officinarum* or *S. robustum* and clones of the wild *S. spontaneum*. In practice, crosses are made freely and the breeding value of the parents is assessed by the performance of the progeny. If a clone is found to contribute desirable characteristics to a series of progenies (*i.e.*, good combining ability) it may be used in a large number of crosses.

Several types of crosses, such as intergeneric, interspecific and intervarietal crosses have been made. Sugarcane has been hybridized with the species of more genera than any other taxon. Eight genera which have been involved intergeneric crosses with *Saccharum* are *Erianthus, Sclerostachya, Sorghum, Narenga, Miscanthus, Miscanthidium, Ripidium* and *Eccoilopus*.

S. officinarum ($2n$ = 80) has been crossed successfully with *Narenga narenga* ($2n$ = 30), *Erianthus sera* ($2n$ = 60), *E. arundinaceous* ($2n$ = 60), *Sorghum durra* ($2n$ = 20), *Sorghum halepense* ($2n$ = 60), *Sorghum durra* ($2n$ = 20), *Sorghum halepense* ($2n$ = 40) and maize ($2n$ = 20 + 2B). Another related species of the genus, *S. spontaneum,* $2n$ = 40 to 128) was also crossed with the above mentioned species of different genera and also to *Bambusa arundinacea.* Besides these, crosses between the hybrids obtained by other genera were made and also hybrids of *S. officinarum* × *S. barberi* were crossed to *Bambusa arundinacea.*

Important varieties of sugarcane under cultivation are listed below:

Co. 213 (P.O.J. 213 × Co 291), Co 205 (Co 213 × POJ, 1410), Co 300 (Co 213 × POJ 1410), Co. 301 (Co. 213 × POJ. 1399). Co. 312, (Co 213 × Co. 244), Co. 313 (Co 213 × Co, 244), Co 331 (Co 213 × Co 213), Co 356 (Sorghum × Sugarcane), Co 541, Co. 602 (Co 421 × Poovan), Co 603 [(Co 421 × Co 440) × Co 419)]. Co. 606 [(Co 231 × POJ. 2725) × Imperata], Co. 612 Co. 419 × Co. 185). Other good varieties are Co. 668, Co. 672, Co. 677, Co. 678, Co. 685, Co. 686, Co. 687, Co. 1148, Co 1158, Co S510 (Co 453 × Co 557), Co 975 (Co. 527 × Co. 617), Co S109, etc.

A new sugarcane variety-COC 671 which is short duration variety with very high sugar content has been released in 1981-82.

Some of the popular varieties developed at Mysore Breeding Station are H.M. 658, HM 544 etc.

Early varieties recommended for cultivation in U.P. are Co S 510, COS 1336, Co S 136, Bo 47, Co 687.

Varieties Resistant to Diseases

The utilization of wild species of *Saccharum* has been an integral part of sugarcane improvement, for incorporating disease resistance genes into cultivated varieties of sugarcane. Wakker made the first interspecific cross between *S. officinarum* and *S. spontaneum* for incorporating disease resistance into noble canes. Varieties resistant to red rot disease caused by *Physalospora tucamanensis* are Co 1082, CoL 5, Co 1158, Co 1336, CoS 611, CoS 630 and CoS 568.

Varieties resistant or tolerant to sugarcane smut caused by *Ustilago scitaminea* are Co 312, Co 527 (Co 349 × Co. 312), CoS 416, and Co 997.

Since no variety stays resistant for long, these should be changed at least after 5 years

Mutation Breeding. Both irradiation and chemical mutagens are used to induce maturation in sugarcane. In sugarcane, nodal buds are exposed to radiation and the mutant buds or tillers are selected in M_1 and M_2 generations.

Co 8152 and Co 8153 varieties of sugarcane (*S. officinarum*) have been developed through mutagenesis. Sugarcane variety Co 8152 is a gamma-ray induced mutant from Co 527. This mutant variety gives 40 per cent higher cane yield than the parent variety Co 527, but it has one undesirable character, that is, its lower leaves dry more quickly than in the parent variety as well as other standard sugarcane varieties. The mutant sugarcane variety Co 8152 has two chromosome less than the parent var. Co 527. This is an example of a change in chromosome number producing desirable mutant phenotype.

Co 8153 is also a gamma-ray induced mutant sugarcane variety from parent var. Co 775 which is highly susceptible to smut. This mutant variety is highly resistant to smut and is as productive as the parent variety Co 775.

Polyploid Breeding. Since sugarcane is already a complex polyploid, possibility of economic improvement by further increase in chromosome number does not appear to offer much opportunity.

BREEDING OF PULSE CROPS

BREEDING OF GRAM OR CHICK PEA OR BENGAL GRAM

Cicer arietinum L

Bengal Gram or Chick Pea is one of the first grain legumes to be domesticated by humans (Vander Maesen 1972). It is third most important pulse crops in the world. In India it is an important source of protein in human diet. Major production of chick pea comes from Central and North India. The area under chick pea crop is 33 per cent of the total area under pulses cultivation which is about 8.41 million hectares and 45 per cent of the total pulses production (6.68 million tons). The major chick pea growing states in India are: Rajasthan, M.P., U.P., Maharashtra, Gujarat, A.P., Karanataka,

Haryana and Bihar. The Indian Institute of Pulses Research of Kanpur has the National Mandate to promote research and development efforts on chick pea. Varieties developed for extra bold seeds, salt tolerance and tolerance to lodging under excessive moisture and fertility condition will revolutionise the chick pea production.

Chick pea is consumed by human being in a variety of ways, such as Dal (split cotyledons), flour (besan), crushed or whole grain, boiled or parched, roasted or cooked, salted or unsalted or sweet preparations. Green foliage and grains are used as vegetable. Germinated seeds are recommended to scurvy patients and diabetics. Malic acid and oxalic acid collected from green leaves are called *Chana Khar* which is prescribed for intestinal disorder and sun stroke. Soaked grains and husks are fed to horses and cattle as concentrate and roughage respectively.

Gram is grown in dry or rain fed areas as well as in irrigated field during *rabi* season. Excessive rain soon after sowing or at flowering stage does great harm to this crop. Similarly serious cold is injurious. Gram is grown alone or mixed with wheat, barley, linseed, safflower or mustard. It is rotated with jowar, bajra, wheat, rice, etc. The crop takes about 120–150 days in maturation.

Origin

The genus *Cicer* is represented by several species. Majority of which are wild. *Cicer arietinum L* is the only cultivated species of this genus. Ladizinsky Adlerl (1976) have suggested *Cicer reticulatum* as a wild progenitor of cultivated chick pea on the basis of electrophoresis of seed proteins. This wild species is found in South-East Turkey and the adjoining regions of Northern Syria. *Cicer arietinum* first originated from that region and subsequently spread to India, Europe, Australia and other regions of the world.

Breeding Approaches

During early phage of development most of the varieties of chick pea were developed either through selection from local germplasm or through introduciton and selection from specific areas. Before the start of All India Coordinated Pulse Improvement Project (AICPIP) in 1966 only 20 varieties were released of which 80 per cent were evolved through pure line selection. After 1996 breeders realised the importance of introducing materials from one agroclimatic zone to the other and started generating breeding materials through hybridization and mutation. The important aspects of breeding varieties of chick pea are as follows:

1. To develop bold seeded high yielding lines.
2. To develop early maturing varieties.
3. To develop input responsive varieties (*i.e.,* varieties responsive to fertilizers and moisture but resistant to lodging).
4. To develop varieties tolerant to abiotic stresses such as salinity, alkalinity, drought and for late sowing conditions.
5. To develop varieties, particularly resistant to *Fusarium* wilt, root rot and *Ascochyta* blight.

Breeding Methods Employed

After inception of All India Coordinated Pulse Improvement Project (AICPIP) in 1966 breeders started realising the importance of generating variability through hybridization and mutation.

As a result, a number of varieties have been released using different breeding methods of plant breeding which are given in the following Table:

Table 39.1 Varieties of chick pea developed through different breeding methods

Breeding methods	*Varieties*
Pure line selection from germplasm	DCP 92-3, Chaffa, Dahod Yellow, BDN 9-3, Annegeri 1, JG 315, RAU52, GNG 146, KWR 108, KGD 1168.
Composite/Mass selection	JG 74, CSG 8962

Breeding methods	Varieties
Hybridization	Radhey, C235, K 850, Pant G 186, BG 256, Phule G 5, ICCC 32, BG 212, BG 267, RSG 44, Gaurav, Hare Chhole 1, L 550, ICCC 4, Vikas, B 108, BG 372, B 115, GNG 469, BG 329, PBG 1, GPF 2, BGD 72, BG 1053, GCP 101, GCP 105, Phule G 12, KPG 59, ICCV 10, Sadabahar, ICCV 2, Vijay, GNG 663, JG 218, JG 11, Vishal, BG 391, H 82-2.
Mutation Breeding	BG 1003 (Kabuli), WCG 2

Types of Chick Pea

Based on seed size and seed coat colour, chick pea varieties can be grouped into the following three distinct classes:

(*a*) *Desi small-seeded types,* which are mostly dark brown, yellowish brown or black-seeded and have 100 seed weight below 20 gm.

(*b*) *Desi bold-seeded types*, which are bold-seeded and have more than 20 gm per 100 seed weight.

(*c*) *Kabuli types*, which have white seed coat and are soft as compared to desi type. Bold-seeded Kabuli types have 100 seed weight more than 25 gm.

The development of bold-seeded desi chick pea varieties has assumed significance particularly in Central and South zone where maturity period limits to 95–100 days. Among bold-seeded varieties BG 256, Phule G5, BG 391, K 850 and Radhey are the prominent ones.

Seeds in some varieties of Kabuli types like L 550, BG 267, Gora, Hisari and ICC B32 are bold but have 100 seed weight below 25 g. Recently limit of 100 seed weight in Kabuli type for table and export purposes have been fixed at 25 g or above. As a result of which, new varieties of Kabuli type BG 1003, BG 1053 have been developed which have more than 25 g/100 seeds.

Disease Resistant/Tolerant Varieties

The important diseases of chick pea are: *Fusarium* wilt, root rot and *Ascochyta* blight. The disease problem has been tackled through resistance breeding which has resulted in several disease resistant/ disease tolerant varieties. The disease resistant chick pea varieties are given below in the following Table:

Table 39.2: Disease resistant/tolerant varieties of chick pea

Disease	Varieties
Fusarium wilt	KWR 108, ICCV 10, H 82-2, DCP 92-3, GCP 101, GCP 105, JG 315, GPF 2, Vijay, KGD 1168, JG 74, GNG 663, K 850, Radhey, BG 256, BG 212, KPG 59, BG 1003, BG 1053, Annergeri 1, Mahamaya 1, Vikas, BGD 72, Gaurav, PBG 1, GNG 469.
Ascochyta blight	GNG 469, Gaurav, PBG 1, GNG 146, C 235, BG 261

Varieties Developed for Specific Situations and Traits

1. **For late sowing condition.** The area under rice gets vacated for late *rabi* planting. In some parts particularly in *Tal* and *Bhal* areas, the land for cultivation becomes available only when water recedes during late *rabi* season. For such situation it becomes imperative to develop chick pea varieties suitable for late sowing where less irrigation facilities are available during *rabi* season. A significant achievement has been made in this direction and varieties like DG 372, KPG 59 and Pant G 186 have been developed for late sowing conditions in Northern India. These varieties can be sown in the first week of December after paddy harvest.
2. **Varieties suited to excessive moisture and high fertilizer input condition.** Under excessive moisture and higher input condition, chick pea genotypes attain more vegetative growth resulting in poor grain production. The crop suffers from lodging under these situations. Variety DCP 92-3, has been developed which is responsive to high input condition and excessive moisture. This variety withstands the adversaries and gives good yield under such situation.

3. **Varieties suited for moderately salt-stress situation.** Chick pea being the crop of normal soil conditions may suffer heavy loss even with small increase of salinity or alkalinity events. Therefore, attention is being paid to develop varieties tolerant to small stress of salts. Accordingly a trial was started under AICPIP to indentify varieties tolerant to salt-stresses. Recently a salt tolerant variety CSG 8962 (Karnal Chana 1) has been developed which will bring an impact on the cultivation of chick pea in areas prone to salinity or alkalinity problems. Karnal Chana 1 is a promising variety for North-West plain for normal soil condition and has an additonal feature of tolerating salt stress. Therefore, it can be grown in saline soils as well.

Some Improved Varieties of Chick Pea and Their Description

Till date about 120 varieties of chick pea have been released for cultivation in different parts of the country. The description of some recently released as well as some old varieties of chick pea which are under cultivation in various parts of the country are given below. These varieties have been recommended/released either by central sub-committee on release of crop varieties or by the state variety release committee.

1. **Alok (KGD 1168).** Developed through selection from K 158 at Kanpur in 1996. Plants are resistant to wilt and root rot diseases with maturity period 140–150 days. Seeds are medium bold, seed yield 19–20 q/ha.
2. **Annegeri 1.** It is selection from local germplasm. Seeds yellowish brown, medium bold, 17–20 g/100 seeds, early maturing (90–100 days), seed yield, 10–15 q/ha. It performs well under rain-fed conditions in Karnataka.
3. **Avarodhi.** It was developed at Kanpur in 1987 through hybridization (T-3 × K315), good for cultivation in U.P. The genotype is semi-spreading, 60–70 cm tall, resistant to wilt, maturity 145–150 days, seed yield 22–23 q/ha. Seeds are bold and brown.
4. **B 108.** It is a semi-erect, wilt tolerant genotype developed through hybridization (N 31 × B 75) in 1980 for cultivation in West Bengal, maturity 130–140 days, seed yield 20–22 q/ha, seeds small brown, 11 g/100 seeds.
5. **Bharti (ICCV 10).** This is a wilt and dry root-rot resistant, semi-erect genotype developed through hybridization (P1231 × P 1265) in 1992 at ICRISAT, Hyderabad for South Zone. It has maturity of 95–100 days and seed yield 18–20 q/ha.
6. **C 235.** This is a medium tall, semi-erect, *Ascochyta* blight resistant genotype developed in 1980 through hybridization (IP 58 × C 2134) at Ludhiana, suited for rain-fed areas. Seeds are small, 30 g/100 seeds, round wrinkled and brownish yellow, seed yield 19–20 q/ha.
7. **Chaffa.** This is a first variety of chick pea released in 1948 through selection from NIPHAD (Maharashtra) and is still under cultivation. It is early maturing, maturity period 105–110 days. Seed yield 10–11 q/ha, seed yellowish brown, medium bold, 100 seeds weight 16–19 g.
8. **Chamatkar (B.G 1053).** This is a tall, semi-erect, wilt tolerant genotype developed through hybridization (ICCV 3 × FLIP 88-20) at I.A.R.I., New Delhi in 1999. Seeds are white and bold (27 g/100 seeds). Seed yield 17–19 q/ha.
9. **Co 3.** Bold-seeded variety developed in 1999 at Coimbatore for cultivation in Tamil Nadu. Seed yield 15–17 q/ha.
10. **Co 4.** This is bold-seeded variety developed at Coimbatore in 1999 for cultivation in Tamil Nadu. Seed yield is 15–18 q/ha.
11. **Dahod yellow.** It is a selection from local germplasm in 1960 for cultivation in Eastern part of Gujarat and adjoining areas of Maharashtra and M.P. Seeds are yellowish brown, medium in size (15 g/100 seeds), maturity period (120–130 days), seed yield 17–20 q/ha.

12. **DCP 92-3.** It is a selection from germplasm at IIPR (Indian Institute of Pulses Research), Kanpur. It was released in 1998 for high fertility and excessive moisture conditions of North-West plain. Genotype is semi-erect, medium tall, lodging resistant, wilt resistant and with maturity of 145–150 days. Seeds yellowish brown, medium bold, 100 seeds weight 17 g. Average grain yield 19–20 q/ha.
13. **Dharwad Pragati (BGD 72).** This genotype was developed through hybridization [(Pusa 256 × E 100 Ym) × Pusa 256] in 1999 at I.A.R.I., New Delhi for cultivation in central zone. It has a maturity period of 115–120 days and seed yield 18–20 q/ha.
14. **Gaurav (H 75–35).** This is a medium tall and erect genotype developed through hybridization (C 235 × E 100 Ym) in 1983 at Hissar. It is resistant to *Ascochyta* blight. Seeds are bold, 100 seeds weight 20 g and yield 18–20 q/ha.
15. **Gora Hisari (Kabuli).** It was released in 1988 for cultivation in Haryana. Plants are semi-erect, seeds are small and white, seed yield 15–16 q/ha.
16. **Gujarat Gram 1 (GCP 101).** It is medium tall, semi-erect, wilt resistant genotype developed through hybridization (GCP × ICCV2) at Junagadh and released in 1999 for central zone. Maturity period of 115–120 days, seeds medium bold and dark brown, seed yield 18–20 q/ha.
17. **Gujarat Gram 4 (GCP 105).** It is a semi-erect, developed through hybridization (ICCL 84224 × Annegeri) at Junagadh and released in 2000 for cultivation in North-East plains. Seeds are dark brown, yield 18–20 q/ha.
18. **Hare Chhole 1.** This is erect genotype developed through hybridization (S 26 × GG Bizapur), matures in 165–170 days. Seeds are green and small, matures in 165–170 days, 100 seeds weight 15 g, suitable for irrigated areas of Punjab, seed yield 12–13 q/ha.
19. **Haryana Chana 1 (H 82–2).** This is a dwarf genotype with long fruiting branches developed through hybridization (F 61 × L 550) in 1990 at Hissar. It is highly resistant to *Fusarium* wilt. Seeds are small, 100 seeds weight 13 g, light in colour, seed yield 22–23 q/ha. It is suitable for timely and late sown situation in North-West plains.
20. **Hirwa Chaffa (AKGS-1).** This genotype is green-seeded developed through selection in 1996. It matures in 105–110 days, resistant to wilt and suitable for cultivation in Maharashtra. Seed yield 15–17 q/ha.
21. **ICCC4.** This is semi-erect, tall genotype with basal branching, developed through hybridization (H 208 × T 3) in 1983 at ICRISAT (International Crops Research Institute for Semi-arid Tropics), Hyderabad. Seeds are yellow to light brown, medium size, 100 seeds weight 15 g, adapted to rain-fed condition Gujarat, seed yield 17–19 q/ha.
22. **ICCC 32 (Kabuli).** Developed as ICCV6 through hybridization (L 550 × L2) at ICRISAT, Hyderabad. Identified in 1984 for central zone and in 1985 for North-West plain, resistant to wilt and tolerant to root rot and pod borer, seeds medium bold, 17.5 g/100 seeds and seed yield 24–26 q/ha.
23. **J.G11.** Developed through hybridization (Phule G 5 × Narsingpur bold × ICCC 37) at ICRISAT, Hyderabad and released in 1999. Genotype is semi-spreading, bold-seeded, maturity 95–100 days, resistant to wilt and moderately resistant to root rot, seed yield 15–17 q/ha, suitable for irrigated and rain-fed areas of south zone.
24. **J.G 218.** Early maturing, wilt tolerant genotype developed through hybridization (ICC4 × P 1353) in 1996 at Jabalpur, suitable for cultivation in MP, seed yield 18–19 q/ha.
25. **J.G 322.** Developed through hybridization (Chaffa × BG1) and released in 1999 for cultivation in MP. It is wilt resistant, matures in 110–115 days, seed yield 18–20 q/ha.
26. **JGG 1.** Selection for land race Gulabi Gram was released in 1997 for cultivation in MP.

27. **K 850.** Released in 1982 and developed through hybridization (Banda local × Etah bold). Seeds reddish brown, round smooth and bold, 100 seeds weight 24 g, seed yield 25–28 q/ha, suitable for cultivation in Northern-plains.
28. **Karnal Channa 1 (CSG 8962).** A selection from GPF 7035 was released in 1998. Plants medium tall, semi-erect, maturity 150–160 days, wilt resistant, suitable for saline and normal soil of North-West plain.
29. **Kranti (ICC 37).** Dwarf bushy genotype with light green narrow leaves developed through hybridization [(P 481 × JG 62) × P 1630] at ACRISAT, Hyderabad in 1981 for Central and South India. Seeds are brownish, small and seed yield 19–20 q/ha.
30. **KWR 108.** Developed through selection from P 108 in 1996 at Kanpur. Seeds are brown and small.
31. **L 550 (Kabuli).** Developed through hybridization (PB 7 × RABAT) at Ludhiana. Seeds small and creamy white, suitable for cultivation in North-West plains.
32. **L 551 (Kabuli).** Small-seeded Kabuli variety developed in 1999 at Ludhiana for cultivation in Punjab.
33. **Mahamaya (B 115).** It is selection from a cross N 31 × B 75, good for cultivation in West Bengal.
34. **Pant G 186.** Developed through hybridization (ILG 613 × Pant G 114) in 1996 at Pant Nagar. It is small-seeded and good for cultivation in U.P.
35. **PBG 1.** Developed through hybridization (GG 578 × NEC 206) at Ludhiana in 1988.
36. **Phule G 12.** Developed through hybridization (GW 517 × Ceylone 2) in 1990 at Rahuri for cultivation in Maharashtra. It is resistant to wilt, matures in 95–100 days and seed yield is 20–22 q/ha.
37. **Pragati (K 3256).** Developed through selection from old cultivar Cyperus local at Kanpur in 1994. It is white-seeded, wilt tolerant with seed yield 17–20 q/ha, and suitable for cultivation in Uttar Pradesh.
38. **Pusa 212 (P 340 × G 130), Pusa 256 or BG 256.** *[(BG 62 × 8503/27) – 1 × (L 550 × H 208)], Pusa 267 (BG 267): (US 613 × BEG 482), Pusa 329 (BG 329):(BG 203 × P 179), Pusa 362 (BG 362);(BG 303 × P 179), Pusa 372 (BG 372):(P 1231 × P 1265), Pusa 391 (BG 391):(ICC 3935 × Pusa 256) and Pusa Kabuli 1003 (BG 1003):* A Kabuli mutant of L 532 were developed at I.A.R.I., New Delhi and they are recommended for North-West, North-East and Central India.
39. **Radhey.** It was developed through hybridization (197 × 79) at Kanpur. Variety tall, late maturing, seeds are light brown and bold, yield 20–25 q/ha.
40. **Sadabahar.** Developed through hybridization (Hima × L 245) in 1992 at Kanpur. It is green-seeded, late maturing variety, suitable for cultivation in Uttar Pradesh.
41. **Sadbhawana (WCG 1) (Kabuli).** Developed at Modipuram, Meerut. It is bold-seeded Kabuli variety released in 1998 for U.P.
42. **Samrat (GNG 469).** Developed through hybridization (Annegeri × H 75–35) in 1996 at Sri Ganga Nagar, suitable for rain-fed and irrigated areas of North-West plains.
43. **Udai (KPG 59).** Developed through hybridization in 1992 at Bharari for cultivation in North-East and North-West plains. It is small-seeded, semi-erect and medium tall variety with seed yield 19–20 q/ha.
44. **Ujjain 21.** A selection from local material and released in 1956 for M.P.
45. **Vardan (GNG 663).** Developed through hybridization (GNG 16 × GNG 146) in 1995 at Sri Ganga Nagar. It is semi-erect, late maturing, wilt resistant genotype with medium dark brown seeds and seed yield 22–25 q/ha, suitable for cultivation in North-West plains.

46. BG 1053. It is an extra bold-seeded (35 g/100 seeds). It is a true Kabuli type recommended for cultivation in the major chick pea growing states of India, Punjab, Haryana, U.P., Rajasthan, Gujarat, M.P. and Maharashtra under irrigated agro-ecosystem. Seeds are attractive, uniform with excellent culinary and table conditions and seed yield 25 q/ha. The variety is resistant to wilt. The variety was released in 1999 by the Central Sub-Committee on Crop Standards.

BREEDING OF PIGEON PEA OR RED GRAM OR ARHAR

Cajanus Cajan (L.) Millsp

Pulses form an important part of human diet. They are important source of proteins (20–26%) and are essential component of predominantly cereal-based diets. Pulses are the best source of proteins for vegetarian population in India. Besides proteins, pulses provide mineral salts like calcium, phosphorus, iron, carbohydrate and to some extent vitamins also.

In India pigeon pea is the second most important pulse crops after chick pea which is grown mainly in Maharashtra, Uttar Pradesh, Karnataka, M.P. and Gujarat. In Uttar Pradesh also the pigeon pea is the second most important pulse crop after chick pea. It is grown either singly or mixed with other crops such as jowar, bajra, urd and cotton. In productivity Uttar Pradesh stands better than other states in India, as well as in total average production in the country.

Pigeon pea is extensively used as dal. Its green pods may be used as vegetable. Green leaves and young branches are fed to cattle. The husks of pods or seeds with parts of the kernal constitute valuable cattle feed. Dry stalks after threshing are used for making baskets or as fuel or as thatching material. It is deep rooted crop so it is planted as soil renovator to bring up hard sub-soil and as a hedge to check erosion.

Pigeon pea is grown in almost every state of India. The average annual area under this crop is about 2.5 million hectares and the annual production is nearly 2 million tons. The crop can be grown in dry and moist climates. In dry condition it requires irrigation. The plants cannot withstand frost. The crop is sometimes irrigated in late December or early January to protect it against frost. Good drainage is essential requirement because this crop cannot withstand water logging. Though perennial, pigeon pea is invariably grown as an annual crop.

Several varieties of this crop are under cultivation. The varieties differ from one another in height, growth habit and the maturity of plants or in colour, size and shape of pods and in their resistance to diseases such as, wilt, stem rot etc. The varieties of Arhar have been classified as:

(*i*) Extra early maturing (maturity period 120 days),

(*ii*) Early maturing (maturity 150 days), and

(*iii*) Late maturing (maturity period more than 200 days).

The development of early maturing pigeon pea varieties was an important land mark in varietal development. In 1960s an early maturing variety type 21 was developed, thereafter several "T21 group" varieties with maturity period of 140 days were developed. These early varieties under rain-fed conditions were very successful as they remain unaffected by water stress during post-rainy season and give good yields. The early varieties made it possible to have pigeon pea-wheat rotation with slight advancement in sowing time. Later on, extra early varieties were developed which brought about revolution in cropping patterns and fitted well in pigeon pea-wheat rotation. Some of the successful varieties are UPAS 120, Pusa Ageti, Pusa 74, Pusa 84, Pusa 33, Manak, Sagar, TT 5, ICPL 87, ICPL 151 etc. In some Northen States like Punjab and Haryana where late varieties were not suited because of frost effect, early maturing varieties have made it possible to cultivate pigeon pea without disturbing wheat cultivation. In traditional pigeon pea growing states where wheat cultivation in *rabi* season is increasing due to availability of irrigation water, early varieties are replacing late varieties.

Pigeon pea varieties have also been developed for pre-*rabi* season (*i.e.,* September/October sowing). Varieties Bahar, WB 20 (105), AS 71–37, DA 11 etc. have shown potential in the Eastern Uttar Pradesh, Bihar, West Bengal, etc. They could be sown in September, October if sowing could not be done due to any reason in June, July or if the *Kharif* crop failed. However, in Bihar region DA 11 should be preferred over Bahar for pre-*rabi* or *rabi* sowing, because of its susceptibility to *Alternaria.*

In chick pea growing areas where farmers are interested in wheat cultivation because of availability of irrigation water, there is need to develop varieties responding to inputs and high level management. Recently identified variety Phule G 5 has shown promise in this regard.

Breeding Methodology

1. **Direct Selection from Land Races/Germplasm.** In the early phase of improvement, selection was practised from natural variability by collecting local cultivars from different regions and selecting the prominent better genotypes. Later on, some good varieties were developed through direct selection from land races/germplasm. The important among them are UPAS 120 (early maturing), BR 65, Bahar (SMV resistant), Gwalior 3, No 148, T 15–15, hy 3C, C 11, B 517, BDN 1, Basant, BR 183, Type 7, Type 17, BR 60, NPWR 15, LRG 30, BDN 2, GS 1, Maruthi, Co 1, Co 2, Co 3, Co 4, B 7 etc.
2. **Hybridization.** Hybridization programme were taken up for obtaining promising recombinants and thus, creating more variability for further use. Incorporation of disease and pest resistance in high yielding varieties with broad agronomic base has been one of the major aims for the last few years. Since 1960 a number of varieties have been developed through hybridization. In pigeon pea an early maturing variety Type 21, a selection from a cross between $T_1 \times T190$ was among the varieties developed in the early phage of hybridization programme. Later on, variety Prabhat was developed from the same cross. Type 21 and Prabhat have been frequently used in hybridization programme to develop early maturing varieties. The other early maturing variety UPAS 120, a selection from P 4758 has been widely used in breeding programme. An exotic line Brazil 1–1 in combination with NP (WR 15), NP41 and NP68 led to the development of Mukta, Sharda and Pusa Ageti respectively.
3. **Mutation Breeding.** Recently Mutation Breeding has been attempted in Arhar with a view to create more variability and to develop superior genotype. A number of mutant varieties have been developed in Arhar through mutagenesis. TT_5 and TT_6 have been developed from Type 21 and Co 3 mutant variety has been developed from Co 1.
4. **Incorporation of Disease and Pest Resistance.** The common diseases affecting the pigeon pea are wilt caused by *Fusarium*, stem blight caused by *Phytophthora* and sterility Mosaic Virus (SMV). The important varieties which are resistant to wilt are JA_3 (C_{11} × No. 148), Pusa 33 (C_{11} × UPAS 120), Mukta [Brazil 1–7 × NP (WR) 15], ICPL 87 (Type 21 × JA 277). The SMV tolerant variety is ICPL 151 (CP 6997 × Prabhat). Bahar, BSMR 736 or resistant to *Alternaria.*

Improved Varieties of Pigeon Pea or Arhar [*Cajanus Cajan (L)*]

Type 7: A selection from a Lucknow land race, released in 1960. It is semi-spreading, tall, maturity period 260–270 days, seeds light brown, medium size, seed weight 7–8 g/100 seeds, yield 18 q/ha.

Type 21: Developed from a cross $T_1 \times T_{190}$, released in 1973, medium size, semi-spreading, maturity medium early, seed yellowish brown, small, seed weight 7 g/100 seeds, average yield 17 q/ha, suitable for pigeon pea-wheat rotation.

Type 17: A selection from Pratapgarh land race, semi-spreading, tall, maturity 260–270 days, seeds light brown, weight 8 g/100 seeds, average yield 18 q/ha, suitable for plains of U.P.

Prabhat: Developed from a cross $T_1 \times T_{190}$. Released in 1975, extra early maturing in 120 days, plants dwarf (110–120 cm), seed small, weight 6.5 g/100 seeds, average yield 12–15 q/ha, suitable for double cropping in North-Western region.

Pant A-3: Released in 1975, semi-erect, dwarf (110–120 cm), seeds brown medium-bold, seed weight 8 g/100 seeds, average yield 12–15 q/ha, resistant to sterility mosaic disease, suitable for double cropping in North plains.

Pusa 84: A cross from Pusa Ageti × T21, released in 1980, early maturity 140–150 days, medium tall, semi-spreading, seeds small brown, weight 7.5 g/100 seeds, average yield 15 q/ha, suitable for North-Western region of U.P.

Manak (H77-216): A cross of T21 × UPAS120, released in 1983, semi-spreading, medium height, maturity 132–166 days, seeds small 6.5 g/100 seeds, average yield 20 q/ha, suitable for Arhar-Wheat areas of North-West plains and Uttar Pradesh.

TT5: It is mutant of T21 released in 1984, semi-spreadng, seeds red, bold, weight 10–11 g/100 seeds, suitable for Western U.P.

Sagar: A cross from T21 × UPAS120, released in 1984, semi-spreading, medium height, small seeded, better yield 20 q/ha, suitable for NPW zone and U.P.

Bahar: A selection from local land race of Motihari (Bihar), released in 1980, semi-erect, seeds brown, round, medium bold, seed weight 10.2 g/100 seeds, maturity 220–250 days, average yield 22 q/ha, tolerant to Sterility Mosaic Virus and Yellow Mosaic Virus.

ICPL 151: A cross from CP 6997 × Prabhat, released in 1985, semi-spreading, seeds cream coloured, round, seed weight 10.6 g/100 seeds, average yield 20 q/ha, suitable for Western U.P. and hills of U.P. under late sown condition.

BR 60: Developed from local germplasm, plant spreading maturity 220 days, seeds reddish brown, weight 10 g/100 seeds, average yield 18–20 q/ha, suitable for Bihar.

Laxmi (Kanak-3): A cross from BR_{183} × local (perennial), medium maturity 180–220 days, seeds cream coloured, bold average yield 20–25 q/ha, resistant to wilt, suitable for Chhotanagpur plateau.

Basant: A selection from local land race, semi-spreading, brown, small 9 g/100 seeds, late maturing, average yield 18–20 q/ha, suitable for Bihar (Tirhut region).

BR 183: A selection from local land race, spreading medium early, maturity 150–170 days, seeds small, weight 8–9 g/100 seeds, yield 12.5 q/ha, suitable for double cropping in Bihar.

Gwalior 3: A selection from local germplasm of Morena district of M.P., released in 1980, late maturing, tall, seeds light brown, small weight 7–8 g/100 seeds, average yield 20 q/ha, suitable for mixed cropping with sorghum, bajra.

Khargone 2: A selection from local cultivar of Nimar region of M.P., developed in early sixties, early maturing 160–170 days, medium tall, small seeds 7–8 q/100 seeds, average yield 15 q/ha, suitable for dry region of Malwa.

JA 9–19: A local selection from Jabalpur developed in 1980. It is tall, semi-spreading, maturity in 180–200 days, seeds brown, weight 8.5 g/100 seeds, average yield 10–15 q/ha.

No. 148: A local selection from Nagpur germplasm, developed in 1975, semi-spreading, medium maturity 180–190 days, medium tall, seeds brown, round, medium bold 9.5 g/100 seeds, average yield 10–15 q/ha, suitable for central zone.

AL 15: Selection from a line P_{8-9} of unknown origin in 1978, developed in 1981, early maturity 130–140 days, small-seeded, tolerant to pod fly, average yield 14 q/ha, suitable for double cropping.

T. 15–15: A selection from local germplasm developed in 1960. It is medium maturing 200–210 days, seeds white, average yield 12–15 q/ha.

NP (WR) 15: A wilt resistant, tall, medium maturing 170–180 days, seeds brown, round medium, weight 8.5 g/100 seeds, average yield 20 q/ha, recommended for central region of M.P. and Delhi.

Mukta (R 60): It is a cross of Brazil 1–1 × NP (WR) 15, released in 1972, medium maturing 180–200 days, medium tall, semi-spreading, seeds brown, medium bold 10 g/100 seeds, average yield 20 q/ha, tolerant to wilt, suitable for Delhi region.

Pusa Ageti (S5): It is cross of Brazil 1–1 × NP 69, released in 1972, early maturing 150–160 days, semi-erect, seeds dark brown, medium 9 g/100 seeds, average yield 15 q/ha, suitable for double cropping in Delhi areas.

Pusa 74: It is developed through cross between Khargone-2 × Pusa Agetic, an early maturing 150–160 days, medium tall, seeds brown, round, weight 8.5 g/100 seeds, average yield 15 q/ha, recommended for double cropping.

Sharda (S 8): A cross of Brazil 1–1 × NP (WR) 15 released in 1971, plant medium tall, maturity period 170–180 days, seeds brown small, weight 8.3 g/100 seeds, average yield 20 q/ha.

BDN 1: A selection from local Bari 192–42, developed in 1978, maturity period 160–180 days, semi-spreading, medium tall, seeds brown, medium bold weight 9 g/100 seeds, wilt tolerant, average yield 15 q/ha, suitable for Maharashtra.

Vishakha 1 (TT_6): It is mutant of T21, developed in 1982, compact and erect plant, maturity period 130–140 days, seeds red, bold, weight 9.3 to 11.4 g/100 seeds, pods long and broad, recommended for Vidharbha region and Western Maharashtra.

PDM 1: Developed from local germplasm, tall, semi-spreading, seeds brown, yield 20–25 q/ha, suitable for coastal region of Andhra Pradesh.

Hy 1: Developed from PL 3704 in 1975 for Peninsular region, suitable for rain-fed area, maturity period 160–170 days, white seeded, seeds weight 9–10 g/100 seeds, average yield 12–15 q/ha.

Hy 2: Developed from PL 4628 in 1978, semi-erect, tall, maturity in 140–150 days, white seeded, bold, weight 11–12 g/100 seeds, average yield 13–19 q/ha, suitable for Andhra Pradesh.

Hy 3a: Developed from 2817–1–A in 1980, Plant erect, tall, green, medium maturing 160–170 days, white seeded, bold seeds weight 18–20 g/100 seeds, average yield 16–20 q/ha.

Hy 3c: Developed from PL 2817–2 in 1980, late maturing (180 days), semi-spreading, large pods, bold, seeds weight 20.3 g/100 seeds, average yield 17–24 q/ha.

Hy 5: Developed from PL 3701 in 1980, medium maturing 160–170 days, semi-spreading, white round medium bold, seed weight 11.2 g/100 seeds, average yield 15–20 q/ha.

LRG 36: Single plant selection from local land race, maturity 170–180 days, seed weight 9.2 g/100 seeds, average yield 20 q/ha.

LRG 30 (Palnadu): A local selection from Lam (Guntur), medium maturing 170 days, semi-spreading, seeds medium bold, weight 9.6 g/100 seeds, average yield 17 q/ha.

C 11: A local selection from Sangareddi (A.P.), developed in 1975 for Peninsular region, medium late maturing 200–220 days, medium tall, seed brown, small, weight 8.5 g/100 seeds, tolerant to wilt, average yield 15–20 q/ha.

ICPL 87: It is cross between T21 × JA 277, suitable for peninsular region, plants semi-spreading, pod large, medium seeds, weight 10.2 g/100 seeds, susceptible to wilt, average yield 12 q/ha.

GS 1: A selection from local land race, maturity period 165–195 days, seeds small, weight 6.5 g/100 seeds, susceptible to wilt disease, average yield 12 q/ha, suitable for Karnataka.

PT 221: A local selection, medium maturing 210–220 days, semi-spreading, average yield 10 q/ha.

Maruthi (ICP 8863): A selection from germplasm, medium maturing 200–220 days, resistant to wilt and sterility mosaic virus (SMV), average yield 10–12 q/ha.

SA 1: A selection from a land race of Tirupathur (Tamil Nadu) developed in 1973 for Peninsular region, maturity 180 days, semi-spreading, medium tall, seed brown, small, weight 7 g (100 seeds), average yield 10–12 q/ha, suitable for Tamil Nadu.

Co 1: A selection from Parambalur local (Tamil Nadu), released in 1970, maturity 135 days, seeds brown, small, weight 7 g/100 seeds, average yield 15 q/ha.

Co 2: A selection from P 4728, released in 1974, maturity period 120–130 days, seeds brown, small, weight 7 g/100 seeds, average yield 12–15 q/ha.

Co 3: A mutant of Co 1, released in 1977, maturity period 130 days, seeds brown, small, weight 7 g/100 seeds, average yield 12–14 q/ha.

Co 4 (S 80): Pure line selection from a germplasm from Hyderabad, maturity period 130 days, seeds brown, medium, weight 8–9 g/100 seeds, average yield 12–15 q/ha.

Sweta (B 7): A pure line selection from land race developed in 1972, maturity 230 days, tall, semi-compact, seeds white with brown dots, medium small, weight 8–10 g/100 seeds, average yield 15 q/ha, most suited for West Bengal.

QUESTIONS

1. Discuss the different methods applied for improvement of wheat crop and give the list of promising varieties of wheat that have been developed recently in India.
2. Give an account of breeding work done in India on sugarcane and name some of the high yielding varieties developed in this country.
3. Discuss the different breeding techniques used for evolving high yielding varieties of paddy and give a list of recently released high yielding varieties of rice which are under cultivation in India.
4. What are the different methods employed for the improvement of Potato? Give the names of high yielding varieties of Potato along with their parentage that have been evolved recently in India.
5. Give an account of breeding work done in India on cotton and list some of the high yielding varieties which have been released recently.
6. Write short notes on:
   (*a*) Central Potato Research Institute, Simla,
   (*b*) Sugarcane Breeding Research Institute, Coimbatore,
   (*c*) M.S. Swaminathan,
   (*d*) Pushkarnath,
   (*e*) K. Ramiah,
   (*f*) Albert Howard,
   (*g*) Normann E. Borlaug, and
   (*h*) Norin genes.
7. Give an account of breeding work done in India on pigeon pea.
8. Discuss the aims and objectives of breeding of Chick pea (*Cicer arietinum*). Give the details of at least ten varieties of gram that have been evolved through different methods of breeding.